Handbook of Rural Studies

Handbook of Rural Studies

Edited by

Paul Cloke, Terry Marsden and Patrick Mooney

Los Angeles | London | New Delhi
Singapore | Washington DC

First published 2006
Reprinted 2009

SAGE Publications Ltd
1 Oliver's Yard
55 City Road
London EC1Y 1SP

SAGE Publications Inc.
2455 Teller Road
Thousand Oaks, California 91320

SAGE Publications India Pvt Ltd
B 1/I 1, Mohan Cooperative Industrial Area
Mathura Road
New Delhi 110 044

SAGE Publications Asia-Pacific Pte Ltd
33 Pekin Street #02-01
Far East Square
Singapore 048763

British Library Cataloguing in Publication data

A catalogue record for this book is available from the British Library

ISBN 978 0 7619 7332 4

Library of Congress Control Number 2005926014

Typeset by C&M Digitals (P) Ltd., Chennai, India
Printed on paper from sustainable resources
Printed in Great Britain by the MPG Books Group

Contents

List of Editors and Contributors

EDITORS

Paul Cloke, Department of Geography, SoGAER, University of Exeter, Exeter, UK

Terry Marsden, Cardiff School of City and Regional Planning, Cardiff University, Cardiff, UK

Patrick H. Mooney, Department of Sociology, University of Kentucky, Lexington, Kentucky, USA

CONTRIBUTORS

David Bell, Sociology Department, Manchester Metropolitan University UK

Alessandro Bonanno, Department of Sociology, Sam Houston State University, Huntsville, Texas, USA

Bruce Braun, Geography Department, University of Minnesota, Twin Cities Campus, Minneapolis, USA

Frederick H. Buttel, a leading scholar in rural sociology, died January 2005.

Noel Castree, Geography, School of Environment and Development, The University of Manchester, Manchester, UK

A.I. (Lex) Chalmers, Department of Geography, Tourism and Environmental Planning, Faculty of Arts and Social Sciences, The University of Waikato, Hamilton, New Zealand

Lynda Cheshire, School of Social Science, The University of Queensland, St Lucia Qld, Australia

David Crouch, Centre of International Tourism, Leisure and Hospitality Management Research, Derby, UK

E. Melanie DuPuis, Sociology Department, University of California, Santa Cruz, California, USA

Tim Edensor, Department of Environmental and Geographical Sciences, Manchester Metropolitan University, Manchester, UK

Mark Goodwin, Department of Geography, SoGAER, University of Exeter, Exeter, UK

Keith Halfacree, Department of Geography, University of Wales Swansea, Swansea, UK

Owain Jones, Department of Geography, SoGAER, University of Exeter, Exeter, UK

Alun E. Joseph, Department of Geography, University of Guelph, Guelph, Ontario, Canada

Lawrence Kitchen, Cardiff School of City and Regional Planning, Cardiff University, Cardiff, UK

Mark B. Lapping, University of Southern Maine, Muskie School, Portland, Maine, USA

Geoffrey Lawrence, School of Social Science, The University of Queensland, St Lucia Qld, Australia

Jo Little, Department of Geography, SoGAER, University of Exeter, Exeter, UK

Stewart Lockie, Faculty of Arts Health & Sciences School of Psychology & Sociology, Central Queensland University, Queensland, Australia

Thomas A. Lyson, Polson Institute for Global Development, Cornell University, Ithaca, New York, USA

Matteo B. Marini, Dipartimento di Economia e Statistica, Università degli Studi della Calabria, Italy

Joan Martinez-Alier, Department of Economics and Economic History, Bellaterra (Barcelona), Spain

Mara Miele, School of City and Regional Planning, Cardiff University, Cardiff, UK

Paul Milbourne, Cardiff School of City and Regional Planning, Cardiff University, Cardiff, UK

Jonathan Murdoch, Cardiff School of City and Regional Planning, Cardiff University, Cardiff

Ruth Panelli, Department of Geography, University of Otago, Dunedin, New Zealand

Harvey C. Perkins, Environment, Society and Design Division, Lincoln University, Canterbury, New Zealand

Jan Douwe van der Ploeg, Rural Sociology Group, Wageningen University, Wageningen, The Netherlands

Christopher Ray, Centre for Rural Economy, School of Agriculture, Food and Rural Development, University of Newcastle upon Tyne, Newcastle upon Tyne, UK

Sonya Salamon, Department of Human and Community Development, University of Illinois at Urbana–Champaign, Urbana, Illinois, USA

Eduardo Sevilla Guzmán, Escuela Tecnica Superior de Inqenieros Agronomos y de Montes de la Universidad de Cordoba, Cordoba, Spain

Brian Short, Department of Geography, University of Sussex, Brighton, UK

David Sibley, University of Leeds, Leeds, UK

Kieron Stanley, Cardiff School of City and Regional Planning, Cardiff University, Cardiff, UK

Ann R. Tickamyer, Department of Sociology and Anthropology, Athens, Ohio, USA

Preface

Until the 1970s, the domain within which contemporary social science was being practised was increasingly urban. The role of the city as political and economic hub, and the construction of the urban as ostensibly the archetypal site of cultural distinction and social deprivation, meant that rural areas were being left behind and often ignored. Since that time, however, there has been something of a revival in rural studies. Not only has the changing nature of the rural been comprehensively charted in different contexts, but there has also been an upsurge in rural theorization and conceptualization. Rurality, we might argue, has been put back on the map through a revitalized rural studies.

Such a picture, however, is only partially accurate. In truth rural studies are often pursued within disciplinary and subdisciplinary boundaries, and the publications of rural researchers tend still to be placed in diverse and dispersed locations. These literatures are also often constrained within subdisciplinary boundaries, leading to poor dialogue, both practically and intellectually, both amongst rural researchers and between rural specialists and other social scientists. Indeed, we face a significant challenge not only to articulate and expose rural research to a wider audience, but also to enhance the interdisciplinary 'stock' of its creative theoretical and empirical endeavours over recent years. We have created this Handbook as an attempt to contribute to both of these goals, to progress a true reflection of the vitality and theoretical innovation displayed in the rural studies of advanced societies.

Our objective for the Handbook then is to encapsulate the intellectual excitement which has arisen from the application of new theorizations of rural life, landscape, work and leisure over the past decade. This has involved engagements both with critical political economy and the 'cultural turn' in social sciences, both of which have led to very significant insights into the assemblages of power, process, practice and change which have (re)produced and (re)encultured rural areas over recent years. The Handbook will reveal, moreover, not just one-way traffic of ideas into rural studies, but also key moments in which the theorizations of culture, nature, politics, agency and space in rural settings are transmitting significant ideas into wider social science debates. The Handbook attempts, then, to deal with both the substantive components of contemporary theorized research in rural areas and the emergent trajectories of theory/research and practice.

In seeking to achieve this objective, it is important to acknowledge the continuing significance of barriers to truly interdisciplinary and internationally relevant studies of the rural. Not all rural areas are the same, even within particular nation-states. Not all rural researchers adopt the same theoretical, philosophical and methodological frameworks for their work, even within particular disciplinary boundaries. Not all substantive issues cropping up in rural areas will show the same face, even where contexts appear similar. In the Handbook we have brought together authors from different national and disciplinary backgrounds but, even so, many of the perspectives offered will appear too narrow and constrained by their contexts. Any such response to the Handbook represents a welcome affirmation of the diversity and differentiated excitement of rural studies, and we hope that the material presented here will act as a springboard for further discussion of those differences and excitements.

As with other Sage Handbooks, the aim here is to present a retrospective and prospective overview of rural studies that focuses on critical discussions of the role of theory in the development and contingency of rural research. In order to engage with the interdisciplinary nature of rural studies, the Handbook emphasizes the theoretical and empirical diversity of research in the field, and the interconnections that may be made between the different movements in this diversity. We have attempted to map out and problematize the development of rural studies, exploring the key terrains of coherence and incoherence in the theorizations which have been brought to rural studies, and the critical pathways taken by new research that seeks to develop a range of theoretical resonances from, and to, wider social sciences. In this sense we hope the Handbook will be read by non-rural researchers as well; those interested in the macro-concerns which emanate now from environmental, economic, social and political uncertainties.

The Handbook is divided into three parts. The first, 'Approaches', presents a retrospective cartography of rural studies. It charts how rural knowledge, and in particular how the notion of rurality, has been informed by different theoretical strands and impulses throughout the latter part of the twentieth century, and through the transitions between modernization and postmodernization. It also deals with key routes into the rural – the spatial, the social, the economic, the resource base and planning problematics. These chapters provide a foundation for the systematic accounts that follow, and present a welcome basis for assessing how far rural studies has progressed over the past three decades.

The second part has been compiled as a series of key theoretical coordinates, each of which is given detailed treatment in three chapters dealing with different forms and foci of research. This is the most extensive part of the Handbook, representing a state-of-the-art survey of different facets of rural studies. Each group of chapters presents a critical assessment of the different 'packages' of theoretical and empirical issues around what we regard as the key research avenues for existing and future development. These are necessarily intersectoral, and they highlight key clusters of creative endeavour in rural research, as well as linking this to wider social science debates – cultural representation, nature, sustainability, new economies, power, new consumerism, identity and exclusion.

The third section, by way of conclusion, attempts to focus on the question of new formulations of rural relations. These we suggest are only beginning to be developed in the rural studies field and are therefore more tentative treatments of imaginative engagement. Nevertheless, it seems crucial to explore new ways of thinking and practising the rural. We do not regard the four contributions as an exhaustive treatment of this new agenda; but they are in our view important new traces, ones which require further theoretical and empirical concern.

Rural studies has reached a stage of maturity which deserves to be mapped out and pored over. It is our hope that this Handbook adds to the resources with which such necessary tasks are undertaken. However, we also appreciate the urgent need for rural studies to remain innovative and dynamically critical. This may involve importing the most recent theoretical and philosophical insights being developed in urban domains, and it may also mean a patient re-excavation of previous approaches to maximize their potential for insights into changing rural relations – insights that are sometimes trampled over in the rush for the fashionable and the supposed cutting-edge. Mostly, however, we hope that rural researchers will increasingly be able to export their theoretical insights, rather than merely accept urban-centred dictates. Were this to be the case, we would soon be needing another Handbook to express another exciting period of innovation.

Acknowledgements

The Editors want to thank Robert Rojek and David Mainwaring, both for believing in the idea of this Handbook and for their enthusiastic efficiency in dealing with it. The process of bringing the Handbook to fruition has been complex and iterative. We want to thank the authors of chapters for their patience and expertise in producing high-quality contributions. Finally, Joek Roex has undertaken very significant editorial support, and has in no small measure been the catalyst for the completion of the manuscript. We are enormously grateful to Joek for his efficiency, diligence and general good humour – he is the unsung hero of the *Handbook of Rural Studies*.

Part I

APPROACHES TO RURAL STUDIES

1

Pathways in the sociology of rural knowledge

Terry Marsden

INTRODUCTION: FROM VACUUM AND HIATUS TOWARDS A THEORETICAL MATURITY

The 1970s and 1980s were in some ways rather frustrating for the rural studies of advanced societies. Like other areas of social science, there was a growing recognition of a 'crisis', not just of substance but of confidence, in the field and a continuous struggle both to release the subject from the restrictive (not least institutional) frameworks of the past, and to adopt a more 'holistic' and renewed focus.

In short, the subject had been significantly retarded and marginalized by the post-war (agricultural) modernization project. This had tended, at its most positive, to render rural spaces as sites for the playing out of a particular type of agricultural modernization based upon increasingly intensive methods and scale economies. This was expressed by rendering anything that did not fit into this model (such as upland farming regions) as more of a 'problem of agricultural adjustment' rather than as a potentially different pathway of rural development. Rural sociology, for instance, by the early 1980s, was still seen as something of a 'side-show' with regard to the main social science drift. As Newby sums up:

> Rural sociology has still not filled the conceptual vacuum left by the demise of the rural–urban continuum. This is reflected in the paucity of problem formulation which continues to afflict the field, and which lies at the heart of rural sociology's current malaise ... To overcome these problems rural sociology could learn from the example of its urban counterpart by beginning from a holistic theory of society within which the rural can be satisfactorily located, and as a corollary, developing theories which link social structure to spatial structure. (1980: 108)

There was a sense at that time that any theoretical or, indeed, scientific justification there may have been for delineating 'the rural' as a distinctive area of study had largely been 'hollowed out', leaving, as Newby believed, 'the rural' as simply a physical and social expression of modernization, rather than a robust and engaging conceptual and scientific assemblage. As a result it was not possible to have a theory of rural society without a theory of society *tout court*; moreover, this demanded a theory that links the spatial with the social, and indeed one that gave the social primacy.

It is interesting that a decade later – a period which brought considerable vibrancy to a large-scale and creative reaction to these circumstances in the form of a new political economy of rural space and agriculture – a similar concern was still prescient. I myself, and colleagues, argued in 1990 that despite the progress made:

> The problems associated with the transformation of rural areas thus need to be rescued from the conceptual hiatus to which they are increasingly assigned ... An appropriate starting point would be to reverse the respective telescopes of agrarian political economy and rural economic restructuring. In other words, to locate the contemporary predicament of rural areas at the intersection of the two major forces transforming them, the reorganisation of the international food system and the social and economic restructuring of rural regions under the pressure of capitalist recombination. (Marsden et al., 1990: 12)

I want to suggest in this chapter, and in introducing this volume as a whole, that considerable progress has been made since this scientific 'call for arms', both in recognizing the need for diverse sociological and spatial theoretical developments, on the one hand, and in bridging the conceptual hiatus potentially created by the bifurcation of effort, vertically along the agro-food supply chain and laterally by addressing the diversity of economic and social restructuring processes affecting rural areas of advanced economies, on the other. I will argue, however, that, despite the deepening and growing plurality of theoretical and conceptual endeavour experienced over the past decade, this still requires a need to consider a revised political economy of rural space. One which foregrounds the distinctive features of rural life, but does so not at the expense of conceptually isolating it from broader social science theoretical and conceptual trends and interpretations. Indeed, one of the features of the most recent decade of research effort, as this Handbook clearly expresses, has been to significantly build a more interdisciplinary critical rural social science and to link this directly to broader societal and restructuring trends. Taking the continued major themes of (the vertical) agro-food regulation and (the lateral process of) regional/rural restructuring, it will be argued here that rural studies (itself now much more of an amalgam of social science disciplines) has now reached a new and more complex terrain, a terrain which can justify a much more confident and mature theoretical and scientific positioning, and which also needs to be anchored at the juxtaposition of the agro-food and rural and regional restructuring dynamics.

In studying how these two bodies of rural knowledge have developed recently and, moreover, how they now contribute to a much more malleable and robust political economy of rural space, it is necessary to identify some key macro features which seem to have stimulated these scholarly trends. Indeed, I tend to consider these, on reflection, as a series of contextual paradoxes. Three seem here to be most pertinent.

Paradox 1 *More intense and diversified social science rural research despite the continual urban cosmopolitanism and globalism of advanced societies and the 'urbanization' of the countryside.* Advanced economies over the past two decades have both urbanized and suburbanized, with traditional rural societies and 'ways of life' being increasingly marginalized.

There has been undoubtedly a growth in the 'consumption countryside' and the highly mobilized networked society. Moreover, traditional rural sectoral interests have given way to more complex environmental and other issue-based concerns (like food quality and animal welfare). Interestingly, however, this has not diminished the needs and priorities for rural-based research. Rather, it has re-invigorated a broader cultural and social understanding of what Keil (2000), in a US context, has called the 'suburban frontier'. This has created a fertile ground for not so much a geographically legitimating social definition of the 'rural', as a true and more diverse and eclectic 'sociology of the rural'. As Mormont in a seminal paper outlining this new intellectual empowerment argued:

> The benefit of a history of rural sociology – yet to be established on an international scale – would be to start from the hypothesis that the rural–urban opposition is socially constructed and that the rural exists primarily as a representation serving to analyse both the social and space–or rather to analyse the social while defining space–borne and interpreted by social agents. The fact that it is a constructed representation and not an ascertained reality does not deplete a sociology of the rural of subject. Its subject may be defined as the set of processes through which agents construct a vision of the rural suited to their circumstances, define themselves in relation to prevailing social cleavages, and thereby find identity, and through identity, make common cause. (1990: 41)

Hence the sociology of the rural now holds a broader 'terrain, map and compass' both in the public's imagination and in the social scientist's repertoire, because of, rather than despite, the onset of global cosmopolitanism.

Paradox 2 *More intense and diversified social science research despite the application of neo-liberal projects and the further de-institutionalization of critical rural (and especially agricultural) research and development.* The social science of the rural has grown as an active interdisciplinary area despite the institutional constraints placed upon disciplinary research institutes and academic departments. While the 'land-grant college' system has remained in the US, budgetary cuts in staffing, reorganization and partial privatization of research have generally failed to reduce the vibrancy of critical rural research. In Europe, the onset of the bio-science revolution in research universities has been a major investment priority, which far outweighs the former support for conventional agricultural research, development and extension. Nevertheless, we witness the growth in critical rural studies. Indeed, as seen in the case of the UK, the point has been long made (see Crow et al., 1990) that the very vibrancy of interdisciplinary rural social science has been facilitated by *not having* a stable disciplinary home. Indeed, the disciplinary 'home', which may have once been recognized, of agricultural economic faculties, have either closed or have been forced to adopt a broader interdisciplinary focus. Rural researchers are found populating geography, sociology, planning and political science schools whilst the traditional (disciplinary) agricultural economics departments have become isolated and moribund places for critical and theoretical rural conversation between the social science disciplines.

Paradox 3 *New processes of modernity and technology are attempting to deny local rural nature and communities, at the same time as both rural actors and researchers identify new, alternative socio-ecological paradigms of local rural development.* Conventional agro-food technologies in particular (not least GM), as well as corporate retail strategies, are currently being applied in ways that attempt to prolong the 'sustainability of the unsustainable' (see Buttel, in this volume). That is, they are continuing to largely deny the embeddedness of rural nature, and reinforce the technological 'treadmill' of production and scale economies of standardization in agro-food. However, much of rural social science is now understanding the significant reactions and contingencies in this process and, in some cases, the emergence of an alternative rural development paradigm (see Marsden, Chapter 14 in this volume). There is, therefore, a widening vector and critical research agenda associated with the social and political development and application of technologies in and through rural space, associated with both GM developments in agro-food and ICT developments in shaping rural development (see Andersson, 2003).

I shall return to the continuity and endurance of these apparent paradoxes in conclusion, after giving some treatment to two major themes in the revised political economy of rural space: the onset of the 'risk society' and new regulation in the conventional food supply chain, and the emergence of diverse regionalized ruralities, within a European context. In both cases, we see contradictory governance and regulatory arrangements between these dynamic forces which are reshaping, in combination, the revised political economies of rural space and society. As such, it is argued that they provide quite rich comparative theoretical and empirical pathways with which to further progress the critical political economy of rural change. In addition, in both of these spheres, we can begin to track what we might regard as a 'postmodernization' tendency, that is, a contextual and political redefinition of rurality which goes beyond the conventional agricultural and industrial modernization process (a 'growth machine') so prevalent in the advanced societies of the late twentieth century. Both cases also demonstrate the growing political and cultural influence of the wider (cosmopolitan) public and consumer realm for the reconstitution of these new critical ruralities.

THE REVISED FOOD CHAIN DIMENSION: REGULATING FOOD RISK AND REGULATING THE DANGEROUS

Governments are continually having to devise innovative strategies to deal with the innumerable risks to health and well-being. The 'successes' of late twentieth century modernization have been marked by new social and environmental risks that are capable of manifesting deleterious impacts at indeterminate points in the future and over indefinite spatial realms. The significant differences, in characteristics and impacts, between contemporary risk and 'age-old' risk have been a feature of recent sociological writing, associated with the 'risk society' (Beck, 1992; see also Giddens, 1994; Cohen, 1997).

According to Beck (1992), postmodern society is facing a new state of human insecurity, characterized not by the desire to satisfy basic and material needs, but by fear of the 'dark side of progress', that is, the tangible and intangible by-products of industrial development. Confidence, therefore, that progress in human development has been synonymous with greater security is challenged by the recognition that modern day 'manufactured dragons', such as bovine spongiform encephalopathy (BSE) in cattle, are actually a product of science and technology, and cannot easily be mitigated by it.

The crisis surrounding BSE (discovered first in British cattle during the mid-1980s) and its links with new variants of the Creutzfeldt–Jakob disease (vCJD) in humans in 1996, has been cited in the food risk literature as a classic illustration of 'manufactured risks'. This image of modernization undermining feelings of confidence and security represents a formidable challenge to politicians, policy-makers, scientists, producers and others to find new ways of not only minimizing these new threats, but to do so without creating hindrances to the conventional modernization process and project itself. In other words, this postmodernization era requires that economic and social development continue within a conventional economic model (Smith, 2002). In the context of exposure to new types of food risks, public concern has extended beyond the mere occurrences of food-borne diseases due to periodic microbial contamination. What has become significantly more important has been the increasing incidence of these microbial episodes, the emergence of new food pathogens, as well as infectious strains of familiar pathogens that bring resistance to customary anti-microbial treatments; *E. coli* 0157 is a notable example. In the case of non-infectious, food-borne diseases, public concern surrounds the uncertain effects of a range of synthetic food additives, pesticide residues, nitrate residues, dioxins, heavy metals, hormones in beef, genetically modified organisms and so on (see Lang et al., 2001; WHO, 1999). Public confidence in food is consequently undermined by the belief that chemical residues and genetic modification may harbour deleterious consequences for human health, such as mutagenic, carcinogenic and teratogenic effects (WHO, 1999).

However, to attribute the erosion of public confidence in food (especially in the UK) during the 1980s and 1990s merely to a steady catalogue of food scares would be to ignore broader societal developments. These gave greater importance to expressions of public disaffection with how the public responsibilities of the state were being discharged. For instance, during the 1980s and into the 1990s the UK experienced a period of pressure for greater openness and transparency in dealings between the public and public institutions. These began to challenge the pre-existing ethos of authority between the executive state and the client public. The ensuing transition from public administration through virtually closed policy avenues of *government* (such as in the Ministry of Agriculture) to the more socially inclusive notion of *governance* gradually provided useful political space for new participants, such as environmental and social non-governmental organizations (NGOs), to contribute to the definition of problems of public significance as well as in devising likely solution options. Consequently, where old-style paternalistic statements concerning the diligence of regulators to reduce the range and number of breaches in food safety, for instance, may have ameliorated public concern previously, this became increasingly inadequate. Hence, issues of accountability within the food chain (particularly of producers) and effectiveness of mechanisms of food safety regulation, began to take on greater importance.

This wider process of food governance, particularly in relation to its contribution to the construction of a new regulatory context for food, has been positively influenced by the parallel demand for participative democracy across Europe. Of particular importance has been the institutionalization of greater openness in matters of public affairs that has been facilitated by the passage of key pieces of EU legislation such as 90/313/EEC,[1] 93/730/EC[2] and 95/46/EC,[3] as well as the Treaty of the European Union itself. These, together with legislation requiring the release of archived information, have opened the way for the public (often through the media) to gain knowledge of the workings of the state, knowledge that previously would have been known only to governments, their advisers and to privileged insider individuals and groups. This change in how, when and what types of information becomes public knowledge meant that issues of public interest, such as food safety and farming practices specifically, began to be reported in the electronic and printed media much more readily, and in much greater detail, than in the past. Easier access to information of public interest has also facilitated NGO campaigning activities in areas such as animal welfare, application of the 'precautionary principle' to matters of food safety, as well as to environmental concerns generally.

Consequently, it is not sufficient to relate attempts at devising new strategies for assuring food safety in the UK and across the EU simply to a raft of recent food scares. Important though this has been, the process has nevertheless been significantly influenced by the wider social, political and economic considerations shown in Figure 1.1.

There is now wider recognition that the factors in Figure 1.1 have been guiding how the public perceive food risks and the rural/farm sector. This has seriously challenged conventional food regulatory practices with their reliance on a combination of science, technology and expert advice to allay public fears. Governments across the EU began to acknowledge that the conventional approach dealt poorly with the crucial issue of scientific uncertainty. And, whilst absolute guarantees about the safety of foods cannot be given or, indeed, may not be expected by the public, the fear that lives and well-being are being exposed to 'manufactured risks' has raised public expectations for improvements in food regulation and particularly the competence of the farm sector. The question for governments becomes: how can new food safety and quality assurances be established given these new food-governance pressures in ways that support the agro-industrial complex?

Beck's (1992) *risk society* thesis elucidates this contemporary risk/governance concern. Scientific research is not always able to provide a full and clear picture of what effects, for instance, widespread consumption of foods containing genetically modified organisms will have on human health. In view of this deficiency in science-based decision-making, states such as the UK have become less resistant to ethical, value and culture-based advice infusing the conventional process of food safety regulation. This has served to pluralize the political context within which the EU's policies of food regulation are being constructed.

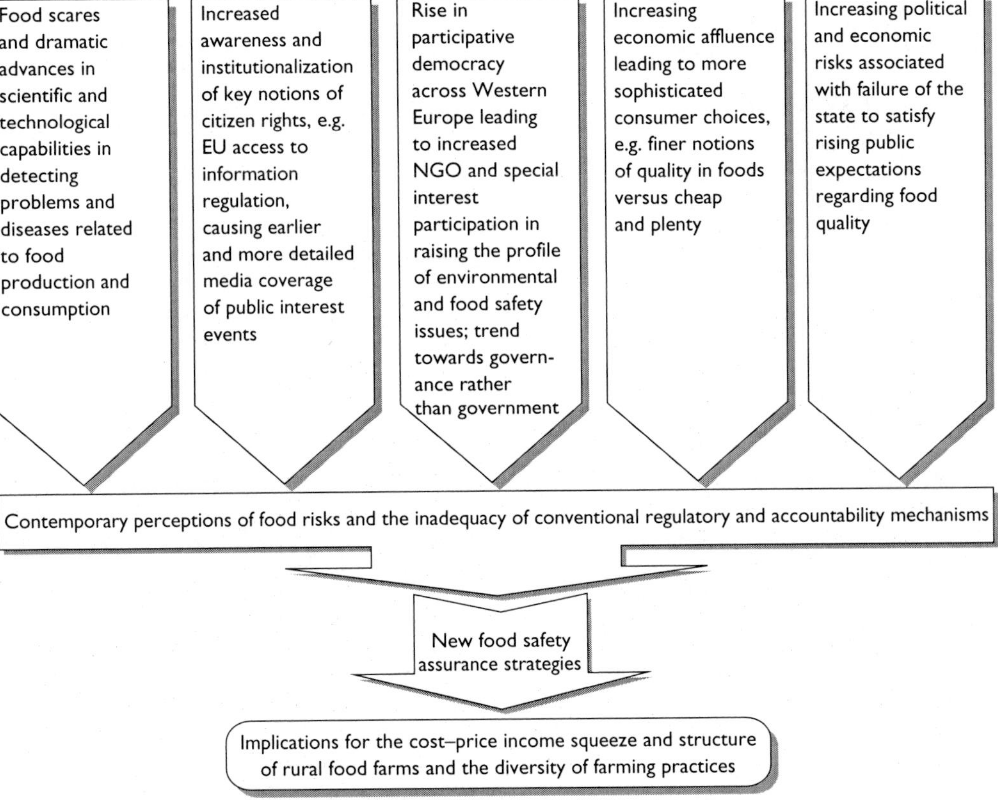

Figure 1.1 ***Factors contributing to contemporary perceptions of food risk and new regulatory approaches***

WHAT ARE SOME OF THE FEATURES OF THIS EMERGENT EU FOOD REGIME? THE MATURING EUROPEANIZATION OF UK FOOD POLICY

In earlier work (see Marsden et al., 2000), we began to show, through both the design and implementation of European food hygiene directives, how corporate retailer-led supply regulation became empowered, and the use of quality control and risk management techniques, such as hazard and critical control point (HACCP), tended to empower the corporate retailer-led forms of food regulation. The failure in reducing food risks and legitimacy concerns since the mid-1990s, and, indeed, their diversification and differential spatial spread across member states (like swine fever in the Netherlands, dioxins in Belgium), tended to further press the EU to develop wider powers and ambitions. This was embodied in the 2000 Food Safety White Paper, which addressed such corporate/public tensions:

> Consumers should be offered a wide range of safe and high quality products coming from all Member States. This is the essential role of the Internal Market. An effective food safety policy must recognise the inter-linked nature of food production. It requires assessment and monitoring of the risks to consumer health associated with raw materials, farming practices and food processing activities; it requires effective regulatory action to manage risk; and it requires the establishment and operation *of control systems to monitor and enforce the operation of these regulations* ... Historically, these measures have mainly been developed on a sectoral basis. However, the increasing integration of national economies within the Single Market, developments in farming and food processing, and new handling and distribution patterns require the new approach. (CEC, 2000: 6–7)

After a decade of economic integration in the European Single Market, the crucial priority of the White Paper was to establish an independent European Food Safety Authority to provide:

> scientific advice on all aspects relating to food safety, operation of rapid alert systems, communication and dialogue with consumers on food safety and health issues, as well as networking with national agencies and scientific bodies. (CEC, 2000: 3)

And, while the White Paper prepared the underlying aims of this more comprehensive European approach to monitoring and enforcing food safety and assurance, a further challenge is the building of multi-interest support for the process of implementing and institutionalizing this new EU-led food safety regime.

Collaboration between DGs in the formation of food safety policies

Although the major elements of food production within the EU still fall within the remit of DG Agriculture, it is DG SANCO (Health and Consumer Affairs) that now has the principal responsibility for overseeing the EU's progression to the new food safety regime that was described in the White Paper. In fact,

> SANCO itself was born out of food crises – the big one being BSE which has a political dimension … many orders of magnitude greater than its public health dimension. The crises and repeated crises that the political machinery responds to, creates [*sic*] a paradox for as policy starts to tighten up at all levels – rules, rule-making, enforcement – you actually discover more and new problems [see Figure 1.1]. Responses [therefore] reflect the conflicts between remediation and party-political objectives and their relationship to economic and private interest concerns. (Interview with DG SANCO senior official)

Despite the new leadership position held by DG SANCO, the view from DG Agriculture is that 'good communication exists between Commissioners David Byrne [DG SANCO] and Franz Fischler [DG Agriculture] because their jobs overlap a lot and they have common views' (interview with DG Agriculture senior official). Similarly, DG Trade, which is primarily responsible for the EU's compliance with World Trade Organization (WTO) rules and in particular the Sanitary and Phytosanitary Agreement (SPS), is satisfied with the opportunities to contribute to the development of the new food safety regime.

In reference to the establishment of the EFSA (European Food Safety Agency) – the centrepiece of the new Commission's new food safety regime – DG Trade is also satisfied that its somewhat long-distance relationship with DG SANCO does not prevent its key concerns from being taken into consideration. According to DG Trade, 'our interest in the EFSA is in ensuring that the right questions are asked, and that third countries have access to it' (interview with DG Trade senior official). This is very important as, despite advocating 'health before trade', the key operating principle of the SPS Agreement is that measures designed to protect health (and the environment) 'have to be based in science'. This science-based approach to food/trade policy is one of the key features of the new food safety regime emerging from Brussels, and is currently being significantly tested with the onset of GMOs through the World Trade Organization.

We can see here how the new, or at least revised processes of Europeanization, consumerization and institution-building are developing in both a contingent and dynamic way. This involves the (differential) empowerment of a wider range of interest groups and consultation procedures in food policy-making. These become more fluid than before in that they are networks based around significant issues rather than simply group- or sector-led. For instance, we see new, issue-based alliances being formed around issues like GM by consumer, environmental, animal-welfare and corporate retailing groups. This acts as a significant counterweight to the 'industry-led', manufacturers and producer interests and alliances.

Also, we see the boundaries between the traditional 'insider-groups' and outside groups becoming much more blurred, with DG SANCO and the EFSA establishing new clientelistic relationships with some key consumer organisations. Indeed, there is significant fluidity both in the means of networking and the content of intermediation, on the one hand, and in terms of variable 'state-capture' and influence on the other.

The second regulatory trend concerns the contemporaneous and co-evolving model of *private interest food regulation*. At the same time as we see a more complex, fluid and networked policy process operating at the EU level, interestingly we also see a re-affirmation of a more complex, yet authoritative *private interest model of food regulation* embedding itself in the conventional food supply chains in Europe. In other words, the stronghold that, for instance, major food retailers in the UK have over their supply chains is being diffused throughout the Union and beyond (see van der Grijp et al., 2005).

The growing, if contested, legitimacy and authority of the more complex private interest model of food regulation

Over time, the private interest model of food regulation in the UK has had to both adjust and Europeanize to rapidly changing circumstances. This has not reduced its relevance, but its character has been modified. It now has to engage with stronger public sector agencies and actions, and a stronger reliance upon independent 'scientific' mitigation and assessment of food risks. Hence, having originally developed out of the limitations of the state, the private interest food regulatory model is now about to be re-engaged by state intervention. In addition, it now has to project itself as an effective, market-based and efficient system for policing the global food chains in a more complex and, as we have seen above, more networked, participatory and fluid policy-making community; one in which some private sector interests – particularly farmers and food manufacturers – have been increasingly questioned as the rightful custodians of the European food system. In doing this it would seem that British retailer practices have been to the fore.

The British influence in shaping the European food safety agenda has thus not only been associated with it being the unique source of BSE and foot and mouth disease. Rather, the now-established practices of the concentrated British retail sector in policing their food chains through arm's-length control mechanisms and developing innovative baskets of own-brand goods has led to this being seen as a preferred model at the European level. Issues of traceability, labelling, implementing a precautionary approach, and the specific debates about the regulation of GM products, are all tending to reinforce the private interest model of food regulation that has been projected and implemented by these retailers. The trick, therefore, becomes how to dovetail the existing private interest model within the newly enhanced public interest regime that is being projected by the EU, and outlined above.

Private interest umbrella organizations, such as Eurocommerce and the Confederation of the Food and Drink Industries (CIAA) at the EU level and the British Retail Consortium in the UK, see and project these public interest developments as a positive European step as long as they remain focused upon assuring the stability of European food markets and allowing the full operation of the European, and indeed the global food market, to operate as freely as possible under these circumstances.

Private interest representative groups at the European level have equipped themselves well to influence the new food policy regime. By demonstrating their technical competence in matters of food safety, the evolving policy for them becomes one of *mutual co-evolution of both public and private systems* with the belief that it is for the public sector to set in place minimum guarantees and risk assessment and assurance mechanisms, and for the private sector, given this more stable risk environment, to compete for the attentions of the more risk-averse and selective European food consumer. Moreover, the costs of compliance in this mutually reinforcing system will tend to always fall more heavily upon the upstream sectors (manufacturers and farmers), for it is they – for instance, in the case of traceability of GM grains – who will have to find the necessary hardware and software to ensure compliance with the private and public systems of regulation.

But, whilst the EU's emerging food safety regime is clearly increasing the grip of retailer-led regulation, these mutually reinforcing private and public interest mechanisms must also ensure that increased globalized and liberalized trade in foods continues as freely as possible. This could create increasingly difficult headaches for the retailers with regard to traceability and quality regulation.

The main aim of this section of the chapter has been to begin to re-conceptualize European food regulation during a continuing period of food risk and insecurity (see Figure 1.1). We can identify some of the key parameters of change at the European level, including: (i) the gradual but significant *Europeanization of policy*, through, not least, the implementation of the recommendations of the European Food Safety White Paper, and the overall commitment to a stronger 'top-down' and standard European approach to both the assessment and the management of risks; (ii) the growing *consumerization and institutionalization* of these policies and the empowerment of a wider set of interest groups; and (iii) the increasing use and legitimacy of a *more complex private interest model of food regulation*. The latter tendency is acting to further entrench retailer and commercially led regulation in and through supply chains at the same time as the public authorities are attempting to further guarantee minimum standards of food safety and quality, and take the bulk of the responsibility as and when major food scares occur.

Hence, we can begin to see, perhaps with hindsight, how the somewhat peculiarly British dual model of food regulation depicted in *Consuming Interests* in the late 1990s (what we termed the *second phase* of food regulation) has been mutated at the European level (to the *third phase*), both as European food markets have become more integrated and developed, and as the diversity and intensity of food risks have tended to grow.

We are now more firmly in a 'post-BSE' phase of European food regulation; one that has established new 'independent' food institutions in which to assess and manage these risks, and one that divests considerable power to selected commercial and consumer organizations in delivering

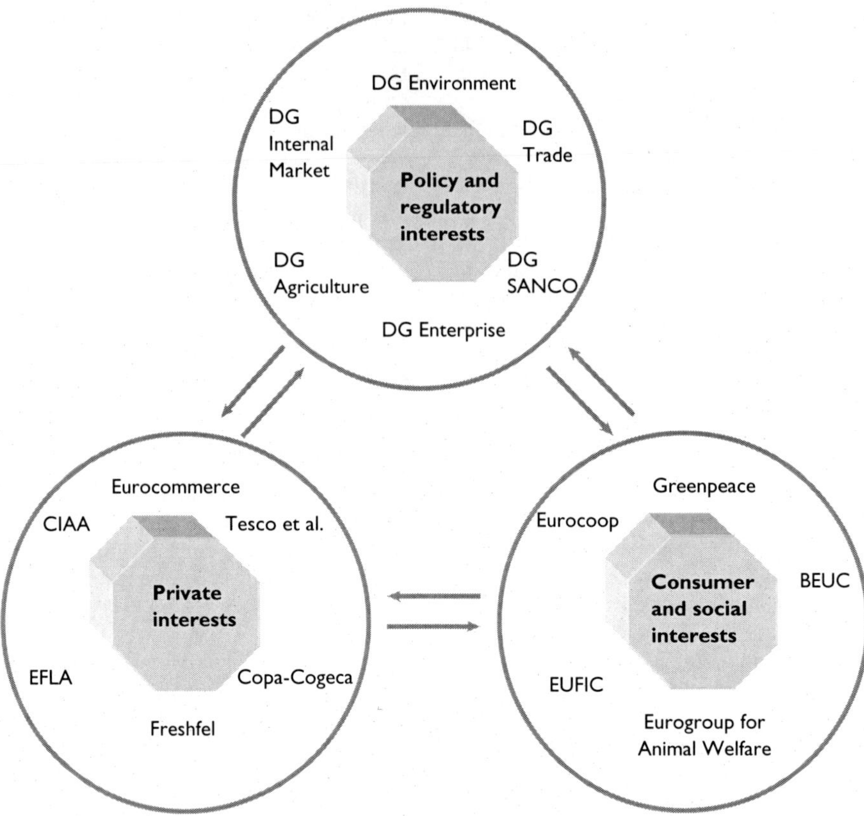

Figure 1.2 ***The emergent and more fluid food policy-formation network at EU level***

food through more accountable supply chains. This is very much then a *state–private sector hybrid model of food regulation*, a particular type of public and private sector response to the five pressures on food regulation (as identified in Figure 1.1). It is important to see this more complex, multi-level system as a response to these pressures in ways that preserve particular notions of the internal (and 'free') market and exchange in food goods within and beyond Europe, at the same time as expressing new controls in the name of the consumer and public interest. In this sense it is also consistent with other branches of the 'European project' which stress supply chain double-dividends and consumerization of public policy (see, for instance, Folkerts and Koehorst, 1998; Skogerbo, 1997). Thus, food regulation is now far more embedded and integrated into the European political project. It carries a far bigger political punch than its actual economic weight in Europe; and perhaps like agriculture and the CAP were viewed over 50 years ago, now food policy is again seen as a major plank for the further overall integration of the European project.

It follows then that as productivist agricultural corporatism has declined in its political and economic power in many of the member states (Murdoch et al., 2003), as well as in European policy-making, this has not weakened the political and economic significance of food in Europe as a major and dynamic regulatory activity. Rather, it has laid a basis for a more comprehensive and commercially led regulatory system – a hybrid model – based upon appeasing consumer *and* private sector concerns. It is increasingly systemic, scientific and standardized, global as well as local in reach, as well as inter-sectoral. And, as we have seen in this analysis, it has emerged out of, and will be maintained by, the interaction of a larger number of actors and policy networks. This now makes the evolution of public policy all the more complex and contingent.

More specifically, these influences and shifts reflect a reconstituted and multi-dimensional set of relationships between private interests, policy and regulatory interests and consumer and social interests. This is illustrated in Figure 1.2, moving from what might be called a formal or corporatist food policy network to a more fluid and participatory model, in which, for instance, key Directorates seek as much themselves to influence what private and consumer groups think as being the focus of their lobbying activities.

This new policy-making landscape and the new and differentiated influences have arisen partly due to the more challenging and complex food risk environment outlined here. In addition, they provide a central challenge or test for the wider European Project in attempting to incorporate consumer and public concerns (that is, uncertainty and risk), at the same time as fostering an integrated and a globally competitive European food market. This, after a decade of operating the European Single Market, makes the requirements for effective transnational regulation and food assurance systems much more pressing than a decade ago. Food risks are now both more mobilized and internationalized; so too are their governance frameworks.

However, from the perspective of the wider European project, major questions remain not only about what types of regulation are required but also how those types of regulation should be formulated and then implemented. In this sense, a focus upon the *contemporary evolutionary governance and regulation of food* tends to expose some of the broader contradictory tendencies of EU policy-making, whereby more regulatory and policy-making fluidity becomes partly a response to regulating uncertainty in a more Europe-wide way.

The very fact that EU policy-making is a collective exercise, involving large numbers of participants, often in intermittent and unpredictable 'relationships', is likely to reinforce both the processes by which national autonomy is being eroded and the capacity for consistent EU-level political leadership. The likelihood of any one government or any one national system of policy actors (for example, governments and interest groups combined) imposing their will on the rest is low. National governments know this. We can, therefore, expect to see the emergence of two apparently contradictory trends. First, the need to construct complex transnational coalitions of actors which will force all actors to become less focused on the nation-states as the 'venue' for policy-making. Just as many large firms have long since abandoned the notion of the nation-state, so will other policy actors. They will seek to create and participate in a multi-layered system of transnational coalitions. Second, the continued 'politics of uncertainty' will lead national governments and national interest groups to try to coordinate their Euro-strategies (for example, DTI, 1993, 1994). In that sense, European policy-making may bring them closer together (Richardson, 2001: 21).

Therefore, food regulation, in this more fluid context, becomes a key 'battlefield of knowledge' and power in the fulcrum of Europe, on the one hand, and in the re-orientation of member states and pan-European policy-making and implementing relationships on the other. The overall tendency, it appears, is for uncertainty and risk to generate more complexity, as well as to re-arrange the power relations between different food supply chain actors (that is, particularly the buying and selling power of corporate retailers and the weakening of farmer unions). This, in a preliminary and more formal manner, is leading to a more complex institutional and policy-making *structure* which is now both multi-level and multi-interest. The drive to minimize food risks for European consumers without creating hindrances to corporate firms and the internal *market is resulting in significant standardization in practice and philosophy across the entire agro-food chain*. And, although the new relationships between the state, corporate and non-corporate private and public interests promise a more viable and accountable food system, such public food safety benefits will be *accompanied by important spatial diversity costs*, such as a threat to regional diversity in food production and preparation. These are food quality concerns which, also, will need to infuse the food policy-making domain. Indeed, the question of diversity, the sanctity and vibrancy of local and short supply chains, and the growing political concerns over ethical and animal welfare issues (see Lucus et al., 2002) are likely to continue to pressure this newly established policy community, which is outlined and conceptualized here. As we see below there are key tensions between a revamped regulatory *standardization* of the conventional agro-industrial complex and the emergence of greater regional/rural and *agro-food diversity*. Thus, the new model of European food regulation traced out here holds significant dynamic and contingent features.

REVISING THE POLITICAL ECONOMY OF RURAL SPACE: THE EMERGENCE OF REGIONAL RURALITIES

In contrast to the somewhat 'virtual' European view taken by recent Brussels food policy developments, we can illustrate a different conception of rural Europe emanating from the EU with regard to local rural development. In an introduction to the recent *Living Countrysides* volume (edited by van der Ploeg et al., 2002), Romano Prodi, the current EU Commissioner, paints quite a different portrait of rural Europe to that referred to above:

> The countrysides of Europe and the diversity and richness of their agricultural systems represent an undeniable social, cultural, ecological and economic patrimony of European society as a whole. In a scenario dominated by the negotiations of the WTO and the enlargement of the Union, rural development constitutes a key defence mechanism for protecting this patrimony. It is therefore highly important to sustain these ongoing rural development processes and to stimulate new ones. (2002: 3)

So far, the analysis and discussion above has concerned itself with the changes in the governmentality of conventional food regulation in the European context. Much of this operates at variable distances from the rural spaces it inevitably affects. Yet such regulatory dynamics are critical for the future trajectory of local and regional food production systems. We have seen that, far from the regulatory state being 'hollowed out', at least in a macro-European context, the past decade has witnessed a significant process of re-regulation, especially at the supra-national level; a process that intends to significantly re-regulate and police hygienic conventional farming systems such that they can service the increasingly integrated European food market, on the one hand, and continue to at least appease the more risk-averse European consumer on the other (Marsden et al., 2001). Overall (and this will be focused upon in more detail in Chapter 14, which deals with rural sustainability), one effect of this re-juvenation in the governance of conventional agro-food is the attempt to continue to justify and legitimate the existing agro-industrial complex in a period of considerable legitimation crisis. This macro-project, a project to maintain the conventional agro-modernization process through its crisis of legitimacy, is, however, what we might see as only half of the story associated with the macro-regulation and political economy of European rural space and resources. For, if we take a more lateral approach, associated with aspects of rural restructuring (as indeed Prodi rhetorically does above), we see equally profound changes and contingencies at hand. These are associated with, first, a process of uneven multi-level governance, and second, a process of regional differentiation. I will deal with these issues here in turn.

The arrival of a problematic rural policy multi-level governance system: strategies beyond the Foucauldian governmentality of Brussels?

Despite the general shifts in the consumerization and Europeanization of food policy outlined above at the EU level, the last decade has also witnessed the re-orientation of agricultural productivist policy through a process of gradual reform and diversification of purpose. Indeed, the gradual reform of the CAP, the development of enhanced regional structural funding, and the uneven growth of regional development agencies in adopting and adapting to rural development objectives, has given growing salience to the concept of multi-level governance of the European rural domain at the same time as, as we see above, the declining powers of agricultural productivist corporatism. The aim of giving more regional and member state subsidiarity with regard to setting rural development objectives has been embodied in the development of the new rural development plans and 'second pillar' of CAP – the rural development regulation. 'Pillar Two' signals more support for sustainable agriculture (see Chapter 14), rural development and other public goods; but as well as a greener CAP, it also signals a more devolved CAP, in the sense that it creates the opportunities whereby regions could design strategies attuned to their own needs. These trends are creating considerable excitement among both the rural academic and policy-making community. But they have yet to fully come to fruition; and it has to be said that funding represents still only a small proportion of the foreseeable CAP total spend. Also, a considerable inert influence remains for the many farmer and producer organizations across Europe, as these tend to see these trends, especially in the context of European enlargement, as a financial zero-sum game; that is, a further threat to and retreat from their traditional sectoral powers at national and EU level, and their abilities to draw down funds for agricultural production *per se*.

Nevertheless, the introduction of 'Pillar Two' at the EU level, the reorientation of Pillar One towards single farm payments based on a variety of food quality and environmental conditions, combined, for instance, with the advent of more devolved regional administrations in the UK (Welsh Assembly, Scottish Parliament, English Regional Development Agencies), is creating a new multi-level policy-making process in the context of a reformed agricultural and rural development policy. And these trends reinforce what Keating (1997) defines as three major regionalization trends which are currently occurring in Europe: 'functional integration', 'institutional re-structuring' and 'political mobilization'. In short, in a variety of ways and for a variety of different reasons, political movements seek greater regional autonomy within or from the supra-national and nation-states:

> there are integrative regionalisms, seeking the full integration of their territories into the nation and the

> destruction of obstacles to their participation in national public life. There are autonomous regionalisms seeking a space for independent action, and there are disintegrative regionalisms, seeking autonomy or separation. (Keating, 1997: 389)

Meanwhile:

> modern developmental policies put more emphasis upon indigenous growth, or the attraction of investment by qualities linked to the region such as environment, the quality of life or trained labour force, rather than on investment incentives provided by the central state. (Keating, 1997: 384)

For Jessop this is leading to a general structural trend towards the 'de-statization' of politics, whereby:

> the overall weight of government and governance has been altered within the overall political system in favour of governance on all levels and, major trans-territorial and inter-local governance mechanisms have also been instituted at regional and local level. Among other phenomena, this is reflected in a growing emphasis on 'partnerships' between governmental, para-governmental and non-governmental organizations in which the state apparatus is often, formally at least, little more than *primus inter pares*. (1997: 304–5)

The steady reform of the CAP (1992–2003) is beginning to stimulate further rural regionalization. This is particularly the case in the UK, where it has developed alongside a process of regional devolution; something which has been a local and regional reaction to the centrist neo-liberal Thatcher years (1979–1997). Ongoing research shows this to be a rich area for comparative rural policy research (see also Chapter 3 in this volume). In very general terms, there are significant differences across rural regions in Europe, as to how this is playing out with regard to governance, philosophy and policy delivery. For instance, the rural policy-making process in Wales has now at its heart the National Assembly Government and a network of Assembly-sponsored bodies. In English regions such as the South-West, the responsibility for agricultural, environmental and rural development issues is largely divided between UK Ministry DEFRA, the Regional Development Agencies and the government Regional Offices. In Italy, by contrast, the regions are historically more autonomous from the central government in the design and implementation of agro-food policy and are responsible for designing regional policy as well as delivery of financial resources. Rather paradoxically, then, just as we witness a stronger move towards levels of European harmonization and integration in areas of food safety and quality policy, almost the opposite is happening with regard to rural and regional spatial policy. The latter is now having to take the notion of the regionally differentiated countryside (Murdoch et al., 2003) seriously as a basis for policy delivery, while the former still clings to a more top-down Euro-centric approach. These new regional mobilizations are becoming important bases for the emergence of alternative agro-food initiatives as a basis for new forms of innovation. This is in many ways despite rather than because of the growing Europeanization of much policy-making and the reliance upon member state national structures to mediate these policies.

The arrival of this more malleable lateral multi-level governance framework for European rural development in some respects becomes more problematic for both central state and corporate interests. The latter become increasingly committed to the advantages of a centralized Brussels bureaucracy channelling legislation which lubricates and ensures the workings of the transnational European internal food market. Agro-food and large retailers support EU food regulations which can transgress the former uneven national forms of regulation; and, as we see with issues of food labelling and the commercialization of GM crops, the corporate interests prefer to articulate with central European agencies (like DG SANCO and Trade), so as to force a common policy which supports their globally competitive ambitions. However, as we see here, this is increasingly problematic, given the emerging differential degrees of subsidiarity unleashed both from the centre, through CAP reform and regional quality labelling etc., and, regionally, by increasingly regionalized politics and new institutional forms. It becomes increasingly unlikely, for instance, that all regions will accept the commercialization of GM crops even if they do not seem to have a choice with regard to the threshold levels being set in Brussels with regard to GM in animal feedstuffs and a range of food products (Carson and Lee, 2005). Hence, the arrival of a more multi-level governance system gives some potential agency to regions, despite the countervailing centrist and harmonizing pressure. But the contradictory centralizing tendencies in other spheres of agro-food EU policy- making make this a highly complex and contingent picture.

The economic and social bases of rural regionalization and differentiation

Rural and regional spatial differentiation is, however, far from simply a matter of politics and the political. As Amin and Thrift (1994) and other 'new regionalists' have recognized, the differentiation in the fortunes of territorial areas in the context of an increasingly globalized economy stems from the emergence of a 'post-Fordist' regime of flexible specialization and accumulation, in which high levels of economic growth derive from clusters of small and medium-sized firms strongly linked to regional institutions (Amin, 1999). Such clusters, it is argued, can achieve rapid growth because they can more easily facilitate learning and innovation than the more traditional hierarchical or atomized production structures. As firms cluster more closely together through rich arrays of network linkages, so the economic region comes into view.

In rural regions, combinations of new private and public sector service industries, many of them tied to the growth in the 'consumption' or 'post-productivist countryside', sit adjacent to more externally managed and traditionally Fordist agro-food processing firms. Nevertheless, regional authorities adopt this logic of new regionalism and increasingly support innovation and R&D on a regional basis. This is perhaps most evident in the highly technologically driven regional policies of Sweden and Finland, but it is also now prevalent across much of North and Western Europe. As Saraceno identifies, this process is reducing the salience of a classic post-war urban–rural spatial division of labour:

> more recently, with diffused industrialization or other forms of decentralization, the rural/urban division of labour did not work in the classical pattern: different kinds of exchange – of labour, entrepreneurial capacities, capital, goods and services – between different sectors of activity, including agriculture, took place in the same geographical space, occasionally concentrating non-agricultural activities in urban centres, but not necessarily doing so. (1993: 451)

In a series of research monographs over the past decade, colleagues and I have researched this process of rural differentiation in the UK (see Marsden et al., 1993; Murdoch and Marsden, 1995; Murdoch et al., 2003). Comparing rural regions like North-East England (Northumberland), South-East England (Buckinghamshire) and South-West England (Devon), this long-term research programme has sought to combine a new regionalist perspective with the more sociological perspectives on regionalism (see Paasi, 1991). This highlights the more historically embedded and contingent nature of regulation, conventions and networks in the shaping of rural regions. What is clear from this research (see Murdoch, Chapter 12 in this volume) is the growing confluence of regional governance, economic and social processes which are mobilizing more differentiated and spatially uneven forms of development. Indeed, regional and intra-regional disparities are growing in the UK generally, and location becomes more important with regard to the 'bundles of rights' rural residents can expect to hold and utilize. Regional authorities also have to increasingly deal with intra-regional variations. In South-West England, for instance, the Government Office of the South-West and the Regional Development Agency are developing agricultural, rural and broader economic development strategies in attempting to connect the internal geographies of the region. Winter (2003) (Figure 1.3) now illustrates this deeper complex and multi-sectoral process of new regional building with regard to agriculture and rural development. New layers of stakeholder complexity are emerging and raising profound questions as to the role of the central UK state in aspects of rural development (see Haskins, 2003).

In this contemporary context, the regional and differentiated countryside is now a *contingent countryside*; one which is unevenly dealing with combinations of internal and external socio-political factors, which in turn affect their ability to compete and perform within the wider national and European governance frameworks. The regionalized ruralities also represent *different compromises* between forms of mobility (see Murdoch, Chapter 12 in this volume) associated with different forms of production and consumption equations. For example, transportation, residential migration, tourism, flexible industrial and service sector strategies, and the socio-natural 'endowments' of given rural spaces (workforces, community buildings, ecological features, reconstituted craft and artisanal skills). These compromises affect the *differential performances of rural places* across Europe as they combine both reflexive and traditional practices and compete for economic activity.

For instance, in parts of the English countryside the modernizing impulses associated with agricultural developments of the second half of the twentieth century have now given way to a much more pragmatic and market-oriented approach which combines agriculture with broader notions of the diversified regional economy. Economic modernization now increasingly applies to rural and regional, rather than agricultural economies. At the same time today's differentiated countrysides are an expression of the seemingly relentless decentralization of (cash- and asset-rich) mobile middle-class populations which are socially as well as economically fuelled by the postmodern and post-urban desire to escape the seeming anomie of urban life (by creating 'new localisms'), even if they still rely upon it in cultural and career terms. As Murdoch et al. conclude with reference to the UK:

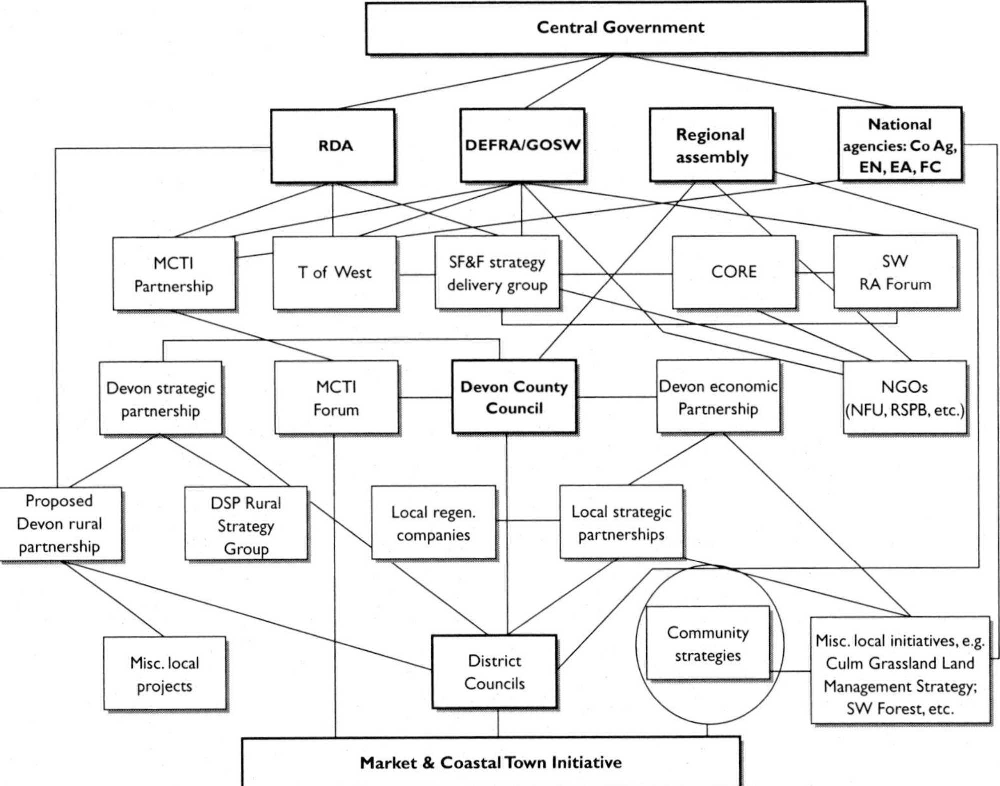

Figure 1.3 Sustainable farming and rural regeneration: a partial representation of the governance network in Devon (Source: Winter, 2003)

> A major driving force in the contemporary processes of change thus concerns the quest on the part of many individuals and families for the rural, both as an experience (in the form of countryside tourism) and as a 'refuge' from modernity (in the form of a house in the country). In this sense the cultural significance of the traditional 'urban–rural dichotomy' is alive and well. However, this broad spatial distinction is also being regionalized. Thus, the differential and multiple imposition of diverse values, codes and conventions in differing social and economic spaces is taking place within new regional contexts. (2003: 152)

Hence there are social, economic as well as political drivers to the recomposition of the diverse regionally based countrysides. As such, there are multitudes of differentiated countrysides all of which show contingent as well as path-dependent developmental features. These become less easy to govern in the traditional Foucauldian sense in that they are less 'regime like' and less prone to classic notions of nationally or internationally based 'governmentality', whereby central control acts at a distance to shape borders and problemizations. Today, in much of Europe's postmodern countryside the social and economic fabrics of rural places are just too fluid and mobile for this to hold. As we have seen above, this does not stop central states or their corporate supporters trying. And so one of the key macro-governance tensions rural areas in Europe face continually lies in the territory between centralist Foucauldian government of conventional agro-food and more regionalist 'consumption countryside' contingencies.

CONCLUSIONS: CONVERGING AND DIVERGING PATHWAYS IN RURAL STUDIES

The sociology of rural knowledge, as illustrated here by these two brief examples, now incorporates a much broader social science agenda. It is as associated with the *social science of risk* (as in the case of food regulation) as it is with understanding the new social, spatial and economic anatomy of regions and localities caught up in a postmodern world. This is being achieved in the context of several broader paradoxes (as described above), particularly if one takes a retrospective perspective with regard to the earlier concerns in rural studies. A refreshing component of these recent developments has been that contemporary rural research has become far more 'de-coupled' from the past institutional, conceptual and empirical rigidities, and now has a confidence to both incorporate and develop broader social science conceptualizations and political and economic frameworks which were hitherto deemed to be outside the ambit of the rural researcher's concern (for example, like food governance), the political economy of risk and the sociology and geographies of consumption.

However, while the earlier three paradoxes suggested in the introduction stand as important drivers for this more empowering re-orientation of rural studies, we can also posit that the two critical dimensions outlined here are influencing the sociological and institutional framing of new rural research questions. In many ways, for instance, we might consider one fulcrum of critical rural studies lying at the very intersection between the macro and geo-political regulation of food systems, on the one hand, and the more spatialized and regionalized regulation of local rural space on the other. This conceptual location – that is, one which critically observes the playing out of complex and often contradictory regulatory tendencies which are re-shaping the very nature of ruralities and rural identities – provides at least one important space for developing further both the critically interpretive as well as normative aspects of the field (see Marsden, 2004). Moreover, with regard to the latter, we begin to see many rural researchers adopting a more critical and more programmatic stance to the question of not only understanding 'what is' but also 'what might be', given the opportunities for continency and potential autonomous social and economic development such a location often suggests.

I have here, however, only related these developments and tendencies to the European context. And we have to recognize that this is increasingly a somewhat partial exercise. North American political economy, as many of the succeeding chapters in this Handbook testify, provides a significantly diverging neo-liberalist formation (for instance, with regard to agricultural, environmental and alternative rural development policy). Increasingly, it will be necessary to conduct geopolitical as well as regional and local comparative analyses in order to capture the divergent projects associated with these neo-liberal and neoregulationist geographies. More divergent pathways of rural development will require further external engagement within and beyond the social sciences.

NOTES

1 Council Directive on public access to environmental information (90/313/EEC, OJ L 158, 26 June 1990, pp. 56–8).
2 Code of Conduct concerning public access to Council and Commission documents (93/730/EC, OJ L 340, 31 December 1993, pp. 41–2).
3 Directive of the European Parliament and of the Council on the protection of individuals with regard to the processing of personal Data and on the free movement of such data (95/46/EC, OJ L 281, 23 November 1995, pp. 31–50).

REFERENCES

Amin, A. (1999) 'An institutionalist perspective on regional economic development', *International Journal of Urban and Regional Research*, 23: 365–78.

Amin, A. and Thrift, N. (1994) 'Living in the global', in A. Amin and N. Thrift (eds), *Globalisation, Institutions and Regional Development*. Oxford: Oxford University Press. pp. 1–11.

Andersson, K. (2003) 'Regional development and Structural Fund measures in two Finnish regions', in K. Andersson, E. Eklund, L. Granberg and T. Marsden (eds), *Rural Development as Policy and Practice: The European Umbrella and the Finnish, British and Norwegian Contexts*, SSKH Skrifter No. 16. Helsinki: University of Helsinki. pp.

Beck, U. (1992) *Risk Society: Towards a New Modernity*. London: Sage.

Carson, L. and Lee, R. (2005) 'Consumer sovereignty and the regulatory history of the European market for genetically modified foods'. BRASS Working Paper, Cardiff University, Cardiff.

CEC (Commission of the European Communities) (2000) *White Paper on Food Safety*, COM (1999) 719 final. Luxembourg: Office for Official Publications of the European Communities.

Cohen, M. (1997) 'Risk society and ecological modernisation: alternative visions for post-industrial nations', *Futures*, 29 (2): 105–19.

Crow, G., Marsden, T. and Winter, M. (1990) 'Recent British rural sociology', in P. Lowe and M. Bodiguel (eds), *Rural Studies in Britain and France*. London: Belhaven Press. pp. 248–62.

DTI (Department of Trade and Industry) (1993) *Review of the Implementation and Enforcement of EC Law in the UK*. London: Department of Trade and Industry.

DTI (Department of Trade and Industry) (1994) *Getting a Good Deal in Europe: Deregulatory Principles in Practice*. London: Department of Trade and Industry.

Folkerts, H. and Koehorst, H. (1998) 'Challenges in international food supply chains: vertical co-ordination in European agribusiness and food industries', *British Food Journal*, 100 (8): 385–8.

Giddens, A. (1994) 'Living in a post-traditional society', in U. Beck, A. Giddens and L. Scott, *Reflexive Modernisation: Politics, Tradition and Aesthetics in the Modern Social Order*. Cambridge: Polity Press. pp. 56–109.

Grijp, X. van der and Marsden, T. and Convalcanti, J.S.B. (2005) 'European retailers as agents of change towards sustainability: the case of fruit production in Brazil', *Environmental Sciences*, 2 (1): 31–46.

Haskins, C. (2003) *Rural Delivery Review: A Report on the Delivery of Government Policies in Rural England*. London: Department for Environment, Food and Rural Affairs.

Jessop, B. (1997) 'Survey article: the regulation approach', *Journal of Political Philosophy*, 5 (3): 287–326.

Keating, M. (1997) 'The innovation of regions: political restructuring and territorial government in Western Europe', *Environment and Planning C: Government and Policy*, 15: 383–98.

Keil, R. (2000) 'Natural spaces? Governing spaces of nature in the new economy'. Paper presented at the Annual Conference of American Geographers, Pittsburgh.

Lang, T., Barling, D. and Caraher, M. (2001) 'Food, social policy and the environment: towards a new model', *Social Policy and Administration*, 35 (5): 538–58.

Lucus, C., Hart, M. and Hines, C. (2002) 'Looking to the local: a better agriculture is possible'. Discussion document, European Parliament, Brussels.

Marsden, T.K. (2004) 'The quest for ecological modernisation: re-spacing rural development and agri-food studies', *Sociologia Ruralis*, 44: 129–46.

Marsden, T.K., Lowe, P. and Whatmore, S. (1990) *Rural Restructuring: Critical Perspectives on Rural Change*, Volume 1. London: Fulton.

Marsden, T., Murdoch, J., Lowe, P., Munton, R. and Flynn, A. (1993) *Constructing the Countryside*. London: UCL Press.

Marsden, T.K., Flynn, A. and Harrison, M. (2000) *Consuming Interests: The Social Provision of Foods*. London: UCL Press.

Marsden, T.K., Renting, H., Banks, J. and Ploeg, J.D. van der (2001) 'The road towards sustainable agricultural and rural development', *Journal of Environmental Policy and Planning*, 3 (2): 75–83.

Mormont, M. (1990) 'Who is rural? Or, how to be rural: towards a sociology of the rural', in T. Marsden, P. Lowe and S. Whatmore (eds), *Rural Restructuring: Global Processes and Their Local Response*. London: Fulton.

Murdoch, J. and Marsden, T. (1995) *Reconstituting Rurality: Class, Community and Power in the Development Process*. London: UCL Press.

Murdoch, J., Lowe, P., Ward, N. and Marsden, T. (2003) *The Differentiated Countryside*. London: Routledge.

Newby, H. (1980) 'Rural sociology: trend report', *Current Sociology*, 28 (1).

Newby, H., Bell, C., Rose, D. and Saunders, P. (1978) *Property, Paternalism and Power*. London: Hutchinson.

Paasi, A. (1991) 'Deconstructing regions: notes on the scales of spatial life', *Environment and Planning A*, 23: 239–56.

Ploeg, J.D. van der, Long, A. and Banks, J. (eds) (2002) *Living Countrysides. Rural Development Process in Europe: The State of the Art*. Doetinchem: Elsevier.

Prodi, R. (2002) 'Foreword', in J.D. van der Ploeg, A. Long and J. Banks (eds), *Living Countrysides. Rural Development Process in Europe: The State of the Art*. Doetinchem: Elsevier. p. 3.

Richardson, J. (2001) 'Policy-making in the EU', in J. Richardson (ed.), *European Union: Power and Policy-making*. London: Routledge.

Saraceno, E. (1993) 'Alternative readings of spatial differentiation: the rural versus the local economy approach in Italy', *European Review of Agricultural Economics*, 21: 451–74.

Skogerbo, E. (1997) 'External constraints and national resources: reflection on the Europeanisation of communications policy', *Telematics and Infomatics*, 14 (4): 383–93.

Smith, E. (2002) 'Ecological modernisation and organic farming in the UK: does it pay to be "green"?'. PhD thesis, Cardiff University, Cardiff.

WHO (World Health Organization) (1999) *Food Safety: Report by the Director-General*, EB105/10. Geneva: World Health Organization.

Winter, M. (2003) 'The changing governance of agriculture and food: a regional perspective'. Inaugural Lecture, 26 March, School of Geography and Archaeology, University of Exeter, UK.

2

Conceptualizing rurality

Paul Cloke

INTRODUCTION: CHANGING RURALITIES

The idea of rurality seems to be firmly entrenched in popular discourses about space, place and society in the Western world. Although the precise nomenclature devoted to the idea is often context-specific – witness the sometimes subtle but always important differences in terms such as rural, countryside, country, wilderness, outback, agricultural and so on – the concept of rurality lives on in the popular imagination and everyday practices of the contemporary world. The rural stands both as a significant imaginative space, connected with all kinds of cultural meanings ranging from the idyllic to the oppressive, and as a material object of lifestyle desire for some people – a place to move to, farm in, visit for a vacation, encounter different forms of nature, and generally practise alternatives to the city. Given the significance of its imaginative and material status, it is surprising how often we seem to lack an adequate understanding of how the concepts that underpin the idea of rurality should be defined and made relevant. It is almost as if the strength of the idea of rurality is in its overarching ability to engage very different situations under a single conceptual banner. Yet as soon as attempts are made to deconstruct the rural metanarrative, much of that conceptual strength dissipates into the nooks and crevices of particular locations, economic processes and social identities. Part of the issue here is that the distinction of rurality is significantly vested in its oppositional positioning to the urban. While cities are usually understood in their own terms, and certainly without any detectable nervousness about defining or justifying that understanding, rural areas represent more of a site of conceptual struggle, where the other-than-urban meets the multifarious conditions of vastly differing scales and styles of living.

In this chapter I want to survey some of the different ways in which rurality has been framed conceptually, signposting along the way some potentially fruitful and imaginative ways of exploring further the mysterious cartographies of the rural. To begin with, however, it is important to emphasize that understandings of rurality are influenced by twin tracks of changing perspectives – not only do we need to survey how different theoretical frames illuminate very different pictures of rurality (and indeed steer rural research down very different pathways) but we also need to be fully aware of the (sometimes rapidly) changing conditions of rural life, rural place and rural political economy which together constitute important shifts in the material manifestation of rurality. The changes occurring in rural areas themselves are such that even a consistent theoretical frame will need to cope with considerable dynamism within its rural subject.

Many of these rural changes are discussed and problematized in the individual chapters of this Handbook, but it warrants emphasis at this point not only that rural change has constituted a blurring of conventional boundaries between country and city, but also that such blurring works in both directions, indicating an urbanization of the rural and (albeit to a lesser extent) a ruralization of the urban. Urbanizing the rural has occurred via an interwoven tapestry of cultural, social and economic trends. The urbanization and indeed globalization of cultural dissemination through broadcast and print media and especially the

Internet, means that most seemingly rural places in the Western world are effectively culturally urbanized. Although distinctive cultural traits are formed in particular globalizations of the local and localization of the global in rural areas, the all-pervading messages of Hollywood, MTV and Google mean that the idea of rurality as an isolated island of cultural specificity and traditionalism has become anachronistic. At the same time, over the past 30 years there has been a hugely significant influx of urban populations into rural locations. In-migrants have been attracted to rural locations because of the perceived advantages of rural lifestyles, yet at the same time they bring with them key attributes of urban living and levels of expectation which often serve to transform the very communities they had been attracted to. Demographic change has both shaped and been shaped by economic change. As the scale of agricultural workforces has diminished, the traditional dominance of the agricultural economy in rural areas has gradually shifted in emphasis from landscapes of production to landscapes of consumption. Economic diversity has been fuelled first by urban-to-rural shifts in manufacturing, and then by new forms of service sector activity, prompted in part by the ability of telecommunications, information technology and increases in personal mobility to 'shrink' the geographic distances between city and country.

These generalizations, of course, mask very considerable variations in and between nations, where different scales and cultures of urban–rural differentiation exert different pressures on these broader processes of blurring. However, such variations notwithstanding, it is easy to agree with Mormont (1990) that the changing relationship between space and society has rendered traditional divisions between rural and urban increasingly indistinct. Rural society and rural space can no longer be seen as welded together. Rather, rurality is characterized by a multiplicity of social spaces overlapping the same geographical area, so while the geographic spaces of the city and the countryside have become blurred it is in the social distinction of rurality that significant differences between the rural and urban remain.

Accounts of the ruralization of the urban have received less emphasis in explaining how town and country distinctions have become blurred. However, two illustrations suffice here to support the argument that rural change does not simply imply a takeover of the rural by urban values and forms. The first is drawn from Wilson's (1992) account of how recent land development in North America has produced suburbs, shopping centres, theme parks, executive estates, tourist development and the like which destabilize ideas about city and country by producing city/country hybrids which owe as much to a bringing-nature-into-the-city as to a spreading-the-city-into-the-country. He illustrates this idea with an account of the West Edmonton Mall in Canada, a 45 hectare suburban indoor shopping centre which includes a one-hectare lake replete with dolphins and sharks, an 18-hole golf course, a water park with six-foot surfing waves and a zoological collection of animals in cages and aquariums. Such developments cannot simply be dismissed as a colonizing commodification of rural nature into urban forms, as the very presence of pseudo-rural landscapes, creatures and practices opens out imaginative spaces of the rural in these hybrid settings.

The second example can be found in the arguments deployed by Urbain (2002) about the ruralizing of the metropolis. Despite the emphasis on how counter-urbanization has urbanized the countryside, Urbain insists that the spread of the city out into the country has effectively ruralized a very significant part of the urban. Given that the nature of the city has been radically changed, both by centralizing tendencies and by decentralizing practices, it can be argued that an important slice of contemporary urbanity can now be found in the village, and that the urban form thereby now encapsulates very strong rural characteristics and influences. Equally, urban managers seem increasingly to be striving for a set of virtues in the city which are more commonly associated with the rural – seemingly fundamental and permanent virtues such as protection, solidarity, community spirit and identity. According to these arguments, then, the blurring of rural–urban distinctions is bringing crucial changes to urbanity as well as to rurality. Such changes present any conceptual framing of the rural with considerable dynamic complexity as to the nature of the subject being framed.

RURALITY: THE SHIFTING THEORETICAL LENS

At least some of the changing narrative of rural studies has been framed around the use of different theoretical perspectives to make sense of and to define the essential characteristics of rurality. The ways in which such theoretical approaches make inroads into rural literatures varies significantly between different social science disciplines. It would not be unfair to suggest that some subject areas have witnessed a reluctance in the main to depart from broadly positivistic approaches (agricultural economists, with some notable exceptions, seem to illustrate this tendency), while others (and here geographers are an obvious example) have tended very readily to embrace alternative theoretical movements as they come along. For some rural researchers, then, the progressive introduction of new theoretical takes on rurality is nothing more than surrendering to the vagaries of conceptual fashion, while for others, fresh theoretical inputs have opened out new channels of inquiry and understanding about the complex nature of rurality.

In broad terms it is possible to recognize three significant theoretical frames which have been influential in constructing conceptualizations of rurality. The first, in some ways not explicitly theoretical at all but in other ways engendering implicitly theoretical assumptions, can be thought of as *functional* concepts of rurality. Here, the search has been to identify functional elements of rural place/landscape/society/existence which together provide an approximation of the overarching concept of rurality. In this manner, rurality can be defined in terms of areas which:

1 are dominated (either currently or recently) by extensive land uses, notably agriculture and forestry;
2 contain small, lower order settlements which demonstrate a strong relationship between buildings and extensive landscape, and which are thought of as rural by most of their residents;
3 engender a way of life which is characterized by a cohesive identity based on respect for the environmental and behavioural qualities of living as part of an extensive landscape (see Cloke and Park, 1984).

Underpinning these functional ideas, however, are a series of assumptions about the relationship between the rural and the urban. For example, some research in rural studies persistently conflates the rural with the agricultural, as if each is somehow interchangeable. There is a suspicion that such conflation merely brings easy settlement to the difficult conceptual issues raised by the blurring of city and countryside discussed above. At least an agricultural focus appears safe conceptually as agricultural production and landscapes (although not markets nor consumption) can confidently be confined to a 'rural' domain. However, these and other devices which assume a functional centrality for rurality often appear to reproduce implicit forms of rural–urban dichotomy or continuum. Tonnies's (1957) theorization of *Gemeinschaft* (community) and *Gesellschaft* (society), which has been taken to reflect the intimate rural and impersonal metropolitan polarities of the dichotomous rural–urban relationship, has generally been superseded by the adoption of continuum models (Pahl, 1965), which suggest a sliding scale of differences between rural and urban poles. Despite strong warnings to the contrary (see, for example, Carlson et al., 1981; Lee and Newby, 1983), these loose concepts continue to underpin aspects of rural studies which see rural areas as functionally different to their urban counterparts. Sadly, empirical work conducted on this basis is often flawed because of arbitrary spatial boundaries of available data, or because of the arbitrary nature of supposed indicators of rurality (see, for example, Cloke, 1977).

A second conceptual landmark is represented by the use of *political-economic* concepts to clarify the nature and position of the rural in terms of the social production of existence (Cloke, 1989). Here, what had been previously recognized as functional rural areas have been increasingly connected into the dynamics of national and international political economy which have often been seen to operate on an aspatial basis. Through this conceptual lens it became apparent that much of what happens within rural areas is caused by factors operating outside the supposed boundaries of these areas. Rurality as an analytical category was desensitized in many of these discussions, and rural researchers were invited to 'do away with rural' (Hoggart, 1990) as an intellectual container and to seek out sectoral research that spanned across previous rural–urban distinctions. This emphasis on the power fields and apparatus of social production effectively sponsored a conceptual blurring of the rural and the urban, and reinforced the concern of rural research with particular sectors, for example, the food sector, which impacted on areas beyond the urban. During the 1980s, the *localities* debate in Britain further destabilized the spatial basis for rural studies, confirming as it did that although certain places achieve a kind of uniqueness associated with local society within broader processes of political economic restructuring, nevertheless rural places did not in general represent distinct localities:

> various critical notions – of different, overlapping spatial divisions of labour, of all localities as sites for the reproduction of labour-power, of variations in local social structures etc. – render problematic the notion that there are distinct 'rural' localities. (Urry, 1984: 198)

The adoption of political economic perspectives did not entirely unplug the research focus on rurality, however. The insistence (see, for example, Dunleavy, 1982) that to study 'rural' anything (and by implication 'urban' anything) was to misrepresent prevailing socio-economic structures was tested out by researchers such as Moseley, who did indeed conclude that

> the inner city and outer rural areas share certain problems relating to a declining or static population and economy and to the selective loss of certain kinds of people and jobs. (1980: 26)

However, he also suggested significantly that structural problems common to urban and rural areas are often manifest differently in rural areas – that there is a 'rural dimension' co-constituted by three basic characteristics:

1 A pleasant environment which will attract the willing or unwilling unemployed.
2 A 'spaced-out' geographical structure which leads to accessibility problems and costly public services.
3 A distinctive local political ideology which favours the market, the volunteer and the self-helper rather than public sector intervention.

So while many rural researchers grasped the political-economy initiative in order to focus on the changing nature of agricultural production, tracing the shift from Fordist to post-Fordist sets of social relations (Kenney et al., 1989, 1991; Sauer, 1990), others deployed regulationist ideas to investigate the idea of 'a rural dimension'. For example, Harvey's (1985) argument that the configuration of mode of regulation and societalization occurs as an ensemble of multi-faceted relations and institutions which produce 'structured coherences' *at particular places* and *at particular times* was taken up in the rural arena to investigate the idea of particular rural structured coherences (Cloke and Goodwin, 1992, 1993). The 'rural dimension', then, may not be incompatible with political economic approaches precisely because the particularities of time and place cannot be disconnected from the social construction of certain time-places as rural. While it is important to emphazise that the changing functions of rural areas are certainly not uniform or predictable, and that it is crucial to avoid over-generalization, there do seem to be grounds for proposing that the idea of a socially constructed rurality need not be incompatible with concepts that locate specific places and people in wider models of changing relations:

> Indeed, it may be crucial that the contracts and strategies of capital in altering institutional forms, networks and norms, the impact of the contesting of change in socio-political spheres, and the role of cultural factors as a glue in establishing locally coherent characteristics are *brought together* in our analysis rather than being regarded as belonging to separate philosophical domains. (Cloke and Goodwin, 1993: 174)

The third theoretical framing of rurality, therefore, involves *social constructions* of rurality, and draws on more postmodern and post-structural ways of thinking, especially about the role of culture in socio-spatial distinctiveness. Regarding rurality as socially constructed suggests that the importance of the 'rural' lies in the fascinating world of social, cultural and moral values which have become associated with rurality, rural spaces and rural life. Such an approach invites study of how practice, behaviour, decision-making and performance are contextualized and influenced by the social and cultural meanings attached to rural places. As a starting point, there has been significant interest in how idyllized meanings of rurality are constructed, negotiated and experienced (Bunce, 1994, 2003), and there is an emerging core of significance in rural studies which focuses on the interconnections between socio-cultural constructs of rurality and nature – which appear to be so important in the reproduction of geographical imaginations of rural space – and the actual lived experiences and practices of lives in these spaces. These practices and lives need to be examined both from the outside looking in (accounting for 'structuring' influences) and from the inside looking out (accounting for difference, identity and embodiment). Researchers have also become increasingly aware of the need to extend beyond the 'obvious' interconnections between social construction and rural practice. In his review of 'neglected' rural geographies in Britain, Philo (1992) contends that most accounts of rural life have viewed the mainstream interconnections between culture and rurality from the perspective of typically white, male, middle-class narratives. He points clearly to a need to explore other windows onto the rural world.

The meanings associated with rurality are not simply derived from differences between individuals and organizations. Other, vital, differences emerge from the divergences of rural society and rural space noted by Mormont above. Drawing significantly on the writings of Baudrillard, Halfacree (1993) identifies three levels of divergence which underpin the multiple meanings of 'rural' in contemporary society. He suggests that the *sign* (rurality) is becoming increasingly detached from the *signification* (meanings of rurality) due to the increasingly diverse social representations of rurality. Moreover, sign and signification are becoming increasingly divorced from their *referent* (rural geographical space). Thus symbolic notions of the rural have become detached from their referential moorings, meaning that socially constructed rural space has become ever more detached from geographically functional rural space, so much so that we might now regard rurality in terms of a 'post-rurality' (Murdoch and Pratt, 1997), in which consumers of the rural realize that rurality represents an inauthentic pastiche of meanings and symbols, but are nevertheless happy to go along with this post-modern condition. One logical outcome of social constructionist approaches to rurality, then, is the prospect of regarding villages, communities and landscapes as hyper-real commodities (Cloke, 1997). According to this view, the rural has become deterritorialized, as the meaningful signs and symbols of rurality have become increasingly detached from their referent geographical spaces, and reterritorialized as more abstract significations begin to define the essential nature of rural space. If at some time in the past some 'real' form of rurality was responsible for cultural mappings of rurality, it may now be the case that cultural mappings precede and direct the recognition of rural space, presenting us with some kind of *virtual* rurality.

CULTURAL TURNS/RURAL TURNS

For many rural researchers the espousal of social constructionism represents a turn to the cultural which has deflected rural studies away from its fundamental core of concern for socio-economic change in rural space. Others, however, would argue that rural studies have yet to embrace the full deconstructionist force of the cultural turn. Either way, significant doubts have been expressed about the intellectual and other kinds of dividends which have resulted from the rural embrace of the cultural. The early 2000s have been a critical period for reassessment of the cultural focus in social science, and such assessment can inform a more specific examination of the interconnections between rural studies and the wider cultural turn.

In terms of a general turn towards the cultural the reassertion of a spatial focus and an increasing engagement with different aspects of social theory have come together to foreground cultural questions of meaning, identity, representation, difference and resistance in social science. However, the detailed outworking of these emphases has produced multiple manifestations of the cultural turn, ranging from the increased use of cultural texts and a heightened reflexivity towards the role of language, meaning and representation in the constitution of 'reality', to the introduction of post-structural epistemologies which emphasize the close relationship of language, power and knowledge, or point to a non-representational engagement with the emergent (see Cloke, 2003). To evaluate *the* cultural turn in rural studies, then, is to attempt to pin down a moving multi-centred target.

There have, however, been notable attempts to provide a critique of the cultural turn (see Barnett, 1998; Cook et al., 2000), and while some critical commentary arises from the desire to promulgate a very different agenda (see, for example, Martin's (2001) dismissal of the cultural turn as intellectual dilettantism), others who are more sympathetic to the cultural project have generated an evaluative mantra which can perhaps be distilled into four principal claims.

1. The cultural turn has *desocialized* social science, withdrawing from the processes which are the stuff of everyday social practices, relations and struggles. The novel concerns with cultural difference and the new identity politics of representation have resulted in a turning away from research into the structures and spatialities and inequality. Gregson has termed this 'an evacuation of the social' (2003: 14), arguing that although the social has not been replaced by the cultural, it is nevertheless increasingly refracted through the cultural. Smith (2000) goes further, suggesting that the cultural has usurped the social as basic social categories of race, class and gender have been recast as subjectivities and identities.
2. The cultural turn has *dematerialized* social science, through its preoccupation with immaterial processes, the constitution of intersubjective meanings and the outworking of identity politics through texts, signs, symbols and emotions. The result, in Philo's terms, is a social science which has become 'less attentive to the more thingy, bump-into-able, stubbornly there-in-the-world kinds of matter, and the diagnosis is a re-emphasis on reclaiming the materiality of the everyday world' (2000: 13).
3. The cultural turn has *depoliticized* social science. Just at a time when the forces of the

political and economic right wing have gained ascendancy, the cultural turn has appeared to reroute research away from the analysis of, and intervention in socio-political struggles. As Mitchell (1995, 2000) argues, much of the post-structuralist debate within the cultural turn has resulted in forms of political quiescence and academic intellectualizing, a move which he regards as a surrender to the forces of reaction and a squandering of intellectual resources.

4 The cultural turn has been insufficiently *deconstructionist*. This fourth claim departs from the previous three in suggesting that there has been an undue conservatism in the cultural foci adopted by social science, which remains dominated by constructionist themes and approaches. Thrift (2000), for example, regards these emphases as 'tired', arguing instead for a closer engagement with non-representational approaches which point to the imminent and the performative in cultural studies as models for wider social science.

How, then, does this fourfold critical commentary map onto rural studies? Has rural research reflected the excitements and achievements of the cultural turn? Have rural researchers participated in the process of development and responding to critiques of the cultural focus. In a 1997 editorial for the *Journal of Rural Studies* I tried to convey some of the excitement and challenge being generated by a resurgent rural studies which had begun to get into the flow of the cultural turn, concluding that,

> I believe that we are now experiencing the most exciting period in rural studies, certainly within the last 20 years of my own engagement with the subject. (Cloke, 1997: 371)

Evidence of this intellectual excitement reflected the potential for rural studies of reconceptualizations of nature–society relations, heightened sensitivity to discourses of rural experience and imagination, incisive reconsiderations of the symbolic texts of rural cultures, and an emergent emphasis on the mobilities and fluidities (rather than the fixities) of rural life and landscape. Seven years later I remain convinced that elements of the cultural turn can be linked with some very significant contributions to rural studies in these and other areas. Witness the focus on nature–society relations in the countryside, with the theoretical and conceptual platforms provided by actor-network theory (ANT) (see, for example, Murdoch, 1997, 1998, 2001) and hybridity (see Whatmore, 2002; Murdoch, 2003) framing innovative insights into the relational contribution of non-human actants to the networks and places of the rural milieux. Clearly one trajectory of ANT conceptualization is its focus on how networks transcend space and time, demonstrating how rural actants are implicated in far-flung and emergent comings-together which can by no means be described as rural in their totality. Another spin-off, however, is the use of *dwelling* concepts (Ingold, 2000; Jones and Cloke, 2002; Wylie, 2003) to present understandings of how animals and plants co-constitute particular places, including rural places. The intellectual excitements of the cultural turn can also be witnessed in other areas of rural research, including both the use of imaginative texts to investigate representations of rurality, and the increasing importance of discursive understandings of rural aesthetics and of rural poetics. Witness also the important new emphases of identities and subjectivities relating to rural masculinity/femininity, sexuality, disability and childhoods, and the broader desire to understand otherness in the rural context (Cloke and Little, 1997). Many of these themes are detailed in subsequent chapters of the Handbook.

Reflecting over the past seven years, however, I want to suggest that my initially very enthusiastic editorial piece failed to appreciate some significant facets of the cultural turn which have emerged over the intervening period. These concerns can be summarized in six propositions. First, it seems potently clear that much of rural studies has carried on as before, effectively untouched by the cultural turn and continuing to rely on untheorized, positivistic or materialist approaches to the understanding of rural issues. Second, with some notable exceptions, the take-up of key tenets of the cultural turn by rural researchers seems to have been relatively half-hearted (in much the same way, for example, as Little (2002) has described the espousal of both feminist perspectives and geographies of embodiment in rural studies). That is, evaluation of the literatures and conferences which constitute the production and display of knowledge in the rural studies arena seems to suggest that rural researchers have been less convinced by, and are less willing to commit to the cultural turn than has been the case in other intellectual arenas. Third, in any case accounts both supportive of and critical of the cultural turn implicitly suggest that the cultural turn has principally been about cities – about re-imagining, re-mapping and re-populating the urban. It can be argued that the cultural turn has only gained purchase on what might be regarded as *rural* issues when such issues intersect with more generic concerns for landscape, nature, environment, leisure, resistance and so on, none of which can be claimed as a specifically rural domain. Again with some notable exceptions, most studies inspired by the cultural turn have taken place quite deliberately *outside* the perceived intellectual boundaries of rural studies, regardless of any incidental or continued overlap with material, social or intellectual spaces for rurality. Fourth, as a result, although post hoc rationalizations by rural researchers might regard the corpus of cultural conceptualization as being intimately bound up with resightings of or in the rural, broader discourses of the cultural turn seem much more likely to place this corpus of work as 'somewhat adjacent to', 'bypassing', or even 'undermining of' rurality as an intellectual or spatial category. Fifth, these constructions of the ambivalence or irrelevance of the cultural turn to rurality have arisen at least in part because rural studies researchers have *both* broadly failed to establish the key wider significances of their work (and may indeed have come not to believe in the idea of rural any more) and been content to deploy the theoretical matrices of the cultural turn within rural settings rather than making theory which perhaps more critically and evenly posits the interconnectedness of society, space and nature with rurality. Lastly, even where the core ideas of the cultural turn have been deployed wholeheartedly in rural arenas, the outcomes remain vulnerable to the critique which perceives the cultural turn to be desocializing, dematerializing, depoliticizing and maybe even insufficiently deconstructionist.

CONCEPTUALIZING RURAL HYBRIDITIES

This critical and somewhat pessimistic review of interconnections between rurality and the conceptual forces of the cultural turn could provide a platform from which to retreat into conventional and uncritical ruralism. To do so in my view would be to miss out on some of the important conceptual opportunities presented to rural studies during the cultural turn. So rather than acceding to a self-fulfilling critical demolition of culturalism and the rural, I prefer to use the experience of the cultural turn to pose two crucial conceptual questions, which are, I believe, central to the continuing liveliness of conceptualizing rurality. First, if we are willing to accept the claims of the cultural turn with regard to reasserting the importance of space, is there anything that can be said to the world outside rural studies about how the hybridities of what we might regard as *rural* space are especially relevant in the interconnection of things, places and people? If so, is it possible to rematerialize, resocialize and repoliticize our understandings of the coming-together of rural space?

In these respects I am optimistic about the prospect of rural researchers seeking out new ways of mapping these rural comings-together. Although there is an obvious requirement here to take full cognisance of the many different rurals, the many different layers of space and the many different reasons why it is appropriate to consider different versions of the post-rural, nevertheless interesting conceptual pathways are emerging by which narratives of hybrid rural spaces can be constructed. One example is Halfacree's deployment of Lefebvrian ideas of representations of space, spaces of representation and spatial practices in order to emphasize rural space as a socially produced set of manifolds (see Chapter 4 in this Handbook). Here, it seems fruitful to bring together material and imaginative conceptions of rural space through their intersections in particular practices. Rather than understanding material, imaginative and practised ruralities as somehow separate, it is possible – indeed seemingly strongly advisable – to see them as intrinsically and dynamically intertwined and embodied with 'flesh and blood' culture and with real life relationships. Part of the task for rural studies, then, is to identify key practices with which to express both internal and external connections between the material and imaginative worlds of the rural.

An alternative conceptual pathway is to follow Deleuzian ideas through which rurality can be expressed in the folded relations between rural reference and rural experience (see Dewsbury, 2003). Rurality can thus be envisaged as a complex interweaving of power relations, social conventions, discursive practices and institutional forces which are constantly combining and recombining. Whatmore's (2002) pioneering work on hybrid geographies has deconstructed nature–culture binaries in its account of how non-human beings, materials, discourses and knowledge combine with human agency in hybrid collectives or relational being and becoming. Although initially played out in the study of nature–society relations, such perspectives on hybridity also allow us to identify overlooked spatialities emerging out of the intersections between culture, economy, biology, planning, governance and so on. As Amin and Thrift (2002) have argued for cities, so we can begin to conceptualize our approach to non-city spaces by seeking to map the intermesh between flesh and stone, humans and non-humans, fixtures and flows and emotions and practices. Part of the task here will be to name neglected spatialities, and to invent new ones which in time help to repopulate the rural; that is to recognize through ideas of hybridity all manner of strange cartographies, networks, fluidities and blank figures. In this way our understandings of rurality can become more open and crosscut by different relations and rationalities, emerging out of the crashing together of myriad practices and performances.

These hybrid approaches seem well capable of rematerializing and even resocializing our cultural understandings or rural spaces, but the question persists as to whether they also permit a repoliticizing of these understandings. Their advocates suggest that such hybridities spawn their own rather innovative cultural ethics, cultural politics and aesthetics of immanent hope – what Thrift has termed 'a politics of the creation of the open dimension of being' (2004: 92), and a 'politics of a generous sensibility that values above all the creation of "joyful encounters" which can boost the powers of all concerned' (2004: 96). Others will be less than fully content with hybridities that are only able to reflect on political power as an affect of relational encounter, preferring to question how particular actors or collectives struggle to impose (explicitly or implicitly) versions of reality on others, for example by establishing problematization, stabilizing identity, enrolment, mobilization and so on. In other words, there will be pressure to take particular interest in hybridities that reveal the ways of the powerful (Murdoch and Pratt, 1993, 1997).

A second crucial conceptual question is this: if we are willing to accept the claim of the cultural turn about social theories of difference and identity, is there a danger that by espousing identity politics we will overlook, trivialize or even reinforce vitally problematic social issues in rural spaces? Here I am less optimistic. Swyngedouw (1995) has argued cogently that French intellectuals (Baudrillard, Foucault, de Certeau, Deleuze and so on) have always been implicated – directly, clearly, actively – in the wider politics of place. By contrast, the deployment of French intellectual thought by British and US academics as part of the cultural turn, he argues, has been solely in the theoretical imagination. As a consequence, it appears that the conceptual core of the cultural turn may have mislaid its constitutive contextualization of politics and place. It may even be that in blowing away the cobwebs of convention, conservatism and prejudice we may inadvertently have turned a commitment to emancipatory social practice and politics into a commitment of the political empowering of pleasure.

In deploying this thought as a litmus test of the potential depoliticizing power of the cultural turn in rural studies, two broad trends emerge. First, in many ways it seems that rural policy and politics have been leading the academic community rather than the other way around. For example, although key members of the rural studies academic community have been drafted into rural policy-making processes in the UK, it seems fair to suggest that the rural policy agenda is responding to the politics of countryside unrest, to crises such as that posed by foot and mouth disease, and to the broader post-productivisms of agriculture. We are in a phase where the policy focus rests on the natural economy and its commodified products and consumptions, and although such a focus is not in theory incompatible with the concerns of the cultural turn, in practice there have been considerable challenges in connecting work inspired by the cultural turn to these particular foci. This is especially so because the current policy agenda relies on traditional epistemologies and fixed binary differentiations, for example between town and county, and between production and consumption. What is clear is that there has been a remarkable lack of interest in the politics of the social. Secondly, where policy *has* connected with concepts that are more familiar in the approaches of the cultural turn, there is a suspicion that the connection has been pragmatic rather than a dynamic reclamation of lost constitutive connections of politics and place. One such example is offered by the adoption of social exclusion as a conceptual tool for understanding rural problematics. Notwithstanding some interesting and informative attempts to map out social exclusion in rural areas (see, for example, Shucksmith and Chapman, 1998) it can be argued that the brand of identity politics more generally represented by social exclusion concepts illustrates how the 'easy' adoption of a concept may actually hinder our grasp of rural problematics. In particular, the relativist positioning of exclusion seems to have replaced important previous understandings, for example as seen through the turns of rural poverty, marking out a prime example of how the re-imagination of neglect can lead to the neglect of what has already been imagined. In other words, we should perhaps be more careful about discarding old ideas unnecessarily just because they are old ideas. In the context of the rural UK, adoption of social exclusion concepts appears to have mystified rather than sharpened the priority needs for policy response, both in its broad focus on a wide range of identity politics and in its overshadowing of problematic inclusions and voluntary exclusions which in each case point to key sectors of rural place politics.

Together, these two broad trends suggest that where the cultural turn is deployed without accompanying critical analysis of power relations, it misses out on the potential impacts of emancipatory social practice and politics. However, when the conceptual fruitfulness of the cultural turn is pursued in conjunction with a more critical analysis of power relations there is a potential to add significantly to the broader understandings of, and critical importance of, rural policy agendas.

CONCLUSION: DOING THEORY IN RURAL STUDIES ...

The intellectual landscape of inquiry that is rural studies is formed of multifarious approaches to, and reactions to theorization. Some will claim with relish to be 'theory-free', while for others the progressive application of a particular theoretical frame provides both continuity and security from the vagaries of social science fashion. Yet others have been open to the challenges of deploying new perspectives as part of an iterative critical process which necessitates the laying down of previous ideas in order to adopt new ones. In reviewing the critical impact of the 'cultural turn' I am aware that for some individuals, academic contexts, and even entire disciplines, the cultural turn will have made little impact on the everyday conceptualization of rurality and rural change. Nevertheless there may be important generic lessons to be learned from the swing-to-the-cultural that has occurred elsewhere. It will be clear from the preceding critique that I believe that the depoliticizing tendencies of a cultural focus do warrant a re-emphasis of the politics of the social, the power relations of policy contexts, the spatialities and practices of ethics and the importance of committed performance and resistance. One route for such a re-emphasis would be to propose yet another 'turn' towards the political/material but an alternative is to find ways of repoliticizing rural studies without sacrificing the insights available from cultural approaches.

The orthodox response to this alternative proposal would be to insist that these different theoretical directions are incompatible – that you cannot hybridize theory. It seems ironic that some of those who are content with theories of hybridity seem unwilling to entertain the ideas of hybridized theory. It seems timely to recover the notion that 'doing theory' does not have to represent the swapping of one complete mindset for another, whether out of the lure of fashionability or out of some kind of totalizing intellectual critique. So maybe rural studies does not require yet more turns. Maybe we need to recognize that we dwell in a palimpsestual theoretical landscape, in which the most recent layers of ideas become eroded down to reveal their integral topographic relations with previous ideas. We may even need to engage in a palingenetic remining of previous theoretical resources so as to reveal their relation with 'new' concepts.

The way forward, then, in conceptualizing rurality may constitute a closer engagement with what Deleuze terms 'minor theory' (see Barnett, 1998; Katz, 1996; Philo, 2000), that is, doing theory in a rather different register which disrupts the binary relations between the theoretical and the empirical, which is far less totalizing, less judgemental, less certain, more fluid. In other words, rural studies would in my view benefit from theoretical reflection that is sufficiently relaxed to be able to recognize theory where it arises in unexpected forms and in unanticipated locations. This is not to advocate sheer pragmatism – conceptualizing rurality still has to be thought through rigorously – but is to suggest that this more relaxed form of 'minor theory' offers scope for easier and more effective theoretical hybridization which can combine, for example, the concerns of the cultural turn with those of political and economic materialism. Perhaps most importantly, such minor theory approaches will also enable rural studies to be a place where unexpected theory in unexpected forms can be *made* rather than simply deployed from other contexts. Conceptual export as well as the current conceptual import would certainly represent a significant marker of maturity in rural studies.

REFERENCES

Amin, A. and Thrift, N. (2002) *Cities: Re-imagining the Urban.* Cambridge: Polity Press.

Barnett, C. (1998) 'The cultural worm turns: fashion or progress in human geography?', *Antipode* 30: 379–94.

Bunce, M. (1994) *The Countryside Ideal.* London: Routledge.

Bunce, M. (2003) 'Reproducing rural idylls', in P. Cloke (ed.), *Country Visions.* Harlow: Pearson.

Carlson, J., Lassey, M. and Lassey, W. (1981) *Rural Society and Environment in America.* New York: McGraw–Hill.

Cloke, P. (1977) 'An index of rurality for England and Wales', *Regional Studies* 11: 31–46.

Cloke, P. (1989) 'Rural geography and political economy', in R. Peet and N. Thrift (eds), *New Models in Geography.* London: Unwin Hyman.

Cloke, P. (1997) 'Country backwater to virtual village? Rural studies and "the cultural turn"', *Journal of Rural Studies* 13: 367–75.

Cloke, P. (ed.) (2003) *Country Visions.* Harlow: Pearson.

Cloke, P. and Goodwin, M. (1992) 'Conceptualising social change: from post-Fordism to rural structured coherence', *Transactions IBG* NS 17: 321–36.

Cloke, P. and Goodwin, M. (1993) 'The changing function and position of rural areas in Europe', *Nederlandse Geografische Studies* 153: 19–36.

Cloke, P. and Little, J. (eds) (1997) *Contested Countryside Cultures.* London: Routledge.

Cloke, P. and Park, C. (1984) *Rural Resource Management.* London: Croom Helm.

Cook, I., Crouch, D., Naylor, S. and Ryan, J. (eds) (2000) *Cultural Turns/Geographical Turns.* Harlow: Prentice Hall.

Dewsbury, J-D. (2003) 'Witnessing space: "knowledge with contemplation"', *Environment and Planning A* 35: 1907–32.

Dunleavy, P. (1982) 'Perspectives on urban studies', in A. Blowers, C. Brook, P. Dunleavy and L. McDowell (eds), *An Interdisciplinary Reader.* London: Harper and Row.

Gregson, N. (2003) 'Reclaiming "the social" in social and cultural geography', in K. Anderson, M. Domosh, S. Pile and N. Thrift (eds), *Handbook of Cultural Geography.* London: Sage.

Halfacree, K. (1993) 'Locality and social representation: space, discourse and alternative definitions of the rural', *Journal of Rural Studies* 9: 1–15.

Harvey, D. (1985) *The Urbanisation of Capital*, Oxford: Blackwell.

Hoggart, K. (1990) 'Let's do away with rural', *Journal of Rural Studies* 6: 245–57.

Ingold, T. (2000) *The Perception of the Environment: Essays in Livelihood, Dwelling and Skill.* London: Routledge.

Jones, O. and Cloke, P. (2002) *Tree Cultures.* Oxford: Berg.

Katz, C. (1996) 'Towards minor theory', *Environment and Planning D: Society and Space* 14: 487–99.

Kenney, M., Laboa, L., Curry, J. and Goe, R. (1989) 'Midwestern agriculture in US Fordism: from new deal to economic restructuring', *Sociologia Ruralis* 29: 131–48.

Kenney, M., Laboa, L., Curry, J. and Goe, R. (1991) 'Agriculture in US Fordism: the integration of the productive consumer', in W. Friedland et al. (eds), *Towards a New Political Economy of Agriculture.* Boulder, CO: Westview Press.

Lee, D. and Newby, H. (1983) *The Problem of Sociology.* London: Hutchinson.

Little, J. (2002) *Gender and Rural Geography.* Harlow: Prentice Hall.

Martin, R. (2001) 'Geography and public policy: the case of the missing agenda', *Progress in Human Geography* 25: 189–210.

Mitchell, D. (1995) 'There's no such thing as culture: towards a reconceptualization of the idea of culture in geography', *Transactions IBG* 20: 102–16.

Mitchell, D. (2000) *Cultural Geography: A Critical Introduction.* Oxford: Blackwell.

Mormont, M. (1990) 'Who is rural? Or, how to be rural: towards a sociology of the rural', in T. Marsden et al. (eds), *Rural Restructuring.* London: Fulton.

Moseley, M. (1980) 'Rural development and in relevance to the inner city debate', Inner Cities Working Party Paper. Social Science Research Council, London.

Murdoch, J. (1997) 'Towards a geography of heterogeneous associates', *Progress in Human Geography* 21: 321–37.

Murdoch, J. (1998) 'The spaces of actor-network theory', *Geoforum* 29: 357–74.

Murdoch, J. (2001) 'Ecologising sociology: actor-network theory, co-construction and the problem of human exceptionalism', *Sociology* 35: 111–33.

Murdoch, J. (2003) 'Co-constructing the countryside: hybrid networks and the extensive self', in P, Cloke (ed.), *Country Visions.* Harlow: Pearson.

Murdoch, J. and Pratt, A. (1993) 'Rural studies: modernism, postmodernism and the post-rural', *Journal of Rural Studies* 9: 411–28.

Murdoch, J. and Pratt, A. (1997) 'From the power of topography is the topography of power', in P. Cloke and J. Little (eds), *Contesting Countryside Cultures.* London: Routledge.

Pahl, R. (1965) *Urbs in Rure.* Geographical Papers 2, London School of Economics.

Philo, C. (1992) 'Neglected rural geographies: a review', *Journal of Rural Studies* 8: 193–207.

Philo, C. (2000) 'More words, more words: reflections on the "cultural turn" and human geography', in I. Cook et al. (eds), *Cultural Turns/Geographical Turns.* Harlow: Prentice Hall.

Sauer, M. (1990) 'Fordist modernization of German agriculture and the future of family farms, *Sociologia Ruralis* 30: 260–79.

Shucksmith, M. and Chapman, P. (1998) 'Rural development and social exclusion', *Sociologia Ruralis* 38: 225–42.

Smith, N. (2000) 'Socialising culture, radicalising the social', *Social and Cultural Geography* 1: 25–8.

Swyngedouw, E. (1995) 'Review of "Geographical Imaginations" by Derek Gregory', *Transactions IBG* 20: 387–9.

Thrift, N. (2000) 'Introduction: dead or alive?', in I. Cook et al. (eds), *Cultural Turns/Geographical Turns*. Harlow: Prentice Hall.

Thrift, N. (2004) 'Summoning life', in P. Cloke, P. Crang and M. Goodwin (eds), *Envisioning Human Geographies.* London: Arnold.

Tonnies, F. (1957) *Community and Society*. New York: Harper.

Urbain, J. (2002) *Paradis Verts: Désirs de Campagne et Passions Résidentielles.* Pans: Payot.

Urry, J. (1984) 'Capitalist restructuring, recomposition and the regions', in T. Bradley and P. Lowe (eds), *Locality and Rurality.* Norwich: GeoBooks.

Whatmore, S. (2002) *Hybrid Geographies: Natures, Cultures, Spaces.* London: Sage.

Wilson, A. (1992) *The Culture of Nature.* London: Routledge.

Wylie, J. (2003) 'Landscape, performance and dwelling: a Glastonbury case study', in P. Cloke (ed.), *Country Visions.* Harlow: Pearson.

3

Reconfiguring rural resource governance: the legacy of neo-liberalism in Australia

Stewart Lockie, Geoffrey Lawrence and Lynda Cheshire

INTRODUCTION

Natural resources have historically been seen in capitalist societies as the key asset of rural spaces, with their utilization being one of the primary roles of rural people (Lockie et al., 2003). In the colonies of the so-called New World, spaces beyond the bounds of settlement were attributed little intrinsic value prior to being 'opened up', 'civilized', and rendered 'productive' in the service of nation and empire building.[1] The contributions that rural people subsequently made to national identities, gross domestic product and export earnings gave them what many believe to be disproportionate levels of political influence (Green, 2001). With the rural conceived – in the Antipodes – as little more than a quarry and a farm, that influence was limited in both magnitude and scope. The economic interests of farmers, foresters and miners have been seen as the interests of all rural people, while groups such as women and Indigenous peoples – together with issues such as rural inequality, economic diversification and environmental decline – have largely been ignored (Lawrence, 1987). Rural resource management in Australia has focused primarily on fostering conditions for the 'development' of natural resources and, to the extent that it has addressed social and environmental issues, managing negative externalities in order to maintain production (Lockie, 1994, 2000). While the state has been, and remains, a key facilitator of this process, changes in the mode of governing in Australia – as elsewhere – have meant that the state is no longer the sole arbiter of legitimate action. Instead, we adopt the concept of rural resource *governance* to refer to the range of institutions and actors that exist, and have quite profound influence over the way Australia's natural resources are managed. In exploring rural resource governance, particular attention is paid to the exercise of political power that occurs 'beyond the state' (Rose and Miller, 1992), and the way in which that power works to contour what is expected (and is possible) of social actors – in this case, farmers, graziers, foresters and miners. Contemporary governance is about extra-state authority as the means of legitimizing action, and of achieving local 'ownership' of natural resource management. This chapter, then, is about the ways in which natural resource governance has been fashioned, within Australia, under neo-liberalism – a policy regime that has had prominence over the past 20 years.

Before and during the twentieth century, Australian governments took a variety of steps to expand and intensify resource utilization. State underwriting was instrumental in providing the infrastructural development (roads, rail and telecommunications) and statutory regimes (price supports, producer boards, subsidies, extension and quarantine services, and a host of taxation 'breaks') that helped to ensure rural prosperity while 'evening out' spatial development (Bolton, 1992; Dovers, 1992). In turn, rural production underpinned national economic growth. When the dustbowls of the 1930s (in both Australia and the US) provided dramatic evidence of the environmental degradation associated with intensive agriculture, governments responded by establishing soil conservation agencies charged with undertaking research and providing advisory services to those farmers and other resource users interested in implementing conservation measures on a voluntary basis (Bolton, 1992; Bradsen, 1988). Soil conservation agencies were small in comparison with production-focused departments of agriculture, forestry and mining, and what few regulatory measures were provided to enforce control of environmental degradation were rarely used (Bolton, 1992; Bradsen, 1988). Not surprisingly, by the late 1970s, these strategies were seen widely to have failed. Massive land degradation (Department of Environment, Housing and Community Development, 1978; Standing Committee on Environment, Recreation and the Arts, 1989) and persistent rural poverty (Lawrence, 1987) contributed to a discourse of 'rural crisis' (Johnston, 1988) that helped justify major changes in the ways governments and producer groups approached rural resource governance. 'Globalization', 'economic rationalism' and more recently 'ecologically sustainable development' provided the conceptual framework for a new approach that linked production, environment and financial management through a variety of strategies designed to replace direct state intervention in production with measures to enhance the capacity of producers to subject themselves to 'market discipline'.

None of this is to say that the governance of rural resources has been uncontested. The question we ask, therefore, is this: are contemporary trends in rural resource governance likely to offer a fundamental challenge to the trajectory of social and environmental decline associated with 'productivism' (that is, the use of high technology inputs such as fertilizers and pesticides to improve the efficiency of agriculture)? While it is impossible to provide a definitive answer to this question, it is possible to discern a number of trends that promise profound effects on the ways we conceptualize and manage rural resources. Critically, many of these effects are likely to be contradictory.

In examining contemporary trends in rural resource governance in this chapter we concentrate upon strategies directed towards agriculture, reflecting the overwhelming dominance of this sector over rural land use. Even so, we are less concerned with the specific details of current policy settings than with the rationalities that underlie the conceptualization and management of resources more broadly. Thus, while empirical examples will draw primarily on the experience of Australian agriculture, much of our discussion will have relevance to broader debates associated with new trends in rural resource governance. The chapter will explore a range of inter-related themes including:

- The roles of science and sustainable development discourses in defining rural resources and the opportunities and threats encountered in relation to them.
- The emergence of neo-liberalism as a political rationality that is supportive of market rule rather than direct state intervention, but which also justifies the evolution of market mechanisms that force producers to internalize the environmental and social costs of production.
- The contradictory imperatives to democratize rural resource management by enhancing the capacity of producers to regulate their own behaviour – through participation in decision-making – and to centralize control over rural resource management in order to secure rapid investment.
- The construction of property rights and obligations, and the implications of these rights and obligations (legal and assumed) for environmental management and Indigenous peoples.

Drawing these threads together, the chapter will argue that the various dimensions of neo-liberalism, scientism and democratization all contribute in important ways to the contemporary governance of rural resources.

PROBLEMATIZING RURAL RESOURCE GOVERNANCE: SCIENCE, SUSTAINABILITY AND THE SOCIAL

Rural Australia continues to face resource degradation on an enormous scale. Estimates put the cost of environmental degradation at more than AU$3.5 billion a year (Standing Committee on Environment, Recreation and Arts, 2001). The main issues for Australia appear to be those of salinization, acidification and erosion of soils, loss of vegetative cover (largely through tree clearing), deteriorating water supplies, weed infestations, and species loss (Price, 1998). Salinization alone affects some 2.5 million hectares of land, with estimates that up to 15.5 million hectares (an area slightly larger than England and Wales combined) could be affected by 2020 unless action is taken (Madden et al., 2000). One report has calculated that the nation must spend something in the vicinity of AU$60 billion simply to manage, let alone overcome, these problems in the first decade of the new century (Virtual Consulting Group, 2000).

However, as compelling as these estimates appear to be, neither their meaning, nor the appropriate manner in which to respond to them, is self-evident. This is not to say that resource degradation estimates are invalid (for irrespective of whether they under- or over-estimate the extent of the biophysical processes involved, there can be little doubt that such processes are production-threatening). Rather, it is to say that the relationships between people, nature and technology implicated in environmental degradation are considerably more complicated than the problem-solution orientation embedded in economic estimates (Callon, 1998). Further, the ways in which human-nature-technology relationships are understood has a major bearing on the labelling of particular processes and behaviours as problematic, and on the identification of appropriate strategies to deal with them. Given that such understandings are likely to be contested, the respective abilities of scientists, farmers and others to account for why people and non-humans (plants, animals, minerals, technologies and so on) alike behave in particular ways is related firmly to the key issue of power in natural resource management.

Although it is more-or-less taken for granted that the ostensible goal of natural resource management is 'sustainability', prior to the late 1980s this was not the case. Rural environmental degradation largely disappeared from political view in post-Second World War Australia as technology-driven productivity increases ameliorated some of the more obvious natural resource management problems and masked others. When degradation first re-emerged as an issue during the crisis-ridden 1980s it was framed either (in the case of agriculture) as a biophysical issue calling for technical solutions or (in the case of mining and forestry) as any infringement on the 'wilderness' value of undeveloped spaces. Even though it is virtually impossible to define either environmental degradation or wilderness without recourse to human values and understandings, social and economic aspects of resource management were framed almost entirely in relation to the apparently objective fact of degradation and its solutions. In agriculture, the social research that was conducted was thus focused on the level of awareness among resource users of degradation; their attitudes towards the environment; and their behavioural responses to information about land degradation (Rickson et al., 1987). Of particular interest were levels of adoption of 'conservation farming' practices designed to minimize soil erosion by replacing mechanical cultivation with chemical weed control (Vanclay, 1986). With few exceptions, social scientists took the technical merits of conservation farming at face value and did little to interrogate the social relations involved in the production of this technology or the reconfiguration of social relationships implicated in its application. Nevertheless, finding that farmers and other resource users had generally positive dispositions towards environmental protection, but often disagreed with scientists over the extent, causes and solutions to environmental degradation, social scientists were forced to confront the limitations of both the technocratic framing of land degradation and the simplistic attitude–stimulus–response model of individual psychology (Barr and Cary, 1992; Peterson, 1991; Vanclay, 1992; Vanclay and Lawrence, 1995). The framing of soil erosion as a problem to be addressed through production-enhancing conservation farming practices, for example, did contribute eventually to reduced levels of erosion from cropping lands. However, it also contributed to increased levels of dependence among farmers on off-farm agribusinesses for material inputs and expertise (Lockie, 1997a, 2001) as well as to community conflict over chemical spraying practices (McHugh, 1996; Short, 1994) and consumer concerns over chemical residues (Short, 1994).

The idea of 'sustainable development' now dominating environmental discourses was popularized in the late 1980s by the so-called Brundtland Report, *Our Common Future* (World Commission on Environment and Development, 1987). Through this, the World Commission on Environment and Development challenged the technocratic framing of environmental degradation by arguing that environmental damage and protection could only be understood and addressed in terms of the relationships between nature conservation, agricultural production, industrial development and social equity, as well as their spatial and temporal dimensions. By framing resource management in a manner that reconciled economic growth with environmental protection and poverty alleviation, 'sustainable development' proved attractive to governments committed to maintaining conditions for capital accumulation while making some accessions to the demands of growing environmental and social justice movements. Conceptually, at least, this represented a major shift from the 'conservation versus development' debates of the 1970s. In

Australia, a series of working groups involving government, industry representatives and environmentalists were established with the task of developing 'ecologically sustainable development' (ESD) plans for each of Australia's major resource-based industries. Consistent with the Brundtland Report, ESD was held to embody principles of intergenerational and intragenerational equity, conservation of biodiversity, a cautious approach to dealing with risk (the 'precautionary principle'), attention to global environmental issues, and economic diversity and resilience (Commonwealth of Australia, 1992).

For our purposes here, one of the most notable features of sustainable development has been its framing of resource management as a social issue – not simply in the sense that humans caused objectively measurable environmental change, but in the sense that the definition of key environmental values and goals was a thoroughly social process that reflects the power relationships and inequities among and between social groups. Environmental degradation cannot, from this perspective, simply be 'fixed'. Rather, better environmental management can only be achieved in the context of more harmonious human relationships. Australia's ESD Working Group on Agriculture thus endorsed an approach to sustainability in agriculture emphasizing 'responsible land management incorporating economic and environmental principles; the formation of community-based self-help groups; and integrated farm planning' (Commonwealth of Australia, 1991b: xx).

The first major manifestation of this more social understanding of resource management in government policy was the establishment in 1989 of the National Landcare Programme (NLP), a programme that sought to address predominantly agricultural land and water degradation through the promotion and support of a national network of autonomous community Landcare groups (Campbell, 1994). With Commonwealth funding provided for assistance for group coordination, training and to establish experimental and demonstration projects, these groups concentrated their activities on farm planning, tree planting and community education (Campbell, 1994). Some 4000 community Landcare groups now involve roughly 40 per cent of Australian farm businesses (Alexander et al., 2000). Despite achieving almost universal political support (Lockie, 1999), the NLP attracted a number of criticisms including questions over the extent to which governments handed over responsibility to deal with environmental degradation to local communities without committing the requisite resources to enable them to fulfil this responsibility (Lockie, 1994, 2000). However, quite apart from the technocratic understanding of land degradation that underlies this criticism, there is a failure to appreciate the rather more fundamental changes that are occurring in rural resource governance and the respective roles that governments, communities and others are playing in these processes. This will be taken up in more detail in the following sections of this chapter. Before closing this section, we wish to point to additional ways in which the more socialized understanding of resource management embodied in 'sustainable development' is currently being enacted through government policy.

Responding to criticisms that the NLP had built community awareness but had not led to significant change 'on the ground' (Lockie, 1997b), the federal government has initiated a process of more far-reaching devolution of authority to regional communities to allocate public monies for natural resource management projects. The National Action Plan for Salinity and Water Quality is a seven-year programme totalling AU$1.4 billion to address soil salinity and poor water quality through the actions of regional bodies which work in partnership with federal and state governments. Under the rubric of the NAP the federal government, in collaboration with the states, is currently:

- Identifying bio-regions and assisting the formation of catchment (watershed) groups that will assume responsibility for the allocation of funds to research, education, on-ground works, and so on, identified as regional priorities.
- Encouraging each catchment group to be inclusive of all groups potentially affected by natural resource management decisions, including representatives of groups traditionally excluded from resource management forums such as tourism, heritage, conservation and Indigenous groups.
- Signing agreements with the catchment groups to provide them with long-term status and funding.
- Imposing systems of accountability to ensure that funding flows only to projects linked directly to regional priorities set by catchment management groups and displayed prominently in each group's sustainability strategy document (see AFFA, 1999; House of Representatives Standing Committee, 2000).

The NLP and NAP incorporate an understanding of the social dimensions of resource management which, in contrast with the individual psychology of technocratic approaches, focuses on the 'capacity' of communities at a number of scales to act in their own interests. This is not to say that the individual has been completely decentred in natural resource management policy and practice. As the next section of this chapter will show, contemporary approaches to resource governance remain very focused on acting on individuals and, in particular, on the ways in which they are likely to construe their own circumstances and respond accordingly.

NEO-LIBERAL POLITICAL RATIONALITIES: REDEFINING RIGHTS AND RESPONSIBILITIES

Before commencing our discussion of neo-liberal approaches to governance it is important that we specify how we intend to use these often loosely defined terms. Following Foucault (1991), governance is treated here as an activity, or social practice, that is not restricted to the agencies or operations of the state. The analysis of governance is concerned with understanding empirically the rationalities, or ways of thinking and doing, that underlie the 'conduct of conduct' in its widest sense. Such rationalities provide ways of doing things that render objects of governance knowable, actionable and contestable. From this perspective, neo-liberalism may be understood as a rationality of governance that – far from providing neo-conservative governments with an ideological smokescreen with which to justify the implementation of *laissez faire* policies – has helped to define the contours of the state, market and civil society in novel ways (Foucault, 1991; Miller and Rose, 1990). More specifically, deregulation and the promotion of market relations have been used to influence both the environment within which people make decisions (Miller and Rose, 1990), and the ways in which they are likely to understand and respond to that environment (Burchell, 1993).

In broad terms, Australian governments and peak industry groups have adopted a neo-liberal rationality of market rule, rather than state rule, in the belief that this will enable Australia to capture a greater share of the perceived benefits of the global economy. This does not imply the 'hollowing out' of the state (Jessop, 1990). Rather, government agencies implement novel strategies to facilitate capitalist expansion through more productive exploitation of rural resources (Moreira, 2003; O'Connor, 1993) and the enrolment of new actors into the network of those deemed responsible for managing the externalities of capitalist production. In other words, the strong support historically shown by governments for productivist resource management remains relatively unchanged. Neo-liberalism (and its populist expression as economic rationality and globalization) has provided the means to conceptualize a shift in state involvement in rural resource management from that of securing national prosperity by protecting Australia's primary industries from the vagaries of the global economy to readily exposing them to the competitive nature of world markets. The result has been the replacement of many of the support structures Australian producers have traditionally come to rely upon, with discourses and practices that extol the virtues of competition, entrepreneurship and efficiency. According to such logic, government subsidies perpetuate dependency and provide farmers and other resource users with little incentive to relinquish uncompetitive production practices. The only solution, it is argued, is for resource users (and, indeed, rural citizens in general) to become less reliant upon state support and to develop a more entrepreneurial attitude to economic and environmental risks (Herbert-Cheshire, 2001; Higgins, 2002). While some have embraced this challenge and established profitable enterprises based on niche marketing and pluriactivity (Gray and Lawrence, 2001), others deemed less viable have been assisted out of primary industries through adjustment strategies and exit payments. The effect is that farm numbers have fallen by 1.3 per cent per annum – that is, 2000 farms per year – since the mid 1950s (see Garnaut and Lim-Applegate, 1998.) While this might mean 'obeying' market signals, it also culminates in the loss of people which, in many of Australia's regions, results in the contraction of rural community viability (Gray and Lawrence, 2001). Yet, for most producers, the only way to remain in business has been to try to increase output by adopting more intensified production practices that, while increasing return in the short term, have simultaneously caused long-term environmental harm.

Associated with the rise of neo-liberal practices are the dual processes of marketization and 'individualization'. Marketization may be understood as the creation and/or extension of market mechanisms, and associated property rights, for the allocation of resources. The most obvious way in which this has occurred has been through the dismantling of government and producer-controlled mechanisms for the collectivization of risk such as statutory marketing boards, production quotas, subsidies and so on (see Gray and Lawrence, 2001). Neo-liberal rationality is also reflected in more proactive moves to create markets through the extension of private property rights. As Reeve (2001) argues, associated with all property rights are sets of legal and normative rights and responsibilities. While these rights and responsibilities are social constructs that are subject to contestation and change over time, freehold property rights in Australia and elsewhere tend to be constructed as absolute rights. What this means is they are understood to provide the owners with the rights to do whatever they like, whenever they like, with their own land, equipment, water and so on, irrespective of any negative impacts these actions may have on the ability of others to exercise their own property rights (for example, through the impact of soil erosion on downstream water quality). Attempts to challenge the absolutist interpretation of private property rights in the rural sector through the imposition of resource-use regulations are typically met with fierce opposition. Actions that appear to support the absolutist interpretation through the extension of private property rights, however, have broad support. The Council of Australian Governments (COAG) policy on water, for example, has, since 1994, provided a framework for water reform based on separating the allocation and trading of water property rights from land property rights, full cost recovery in the provision of water and greater user involvement in water management (Cullen et al., 2000). By treating water as an unsubsidized, but freely exchangeable, commodity (backed up by appropriate institutional arrangements for the registration of rights and transactions), it is believed that competition for water allocations will see them move to their most economically efficient uses while avoiding ecological damage caused by inefficient use and wastage. Ignored in this logic, however, are the contradictions of market-based production and exchange and the opportunities afforded by entirely different forms of property rights, such as common property (Williamson et al., 2003), to promote better environmental and social outcomes.

According to O'Connor (1993), there are two basic contradictions within capitalist systems that ultimately undermine their economic and ecological sustainability. The first is based on the tendency for individual enterprises to attempt to increase their market share by raising productivity and lowering costs (for example, by shedding labour). When multiple enterprises do this, the result is overproduction relative to consumer demand, leading to declining prices. The second contradiction derives from the need for individual enterprises to reduce costs so as to remain competitive relative to other producers. Where social and environmental costs of production can be externalized there is no incentive for individual enterprises to pay them, even though the long-term effect may be to undermine the resource base on which production depends and thus to raise average costs. Capitalist enterprises are not bound by any inherent structural logic to externalize the environmental costs of production in this manner and many producers do absorb some of these costs by acting to minimize the off-site impacts of their activities. However, the internalization of social and environmental costs is far more likely in circumstances where clear responsibilities are associated with property rights either through government regulation or through other institutional arrangements (Reeve, 2001). This creates a dilemma for political authorities in their desire to maintain high levels of economic growth as well as political legitimacy. While both objectives are seemingly at odds with one another, the formation of new forms of resource governance, based on principles of participative democracy, state-community partnerships and individual and community self-help, have gone some way to addressing these contradictions (Lockie, 2000).

Individualization may be seen as one way to resolve the contradictions of marketization through the implementation of what Ulrich Beck calls 'biographical solutions to systemic contradictions' (Beck and Beck-Gernsheim, 2002: xxii). In the Beck-ian sense, individualization refers to the emergence of institutional settings that promote and endorse opportunities for the person, rather than for the group (see Beck and Beck-Gernsheim, 2002). In practice, however, individualization is also consistent with more collective forms of self-interest where rural people are required to establish community-based or catchment-based groups and compete, against others, for scarce markets and funding. In this scenario, the competitive individual becomes the competitive community – succeeding or failing according to the dictates of the market place. The first stage of this individualization process, therefore, is the redefinition of the social and environmental problems facing rural areas in terms of the attributes and supposed deficiencies of rural people and their communities, rather than as outcomes of broader social relationships (Lawrence and Herbert-Cheshire, 2003). Such thinking has been reflected in a number of reports published since the early 1990s that relate the poor economic performance of regional Australia to the lack of skills and 'inappropriate' attitudes of the people living there (Hilmer, 1993; McKinsey and Company, 1994; Industry Taskforce on Leadership and Management Skills, 1995). The effects of this process are twofold. First, in seeking solutions to problems such as environmental degradation, the personal characteristics of farmers and communities, rather than the actions of governments or large business enterprises, come to be viewed as the most appropriate sites of reform. It is for this reason that many of the so-called solutions to rural decline articulated by Australian governments involve leadership training, individual and community capacity building or farm financial counseling programmes which are designed to address the perceived inadequacies of rural people (Herbert-Cheshire, 2000). Not only does this shift the blame for environmental degradation firmly onto the shoulders of individual producers and/or communities, but it also leaves unchallenged the structural imperatives of productivism that induce farmers and others to continue exploiting natural resources. Second, this, in turn, helps to resolve the accumulation/legitimacy contradictions facing contemporary governments by enabling them to be seen to be addressing environmental degradation, but actually doing so in a way that secures conditions for capitalist accumulation (Lockie, 2000). The downside is a worsening environmental crisis as the pressure to over-use the land for smaller returns continues unabated.

Individualization has been operationalized in Australia through the devolution of responsibility for addressing the environmental and social costs of productivist agriculture to individual producers and rural communities. What were once seen as collective problems for the state or society are now regarded as individual problems to be addressed by rational, self-governing producers or community groups at farm or community levels. Such moves away from state assistance towards strategies of self-help/self-reliance (Herbert-Cheshire, 2000; Higgins, 2002) are evident in a range of governmental policies and programmes relating to rural resource management that have been established in the past decade. In the realm of natural resource management, the Australian National Landcare Programme discussed above is a much-cited example of attempts by Australian governments to restore degraded rural environments through the voluntary actions of landholders and catchment groups supposedly 'empowered' by their new partnership with the state (Higgins and Lockie, 2002). The Rural Adjustment Scheme of 1992, established by the former Commonwealth Department of Primary Industries and Energy (DPIE) to improve farm business management practices, was similarly underpinned by a rationality of self-reliance in which effective farm business managers were seen to be those who, with the correct training and skills, could manage their own properties in a planned, productive and rational manner (Higgins, 2002). Finally, regional and rural development policies over the past decade have also privileged self-help and local leadership over state involvement as essential ingredients for the long-term sustainability of Australia's inland regions (Herbert-Cheshire, 2000).

The key point to make here is not that neo-liberalism allows governments and other actors to trick citizens into accepting or taking responsibility for the contradictory effects of capitalist accumulation, but that the ways in which they seek to influence the conduct of others may be subtle and indirect – a process Latour (1987) terms 'action at a distance'. It may also lead to contradictory outcomes. However, while the propensity for governments to displace responsibility for dealing with such costs may be seen to involve tokenistic gestures towards more participatory models of democracy, such gestures also open avenues through which to challenge neo-liberal governance (Dryzek, 1992) and the construction of rural spaces exclusively in terms of production.

DEMOCRATIZATION: DEVOLUTION VERSUS CENTRALIZATION

It has been argued here, as elsewhere, that the inclusion of rural people in structures and processes of rural resource management provides contemporary governments with a range of opportunities to manage the contradictions of global capitalism and civil society (Lockie, 2000; Marsden, 2003). On the one hand, this supports the capitalist agenda by individualizing the social and environmental impacts of productivist agriculture; on the other, it has democratic appeal by virtue of the discourses of partnerships, self-help and local action that accompany these mechanisms. As this dual objective implies, however, the democratization of rural resource management does not constitute a shift towards a *laissez-faire* mode of government in which rural people are now 'free' to manage local resources any way they wish. Centralized influence over Australian rural resource management remains significant in two broad ways. The first is through the articulation of boundaries around those issues that are deemed suitable for various forms of devolved responsibility and those regarded as more appropriate to centralized decision-making in the state or national interest. The second is through a variety of mechanisms to exert influence indirectly, or 'at a distance', over those issues deemed suitable for devolution to individuals and community groups.

Devolution of responsibility has undoubtedly been the dominant trend in rural resource governance over the past decade and a half (Gray and Lawrence, 2001). This chapter has discussed a range of programmes through which the devolutionist agenda has been pursued, including the National Landcare Programme, National Action Plan for Salinity and Water Quality (NAP) and COAG water reforms. Greater levels of public participation have been pursued in a host of other natural resource planning and management processes, ranging from Regional Forestry Agreements to the preparation of environmental impact assessments and local government land-use plans. But in many of these cases, decision-making powers continue to rest with government agencies and elected officials. This is particularly the case in relation to large resource development projects such as mines, dams, power stations and so on, and in circumstances where devolutionist programmes are seen to have failed in changing the behaviour of resource users. Irrespective of the particular arguments for centralized control, rural people who believe themselves to have participated in devolutionist and other participatory processes in good faith often perceive such centralized intervention as a breach of trust. Recent attempts by state governments to introduce moratoriums on tree clearing following a decade of Landcare offer a case in point. The inability of catchment groups charged under the NAP with developing regional natural resource management plans to assume responsibility for approving major resource development projects – despite the huge implications of such projects on a catchment scale – is another. While, therefore, neo-liberal strategies of individualization and devolution offer state agencies a range of opportunities to maintain political legitimacy, despite the contradictory tendencies of capitalist production, new sets of contradictions between centralized and devolved approaches to governance potentially challenge that legitimacy. The outcome is not necessarily the complete breakdown of political legitimacy, but continuing contestation over the appropriate goals, boundaries and modes of state action between and among government agencies, rural communities, environmental movements and private capital.

Indirect forms of influence over devolved issues take a number of forms oriented to shaping the self-calculating and self-regulating behaviour of individuals and community groups. Perhaps the most obvious manifestations of 'action at a distance' are the accountability and reporting requirements placed on supposedly autonomous groups and individuals in exchange for access to government funding and other resources. Muetzelfeldt (1992a, 1992b) interprets such requirements as mechanisms for the maintenance of strategic control at the same time that responsibility for day-to-day decision-making is delegated. Regional planning arrangements under the National Action Plan for Salinity and Water Quality is a good example. While no state agency exists to dictate to catchment groups the projects and strategies to which they should direct funds, access to those funds is contingent on the development of a regional sustainability strategy and investment plan that can be demonstrated to meet a number of minimum requirements as stipulated by the federal government. These minimum requirements relate both to the processes followed in the development of regional strategies and the biophysical resource condition targets they must meet (AFFA, 1999; House of Representatives Standing Committee, 2000). Similar accountability and reporting requirements have been evident in other devolutionary initiatives resulting, in the case of the National Landcare Programme, in a remarkable degree of uniformity in the activities of community Landcare groups across the nation, despite the enormous variability to be found in agroecological conditions and resource uses (see Campbell, 1994). In the case of the Rural Adjustment Scheme of 1992, access to financial assistance due to unfavourable market or climatic conditions is contingent on the adoption of a range of financial and land-use planning techniques (Higgins and Lockie, 2001).

The increasing prevalence of requirements for financial and environmental accountability that these examples illustrate has led Argent (2002) to conclude that the once dominant position of the farm sector over rural resource use policy has effectively been over-ridden. Interestingly, this has been achieved without the need to confront directly the absolutist interpretation of private property rights that resource users typically use to defend themselves against more overt forms of centralized control. This does not mean that governments have fooled farmers into now 'sharing' pre-existing rights, but that they have sought ways of ensuring that farmers and other resource users make the 'correct' decisions based on a rational interpretation of the environment with which they are faced. For governments, the assumption is that better market-based tools will enable producers to utilize their resources more sustainably without having to resort to regulatory measures that, due both to resistance and the complexity of natural resource management, often generate unintended and perverse results (something at the heart of the contradictory discourses of sustainable development). Yet, such an approach is consistent with neo-liberalist ideals such as those of 'abuser-pays' (externalities must be measured, costed and attributed to individuals and firms implicated in pollution).

The less obvious forms of 'action at a distance' are those that relate to the knowledge on which resource users base their decisions and thus regulate their own behaviour. These are evident in many of the accounting procedures referred to above, but extend beyond these to include the scientific and other knowledge on which natural resource management decisions are based (Lockie, 1997a, 1999). For example, Landcare and catchment groups have played key roles in the continued promotion of technical solutions to environmental problems in rural areas that support productivism, not simply because they are consistent with a particular ideological agenda, but because, on the basis of a variety of technologies of knowledge, they appear to be the most rational solutions available. Conservation farming, as mentioned previously, is probably the most widespread example, with many believing chemical use to hold the key to reduced soil erosion, improved pest control and increased production from crops and pastures, not to mention more successful tree establishment in shelter belts and riparian zones (streambanks). The apparent rationality of such practices rests on technologies of knowledge, including soil and leaf tissue testing, that are dependent for their application on interpretive frameworks established by agri-science agencies committed to a particular vision of sustainability – one in which high levels of production are maintained through equally high levels of input use while minimizing environmental leakage. Interpretive frameworks used to give meaning to the apparently objective data provided by soil and leaf tissue testing are established often through 'input-requirement' trials that ignore the innumerable possible alternative strategies. Advice subsequently given to farmers almost invariably, therefore, suggests that problems can be rectified through the addition of farm inputs. Other constructions of sustainability are not only possible, but suggest entirely different interpretations of the meaning of 'objective' test results and an array of alternative strategies with which to respond to them (Pretty, 1998).

As we have seen, the apparent trend toward democratization of rural resource governance raises a number of new contradictions at the same time that it helps state agencies to manage others. In addition to conflicting models of democracy and tensions between centralization and devolution, democratization has provided avenues for a wider variety of actors to participate in rural resource governance (and at the very time of the intensification of a market-driven logic in wider neo-liberal discourse). Two of the most active campaigners for increased participation have been urban environmental movements seeking to protect endangered species and ecosystems, and Indigenous groups attempting to negotiate rights over Native Title (Holmes, 2002). Other, previously excluded, groups, such as local government, industry bodies, tourist operators and regional economic development organizations, are also demanding a say in how rural resources are accessed, used and managed. With the incorporation of nonfarming and nonrural groups into the decision-making arena, the potential for new resource values to emerge that conflict with the productivist regime is increased (Holmes, 2002: 379; see also Woods, 2003). Examples of such conflict have arisen in the formation and functioning of catchment management groups (Ewing, 2003), particularly under the NAP, where representatives of a range of competing stakeholder groups have been brought together into one regional body.

POST-PRODUCTIVISM AND ALTERNATIVE CONSTRUCTIONS OF RURAL SPACE

Our discussion thus far has focused on rural resource management as it occurs largely within a productivist paradigm. However, writers such as Ward (1993), Marsden et al. (1993), Lowe et al. (1993), Argent (2002), Holmes (2002) and Marsden (2003) have assembled considerable evidence to suggest that, in relation to agriculture in particular, there have been discernible moves from the strongly petrochemically driven production methods that emerged after the Second World War to a somewhat more environmentally aware focus. According to Argent (2002: 99, see also Ilbery and Bowler, 1998), where the key features of productivist agriculture are those of *intensification*, *concentration* of production and *specialization* (both of farms and regions), post-productivist forms of agriculture are generally characterized by *extensification* (decreases in external inputs and the use of land); *dispersion* (access by a variety of stakeholders to the land); and, *diversification* (heterogeneity in agricultural, and other, pursuits in rural space). Accompanying post-productivist changes in agriculture, forestry, mining and other rural sectors have been alterations in the ways in which the roles of primary production and producers in resource management are understood (Wilson, 2001). The privileged ideological position of farmers as the 'backbone of the nation' is challenged as they are increasingly implicated in resource depletion and as new social actors acquire an increasingly important role in defining the meaning and use of rural space.

The work of Australian geographer John Holmes (2002) is of interest here. Holmes is critical of the notion that there is a clearcut move from productivism to post-productivism in Australia. He posits that any such move will be locationally specific and determined by the extent to which any region within the landscape is dominated by production, consumption or protection. He views the nation as a combination of various landscape occupance modes, each having identifiable trajectories. The five main ones he identifies are: productivist agricultural occupance (where production values dominate); amenity-oriented rural occupance (where consumption values dominate); small farm or pluriactive rural occupance (where there is a mix of production and consumption values); redundant agricultural/pastoral occupance (where there is potential integration of production and protection values); and conservation and indigenous occupance (where protection values dominate). The usefulness of such a typology is that it allows us to identify the two occupance modes which are putting the landscape at ecological risk (for Australia, productivist agricultural occupance and redundant agricultural/pastoral occupance) and to develop specific policy settings for those occupance modes. By having a more subtle way of describing the landscape, we are at one and the same time aware of the occupance modes that must be altered if sustainability is to be improved, and have the opportunity of developing new policy settings that relate specifically to farming practices in those areas (and thereby avoiding 'blanket' policies that are deemed to have largely failed in previous decades).

Even more radical transformations in the management of rural resources are mooted by those, such as Marsden (2003), who prefer to see post-productivism as a transitional phase on the path to sustainable rural development, with the latter characterized by an expanded focus upon rural livelihoods, agroecology, new institutional arrangements (including new structures of support for sustainable agriculture), food supply chains that are 're-embedded' in the regions, and the integration of farming and other rural activities (see also Lawrence, 1998; Dore and Woodhill, 1999). In such a model, environmental and social policy cannot simply be 'bolted on' to agricultural policy, as it has been in the past (see Drummond et al., 2000), but must bring new sets of organizing principles, or forms of governance, to guarantee ecosystem health over the long term (Marsden, 2003). Yet, what we have seen to date in Australia appears to be of this 'bolted on' character. The emerging promise in overcoming this with one of 'triple bottom line' calculation (Pritchard et al., 2003) is consistent with the move toward what has been identified as 'sustainable regional development' (Lawrence, 1998; Dore and Woodhill, 1999).

In Australia, the idea of a more integrated approach to sustainable regional development and resource governance has been taken up, in part, through the devolution of planning and decision-making associated with the National Action Plan for Salinity and Water Quality and other programmes. These programmes reflect very strongly the framing of resource management as a social problem (and one that must be addressed through more individualist 'solutions'). The question is, will such initiatives re-direct farmers and graziers away from productivist approaches to resource use? As our brief discussion of the National Landcare Programme in the previous section demonstrates, devolved natural resource management instruments may, in fact, reinforce production-driven resource management due to their reliance on technologies of knowledge that construct as rational the application of productivist practices in response to environmental problems (Lockie, 1997a, 1999). While many argue that practices, such as conservation farming, fall short of the ecologically and socially regenerative requirements of a genuinely sustainable agriculture they have, nevertheless, addressed some of productivism's most visible environmental excesses (Pretty, 1998). The important point is that the technocratic application of productivist practices is not the only possible outcome. Ultimately, it might be argued that the long-term success of Landcare and similar programmes that draw on discourses of participatory democracy will rest on the extent to which they encourage resource users to abandon highly individualist and absolutist constructions of private property rights and accept instead responsibility for the impacts of their activities on the property rights of others and on other non-market values (Reeve, 2001). But, at the same time that Landcare and similar initiatives encourage resource users to associate responsibility to reduce the off-site impacts of their activities as part of their total package of property rights, other market-based solutions to environmental problems entrench an individualist notion through their assumption that clearly defined private property rights are essential to the maintenance of the stable investment environment necessary for the long-term orientation to decision-making implied by sustainable resource management. This leads Reeve (2001) to conclude that progress on sustainable rural development is likely to be stilted until serious moves are made to open political and legal debates on the full array of rights and responsibilities inhering in private property and on alternative legal and cultural regimes for their regulation – something which is now occurring with more vigour (Orchard et al., 2003).

In many senses, far more profound challenges to productivism and individualist constructions of private property have come from those actors traditionally excluded from rural resource management. Holmes (2002: 362), for example, argues that a combination of pressure to institutionalize Aboriginal land rights, the influence of an urban-based green protest movement, together with a minimally regulated system of agriculture, have had the effect of 'dissolving the productivist/pastoral hegemony' and replacing them with new, 'amenity-oriented forms of land use'. While this oversimplifies the impact of Native Title (the legal recognition of traditional rights in relation to land and water) on resource use, it is clear that opportunities have been opened up to consider, within regional natural resource management plans and other governance processes, a range of non-market values. Legislative recognition and definition of Native Title and cultural heritage at state and federal levels has provided Australia's Indigenous peoples with rights to negotiation over a wide variety of rural resource management decisions and, in a growing number of cases, with resources to implement alternative land uses on a variety of scales (Godwin, 2001). This represents a challenge to – and, indeed, diminution of – the older, individualist, productivist-based property rights claims.

CONCLUSION

In this chapter it has been argued that global processes both fostered by, and in turn fostering, neo-liberal policy settings in Australia have had major implications for the governance of rural space and resources. The increased exposure of agriculture and other rural industries to global market forces has tended to have two effects. The first has been for the productivity and efficiency imperatives of productivism to become more important in determining the location and intensity of production (in other words, productivism has become entrenched in many regions as producers seek to compete on international markets). The second, more recent, impact has been for post-productivist options to be embraced by those whose economic existence in primary industry has become marginalized (those seeking to remain in rural areas yet lacking any income derived from natural resource-based industries), as well as by those who are living in regions where cultural, scenic, historical advantages have led to the influx of foreign and domestic tourists. Both dynamics are of importance in setting the boundaries of what is acceptable or unacceptable natural resource management practice. And, while a more conservationist ethos has tended to accompany post-productivism, it remains that both productivism and post-productivism have been guided by principles of market rule, individual and community self-help, and limited government involvement in natural resource management. As a consequence, Australia's environmental problems have worsened.

In returning to the four 'themes' alluded to at the beginning of the chapter and discussed throughout, it is clear that new voices have emerged to help shape debates about Australia's resource future. The previously hegemonic productivist discourses of the agricultural scientists and economists are being challenged by those of ecologists, environmentalists, Indigenous groups and a host of other actors. Similarly, while freehold property rights have traditionally been supported by state and market imperatives, the previously taken-for-granted rights of property holders to manage their resources in whatever way they wish is also being undermined. In effect, a new form of government/community participation, guided by a different set of principles and policies is considered to be emerging as Australia struggles with ways of addressing environmental degradation. What has variously been termed 'the rural development dynamic' and 'sustainable regional development' goes beyond post-productivism in seeking to understand the social and economic health of regions in relation to environmental security (Everingham et al., 2003). Sustainable regional development looks to different actors and different mechanisms to bring about change. It is premised upon better coordination and integration of government services at the regional level; partnerships between government and community within bio-physically defined catchments; a blend of market-based and non-market-based incentives to foster change; and the injection of considerable funding from state and federal governments – well beyond the normal cycle of parliamentary elections (Lawrence, 2003).

Alongside this change, however, has been the continued acceptance that market rule provides the best mechanisms for ensuring producers internalize, and pay for, the costs of environmental degradation. This conforms with, rather than challenges, neo-liberalist rationalities regarding the importance of the individual – rather than the social – in dealing with widespread destruction. It also means that the broader kinds of changes required to deal with this destruction are unlikely to occur. Are governments unaware of the fundamental contradiction between seeking sustainable development within a competitive global market regime that rewards productivism? Can the unsustainable path of global marketization really be made sustainable in the manner currently being promoted? It is too early to decide how profoundly the new forms of natural resource governance in Australia will challenge this path of productivism. What is of interest is that the new form of governance (community-based catchment management) not only holds a democratic promise, but also blurs the boundaries between the market, state and civil society with the potential for greater flexibility and inclusion. Governments are now investing in community capacity, rather than in roads and dams, hoping that decision-making at the local level will foster both ownership and change at that level. But, might trust in a community's ability to identify and solve natural resource management problems locally be misplaced? Perhaps the partnership between government and community that keeps the distribution of federal dollars closely aligned with stated and approved catchment priorities will be the mechanism to ensure accountability at the local level. Australia is embarking on a huge experiment in governance: we can only hope that this time, the key to better environmental management has been discovered.

As a final point, it might be useful to consider what new 'threads' are emerging in the analysis of rural space that might assist in identifying ways we can move beyond the current settings that 'sustain the unsustainable' in rural regions of the advanced societies. Holmes's work on modes of rural occupance is still in the development phase – and comes close to the already-existing 'bioregional planning' (see Brunkhorst and Coop, 2001) and 'agroecosystems' (see Llambi and Llambi, 2001) approaches. Nevertheless, it does offer rural social researchers an opportunity to apply and assess multiple governance forms that co-exist and interact at different spatial scales and in different regions. This will inevitably add to the complexity of evaluation of 'rural governance', while providing an opportunity to ascertain the extent to which various combinations of structures/actions/programmes do, or do not, work in specific settings. With stronger theoretical and conceptual development, the occupance modes approach is one, among a number, of possible ways forward in the more advanced study of regional governance.

NOTE

1 This is reflected in Brulle's (1996) analysis of the concept of 'manifest destiny' that dominated environmental discourses in the United States from the time of Puritan settlement in 1620, and through which exploitation of the natural environment was constructed as a duty.

REFERENCES AND FURTHER READING

AFFA (Agriculture, Forestry and Fisheries – Australia) (1999) *Managing Natural Resources in Australia for a Sustainable Future: A Discussion Paper for Developing National Policy*. Canberra: AFFA, July.

Alexander, F., Brittle, S., Ho, A., Gleeson, T. and Riley, C. (2000) *Landcare and Farm Forestry: Providing a Basin for Better Resource Management on Australian Farms*. Canberra: Australian Bureau of Agricultural and Resource Economics.

Argent, N. (2002) 'From Pillar to Post? In Search of the Post-productivist Countryside in Australia', *Australian Geographer* 33 (1): 97–114.

Barr, N. and Cary, J. (1992) *Greening a Brown Land: The Australian Search for Sustainable Land Use*. Melbourne: Macmillan.

Beck, U. and Beck-Gernsheim, E. (2002) *Individualization*. London: Sage.

Bolton, G. (1992) *Spoils and Spoilers: A History of Australians Shaping Their Environment*, 2nd edn. Sydney: Allen and Unwin.

Bradsen, J. (1988) *Soil Conservation Legislation in Australia: Report to the National Soil Conservation Program*. Adelaide: University of Adelaide.

Brulle, R. (1996) 'Environmental discourse and social movement organizations: a historical and rhetorical perspective on the development of U.S. environmental organizations', *Sociological Inquiry* 66 (1): 58–83.

Brunkhorst, D. and Coop, P. (2001) 'The influence of social eco-logics in shaping novel resource governance frameworks', in G. Lawrence, V. Higgins and S. Lockie (eds), *Environment, Society and Natural Resource Management: Theoretical Perspectives from Australasia and the Americas*. Cheltenham: Edward Elgar. pp. 84–103.

Burchell, G. (1993) 'Liberal government and techniques of the self', *Economy and Society* 22 (3): 267–82.

Callon, M. (1998) 'Introduction: the embeddedness of economic markets in economics', in M. Callon (ed.), *The Laws of the Markets*. Oxford: Blackwell.

Campbell, A. (1994) *Landcare: Communities Shaping the Land and the Future: With Case Studies by Greg Siepen*. Sydney: Allen and Unwin.

Commonwealth of Australia (1991) *Ecologically Sustainable Development Working Groups: Final Report – Agriculture*. Canberra: Australian Government Publishing Service.

Commonwealth of Australia (1992) *National Strategy for Ecologically Sustainable Development*. Canberra: Commonwealth Government.

Cullen, P., Whittington, J. and Fraser, G. (2000) *Likely Ecological Outcomes of the COAG Water Reforms*. Canberra: Cooperative Research Centre for Freshwater Ecology.

Department of Environment, Housing and Community Development (1978) *A Basis for Soil Conservation Policy in Australia*. Commonwealth and State Government Collaborative Soil Conservation Study 1975–77, Report 1. Canberra: Australian Government Publishing Service.

Dore, J. and Woodhill, J. (1999) *Sustainable Regional Development: Executive Summary of the Final Report*. Canberra: Greening Australia.

Dovers, S. (1992) 'The history of natural resource use in rural Australia: practicalities and ideologies', in G. Lawrence, F. Vanclay and B. Furze (eds), *Agriculture, Environment and Society: Contemporary Issues for Australia*. Melbourne: Macmillan. pp. 1–18.

Drummond, I., Campbell, H., Lawrence, G. and Symes, D. (2000) 'Contingent and structural crisis in British agriculture', *Sociologia Ruralis* 40: 111–28.

Dryzek, J. (1992) 'Ecology and discursive democracy: beyond liberal capitalism and the administrative state', in M. O'Connor (ed.), *Is Capitalism Sustainable? Political Economy and the Politics of Ecology*. New York: The Guilford Press. pp. 176–97.

Everingham, J., Herbert-Cheshire, L. and Lawrence, G. (2003) 'Regional renaissance? New forms of governance in non-metropolitan Australia'. Paper presented at the Conference of the Australian Sociological Association, University of New England, Armidale, NSW, 4–6 December.

Ewing, S. (2003) 'Catchment management arrangements', in S. Dovers and S. Wild River (eds), *Managing Australia's Environment*. Annandale, NSW: Federation Press. pp. 393–412.

Foucault, M. (1991) 'Governmentality', in Graham Burchell, Colin Gordon and Peter Miller (eds), *The Foucault Effect: Studies in Governmentality*. Hemel Hempstead: Harvester-Wheatsheaf. pp. 87–104.

Garnaut, J. and Lim-Applegate, H. (1998) 'People in Farming', ABARE Research Report 98 (6). Canberra: ABARE.

Godwin, L. (2001) 'Indigenous natural resource management in Central Queensland', in M. Alderton, J. Norton and L. Godwin. *Organics, Biotechnology and Indigenous Natural Resource Management in Queensland*. Rockhampton: Institute for Sustainable Regional Development.

Gray, I. and Lawrence, G. (2001) *A Future for Regional Australia: Escaping Global Misfortune*. Cambridge: Cambridge University Press.

Green, A. (2001) 'Bush politics: the rise and fall of the country/national party', in S. Lockie and L. Bourke (eds), *Rurality Bites: The Social and Environmental Transformation of Rural Australia*. Sydney: Pluto Press.

Herbert-Cheshire, L. (2000) 'Contemporary strategies for rural community development in Australia: a governmentality perspective', *Journal of Rural Studies* 16 (2): 203–15.

Herbert-Cheshire, L. (2001) '"Changing people to change things": building capacity for self-help in natural resource management – a governmentality perspective', in G. Lawrence, V. Higgins and S. Lockie (eds), *Environment, Society and Natural Resource Management: Theoretical Perspectives from Australasia and the Americas*. Cheltenham: Edward Elgar. pp. 270–82.

Higgins, V. (2002) 'Self-reliant citizens and targeted populations: the case of Australian agriculture in the 1990s', *Arena* 19: 161–77.

Higgins, V. and Lockie, S. (2001) 'Getting big and getting out: government policy, self-reliance and farm adjustment', in S. Lockie and L. Bourke (eds), *Rurality Bites: The Social and Environmental Transformation of Rural Australia*. Sydney: Pluto Press.

Higgins, V. and Lockie, S. (2002) 'Re-discovering the social: neoliberalism and hybrid practices of governing in rural natural resource management', *Journal of Rural Studies* 18 (4): 419–28.

Hilmer, F. (1993) *National Competition Policy*. Canberra: Australian Government Publishing Service.

Hindess, B. (1996) *Discourses of Power: From Hobbes to Foucault*. Oxford: Blackwell.

Holmes, J. (2002) 'Diversity and change in Australia's rangelands: a post-productivist transition with a difference?', *Transactions of the Institute of British Geographers* 27: 362–84.

House of Representatives Standing Committee on Primary Industries and Regional Services (2000) *Time Running Out: Shaping Regional Australia's Future*. House of Representatives, Canberra.

Ilbery, B. and Bowler, I. (1998) 'From agricultural productivism to post-productivism', in B. Ilbery (ed.), *The Geography of Rural Change*. London: Longman. pp. 57–84.

Industry Taskforce on Leadership and Management Skills (1995) *Enterprising Nation: Renewing Australia's Managers to Meet the Challenges of the Asia-Pacific Century*. Canberra: Commonwealth of Australia.

Jessop, B. (1990) *State Theory: Putting the Capitalist State in its Place*. Cambridge: Polity Press.

Johnston, M. (1988) *Salt of the Earth: Stories of Families Surviving the Rural Crisis in Australia*. Melbourne: Collins Dove.

Latour, B. (1987) *Science in Action: How to Follow Scientists and Engineers Through Society*. Cambridge, MA: Harvard University Press.

Lawrence, G. (1987) *Capitalism and the Countryside: The Rural Crisis in Australia*. Sydney: Pluto Press.

Lawrence, G. (1998) 'The institute for sustainable regional development', in J. Grimes, G. Lawrence and D. Stehlik (eds), *Sustainable Futures: Towards a Catchment Management Strategy for the Central Queensland Region*. Rockhampton: Institute for Sustainable Regional Development. pp. 6–8.

Lawrence, G. (2003) 'Sustainable regional development: recovering lost ground'. Paper presented at the Social Dimensions of the Triple Bottom Line in Rural Australia One Day Seminar, Bureau of Rural Sciences, ACT, 26 February.

Lawrence, G. and Herbert-Cheshire, L. (2003) 'Regional restructuring, neoliberalism, individualisation and community: the recent Australian experience'. Paper presented at the European Society for Rural Sociology Conference, Sligo, Ireland, 18–23 August.

Lockie, S. (1994) 'Farmers and the state: local knowledge and self-help in rural environmental management', *Regional Journal of Social Issues* 28: 24–36.

Lockie, S. (1997a) 'Chemical risk and the self-calculating farmer: diffuse chemical use in Australian broadacre farming systems', *Current Sociology* 45 (3): 81–97.

Lockie, S. (1997b) 'What future landcare? New directions under provisional funding', in S. Lockie and F. Vanclay (eds), *Critical Landcare*. Wagga Wagga, NSW: Centre for Rural Social Research Key Papers No. 5, Charles Stuart University.

Lockie, S. (1999) 'The state, rural environments and globalisation: "Action at a Distance" via the Australian Landcare Program', *Environment and Planning A* 31 (4): 597–611.

Lockie, S. (2000) 'Environmental governance and legitimation: state-community interactions and agricultural land degradation in Australia', *Capitalism, Nature, Socialism* 11 (2): 41–58.

Lockie, S. (2001) '"Name Your Poison": the discursive construction of chemical use as everyday farming practice', in S. Lockie and B. Pritchard (eds), *Consuming Foods, Sustaining Environments*. Brisbane: Australian Academic Press.

Lockie, S., Herbert-Cheshire, L. and Lawrence, G. (2003) 'Rural sociology', in I. McAllister, S. Dowrick and R. Hassan (eds), *Cambridge Handbook of the Social Sciences*. Cambridge: Cambridge University Press.

Lowe, P., Murdoch, J., Marsden, T., Munton, R. and Flynn, A. (1993) 'Regulating the new rural spaces: the uneven development of land', *Journal of Rural Studies* 9: 205–22.

Llambi, L. and Llambi, D. (2001) 'A transdisciplinary framework for the analysis of tropical agroecosystem transformations', in G. Lawrence, V. Higgins and S. Lockie (eds), *Environment, Society and Natural Resource Management: Theoretical Perspectives from Australasia and the Americas*. Cheltenham: Edward Elgar. pp. 53–69.

Madden, B., Hayes, G. and Duggan, K. (2000) *National Investment in Rural Landscapes: An Investment Scenario for National Farmers' Federation and Australian Conservation Foundation with the Assistance of Land and Water Resources Research and Development Corporation*. Melbourne: Australian Conservation Foundation and National Farmers' Federation.

Marsden, T. (2003) *The Condition of Rural Sustainability*. Assen, The Netherlands: Van Gorcum.

Marsden, T., Murdoch, J., Lowe, P., Munton, R. and Flynn, A. (1993) *Constructing the Countryside*. London: University College of London Press.

McHugh, S. (1996) *Cottoning On: Stories of Australian Cotton-Growing*. Sydney: Hale and Iremonger.

McKinsey and Company (1994) *Lead Local Compete Global: Unlocking the Growth Potential of Australia's Regions*. Canberra: Commonwealth Department of Housing and Regional Development.

Miller, P. and Rose, N. (1990) 'Governing economic life', *Economy and Society*, 19 (1): 1–31.

Moreira, M. (2003) 'Local consequences and responses to global integration: the role of the state in the less favoured zones', in R. Almas and G. Lawrence (eds), *Globalization, Localization and Sustainable Livelihoods*. Aldershot: Ashgate. pp. 189–203.

Muetzelfeldt, M. (1992a) 'Organisational restructuring and devolutionist doctrine: organisation as strategic control', in J. Marceau (ed.), *Reworking the World: Organisations, Technologies and Cultures in Comparative Perspective*. Berlin: Walter de Gruyter. pp. 295–316.

Muetzelfeldt, M. (1992b) 'Economic rationalism in its social context', in M. Muetzelfeldt (ed.), *Society, State and Politics in Australia*. Sydney: Pluto. pp. 187–215.

O'Connor, J. (1993) 'Is sustainable capitalism possible?', in P. Allen (ed.), *Food for the Future: Conditions and Contradictions of Sustainability*. New York: John Wiley and Sons. pp. 125–37.

Orchard, K., Ross, H. and Young, E. (2003) 'Institutions and processes for resource and environmental management in the indigenous domain', in S. Dovers and S. Wild River (eds), *Managing Australia's Environment*. Annandale, NSW: Federation Press. pp. 413–41.

Peterson, T. (1991) 'Telling the farmers' story: competing responses to soil conservation rhetoric', *Quarterly Journal of Speech* 77: 289–308.

Pretty, J. (1998) *The Living Land: Agriculture, Food and Community Regeneration in Rural Europe*. London: Earthscan.

Price, P. (1998) 'Sustainability – what does it mean, and how should it be applied in a regional context', in J. Grimes, G. Lawrence and D. Stehlik (eds), *Sustainable Futures: Towards a Catchment Management Strategy for the Central Queensland Region*. Rockhampton: Institute for Sustainable Regional Development. pp. 15–27.

Pritchard, B., Curtis, A., Spriggs, J. and Le Heron, R. (eds) (2003) *Social Dimensions of the Triple Bottom Line in Rural Australia*. Canberra: Bureau of Rural Science.

Reeve, I. (2001) 'Property rights and natural resource management: tiptoeing round the slumbering dragon', in S. Lockie and L. Bourke (eds), *Rurality Bites: The Social and Environmental Transformation of Rural Australia*. Sydney: Pluto Press. pp. 257–69.

Rickson, R., Saffigna, P., Vanclay, F. and McTainsh, G. (1987) 'Social bases of farmers' responses to land degradation', in A. Chisholm and R. Dumsday (eds), *Land Degradation: Problems and Policies*. Cambridge: Cambridge University Press. pp. 187–200.

Rose, N. and Miller, P. (1992) 'Political power beyond the state: problematics of government', *British Journal of Sociology*, 42 (2): 173–205.

Short, K. (1994) *Quick Poison, Slow Poison: Pesticide Risk in the Lucky Country*. Sydney: Envirobook.

Standing Committee on Environment, Recreation and the Arts (1989) *Effectiveness of Land Degradation Policies and Programs Parliamentary paper* 285/89. Canberra: Commonwealth Government.

Vanclay, F. (1986) 'Socio-economic correlates of adoption of soil conservation technology'. M.Soc.Sc. Thesis. St Lucia: University of Queensland.

Vanclay, F. (1992) 'The social context of farmers' adoption of environmentally sound farming practices', in G. Lawrence, F. Vanclay and B. Furze (eds), *Agriculture, Environment and Society: Contemporary Issues for Australia*. Melbourne: Macmillan. pp. 94–121.

Vanclay, F. and Lawrence, G. (1995) *The Environmental Imperative: Eco-social concerns for Australian Agriculture*. Rockhampton: Central Queensland University Press.

Virtual Consulting Group (2000) *Repairing the Country: A National Scenario for Strategic Investment*. Victoria: Australian Conservation Foundation/National Farmers' Federation.

Ward, N. (1993) 'The agricultural treadmill and the rural environment in the post-productivist era', *Sociologia Ruralis* 33: 348–64.

Williamson, S., Brunckhorst, D. and Kelly, G. (2003) *Reinventing the Common: Cross-Boundary Farming for a Sustainable Future*. Sydney: The Federation Press.

Wilson, G. (2001) 'From productivism to post-productivism ... and back again? Exploring the (un)changed natural and mental landscapes of European agriculture', *Transactions of the Institute of British Geographers* 26: 77–102.

Woods, M. (2003) 'Globalisation, citizenship and the strategies of the rural movement: contesting the global from the local'. Paper presented at the European Society for Rural Sociology Conference, Sligo, Ireland, 18–23 August.

World Commission on Environment and Development (1987) *Our Common Future*. Oxford: Oxford University Press.

4

Rural space: constructing a three-fold architecture

Keith Halfacree

Building a map in order to find,
What's not lost but left behind

(Beth Orton, *Tangent*, 1996)

INTRODUCTION

This chapter develops a framework for exploring the present day character and status of 'rural space'. Reflecting the present era, where fragmentation within all aspects of life appears as a key *leitmotif* (Harvey, 1989), the chapter seeks to construct a heuristic device – a 'map' – with which to interrogate rural space. Such a strategy aims to bring together the dispersed elements of what we already know about rural space more than to reveal some hidden or 'lost' aspect of this space. As such – and as seems appropriate within a Handbook of rural studies – the chapter aims to provide a resource to be drawn upon by those in search of a better understanding of the character of rural space throughout the world today. Thus, whilst clearly written from a British vantage point, the chapter will hopefully resonate much further afield.

At the outset, the chapter is not going to rehearse yet again the debates that have raged within geography and social theory concerning the ontological and epistemological status of 'space' (cf. Crang and Thrift, 2000a; Gregory and Urry, 1985; Peet, 1998). Instead, it takes as a starting point the position that space – and anything that we might call 'rural space' – is not 'a practico-inert container of action' but 'a socially produced set of manifolds' (Crang and Thrift, 2000b: 2). Space does not somehow 'just exist', waiting passively to be discovered and mapped, but is something created in a whole series of forms and at a whole series of scales by social individuals. We thus have a great diversity of 'species of space' (2000b: 3) implicated in every aspect of life, some touched on within this chapter.

Much of what follows draws on the ideas of the late Henri Lefebvre. This material, not yet widely deployed within rural studies, seeks to broaden and to enrich our understanding of space, and to draw out both its mundane everyday significance and its highly abstract character under capitalism. Through developing a Lefebvrian model of (rural) space, the chapter argues that far from disappearing as a significant conceptual category, 'rural space' does indeed retain what Sarah Whatmore (1993: 605) termed an 'unruly and intractable … significance', both within everyday life and for us academics.

Although the issue of defining the rural has its own chapter in this collection (Cloke, Chapter 2 in this volume), it is with reference to this debate that we start. This is because the very idea of 'rural space' is pleonastic.[1] The redundancy in the term comes from the fact that the concept 'rural' is inherently spatial, with 'space' understood in the broad sense implied above. Any attempt to separate rural from space runs the risk of reproducing the unhelpful dualism of society versus space. From the brief engagement with defining rural (space), the second section of the chapter develops a three-fold understanding of space, and then of rural space. Finally, the third section illustrates an application of this model through a brief account of the two key phases of rural spatiality in post-1945 Britain.

DEFINING THE RURAL

In a paper from 1993, I argued that the rural is best understood in two ways (Halfacree, 1993). Both attempted to provide a definition rooted in 'rational abstraction' or the identification of 'a significant element of the world which has some unity or autonomous force' (Sayer, 1984: 126). Unity could be attained, albeit through a degree of time–space bracketing, by regarding the rural as either 'locality' or 'social representation'. Crucially, these two conceptions of rural were seen as intrinsically interwoven and co-existent rather than mutually exclusive. The significance of this point will be developed below. Before this, however, I must say a little about the geographical specificity of the term 'rural' itself (see also Chapters 1 and 2).

The geographical specificity of 'rural'

The problematic issue of what different peoples in different places mean by 'rural' goes right to the heart of the theoretical argument developed below. Quite simply, neither at the official (Halfacree et al., 2002) nor at the cultural or popular level is there consensus on the delineation of the 'non-urban' spaces that the term 'rural' seeks to encapsulate. Even within Europe, Hoggart et al. observe:

> there is little chance of reaching consensus on what is meant by 'rural' … It is both more straightforward and more convenient to establish definitions of urban areas [but cf. Champion and Hugo, 2004], based on population size or building density, than to attempt to identify the defining parameters of rural space. (1995: 21)

A similar problem faces us when talking about the rural in developing countries (Barke and O'Hare, 1991). For example, within the Arab world, 'rural landscapes … are as diverse and complex as those of any other major culture region' (Findlay, 1994: 126).

Unsurprisingly given the 'manifold' character of space, 'rural' can conjure up a huge range of spatial imaginaries. A list of these could include: countryside, wilderness, outback, periphery, farm belt, village, hamlet, bush, peasant society, pastoral, garden, unincorporated territory, open space … (for an Anglo-US perspective, see Bunce, 1994; Marx, 1964; Short, 1991). In part, this variation reflects geographical context. A good example comes through in research on counter-urbanization. In the small, densely populated island of Great Britain, the rural side of this migration trend is typically defined at the scale of the local government District (for example, Champion, 1998). In contrast, in the USA and Australia, rurality is depicted at the more macro scale of the Non-Metropolitan Region (for example, Frey and Johnson, 1998; Hugo and Bell, 1998). None the less, it is much more than a question of scale that shapes 'rural', since each of these diverse spatial imaginaries also bears the imprint of practices of culture, contestation, commodification, etc. The role of these multiple imprints is clear in the rest of this chapter.

Returning to the specificity of the rural, in the light of this chapter's British bias it is important to note that 'in an international context, the English[2] notions of 'rural' and the 'countryside' are strongly contested and often non-existent' (Wilson, 2001: 90). For example, in stark contrast with Britain, Laschewski et al. (2002) regard the rural in Germany as a 'secondary concept', subordinate to other spatialized terms such as 'region', 'peasant' or 'periphery'. This is reflected in the problem of finding any 'direct translations of "rural" (ländlich?) or "countryside" (no equivalent word)' (Wilson, 2001: 90). Any talk or analysis of rural or rural space must always be sensitive to this issue of geographical specificity.

Rural space as material

Defining rural as a 'locality' is inspired by structuralist and broader materialist concerns not to fetishize space through the metaphor of a 'container' but to see it as being constantly produced, reproduced and (potentially) transformed. As Smith sharply observes, following Marx, 'a geographical space [*sic*] … abstracted from society is a philosophical amputee … we do not live, act and work "in" space so much as by living, acting and working we produce space' (1984: 77, 85).

More specifically, localities are relatively enduring spaces inscribed by social processes or, less passively, both inscribed and used by social processes; product and means of production. They can be '*visualized* as islands of absolute space in a sea of relative space' (Smith, 1984: 87; emphasis added). This understanding of locality is more theoretically reflexive and, as a consequence, more restrictive than that promoted through the 'localities research' that emerged in the 1980s (for example, Cooke, 1989). The latter ran the danger of spatial (environmental) determinism, a risk commonplace whenever specific places are studied 'in and of themselves' (Smith, 1987: 62).

To identify rural localities, at least two conditions must be met (Hoggart, 1990). First, we must show *significant* processes in operation that are delineated at a local spatial scale. Second, the resulting spatial inscriptions must enable us to distinguish 'rural' from one or more 'non-rural' environments. In summary, we must 'identify locations with distinctive causal forces' (1990: 248) we can label 'rural'.

Many authors doubt whether rural localities, or at least socially significant rural localities, can be identified today, especially in the so-called developed world (Cloke, 1999). Such doubt reflects, in particular, the influence of a political economy perspective on rural studies (Cloke, 1989). Quite simply, the spatiality of contemporary capitalism has tended to do away with the old geographical demarcations and borders. There has been a 'spatial loosening of the elements once considered indicative of ... rural and urban' (Lobao, 1996: 89). The scale of operations within capitalist society is constantly being re-written and the significance of a 'rural' scale has been incessantly and plurally undermined. From local, national and international (global) perspectives, the rural often seems anachronistic and overtaken by events (Mormont, 1990), at best a minor player in the line-up of localities.

Reflecting this effacement of erstwhile rural space, in a Presidential address to the Rural Sociological Society in 1972, the US sociologist James Copp argued forcefully that:

> There is no rural and there is no rural economy. It is merely our analytic distinction, our rhetorical device. Unfortunately we tend to be victims of our own terminological duplicity. We tend to ignore the import of what happens in the total economy and society as it affects the rural sector. We tend to think of the rural sector as a separate entity ... (1972: 519)

The British geographer Keith Hoggart reiterated the essence of this argument 18 years later:

> undifferentiated use of 'rural' in a research context is detrimental to the advancement of social theory ... The broad category 'rural' is obfuscatory ... since intra-rural differences can be enormous and rural–urban similarities can be sharp. (1990: 245)

From this perspective we must 'do away with rural' (p. 245) for theoretical progress. Indeed, continued belief in its inherent salience may be seen as ideological, in that it denies and confuses our picture of the spatiality of contemporary capitalism.

Rural space as imaginative

In contrast to exploring and evaluating the existence of a socially significant materially expressed rural space, an alternative strategy is to look further at Copp's argument that the rural today is 'merely' an analytic distinction or rhetorical device. Here, the predominant character of rural space moves away from being materially immanent on the ground to being first and foremost imaginative (Mormont, 1990). It may have material expression, of course, but such embodied traces are not its primary reality. The importance of such a perspective has already been suggested in the discussion of the geographical variability of the term 'rural' internationally.

As a concept utilized in everyday life, the rural is a part of what Sayer (1989) terms a 'lay narrative'. Typically, as reflected in Copp's quote, such narratives have been rather dismissed by academics because of their relatively loose and unexamined character as compared to the supposedly rigorous and reflexive theory-saturated concepts favoured by 'theorists'. 'Conversational realities' must yield to 'academic discourses' (Shotter, 1993) or 'academic narratives' (Sayer, 1989). However, this is a problematic positioning, with its elevation of the academic stance over that used by everybody – including academics – most of the time. Its limitations are immediately apparent when the rural as a lay narrative is examined further.

This examination can be undertaken by drawing on the work of Serge Moscovici and the insights of symbolic interactionism. Moscovici (1984) argued that in order to deal with the perpetual complexity of the world around us we are forced to simplify it into a series of 'social representations'. These are understood as:

> organizational mental constructs which guide us towards what is 'visible' and must be responded to, relate appearance and reality, and even define reality itself. The world is organized, understood and mediated through these basic cognitive units. Social representations consist of both concrete images and abstract concepts, organized around 'figurative nuclei'. (Halfacree, 1993: 29)

Applying this theory to the definition of 'rural' we can say that this 'mere' rhetorical device refers to a 'social representation of space' (Halfacree, 1993, 1995; Jones, 1995). Rural space's cognitive representation rather than its appearance in the social and physical landscape thus becomes the entry point of our interest. This is irrespective of whether or not rural localities are acknowledged. Indeed, even if one doubts the continued existence of any distinct rural locality:

> The rural is a category of thought. … The category [is] not only empirical or descriptive; but it also [carries] a representation or set of meanings, in that it [connotes] a more or less explicit discourse ascribing a certain number of characteristics or attributes to those to whom it [applies]. (Mormont, 1990: 40, 22)

The idea that the rural is a social representation of space can, however, be dematerialized further. This comes from a critique of both elements of the term 'social representation'. First, we can take issue with its *social* character, which assumes a degree of group-specific consensus in composition. Whilst this clearly facilitates both communication and understanding (Potter and Wetherell, 1987; Halfacree, 2001a), it is hard to know where to delineate the boundary between social and more 'individual' representations (Potter and Wetherell, 1987). Second, we can question the *representational* or cognitive character of the concept. For example, Shotter suggests that our everyday 'conversational realities' attain their fullness through playing themselves out within specific discursive situations, without there being a need for them to be grounded through 'any reference to any inner mental representations' (1993: 142).

Following both lines of critique, Potter and Wetherell (1987) proposed an alternative concept to social representation, namely the 'interpretative repertoire'. This fundamentally post-structuralist and linguistic concept is located within discourse alone and comprises 'a lexicon or register of terms and metaphors drawn upon to characterize and evaluate actions and events' (Potter and Wetherell, 1987: 138; also Shotter, 1993). Within debates about defining rural, Pratt has taken up this interpretative repertoire direction most directly. He considers that it is paramount to stress 'the variability, contradiction and variety of representation and articulation of rural discourses' (1996: 76). Indeed, he called earlier for the replacement of the term rural with that of 'post-rural'. This would highlight the rural's constant 'reflexive deployment' and indicate how '*the point is there is not one* [rural] *but there are many*' (Murdoch and Pratt, 1993: 425; emphasis in original).

Such a fundamental critique again leads us to question the contemporary validity of any notions of rural space. The concept of social representation, because it is representational, does allow us to retain a notion of rural space, albeit one that is much more 'virtual' than that implicated in the locality definition. However, by denying the representational element, with its implied delineation and boundedness, regarding the rural as an interpretative repertoire fundamentally destabilizes what is left of any fixity. Any linguistic rural space that is produced through an interpretative repertoire becomes fundamentally and irreducibly contextual and thus highly transient. A ghostly ephemerality is suggested.

Rural space as material and ideational

Cloke and Park (1984) presented a range of definitions that depicted the rural (countryside) in terms of areas dominated by extensive land uses, containing small, low order settlements and/or engendering a cohesive sense of community identity. These emphases are also apparent in the probable content of the two definitions outlined above. The attempt to understand rural space through the locality definition is likely to draw upon the distinctiveness of one or more of the following: agriculture and other primary productive activities, low population density and physical inaccessibility, and consumption behaviour (Halfacree, 1993; Moseley, 1984). In contrast, the social representation of space approach is likely to contrast an imagined rural geography of landscape aesthetics and 'community' with that of other spaces, notably the city/urban and the suburb (Halfacree, 1995; Murdoch and Day, 1998).[3]

As noted above, and as suggested by their likely content, both definitions should be seen as interwoven rather than mutually exclusive. In other words, the material and ideational rural spaces they refer to intersect in practice. For example, a belief (social representation) in the distinctiveness of rural space (locality), in terms of a people's way of life and priorities and in terms of general geography, lies behind the politics of the British organization the Countryside Alliance (Woods, 2003). Specifically, the Alliance feels that the 'real rural agenda' (locality) (Countryside Alliance, 2002) has not been recognized, or has been ignored or overwhelmed by an 'urban' perspective (social representation) on the countryside. Whilst one can be critical of such groups, not least through recognizing their ideological agenda, dissection of their rurality-in-discursive-motion exposes the contours of a space that is both material and ideational. This, of course, should come as no surprise to those who hold a 'socially produced manifold' model of space.

Another excellent example of these two dimensions of rural space, plus their interwoven character, is given by Gray's (2000) examination of how the European Union (EU) has defined rural space. He argued that the EU has mixed the 'two modes of conceiving rurality' (p. 32) in formulating, applying and appraising its Common Agricultural Policy (CAP). Four phases, alternating between the two modes, are teased out:

1 *Representation.* When conceiving the CAP, the EU initially understood the rural as a space defined by agriculture. Specifically, when the EU thought of rurality it thought of a certain form of agriculture, namely small family farms. Such farms were seen as the bedrock of rural society – everything 'rural' was ultimately built up upon them. There was also a strong moral element implicated within this vision, with these farms and their resulting society being based on hard, honest work.
2 *Locality.* The EU then sought to define the rural as a locality informed by this idea of what made the rural meaningful. This was so as to be able to implement the CAP. Boundaries were drawn on the maps to delineate rural spaces based upon agriculturally related characteristics: topography, resources, farm development potential, etc. It was thought that a distinct set of processes acted within these localities. Thus, a '*rural* problem' was defined where there were threats to small family *farming* and, thus, to *rural* society. Policies were put in place trying to sustain these delineated landscapes.
3 *Representation.* Since the 1970s, the mental construct of the rural held by the EU has altered, largely as a result of the changes and trends occurring (if not necessarily originating) within rural places themselves. Consequently, the spatial imagination of the CAP was revised. In European Commission documents, such as *The Future of Rural Society* (CEC, 1988) or *The Cork Declaration* (CEC, 1996), the rural is seen as being much more autonomous from agriculture. Rural space is recognized as comprising a heterogeneity of activities – consumption as well as production.
4 *Locality.* Drawing boundaries around rural space can no longer be guided largely by agricultural criteria. Significant processes that are felt to inscribe rural localities are no longer just agriculturally related. The revised CAP thus seeks to delineate a more internally diversified rurality, drawing on the imprint of processes such as those represented through local cultures. For example, in Highland Scotland, effort has been made to recognize rural localities according to Gaelic culture and crofting (see Black and Conway, 1995).

Rural space as practised

Gray's example draws attention to a further crucial point, namely that ideas of space cannot be separated from ideas of time. Time does not exist on the metaphorical head of a pin/black hole and space is always temporal: 'all space is anthropological, all space is practised, all space is place' (Thrift, 1996: 47). In short, moving away from the Kantian absolute, space and time are inseparable and we have '*space as process* and *in process* (that is space and time combined in becoming)' (Crang and Thrift, 2000b: 3; emphasis in original).

It follows that when we think about the possibility of rural space we must also think about the possibility of rural time. Overall, we should consider rural space–time, although my terminology will stick to the expression 'rural space' in this chapter. Thus, we must note how the material space of the rural locality only exists through the *practices* of structural processes, and how the ideational space of rural social representations only exists through the *practices* of discursive interaction. In this respect, therefore, investigation into rural space(–time) requires a strongly contextual approach to enable us to tease out the entanglements at play. This immediately begins to question the ultimate usefulness of exploring the totality of rural space through the somewhat arid, distanced and partial categories (Halfacree, 2001a) used in 'defining the rural'. Instead, we need to think of (rural) space synergistically, as 'more than simply the sum of separate relations that comprise its parts' (Smith, 1984: 83).

TOWARDS A THREE-FOLD UNDERSTANDING OF RURAL SPACE

This section works towards the development of such a more fully contextual process-rooted synergistic understanding of rural space. It aims to produce an architecture within which the totality – its diversity, even 'duplicity' (Halfacree, 2003) – of rural space can be better appreciated. This is a *model* that can be applied to all rural places, although its *content* will be extremely diverse.

Problems with the locality and representational definitions

My stress on the interwoven character of rural space in terms of material and ideational elements sought to overcome any dualism between the definitions, which would force us to choose one or the other as the 'right' perspective. None the less, dualistic ways of thinking still haunt this model. For example, when the two definitions were first laid out, it was suggested that the locality approach was rooted within the distanced, disinterested perspective that has traditionally been the *modus operandi* of academia. In contrast, the social representation approach was an attempt to tap into 'ordinary' people's lay narratives. However, as Jones (1995) has pointed out, we need to see not just two types of discourse but a spectrum of discourses between the lay and the academic, such as what he terms 'popular' and 'professional' discourses. This speaks of a mixed, hybrid character to rural space or even an infinity of spatialities.

As noted at the start of the last section, one of the aims of the attempt to define the rural as a locality and/or a social representation was to consolidate rurality within a rational abstraction. However, as we also saw above, it remains a moot point just how rationally abstract both definitions are. Instead, they may be seen as 'chaotic conceptions' that 'arbitrarily divide the indivisible and/or [lump] together the unrelated and the inessential' (Sayer, 1984: 127). In both cases, therefore, the current theoretical saliency of rural space is questioned, which suggests that to try to maintain a foundational idea of rural space – material *or* ideational – is fetishistic. Only through a focus on contextual practice is the 'truth' of rural space revealed.

An essentialism to rural space also haunts the model through the ways in which each definition appears practised. In the locality version, rural space is practised as a set of distinct social, political and/or economic actions, whereas in the social representation model, the practice is largely symbolic or linguistic. Although both the Countryside Alliance and the EU example showed the insufficiency of these understandings, focusing on the two ways of defining rural thus seems to break up the totality of rural space. Related to this point, the two approaches also run the risk of mis-specifying 'space'. For example, by stressing the rural's status as a 'constructed representation' at the expense of any existence as an 'ascertained reality', Mormont (1990: 41) reproduces a society–space dualism that, as a consequence, reproduces a much more limited sense of space than the present chapter has assumed from the outset.

Rural space as a relative 'permanence'

The fuller concept of rural space that I wish to present is rooted in some of the core ideas contained within Henri Lefebvre's *The Production of Space* (1991a). Although this book was originally published in France in 1974, it was not translated into English until 1991. Its appearance in that year has been described by Merrifield (2000: 170) as '*the* event within critical human geography over the 1990s' and, although there have been some detractors (for example, Unwin, 2000), it has become extremely influential within geography (and beyond). However, its place within rural geography has been much less marked (Phillips, 2002), perhaps in part reflecting a perception of Lefebvre as having had 'very little to say about rural life' (Unwin, 2000: 15).[4]

Lefebvre's work is typically written in a fairly loose and indeterminate – and, admittedly, sometimes impenetrable – style, which for critics such as Unwin (2000) is one of its fundamental flaws. Alternatively, one can see this imprecision and fluidity as facilitating the resourcefulness of his ideas. As with his concept of space discussed below, he can be said to have left 'us to add our own flesh and to re-write it as part of our own chapter or research agenda' (Merrifield, 2000: 173). This is what those inspired by Lefebvre mostly appear to have done and it is also my intention here.

Lefebvre's relative neglect of the rural within his work comes in part from a combination of showing the universality of the production of a particular kind of space – urban and rural – under capitalism, and from his dialectical attempt to resist binaries or dualisms (Shields, 1999). For example, he resists the idea that the principal spatial contradiction of capitalism lies within the dualism 'town' versus 'country', locating it instead within the urban (Gregory, 1994; also Harvey, 1985; Lefebvre, 1996: 118–21). However, the implications of Lefebvre's ideas do not result in a simple rejection of the rural/rural space. This is because, for Lefebvre (after Hegel and Marx), the concept of *production* that underlies capitalist spatiality is a 'concrete universal'. It is 'a social practice [that] suffuses all societies but its concrete form is differentiated from one to the other' (Smith, 1998: 278). The rural can be a significant category that emerges – and not necessarily just as a dualistic 'response' to the urban – within this differentiation.

One of the core tasks of *The Production of Space* (Lefebvre, 1991a), was to draw attention to the way in which space is now intrinsically linked with capitalism and exchange, rather than being somehow relatively independent of society's mode of production. Today, reflecting the lasting legacy of critical geography, this may seem a fairly obvious statement but its implications remain profound. Under capitalism, space becomes *produced*:

> the production of space can be likened to the production of any other sort of merchandise, to any other sort of commodity ... [S]pace ... isn't just the staging of reproductive requirements, but part of the cast, and a vital, productive member of the cast at that ... It is a phenomenon which is colonized and commodified, bought and sold, created and torn down, used and abused, speculated on and fought over. (Merrifield, 2000: 172–3)

More specifically, Lefebvre characterizes the space produced through capitalism as 'abstract'. Constantly being moulded by the pressures and demands of the market and social reproduction, this abstract space transcends earlier 'modes of production' of space (Shields, 1999: 170–85). Abstract space's core feature is its simultaneous homogeneity and fragmentation. As with production, it is a concrete universal, or a concrete abstraction:

> Space ... is both *abstract* and *concrete* in character: abstract inasmuch as it has no existence save by virtue of the exchangeability of all its component parts, and concrete inasmuch as it is socially real and as such localized. This is a space, therefore, that is *homogenous yet at the same time broken up into fragments*. (Lefebvre, 1991a: 341–2; emphasis in original)

This dual character suggests how space both subsumes place, with the loss of any ingrained meanings – in Marx's famous dictum, 'all that is solid melts into air' – and reconfigures places as relative 'permanences' carved out through the flow of processes producing space (Harvey, 1996: 261). These are very much temporary places due to the inherent dynamism of capitalist spatiality, as drawn out so strongly in David Harvey's work. In summary: 'The geographical and technological landscape of capitalism is torn between a stable but stagnant calm incompatible with accumulation and disruptive processes of devaluation and "creative destruction"' (Harvey, 1985: 138).

Moreover, reflecting this inherent dynamism and as Merrifield's previous quote implied, as a vital member of the productive cast, the significance of space for Lefebvre is not ended when we appreciate that it is produced. Space is not simply a product but also a medium through which production occurs (Gottdiener, 1985). To appreciate space, therefore, we must not only note the geography and temporality of its production but also how it operates as a *means of production*. For it to function as the latter, attention must be given to space's more humanistic dimensions. Thus, 'the production of space also implies the production of the meaning, concepts and consciousness of space which are inseparably linked to its physical production' (Smith, 1984: 77). These social constructions can 'operate with the full force of objective facts' (Harvey, 1996: 211).

Rural space, then, suggests one of the productive permanences of capitalist spatiality. However, to avoid 'the grossest of fetishisms' (Harvey, 1996: 320), we need to show how its multi-faceted character comes into being within the context of abstract space. To do this requires us to develop our more complete picture of the architecture of this space. This will 'bring the various kinds of space and the modalities of their genesis within a single theory' (Peet, 1998: 102–3).

A three-fold model of space

> Materiality, representation, and imagination are not separate worlds. (Harvey, 1996: 322)

The complex model of rural space developed below is based on Lefebvre's (1991a: 33, 38–39, *et seq.*) seminal three-fold understanding of spatiality; his 'conceptual triad'. This triad has been outlined and developed in various locations (for example, Gregory, 1994: 401–6; Merrifield, 1993: 522–7; 2000; 2002: 173–5; Shields, 1999: 160–70). Let us consider each element in terms of my interpretation of them.

First, there are *spatial practices*. These are the actions – flows, transfers, interactions – that 'secrete' a particular society's space, facilitating both material expression of permanences and societal reproduction. Bearing in mind the inherent instability of space, they can secrete contradiction as well as stability (Merrifield, 2002: 90). Spatial practices are inscribed routine activities and their expression bears similarities with the concept of the locality outlined earlier. They are associated with everyday *perceptions* of space. They structure our everyday reality, whilst at the same time being rooted within that reality. As such, spatial practices can also be traced to rules and norms, and to space as lived.

Second, there are *representations of space*. These are formal conceptions of space, as articulated by capitalists, developers, planners, scientists and academics. They have much in common with Sayer's (1989) idea of an 'academic narrative' that is the tool of the 'specialist'. Representations of space are *conceived* and abstract and expressed through 'arcane signs, jargon, codifications' (Merrifield, 2000: 174). They also find 'objective expression' directly in such things as monuments, factories, housing estates, workplace and bureaucratic rules, and in the more general rules and norms of everyday life that are operative in a given place. Thus, representations of space will to some extent be perceived, appropriated and perhaps even subverted within daily life. They are always 'shot through with a ... mixture of understanding ... and ideology' (Lefebvre, 1991a: 41).

Third, there are *spaces of representation*. These diverse and often incoherent images and symbols are associated with the tumults and passions of space as *directly lived*. This was a major area of interest for Lefebvre (for example, Lefebvre, 1991b), with the turmoil of the everyday making his 'heart soar' (Merrifield, 2002: 90). Although clearly with links to perceived space, spaces of representation refer to vernacular space symbolically appropriated by its users. Such a 'social imaginary' (Shields, 1999: 164) can have a subversive aspect when it results in space being (re)appropriated from the interests of the dominant. Partly in response to this threat, representations of space will seek to dominate and control the potentially 'hot' spaces of representation. They are thus key sites of political and ideological struggle.

As I have tried to suggest in this outline of Lefebvre's spatiality, each of the three facets cannot be understood in isolation from the other two. Each forms an element of a three-part dialectic and thus each facet is always 'in a relationship with the other two' (Shields, 1999: 161). In any given permanence, however, one or more may be dominant, and one or more may appear almost indistinct; likewise, the individual facets may serve to reinforce, contradict or be relatively neutral with respect to one another. For example, under the regime of abstract space, the abstraction of exchange value elevates representations of space that, through commodification and bureaucratization, seek to dominate the use value concrete realm of spaces of representation (Gregory, 1994: 401). None the less, ultimately, all three facets together comprise space's 'triple determination'.

Moreover, again reflecting Lefebvre's dialectical approach, these three facets of space are seen as *intrinsically dynamic*, as are the relations between them. Lefebvre's spatiality is inherently 'turbulent' (Gregory, 1994: 356). Space as some kind of frozen category has no meaning. As Merrifield argues, the spatial triad must always 'be *embodied* with actual flesh and blood and culture, with real life relationships and events' (2000: 175; emphasis in original). Thus, contra Unwin (2000: 21, 24), Lefebvre does *not* 'dangerously [reduce] the significance of time' and does *not* 'dehumanize' space; quite the opposite.

A three-fold model of rural space

Combining Lefebvre's ideas with the definitional debates outlined earlier enables me to suggest a more complex model of *rural* space. This is outlined in Figure 4.1. Again, it has three facets:

- *Rural localities* inscribed through relatively distinctive spatial practices. These practices may be linked to either production or consumption activities.
- *Formal representations of the rural* such as those expressed by capitalist interests, bureaucrats or politicians. Crucially, these representations refer to the way the rural is framed within the (capitalist) production process; specifically, how the rural is commodified in exchange value terms. Procedures of signification and legitimation are vital here.
- *Everyday lives of the rural*, which are inevitably incoherent and fractured. These incorporate individual and social elements ('culture') in their cognitive interpretation and negotiation. Formal representations of the rural strive to dominate these experiences, as they will rural localities.

All three facets together comprise rural space, an understanding that dissolves the potential dualism between locality and social representation that troubled me above. As was suggested by the quote that starts this chapter, this three-fold architecture for rural space is less about establishing a new understanding than about realizing what we already have. For example, this structure of everyday life, locality and formal representation bears a fair resemblance to the definition of 'rural' given in the fifth edition of the *Concise Oxford Dictionary*: '*In, of, suggesting*, the country' (emphases added).

The extent to which an individual *place* (at whatever scale) can be said to merit the label 'rural' depends on the extent to which the totality of rural space dominates that place relative to other spatialities. Clearly, as recognized long ago (see above), this is a moot point. Places represent the meeting points of networks, 'constructed out of a particular constellation of social relations, meeting and weaving together at a particular locus' (Massey, 1996: 244). The rural status of any place is thus an issue that always must be determined on the ground/in place to avoid rural fetishism. A further issue here will be the extent to which the three facets of rural space cohere in a united front within that place. As noted above, this is inherently problematic. For example, formal representations never completely overwhelm

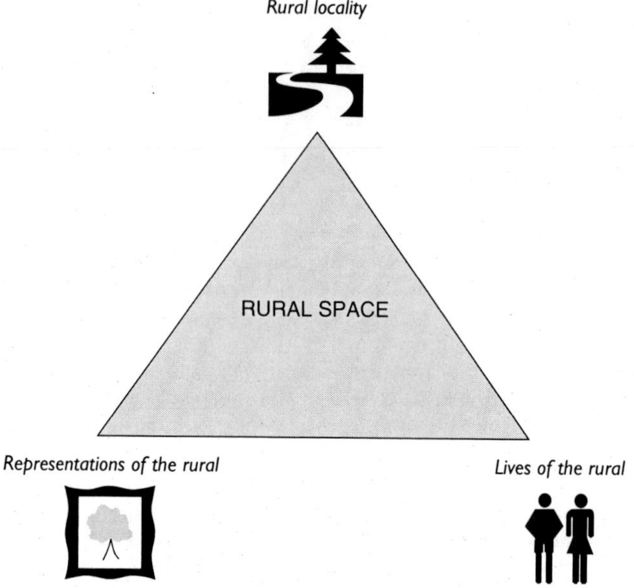

Figure 4.1 ***The totality of rural space***

the experience of everyday life – although they may come close – and the extent to which formal representations and local spatial practices are unified is also uneven. The tension within abstract space between permanence and flow is also apparent throughout.

Following Cloke and Goodwin (1992), the 'middle-ground regulationist concept' (Cloke, 1994: 166) of 'structured coherence' (after Harvey, 1985: 139ff.) is a useful device for exploring the internal consistency of any permanence that is rural place. This concept refers to the extent to which economy, state and civil society mesh together in a *relatively* stable fashion at the local level (Cloke and Goodwin, 1992). Each of these three societal elements will be shaped by and internalize the three facets of rural space, again to varying degrees. Overall, three characterizations of a rural structured coherence under the more meta-level purview of abstract capitalist spatiality are suggested:

- *Congruent and unified.* All elements of rural space cohere in a relatively smooth, consistent manner. Formal representations of the rural are unified, overwhelming and hegemonic.
- *Contradictory and disjointed.* There is tension and contradiction between or within the elements of rural space but an overall coherence does hold, best appreciated at a more meta-level. Differences are more 'induced' than 'produced' (Lefebvre, 1991a: 372). Formal representations of the rural are less hegemonic, and often-diverse spatial practices come to the fore.
- *Chaotic and incoherent.* There is a lack of local structured coherence brought about by more fundamental contradictions between or within the elements of rural space. Rural space is dominated by everyday lives of the rural, as it holds together at neither the perceptual nor conceptual level. This characterization represents a potentially subversive alternative within the overall logic of abstract spatiality as, for example, it can disrupt societal reproduction. Some difference this time is 'produced' and 'presupposes the shattering of a system' (Lefebvre, 1991a: 372) if that system is unable to recuperate it.

RURAL SPACE TODAY: A BRITISH ILLUSTRATION

I will now illustrate the model of rural space in the contemporary British context. Clearly, the international relevance of this illustration will vary enormously. However, the core aim here is elucidatory, to indicate the application of the theoretical ideas developed in the last section. I make this brief sketch under two sub-headings: productivism and post-productivism. Attention will also be paid to the type of structured coherence attained within these spatialities. First, though, consideration must be given to the aforementioned international context.

Rural change in an era of 'globalization'

The debate about changing rural space detailed below takes place within an international context of ongoing rural and urban change (Champion and Hugo, 2004). Two key and interrelated dynamics appear to be at work. First, and of significance to every country in the world, is the changing and intensifying influence of capitalist globalization, not least as manifested in a 'globalized food system' (Symes and Jansen, 1994a: 5; compare Goodman and Watts, 1997; Le Heron, 1993; McMichael, 1994; Wallace, 1992). In the wake of global agreements at the GATT level (now World Trade Organization) on agricultural trade (McMichael, 1993; Winter, 1996), this manifestation is now increasingly characterized by a freer rein for 'market forces', as direct state engagement with agriculture retreats. Of course, as any glance at the news tells us almost daily, this is proving a painful and grudging retreat, seen in ongoing agricultural trade disputes between and within the EU, United States and other large food exporters, such as the Cairns Group (Denny and Elliott, 2002). None the less, it is one that is pushing developing countries towards an export-led agriculture, often to the detriment of more domestic food priorities (Shiva, 2000), of typically 'non-traditional' products (for example, Llambi, 1994; Raynolds, 1994). For developed countries, this restructuring is prompting a questioning of the previously hegemonic position of agriculture within rural society, *ensemble*, whilst also promoting intensification and industrialization within that industry, from hogs in Iowa (Page, 1997) to genetically modified rapeseed in Oxfordshire (Monbiot, 2002: ch. 7).

The second key dynamic powering rural change is the recognition of the increasing consumption role and potential of rural places. This is apparent within developing countries in, for example, the sequestration of former commons for National Parks, where a primary objective is to attract big-spending tourists and trophy hunters (Monbiot, 1994; also Pretty, 2002), but it is most noted in the rich world. Here, 'agrarian marginalization' (Buttel, 1994: 16) and rising consumption concerns are reflected in everything from the increasing weight given to 'environmental' considerations within some agriculture (Bowler et al., 1992; Harper, 1993; Potter, 1998) to the replacement of agriculture by other land uses in many locations, typically those attempting to service the 'external' (Marsden, 1999) demands of urban residents. Consequently, developed countries are seeing the emergence of a multifunctional *rural* regime (cf. Wilson, 2001), an increasingly regionalized rural (Halfacree et al., 2002; Murdoch et al., 2003). This latter trend is especially evident within Britain, to which I now turn.[5]

The spaces of productivism

Eager to overcome the boom–bust cycles to which the agricultural industry had been exceptionally prone, and to reward the efforts of the farming community during wartime, the post-1945 British government sought, through legislatively driven reforms, to provide a more secure base for the country's farmers – to 'make two blades of grass grow where one grew before' (Shucksmith, 1993: 466). This heralded the start of the productivist era, which lasted until at least the late 1970s.

Productivism positioned agriculture as a production maximizer, a progressive and expanding food production-orientated industry (Marsden et al., 1993) but also as an industry with its roots in an established agricultural landscape. As we shall see below, this was a contradiction that has increasingly proved problematic. Additionally and crucially, productivism was not just experienced by the farming community but filtered into every corner of British rural life and even beyond. It was a cornerstone of rural local structured coherences and can be seen almost as shorthand for British rural life between 1945 and 1980. Let me now consider its three facets of rural space in more detail.

First, a *rural locality* was inscribed through the predominance of particular agricultural practices. These encapsulated the daily and seasonal activities of the farmers themselves, plus their multi-faceted and increasingly specialized support services. These practices are outlined, for example, in classic textbooks of the time (for example, Morgan and Munton, 1971; Tarrant, 1974). Thus, in 1967's *Agricultural Geography*, Symons characterized the 'agricultural activities' to be studied 'in a spatial context' as:

> the constituent activities of cropping and livestock rearing, and … the outputs (crops and livestock) and … the farms, fields, labour, machinery and all other inputs required for production. (1978 [1967]: 1)

Overall, these agricultural practices focused on increasingly industrialized modes of food production and of increasing both the output and the profitability from the land. This is also to some extent implied by the interrogation of agriculture almost exclusively through neoclassical economics within these same textbooks.

Beyond the farm, predominant practices also had an agricultural flavour, from the harvest festival and other village celebrations, to the imprints of the local landed elites, to the service roles played by market towns. The whole rural locality appeared to radiate out positively from agriculture. Illustrating this, in an upbeat account of dairy farming in Wales, Bowen described how:

> the daily collection of milk and its transportation to the local factory ... means employment for lorry drivers, milk-factory workers, mechanics and garage-hands in the countryside, a whole range of supplementary occupations which keep more and more people in rural areas. In this way the peripheral areas of the Welsh massif have not only been able to arrest the seepage of population ... but ... to record slight increases ... Nowhere has this been more marked than in the small market towns which act as service centres for the countryside ... [A]part from agriculture no other industry is now sufficiently prominent to leave its mark on the general distribution of population [of rural Wales]. (1962: 257)

The *formal representations of the rural* which underpinned this (re)focusing of rural life were outlined most clearly in official government statements and publications. Early on, these included the Scott Report [Land Utilization in Rural Areas Survey] of 1942, with its moralistic expression of farming's 'prescriptive right' (Clout, 1972: 83) over land, and the 1947 Agriculture Act, with its more prosaic institutionalization of assured markets and guaranteed prices. Even as late as the 1970s, the White Papers *Food from Our Own Resources* (1975) and *Farming and the Nation* (1979) saw continued expansion of agriculture as an urgent priority (Bowers, 1985). All these documents, reflecting in part an entrenchment of a corporatist relationship between the agricultural industry and the state (Winter, 1996), clearly nominated, normalized and nurtured the countryside as first and foremost a food production resource. Moreover, as we have already seen from Gray's (2000) work, this attitude extended well beyond Britain.

Once again, beyond the farm gate this representation was sustained and reinforced. For example, non-agricultural institutions concerned with rural issues tended both to acknowledge and to accept a leadership role for agriculture, as witnessed by the persistence of 'agricultural exceptionalism' (Newby, 1987: 216) within otherwise strict planning controls on development. In national (and EU) policy terms, 'rural' was synonymous with 'agricultural' (Woods, 2003), just as 'rural geography' as an academic subdiscipline was largely held in the thrall of 'agricultural economics', at least until Clout's 1972 textbook (Cloke, 1980).

Third, not least through the connections of productivist agriculture to the wider civil society of rural places, *everyday lives of the rural* existed largely through this productivist vision. With a clear state-endorsed place within the Atlanticist food order (Goodman and Watts, 1997; Le Heron, 1993), farmers themselves felt this intended sense of security in a variety of dimensions: land rights, land use, finance, politics, ideology (Marsden et al., 1993: 59–61). Such an experience was reinforced by local landed elites (Cloke and Goodwin, 1992), as outlined so effectively in Newby's (1977; Newby et al., 1978) accounts of how and why they prevented the construction of much local authority housing in rural areas and more generally structured everyday rural life in their own (productivist) interest.

Overall, therefore, productivism was the glue that consolidated a structured coherence for rural Britain. Moreover, this was largely a congruent and unified coherence, with the three facets of spatiality meshing together well. Within the rural totality, the formal representation of British rurality as productivist agriculture was strongly unified, quite overwhelming and fairly hegemonic. This is illustrated, for example, by the lack of a 'political' profile or of general public interest in agriculture or the rural at this time. 'Rural politics' was almost oxymoronic, characterized by conservative, stable voting traditions (Grant, 1990) and a Ministry of Agriculture, Fisheries and Food working largely outside the public view, marginalizing rural issues far from the political mainstream (Woods, 2003).

None the less, especially in some parts of Britain, each facet of this productivist rural space *was* contested by other spaces, rural and non-rural. For example, productivist agricultural practices were tempered by the persistence of less capitalistically rational farming, often buttressed by state welfare payments. From the 1946 Hill Farming Act onwards (Morgan and Munton, 1971), these payments have been crucial to the survival of sheep farmers in the uplands, in particular. Another key spatial practice of rural localities was population out-migration, as revealed so clearly by the *Beacham Report* on mid-Wales in 1964. Contra the impression given by many of the Welsh 'community studies' (for example, Rees, 1950), 'migration out of the area is ... a very important feature in the life of the community'

(Bowen, 1962: 257). This practice could have fitted in with a 'pure' productivist representation, whereby 'surplus' population would be expelled for 'efficiency' gains. Indeed, this was suggested, for example, in the 1954 report of the Welsh Agricultural Land Sub-Commission (1962: 258). However, the formal representation never went this far. Throughout productivism's currency, a contradictory romantic, communitarian and ideological moral stewardship vision of farmers and landowners overlay its representation of rural space. In Murdoch et al.'s (2003) terms, modernism was challenged by pastoralism. Again, the Scott Report epitomized this contradiction. Finally, as might be expected from this tension, everyday lives of the rural were also not always in tune with productivism. Remnants of less productivist agriculture and other economic activities remained, and other rural practices impinged on people's daily lives. This facilitated the broadening of their 'social imaginary' away from any singular productivist vision.

Undermining the spaces of productivism

> Changes in the political economy of agriculture compel a need to reappraise the future of rural space. The late 1980s and early 1990s have marked the beginning of the end for the productivist rationale which had given the agricultural industry a dominant, almost exclusive, role in determining the shape and function of the countryside. (Symes and Jansen, 1994a: 2)

Over the past couple of decades, the productivist agricultural rural spatiality has come under increasingly intense and probably fatal strain.[6] Bluntly, productivist British agriculture is now in 'structural crisis' and cannot be cured by the kinds of 'technical fixes' that rescued it from previous more minor 'conjunctural crises' (Drummond et al., 2000). All three facets of productivist rural space have been undermined. Its representation of rural was ultimately unable to achieve total dominion over either rural localities or everyday lives of the rural. Nor has it even been able to retain its dominance within the realm of representation, as rivals have grown more significant. Rural change is not being driven *primarily* through some idealist desire to change the landscapes of rural Britain but due to the totalizing failure of productivism.

Rural localities have been profoundly affected by a different rural spatiality expressed through economic restructuring and social recomposition (Cloke and Goodwin, 1992). The former is expressed through practices associated with urban workplaces (commuting), leisure-related commodification, industrialization and exploitation of marginality (waste dumps, mining), whilst the latter's practices are especially associated with counter-urban in-migration and the appeal of the 'rural idyll' (Halfacree, 1994). Farming practices, too, have been forced to adjust in response to the varied and numerous economic and other contradictions of productivist agriculture, from having to cope with surpluses and over-production, to dealing with the legacy of publicly acknowledged environmental destruction (for example, Harvey, 1997).

In response to these challenges, the lives of the rural for individual farmers and their families have increasingly been characterized by insecurity and uncertainty, expressed most acutely through high levels of debt and depression (Derounian, 1993; Simmons, 1997). This situation is enhanced by a general public – and a new rural population – that increasingly questions both the role of farmers as 'guardians of the countryside' and the financial and other implications of agricultural support (Harvey, 1997; Seymour et al., 1997).

Finally, formal representations of the rural are also no longer so dominated by the productivist vision. Instead, other ways of commodifying the rural have come to the fore (Cloke, 1999). For example, Cloke has noted how a push towards a degree of deregulation in rural planning has provided an opening for new economic opportunities to be exploited. This went hand-in-hand with:

> new markets for countryside commodities: the countryside as an exclusive place to be lived in; rural communities as a context to be bought and sold; rural lifestyles which can be colonised; icons of rural culture which can be crafted, packed and marketed; rural landscapes with a new range of potential ... (1992: 293)

Illustrating this in respect to rural leisure, in 1985 the then Conservative government released a report entitled *Pleasure, Leisure and Jobs: The Business of Tourism*. This heralded the start of what Veal (1993) terms its 'enterprise phase', where emphasis shifted from social welfare concerns towards market demand being met via the private sector. This more diversely commodified representation of rural impinged on agriculture as well. For example, England's Rural Development Programme:

> underpins [the] Government's New Direction for Agriculture by helping farmers and foresters to respond better to consumer requirements and become more competitive, diverse, flexible and environmentally responsible. (DEFRA, 2003: no pagination)

Likewise, the EU has a broader vision of rurality within its CAP reforms (Gray, 2000; see earlier).

A recent series of devastating agricultural crises have also directly challenged all three facets of British productivist spatiality. In particular, the Foot and Mouth epidemic of 2001–2 brought the twin but related issues of rural crisis and an increasingly uncertain rural future to the fore (Bennett et al., 2002). Practices of the rural became embodied in mass livestock slaughter, cremation and burial, and in 'footpath closed' notices; lives of the rural became dominated by loss of livelihoods (farmers, tourism providers, support services) and the harsh light of media attention; and formal representations of the rural seemed simply overwhelmed. All established rural 'certainties' – even as far as producing food at all – were sharply questioned. For example, government sources suggested that around a quarter of British farms – almost all of them small ones – would have closed or merged by 2005, with 50,000 people forced to leave the industry (*Guardian*, 2001). Constitutionally, the replacement of the Ministry of Agriculture, Fisheries and Food by a new Department for the Environment, Food and Rural Affairs in 2001 removed, at least nominally, 'agriculture' from its assumed pre-eminence within rural policy-making. The rural world of 1950s' productivism seemed as dead and buried as much of the country's livestock …

All these disruptions to rural Britain over the past decade throw sharply into focus Ilbery and Bowler's (1998) talk of a 'post-productivist transition' for agriculture, characterized by reduced food production and state support, and the internationalization of the food industry within a more free market global economy, albeit with increased environmental regulation placed on the industry. Three 'bipolar dimensions of change are recognized' (Ilbery and Bowler, 1998: 70): intensification to extensification, concentration to dispersion, and specialization to diversification. However, Ilbery and Bowler are careful not to homogenize and essentialize these trends, recognizing diverse 'pathways of farm business development' emerging (Bowler, 1992).

This emphasis on diversity has extended beyond the agricultural sector and has prompted talk not just of a post-productivist transition but of a 'post-productivist countryside' (Halfacree, 1997, 1998, 1999; Ward, 1993). Within such a vision, as laid out prophetically in the European Commission's 1988 report *The Future of Rural Society* (CEC, 1988), 'Agriculture exists within and is encompassed by rural space and society rather than the other way around' (Gray, 2000: 42). Consequently, the EU now seeks to '[transpose] the image of diversified rurality into tangible localities' (2000: 44). The spatiality of the post-productivist countryside is, in short, fundamentally *heterogenous*.

Spaces of post-productivism

Referring to the earlier account of the spatiality of capitalism, abstract space is ever searching to maximize exchange values through creating the most economically 'successful' productive permanences. Under productivism, its rural grounding was rooted in a specific model of agricultural production. Given that such groundings are never truly 'permanent' – and reflecting the danger of putting all your eggs in one basket – it is perhaps unsurprising that this permanence has now become increasingly untenable. A search is on for alternatives that promise greater profitability. A key way of doing this, given that the spatial frontier of 'rurality' has by now largely been 'colonized' by capitalism, is through 'involution', whereby the production of space reworks its internal subdivisions (Katz, 1998). This seems to be what is happening in the post-productivist countryside. Hence, the emphasis on heterogeneity.

A number of authors have suggested models of the possible ongoing production of the post-productivist countryside. For example, Marsden (1998a; Marsden et al., 1993) has suggested the emergence of four ideal types within a 'differentiated countryside', as outlined in Table 4.1 and illustrated further in Murdoch et al. (2003). Marsden goes on to argue that the new territories of this differentiated countryside will comprise varying proportions of at least the four dimensions (Marsden, 1998b), described in Table 4.2. All four involve processes of commoditization, an emphasis which takes us 'beyond agriculture' (Marsden, 1995, 1999) to explore the varying arenas in which varied social and political processes construct and attribute commodity values to a range of rural objects, artefacts, people and places (Marsden, 1998a).

Marsden's model is very much an incremental expression of rural change. Alternatively, in the spirit of Lefebvre's more 'open' notion of the production of space, such models can be extended to incorporate more *potential* rural futures. This is the intention of the four spatial scenarios sketched below, which is developed from Halfacree (1999).

Super-productivism This vision re-states the spatiality of productivism, but this time in

Table 4.1 ***Ideal types for the post-productivist countryside***

- *Preserved countryside*: areas dominated by anti-development and preservationist attitudes; middle-class pressure groups active and powerful. For example, much of South-East England
- *Contested countryside*: areas beyond the major commuter areas where strongly local agricultural and small business interests hold sway but are increasingly challenged by in-migrants. For example, much of South-West England
- *Paternalistic countryside*: areas where large private estates remain important and/or where large farms predominate; their owners still feel some sort of obligation to their constituents. For example, parts of the Highlands and Islands of Scotland
- *Clientalist countryside*: remote areas where agricultural productivism still dominates the landscape but where considerable reliance is placed on external finances, such as EU monies. For example, West Wales and much of Scotland

Source: Marsden, 1998a: 17–18

Table 4.2 ***Components of the differentiated countryside***

- *Mass food markets*: building on a post-CAP reform and a more free market attitude to agriculture, this is the highly productive agricultural land of agribusiness, high capital-intensive inputs and genetically modified crops
- *Quality food markets*: farming influenced by added value that can be obtained by 'quality'. This landscape will include organic farming and the production and marketing of locally identified products, e.g. cheeses, livestock
- *Agriculturally related changes*: diversification-led changes on the farm. Includes the re-commodification of formerly disused buildings (e.g. for accommodation) and defunct objects (e.g. old implements) and the marketing of the rural for recreation and tourism
- *Rural restructuring (non-agricultural)*: from the use of the countryside for nasties (waste dumps, industry, open cast mines) to residence

Source: Marsden, 1998b: 109–13

a much less moderated form, shorn of its moral dimension. The capitalist 'logic' of abstract space is fully released. An emergent super-productivism is readily apparent in the practices of agribusiness, the genetic modification of plants and animals, and biotechnology generally. Here, formal representations position land solely as a productive resource linked to profit maximization. 'Nature' is very much seen as 'an accumulation strategy' (Katz, 1998; see Goodman and Redclift, 1991). Indeed, such is the physical impact of super-productivism, with its 'monoscape' of sameness (Pretty, 2002), that everyday lived rurality has little scope to diverge from the representation. Overall, what is suggested is a potentially congruent and unified local structured coherence, albeit one that is unpalatable to many.

Consuming idylls This alternative directly opposes – in terms of rural spatiality – super-productivism. It takes and develops the erstwhile moderating element of the old productivist representation, namely its moral 'pastoralist' (Murdoch et al., 2003) angle, expressed most strongly through ideas of 'community' (Murdoch and Day, 1998). The rural locality may have agriculture as a backdrop but its key spatial practices are consumption-orientated, notably leisure, residence and attendant migration (counter-urbanization). The formal representation underlying these practices tends to be that of the '*rural idyll*' (Halfacree, 1994, 2003), upheld by rules such as those of the planning system. However, unlike super-productivism, the extent to which everyday lives in the rural conform to this spatial imagination vary (*cf.* Cloke et al., 1995; Halfacree, 1995), reflected in battles within the planning arena over what are seen as unsuitable developments in rural locations (Ambrose, 1992). Whilst in some places there may well be an 'ascendance of certain aesthetic representations of the countryside over previous economic ones' (Murdoch and Marsden, 1994: 215–16), this is highly uneven. Overall, the local structured coherence is likely to be contradictory and disjointed, although differences between co-existent rural spatialities are more induced than produced.

Effaced rurality As noted above, many academics have sought to emphasize just how lacking in distinctiveness the categories 'rural' and 'urban' really are. The rural has, in effect, been effaced by the geographical development of late capitalism. Thus (formerly) rural places may be seen as dominated by distinctly non-rural spatialities, leaving rural space only as a ghostly presence, experienced through folk memory, nostalgia, hearsay, etc. Here, locality, formal representations and daily lives will have little significant 'rural' content. Where rurality does still come through, however, we might again expect to see a contradictory and disjointed local structured coherence.

Radical visions Radical visions for rural space have been neglected to date in discussions of the topography of the post-productivist countryside (Halfacree, 1999, 2001b). By 'radical' I mean here rural spatialities that try to express produced rather than induced difference and thus challenge the spatial logic of capitalism. Rooted in the social imaginary of everyday lives, Mormont captures the thrust of my argument thus:

> Rurality is claimed not only as a space to be appropriated ... but as a way of life, or a model of an alternative society inspiring a social project that challenges contemporary social and economic ill ... Peasant autarky, village community and ancient technique are no longer relics, but *images which legitimize this social project of a society which would be ruralized* ... The aim is not to recreate a past way of life but to develop forms of social and economic life different from those prevailing at present ... (1987: 18; emphasis in original)

Thinking of the content of this 'radical' rural space, we can imagine a locality revolving around decentralized and relatively self-sufficient living patterns, representations that imagine the countryside as a diverse home accessible to all, and everyday experiences celebrating the local and the individually meaningful (Juckes Maxey, 2002; Schwarz and Schwarz, 1998). These alternatives may be imagined through the efflorescence of 'festival' (Lefebvre, 1991b) or 'temporary autonomous zone' (Bey, 1991), or through the more sober practices of 'low impact development' (Fairlie, 1996) or 'agri-culture' (Pretty, 2002). Where they emerge, however, they tend to result in chaotic and incoherent local structured coherences. This is because they feature centrally a struggle between these nascent produced differences and the existing spatialities of (rural) capitalism. The latter represent 'the existing centre and the forces of homogenization [that] must seek to absorb all such differences ... [so as] to integrate, to recuperate, or to destroy whatever has transgressed' (Lefebvre, 1991a: 373). The challenge for radical politics, of course, is somehow to make these temporary, provocative spatialities more lasting and permanent.

CONCLUSION

Michael Woods (2003) has recently discussed how 'rural politics' has ceased to be an oxymoron and has become a key issue for struggle within the developed world, detailing examples of rural protest as a 'new social movement' in Britain, France and the United States. Such a positioning is not to fetishize the rural but to place it within the 'socially produced [and contested] set of manifolds' (Crang and Thrift, 2000b: 2) that is space. Indeed, internationally, rurality's obituary must not be prematurely written, it being a global site of struggle and radicalism (Halfacree, 2001c). These struggles are struggles over the practices, representations and everyday lives that inscribe the rural totality. They are struggles about rural space.

Using as a backdrop the abstract space of capitalism, this chapter has sought to outline and then to illustrate briefly an architecture through which we can come to analyse this totality of rural space. The resulting general model must be seen, finally, in the spirit of the 'concrete universal'. In a *global* capitalist era, such a general model has 'universal' applicability and it is hoped that future work will fill in its 'concrete' contours in a wide range of different places.

NOTES

1 Pleonasm: 'redundancy of expression' (*Concise Oxford Dictionary*).

2 Within Great Britain we can recognize a range of rurals, again linked to but not reducible to scale. For example, rurality arguably varies between England, Scotland and Wales (for example, Cloke et al., 1995; Jedrej and Nuttall, 1996), and Cloke et al. (1998) note national-, regional- and local-level constructs of rural within England.

3 Once again, we might reflect that these contents represent a developed world, specifically British, bias. None the less, the content of a more developing world rural locality is also likely to reflect, at least, agriculture and primary production, low population and inaccessibility, plus additional features such as the challenge of obtaining key resources (food, water, etc.) (Barke and O'Hare, 1991; Findlay, 1994). The social representation perspective, though, is likely to be much more geographically specific, as already noted.

4 This is in spite of Lefebvre's early career as a rural sociologist (Shields, 1999) and the insights he gained from small French towns such as Navarrenx (Merrifield, 2002; Lefebvre, 1991b).

5 The productivist/post-productivist framework illustrated here is not one that has been completely accepted, even for Great Britain. Controversy shadows this model, although critique has been developed most with respect to its agricultural dimension (see Evans et al., 2002; Wilson, 2001). There is not the space here to go into this debate, suffice it to note that the model's general applicability seems to dissipate as we go away from Britain, through northern Europe (Buller et al., 2000; Hoggart et al., 1995; Symes and Jansen, 1994b), Japan (Takahashi, 2001), North America (Curry-Roper, 1992) and Australia (Argent, 2002; Panelli, 2001), where there are clear resonances, to a Mediterranean

Europe that is still not even fully productivist (Wilson, 2001), to developing nations, where the model presently has little relevance (Barke and O'Hare, 1991: ch. 4; Madeley, 2002). Such geographical diversity (Marsden, 1999) is, however, unsurprising if we follow neither an overly structuralist line, allowing endogenous change (van der Ploeg and Long, 1994), nor a 'developmentalist' (Taylor, 1993) assumption of linear change. None the less, we must take care not to overlook the British bias and general lack of meaningful international comparison that still bedevils the productivist/post-productivist debate (Wilson, 2001): national traditions remain very strong within rural studies generally (Symes and Jansen, 1994a).

6 This argument to some extent parallels that over the fate of Fordism; indeed, some scholars have linked the two (Buttell, 1994; Marsden, 1992). This suggests the pairings of Fordism/productivism and post-Fordism/post-productivism, although one needs to exercise caution here, as with all such schema.

REFERENCES

Ambrose, P. (1992) 'The rural/urban fringe as battleground', in B. Short (ed.), *The English Rural Community*. Cambridge: Cambridge University Press. pp. 175–94.

Argent, N. (2002) 'From pillar to post? In search of the post-productivist countryside in Australia', *Australian Geographer*, 33: 97–114.

Barke, M. and O'Hare, G. (1991) *The Third World*, 2nd edn. Harlow: Oliver and Boyd.

Bennett, K., Carroll, T., Lowe, P. and Phillipson, J. (2002) *Coping With Crisis in Cumbria: Consequences of Foot and Mouth Disease*. University of Newcastle: Centre for Rural Economy.

Bey, H. (1991) *TAZ: The Temporary Autonomous Zone, Ontological Anarchy, Poetical Terrorism*. New York: Autonomedia.

Black, J. and Conway, E. (1995) 'Community-led development policies in the Highlands and Islands: the European Community's LEADER programme', *Local Economy*, 10: 229–45.

Bowen, E. (1962) 'Rural Wales', in J. Mitchell (ed.), *Great Britain: Geographical Essays*. Cambridge: Cambridge University Press. pp. 247–64.

Bowers, J. (1985) 'British agricultural policy since the Second World War', *Agricultural History Review*, 33: 66–76.

Bowler, I. (1992) 'Sustainable agriculture as an alternative path of farm business development', in I. Bowler, C. Bryant and M. Nellis (eds), *Contemporary Rural Systems in Transition. Volume 1. Agriculture and Environment*. Wallingford: CAB International. pp. 237–53.

Bowler, I., Bryant, C. and Nellis, M. (eds) (1992) *Contemporary Rural Systems in Transition. Volume 1. Agriculture and Environment*. Wallingford: CAB International.

Buller, H., Wilson, G. and Höll, A. (eds) (2000) *Agri-environmental Policy in the European Union*. Aldershot: Ashgate.

Bunce, M. (1994) *The Countryside Ideal. Anglo-American Images of Landscape*. London: Routledge.

Buttel, F. (1994) 'Agricultural change, rural society, and the state in the late twentieth century: some theoretical observations', in D. Symes and A. Jansen (eds), *Agricultural Restructuring and Rural Change in Europe*. Wageningen: Agricultural University. pp. 13–31.

CEC (Commission of the European Communities) (1988) *The Future of Rural Society*. Brussels: CEC.

CEC (Commission of the European Communities) (1996) *The Cork Declaration: A Living Countryside*. Brussels: CEC.

Champion, A. (1998) 'Studying counterurbanisation and the rural population turnaround', in P. Boyle and K. Halfacree (eds), *Migration into Rural Areas. Theories and Issues*. Chichester: Wiley. pp. 21–40.

Champion, T. and Hugo, G. (eds) (2004) *New Forms of Urbanization: Beyond the Urban–Rural Dichotomy*. Aldershot: Ashgate.

Cloke, P. (1980) 'New emphases for applied rural geography', *Progress in Human Geography*, 4: 181–217.

Cloke, P. (1989) 'Rural geography and political economy', in R. Peet and N. Thrift (eds), *New Models in Geography. Volume 1*. London: Unwin Hyman. pp. 164–97.

Cloke, P. (1992) '"The countryside": development, conservation and an increasingly marketable commodity', in P. Cloke (ed.), *Policy and Change in Thatcher's Britain*. Oxford: Pergamon. pp. 269–95.

Cloke, P. (1994) '(En)culturing political economy: a life in a day of a "rural geographer"', in P. Cloke, M. Doel, D. Matless, M. Phillips and N. Thrift, *Writing the Rural: Five Cultural Geographies*. London: Paul Chapman. pp. 149–90.

Cloke, P. (1999) 'The country', in P. Cloke, P. Crang and M. Goodwin (eds), *Introducing Human Geographies*. London: Edward Arnold. pp. 256–67.

Cloke, P. and Goodwin, M. (1992) 'Conceptualizing countryside change: from post-Fordism to rural structured coherence', *Transactions of the Institute of British Geographers*, 17: 321–36.

Cloke, P. and Park, C. (1984) *Rural Resource Management*. London: Croom Helm.

Cloke, P., Goodwin, M. and Milbourne, P. (1995) '"There's so many strangers in the village now": marginalization and change in 1990s Welsh rural life-styles', *Contemporary Wales*, 8: 47–74.

Cloke, P., Goodwin, M. and Milbourne, P. (1998) 'Inside looking out; outside looking in. Different experiences of cultural competence in rural lifestyles', in P. Boyle and K. Halfacree (eds), *Migration into Rural Areas: Theories and Issues*. Chichester: Wiley. pp. 134–50.

Clout, H. (1972) *Rural Geography*. Oxford: Pergamon.

Cooke, P. (ed.) (1989) *Localities: the Changing Face of Urban Britain*. London: Unwin Hyman.

Copp, J. (1972) 'Rural sociology and rural development', *Rural Sociology*, 37: 515–33.

Countryside Alliance (2002) Homepage at <http://www.countryside-alliance.org/> (accessed June).

Crang, M. and Thrift, N. (eds) (2000a) *Thinking Space*. London: Routledge.

Crang, M. and Thrift, N. (2000b) 'Introduction', in M. Crang and N. Thrift (eds), *Thinking Space*. London: Routledge. pp. 1–30.

Curry-Roper, J. (1992) 'Alternative agriculture and conventional paradigms in US agriculture', in I. Bowler, C. Bryant and M. Nellis (eds), *Contemporary Rural Systems in Transition. Volume 1. Agriculture and Environment*. Wallingford: CAB International. pp. 254–64.

DEFRA (Department for Environment, Food and Rural Affairs) (2003) 'Welcome to the England Rural Development Programme', at <http://www.defra.gov.uk/erdp/> (accessed June).

Denny, C. and Elliott, L. (2002) 'Shaping up for Seattle at the beach', *Guardian*, 4 September at <http://www.guardian.co.uk/business/story/0,3604,785702,00.html> (accessed June 2003).

Derounian J. (1993) *Another Country. Real Life Beyond Rose Cottage*. London: NCVO Publications.

Drummond, I., Campbell, H., Lawrence, G. and Symes, D. (2000) 'Contingent or structural crisis in British agriculture?', *Sociologia Ruralis*, 40: 111–27.

Evans, N., Morris, C. and Winter, M. (2002) 'Conceptualizing agriculture: a critique of post-productivism as the new orthodoxy', *Progress in Human Geography*, 26: 313–32.

Fairlie, S. (1996) *Low Impact Development*. Charlbury, Oxfordshire: Jon Carpenter.

Findlay, A. (1994) *The Arab World*. London: Routledge.

Frey, W. and Johnson, K. (1998) 'Concentrated immigration, restructuring and the "selective" deconcentration of the United States population', in P. Boyle and K. Halfacree (eds), *Migration into Rural Areas: Theories and Issues*. Chichester: Wiley. pp. 79–106.

Goodman, D. and Redclift, M. (1991) *Refashioning Nature: Food, Ecology and Culture*. London: Routledge.

Goodman, D. and Watts, M. (eds) (1997) *Globalising Food: Agrarian Questions and Global Restructuring*. London: Routledge.

Gottdiener, M. (1985) *The Social Production of Urban Space*. Austin: University of Texas Press.

Grant, W. (1990) 'Rural politics in Britain', in P. Lowe and M. Bodiguel (eds), *Rural Studies in Britain and France*. London: Belhaven. pp. 286–98.

Gray, J. (2000) 'The Common Agricultural Policy and the re-invention of the rural in the European Community', *Sociologia Ruralis*, 40: 30–52.

Gregory, D. (1994) *Geographical Imaginations*. Oxford: Blackwell.

Gregory, D. and Urry, J. (1985) *Social Relations and Spatial Structures*. Basingstoke: Macmillan.

Guardian (2001) 'Half of farms "will close by 2020"', 13 August.

Halfacree, K. (1993) 'Locality and social representation: space, discourse and alternative definitions of the rural', *Journal of Rural Studies*, 9: 23–37.

Halfacree, K. (1994) 'The importance of "the rural" in the constitution of counterurbanization: evidence from England in the 1980s', *Sociologia Ruralis*, 34: 164–89.

Halfacree, K. (1995) 'Talking about rurality: social representations of the rural as expressed by residents of six English parishes', *Journal of Rural Studies*, 11: 1–20.

Halfacree, K. (1997) 'Contrasting roles for the post-productivist countryside: a post-modern perspective on counterurbanisation', in P. Cloke and J. Little (eds), *Contested Countryside Cultures*. London: Routledge. pp. 70–93.

Halfacree, K. (1998) 'Neo-tribes, migration and the post-productivist countryside', in P. Boyle and K. Halfacree (eds), *Migration into Rural Areas: Theories and Issues*. Chichester: Wiley. pp. 200–14.

Halfacree, K. (1999) 'A new space or spatial effacement? Alternative futures for the post-productivist countryside', in N. Walford, J. Everitt and D. Napton (eds), *Reshaping the Countryside: Perceptions and Processes of Rural Change*. Wallingford: CAB International. pp. 67–76.

Halfacree, K. (2001a) 'Constructing the object: taxonomic practices, "counterurbanisation" and positioning marginal rural settlement', *International Journal of Population Geography*, 7: 395–411.

Halfacree, K. (2001b) 'Going "back-to-the-land" again: extending the scope of counterurbanisation', *Espace, Populations, Sociétés*, 2001–1–2: 161–70.

Halfacree, K. (2001c) 'A place for 'nature'?: new radicalism's rural contribution'. Paper presented at 19th European Congress for Rural Sociology, Dijon, France, September.

Halfacree, K. (2003) 'Landscapes of rurality: rural others/other rurals', in I. Roberston and P. Richards (eds), *Studying Cultural Landscapes*. London: Arnold. pp. 141–69.

Halfacree, K., Kovach, I. and Woodward, R. (eds) (2002) *Leadership and Local Power in European Rural Development*. Aldershot: Ashgate.

Harper, S. (ed.) (1993) *The Greening of Rural Policy: International Perspectives*. London: Belhaven.

Harvey, D. (1985) *The Urbanization of Capital*. Oxford: Blackwell.

Harvey, D. (1989) *The Condition of Postmodernity*. Oxford: Blackwell.

Harvey, D. (1996) *Justice, Nature and the Geography of Difference*. Oxford: Blackwell.

Harvey, G. (1997) *The Killing of the Countryside*. London: Jonathan Cape.

Hoggart, K. (1990) 'Let's do away with rural', *Journal of Rural Studies*, 6: 245–57.

Hoggart, K., Buller, H. and Black, R. (1995) *Rural Europe: Identity and Change*. London: Arnold.

Hugo, G. and Bell, M. (1998) 'The hypothesis of welfare-led migration to rural areas: the Australian case', in P. Boyle and K. Halfacree (eds), *Migration into Rural Areas: Theories and Issues*. Chichester: Wiley. pp. 107–33.

Ilbery, B. and Bowler, I. (1998) 'From agricultural productivism to post-productivism', in B. Ilbery (ed.), *The Geography of Rural Change*. Harlow: Longman. pp. 57–84.

Jedrej, C. and Nuttall, M. (1996) *White Settlers. The Impact of Rural Repopulation in Scotland*. Luxembourg: Harwood Academic Publishers.

Jones, O. (1995) 'Lay discourses of the rural: developments and implications for rural studies', *Journal of Rural Studies*, 11: 35–49.

Juckes Maxey, L. (2002) *One Path Forward? Three Sustainable Communities in England and Wales*. Unpublished PhD thesis, Department of Geography, University of Wales, Swansea.

Katz, C. (1998) 'Whose nature, whose culture?: private productions of space and the "preservation", of nature', in B. Braun and N. Castree (eds), *Remaking Reality. Nature at the Millennium*. London: Routledge. pp. 46–63.

Laschewski, L., Teherani-Krönner, P. and Bahner, T. (2002) 'Recent rural restructuring in East and West Germany: experiences and backgrounds', in K. Halfacree, I. Kovach and R. Woodward (eds), *Leadership and Local Power in European Rural Development*. Aldershot: Ashgate. pp. 145–72.

Le Heron, R. (1993) *Globalized Agriculture: Political Choice*. Oxford: Pergamon.

Lefebvre, H. (1991a) *The Production of Space*. Oxford: Blackwell.

Lefebvre, H. (1991b) *Critique of Everyday Life. Volume 1*. London: Verso.

Lefebvre, H. (1996) *Writings on Cities*. Oxford: Blackwell.

Llambi, L. (1994) 'Comparative advantages and disadvantages in Latin American nontraditional fruit and vegetable exports', in P. McMichael (ed.), *The Global Restructuring of Agro-food Systems*. Ithaca, NY: Cornell University Press. pp. 190–213.

Lobao, L. (1996) 'A sociology of the periphery versus a peripheral sociology: rural sociology and the dimension of space', *Rural Sociology*, 61: 77–102.

Madeley, J. (2002) *Food For All: The Need For a New Agriculture*. London: Zed Books.

Marsden, T. (1992) 'Exploring a rural sociology for the Fordist transition: incorporating social relations into economic restructuring', *Sociologia Ruralis*, 32: 209–30.

Marsden, T. (1995) 'Beyond agriculture? Regulating the new rural spaces', *Journal of Rural Studies*, 11: 285–96.

Marsden, T. (1998a) 'Economic perspectives', in B. Ilbery (ed.), *The Geography of Rural Change*. Harlow: Longman. pp. 13–30.

Marsden, T. (1998b) 'New rural territories: regulating the differentiated rural spaces', *Journal of Rural Studies*, 14: 107–17.

Marsden, T. (1999) 'Rural futures: the consumption countryside and its regulation', *Sociologia Ruralis*, 39: 501–20.

Marsden, T., Murdoch, J., Lowe, P., Munton, R. and Flynn, A. (1993) *Constructing the Countryside*. London: UCL Press.

Marx, L. (1964) *The Machine in the Garden*. New York: Oxford University Press.

Massey, D. (1996) 'A global sense of place', in S. Daniels and R. Lee (eds), *Exploring Human Geography. A Reader*. London: Arnold. pp. 237–45.

McMichael, P. (1993) 'World food system restructuring under a GATT regime', *Political Geography*, 12: 198–214.

McMichael, P. (ed.) (1994) *The Global Restructuring of Agro-food Systems*. Ithaca, NY: Cornell University Press.

Merrifield, A. (1993) 'Place and space: a Lefebvrian reconciliation', *Transactions of the Institute of British Geographers*, 18: 516–31.

Merrifield, A. (2000) 'Henri Lefebvre: a socialist in space', in M. Crang and N. Thrift (eds), *Thinking Space*. London: Routledge. pp. 167–82.

Merrifield, A. (2002) *Metromarxism*. New York: Routledge.

Monbiot, G. (1994) *No Man's Land*. London: Macmillan.

Monbiot, G. (2002) *Captive State. The Corporate Takeover of Britain*. London: Macmillan.

Morgan, W. and Munton, R. (1971) *Agricultural Geography*. London: Methuen.

Mormont, M. (1987) 'Rural nature and urban natures', *Sociologia Ruralis*, 27: 3–20.

Mormont, M. (1990) 'Who is rural? Or, how to be rural: towards a sociology of the rural', in T. Marsden, P. Lowe and S. Whatmore (eds), *Rural Restructuring*. London: David Fulton. pp. 21–44.

Moscovici, S. (1984) 'The phenomenon of social representations', in R. Farr and S. Moscovici (eds), *Social Representations*. Cambridge: Cambridge University Press. pp. 3–69.

Moseley, M. (1984) 'The revival of rural areas in advanced economies: a review of some causes and consequences', *Geoforum*, 15: 447–56.

Murdoch, J. and Day, G. (1998) 'Middle class mobility, rural communities and the politics of exclusion', in P. Boyle and K. Halfacree (eds), *Migration into Rural Areas. Theories and Issues*. Chichester: Wiley. pp. 186–99.

Murdoch, J. and Marsden, T. (1994) *Reconstituting Rurality*. London: UCL Press.

Murdoch, J. and Pratt, A. (1993) 'Rural studies: modernism, postmodernism and the "post-rural"', *Journal of Rural Studies*, 9: 411–27.

Murdoch, J., Lowe, P., Ward, N. and Marsden, T. (2003) *The Differentiated Countryside*. London: Routledge.

Newby, H. (1977) *The Deferential Worker*. London: Allen Lane.

Newby, H. (1987) *Country Life*. London: Weidenfeld and Nicolson.

Newby, H., Bell, C., Rose, D. and Saunders, P. (1978) *Property, Paternalism and Power*. London: Hutchinson.

Orton, B. (1996) 'Tangent', from *Trailer Park*. Heavenly Records HVN LP17CD.

Page, B. (1997) 'Restructuring pork production, remaking rural Iowa', in D. Goodman and M. Watts (eds), *Globalising Food: Agrarian Questions and Global Restructuring*. London: Routledge. pp. 133–57.

Panelli, R. (2001) 'Narratives of community and change in a contemporary rural setting: the case of Duaringa, Queensland', *Australian Geographical Studies*, 39: 156–66.

Peet, R. (1998) *Modern Geographical Thought*. Oxford: Blackwell.

Phillips, M. (2002) 'The production, symbolization and socialization of gentrification: impressions from two Berkshire villages', *Transactions of the Institute of British Geographers*, 27: 282–308.

Ploeg, J. D. van der and Long, A. (eds) (1994) *Born From Within: Practice and Perspectives of Endogenous Rural Development*. Assen: Van Gorcum.

Potter, C. (1998) *Against the Grain: Agri-environmental Reform in the United States and the European Union*. Wallingford: CAB International.

Potter, J. and Wetherell, M. (1987) *Discourse and Social Psychology*. London: Sage.

Pratt, A. (1996) 'Discourses of rurality: loose talk or social struggle?', *Journal of Rural Studies*, 12: 69–78.

Pretty, J. (2002) *Agri-culture. Reconnecting People, Land and Nature*. London: Earthscan.

Raynolds, L. (1994) 'The restructuring of Third World agro-exports: changing production relations in the Dominican Republic', in P. McMichael (ed.), *The Global Restructuring of Agro-food Systems*. Ithaca, NY: Cornell University Press. pp. 214–37.

Rees, A. (1950) *Life in a Welsh Countryside*. Cardiff: University of Wales Press.

Sayer, A. (1984) *Method in Social Science*. London: Hutchinson.

Sayer, A. (1989) 'The "new" regional geography and problems of narrative', *Society and Space*, 7: 253–76.

Schwarz, W. and Schwarz, D. (1998) *Living Lightly. Travels in Post-Consumer Society*. Charlbury, Oxfordshire: Jon Carpenter.

Seymour, S., Lowe, P., Ward, N. and Clark, J. (1997) 'Environmental "others" and "elites": rural pollution and changing power relations in the countryside', in P. Milbourne (ed.), *Revealing Rural 'Others'*. London: Pinter. pp. 57–74.

Shields, R. (1999) *Lefebvre, Love and Struggle. Spatial Dialectics*. London: Routledge.

Shiva, V. (2000) *Stolen Harvest*. Cambridge, MA: South End Press.

Short, J. (1991) *Imagined Country*. London: Routledge.

Shotter J. (1993) *Cultural Politics of Everyday Life*. Buckingham: Open University Press.

Shucksmith, M. (1993) 'Farm household behaviour and the transition to post-productivism', *Journal of Agricultural Economics*, 44: 466–78.

Simmons, M. (1997) *Landscapes of Poverty*. London: Lemos and Crane.

Smith, N. (1984) *Uneven Development*. Oxford: Blackwell.

Smith, N. (1987) 'Dangers of the empirical turn: some comments on the CURS initiative', *Antipode*, 19: 59–68.

Smith, N. (1998) 'Nature at the millennium: production and re-enchantment', in B. Braun and N. Castree (eds), *Remaking Reality. Nature at the Millennium*. London: Routledge. pp. 271–85.

Symes, D. and Jansen, A. (1994a) 'Introduction', in D. Symes and A. Jansen (eds), *Agricultural Restructuring and Rural Change in Europe*. Wageningen: Agricultural University. pp. 1–12.

Symes, D. and Jansen, A. (eds) (1994b) *Agricultural Restructuring and Rural Change in Europe*. Wageningen: Agricultural University.

Takahashi, M. (2001) 'Changing ruralities and the post-productivist countryside of Japan: policy changes of the central government in the 1990s', in K. Kim, I. Bowler and C. Bryant (eds), *Developing Sustainable Rural Systems*. Pusan: Pusan National University Press. pp. 163–74.

Tarrant, J. (1974) *Agricultural Geography*, Newton Abbot: David and Charles.

Taylor, P. (1993) *Political Geography*, 3rd edn. Harlow: Longman.

Thrift, N. (1996) *Spatial Formations*. London: Sage.

Unwin, T. (2000) 'A waste of space? Towards a critique of the social production of space …', *Transactions of the Institute of British Geographers*, 25: 11–29.

Veal, A. (1993) 'Planning for leisure: past, present and future', in S. Glyptis (ed.), *Leisure and the Environment*. London: Belhaven. pp. 53–67.

Wallace, I. (1992) 'International restructuring of the agri-food chain', in I. Bowler, C. Bryant and M. Nellis (eds), *Contemporary Rural Systems in Transition. Volume 1. Agriculture and Environment*. Wallingford: CAB International. pp. 15–28.

Ward, N. (1993) 'The agricultural treadmill and the rural environment in the post-productivist era', *Sociologia Ruralis*, 27: 21–37.

Whatmore, S. (1993) 'On doing rural research (or breaking the boundaries)', *Environment and Planning A*, 25: 605–7.

Wilson, G. (2001) 'From productivism to post-productivism … and back again? Exploring the (un)changed natural and mental landscapes of European agriculture', *Transactions of the Institute of British Geographers*, 26: 313–32.

Winter, M. (1996) *Rural Politics*. London: Routledge.

Woods, M. (2003) 'Deconstructing rural protest: the emergence of a new social movement', *Journal of Rural Studies*, 19: 309–25.

5

Rural society

Ruth Panelli

INTRODUCTION

Rural studies have long sought to understand how people experience and organize rural life; how families operate farms; how communities construct cultural meanings and control space; and how marginal groups negotiate inequalities and sometimes contest social relations and structures. Collectively, these types of endeavour represent studies of rural society. Products of this work include the conceptualization of social formations or systems, and the identification of processes and relations that influence – and are shaped by – the practice of rural life. Using modernist and social scientific epistemologies, some studies have approached rural society as a given; a naturally existing phenomenon that can be identified, mapped – and sometimes critiqued – according to its complex but uneven structure and processes. Alternatively, postmodern and poststructural scholarship has approached rural societies as socio-cultural constructions, landscapes and texts that can be read for the meanings, values and politics associated with rural identities and the diversity of formations and change occurring in different rural societies.

In reviewing scholarship that addresses rural society, this chapter draws on the work of anthropologists, geographers and sociologists to outline how knowledge of rural societies has developed in recent times. Accounts of this kind face the highly contestable politics of representation since such reviews are always reconstructions of history and/or partial accounts of diverse future horizons (Livingstone, 1992; Rose, 1995). Phillips (1998) notes that exclusions that are likely to occur both through authors' positionality[1] and intent. In acknowledging these issues, the scope of this chapter is delimited by a focus on literature concerning Western rural societies, especially those in Europe, North America, Australia and New Zealand. It is also shaped by a principal consideration of works written in the past three decades, although some reference is made to earlier scholarship.

After initially sketching the heritage of this literature and the contrasting theoretical perspectives that have grounded studies of rural society, the second section identifies key units of analysis: families, communities and localities. A third section highlights analytic themes (including change, difference and power in rural societies). Then, a fourth section records how these themes continue and are joined by new analytical dimensions to constitute contemporary foci in studies of rural society. Finally, a concluding section synthesizes work to date and suggests future directions for the study of rural societies in an increasingly postindustrial, if not post-rural era.

BACKGROUND

A chapter of this size cannot hope to review the enormous and multi-faceted corpus of academic work on rural society. Instead, it is possible to acknowledge some of the broad differences and clusters that characterize this literature and

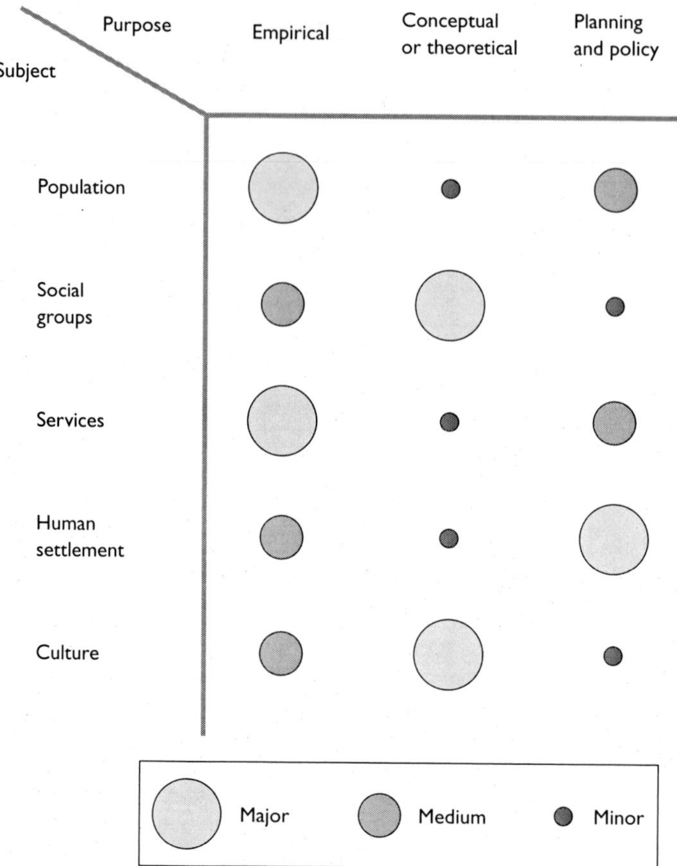

Figure 5.1 ***Emphasis of research into rural societies by purpose and subject***

provide a background and framework for the following discussion. Studies of rural society can initially be recognized for their heritage and macro foci that are grounded in different disciplines, particularly anthropology, sociology and geography. These cultural, social and spatial foci have all provided conceptual strategies (and influenced research methodologies) that have resulted in diverse classifications and interpretations of rural society including the beliefs, structures, processes and spaces involved.

These different disciplinary roots have sustained twentieth century studies of rural society that can be described both in the traditional macro terminology of 'social systems' (e.g. Loomis and Beegle, 1950) and via the focus on specific subject areas, such as: population structure and dynamics; social groups; human settlements and regions; cultural values and practices; and service and resource access. As shown in Figure 5.1, these subjects have also been concentrated according to different purposes of inquiry. For instance, studies of service access have most directly fed into planning and policy oriented research, while theoretical investigations have most often concentrated on conceptualization and explanation of social groups (e.g. stratification and interaction) and cultures (e.g. meanings and practices).

Alongside these functional differences, it should be noted that studies of rural society also draw on a variety of epistemological and scientific traditions. A long tradition of empirical works in all three disciplines has applied either ethnographic or logical- or neo-positivist approaches to the construction of knowledge about rural societies. In some cases, such perspectives have been grounded in broad social scientific endeavours set within wider science-state relations and institutional structures of extension or economic development (e.g. Larson, 1965; Marsden, 1999; Working Party on Rural Sociological Problems, 1965). Since the 1970s, critical political economy and feminist approaches have sought to identify the inequalities, the political struggles and the alternative possibilities in rural society and its organization. Finally, and in contrast, phenomenological and later post-structural approaches have employed interpretative or deconstructive strategies when recounting the experience and meaning of rural life or the organization of society. These approaches have used different philosophical and analytic repertoires to investigate, explain, theorize or read the diversity of individual and/or local experiences as they occur within wider structures, discourses and power relations. These contrasting approaches are summarized in Table 5.1. The table highlights the interconnection between philosophical foci, contextual influences and characteristics of practice that have occurred when scholars have undertaken each type of work. The elements in Table 5.1 provide a framework for reading the unfurling landscape of rural social scholarship as it has cumulatively established these repertoires over time. As will be shown in the third and fourth sections of this chapter, each approach has enjoyed considerable popularity at various times. But rather than occurring as discrete and separate endeavours (as is depicted for simplicity in the five rows of Table 5.1), these contrasting approaches have each dominated rural scholarship yet also have drawn upon past perspectives and have laid down features that future approaches have used to chart new complementary directions or critical alternatives. Drawing a geomorphologic metaphor, these philosophical and theoretical differences produce a palimpsestual landscape of rural studies where contrasting conceptions of rural research overlay each other but underlying textures and practices continue to have some influence on newer strata or approaches (for discussions of these palimpsestual or cumulative philosophical interconnections in different settings, see Cloke, 1997a; Crang, 1998; McCormack, 1998; Panelli, 2004; Phillips, 1998). These layered and interconnected approaches are highlighted through the following discussion in order to show the diversity of effort and outcomes occurring in rural social research.

UNITS OF ANALYSIS

Whether studies of rural society have been conducted for applied, planning and policy purposes, or for more conceptual and theoretical reasons, all researchers have faced the need to define the scope of their work. Some of these decisions delimit studies by time and place; however, the selection of units of analysis is also important. Rural society has been investigated through reference to many different units. Some of these are aggregate concepts such as social formations, social systems and countryside types (Hoggart et al., 1995; Marsden et al., 1993; Stacey, 1969). However, smaller units of analysis are more common and three key ones are addressed in this section, namely: family, community and locality. Farm and village are complementary units, but they are not considered in detail here since 'the farm' is often treated as a foundational unit for many economic studies, and 'the village' has been less frequently addressed (and in a more descriptive and case-specific form) than the generic concepts of family, community or locality. Nevertheless the following discussion will note how consideration of each analytic unit has involved considerable overlap with the others, for example, family and farm in studies of the social organization of agriculture.

'The family'

The family has long been a unit of inquiry and analysis in rural studies, especially through investigations of farming or community life (e.g. Arensberg and Kimball, 1940; Coward and Smith, 1981; Williams, 1963a). Different positivist, Marxist and feminist approaches have employed or interrogated the notion of 'family' in rural studies, and it remains a commonly cited unit when investigating kinship, conjugal, generational or household structures and relations.

Post-war quantitative studies of rural and peri-urban life, illustrate the use of 'family' as a pre-given social category. For instance, early studies of rural social change resulting from industrialization adopted 'the family' as an analytical unit for making rural–urban or cross-cultural comparisons (Lupri, 1965; Wilkening and Lupri, 1965), while later analyses of housing provision and needs drew on implicitly defined notions of 'family' (Dunn et al., 1981; Phillips and Williams, 1992). Most recently, Beesley (1999) and Elder and

Table 5.1 ***Contrasting philosophical and theoretical practices in the study of rural society***

	Contextual influences within/beyond rural society	**Philosophical foci applied to rural social scholarship**	**Practice of rural social research**
Positivist and quantitative	Post-WWII drive for economic development and modernization Industrialization of agriculture Growth of rural services, infrastructure, policy and planning	Investigation of 'positive' phenomena (e.g. population) Aim to establish generic patterns and relations that could be systematically tested Potential to model and predict variables of rural society (e.g. population growth and transport needs)	Attention to observation and measurement of material/ tangible social variables Commitment to objective documentation of rural social phenomena Production of maps, statistics and models of rural society (e.g. population change and settlement hierarchies)
Hermeneutic	Dissatisfaction with positivist and quantitative approaches Popularity of humanist and field/real-life stories in contemporary society Growth in local symbols and identities for rural (and other) places/ communities	Investigation of material and symbolic dimensions of everyday life (e.g. work patterns and community events) Aim to document accounts and meanings of rural experiences (that are particular rather than generic)	Attention to in-depth observation, description and interpretation of rural experiences and meanings Commitment to collect authentic and unique records of everyday lives, social practices, values and meanings Production of ethnographies, textual records and symbolic analysis (e.g. community and landscape studies)
Marxist	Widespread social and political protests in developed societies (1960s–1970s) Industrialization of rural industries leads to globalization and restructuring (of farming etc.)	Historical materialist critique of political economies (e.g. capitalism) Aim to highlight and challenge unequal capitalist power relations in and beyond rural societies Use of dialectic reasoning to document and challenge inequalities associated with the rural social reproduction of capitalist industries (e.g. farming and mining)	Attention to conceptualization of the economic and social relations involved in different modes of production (e.g. class relations in industrial capitalism) Commitment to radical critiques of Western rural economies and societies Production of commentary, statistical analysis and case studies (e.g. regional employment differences, social impact of restructuring)
Feminist	Women's liberation and third wave feminist movements Institutionalization of gender equity in public sector Recognition of gender differences and in some cases inequalities	Study of rural women's social positions and experiences Aim to critique patriarchal gender relations underpinning the structure of rural society Use of feminist theory and emancipatory methodologies to incorporate women's experiences into rural literature	Attention to women's own accounts of rural life Commitment to document and challenge the unequal gender relations and implications of gender identities Production of case studies and critical commentaries of gendered rural life (e.g. under-recognition of women's contribution to agriculture)

Table 5.1 (Continued)

	Contextual influences within/beyond rural society	Philosophical foci applied to rural social scholarship	Practice of rural social research
Postmodern and post-structuralist	Apathy and scepticism surrounding past progressive and modernist hopes Increasing importance of cultural attributes and revaluing of cultural texts alongside scientific knowledge Increased pressure to conserve and consume particular rural environments and lifestyles	Deconstruction of key concepts and metanarrative accounts and of rural society (e.g.'community','rural') Aim to highlight diversity of society and the constitutive importance of language Use of deconstructive and archaeological strategies to document the various interconnected discursive and political dimensions of rural knowledge and social relations	Attention to social diversity and the discursive construction of rural societies and imagination Commitment to deconstruct dominant narratives of rural life (e.g. as productive or idyllic or harmonious/safe) Production of multiple readings of rurality and the diversity of rural groups and their uneven experiences (e.g. records of rural lives of children, homeless and mentally ill)

Conger (2000) have used 'family life' as a natural and homogenous indicator of 'life satisfaction' or children's 'life success' and implicitly present families as synonymous with the presence of two parents and children.

Studies of agriculture have also recognized the notion of family as a key analytical unit through which the organization of farming can be understood (Marsden, 1984; Moran et al., 1993; Williams, 1963b). This has been a common occurrence since most capitalist agricultural systems remain structured around family units (for example, Alston, 1995a notes that 90 per cent of Australian farms are family units). Most commonly, detailed critiques of capitalism and the industrialization of agriculture have debated the position of 'the family' as a social and economic unit through which property, labour and kinship relations are mediated in different ways (Moran et al., 1993; and see also the subsumption and survival debates concerning family farming: Goodman and Redclift, 1985; Whatmore et al., 1987). In farming, diverse and complex family forms have been acknowledged as contrary to smaller urban units where industrial capitalist economies have seen the spatial division between productive and reproductive labour (Alston, 1995b; Toynbee and Jamieson, 1989)[2]. Further, the influence of feminist approaches has seen the development of useful gender-sensitive analyses of farm labour, production, authority and decision-making across a variety of family forms (Toynbee and Jamieson, 1989; Wallace et al., 1994; Whatmore, 1990). Similarly, a smaller number of generational studies have been important for recognizing farm-related negotiations between parents and children in terms of farm labour and transference (e.g. Hutson, 1987; Wallace et al., 1994).

While the family has been a common unit of rural social research, the 1980s and 1990s saw feminist critiques of the 'family' unit make a radical and invigorating contribution to studies of farming and community life (Berlan-Darque and Gasson, 1991; Little, 1987; Whatmore, 1990, 1991). Early studies indicated the spatial and labour-related gender divisions of labour and expectations associated with families (e.g. Bouquet, 1982; Stebbing, 1984). Later investigations have adopted theories of gender relations and identities to critique the constructions of 'family' as a unit and place primarily for rural women. These works have also highlighted the effect of beliefs about family, and domestic ideologies, on the social and employment experiences of women (Hughes, 1997a; Little, 2002; Little and Austin, 1996).

Most recently, poststructural approaches to rural studies have recognized the family as a social construction that is established and/or contested in a variety of cultures and discourses (Liepins, 1998c; Mackenzie, 1992; Philo, 1997).

The historical and spatial implications of such constructions are emerging as a new way to recognize the diversity of rural societies and political struggles. For instance Philo's (1997) attention to the alternative family form of the North American Shakers of the mid-eighteenth century reminds us of how geographies of alternative family units have established different rural spaces. Additionally, Mackenzie (1992, 1994) and Liepins (1998c) have argued that farm women's strategic use of their multiple subjectivities (including those of wives and mothers *in families*) enabled them to act politically for their farm business and kin.

In sum, the family has been widely used as an analytic unit by scholars from many contrasting backgrounds. While some under-theorized or implicitly defined usage continues, feminist and poststructural critiques have highlighted the diverse forms, constructions, and cultural and power relations involved in family units.

'The community'

Community has a long and variable history as a unit of analysis in studies of rural society but has dominated much of the English-language literature.[3] As a mid-scale concept, community has been used to address the structural, relational (and in some cases spatial) dimensions of a social grouping that is popularly and politically recognized between family and regional or national groups.[4] Notions of community have varied enormously (Hillery, 1955) but have provided the framework for many studies of rural society based on the interaction of people in specific locations.[5]

Drawing on North American sociologists' interest in social ecology, ideas about 'natural communities' in urban studies were to influence rural research, as was the German work of Tonnies (1963), which established a distinction between *Gemeinschaft* (community) and *Geselleschaft* (association/society). In the latter case, community emerged as a positive and relatively stable state in contrast with (urban) *Geselleschaft* association or interaction. The enormous range of British and North American community studies written in the second half of last century drew on Tonnies's work[6] and focused on village-based communities as indicative of a rural–urban distinction, or later as part way along a rural–urban continuum (Frankenberg, 1966). These works concentrated on the structure of place-based or territorial communities but were criticized during the 1960s and 1970s for being descriptive, static, homogenizing, traditional, unscientific, abstractly empiricist and even premodern (Day and Murdoch, 1993; Harper, 1989; Newby, 1980; Saunders et al., 1978). The influence of modernization and the industrialization of Western rural societies led to fundamental changes in many rural industries and settlements. Consequently, notions of community were challenged by exogenous processes. Further theoretical criticism greatly reduced the influence and popularity of this term when Pahl (1968) argued that rural and urban categories were not explanatory in and of themselves. Since community had become so intimately conflated with rural areas the move away from comparative rural–urban explanations saw scholars eschew community investigations.

In the face of such criticism, community-level analysis declined. In some quarters it was still recognized as relevant to applied and policy studies. For instance, studies drawing on the positivist and quantitative philosophical foci and practices (noted in Table 5.1) took community as a measurable indicator of 'rural well-being' or 'life satisfaction' (Wilkinson, 1986; Beesley, 1999) and as a key arena for development and policy work (e.g. Henderson, 1991). However, anthropologists were the only scholars to remain fully engaged with the value of a community approach to the analysis of rural societies through the 1980s. Anthropology served as a conceptual reservoir for the study and theorization of community. Drawing on hermeneutic philosophies and interpretative practices, rural ethnographic studies took the community as a unit of inquiry and the 1980s saw British anthropologists produce and edit an important body of work (Cohen, 1982, 1983, 1985, 1986; Strathern, 1982). These ethnographies and 'immaterial' emphases on community were crucial for highlighting the cultural meanings and symbols of rural communities (even when economic and social changes meant a material blurring of rural and urban environments/societies). In particular, Cohen countered structural-functionalist approaches with attention to symbolic interactionism when he argued:

> the 'community' as experienced by its members does not consist in social structure or in 'the doing' of social behaviour. It inheres, rather in 'the thinking' about it. It is in this sense that we can speak of the 'community' as a symbolic, rather than a structural, construct. (1985: 98)

This primarily British trend has been registered elsewhere (e.g. Poiner, 1990 – Australia; and Scott et al., 2000 – New Zealand). Consequently these perspectives were to provide core contributions that would resonate in later cultural and post-structural re-engagements with community on the part of sociologists and geographers (Day, 1998; Liepins, 2000b). Following philosophical critiques of the exclusionary implications in employing homogenizing and reifying notions of community (Young, 1990), and analyses of the gendered inequalities within such social units (e.g. Dempsey, 1992; Little and Austin, 1996; Poiner, 1990; Porter, 1991), these later scholars sought to incorporate both the material and the spatial mediums through which community was established with the cultural meanings and practices that reproduced senses of community and identity in contrasting situations (Day, 1998; Day and Murdoch, 1993; Liepins, 2000a). In measure these works have picked up on what Newby (1980: 80) called a 'revival in "community" values [following] ... the re-assertion of a Romantic anti-urbanism and anti-industrialism in Western societies'. While the blurring of urban and rural lifestyles and economies has continued into postindustrial societies, scholars are continuing to recognize and study community as both an important social scale of analysis and a cultural unit in the discourses and social relations that shape people's experiences. The sites and spaces of 'community' (Hall et al., 1983; Liepins, 2000a) have been recognized as interdependent with the socially differentiated struggles and cultural practices that continually rework ideals and day-to-day experiences of community (Day, 1998; Panelli, 2001; Scott et al., 2000).

There remains much to be explored in terms of the power of 'community' as a unit linking class, property and change to post-agricultural communities (e.g. Salamon and Tornatore, 1994), and as a cultural notion, even: 'a memory of the past and a hope for the future' (Connell, 1978: 214). As long as the term remains common to both popular discourse and political rhetoric it is likely that rural scholars will need to maintain their engagement and critique with it.

'Locality'

In comparison with the notion of community, locality has a relatively short and modest history in studies of rural society. Sociological foundations for locality studies rest in Pahl's (1968) call for attention to the locally specific implications of national processes, and in Stacey's (1969) conceptualization of the 'local social system'. Scholars drew upon these proposals as community studies lost favour in the 1960s and 1970s. A 'local social systems' approach enabled locally specific studies to be conducted with reference to wider society via vertical institutional links. Stacey's own development of a local social systems approach occurred within the dominant functionalist approaches of her day. However, intermittent sociological considerations of locality have continued (Bradley, 1985), including the relatively recent analysis of cultural change in Dutch and French villages as two local systems displaying differences that could be understood in terms of locality and identity (de Haan, 1997).

Within anthropological circles, interests in locality as a unit of inquiry have been established through hermeneutic and interpretative strategies (Table 5.1) that produced varied ethnographies of local meanings. Cohen (1982) and Strathern (1984) illustrate the types of work that recorded the social meanings of local-ness or localism. These primarily village-based studies served to complement the class analyses that became important as migration and economic restructuring stimulated widespread social changes and tensions between 'locals' and 'newcomers' (Quale, 1984; Strathern, 1982).

While hermeneutic approaches were an important alternative to the science-inspired positivist and quantitative accounts of post-war rural development, it was not until the growth of radical political economy analyses, however, that locality studies became a popular unit of analysis in rural studies. As summarized in Table 5.1, Marxist approaches drew on dialectic reasoning to critique capitalist production and social reproduction in rural areas. Through this process, localities became a popular unit for study in the 1980s. Following critiques about the overly macro-economic perspectives being written of capitalist economies and societies, attention was increasingly given to spatial inequalities, local difference and restructuring occurring in specific localities (Urry, 1981; Newby, 1986; Massey, 1984).[7] Cooke's (1989) attention to urban and regional change led him to argue that locality was a superior concept to community, being more appropriate for a nuanced study of capitalism and modernity.[8] In this respect we can trace a philosophical and conceptual concentration and division that occurred during the 1980s between hermeneutic-inspired ethnographies of community and the Marxist-inspired critiques of capitalist relations in localities.

Parallel with the general popularity of political economic accounts of agriculture, British locality

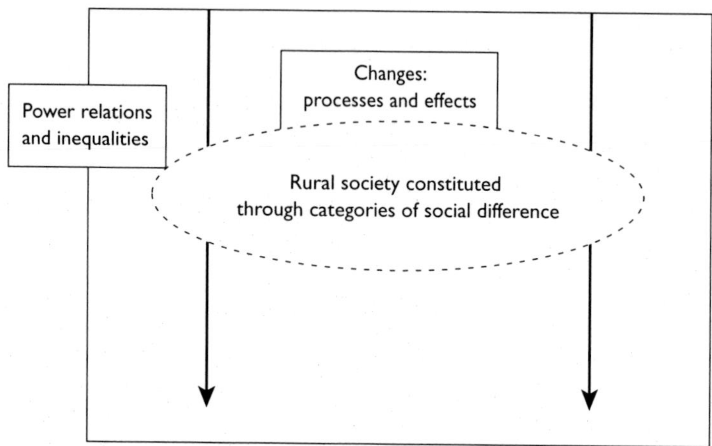

Figure 5.2 ***Analytic themes in the study of rural society***

studies of rural labour and property provided detailed accounts of local relations that were also intimately connected to wider economic processes (e.g. Bradley, 1985; Gilligan, 1984; Marsden, 1984; Munton, 1984). And the influence of these studies continued to resonate in later local case studies of actor network relations (Murdoch and Marsden, 1995). The most numerous of these analyses have had an economic focus, however, contemporary approaches (discussed in section four) have continued to include locality as an analytic unit. It has been recognized as a discursive construction (Pratt, 1991) that can be studied when seeking to interpret how economic and cultural meanings are invested in different ways in different places (de Haan, 1997).

In sum, analyses of locality, while not so prolific as community studies, have offered a strategy for enabling rural society (and economic systems such as agriculture) to be understood as linked to wider national and global processes in a range of mutually constituting relationships. Locality studies simultaneously encouraged an emphasis on vertical relations with wider social and economic processes, and an attention to the local specificities that ensured broad processes and changes were never uniform in their affect on, or implications for, rural societies.

ANALYTIC THEMES: CHANGE, DIFFERENCE AND POWER

While studies of rural society are diverse, they do include complex iterations of key themes. This section sketches out some of the most influential themes permeating the literature. It shows how, beyond descriptive accounts of key units such as family and community, rural scholars have concentrated primarily on three powerful foci: change, difference and power. As will be demonstrated below, these themes have been conceptualized to build explanations, interpretations or readings of rural society as a dynamic and heterogeneous phenomenon. These outcomes vary because of the contrasting philosophical and theoretical approaches scholars have taken. As outlined earlier, each of the five main approaches shown in Table 5.1 have enjoyed (or continue to enjoy) academic popularity at different times and thus emphases have varied.

Despite philosophical differences, socially differentiated rural society is the central object of study, as shown in Figure 5.2. This diagram indicates that rural society undergoes different processes of change, all of which are articulated through different systems of power relations, institutions and inequalities. Rural societies are clearly influenced by many other phenomena and relations (for example, economic and environmental processes) operating at different scales; however, Figure 5.2 and this section focus upon those motifs that have been most dominant in constructing accounts of rural society, namely: change, difference and power.

Rural change: processes and effects

Rural studies have historically been written with either an implicit or explicit acknowledgement of

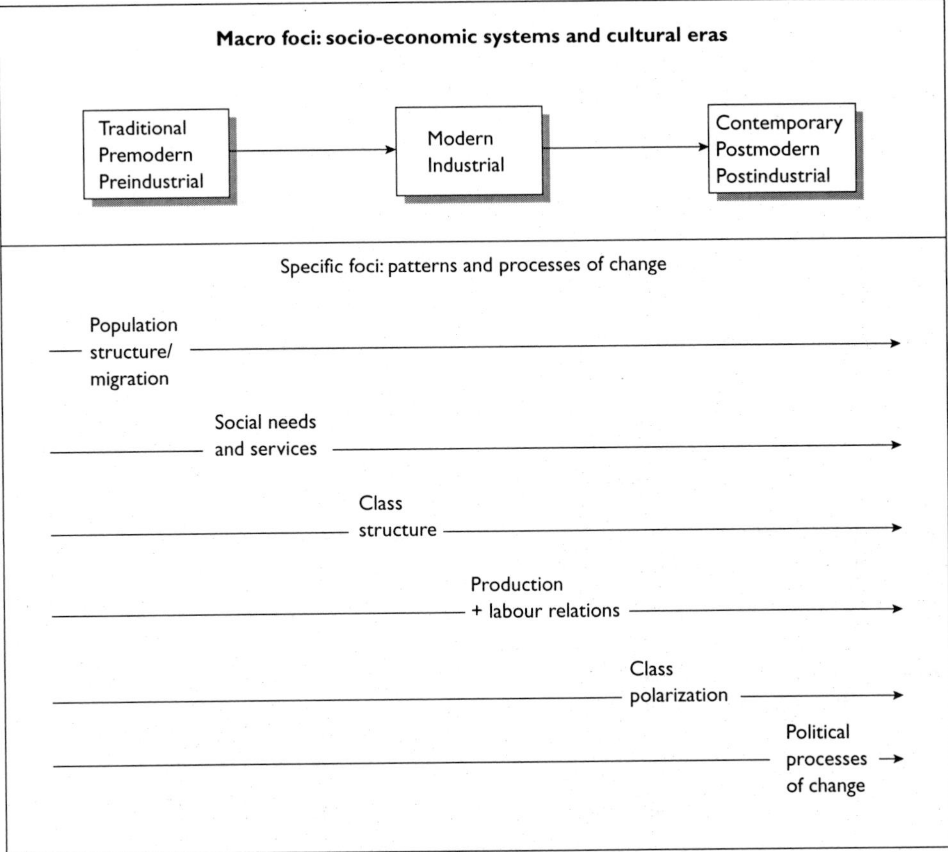

Figure 5.3 ***Foci in the accounts of change in rural societies***

rural areas and societies as dynamic entities. The diversity of areas and contexts has meant that some studies have emphasized rural stability (in a changing world) while others have highlighted dynamics of decline or growth. Nevertheless, accounts of social change arguably constitute the most enduring theme in the study of rural society. In each discipline and at all scales of inquiry, scholars have identified patterns and experiences of change and sought to explain these in terms of specific processes or macro socio-economic systems and eras (see Figure 5.3). While Figure 5.3 highlights the content of these foci, they also represent different interests associated with the five theoretical approaches previously depicted in Table 5.1, for example, population and 'needs' studies have frequently been generated from quantitative approaches while analyses of labour relations and class polarization have been part of the later popular Marxist approaches.

Work during the 1960s and 1970s was important for documenting the modernization, industrialization and urbanization of rural societies from a traditional and, in some cases, preindustrial, state (e.g. Bergman, 1975; Connell, 1978; Fleigel, 1976; Lewis and Maund, 1976; Pahl, 1965; Sachs, 1965; Working Party on Rural Sociological Problems, 1965). Alongside quantitative approaches documenting the changes, theories of 'integration' were commonly employed to argue that rural societies would be increasingly integrated into broader (urbanized and industrialized) social and economic systems (Munters, 1975; Photiadis and Ball, 1976). Evidence of mechanization of agriculture, increasing productivity, changing labour relations, depopulation and changing values were all used to substantiate arguments about major changes in rural society in a modern/industrial era (Capo and Fonti, 1965; Fleigel, 1976; Lewis, 1979; Newby, 1977; Sachs, 1965), although some scholars were quick to point out the complexity and unevenness of these trends (Brandes, 1975; Bowler, 1986; Thieme, 1983; Took, 1986).

In studying these changes over the past 50 years, descriptions have abounded regarding the recomposition of social groups – in terms of both units such as families, communities and farms, and categories of social difference such as class and gender (discussed below). The theoretical differences between scholars have been marked. A dominant approach involved the philosophies and practices underpinning positivist and quantitative perspectives (Table 5.1). For instance, studies as diverse as changing population structure and cultural values relied on logical positivism. These works concentrated on modelling of population stratification and hypothesis testing of value changes (e.g. Fleigel, 1976). Others have pointed to the difficulty of single-theory accounts of change and have called for methods that address the diversity of processes and social experiences (for example, in explaining population changes: Bolton and Chalkley, 1990). In wider contrast, ethnographies of community change and counter-urbanization have built up field-based accounts of how changing rural societies are diversified and in some cases must negotiate alternative social and cultural values that influence collective units such as communities and villages (e.g. Richling, 1985; Scott et al., 2000). Such examples show how even when one theoretical approach might dominate, as in the case of early quantitative accounts of change, still other markedly different approaches will complement inquiries by providing in-depth individual accounts of the processes and values involved in such changes.

Marxist and feminist analyses of change have been more common in studies of agricultural production and social relations (Bouquet, 1982; Friedmann, 1978; Marsden, 1984; Whatmore et al., 1987). Both these approaches draw on a radical heritage to shape their theoretical focus and methodological practice (Table 5.1). Consequently, their results have often provided critical commentary and cases of the social impact of farming and agricultural restructuring. More recently, geographers have applied regulation theory and actor-network theory to provide detailed, multidimensional analyses of how rural societies and economies are being incorporated into wider economic and social systems whilst still operating in locally specific and differentiated ways. (For British studies see Cloke and Goodwin, 1992; Goodwin et al., 1995; Marsden et al., 1993; Murdoch and Marsden, 1995. For New Zealand examples see Cocklin and Wall, 1997; Conradson and Pawson, 1997; Moran et al., 1996.)

Overall, studies of change have remained a constant theme in analyses of rural society. This macro focus on the dynamics of social systems and relations has enabled scholars to build explanations of the material and cultural changes that have continually reshaped rural lives, institutions, cultures and landscapes. This brief discussion has shown the interests of scholars working from different philosophical positions, with more quantitative studies in the 1970s/1980s and more radical ones in the 1980s/1990s. Most recently, attention to change has concentrated on post-structural and postmodern approaches to rural society (Mitchell, 1998; Murdoch and Pratt, 1993; Philo, 1993) with mixed conclusions – yet the endurance of the theme of change remains steady. It provides researchers from many different perspectives with the opportunity to develop their interests in, or readings of complex social systems as dynamic entities.

Social difference

A second core analytic theme that can be traced through many studies of rural society involves accounts of social difference. A wide range of empirical and theoretical work has documented how rural societies differ in terms of social groups, including their economic, cultural and political activities, experiences and interests. Early axes of difference related to analyses of people's social positions as mediated through property and labour relations, thus studies of social class have formed an early and enduring feature within this theme. They have been supported primarily by radical historical materialist approaches undertaken by Marxist scholars, critiquing the social consequences of capitalist rural economies from farm to global industry scales (see Table 5.1). More recently, and shaped by wider feminist movements, the significance of gender has been recognized as an important cleavage distinguishing people's experiences of labour, community life and rural spaces. The critical politics underpinning theory and practice of feminist research has led to increasing studies of gender inequalities in rural settings (Table 5.1). These two categories of difference have been the focus of most work until recent times, and the following discussion reviews some of the advances made, while broader acknowledgements of difference are also made in the fourth section of the chapter, devoted to contemporary works.

Class

Class analyses of rural society have seen traditional, descriptive and graduated typologies of class (using pluralist perspectives) critiqued and replaced with more radical and relational analyses, using Marxist and neo-Marxist perspectives on class and the political-economic organization of production (and more latterly consumption). While this has been the dominant approach to class, recent accounts have been complemented by more process-oriented and cultural readings of class groups. This section reviews these developments and suggests that class studies have lost some political edge, although potential for future insightful critiques remains.

Over the past 40 years, analyses of social difference in rural areas have included descriptions of class categories and accounts of the relations and conflicts between and within such groupings. Past American sociology and geography argued that social differences could be usefully understood in terms of class (e.g. Stinchcombe 1961). This allowed scholars to balance many of the descriptions of farm and village life with generic conceptualizations and explanations regarding status and social differentiation. Drawing on Wissink's (1962) identification of social heterogeneity on the rural–urban fringe, Pahl (1965a) became a key figure in rural class analysis. He established a series of works documenting class-based structures and changes (related to occupation, income, property, housing and education: see Pahl, 1963, 1965a, 1965b). Pahl's writings identified complexity and variation within various classes that were primarily recognized as providing a foundation for investigations of class conflict and change. Building on from his accounts of a (traditional, settled) working class and a (urban- and modern-oriented, newcomer) middle class in commuter villages surrounding London, other scholars developed a literature that explored class differences, migration-based social change and subsequent class polarization (Connell, 1978). For example, Newby's analysis of East Anglia, concluded that:

> The social polarization of the countryside has been a slow but inexorable process since the end of the Second World War. Within agriculture the large-scale landowner and farmer has generally benefited at the expense of the small-marginal producer and the farm worker. At the same time, a stark contrast has arisen in most villages between a comparatively affluent, immigrant, ex-urban middle class and the remnants of the former agricultural population tied to the locality by their (low-paid) employment, by old age and by lack of resources to undertake a move. (1979: 494)

A good deal of research through the 1970s and 1980s employed relational theories of class (either Weberian or Marxist). This enabled twin foci on class structures in agriculture and analyses of the uneven property and/or production relations, maintaining and exacerbating class difference between farm owners and/or middle classes on the one hand and (farming and nonfarming) working classes on the other. (See Bell and Newby, 1974 and Bradley, 1985 for British cases; Hall et al., 1983 (a New Zealand case); Newby, 1972 (a further British example); and Wild, 1974 (an Australian case).) Across such works, the relationship between (rural) property and class emerged as a key dimension (for example, for agricultural production, display of status, or political leverage). For instance, Saunders et al. reviewed American, Carribean, Asiatic, European and British cases over time and argue that:

> the landowning class in each case operates an elaborate web of paternalistic relationships because of the stability which this confers on an otherwise highly inegalitarian social system, and … the exercise of traditional authority essentially involves a handling of the contradictions that arise from … the social differentiation which a hierarchical structure engenders, and … the identification which must be inculcated if stability is to be achieved. (1978: 62)

As traditional agrarian societies became industrialized and urban–rural migration increased, further attention was afforded the growing middle class groupings, especially in British scholarship (e.g. Cloke et al., 1995c; Savage et al., 1992; Thrift, 1987b; Urry, 1995a). Employing the aims and commitments of Marxist approaches (Table 5.1), an upsurge in studies recorded the social change and struggle occurring as capitalist and class structures altered. Cloke and Thrift's (1987) investigation of intra-class conflict provided a critique and advance on the homogenizing tendencies of past class studies – pointing to the class factions (after Wright, 1985) and intra-class struggles occurring within a fraction of the middle class (that is, the service class). More latterly scholars have acknowledged that patterns and politics of consumption are equally important to class analyses. Of the service class within the middle class fractions, Thrift argued:

> the service class lives in a series of milieux bent towards tasteful consumption. These are designer civil societies, the consumption cultures. In them, the consumption-cum-reproduction preferences of the service class are made particularly clear. (1987a: 242)

Later studies of housing, recreation, tourism and local politics explored class relations based on consumption (rather than previous emphases on production) more fully (e.g. Phillips, 1993; Savage et al., 1992; Urry, 1995a; see also the analysis of different class actors by Murdoch and Marsden, 1995). These works have indicated how such middle class interests and consumption practices draw economic and cultural relations together as rural space and society are reconstructed (and sometimes regulated) to realize notions of a rural idyll. Cloke et al. (1995c: 233) contend that 'class fractions ... [invest] not only economically but also culturally and psychologically in the countryside', consuming idyllic discourses of rurality and mobilizing class interests to colonize the countryside in particular ways.

The British scholarship on rural classes has been most prolific and influential. While it is firmly rooted in Marxist approaches, popular in the 1980s, a trend can be discerned from a material (and increasingly critical) analysis of property and labour relations in the 1980s to one focused more on relations of consumption and cultural meaning in the 1990s. These latter works have been important for identifying how powerful class processes may continue to shape rural areas (even as definitions of rurality and urbanity become problematic); for rural societies and spaces are noted to carry class-significant cultural meanings and material and aesthetic resources. Geographers have been particularly important for showing how class interests enable the control of property and access to political networks and processes. The variations in such interests are shown to differentiate people's ability to value and control rural space, local housing property and processes of landscape management (Cloke and Thrift, 1987; Cloke et al., 1995c; Lowe et al., 1993; Marsden et al., 1993).[9]

These works have nevertheless lost some of the critical edge of earlier studies, for class has been somewhat overtaken by culture, while the material and political inequalities experienced by significant lower and/or working class groups have fallen almost from view. Newby's material and political note of the less affluent classes echoes a critical note for recasting future academic class analyses:

> The rural poor find, somewhat disconcertingly, that they and their needs are increasingly regarded as residual – or even unacknowledged ... They find that more attention seems to be given to the visual appearance of the countryside than to the standard of living of those who are employed in maintaining it. (1979: 497)

While important studies of poverty and deprivation have been undertaken by British scholars more recently (see section four of the chapter below), future class-specific analyses could more critically re-engage with the continuing struggles over labour, resources (including but not confined to property) and decision-making as rural cultures and spaces change.

The focus on cultural and middle classes has not been so widespread beyond British scholarship; however, the critical implications of class struggles and social polarization have resonated through other inquiries. (See, for example, Poiner's (1990) class-sensitive gender analysis of an Australian community; Naples's (1994) reading of class, gender and ethnicity through economic change in rural USA; and Salamon and Tornatore's (1994) identification of class tensions surrounding property and community change in rural USA.) More recently, Gibson-Graham's (1996) anti-essentialist conceptualization of class as a *process* has drawn on detailed post-structuralist readings of class as an overdetermined social category which is integrally connected through constantly negotiated processes with other relations and axes of identity (such as gender and ethnicity). Gibson-Graham's study reflects the post-structural commitment to deconstructing past, stable notions (Table 5.1), to show the complex and multiple class identities and relations in contrasting Australian rural mining communities. It illustrates how traditional dimensions of property and labour relations are deconstructed and reworked through diverse variations in place, gender and ethnicity. This work provides a further theoretical 'boost' to rural class investigations, for it provides a conceptually rigorous framework for considering the multiplicity of class identification and power relations involved in the class processes in contemporary societies.

Gender

Gender is a second key means by which rural difference has been studied and theorized in recent decades. Academic investigations and theories of rural society have increasingly acknowledged gender difference in all aspects of rural life. As with class, gender has been recognized as an analytical axis around which social groups and rural life is differentiated, for example, the organization of agricultural production; the social relations and spatial politics occurring in rural communities; or the cultural meanings and power associated with different notions of agriculture or rurality. Major inquiries centred primarily on farming, but increasingly broadened as explanations of gender difference developed away from early sex-role typologies to grapple with the power relations and socio-cultural identity processes that constitute gender in different societies, places and times. This section reviews these changes and

Table 5.2 ***Examples of gendered farm role analysis in different agricultural systems***

Context	Year of publication	Publication
USA: capitalist farming (women)	1978	Flora and Johnson (1978)
Yugoslavia: socialist and 'private' (women)	1978	First-Dilic (1978)
Australia: capitalist (women)	1979	Craig (1979)
England: capitalist (women)	1980	Gasson (1980)
USA: capitalist (young men and women)	1980	Dunne (1980)
England: capitalist (women)	1983	Symes and Marsden (1983)
England: productionist and post-productionist (men and women)	1991	Symes (1991)
Norway: productionist (men and women)	1991	Almas and Haugen (1991)
New Zealand: productionist (women)	1994	Keating and Little (1994)

points to the further concentration on identity and masculinities that is emerging in contemporary work.

Historical works in Britain and the United States have highlighted the gendered organization of farming and community life. For instance Davidoff et al. (1976) and Adams (1994) have noted how gender differences occurred in household and agricultural labour, as did the varying cultural responsibilities and expectations surrounding the reproduction of 'home', 'family' and 'community'. They argue that notions of domesticity, modernity and rurality were powerful cultural frameworks through which gendered social and economic relations were negotiated as rural societies have been 'modernized'.

Beyond social histories, gendered analyses of rural society commenced primarily through recognition of men's and women's different experiences and responsibilities in farming (Wilkening and Lupri, 1965). Drawing on the dominance of role theory in the 1970s (e.g. Albrecht et al., 1977; Burke and Tully, 1977), men's and women's contrasting experiences of labour and decision-making on farms were categorized through various occupational and social sex roles. Typologies of women's roles were most common (for example, producer, housekeeper, socializer – First-Dilic, 1978; independent operator, matriarch, partner, helper, homemaker – Craig, 1979), and a widespread application of role descriptions occurred in a range of contrasting agricultural societies (see Table 5.2).

These early accounts of gender, while reflecting some feminist interests, were primarily working from quantitative and descriptive traditions to document rather than explain gender patters and models. Then, through the 1980s the upsurge in feminist scholarship saw a growth in theoretical and empirical accounts of rural gender relations based on both feminist theories of patriarchy and methodological commitment to include women's voices and experiences more fully in academic work (Table 5.1). Critiques of the limited explanatory power of descriptive, role-based accounts of gender differences led, through attention to divisions of labour (Berlan-Darque, 1988), to the adoption of gender relations theory. Based on wider considerations of gender relations (in association with capitalism: Foord and Gregson, 1986, McDowell, 1986; and rural society: Little, 1986, 1987), Whatmore (1991) argued that analyses of capitalist agriculture would always be limited unless a full investigation occurred of the power relations contested between men and women. Based on her extensive study of productive and reproductive labour on varying types of British farms, Whatmore's (1990) conceptualization of a domestic political economy was a key feminist contribution to theoretical explanations of both the social and economic organization of agriculture. While the conceptually detailed framework of an agricultural and domestic political economy has not been addressed in detail elsewhere, gender relations theory has been more widely adopted in studies of farm pluriactivity (Blanc and MacKinnon, 1990) and wider non-agricultural rural studies (discussed shortly).

While the majority of works have focused on women's position in farming, a small number have considered both men's and women's experiences (e.g. Buttel and Gillespie, 1984; Coughenour and Swanson, 1983). Consideration of men and women's farm involvement stimulated a debate over the increasing feminization or masculinization of agriculture and various explanations were proffered. These included accounts of gender changes following industrialization (First-Dilic, 1978); crises in capitalist agriculture (Alston, 1995b); or the growth of pluriactivity and/or post-productive enterprises (Blanc and MacKinnon, 1990; Evans and Ilbery, 1996); and broad social and demographic changes for rural women (Berlan-Darque, 1988). These works began to hint at the complexity of gender differences. Simultaneously with wider gender studies, rural scholarship began to acknowledge the heterogeneity of gender categories and the diversity of men's and women's diverse experiences. These advances paralleled the influence of cultural studies of gender that also addressed the multiple ways gender was experienced and constituted in society. In this way we see a palimpsestual impact occurring between the feminist scholarship that had been established and the more postmodern and post-structural attention to diversity that has been a key characteristic in these later approaches (Table 5.1).

In rural studies these works appeared first in feminist critiques of agriculture in the 1990s. Scholars began to extend their focus from labour and property relations (common in socialist and Marxist traditions) to consider cumulative gender differences occurring through multiple spheres (from home and farm spaces through farming cultures to political arenas in agriculture). Studies through European, North American and Australasian settings documented how identities and discursive construction of a gendered agriculture occurred in language, cultural practices, rural organizations and industry politics (Brandth, 1995; Liepins, 1996, 1998b; Mackenzie, 1992, 1994; Shortall, 1992, 1994; Teather, 1996; Walter and Wilson, 1996). Many of these works illustrate the post-structural foci on the constitutive power of discourses and the opportunities that exist to deconstruct notions of gender, thereby challenging related power relations.

These works resonate with the contemporary cultural focus of much rural scholarship and they were important for complementing the material analyses and explanations of gender inequality in farming. The newer works demonstrated that even if alternative or more egalitarian gender relations were to be established on particular farms, wider social practices and discursive processes were maintaining dominant gender identities – even while providing some opportunities for resistance and alternative discourses.

Gender analysis of rural society beyond farming has been less voluminous and more ethnographic, drawing on hermeneutic and interpretive strategies (summarized in Table 5.1). However, it has also followed a similar trajectory, moving from a focus on the description of gender roles, through to more explanatory endeavours concerning gender relations and identity construction, including the contestation of rural femininities and masculinities. Early work is illustrated by Stebbing, who employed historical commentary on rurality and domesticity to establish a foundation for her gender role analysis of British women in rural communities:

> in the nineteenth century, idealization of rural life became associated with an idealization of home and family ... The woman's role was almost entirely domestic, at the hub of the family and hence at the hub of the idealized community. (1984: 202)

Middleton's anthropological ethnography of a Yorkshire village, supports this perspective and is important for the geographies that followed for she recorded the spatial implications of gender differences and inequalities. She concluded:

> Women ... are seen as 'out of place' in most public space in the village. They spend most of their lives in the home, enmeshed in family activities, whilst men of the village fraternize with whomever they find to talk to in public space. (1986: 132)

Dempsey's (1987, 1988) Australian work corroborated these themes, but focused more on gender relations in terms of both social and labour relations. He showed men's employment and social status were privileged in a rural community while women were disadvantaged economically and socially. Both his analyses of employment and domestic labour concluded that women experienced inequality and exploitation because customs and expectations supported relations that left women primarily as carers and nurturers whose labour could be exploited and appropriated by men. The dominant view in Dempsey's study was that: 'women should be able to find all the happiness they need through the family and home life' (Dempsey, 1988: 434).

Parallel to Dempsey's argument regarding gender exploitation, Little (1987: 304), working in the United Kingdom, was to argue that women's unpaid labour and caring skills support voluntary activities in communities. Inquiries into the positioning of women in disadvantaged positions saw a re-engagement between theories of gender and rurality. By the 1990s, understandings of gender were also influenced by the emerging postmodern and post-structural approaches and linked to the discursive construction of rural idylls (in different historic and spatial contexts). Rose's (1993: 95) arguments thus resonate with the early studies of gender in rural communities, while taking a post-structural reading of rural discourses: 'The rural idyll was envisioned as a village community. Everyone knew their place, and the harmony of such a community

Table 5.3 ***Beyond class and gender: established accounts of social difference***

Categories of social difference	Selected authors and studies
Youth	Dunne (1980) Occupational sex-stereotyping among rural young women and men Hutson (1987) Family farms, family businesses and the farming industry
Elderly	Smith and Gant (1982) The elderly's travel in the Cotswolds Wenger (1982) Ageing in rural communities
Deprivation/poverty	Knox and Cottam (1981) Rural deprivation in Scotland McLaughlin (1986) The rhetoric and reality of rural deprivation Moseley (1980) Is rural deprivation really rural?
Ethnicity	Salamon (1980) Ethnic difference in farm family land transfers Salamon (1985) Ethnic communities and the structure of agriculture

was centrally represented through "natural" gender differences ... with women naturally natural mothers'. Little's (1994, and Little and Austin, 1996) later study of women's employment continued these themes, noting that such constructions of rural women ensured that their employment patterns and career options were maintained in secondary positions behind commitments as carers. In a complementary fashion, Macklin (1995) and Hughes (1997b) note the influence of popular discourses on the unequal constructions of femininity and masculinity in specific communities.

Later gender analyses of rural communities have continued to study women's positions in domestic and paid labour, as well as community activities and voluntary work (Fisher, 1997; Hughes, 1997a, 1997b; Little and Austin, 1996; Poiner, 1990). However, an increasing focus on gender *identities* has come from the post-structural approaches that have theorized gender as constituted in discursive and cultural practices, for example, surrounding domesticity and femininity. Investigations into the construction of rural womanhood paralleled discursive analyses of gender identities in farming and complement developing accounts of rural masculinity – see the fourth section of the chapter (Agg and Phillips, 1998; Hughes, 1997a; Macklin, 1995).

Additional differences

The preceding discussion has illustrated how studies of class and gender have been established as core explanatory axes for understanding how rural society is differentiated. While these foci dominated a good deal of rural scholarship in certain quarters, it is important to acknowledge that other categories of social difference have been recognized for some time, with scholars addressing various applied and policy goals, established accounts of rural social difference based on age and generational relations, poverty and deprivation, and to a lesser extent, ethnicity and disability (see Table 5.3).

The quantitative and/or applied nature of many of these works meant that the descriptive quality of the pieces, or their needs-assessment character reduced the degree to which academic dialogue occurred. Those scholars more interested in conceptualizing rural society in terms of class within capitalist modes of production, or divisions of labour or ideologies of womanhood within patriarchal gender relations rarely acknowledge the more practical and planning-oriented research of those identifying material needs of different groups. This illustrates some of the gaps that exist between approaches to rural society. Even though this chapter has documented considerable overlap and cross-fertilization between the five core approaches depicted in Table 5.1, the contrasts in philosophy and practice can lead, as this example shows, to major chasms and silences between potential dialogue partners. Nevertheless, these initial accounts of age, deprivation and so forth, laid a foundation for a more critical assessment of rural social research in the 1990s, as will be shown later. The important foci on otherness, marginalization and exclusion were to result.

POWER: RESOURCES, RELATIONS AND INEQUALITIES

A third analytical theme that permeates some studies of rural society involves a consideration of power. Notions of power and political systems are recognized to exist beyond rural society (as shown in Figure 5.2) and relatively few research programmes or publications have focused on power exclusively. However, anthropologists, geographers and sociologists have all contributed to the identification of particular local and regional manifestations of power and political processes that can be analysed in terms of their rural contexts and implications. Different theoretical perspectives have influenced analyses of power in rural society and can be seen within the following account which outlines the attention that has been given to resources, relations and inequalities; however, it was the critical philosophies and practices of Marxist and feminist scholars that most have advanced our understanding of power.

Traditional studies of power in rural society have concentrated upon formal political topics and pluralist conceptions of power and political interests. Consequently, analyses of agrarian political interests, political consciousness and local state processes have been most common (e.g. Bergman, 1975; Bokemeier and Tait, 1980). Across these works, politics has been assumed to be democratic and struggles have been normalized as competing interests; with the occasional acknowledgement of 'elites' (see British and Australian commentaries in Gray, 1991; Saunders et al., 1978).

More critical resource and process-based accounts of power have shown how power is unevenly held or mobilized by dominant groups, classes or interests to 'mask, repress, or preempt' interests and groups who may challenge dominance (Saunders, 1978: 482). These works have focused on resources such as property and capital, and processes such as labour relations, patronage and philanthropy that maintain – and at times mask – the legitimacy of dominant minorities (Newby, 1977; Rose et al., 1976). For instance, works such as Saunders et al. (1978) pointed to the importance of class- and property-based structures in rural communities whereby power – in the form of 'traditional authority' – could be maintained through paternalistic personal relations that managed the tensions surrounding inequalities and therefore maintained political stability. These types of works were important for highlighting the relational aspects of power that can be mediated through institutions as diverse as local government units and family farms (e.g. Gray, 1991; Sachs, 1983). In contrast and with the benefit of Marxist and feminist approaches, critical analysis of power in the social organization of agriculture shows the classed and gendered nature of property and operations as access to material resources and decision-making is mediated through gender relations and divisions both on and beyond the farm (e.g. Alston, 1995b; Shortall, 1992). In particular they highlighted economic and social power relations that bind material and immaterial issues together through unequal institutions: for example, 'the family farm'; and practices: for example, 'division of labour' and farm transfer (Berlan-Darque, 1988; Voyce, 1994; Whatmore, 1991). Similarly, community studies have noted the institutions and relations that maintain social differences (such as class and gender) while ensuring particular interests are maintained (Dempsey, 1992; Poiner, 1990; Stebbing, 1984). Overlaps with studies of change have also shown how power is integral to an understanding of rural change, planning and political struggle of space and community (Barlow and Savage, 1986; Cloke and Thrift, 1990; Murdoch and Marsden, 1994). Across both farming and community studies, these analyses indicated that power was invested or mobilized not only in resources or possessions (property, for example), but in the processes and interactions that maintained unequal relations and enable some groups and interests to control and structure practices in rural societies.

Most recent approaches have continued to consider relational and process-based studies of power but have done so as an underlying theme rather than a prime focus of inquiry. For instance, studies of environmental regulation, planning and development, social service provision, or governance and policy have been thoroughly investigated while different conceptions of power have more or less explicitly supported the activity (Lowe et al., 1993; Murdoch and Marsden, 1994). Similarly, studies of gender and other social differences have addressed power relations between dominant groupings and those social groups (for example, women, children, elderly) that are positioned less powerfully.

Across these endeavours two newer and popular perspectives on power have emerged. First, the discursive constitution and negotiation of power, after Foucault, has been widely adopted in some circles. This perspective has spawned widely divergent studies inspired by post-structural philosophical foci and practices (Table 5.1). These studies have produced readings of the discursive construction of meanings and practices by which knowledge and power is established and articulated (for example, of agriculture or development or community). In some cases, certain groups are constructed as marginal or totally absent from rural society (for example, see constructions: of poverty – Cloke, 1997b; of community – Philo,

1997; of agricultural policy – Higgins, 2001; Liepins and Bradshaw, 1999; and of the gendering of agricultural politics – Liepins, 1998b; Mackenzie, 1992, 1994).

Second, an equally popular perspective on power in rural society has concentrated on the mobilization and articulation of assemblages of power through networks of association. Drawing on theorists such as Latour, scholars predominantly from the United Kingdom have established studies of local relations or specific issues as significantly influenced by – and connected to – wider economic, social and political networks (Murdoch, 2000; Murdoch and Marsden, 1995; Murdoch and Pratt, 1997; and see articles in *Journal of Rural Studies,* 1998 vol. 14, no. 1).

These two recent developments are influential but their adoption in studies of rural society is still evolving. However, they provide contrasting sets of conceptual resources and strategies for the rural social studies – both where conservative social formations continue (often agricultural or modern/albeit restructured) and also those societies where post-productive, socially diversifying trends are reconstituting power relations and practices through new structures and in more fluid ways (Murdoch and Pratt, 1997). Both approaches encourage scholars to look beyond resources and institutions of power (measured in property or votes or electoral boundaries or local organizations). Indeed, they support study of the spectrum of power relations in rural society from continuing attempts to understand the conservative or historical hegemonies that may exist (Murdoch and Marsden, 1994; Tonts, 2001), through to new political units, such as neo-tribal groupings and new social movements (Halfacree, 1997; Liepins, 1998a; Mertig and Dunlop, 2001). They also provide frameworks for the critique of marginalization and exclusion that forms one of the key contemporary themes addressed in the following section.

CONTEMPORARY STUDIES

Drawing a distinction between 'contemporary' and 'past' literature is an artificial construct since both empirical and theoretical work most often evolves in a series of ebbs and flows. Distinct initiatives have occurred in studies of rural society but have never been exclusively without reference to existing approaches, nor have they totally replaced existing work. Thus while this section discusses works primarily published in the past ten years showing both continuities and new directions in the literatures, the 'contemporary' notion is primarily a chronological one.

Continuities

In terms of continuity, substantial empirical efforts remain focused on the measurement, assessment – and sometimes modelling – of contemporary rural societies; including their structures, relations and needs (e.g. Berry et al., 1990; Champion and Watkins, 1991; Gant, 1999; Goudy, 1990; Phillips and Williams, 1992; Reimer et al., 1992; Shucksmith et al., 1994; Smith et al., 2001). This emphasizes the point made initially in the first section of this chapter, that while newer philosophical and theoretical energies have developed in rural studies, early approaches have continued to be important. The ongoing use of positivist and quantitative approaches provides important applied and practical documentation of key components and needs in contemporary rural societies. Some of the stability in the literature also rests in the continuing adoption of family, community and locality as core analytic units. In analyses of social conditions and livelihoods, family units remain an important basis of analysis and explanation; even if the units themselves are thought to be changing (Djurfeldt, 1996; Gonzalez and Benito, 2001).[10]

Parallel to family-based work, contemporary studies continue to highlight the importance of communities and localities as relevant scales and units of inquiry. Notions of community continue to be employed (for example, regarding social relations and networks: Liepins, 2000a; Sharp, 2001; Tonts, 2001; socio-economic change and/or 'disruption': Naples, 1994; Panelli, 2001; Smith et al., 2001; and planning and development: Beyers and Nelson, 2000; Herbert-Cheshire, 2000). And through resonance with past works, notions of locality have been employed to emphasize the connections between local particularities and wider (often economic) processes – including the differentiation between 'powerful' and 'dependent' localities (Marsden, 1996; see also Kalantaridis and Barianidis, 1999; Kneafsey et al., 2001).

A further constancy involves the perennial themes of change. Accounts continue of rural decline, population contraction, loss of services and government restructuring (Conradson and Pawson, 1997; Mayer and Greenberg, 2000; Panelli, 2001), especially in relation to questions of sustainability (Copus and Crabtree, 1996;

Table 5.4 ***Contemporary approaches to the study of change in rural societies***

Approach	Characteristics	Recent examples
Socio-material	Studies of population change and/or economic implications on both long-term and newer residents of rural centres	Beyer and Nelson, 2000 Broadway, 2000 Smith et al., 2001 Stockdale et al., 2000
Social/cultural capital	Studies of social and/or cultural knowledge, values and relations that are eroded or enabled through change	Brunori and Rossi, 2000 Israel et al., 2001 Putnam, 2000 Schulman and Anderson, 199[illegible] Warner, 1999
Cultural-economic	Studies of how rural societies/units may mobilize cultural 'resources' through local economies	Ekman, 1999 Kneafsey et al., 2001 Marsden, 1999 Ray, 1998
Networks	Studies of how networks (including political and economic ones) are mobilized as rural societies, communities or economic sectors are reorganized	Kneafsey et al., 2001 Lockie and Kitto, 2000 Murdoch, 2000

Scott et al., 2000). However, new accounts of rural growth have also emerged, both emphasizing the economic and cultural complexities of migration (Halfacree, 1997; Mitchell, forthcoming; Stockdale et al., 2000) as well as the social stability, cultural links and implications involved in economic change (Beyers and Nelson, 2000; Broadway, 2000; Mitchell, 1998; Smith et al., 2001).[11]

Across the now diverse rural change literature we can track different but overlapping conceptual and methodological influences (see Table 5.4). In one grouping, analyses of population change and the various social and economic implications occurring between long-term residents and newcomers are primarily socio-demographic (Row 1, Table 5.4). A second approach to change spans multiple conceptions of social and cultural capital as a way to analyse rural society in terms of participation, linkages and cultural resources. Drawing on differing perspectives of Bourdieu (1977), Coleman (1988), Putnam (2000) and Throsby (1999), rural scholars have begun to pay increasing attention to various indicators of social cooperation, trust, participation and cultural knowledge that occur – or are eroded – as changes occur in different household, production, community or local state settings (see examples in Row 2, Table 5.4).

A third approach draws economic and cultural notions of consumption together (Marsden, 1996; Urry, 1995b). These types of work show how local cultural resources and histories can be employed to reconstruct rural spaces and links between activities, for example, cultural festivals or wine and food production (Brunori and Rossi, 2000; Ekman, 1999). Cultural meanings and assets can be mobilized for rural and regional development, and cultural identities become important for local economies (see Row 3, Table 5.4).

A final perspective that is closely connected to studies of culture and economy has involved scholars focusing on the relations or linkages that enable such cultural and economic developments to occur (Row 4, Table 4). Drawing upon actor-network theories, these writers produce analyses of the associations (and sometimes politics) underpinning changes and reorganization of rural societies.

Overall, contemporary works that retain common themes are drawing on the heritage of past studies while recognizing new configurations and values (or relegation) of rural areas and societies. While some empirical energies focus most on quantitative approaches and on policy and planning-related goals, other critical and cultural studies draw on the recent, poststructurally inspired approaches to power (see above) to develop what has been called the 'cultural turn'.

The 'cultural turn', diversity and otherness

Although a significant number of motifs recur in contemporary studies of rural society, the influence of cultural, and post-structural, thought can be seen as a major change across the spectrum of rural studies. Just as attention to pluralist and positivist social scientific epistemologies or Marxist critiques influenced scholarship between he 1960s and 1980s, so too, cultural theory and post-structural and postmodern approaches to society, knowledge and research have shaped many of the styles and emphases in recent work. This development saw the emergence of philosophical foci and scholarly practice associated with postmodern/post-structural thought, especially in terms of the critical deconstruction of previous power concepts, and a heightened recognition of social diversity and the social and methodological implications of this recognition (Table 5.1). This expansion of a new philosophical and theoretical current within rural studies had a number of origins; however, a key moment occurred with Philo's publication of 'neglected rural geographies' which alerted many to the previous dominance of perspectives from 'white, middle-class, middle-aged, able-bodied, sound-minded, heterosexual men' (1992: 193). Cloke (1997a) and others refer to subsequent proliferation of more nuanced studies as the 'cultural turn' with its heightened awareness of the constructed and contested notions of rurality, nature, landscape, difference, identity and otherness – including their constructions in lay, popular and academic discourse (Halfacree, 1993; Jones, 1995). Phillips (1998: 139) notes that geographers have developed a 'willingness to tackle the intangibilities of meaning and understanding'. This has enabled an important growth in the study of rural society, especially in terms of how people's lives and rural spaces are shaped; not only by material and economic conditions but by the meanings and power attributed to images and myths of rural life (e.g. Little and Austin, 1996; Phillips, 1998).

Parallel with a postmodern attitude (Philo, 1993) and awareness of the dominances and marginal subjects in academic scholarship (Philo, 1992), a major part of this new cultural era involves attention to diversity (be it economic or social). This extends and invigorates the established practice of acknowledging social difference (noted in the third section of the chapter) and has resulted in the study of how cultural processes and constructions are crucial to the politics of difference, marginalization and otherness. Beginning with established categories of social difference such as gender, attention to the heterogeneous nature of such differences has grown (that is, new and emerging studies of plural femininities and masculinities: Brandth, 1995; Brandth and Haugen, 2000; Campbell and Bell, 2000; Liepins, 1998c, 2000c; Watkins, 1997). Additionally, a focus on identity formation as a cultural and political process has become well established in gender analyses (Bryant, 1999 – Australia; Oldrup, 1999 – Denmark; Brandth, 1994 – Norway; Hughes, 1997b and Little, 1997 – UK; see also Chapter 26 this volume). Notions of identity formation and the social construction of economic identities have also been important for expanding the social and cultural analyses of local and regional economies (Brunori and Rossi, 2000; Kneafsey et al., 2001; Roest and Menghi, 2000) and specific socioprofessional identities (for example, farming: Gonzalez and Benito, 2001; and forestry: Brandth and Haugen, 2000).

Beyond the popularity of identity analyses, a further major development has emerged in contemporary foci on diversity, namely attention to: otherness, marginalization and exclusion. While this work resonates with earlier studies of social exclusion, poverty and deprivation, geographers in particular have been prolific in responding to Philo's (1992) critique of the narrow mainstream spectrum of rural inquiries. Studies of rural society have consequently expanded in a rich array of descriptions and critiques surrounding Othered groups; their living conditions, use of space and experiences of exclusion. Table 5.5 depicts a sample of this type of scholarship and represents some of the major attempts being made to more fully acknowledge and include voices and experiences that have been predominantly hidden, marginalized or neglected in past rural studies. Also, see more recent material in Chapters 26–31 of this volume. This work often illustrates not only the philosophical foci but also the practice of post-structural approaches (Table 5.1), where readings (rather than definitive modelling or singular critiques) are made of the complexity of rural society. Nevertheless, the palimpsestual processes of overlapping approaches are appropriate as the rural post-structural scholarship has also drawn much from preceding critical Marxist and feminist viewpoints and thus authors are attentive to debates and questions that remain over the aims, politics and ethics of such work (Cloke and Little, 1997; Murdoch and Pratt, 1993).

SYNTHESIS AND FUTURE DIRECTIONS

This chapter has canvassed the breadth of approaches various scholars have adopted when taking a social route into rural studies. It has documented how investigations of rural society have frequently concentrated on common units (family, community, locality) while attempting to

Table 5.5 *'Other' studies of rural society*

Forms of otherness	Examples
Poverty and homelessness Particularly UK-based studies of people's experience of rural life when their income is low or they are without 'standard' secure housing	Cloke et al. (1995a) Poverty in the countryside; plus Cloke et al., 2000a, 2000b, 2000c, 2001 Woodward (1996) Deprivation and the rural
Ethnicity Including both UK and US studies of ethnicities that are seen as minorities or Other to white ethnicities	Agyeman and Spooner (1997) Ethnicity and the rural environment Kimmel and Ferber (2000) White men are this nation
Sexuality Studies of Australian, US and UK experiences of homosexuality as Other to the 'normality' of heterosexual rural society	Bell (2000) rurality, masculinity and homosexuality Little (forthcoming) Heterosexuality and the rural community Watkins (1997) The cultural construction of rurality
Transience and nomadism Accounts of how gypsies and travellers challenge the social, cultural and spatial norms of 'settled' rural societies; using space in different ways and appearing 'out of place'	Halfacree (1996) Out of place in the country Hetherington (2000) New Age travellers McLaughlin (1998) Anti-traveller racism in Ireland
Age Including UK and NZ studies of elderly experiences and marginalization in rural settings	Chalmers and Joseph (1998) Rural change and the elderly in rural places Harper (1997) Contesting later life
Youth Including international studies of children and young people as distinct, heterogeneous, competent social actors in rural environments and societies	*Journal of Rural Studies* (2002) Special Issue: Young Rural Lives Matthews et al. (2000) Growing up in the countryside Panelli et al. (2002) 'We make our own fun'

understand both the intricacies of specific local examples and the more generic processes and cultural frames that shape them (for example, ideologies of family and community; or the political and economic processes linking localities and wider national and global systems). The largest, central part of the chapter then concentrated on three recurring themes that have gained greatest attention in studies of rural society. Analyses of change, difference and power have all provided fruitful lines of inquiry, enabling scholars to document and theorize the dynamic but differentiated nature of rural societies. As noted when introducing Table 5.1, the literatures vary enormously, based on the contrasting approaches that have dominated different disciplines over time. Positivist and quantitative interests in measuring and modelling rural societies, the needs and changes had been particularly important in post-war modernization while hermeneutic viewpoints produced contrasting attention to lived experiences and meanings attributed to rural life. Each of these has since been challenged in new ways by theoretical and practical changes in the 1980s and 1990s. However, while research approaches have varied, many rural social units and themes of analysis continue to the present day. In addition, the final section of this chapter noted the crucial significance of the 'cultural turn' in rural studies. This latest trend has influenced both theoretical and empirical work, more closely incorporating cultural meanings into studies of societies and economies and widening the range of research subjects and foci to include the previously hidden or marginal experiences of groups such as children and travellers.

Cumulatively these works produce a repertoire of contrasting academic effort and a palimpsestual landscape of scholarly thought and interpretation of rural societies. As noted in the first section, effort is divided between empirical, conceptual and planning/policy goals, and this appears likely to continue. Nevertheless, studies of rural society now constitute a multi-layered landscape where although disciplinary and epistemological differences leave striking cleavages, at times there exists a series of exciting new terrains and directions that both established and emerging researchers may adopt.

First, engagements between statistics, policy and socio-cultural theory provide a possible way to move beyond the entrenched differences and distances that usually exist between applied and theory-focused scholars and the institutions they might wish to influence. Cloke and colleagues illustrate one example of the frustrations and possibilities that exist when dialogue and socially significant rural research are attempted across agency, policy and academic lines (Cloke et al., 2000a, 2000b). Learning from these experiences and creating alternative examples in different cultural, political and economic settings may enable studies of rural society to further advance both policy-relevant and theoretically robust accounts of rural conditions and dynamics in contemporary societies.

Second, support for integrating or multi-disciplinary scholarship may provide the opportunity to further benefit from the already substantial body of detailed issue- and location-specific studies of rural society. Such work would be logistically difficult and challenging to fund. However, comparative activities across different social foci (for example, class, age, gender, mobility etc.) and settings (for example, Europe, Antipodes, North America), would helpfully challenge national academic cultures and assumptions and provide further energies for identifying generic social concepts and relations as well as highlighting the particularities that differentiate rural societies.

Third while rural and other social formations become increasingly blurred and influenced by so-called processes of consumption and globalization, creative and risk-taking critiques of the ongoing relevance of rurality may forge new paths for the study of rural societies. While much of the initial fervour of radical approaches has been tempered by post-structural challenges, there is space for compassionate critiques of ongoing material conditions within and between different rural and other settings. There is opportunity for assessing the importance of cultural diversity in (material or symbolic) constructions of rurality and rural society. And there is potential yet to be realized in analysing the politics and psycho-social value of rural life as an assemblage of power relations, cultural meanings and moral values – whether this be in a 'family', 'community' or 'locality' or other setting.

Finally, recent post-structural critiques of social science provide the opportunity to establish reflective practices in the academic study of rural societies. Challenging ourselves in terms of the purpose and positionality of our work is important. But so too will be the opportunity we may take to converse beyond theoretical and empirical strongholds (e.g. Djurfeldt, 1996; Murdoch and Pratt, 1993) and mobilize scholarly diversity for the benefit of those rural societies we value.

NOTES

1 In this case my positionality includes working as a social geographer with an Antipodean, Western and anglocentric background and with contemporary post-structural and feminist interests.
2 But also see Gibson-Graham's (1996) post-structural reading of post-industrial economic and domestic relations that unravel the dualisms and divisions presumed to exist so clearly in modern capitalist systems.
3 The notion of community is differently understood in various cultures and languages, see, for example, Mendras's (1982) discussion of French rural analyses.
4 While informative debate and empirical work is providing wider studies of community at national, hyper-spatial and international scales, most rural scholarship continues to employ the term 'community' in place-based or territorially defined ways.
5 For instance, community studies are often most synonymous with village analysis in the British literature.
6 It is important to remember that Tonnies identified forms of association while geographers and sociologists applied this to rural and urban settlements and established a 'taxonomy of settlement patterns' (Newby, 1977: 95).
7 Note Urry's (1981: 464) contention that: 'important changes in contemporary capitalism are at present heightening the economic, social and political significance of each locality'; and Massey's (1984: 59) acknowledgement that: 'local histories and local distinctiveness are integral to the social nature of production relations'.
8 Note though, that Pahl had argued in 1968 that local level impacts should be an object of study when seeking to understand the implication of broader processes.
9 In addition to the geographies of class difference, scholars have also registered the cultural attachments to rurality that may be partly shared across classes (Cloke et al., 1995c).
10 However, feminist household-oriented works have also been constructed as critical alternatives (e.g. Fisher, 1997).
11 Note that commentaries of change often differ greatly. They range from traditional notions of change as a disruption to the long-term stability of community units (e.g. Smith et al., 2001) through to 'postmodern' contentions that communities are being fundamentally reconstituted (e.g. Mitchell et al., 1998).

BIBLIOGRAPHY

Adams, J. (1994) *The Transformation of Rural Life.* Chapel Hill, NC: University of North Carolina Press.

Agg, J. and Phillips, M. (1998) 'Neglected gender dimensions of rural social restructuring', in P. Boyle and K. Halfacree (eds), *Migration Issues in Rural Areas.* London: Wiley. pp. 252–279.

Agyeman, J. and R. Spooner (1997) 'Ethnicity and the rural environment', In P. Cloke and J. Little (eds), *Contested Countryside Cultures: Otherness, Marginalisation and Rurality.* London: Routledge. pp. 197–217.

Albrecht, S.L., Bahr, H.M. and Chadwick B. (1977) 'Stereotyping of sex roles, personality characteristics and occupations', *Sociology and Social Research*, 61: 223–240.

Almas, R. and Haugen, M. (1991) 'Norwegian gender roles in transition: the masculinization hypothesis in the past and in the future', *Journal of Rural Studies*, 7: 79–83.

Alston, M. (1995a) 'Women and their work on Australian farms', *Rural Sociology*, 60: 521–532.

Alston, M. (1995b) *Women on the Land: The Hidden Heart of Rural Australia.* Sydney: University of New South Wales Press.

Arensberg, C.M. and Kimball, S.T. (1940) *Family and Community in Ireland.* Cambridge, MA: Harvard University Press.

Barlow, J. and Savage, M. (1986) 'The politics of growth: cleavage and conflict in a Tory heartland', *Capital and Class*, 31: 156–181.

Beesley, K.B. (1999) 'Living in the rural–urban fringe: toward an understanding of life and scale', in N. Walford, J. Everitt and D. Napton (eds), *Reshaping the Countryside: Perceptions and Processes of Rural Change.* London: CAB International. pp. 91–104.

Bell, C. and Newby, H. (1974) 'Capitalist farmers in the British class structure', *Sociologia Ruralis*, 14: 86–107.

Bell, D. (2000) 'Farm boys and wild men: rurality and homosexuality', *Rural Sociology*, 65: 547–561.

Bergman, T. (1975) 'Change processes in farming and political consciousness and attitudes of peasants and worker-peasants', *Sociologia Ruralis*, 15: 73–89.

Berlan-Darque, M. (1988) 'The division of labour and decision-making in farming couples: power and negotiation', *Sociologia Ruralis*, 28: 271–292.

Berlan-Darque, M. and Gasson, R. (1991) 'Changing gender relations in agriculture: an international perspective', *Journal of Rural Studies*, 7: 1–2.

Berry, E.H., Krannich, R.S. and Greider, T. (1990) 'A longitudinal analysis of neighbouring in rapidly changing rural places', *Journal of Rural Studies*, 6: 175–186.

Beyers, W.B. and Nelson, P.B. (2000) 'Contemporary development forces in the non-metropolitan west: new insights from rapidly growing communities', *Journal of Rural Studies*, 16: 459–474.

Blanc, M. and MacKinnon, N. (1990) 'Gender relations and the family farm in Western Europe', *Journal of Rural Studies*, 6: 401–405.

Bokemeier, J. and Tait, J.L. (1980) 'Women as power actors: a comparative study of rural communities', *Rural Sociology*, 45: 238–255.

Bolton, N. and Chalkley, B. (1990) 'The rural population turnround: A case-study of North Devon', *Journal of Rural Studies*, 6: 29–43.

Bourdieu, P. (1977) 'Cultural reproduction and social reproduction in J. Karabel and A.H. Halsey (eds), *Power and Ideology in Eduction.* New York, Oxford University Press.

Bouquet, M. (1982) 'Production and reproduction of family farms in south-west England', *Sociologia Ruralis*, 22: 227–244.

Bowler, I. (1986) 'Intensification, concentration and specialization in agriculture: the case of the European Community', *Geography*, 71: 14–24.

Bradley, T. (1985) 'Reworking the quiet revolution: industrial and labour market restructuring in village England', *Sociologia Ruralis*, 25: 40–59.

Brandes, S.H. (1975) *Migration, Kinship and Community: Tradition and Transition in a Spanish Village.* New York: Academic Press.

Brandth, B. (1994) 'Changing femininity: the social construction of women farmers in Norway', *Sociologia Ruralis*, 34: 127–149.

Brandth, B. (1995) 'Rural masculinity in transition: gender images in tractor advertisements', *Journal of Rural Studies*, 11: 123–133.

Brandth, B. and Haugen, H. (2000) 'From lumberjack to business manager: masculinity in the Norwegian forestry press', *Journal of Rural Studies*, 16: 343–355.

Broadway, M.J. (2000) 'Planning for change in small towns or trying to avoid the slaughterhouse blues', *Journal of Rural Studies*, 16: 37–46.

Brunori, G. and Rossi, A. (2000) 'Synergy and coherence through collective action: some insights from wine routes in Tuscany', *Sociologia Ruralis*, 40: 409–423.

Bryant, Lia (1999) "The detraditionalization of occupational identities in farming in South Australia." *Sociologia Ruralis*, 39: 36–261.

Burke, P.J. and Tully, J. (1977) 'The measurement of role identity', *Social Forces*, 55: 881–897.

Buttel F.H. and Gillespie G.W. (1984) 'The sexual division of farm household labor: an exploratory study of the structure of on-farm and off-farm labor allocation among farm men and women', *Rural Sociology*, 49: 182–209.

Campbell, H. and Bell, M.M. (2000) 'The question of rural masculinities', *Rural Sociology*, 65: 532–546.

Capo, E. and Fonti, G.M. (1965) 'L'exode rural vers les grandes villes', *Sociologia Ruralis*, 5: 267–287.

Chalmers, A.I. and Joseph, A.E. (1998) 'Rural change and the elderly in rural places: commentaries from New Zealand', *Journal of Rural Studies*, 14: 155–165.

Champion, T. and Watkins, C. (eds) (1991) *People in the Countryside: Studies of Social Change in Rural Britain.* London: Paul Chapman.

Cloke, P. (1997a) 'Country backwater to virtual village? Rural studies and "the cultural turn"', *Journal of Rural Studies*, 13: 367–375.

Cloke, P. (1997b) 'Poor country: marginalization, poverty and rurality', in P. Cloke and J. Little (eds), *Contested Countryside Cultures: Otherness, Marginalization and Rurality.* London: Routledge. pp. 252–271.

Cloke, P. and Goodwin, M. (1992) 'Conceptualizing countryside change: from post-fordism to rural structured coherence', *Transactions of the Institute of British Geographers*, 17: 321–336.

Cloke, P. and J. Little (1997) 'Introduction: other countrysides?', in P. Cloke and J. Little (eds), *Contested Countryside Cultures: Otherness, Marginalisation and Rurality.* London: Routledge.

Cloke, P. and Thrift, N. (1987) 'Intra-class conflict in rural areas', *Journal of Rural Studies*, 3: 321–33.

Cloke, P. and Thrift, N. (1990) 'Class and change in rural Britain', in T. Marsden, P. Lowe and S. Whatmore (eds), *Rural Restructuring.* London: David Fulton. pp. 165–181.

Cloke, P., Goodwin, M. and Milbourne, P. (1995a) 'Poverty in the countryside: out of sight and out of mind', in C. Philo (ed.), *Off the Map: The Social Geography of Poverty in the U.K.* London: Child Poverty Action Group.

Cloke, P., Goodwin, M., Milbourne, P. and Thomas, C. (1995b) 'Deprivation, poverty and marginalization in rural lifestyles in England and Wales', *Journal of Rural Studies*, 8: 351–365.

Cloke, P., Milbourne, P. and Widdowfield, R. (2000a) 'Partnership and policy networks in rural local governance: homeless in Taunton', *Public Administration*, 78 (1): 111–133.

Cloke, P., Milbourne, P. and Widdowfield, R. (2000b) 'The hidden and emerging spaces of rural homelessness', *Environment and Planning A*, 32: 77–90.

Cloke, P., Milbourne, P. and Widdowfield, R. (2000c) 'Homelessness and rurality: "out-of-place" in purified space?', *Environment and Planning D: Society and Space*, 18 (6): 715–735.

Cloke, P., Milbourne, P. and Widdowfield, R. (2001) 'The geographies of homelessness in rural England', *Regional Studies*, 35 (1): 23–37.

Cloke, P., Phillips, M. and Thrift, N. (1995c) 'The new middle classes and the social constructs of rural living', in T. Butler and M. Savage (eds), *Social Change and the Middle Classes.* London: UCL Press. pp. 220–238.

Cocklin, C. and Wall, M. (1997) 'Contested rural futures: New Zealand's East Coast forestry project', *Journal of Rural Studies*, 13: 149–162.

Cohen, A. (ed.) (1982) *Belonging: Identity and Social Organization in British Rural Cultures.* Manchester: Manchester University Press.

Cohen, A. (1983) *Anthropological Studies in Rural Britain, 1968–1983.* London: Social Science Research Council.

Cohen, A. (1985) *The Symbolic Construction of Community.* London: Ellis Horwood.

Cohen, A. (ed.) (1986) *Symbolizing Boundaries: Identity and Diversity in British Cultures.* Manchester: Manchester University Press.

Connell, J. (1978) *The End of Tradition: Country Life in Central Surrey.* London: Routledge & Kegan Paul.

Conradson, D. and Pawson, E. (1997) 'Reworking the geography of the long boom: the small town experience of restructuring in Reefton, New Zealand', *Environment and Planning A*, 29: 1281–1397.

Cooke, P. (1989) *Localities.* London: Unwin Hyman.

Copus, A.K. and Crabtree, J.R. (1996) 'Indicators of socio-economic sustainability: an application to remote rural Scotland', *Journal of Rural Studies*, 12: 41–54.

Coughenour C.M. and Swanson L.E. (1983) 'Working statuses and occupations of men and women in farm families and the structure of farms', *Rural Sociology*, 48: 23–43.

Coward, R.T. and Smith, W.M. (eds) (1981) *The Family in Rural Society.* Boulder, CO: Westview Press.

Craig, R. (1979) 'Down on the farm: role conflicts of Australian farm-women', in *The Woman in Country Australia Looks Ahead; Conference Proceedings.* Melbourne: LaTrobe University. pp. 1–12.

Crang, M. (1998) *Cultural Geography.* London and New York: Routledge.

Davidoff, L., L'Esperance, J. and Newby, H. (1976) 'Landscape with figures: home and community in English society', in J. Mitchell and A. Oakley (eds), *The Rights and Wrongs of Women*, London: Penguin.

Day, G. (1998) 'A community of communities? Similarity and difference in Welsh rural community studies', *The Economic and Social Review*, 29: 233–257.

Day, G. and Murdoch, J. (1993) 'Locality and community: coming to terms with place', *Sociological Review*, 41: 82–111.

Dempsey, K. (1987) 'Economic inequality between men and women in an Australia rural community', *Australian and New Zealand Journal of Sociology*, 23: 358–374.

Dempsey, K. (1988) 'Exploitation in the domestic division of labour: an Australian case study', *Australian and New Zealand Journal of Sociology*, 24: 420–436.

Dempsey, K. (1992) *A Man's Town: Inequality between Women and Men in Rural Australia.* Melbourne: Oxford University Press.

Djurfeldt, G. (1996) 'Defining and operationalizing family farming from a sociological perspective', *Sociologia Ruralis*, 36: 340–351.

Duncan, C.M. (1999) *Worlds Apart: Why Poverty Persists in Rural America.* New Haven, CT: Yale University Press.

Dunn, M., Rawong, M. and Rogers, A. (1981) *Rural Housing: Competition and Choice.* London: Allen and Unwin.

Dunne, F. (1980) 'Occupational sex-stereotyping among rural young women and men', *Rural Sociology*, 45: 396–415.

Ekman, A. (1999) 'The revival of cultural celebrations in regional Sweden: aspects of tradition and transition', *Sociologia Ruralis*, 39 (3): 280–293.

Elder, G.H. Jr, and Conger, R.D. (2000) *Children of the Land: Adversity and Success in Rural America.* Chicago: University of Chicago Press.

Evans, N. and Ilbery, B. (1996) 'Exploring the influence of farm-based pluri-activity on gender relations in capitalist agriculture', *Sociology Ruralis*, 36: 74–93.

First-Dilic, R. (1978) 'The productive roles of farm women in Yugoslavia', *Sociologia Ruralis*, 18: 124–139.

Fisher, C. (1997) '"I bought my first saw with my maternity benefit": craft production in west Wales and the home as the space of (re)production', in P. Cloke and J. Little (eds), *Contested Countryside Cultures: Otherness, Marginalization and Rurality*. London: Routledge. pp. 232–251.

Fleigel, F.C. (1976) 'A comparative analysis of the impact of industrialism on traditional values', *Rural Sociology*, 41: 431–451.

Flora, C.B. and Johnson, S. (1978) 'Discarding the distaff: new roles for rural women', in T.R. Ford (ed.), *Rural U.S.A. Persistence and Change*. Ames, IO: Iowa State University Press.

Foord, J. and Gregson, N. (1986) 'Patriarchy: towards a reconceptualisation', *Antipode*, 18: 186–211.

Frankenberg, R. (1966) *Communities in Britain: Social Life in Town and Country*. Harmondsworth: Penguin.

Friedmann, H. (1978) 'Simple commodity production and wage labour in the American plains', *Journal of Peasant Studies*, 6: 71–100.

Gant, R. (1999) '"Enabling" technology for disabled people. Telecommunications for community care in rural Britain', in N. Walford, J. Everitt and D. Napton (eds), *Reshaping the Countryside: Perceptions and Processes of Rural Change*. London: CAB International. pp. 135–145.

Gasson, R. (1980) 'Roles of farm women in England', *Sociologia Ruralis*, 20: 165–180.

Gibson-Graham, J.K. (1996) *The End of Capitalism (as we knew it): A Feminist Critique of Political Economy*. Cambridge MA: Blackwell.

Gilligan, J.H. (1984) 'The rural labour process: a case study of a Cornish town', in T. Bradley and P. Lowe (eds), *Locality and Rurality: Economy and Society in Rural Regions*. Norwich: Geobooks. pp. 91–112.

Gonzalez, J.J. and Benito, C.G. (2001) 'Profession and identity: the case of family farming in Spain', *Sociologia Ruralis*, 41: 343–357.

Goodman, D. and Redclift, M. (1985) 'Capitalism, petty commodity production and the farm enterprise', *Sociologia Ruralis*, 25: 231–247.

Goodwin, M., Cloke, P. and Milbourne, P. (1995) 'Regulation theory and rural research: theorising contemporary rural change', *Environment and Planning A*, 27: 1245–1260.

Goudy, W.J. (1990) 'Community attachment in a rural region', *Rural Sociology*, 55: 178–198.

Gray, I. (1991) *Politics in Place: Social Power Relations in an Australian Country Town*. Cambridge: Cambridge University Press.

Haan, H. de (1997) 'Locality, identity and the reshaping of modernity: an analysis of cultural confrontations in two villages', in H. de Haan and N. Long (eds), *Images and Realities of Rural Life*. Assen: Van Gorcum. pp. 153–177.

Halfacree, K.H. (1993) 'Locality and social representation: place, discourse and alternative definitions of the rural', *Journal of Rural Studies*, 9: 23–37.

Halfacree, K. (1996) 'Out of place in the country: travellers and the "rural idyll"', *Antipode*, 28: 42–71.

Halfacree, K.H. (1997) 'Contrasting roles for the post-productivist countryside: a postmodern perspective on counter-urbanization', in P. Cloke and J. Little, *Contested Countryside Cultures: Otherness, Marginalization and Rurality*. London: Routledge. pp. 70–93.

Hall, B., Thorns, D. and Willmott, B. (1983) 'Class, locality and family: bases of communion in a locality', in *Community Formation and Change: A Study of Rural and Urban Localities in New Zealand*. Working Paper Number 4. Christchurch: Department of Sociology, University of Canterbury. pp. 171–190.

Harper, S. (1989) 'The British rural "community": an overview of perspectives', *Journal of Rural Studies*, 5: 161–184.

Harper, S. (1997) 'Contesting later life', in P. Cloke and J. Little (eds), *Contested Countryside Cultures: Otherness, Marginalization and Rurality*. London: Routledge. pp. 180–196.

Henderson, P. (1991) 'For a rural people's Europe', *Community Development Journal*, 26: 118–123.

Herbert-Cheshire, L. (2000) 'Contemporary strategies for rural community development in Australia: a governmentality perspective', *Journal of Rural Studies*, 16: 203–215.

Hetherington, K. (2000) *New Age Travellers: Vanloads of Uproarious Humanity*. London: Cassell.

Higgins, V. (2001) 'Governing the boundaries of viability: economic expertise and the production of the "low-income farm problem" in Australia', *Sociologia Ruralis*, 41: 358–375.

Hillery, G.A. (1955) 'Definition of community: areas of agreement', *Rural Sociology*, 20: 94–119.

Hoggart, K., Buller, H. and Black, R. (1995) *Rural Europe: Identity and Change*. London: Arnold.

Hughes, A. (1997a) 'Rurality and "cultures of womanhood": domestic identities and the moral order in village life', in P. Cloke and J. Little (eds), *Contested Countryside Cultures: Otherness, Marginalization and Rurality*. London: Routledge. pp. 123–137.

Hughes, A. (1997b) 'Women and rurality: gendered experiences of "community" in village life', in P. Milbourne (ed.), *Revealing Rural 'Others': Representation, Power and Identity in the British Countryside*. London: Pinter. pp. 167–188.

Hutson, J. (1987) 'Fathers and sons: family farms, family businesses and the farming industry', *Sociology*, 21: 215–229.

Israel, G.D., Beaulieu, L.J. and Hartless, G. (2001) 'The influence of family and community social capital on educational achievement', *Rural Sociology*, 66: 43–68.

Jones, O. (1995) 'Lay discourses of the rural: development and implications for rural studies', *Journal of Rural Studies*, 11: 35–49.

Kalantaridis, C. and Barianidis, L. (1999) 'Family production and the global marketplace: rural industrialization in Greece', *Sociologia Ruralis*, 39: 146–164.

Keating, N.C. and Little, H.M. (1994) 'Getting into it: farm roles and careers of New Zealand women', *Rural Sociology*, 59: 720–736.

Kimmel, M. and A.L. Ferber (2000) 'White men are this nation: right wing militias and the restoration of rural American masculinity', *Rural Sociology*, (65): 582–604.

Kneafsey, M., Ilbery, B. and Jenkins, T. (2001) 'Exploring the dimensions of culture economies in rural West Wales', *Sociologia Ruralis*, 41: 296–310.

Knox, P.L. and Cottam, M.B. (1981) 'Rural deprivation in Scotland: a preliminary assessment', *Tijdschrift voor Economische en Sociale Geografie*, 72: 162–175.

Larson, O. (1965) 'Contributions of rural sociology research and evaluation to extension development in the United States', *Sociologia Ruralis*, 5: 308–328.

Lewis, G. (1979) *Rural Communities: A Social Geography.* London: David and Charles.

Lewis, G. and Maund, D. (1976) 'The urbanization of the countryside: a framework for analysis', *Geografiska Annaler*, 58B: 17–27.

Liepins, R. (1996) 'Reading agricultural power: media as sites and processes in the construction of meaning', *New Zealand Geographer*, 52: 3–10.

Liepins, R. (1998a) 'Fields of action: Australian women's agricultural activism in the 1990s', *Rural Sociology*, 63: 128–156.

Liepins, R. (1998b) 'The gendering of farming and agricultural politics: a matter of discourse and power', *Australian Geographer*, 29: 371–388.

Liepins, R. (1998c) '"Women of broad vision": nature and gender in the environmental activism of Australia's "Women in Agriculture"', *Environment and Planning A*, 30: 1179–1196.

Liepins, R. (2000a) 'Exploring rurality through "community": discourses, practices and spaces shaping Australian and New Zealand rural "communities"', *Journal of Rural Studies*, 16: 83–99.

Liepins, R. (2000b) 'New energies for an old idea: reworking approaches to "community" in contemporary rural studies', *Journal of Rural Studies*, 16: 23–35.

Liepins, R. (2000c) 'Making men: the construction and representation of agriculture-based masculinities in Australia and New Zealand', *Rural Sociology*, 65: 605–620.

Liepins, R. and Bradshaw, B. (1999) 'Neo-liberal agricultural discourse in New Zealand: economy, culture and politics linked', *Sociologia Ruralis*, 39: 563–582.

Little, J. (1987) 'Gender relations in rural areas: the importance of women's domestic role', *Journal of Rural Studies*, 3: 335–342.

Little, J. (1994) 'Gender relations and the rural labour process', in S. Whatmore, T. Marsden and P. Lowe (eds), *Gender and Rurality.* London: David Fulton. pp. 11–29.

Little, J. (1997) 'Employment, marginality and women's self-identity', in P. Cloke and J. Little (eds), *Contested Countryside Cultures: Otherness, Marginalization and Rurality.* London: Routledge. pp. 138–157.

Little, J. (2002) *Gender and Rural Geography.* London: Pearson.

Little, J. (2003) 'Riding the rural love train, heterosexuality and the rural community', 43: 401–417.

Little, J. and Austin, P. (1996) 'Women and the rural idyll', *Journal of Rural Studies*, 12: 101–111.

Livingstone, D. (1992) *The Geographical Tradition: Episodes in the History of a Contested Enterprise.* Oxford: Blackwell.

Lockie, S. and Kitto, S. (2000) 'Beyond the farm gate: production-consumption networks and agri-food research', *Sociologia Ruralis*, 40 (1): 3–19.

Loomis, C. and Beegle, J.A. (1950) *Rural Social Systems.* New York: Prentice Hall.

Lowe, P., Murdoch, J., Marsden, T., Munton, R. and Flynn, A. (1993) 'Regulating the new rural spaces: the uneven development of land', *Journal of Rural Studies*, 9: 202–222.

Lupri, E. (1965) 'Urbanisierung and Familienstruktur: ein Beitrag zur interkulturellen Forschung', *Sociologia Ruralis*, 5: 57–76.

McDowell, L. (1986) 'Beyond patriarchy: a class-based explanation of women's subordination', *Antipode*, 18: 311–321.

Mackenzie, F. (1992) '"The worse it got the more we laughed": a discourse of resistance among farmers in Eastern Ontario', *Environment and Planning D: Society and Space*, 10: 691–713.

Mackenzie, F. (1994) '"Is where I sit, where I stand?" The Ontario Farm Women's Network, politics and difference', *Journal of Rural Studies*, 10: 101–115.

MacLaughlin, J. (1998) 'The political geography of anti-traveller racism in Ireland: the politics of exclusion and the geography of closure', *Political Geography*, 17 (4): 417–435.

Macklin, M. (1995) 'Local media and gender relations in a rural community', in P. Share (ed.), *Communication and Culture in Rural Areas.* Wagga Wagga: Centre for Rural Social Research, Charles Stuart University. pp. 291–303.

Marsden, T. (1984) 'Capitalist farming and the farm family: a case study', *Sociology*, 18: 205–224.

Marsden, T. (1996) 'Rural geography trend report: the social and political bases of rural restructuring', *Progress in Human Geography*, 20: 246–258.

Marsden, T. (1999) 'Rural futures: the consumption countryside and its regulation', *Sociologia Ruralis*, 39: 501–520.

Marsden, T., Murdoch, J., Lowe, P., Munton, R. and Flynn, A. (1993) *Constructing the Countryside.* London: UCL Press.

Massey, D. (1984) *Spatial Divisions of Labour: Social Structures and the Geography of Production.* London: Macmillan.

Matthews, H., Taylor, M., Sherwood, K., Tucker, F. and Limb, M. (2000) 'Growing-up in the countryside: children and the rural idyll', *Journal of Rural Studies*, 16: 141–153.

Mayer, H.J. and Greenberg, M. (2000) 'Responding to economic change in remote, rural regions: federal installations in Idaho and Washington', *Journal of Rural Studies*, 16: 421–432.

McCormack, J. (1998) Constructing 'rurality': a study of lay and academic discourses of the 'rural'. Unpublished BA(Hons) thesis, Dunedin: University of Otago, Department of Geography.

McLaughlin, B. (1986) 'The rhetoric and reality of rural deprivation', *Journal of Rural Studies*, 2: 291–307.

McLaughlin, B. (1998) 'The political geography of anti-Traveller racism in Ireland',

Mendras, H. (1982) 'Outline for an analysis of the French peasantry', in H. Mendras and I. Mihailescu (eds), *Theories and Methods in Rural Community Studies.* Oxford: Pergamon Press. pp. 105–125.

Mertig, A.G. and Dunlop, R.E. (2001) 'Environmentalism, new social movements, and the new class: a cross-national investigation', *Rural Sociology*, 66: 113–136.

Middleton, A. (1986) 'Marking boundaries: men's space and women's space in a Yorkshire village', in P. Lowe, T. Bradley, S. Wright (eds), *Deprivation and Welfare in Rural Areas.* Norwich: Geobooks. pp. 121–134.

Mitchell, C.J.A. (1998) 'Entrepreneuralism, commodification and creative destruction: a model of post-modern community development', *Journal of Rural Studies*, 14: 273–286.

Mitchell, C.J.A (forthcoming) 'Making sense of counterurbanization', *Journal of Rural Studies*, in press.

Moran, W., Blunden, G. and Greenwood, J. (1993) 'The role of family farming in agrarian change', *Progress in Human Geography*, 17: 22–42.

Moran, W., Blunden, G., Workman, M. and Bradley, A. (1996) 'Family farmers, real regulation and the experience of food regimes', *Journal of Rural Studies*, 12: 245–258.

Moseley, M.J. (1980) 'Is rural deprivation really rural?', *The Planner*, 66: 97.

Munters, Q. (1975) 'Some remarks on the opening up of rural social systems', *Sociologia Ruralis*, 15: 34–45.

Munton, R. (1984) 'The politics of rural land ownership: institutional investors and the Northfield enquiry', in T. Bradley and P. Lowe (eds), *Locality and Rurality: Economy and Society in Rural Regions.* London: Geobooks. pp. 167–178.

Murdoch, J. (2000) 'Networks – a new paradigm of rural development', *Journal of Rural Studies*, 16: 407–420.

Murdoch, J. and Marsden, T. (1994) *Reconstituting Rurality: Class, Community and Power in the Development Process.* London: University College London Press.

Murdoch, J. and Marsden, T. (1995) 'The spatialization of politics: local and national actor-spaces in environmental conflict', *Transactions of the Institute of British Geographers*, 20: 368–380.

Murdoch, J. and Pratt, A. (1993) 'Modernism, powermodernism and the post-rural', *Journal of Rural Studies*, 10: 83–87.

Murdoch, J. and Pratt, A. (1997) 'From the power of topography to the topography of power: a discourse of strange ruralities', in P. Cloke and J. Little (eds), *Contested Countryside Cultures: Otherness, Marginalization and Rurality.* London: Routledge. pp. 51–69.

Naples, N. (1994) 'Contradictions in agrarian ideology: restructuring gender, race-ethnicity, and class', *Rural Sociology*, 59: 110–135.

Newby, H. (1972) 'Agricultural workers in the class structure', *Sociological Review*, 20: 413–439.

Newby, H. (1977) *The Deferential Worker: A Study of Farm Workers in East Anglia.* London: Allen Lane.

Newby, H. (1980) 'Rural sociology', *Current Sociology*, 28: 3–141.

Newby, H. (1986) 'Locality and rurality: the restructuring of rural social relations', *Regional Studies*, 20: 209–215.

Pahl, R. (1963) 'Education and social class in commuter villages', *Sociological Review*, 11: 241–246.

Pahl, R. (1965a) 'Class and community in English commuter villages', *Sociologia Ruralis*, 5: 5–23.

Pahl, R. (1965b) *Urbs in rure: The Metropolitan Fringe in Hertfordshire.* London: Weidenfeld & Nicolson.

Pahl, R. (1968) 'The rural–urban continuum', in R. Pahl (ed.), *Readings in Urban Sociology.* Oxford: Pergamon Press. pp. 263–297.

Panelli, R. (2001) 'Narratives of community and change in a contemporary rural setting: the case of Duaringa, Queensland', *Australian Geographical Studies*, 39: 156–166.

Panelli, R. (2004) *Social Geographies: From Difference to Action.* London: Sage.

Panelli, R., Nairn, K. and McCormack, J. (2002) '"We make our own fun": reading the politics of youth with (in) community', *Sociologia Ruralis*, 42: 106–130.

Phillips, D. and Williams, A. (1992) *Rural Britain: A Social Geography.* Oxford: Blackwell.

Phillips, M. (1993) 'Rural gentrification and the processes of class colonization', *Journal of Rural Studies*, 9: 123–140.

Phillips, M. (1998) 'The restructuring of social imaginations in rural geography', *Journal of Rural Studies*, 14: 121–153.

Philo, C. (1992) 'Neglected rural geographies: a review', *Journal of Rural Studies*, 8: 193–207.

Philo, C. (1993) 'Postmodern rural geography? A reply to Murdoch and Pratt', *Journal of Rural Studies*, 9: 427–436.

Philo, C. (1997) 'Of other rurals?', in P. Cloke and J. Little (eds), *Contested Countryside Cultures: Otherness, Marginalization and Rurality.* London: Routledge. pp. 19–50.

Photiadis, J.D. and Ball, R.A. (1976) 'Patterns of change in rural normative structure', *Rural Sociology*, 41: 60–75.

Poiner, G. (1990) *The Good Old Rule: Gender and Other Power Relationships in a Rural 'Community'.* Sydney: Sydney University Press.

Porter, M. (1991) 'Time, the life course and work in women's lives: reflections from Newfoundland', *Women's Studies International Forum*, 14: 1–13.

Pratt, A. (1991) 'Discourses of locality', *Environment and Planning A*, 23: 257–266.

Putnam, R.D. (2000) *Bowling Alone: The Collapse and Revival of American Community.* New York: Simon and Schuster.

Quale, B. (1984) 'Images of place in a Northumbrian dale', in T. Bradley and P. Lowe (eds), *Locality and Rurality: Economy and Society in Rural Regions.* London: Geobooks Elsevier. pp. 225–241.

Ray, C. (1998) 'Culture, intellectual property and territorial rural development', *Sociologia Ruralis*, 38 (1): 3–20.

Reimer, B., Ricard, I. and Shaver, F.M. (1992) 'Rural deprivation: a preliminary analysis of census and tax family data', in R.D. Bollman (ed.), *Rural and Small Town Canada*. Toronto: Thompson Educational Publishing. pp. 319–335.

Richling, B. (1985) '"You'd never starve here": return migration to rural Newfoundland', *Canadian Review of Sociology and Anthropology*, 22: 236–249.

Roest, K. de and Menghi, A. (2000) 'Reconsidering "traditional" food: the case of Parmigiano Reggiano cheese', *Sociologia Ruralis*, 40: 439–451.

Rose, D., Saunders, P., Newby, H. and Bell, C. (1976) 'Ideologies of property ownership: a case study', *Sociological Review*, 24: 699–730.

Rose, G. (1993) *Feminism and Geography: The Limits of Geographical Knowledge*. Cambridge: Polity Press.

Rose, G. (1995) 'Tradition and paternity: same difference?', *Transactions of the Institute of British Geographers*, 20: 414–416.

Sachs, C.E. (1983) *The Invisible Farmers: Women in Agricultural Production*. Totowa, NJ: Rowman and Allanfield.

Sachs, R. (1965) 'Wandlungen des Ziel-und Wertsystems', *Sociologia Ruralis*, 5: 133–148.

Salamon, S. (1980) 'Ethnic difference in farm family land transfers', *Rural Sociology*, 45: 290–308.

Salamon, S. (1985) 'Ethnic communities and the structure of agriculture', *Rural Sociology*, 50: 323–340.

Salamon, S. and Tornatore, J.B. (1994) 'Territory contested through property in a midwestern post-agricultural community', *Rural Sociology*, 59: 636–654.

Saunders, P., Newby, H., Bell, C. and Rose, D. (1978) 'Rural community and rural community power', in H. Newby (ed.), *International Perspectives in Rural Sociology*. London: Wiley. pp. 55–85.

Savage, M., Barlow, J., Dicken, P. and Fielding, T. (1992) *Property, Bureaucracy and Culture: Middle Class Formation in Contemporary Britain*. London: Routledge.

Schulman, M.D. and Anderson, C. (1999) 'The dark side of the force: a case study of restructuring and social capital', *Rural Sociology*, 64: 351–372.

Scott, K., Park, J. and Cocklin, C. (2000) 'From "sustainable rural communities" to "social sustainability": giving voice to diversity in Mangakahia Valley, New Zealand', *Journal of Rural Studies*, 16: 433–446.

Sharp, J.S. (2001) 'Locating the community field: a study of interorganizational network structure and capacity for community action', *Rural Sociology*, 66: 403–424.

Shortall, S. (1992) 'Power analysis and farm wives: an empirical study of the power relationships affecting women on Irish farms', *Sociologia Ruralis*, 32: 431–451.

Shortall, S. (1994) 'Farm women's groups: feminist or farming or community groups or new social movements', *Sociology*, 28: 279–291.

Shucksmith, M., Chapman, P., Clark, G. and Black, S. (1994) 'Social welfare in rural Europe', *Journal of Rural Studies*, 10: 343–356.

Smith, J. and Gant, R. (1982) 'The elderly's travel in the Cotswolds', in A.M. Warnes (ed.), *Geographical Perspectives on the Elderly*. Chichester: Wiley.

Smith, M.D., Krannich, R.S. and Hunter, L.M. (2001) 'Growth, decline, stability and disruption: a longitudinal analysis of social well-being in four western rural communities', *Rural Sociology*, 66: 425–450.

Stacey, M. (1969) 'The myth of community studies', *British Journal of Sociology*, 20: 34–47.

Stebbing, S. (1984) 'Women's roles and rural society', in T. Bradley and P. Lowe (eds), *Locality and Rurality: Economy and Society in Rural Regions*. Norwich: Geobooks. pp. 199–208.

Stinchcombe, A.L. (1961) 'Agricultural enterprise and rural class relations', *American Journal of Sociology*, 67: 169–176.

Stockdale, A., Findlay, A. and Short, D. (2000) 'The repopulation of rural Scotland: opportunity and threat', *Journal of Rural Studies*, 16: 243–257.

Strathern, M. (1982) 'The village as an idea: constructs of villageness in Elmdon, Essex', in A. Cohen (ed.), *Belonging*. Manchester: Manchester University Press.

Strathern, M. (1984) 'The social meaning of localism', in T. Bradley and P. Lowe (eds), *Locality and Rurality: Economy and Society in Rural Regions*. London: Geobooks Elsevier. pp. 181–198.

Symes, D. (1991) 'Changing gender roles in productionist and post-productionist capitalist agriculture', *Journal of Rural Studies*, 7: 85–90.

Symes, D. and Marsden T. (1983) 'Complementary roles and asymmetrical lives: farmers' wives in a large farm environment', *Sociologia Ruralis*, 23: 229–241.

Teather, E.K. (1996) 'Farm women in Canada, New Zealand and Australia redefine their rurality', *Journal of Rural Studies*, 12: 1–14.

Thieme, G. (1983) 'Agricultural change and its impact in rural areas', in M.T. Wild (ed.), *Urban and Rural Change in West Germany*. London: Croom Helm. pp. 220–247.

Thrift, N.J. (1987a) 'The geography of late twentieth century class formation', in N.J. Thrift and P. Williams, *Class and Space*. London: Routledge & Kegan Paul. pp. 207–253.

Thrift, N.J. (1987b) 'Manufacturing rural geography', *Journal of Rural Studies*, 3: 77–81.

Throsby, D. (1999) 'Cultural capital', *Journal of Cultural Economics*, 23: 3–12.

Tonnies, F. (1963) *Community and Society*. New York: Harper & Row.

Tonts, M. (2001) 'The exclusive brethren and an Australian rural community', *Journal of Rural Studies*, 17: 309–322.

Took, L. (1986) 'Land tenure, return migration and rural change in the Italian province of Chiete', in R.L. King (ed.), *Return Migration and Regional Economic Problems*. London: Croom Helm. pp. 79–99.

Toynbee, C. and Jamieson, L. (1989) 'Some responses to economic change in Scottish farming and crofting family life, 1900–1925', *Sociological Review*, 37: 706–732.

Urry, J. (1995a) 'A middle-class countryside?', in T. Butler and M. Savage (eds), *Social Change and the Middle Classes*. London: UCL Press. pp. 205–219.

Urry, J. (1995b) *Consuming Places*. London: Routledge.

Urry, J. (1981) 'Localities, regions and social classes', *International Journal of Urban and Regional Research*, 5: 455–474.

Voyce, M. (1994) 'Testamentary freedom patriarchy and inheritance of the family farm in Australia', *Sociologia Ruralis*, 34: 71–83.

Wallace, C., Dunerley, D., Cheal, B. and Warren, M. (1994) 'Young people and the division of labour in farming families', *Sociological Review*, 42: 501–530.

Walter, G. and Wilson, S. (1996) 'Silent partners: women in farm magazine success stories, 1934–1991', *Rural Sociology*, 61: 227–248.

Warner, M. (1999) 'Social capital construction and the role of the local state', *Rural Sociology*, 64: 373–393.

Watkins, F. (1997) 'The cultural construction of rurality: gender identities and the rural idyll', in J.P. Jones, H. Nast and S. Roberts (eds), *Thresholds in Feminist Geography*. Oxford: Rowman and Littlefield. pp. 383–392.

Wenger, G.C. (1982) 'Ageing in rural communities: family contacts and community integration', *Ageing and Society*, 2: 211–229.

Whatmore, S. (1990) *Farming Women: Gender Work and Family Enterprise*. London: Macmillan.

Whatmore, S. (1991) 'Life cycle or patriarchy? Gender divisions in family farming', *Journal of Rural Studies*, 8: 387–397.

Whatmore, S., Munton, R., Marsden, T. and Little, J. (1987 'Towards a typology of farm businesses in contemporary British agriculture', *Sociologia Ruralis*, 27: 21–37.

Wild, R. (1974) *Bradstow: A Study of Class Status and Power in a Small Australian Town*. Sydney: Angus and Robertson

Wilkening, E. and Lupri, E. (1965) 'Decision-making in German and American farm families', *Sociologic Ruralis*, 5: 366–385.

Wilkinson, K.P. (1986) 'In search of the community in the changing countryside', *Rural Sociology*, 51: 1–17.

Wilkinson K.P. (1991) *The Community in Rural America*. New York: Greenwood Press.

Williams, W.M. (1963a) *A West Country Village: Ashworthy: Family, Kinship and Land*. London: Routledge and Kegan Paul.

Williams, W.M. (1963b) 'The social study of family farming', *Geographical Journal*, 129: 63–75.

Wissink, G.A. (1962) *American Cities in Perspective, with Special Reference to the Development of their Fringe Areas*. Assen: Van Gorcum.

Woodward, R. (1996) 'Deprivation and the rural: an investigation into contradictory discourses', *Journal of Rural Studies*, 12: 55–67.

Working Party on Rural Sociological Problems in Europe (1965) 'Report of the First Meeting', *Sociolgia Ruralis*, 5: 207–237.

Wright, E.O. (1985) *Classes*. London: Verso.

Young, I.M. (1990) 'The ideal of community and the politics of difference', in L. Nicholson (ed.), *Feminism/ Postmodernism*. New York: Routledge. pp. 300–323.

6

Rural economies

*Matteo B. Marini and Patrick H. Mooney**

INTRODUCTION

We must begin with the specification of our assumptions regarding the term 'rural economies'; we do not view the rural as strictly separable from the urban, the suburban, or any other spatial form of the *'non-rural'*. Neither can we view the economy as entirely separable from the political, the cultural, the social, etc. Hence, we begin by recognizing that the notion of rural economy is itself only an ideal type. This analytical construction is useful, however, in that it permits the generation of a number of fundamental questions: Are rural areas moving toward increased diversity, ever more homogeneity, or some simultaneous and contradictory combination? How does globalization effect these trends? Can types of rural economies be formulated to better understand the remaining or developing diversity? What roles do states play in rural economies? What role does social class play in rural economies? How do these factors facilitate or subvert the sustainability of rural areas?

Answers to these questions require a historical and comparative perspective. However, while maintaining such a perspective, the focus of our attention is on rural areas within relatively developed societies. A discussion of rural economies in less developed societies would necessitate far greater levels of abstraction and generality due to their distinctive dynamics. Though globalization may generate some tendencies toward convergence (Dollar and Kraay, 2002), at the moment, the less developed economies must be seen as having somewhat unique characteristics.

RURAL ECONOMIES: DEFINITIONS

Social scientists have long debated the nature of the rural–urban relationship. In noting the difficulty of developing a useful definition of 'the city', Max Weber apparently left the 'rural' to be merely the residual in various types of cities (e.g., the 'consumer city', the 'merchant city' or the 'producer city'). His colleague Georg Simmel (1978) argued that the urban environment produced a distinctive 'mentality' and, significantly, that such a mentality was tied to the predominance of the 'money economy' in urban areas. This economy was, in turn, associated with the insensitive and impersonal 'dominance of the intellect', implying that the rural was characterized by the subordination of the intellect. Kolb and Brunner (1946: 6–7) contended that an immediate environment that is closer 'to nature' has many implications for a different social life than urban places where '... nature is, so to speak, artificially on exhibit'. There is, in the rural studies literature, a long history of the development and use of typologies that either explicitly denote a difference between urban and rural places or imply this distinction indirectly through a focus on dichotomies such as traditional and modern, folk and urban or *Gemeinschaft* and *Gesellschaft*.

More recently, however, some have argued that in advanced capitalist society there is no longer any space or place that can or should be understood, meaningfully, as distinctively rural. Friedland (1982), for instance, sees 'the end of rural society' and thus, an end to rural sociology. This would also, of course, imply an end to rural

* While this chapter is a collaborative product of the two authors, Matteo B. Marini drafted and has primary responsibility for the sections 'Historical Background' and 'The Contemporary Patchwork'. Patrick H. Mooney drafted and has primary responsibility for the 'Introduction', 'Conclusion' and the section of 'Rural Economies: Definitions'.

studies. Such a position often derives from a Marxism focused on the eventual colonization of all space by capital that levels any difference between rural and urban. Indeed, today, transportation and communication technology permits the location of industrialization and service sector development (especially, information technology services) in rural places that may have little or nothing in their 'natural' environment associated with the particular product or service, other than the availability of cheaper labour. Yet, we contend, in common with other authors in the present text, that there is still, in the fundamental demographic fact of low population density, both a material as well as a socially constructed and meaningful difference associated with the rural in general and with rural economies more specifically. Of course, nearly all modern states recognize the existence of rural places and design some policy in accord with specific definitions or understanding of the rural. Further, most people in these societies carry some idea, however formed, of a distinct rural environment (see Bodenstadt, 1990; Jacob and Luloff, 1995; Singleman, 1996).

At the most fundamental or micro levels of interaction, the multiplex character of status and role structures in rural places lays a foundation of overlapping norms associated with these various roles. This diminishes the capacity for rural people to interact in accordance with only economic role expectations. Though urban ethnic enclaves may be somewhat similar, this is in contrast to the relatively simple series of discrete bilateral role interactions characteristic of most economic transactions in urban places. This is Simmel's point when he notes the correspondence, if not reduction, of urban social interaction to exchange value and its calculating anonymity. Economic interaction within rural places is more likely to take place in the context of 'other than economic' relationships (kinship, cohort, neighbour, friendships, etc.) that bring distinctive but overlapping normative expectations and obligations to bear on the economic transaction. Thus, urban and rural network structures differ in both form as well as substance, in turn, giving rise to distinctive social capital formation (Beggs et al., 1996). Whether this is a curse (to the neo-classical economist, for example) or a blessing (to the social capital analyst) is debatable. Our point here is only that the economic actor cannot, under such circumstances enjoy the normative autonomy of singular rationality that exists in the relative anonymity of the urban economy and that this may generate a fundamental difference in rural micro-economic behaviour and institutions.

Globalization raises the question of whether or not we can continue meaningfully to discuss place-specific economies, as implied in notions of a 'rural economy', an urban economy, or even a national economy. From some points of view globalization involves the homogenization of places, cultures, values, etc. Another view is the paradoxical renewed emphasis on the very distinctiveness of locale as each place finds its niche in the global division of labour. This chapter will argue that there are still, even in the most advanced industrialized societies, distinguishable rural places that differ in significant ways from 'other than rural' places. Further, it is argued that considerable variability among rural places not only exists as remnants of the past but that such difference continues to be preserved and, sometimes even newly constructed, by the processes of modernization and globalization.

We view this conceptual distinction of rural economies as referring to a dialectical tension within a unified whole where the rural and urban (and all that lies between) define, structure and transform each other. Marsden (2003: 151) argues explicitly for a need 'for both analytical and policy reasons to move away from the strictly geographically defined notion of "local rural area"' and to recognize the differentiation of rural spaces as 'caught up in different webs of local, regional, national and international supply chains, networks and regulatory dynamics'. In this sense, understanding of the economies of rural places demands the recognition of more than the influence of adjacent urban places but of other places both near and far; both 'other rural' places as well as 'other than rural' places. Here, again, Marsden's (2003: 142) conception of 'rural spaces as ensembles of local and non-local connections, of combinations of local actions and actions "at a distance"' is useful.

Defining the rural

A rural area is usually defined as a less densely populated area. In fact the city – the opposite pole of rurality – is by definition a concentration of people and activities for commercial and institutional purposes (Sorokin et al., 1930). The significance of the rural in US policy has led to sometimes complicated means for defining rural areas. On one hand, the US federal government (USDA, 2004) defines 'rural areas' as 'places (incorporated or unincorporated) with fewer than 2,500 residents and open territory'. On the other hand, there is also a 10-category scale measuring a continuum of 'urban influence' among counties.

Table 6.1 ***Characterestics of rural versus urban economies***

Question	Urban	Rural
What goods and services are produced and in what quantities?	Manufacturing goods, technological services, administrative services	Extractive-based goods, agricultural goods, environmental services
How are goods and services produced?	By advanced technologies	By advanced and traditional technologies
When are goods and services produced?	Continuous cycle; all year long	Greater seasonality; greater relative impacts of climate
Where are goods and services produced?	In controlled environment	In open air
Who consumes the goods and services that are produced?	Distant and local markets	Local and distant markets

The first four categories are various types of 'metro counties'. Only two of the six nonmetro categories in this scheme are labelled rural. These counties are distinguished as either adjacent to or not adjacent to metropolitan counties and are otherwise 'completely rural or less than 2,500 urban population'. The 'least urban' or 'most rural, urban' category is composed of counties with urban populations of 2,500 to 19,999 that are not adjacent to metropolitan areas. In these two schemes, 'rural' areas and 'nonmetro' areas are not necessarily identical.

European level statistical data on rural populations are both less developed and more variable (by nation) than in the American case, though rural areas are defined for specific EU policy purposes.[1] For example, rural areas targeted by EU structural funds, must comply with at least two of the four following conditions: population less than 100 inhabitants per square kilometre (or employed in agriculture at greater than twice the European average) and unemployment rate greater than European average (or decrease of population greater than European average).

Distinguishing rural economy

Parkin (1998) raises five questions that point toward an analysis of any economy. What goods and services are produced and in what quantities? How are goods and services produced? When are goods and services produced? Where are goods and services produced? Who consumes the goods and services that are produced? To the extent that the answers to these questions vary between rural and urban places, we can recognize the existence of *tendencies* toward distinctive rural economies, as shown in Table 6.1. If we accept the position of Kolb and Brunner (1946), that the economy of rural places is, more so than urban places, contingent on the unique mix of factors given in the natural environment, then there is no single form of 'rural economy'. Rural economies are quite varied insofar as they are grounded in such different bases as agriculture, fishing, forestry, mining, tourism, etc. Even within agriculturally dependent regions, for example, the broadest contours of a rural economy will vary with the specific commodity mix of that rural place.

Thus, the US Department of Agriculture distinguishes nonmetro counties on the basis of their *primary economic characteristics*. There are six discrete county categories:

- about 25 per cent of US nonmetro counties are *farming-based* counties;
- 7 per cent are *mining-based* (coal, gas, oil, metals);
- 23 per cent are *manufacturing-based*;
- 11 per cent are *government-based* (75 per cent from state and local jobs, 25 per cent from federal jobs);
- 14 per cent are *service sector-based*;
- the remainder are classified as *nonspecialized* in their economy.

Certain other characteristics are also used to distinguish counties in relation to ***policy*** concerns. These overlapping categories are:

- counties that are major *retirement destinations*;
- those dominated by *federal government land ownership*;
- counties whose *residents commute to work* in other counties;
- counties characterized by *persistent poverty*; and
- counties *dependent* on transfers of 'unearned income', primarily from various government agencies.

We are unaware of any such comparable classification schemes for the collection of statistical data on rural Europe.

HISTORICAL BACKGROUND

The distinction of rural and urban is not one in which the two entities can be understood as independent from one another but as integrally or dialectically tied to one another as two parts of a whole. Max Weber (1978: 1217) recognized this when he noted that: 'Historically, the relation of the city to agriculture has in no way been unambiguous and simple.' The differentiation between urban and rural is usually argued to have emerged with the transition from nomadic to settled agriculture and the creation of an agricultural surplus (e.g., Henslin, 2002; Palen, 1997). The appearance and exacerbation of various forms of social inequality were aspects of struggles for control of that surplus. Urban development involved the flow of surplus from the countryside, often along with the owners of that surplus, to the city to better manage their investments in commercial and, eventually, industrial activities. Ironically, this spatial division of labour, which depended upon the surplus produced in rural places, simultaneously also established a hierarchical structure of power, with the city dominating the countryside economically and politically.

The dominant economic explanation for the concentration of economic as well as political, religious and educational activity in urban places is that distance is a cost for transactions of material goods or services. The spatial concentration or centralization of activity is expected to generate economic efficiencies in the total cost of production (Weber, A., 1899). Such cost reductions are referred to as *economies of agglomeration*. Rural economies were usually, then, originally based upon extensive agriculture and/or extractive industry (for example, agriculture, forestry and fishing), primary sector activities with high ratios of space to population. Commerce and public affairs constituted the core of urban economies because of the existence of economies of agglomeration. In terms of consumption, twentieth century Fordist mass production demanded sales in huge quantities, a function which urban populations more readily satisfied. Cities also functioned as network nodes for transportation to reach distant markets.[2]

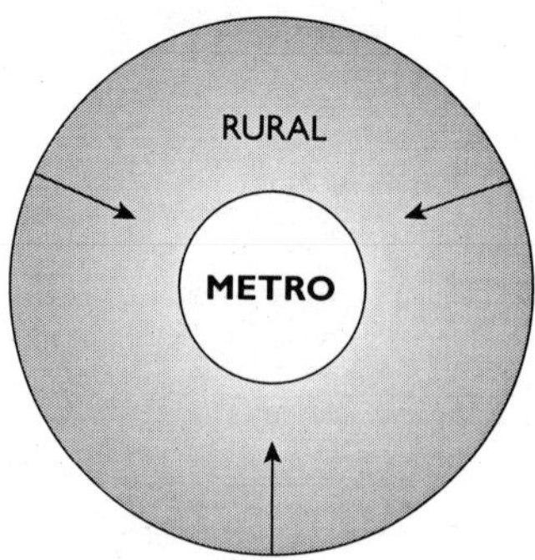

Figure 6.1 ***The exodus from rural to metropolitan areas***

Modernization: escape from the countryside

Some towns and cities grew quickly into metropolitan areas, leaving many rural economies behind in per capita income. Most rural economies became synonymous with lagging or backward economies, as if frozen at an early stage of development, producing mainly primary goods. On the other hand, urban economies were identified with advanced economies, characterized by manufacturing (secondary sector) and service provision (tertiary sector). 'Rural' was often associated with poverty, lack of opportunities, traditionalism and isolation, while 'urban' was associated with wealth, opportunity, modernity and concentration. The modernization process primed an out-migration flow from traditional rural areas to metropolitan areas. By the second half of the twentieth century warnings were put forward that the exodus from the countryside – attracted by wage differentials and opportunity expectations – would soon empty traditional rural areas while congesting metropolitan areas.

Desertion of rural areas (often deprived of their most valuable human capital) was often forecast as the biggest threat to local rural economies. In many cases, the pace and intensity of urbanization facilitated the perception, if not the reality, of an association of the urban with overcrowding, class conflict, criminal behaviour, moral decay and various forms of environmental pollution. In this context, nostalgia for idealized rural settings began to take root in the minds of city dwellers. Bell's account of the 'rural idyll' (Chapter 10 in this volume) provides an excellent discussion of this complex and contradictory framing of the rural (see also Cloke and Milbourne, 1992).

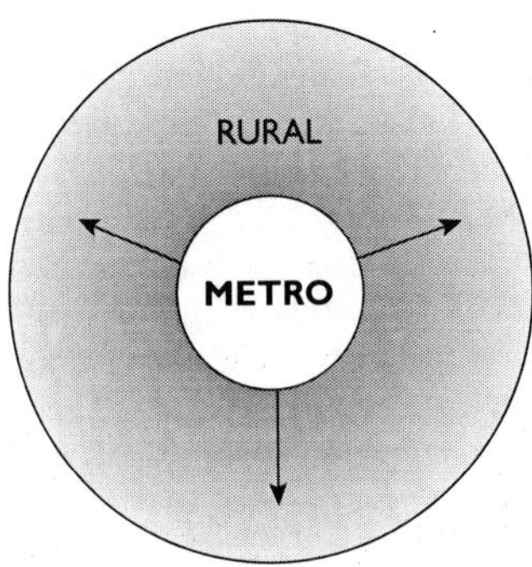

Figure 6.2 ***The population turnaround***

The reversal of the trend: the population turnaround

During the last quarter of the twentieth century, a reversal took place as a new flow of capital, goods and opportunities headed toward many rural areas throughout Europe and North America (see Champion, 1989; Elliott and Perry, 1996; Fulton et al., 1997; Halfacree, 1994; Serow, 1991; Stockdale et al., 2000). As Murdoch notes regarding Britain (Chapter 12 in this volume), Johnson and Beale (1999) find that there is continued population growth in nonmetropolitan America. The 'population turnaround' that took place in the 1970s is now argued to be a long-term phenomenon, with the slight downturn in the 1980s being merely a short-term function of the economy. In rural America, Johnson and Beale found that 75 per cent of nonmetropolitan counties gained population from 1990 to 1996 with migration patterns tied to certain economic types of rural places.

Among the reasons for this reversal were capital's search for cheaper labour, the creation of new markets, and better places for living and raising children (Brown et al., 1997; see also the special issue of *Sociologia Ruralis*, 31: 1, 1991). The means were decisive: the improvement in infrastructures and telecommunication technology reduced the space in which transactions take place. Electronic information and financial exchanges have practically abolished the obstacle of distance (Grimes, 2000), while the cost of transporting material goods has been considerably lowered. These trends are, of course, part of the reason why some scholars increasingly call into question the very differentiation of urban and rural (Dahms and McComb, 1999; Friedland, 1982; Thomson and Mitchell, 1998). However, short of any absolute dismissal of the concept of rural, this reduction of the significance of time and space as costs of production certainly has many implications.

In agriculture, for instance, much scientific research and development is devoted to developing plants that can grow independently of the soil or climatic conditions of specific places, or even, for example, hydroponically (independent of soil altogether). Food markets in advanced societies provide fresh fruits and vegetables from halfway around the world. We are as likely to find apples from New Zealand in our Italian or American grocery store as apples from our own regions. Shopping malls and hypermarkets eliminate the distinctiveness of locale (urban or rural). Similarly, the process of rural industrialization means that factories are now as likely to be found in rural places as in urban places. In Giddens's (1990: 19) terms: 'locales are thoroughly penetrated by and shaped in terms of social influences quite distant from them'. The globalization of this process gives rise to questions about the possible homogenization of all space, not merely the obliteration of difference between urban and rural. Yet a contradiction lies beneath this process for it is often precisely the uniqueness of a rural place that attracts the exurban in-migrant.

For example, those rural counties growing most quickly in the US are those with unique natural amenities that attract retirees, tourists, as well as manufacturing. Those counties dependent upon traditional economic activities such as agriculture or mining were the least likely to gain population in the 1990s (Johnson and Beale, 1999). On the other hand, the very immigration and economic transformation may eclipse the amenities base that provided the initial attraction of migrants. In this sense rural economic development may create 'its own gravediggers'. Nevertheless, there is tremendous pressure on rural locales to construct their own unique 'niche' to attract development.

What Marsden calls the 'consumption countryside' demands the creation of a heterogeneous product differentiation of place while simultaneously being subjected to forces of globalization and rationalization that demand homogenization of space. This contradictory process is full of interesting and provocative lines of research. Salamon observes (Chapter 23 in this volume) the blurring of suburb and rural in relation to home furnishing, in which suburbanites appropriate rural symbols that are, in turn, re-appropriated by rural residents who, in many instances are, ironically, transforming what were once means of farm production (butter churns, saws, etc.) into household consumption items that now merely decorate the home in the fashion of suburbia.

Thus, the globalization process can, somewhat ironically, be seen as a force countering the agglomeration forces. While the latter had been the primary determinant of industrial location in the past, this is now a function of a contradictory tension between both centripetal (agglomeration) and centrifugal (decentralizing) forces (Krugmann, 1995). For many rural areas that run the risk of abandonment through the modernization process, the centrifugal force that pushes capital away from the center and toward the periphery has been a salvation. The counter-movement of capital and opportunities flowing to the countryside has deeply transformed rural economic and social structures. Clearly, it can no longer be described merely with the 'traditional/modern' dichotomy, but rather, as a 'patchwork' of diverse, local economies.

To develop ideal types that reflect the composition of this emergent patchwork, we need to identify the internal characteristics of local places that play a role in responding to the external stimuli. An increasing body of literature (Harrison and Huntington, 2000) points to local cultures as the major determinant of the pattern pursued by local economies. By 'culture' we mean the transmission of values, beliefs (unexamined assumptions) and norms (standard operating procedures) that reflect 'what has worked' in the history of a region's population (Cloke, 1997; Triandis, 1996; for the linkage between economic development and rural culture, see the special issue of *Sociologia Ruralis*, 38 (1), 1998). Differences between, as well as within, both Europe and the Americas that cannot be explained merely by variable factors of production (land, labour, capital) exemplify these cultural factors.

THE CONTEMPORARY PATCHWORK

Following Marsden (2003: 103), the analysis of rural change and differentiation necessitates 'the development and refinement of typologies'. As we stated at the end of the previous section, the construction of contemporary ideal typical rural economies demands an account of both global forces and local responses in terms of economic strategies. The former may consist of capital supply or demand for goods. The latter will depend on the local endowment of resources (that is, physical, financial, human and social capital) as well as on the culturally rooted values system of that specific society, which may facilitate or hinder the local economy's capacity to take advantage of global opportunities.

Marsden (2003) lays out four ideal types that characterize socio-political relations in the British countryside. In the following, we will elaborate three other ideal types that characterize the economic relations of rural regions. These types of rural economies (rent-seeking, dependent, entrepreneurial) can be associated with Marsden's types for a more nuanced analytical framework. Marsden characterizes the *preserved countryside* as 'attractive' rural regions in which recent growth has stimulated a new middle class increasingly inclined toward preservationist political regulation of the local economy focused on the creation of a service sector and clean industry. His *contested countryside* tends to lie outside the core commuter corridors and 'as yet may be of no special environmental quality'. It is characterized by an incomplete usurpation of political regulation by the newcomers who find themselves in conflict with the long-standing landowning resident farmers over many issues associated with the local economy and development. The *paternalistic countryside* is characterized as still being under the sway of established large landholders and farmers who exert relatively unchallenged regulatory control over minimal economic development. The *clientelistic countryside* is associated with rural regions that are economically dependent on transfer payments associated with political institutions and subsidy of agricultural production. However, Marsden's typology has much more to do with social relationships and political governance rather than with the local economy *per se*, the theme of the present chapter. Marsden accomplishes this with respect to the conditions of contemporary Britain, our typology seeks a broader scope. Nevertheless, we will note points of convergence with Marsden's typology. In the following section, we present three ideal types of rural economies. The first two of these, the rent-seeking economy and the dependent economy, are argued to pose impediments that hinder economic development consistent with the opportunities presented by globalization. The latter type, the entrepreneurial economy, on the other hand, is argued to tend toward the facilitation of rural economic development in the context of globalization.

Rent-seeking economy

Rent-seeking economies refer to those rural areas whose resources are mainly based on agriculture and extractive industry. As we have seen in the previous section concerning the historical background, these are typical features of rural areas, insofar as the original division of labour between urban and rural economies localized industrial and administrative activities in the city, leaving the countryside to space-consuming activities such as farming and mining. Moreover, these natural resources are deeply embedded in a specific locality, and cannot be reproduced elsewhere. It is the non-replicable nature of monopolized goods which forms the basis of this concept of 'rent'. According to Ricardo, land has differential productivity that accounts for surplus falling into the hands of landlords without the need for investments, due to the monopolization of non-replicable resources. Profit needs investment, rent does not. However, investment is the most productive factor in economic growth (Levine and Renelt, 1992). Consequently, rent-seeking strategies tend to constitute an obstacle for economic development.

Contemporary research confirms this theoretical account, showing that 'nations having the greatest abundance of natural resources tend to perform more poorly than those that do not have an abundance of natural resources' (G. Sachs, quoted by Lindsay, 2000). Besides the above consideration, claiming that rent-seeking strategies hinder economic growth because of an associated lack of investment, Lindsay explains that poor economic performance is due to the commodity nature of rent-seeking products. As is often noted in discussions of the shift from a production-driven to a consumer-driven agriculture, commodity prices in recent decades have been declining because producers have less control over them. Although many nations are exporting a greater amount of raw materials in these days, they are earning less money in real terms. 'In today's global economy, a comparative advantage in natural resources does not assure economic prosperity' (Lindsay, 2000: 285).

Rent-seeking economies are thus trapped in a low income status that is not easy to escape, given two primary constraints: the social structure of economic power and the character of local values systems (Freudenburg, 1992). Rent-seeking economies usually flourish in marginal areas, where class structure is polarized and local culture is deeply affected by isolation. Within the class structure of the rent-seeking economy, a few families often control the majority of a valuable natural resource (be it land, oil, coal, forest, coastline, etc.) The monopolistic or oligopolistic structure of ownership, diminishes the need for landlords to diversify the local economy. This, in turn, often lays the material basis for the social relations that Marsden refers to as the 'paternalistic' countryside. Even when opportunities arise from outside, as with the recent population turnaround, the economic strategy of landlords does not diverge from a rent-seeking one, as in the re-utilization of existing buildings for newcomers (Spencer, 1997).

Rent-seeking economy is usually associated with a local culture that is averse to any change, even the smallest, because it considers change as a threat to its own entire value system. In these 'tight' cultures, the cost to elites of cultural change necessary to economic development is perceived as greater than the benefits that the latter may bring, and is thus opposed. Scholars find examples of such 'tightness' of local culture in some geographical areas more than in others. For instance, among industrialized countries, Southern Europe is one of these regions where rural areas, despite their change in appearance, are often described as firmly rooted in traditional customs (Hogart and Paniagua, 2001; Jansen, 1991).

Moreover, rent-seeking economies may result in highly unstable settings from a political and institutional point of view because the hegemonic construction of reality is based on the 'image of the limited good' (Foster, 1965). According to this perspective, 'goods are limited' given their natural and irreproducible character. While this rationality is often associated with a 'peasant' perspective, it is also fundamental to a local elite worldview, with significant subsequent impacts on local resource utilization. Constructed as a 'zero-sum game', the only way to climb the social ladder is by appropriating the limited good itself, whether it is land, oil, coal, or political power. One consequence is that it does not lure foreign investments from outside, even in a time of global fluid capital movement such as the current one. Rather, foreign investments are more likely to be directed towards rural areas with different structural and cultural characteristics.

Dependent economy

By 'dependent economies' we refer to localities whose income is primarily derived from external sources. The population turnaround witnessed by rural areas from the 1970s on, is indeed mainly due to external sources. Such sources may be of a private as well as of a public nature: for example, a large factory built by a multinational corporation belongs to the first category, while the extension of public services such as most schools and public or state supported hospitals would exemplify the second category.

Dependence on the private sector is usually grounded in a locally high ratio of labour to capital, in which the former factor of production is relatively cheap, thus attracting global flows of capital in search of lower costs of production. In rural Europe, this is the situation in many of the former state socialist countries. However, the capacity of these areas to attract resources from outside is not due merely to material factors (that is, the labour/capital ratio). It also demands a specific attitude on the part of the population: private corporations must find a 'friendly' environment to make investments. Recently, Ireland has exemplified such an environment attracting global business interests. Irish rural areas have been among the most successful in Europe attracting industrial plants from abroad (Curtin and Varley, 1986).

On the other hand, if the population is hostile (for example, high rates of crime or corruption, worker absenteeism, political instability), as in the case of Southern Italy (Leonardi, 1995), investments will be made elsewhere. In such hostile environments, dependence on the public sector rather than from the private sector is more likely. Where local areas have not been able to attract foreign investments, the state has often compensated through Keynesian policies aimed at building infrastructure (such as roads, land reclamation, afforestation) and services (education and public health, for example), in order to create jobs and raise the standard of living of these populations (Mencken, 2000). There are similar tendencies in the former socialist countries, where many rural families are still dependent on social welfare programmes (Brown and Kulcsar, 2002). Within the formerly state socialist areas, that will soon be integrated into the European Union, there has been a somewhat uneven benefit from EU funds for infrastructural development, based on the vigour with which different regions and communities within these nations have pursued such funds.

A recent form of rural dependence on the public sector involves the attraction of waste facilities, penitentiaries and the like (Albrecht et al., 1996). Since most communities refuse to comply with mandatory decisions related to the location of these facilities, those that accept them, do so in exchange for financial compensation (Bourke, 1994). In the long term, dependent economies – based on either public or private resources – are vulnerable, since the source of investment is outside the control of the local population.

Dependence on public expenditures is fragile because the latter depend upon the performance of the real economy. For example, most governments are no longer willing or able to afford the huge budget deficits experienced during the 1970s and the 1980s. In fact, public institutions are aware of the vulnerability of those rural economies that are based on public expenditures (OECD, 2003).

Just as we suggested that rent-seeking economies lay the material basis for the prevalence of what Marsden calls paternalistic social relations, dependent economies facilitate the development of what he calls the 'clientelistic countryside'. Further, where public expenditures are allocated through clientelistic practices, the impact on economic growth potentials is often negative. Littlewood (1981) argues that this is because clientelism contributes to a reduction of the local population's collective self-esteem.

Dependence on private investment, on the other hand, is often even more volatile, because industrial plants may always flee toward locations where the cost of labour is even lower. Through the tendencies of the globalization process, the more remote the region, the lower will be the average cost of labour; thus, the greater the possibility for capital to flee the earlier settlements and to relocate in more remote areas. However, the fragility of the dependent pattern of development, public or private, can be moderated by spin-off effects produced by the presence of a large investment in physical assets. According to some research, local workforces may learn entrepreneurial skills and attitudes, and thus the local economy moves toward an 'entrepreneurial economy' (Hirschman, 1977).

Entrepreneurial economy

Entrepreneurial economies draw their incomes mainly from the valorization of local resources. Thus, there is a rough overlap with Marsden's preservationist countryside, although entrepreneurial economies reflect a broader scope of action. Rather than trying to attract external capital investment, whether private or public, they fill the demand for high-quality goods promoted by the globalization process through their local, but socially widespread, tacit knowledge. Paradoxically, most of these goods (local cuisine, furniture, rural tourism and the like) are 'traditional' and endangered by the standardization associated with industrialization. Insofar as these goods have traditional features but are integrated into modern marketing structures, they are sometimes referred to as 'postmodern' (Brunori and Rossi, 2000; Buller and Hokkart, 1994; Dahms, 1995; Ehrentraut, 1996). Again, Salamon's discussion (Chapter 23 in this volume) of the contemporary consumption of what were once means of rural production as home decor is an interesting example of this feature.

In entrepreneurial economies, labour is not as cheap as in the more remote areas, because modernization has brought relatively high standards of living. By the same token, these areas are not particularly attractive for foreign industries in search of cheaper labor. On the contrary, some of them have lost industrial plants, which have relocated to areas where the cost of labour is cheaper. The rural characteristics of these places, along with the endowment of modern infrastructures (highways, broadband access, etc.) may, however, attract industries fleeing the city, not in search of cheaper labor, but better residential places for their employees (Beyers and Nelson, 2000; Goe, 2002; Luloff and Swanson, 1990).

In all these cases – whether they sell goods or places – the cultural factor which characterizes these economies is the entrepreneurial capability (Anderson and Eklund, 1999; Terluin, 2003). These local businesses tend to be small, completely different from the vertically integrated corporate firm which represents the main outcome of the modernization process in metropolitan areas. Rather, these small firms reach their economies of scale through horizontal networks (Piore and Sabel, 1984; Putnam, 1993) in which cooperation, more than hierarchy, is the functional value lying at the top of their cultural system. Social capital has often been invoked to explain the main features of these entrepreneurial communities (Jóhannesson et al., 2003; Sharp et al., 2002; Zeckeri et al., 1994). Indeed, in some cases, this cooperation of entrepreneurs has been formalized into an increasingly diverse array of cooperative organizations that facilitate networking potentialities among enterprises that might otherwise conflict with one another in a competitive marketplace and simultaneously functioning to tie these firms to place, diminishing both dependency and capital flight (Mooney, 2004; Mooney et al., 1996).

Generally speaking, the paths of development experienced by entrepreneurial economies are significantly different from one another, since they are culturally rooted. The literature on regional economy speaks of a 'Rhinean capitalism', from the name of the Rhine river cutting across Germany and France, as distinctive from an 'Asian capitalism' and from the early model of the 'Anglo-Saxon capitalism' (Berger and Dore, 1996). Differences concern the relationships between managers and shareholders, entrepreneurs and workers, firms and financial institutions. Without entering into the details of such differences, the literature underscores the idea that each regional economy is profoundly influenced by its own cultural environment (Fukuyama, 1995).[3]

The existing literature does not adequately assess the viability of entrepreneurial communities as compared with large metropolitan corporations. However, compared with dependent economies, these rural populations exhibit much more control over their own destiny. These communities seem to express the best of the two worlds (the entrepreneur capability of the city with the communitarian spirit of rurality).

CONCLUSION

Figure 6.3 is a spatial representation of our rural economies' typology. Metropolitan areas are situated at the core of territorial space. Metro and nonmetro areas are connected in a web of economic transactions, where metropolitan areas are the sites of technological, economic and administrative power. They draw raw materials and commodities from rent-seeking economies, which are located in the most peripheral areas. They also draw manufactured goods from dependent economies, where industrial plants are increasingly located since the cost of labor is cheaper. Eventually, they draw high quality goods from entrepreneurial economies, where entrepreneurial skills are associated with a preserved, rural environment that provides a better quality of life and an escape route for stressed metropolitan newcomers. Of course, this model is itself of typology. The variable articulation of types leads, as Murdoch notes elsewhere in this volume, to an increasingly differentiated regionalization of rural space.

From a historical perspective, it can be said that nonmetro areas, which experienced a heavy loss of human capital to the city at the beginning of the modernization era, are now witnessing the reversal of the trend that brings opportunities to them. Remote, rural areas, once a site specialized in raw materials, may upgrade to a dependent economy hosting an industrial plant from a multinational corporation, and eventually become an entrepreneurial economy, if spin-off effects produce the emergence of a local capitalist class.

The process is neither mechanical, nor deterministic, of course, but it displays a sort of virtuous cycle promoting the integration of remote rural areas into the globalization process. Due to industry attraction or to public jobs, rural population may acquire those skills and the mentality needed to move towards the next category, which is the entrepreneurial economy. By the same token, the latter may follow a specific path of

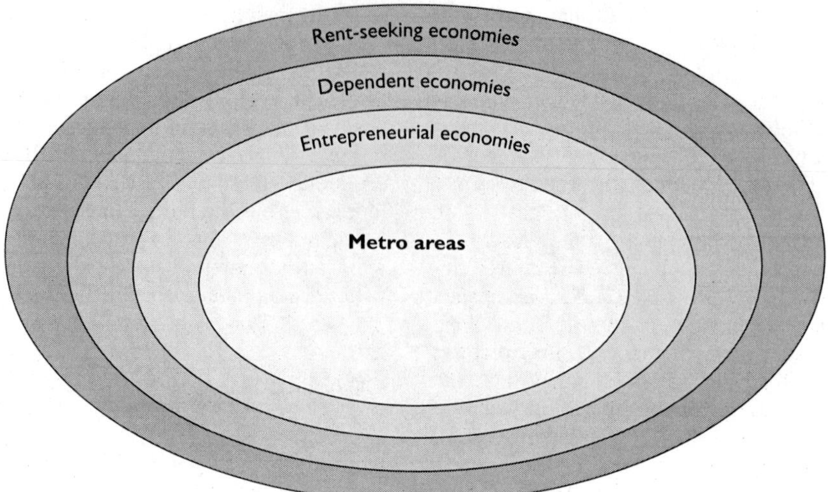

Figure 6.3 ***The contemporary patchwork of rural economies***

development of its own, linked to the local culture and able to escape the flaws of the early modernization and industrialization process. On the other hand, upgrading cannot be taken for granted. The opposite can also occur, with dependent economies being downsized to the level of rent-seeking economies, or entrepreneurial communities downscaled to the level of dependent economies. In the end, all depends on the response that the local community is willing, or able to give, to the increasing opportunities emerging from the globalization process.

Against the all-too easy contention that globalization processes are simply levelling all rural difference, this analysis suggests that the future of rural places is not given. Such places are continuously (re)constructing the social, economic, political and cultural structures, and even we might say, the nature of, their locale. Both the substance of that restructuring and the mechanisms by which it takes place vary with the types of economies that we have outlined in the above discussion. While this restructuring is not 'path dependent' in the strict sense (Stark, 1992), the effects of these historical economic structures impact developmental possibilities. These analytical constructs of rent-seeking, dependent, and entrepreneurial economies reflect the material context in which Marsden's sociological types develop. As we have argued: paternalism is likely grounded in rent-seeking economies; clientalistic social relations flourishes in dependent economies; preservationist strategies are often cause and/or consequence of entrepreneurial economies; and the contested countryside reflects struggles that are grounded in conflicts between these types of economies, perhaps especially when entrepreneurial aspirations arise to undermine traditional forms of dependency and clientelism, or threaten to diversify a rent-seeking economy's singular resource base, fragmenting elite control.

Changes within and between each of these types of rural economies may also be further specified with Marsden's analytical focus on transformations associated with regulation, commodification and spatialization processes. This scheme permits a historical and comparative examination of rural economies and the contradictory tensions between homogeneity and heterogeneity associated with globalization. First, an examination of the forms and functions of *regulation* can show that rent-seeking, dependent and entrepreneurial economies each entail specific patterns associated with globalization's breaking down *(deregulation)* of existing regulatory mechanisms and reveal the embryonic forms of re-regulation, that are shaped by the type of economy that currently characterizes the region.

Second, Marsden's focus on the commodification and resistance to commodification of a region will vary with the types of economies we have specified. Further, this commodification may reflect an even broader range of regional qualities; from the commodification of natural amenities such as fertile soil, mineral resources or aesthetics to the commodification of the residents themselves who might be 'on display' as quaint and picturesque villagers or peasants. Resistance to such commodification might range from opposition to 'surface mining' of coal to refusal to have one's image photographed by tourists (as in the case of many Amish, for instance). Rent-seeking economies, dependent economies and entrepreneurial economies each have specific interests in the substance (what gets commodified) of *commodification* and *decommodification*. Entrepreneurs, of course, will have immediate interests in *decommodifying* some regional resources while *recommodifying* others.

Finally, Marsden points to the need to examine the spaces within which actors act through the spatial character of their networks. Sharp (2001), for instance, has provided an exemplary analysis of networks associated with economic development of Iowa rural communities. However, that analysis focused only on networks *within* communities. Marsden would point us to the embeddedness of actors within networks that reach beyond the local community. This focus on *spatialization* allows us to direct attention at the interesting contradictory tension discussed above concerning the spatial homogenization, or *despatialization,* associated with globalization as well as the opposing force of what is often called 'localization' or the process of *respatialization* as locales seek to reconstruct new identities that differentiate them from both the urban as well as from other rural places.

Again, the dynamics associated with spatial relations will play out differently in the different types of rural economies that we have discussed above. This is most clear in the tendency of rent-seeking economies toward isolation and the dependent economy's inherent reliance on exogenous resources through variable networks. Entrepreneurial economies may be particularly prone to be agents of respatialization as they seek to construct new forms of regulation around local-specific commodities that remain oriented toward global forms of investment and consumer capital flows.

To ignore these differential effects on each type of rural economy is to sabotage effective planning or policy development. However, bureaucratic structures tend to treat any and all places as already homogenized or soon-to-be homogenized. The push for the development of a multifunctional village or region (especially pronounced in the EU) only enhances the need to recognize the unique mix of factors that come together in any rural community. Rent-seeking and dependent economy elites, for instance, would have clear interests in disrupting the development of multi-functional economies, while entrepreneurial economies would likely embrace this development strategy. Marsden's concept of the *economies of synergy* is particularly useful for recognizing the strength or weakness of interaction effects of various combinations of development projects within a community or region. Successful rural economies may now and in the future need to focus more on these economies of synergy and *economies of scope* than on *economies of scale.* The latter dominated policy and economic decision-making in the development of the increasingly unsustainable agro-industrial model that is now being eclipsed in the construction of new rural economies and countrysides. Sustainable rural development demands privileging economies of scope and synergy.

Finally, this brings us back to the need to remember that, at the outset of this discussion, we recognized that, in fact, economies do not actually operate independently of other institutions. The concept of *social economy* (Marsden, 2003) explicitly refers to the fact that economic relations are deeply embedded in 'other than economic' institutions. There is a pragmatic need to recognize the impact of social, political, regulatory, scientific and cultural spheres in concrete rural economic development. This may be even more significant in rural development since the multiplex role complex of the rural actor demands that they bring the norms and practices associated with these other overlapping institutions in which they are embedded to bear on their economic actions. Attempts to separate these spheres in accord with disciplinary interests or the boundaries of bureaucratic governmental agencies can only have a disorganizing, 'Tower of Babel' effect on the sustainability project. Failure to remember that our notion of 'rural economy' is an ideal type and that the rural cannot, in practice, be fully separated from the 'other than rural' or that the economy cannot be fully separated from the political, the social, the cultural, the scientific etc. may result in the continued fragmented development that is increasingly recognized as unsustainable.

NOTES

1 The minutes of an official EU meeting held on 17 November 2003 in Luxembourg by a Working Group entitled RURAL DEVELOPMENT STATISTICS note that during the discussion of the working paper that was presented, different approaches to several issues such as the level of geographical detail to be used, the criteria to be used and the subtypologies to be introduced were

discussed. None the less, no common position could be reached. Moreover, several member states would favour creating different sets of typologies for rural areas rather than trying to find a common definition for them.

2 Not all scholars agree with such an economic explanation of the origin of the rural–urban divide. Others stress political or religious factors, like the rise of the city-state and the location of educational centres (Sorokin et al., 1930). Whatever the reasons, most interpretations share a conviction that the spatial concentration of activities (agglomeration economies) has been the result of a long-term evolutionary process which left behind the most remote and isolated areas.

3 Of course, cultural capitalism is not a specific matter of rural areas. The above definitions of 'Rhinean', 'Asian' and 'Anglo-Saxon' apply to metropolitan as well as nonmetropolitan areas. But it is reasonable to believe that among nonmetro areas, entrepreneurial economies will bring more of the local cultural mark than dependent economies, for example, the latter being a byproduct of external multinational corporations.

REFERENCES

Albrecht, S.L., Arney, R.G. and Amir, S. (1996) 'The siting of radioactive waste facilities: what are the effects on communities?', *Rural Sociology*, 61 (4): 649–673.

Anderson, K. and Eklund, E. (1999) 'Tradition and innovation in coastal Finland: the transformation of the Archipelago Sea Region' *Sociologia Ruralis*, 39 (3): 377–393.

Beggs, John J., Haines, Valerie A. and Hurlbert, Jeanne S. (1996), 'Revisiting the rural–urban contrast: personal networks in nonmetropolitan and metropolitan settings', *Rural Sociology*, 612: 306–325.

Berger, S. and Dore, R. (eds) (1996) *National Diversity and Global Capitalism.* Ithaca, NY: Cornell University Press.

Beyers, W.B. and Nelson, P.B. (2000) 'Contemporary development forces in the nonmetropolitan West: new insights from rapidly growing communities', *Journal of Rural Studies*, 16 (4): 459–474.

Bodenstedt, A.A. (1990)'Rural culture – a new concept', *Sociologia Ruralis*, XXX (1): 34–47.

Bourke, L. (1994) 'Economic attitudes and responses to siting hazardous waste facilities in rural Utah', *Rural Sociology*, 59 (3): 485–496.

Brown, D.L. and Kulcsar, L. (2002) 'Household economic behavior in post-socialist rural Hungary', *Rural Sociology*, 66 (2): 157–180.

Brown, D.L., Fuguitt, G.V. Heaton, T.B. and Waseem, S. (1997) 'Continuities in size of place preferences in the United States, 1972–1992', *Rural Sociology*, 62 (4): 408–428.

Brunori, G. and Rossi, A. (2000) 'Synergy and coherence through collective action: some insights from wine routes in Tuscany', *Sociologia Ruralis*, 40 (4): 409–423.

Buller, H. and Hoggart, K. (1994) 'The social integration of British home owners into French rural communities', *Journal of Rural Studies*, 10 (2): 197–210.

Champion, A.G. (ed.) (1989) *Counterurbanization: The Changing Pace and Nature of Population Deconcentration.* London: Edward Arnold.

Cloke, P. (1997) 'Country backwater to virtual village? Rural studies and "The Cultural Turn"', *Journal of Rural Studies*, 13 (4): 367–375.

Cloke P. and Milbourne, P. (1992) 'Deprivation and lifestyles in rural Wales: rurality and the cultural dimension', *Journal of Rural Studies*, 8 (4): 359–371.

Curtin, C. and Varley, A. (1986) 'Bringing industry to a small town in the West of Ireland', *Sociologia Ruralis*, XXVI (2): 170–185.

Dahms, F.A. (1995) '"Dying Villages", "Counterurbanization" and the urban field – a Canadian perspective', *Journal of Rural Studies*, 11 (1): 21–33.

Dahms, Fred and McComb, Janine (1999) '"Counterurbanization", interaction and functional change in a rural amenity area – a Canadian example', *Journal of Rural Studies*, 15 (2): 129–146.

Dollar, David and Kraay, Aart (2002) 'Spreading the wealth', *Foreign Affairs*, 81 (1).

Ehrentraut, A. (1996) 'Globalization and the representation of rurality: Alpine open-air museums in advanced industrial societies', *Sociologia Ruralis*, 36 (1): 4–26.

Elliott, J.R. and Perry, M.J. (1996) 'Metropolitanizing nonmetro space: population redistribution and emergent metropolitan areas, 1965–90', *Rural Sociology*, 61 (3): 497–512.

Foster, G.M. (1965) 'Peasant society and the image of limited good', *American Anthropologist*, 67: 293–315.

Friedland, William H. (1982) 'The end of rural society and the future of rural sociology', *Rural Sociology*, 47 (Winter): 589–608.

Fukuyama, F. (1995) *Trust: The Social Virtues and the Creation of Prosperity.* New York: The Free Press.

Fulton, I.A., Fuguitt, G.V. and Gibson, R.M. (1997) 'Recent changes in metropolitan–nonmetropolitan migration streams', *Rural Sociology*, 62 (3): 363–384.

Freudenburg, W.R. (1992) 'Addictive economies: extractive industries and vulnerable localities in a changing world economy', *Rural Sociology*, 57 (3): 305–332.

Giddens, Anthony (1990) *The Consequences of Modernity.* Cambridge: Polity Press.

Goe, W.R. (2002) 'Factors associated with the development of nonmetropolitan growth nodes in producer services industries, 1980–1990', *Rural Sociology*, 67 (3): 416–441.

Grimes, S. (2000) 'Rural areas in the information society: diminishing distance or increasing learning capacity?', *Journal of Rural Studies*, 16 (1): 13–21.

Halfacree, K.H. (1994) 'The importance of "the rural" in the constitution of counterurbanization: evidence from England in the 1980s', *Sociologia Ruralis*, 34 (2–3): 165–189.

Harrison, L.E. and Huntington, S.P. (eds) (2000) *Culture Matters: How Values Shape Human Progress.* New York: Basic Books.

Henslin, James M. (2002) *Essentials of Sociology: A Down-to-Earth Approach*, 4th edn. Boston, MA: Allyn and Bacon.

Hirschman, A.O. (1977) 'A generalized linkage approach to development, with special reference to staples', *Economic Development and Cultural Change*, 25, Supplement.

Hoggart, K. and Paniagua, A. (2001) 'The restructuring of rural Spain?', *Journal of Rural Studies*, 17 (1): 63–80.

Jacob, S. and Luloff, A.E. (1995) 'Exploring the meaning of rural through cognitive maps', *Rural Sociology*, 60 (2): 260–273.

Jansen, A.J. (1991) 'The future of the periphery of the periphery', *Sociologia Ruralis*, 31 (2–3): 122–139.

Jóhannesson, G., Skaptadóttir, V.D. and Benediktsson, K. (2003) 'Coping with social capital? The cultural economy of tourism in the North', *Sociologia Ruralis*, 43 (1) 3–16.

Johnson, Kenneth M. and Beale, Calvin L. (1999) 'The continuing population rebound in nonmetro America', *Rural Development Perspectives*, 13 (3): 2–10.

Kolb, J.H. and Brunner, Edmund de S. (1946) *A Study of Rural Society*. Boston, MA: Houghton Mifflin.

Krugmann, P.R. (1995) *Development, Geography, and Economic Theory*. Cambridge, MA: MIT Press.

Leonardi, R. (1995) 'Regional development in Italy: social capital and the Mezzogiorno', *Oxford Review of Economic Policy*, 11 (2): 165–179.

Levine, R. and Renelt, D. (1992) 'A sensitivity analysis of cross-country growth regressions', *American Economic Review*, 82 (4): 942–963.

Lindsay, S. (2000) 'Culture, mental models, and national prosperity', in L.E. Harrison and S.P. Huntington (eds), *Culture Matters: How Values Shape Human Progress*. New York: Basic Books.

Littlewood, P. (1981) 'Patrons or bigshots? Paternalism, patronage and clientelist welfare in Southern Italy', *Sociologia Ruralis*, 21 (1): 1–18.

Luloff, A.E. and Swanson, L.E. (1990) *American Rural Communities*. Boulder, CO: Westview Press.

Marsden, Terry (2003) *The Condition of Rural Sustainability*. Assen, Netherlands: Royal Van Gorcum.

Mencken, F.C. (2000) 'Federal spending and economic growth in Appalachian counties', *Rural Sociology*, 65 (1): 126–147.

Mooney, Patrick H. (2004) 'Democratizing rural economy: institutional friction, sustainable struggle, and the cooperative movement', *Rural Sociology*, 69 (1): 76–98.

Mooney, Patrick H., Roahrig, Jerry and W. Thomas Gray (1996) 'The de/repoliticization of cooperation and the discourse of conversion', *Rural Sociology*, 61 (4): 559–576.

OECD (2003) Introduction to *The Future of Rural Policy*. Paris: OECD Publications. pp. 11–30.

Palen, V. John (1997) *The Urban World*, 5th edn. New York: McGraw-Hill.

Piore, M.J. and Sabel, C. (1984) *The Second Industrial Divide: Possibilities for Prosperity*. New York: Basic Books.

Putnam, R. (1993) *Making Democracy Work: Civic Traditions in Modern Italy*. Princeton, NJ: Princeton University Press.

Serow, W.J. (1991) 'Recent trends and future prospects for urban–rural migration in Europe', *Sociologia Ruralis*, 31 (4): 269–280.

Sharp, Jeff S. (2001) 'Locating the community field: a study of interorganizational network structure and capacity for community action', *Rural Sociology*, 66 (3): 403–424.

Sharp, J.S., Agnitsch, K., Ryan, V. and Flora, J. (2002) 'Social infrastructure and community economic development strategies: the case of self-development and industrial recruitment in rural Iowa', *Journal of Rural Studies*, 18 (4): 405–417.

Simmel, Georg (1978) 'The metropolis and mental life', from *The Sociology of Georg Simmel*. New York: The Free Press.

Singelmann, J. (1996) 'Will rural areas still matter in the 21st century? or can rural sociology remain relevant?', *Rural Sociology*, 61 (1): 143–158.

Sorokin, P.A., Zimmerman, C.C., Galpin. C.J. (1930) *A Systematic Sourcebook in Rural Sociology*. Minneapolis, MN: University of Minnesota Press.

Spencer, D. (1997) 'Counterurbanization and rural depopulation revisited: landowners, planners and the rural development process', *Journal of Rural Studies*, 13 (1): 75–92.

Stark, David (1992) 'Path dependence and privatization strategies in East Central Europe', *East European Politics and Society*, 6: 17–54.

Stockdale, A., Findlay, A. and Short, D. (2000) 'The repopulation of rural Scotland: opportunity and threat', *Journal of Rural Studies*, 16 (2): 243–257.

Terluin, I.J. (2003) 'Differences in economic development in rural regions of advanced countries: an overview and critical analysis of theories', *Journal of Rural Studies*, 19 (3): 327–344.

Thomson, M.L. and Mitchell, C.J.A. (1998) 'Residents of the urban field: a study of Wilmot Township, Ontario, Canada', *Journal of Rural Studies*, 14 (2): 185–201.

Triandis, Harry C. (1996) 'The psychological measurement of cultural syndromes', *American Psychologist*, 51: 407–415.

USDA (2004) 'Measuring rurality: what is rural?' United States Department of Agriculture Economic Review Service Website: http://www.ers.USDA.gov/Briefing/Rurality/WhatisRural

Weber, Adna F. (1899) *The Growth of Cities in the Nineteenth Century: A Study in Statistics*. New York: Macmillan.

Weber, Max (1978) *Economy and Society*, edited by Guenther Roth and Claus Wittich. Berkeley, CA: University of California Press.

Zeckeri, A.A., Wilkinson, K.P. and Humphrey, C.R. (1994) 'Past activeness, solidarity, and local development efforts', *Rural Sociology*, 59 (2): 216–235.

7

Rural policy and planning

Mark B. Lapping

INTRODUCTION: ROOTS AND BRANCHES

Because rural planning and policy can cover an array of issues and problems, such as housing, poverty, health care delivery, transportation, gender, land use, conservation and much else – indeed it is such a broad term as to almost lose meaning – it is perhaps best to recognize that the very notions of 'planning' and 'planned change' reflect the optimism, the belief in the power of empirical science, and the desire of people and states to shape both nature and society which accompanied the Enlightenment. As Isaiah Berlin pointed out throughout his long career as a student of ideas, many of these Enlightenment and 'modernist' impulses actually found their origins in utopian thinking (see, generally, Berlin, 1991).

The utopian tradition in Western thought proved to be particularly powerful for rural planning in that it so often reinforced positive representations and notions of rural life – conviviality, community, coherence, simplicity and organic integrity – at the very time when 'modern' was coming to be defined by city life and cosmopolitan ideas and lifestyles. The utopian impulse had a profound impact upon the rise of planning and community organization theory, as Thomas Reiner, amongst others, has pointed out (Reiner, 1963). In North America utopianism was further reinforced by a religious pluralism which encouraged rural communitarianism and experimentation among the Shakers, the Hutterites, the Amana Colonists, the Amish and other Mennonites, and similar groups which all saw in agrarianism both a positive statement framed by faith and a pattern of sustainable living (Lapping, 1997, 2001). In this way utopianism was for some a response to the excesses of industrialization, urbanization, secularism and alienation from tradition, work and family, which modernism seemed to represent.

In the United Kingdom, as elsewhere in Europe where 'national romanticism' was a major artistic and intellectual force, some of these ideas found further amplification in the 'Arts and Crafts' movement of William Morris and others, which had the village and the tradition of premodern craftsmanship at the centre of their critique of the modern world and their conception of the good society (Bunce, 1994; Heathorn, 2000). Rural simplification also developed deep roots in North America as well (see Schmitt, 1969; Shi, 1985). In understanding things in this way some early planners came to see the rural village, bucolic simplicity and organicism as essential counterpoints to the social chaos, alienation, class warfare and rootlessness so many came to see in the all-powerful forces of industrial capitalism and urbanization (Lears, 1981).

The first generation of planners who held an avowed interest in rural people and places included Raymond Unwin and Ebenezer Howard. They led the English 'Garden City' movement which sought to limit British urbanization by connecting cities – whose growth would be confined within specified physical limits through the creation of 'green belts' and other tools – with new smaller settlements nested in productive rural hinterlands and connected through rapid transportation. Howard argued that through controlling the use of land and its ownership the culture and the amenities of urban places could be made accessible to rural people, thereby improving the life of those who lived in the countryside. And the lives of those in cities would also be enhanced through the imposition of limits on urban expansion and by providing greater proximity to green space and natural landscapes. The noted planner Peter Hall aptly summed up what Ebenezer Howard attempted to do in his work and writing, most importantly in *Garden Cities for To-morrow* (1902), when he wrote that 'Howard was saying here that both existing cities and the existing countryside had an indissoluble mixture of advantages and disadvantages' (Hall, 1992: 36). Some, including the great urbanist Jane Jacobs, have come to see these ideas as fundamentally 'anti-urban' (Jacobs, 1961), while Stephen Heathorn has argued that Howard was the leader of a loose intellectual movement whose sole 'binding element' was 'a shared idealization of English rusticity' (Heathorn, 2000: 120). Nevertheless, the 'Garden City' movement genuinely reflected an expression of the need for limits to and boundaries on the very social forces and technological processes – largely uncontested and rarely debated in the larger society – which were generating irreversible changes in people's lives and the ordering of social and economic life (Richert and Lapping, 1998).

In North America these and related ideas found resonance among decentralists, regionalist thinkers and, most importantly, a small group of planners known as the Regional Planning Association of America (RPAA). Composed of some of America's truly most seminal planning intellectuals, Clarence Stein, Lewis Mumford, Catherine Bauer, Henry Wright, and Benton MacKaye, the RPAA implemented its own conception of the 'Garden City' ideal in the establishment of the new community of Radburn, New Jersey (Lapping, 1977) and through the late 1920s and 1930s continued to generate important plans and proposals, such as the introduction of comprehensive regional planning for the entirety of New York State, the elaboration of extensive systems of rural highways, rural electrification, new towns developed along the lines of the original 'Garden City' ideal, and regional natural resources conservation.

These ideas struck a responsive chord with other architects and planners, academics, social critics and 'progressives' across the country. This was especially true among a group of thinkers centred around the person of Howard Odum, a sociologist at the University of North Carolina, who, during the 1930s, focused on the problem of regional backwardness and peripheral regions, most especially the American South. Odum and his colleagues, labelled the 'Regionalists', sought to understand the underpinnings of rural backwardness, dependency and poverty. As Friedmann and Weaver have rightly noted of them, 'they wanted to fend off the attack of northern industrial interests and metropolitan culture on southern rural values; they wanted to alleviate agrarian poverty and racism' (Friedmann and Weaver, 1980: 35; see, more generally, Dorman, 1993). Odum and the reformers advocated land reforms to break the hold of tenancy and to favour small and family farmers, both black and white, a transformation in southern agriculture away from export and cash cropping systems, like 'King Cotton' and peanuts, toward a more diversified food production model, the introduction of critically important physical infrastructure into rural areas, the provision of new rural credit systems and other financial institutions, the conservation of marginal lands and forests, and the establishment of a broad system of cooperatives, amongst other programmes and policies (Singal, 1982). All of this was to be done through a national commitment to social planning implemented on a regional basis and locally controlled industrialization (O'Connor, 1992; Odum, 1934).

Together these 'roots and branches' form what Hall has labelled the 'Anglo-American tradition' in planning (Hall, 1992: 31). Though often neglected by the overwhelming urban orientation and bias among planners and policy-makers, many of the core and most seminal ideas about planning really sought to speak to the issues and problems facing rural people and rural places.

THE MODERNIST POSE

If there were those who advocated rural policies that would substantially alter the nature of society and contest the urban–industrial model of development – which invariably required the continued underdevelopment of rural areas and places – there were others who struck a different ideological pose. Accommodation rather than resistance to the unfolding urban–industrial hegemony was the strategy which the Country Life movement adopted (see, generally, Bowers, 1974; Danbom, 1979). The movement was most active through the United States and Canada, although its impulses were extended to much of Europe and Australia through the promulgation of similar 'modernization' efforts (on the Country Life movement in Canada, see Jones, 1982).

Country Lifers understood the impact of emerging technologies and aimed to secure for rural people an appropriate niche in the unfolding urban–industrial order, through the full adoption of technology and modern business principles to farming as well as the reform of many rural institutions. The leader of the movement, and head of the US Country Life Commission, was Liberty Hyde Bailey, Dean of the College of Agriculture at Cornell University, arguably then America's foremost agricultural educator.

Bailey reflected the rising pre-eminence of the land-grant university model of public higher education in the United States. Created under the terms of the 1862 Morrill Act, nearly every state in the country accepted a federal authorization which permitted states to sell off federal lands with the receipts directed to the establishment of state colleges where agriculture, the mechanical arts, as engineering was then known, and home economics were to be taught. These institutions came to form the most significant 'human capital' engine in existence in rural America, combining teaching with applied research, through federally supported agricultural and forestry research stations, and direct assistance to producers in the field, via the jointly federally and locally sponsored Cooperative Extension Service, which maintained 'agents' in many rural counties throughout the nation. Cooperative Extension became the backbone of a unique system of technology transfer which, at every step, assisted farmers and country people to adopt new technologies to production, to bring scientific discoveries and innovations into the countryside, and to shape a new and more modern rural sector. Extension agents, along with management experts in the banking system, were, as Deborah Fitzgerald has called them, the 'shock troops of rural transformation' (Fitzgerald, 2001: 194). Together with agricultural economists in the newly organized Bureau of Agricultural Economics in the USDA, farm management experts were busy advocating a fundamentally different model for American agriculture, one which would see farmers 'industrialize their farms' (Fitzgerald, 2001: 100; see, too, Busch and Lacy, 1983; Taylor and Taylor, 1952). As one prominent Country Lifer put it, 'the small farm of today is similar in its organization to the shop of yesterday, and must surely give way' (Holmes, 1912: 523).

While the Country Life commissions continued to study, research and advocate for change in the rural sector, it was the First World War which made its agenda national policy throughout North America and much of Europe. As David Danbom has noted:

> the war, then, was the pivotal occurrence in the industrialization of agriculture. It unnaturally speeded the attainment of agricultural efficiency to the point where depression resulted, which in turn decreed that the gain of the war would not be lost. Not only did the depression (1919–1920) assure the cities the cheap food that had been their primary goal all along, it also stimulated the sharp competition in agriculture which assured that it would become still more productive. And finally, the war accelerated those trends which were breaking down rural social institutions leading to increased realization of the Country Lifers' social goals. (Danbom, 1979: 104)

THE POWER OF POWER

The place where these two seemingly oppositional threads of thinking came together lies in the experience of numerous national governments in utilizing mega-projects to modernize their rural sectors. And the medium for doing this, for transforming, modernizing and radically altering the entire nature of rural living was electrification. By the 1920s more than two-thirds of the farmers in Germany, France, the Netherlands and Scandinavia had electricity while 90 per cent of American farmers did not.

American farmers and others would have to await the New Deal during the Great Depression before the Rural Electrification Administration (REA) would create a system of cooperatives which would bring power to rural Americans. Under the dynamic leadership of Morris Llewellyn Cooke, the REA made low-cost loans available to private utilities to bring power to farms, rural homes and local businesses. With electricity, farmers increased their productivity, the healthfulness of foodstuffs was enhanced through refrigeration and better sanitation, and the quality of life in rural America more closely approximated the standards of cities. But even before the creation of the REA, the Tennessee Valley Authority (TVA) brought power to an important part of rural America, the watershed of the Tennessee River which encompassed parts of seven states of the South. The TVA was to become the model for rural development mega-projects throughout the world.

Ostensibly a flood prevention project, the TVA built a number of major dam projects on the river which developed its extensive hydroelectric capacity. The TVA used its control over electric rate structures to both modernize agriculture in the region – one of America's poorest rural areas – and also to industrialize the Tennessee Valley.

Textiles, chemical plants, pulp and paper mills and aluminium production all established themselves in the area (see, as examples, Chandler, 1984; Creese, 1990; Hargrove and Conkin, 1983; Martin, 1967; Selznick, 1949). The TVA represents a milestone in the evolution of rural policy in a number of significant ways. The TVA became the model for rural regional development throughout much of the world by encouraging a concentration on the development of an entire river basin and watershed and by fostering a large-scale or a mega-project approach which invariably required dams which, most importantly, produced electricity. Electricity became the currency of the TVA and all subsequent river basin development projects. It was heavily capital-intensive, centralized and promoted rural industrialization, the growth of settlements and the modernization and standardization of agriculture. The TVA development model used electricity and a relatively cheap rate structure to attract and to relocate development into a rural region from areas outside of it. In this way it established something of a precedent for the use of growth incentive policies which would redirect development from one place to another. By permitting economic opportunity to essentially 'trickle down' from these newly emerging metropolitan areas to rural places, rural people would be helped, as measured by the growth in personal income.

The TVA experience substituted the region for the local as the scale of analysis for rural planning and policy development. Hereafter rural development came to be expressed very largely as regional policy. To a great extent, then, the TVA model of rural development, which would also become the backbone of the plans for the Mekong River basin, the Aswan High Dam on the Egyptian Nile, the James Bay project in Canada and the Three Gorges Project in China, was fundamentally about altering rural areas, stimulating industrialization and urbanization, and changing the focus of planning and policy from the local to the regional scale of description and analysis.

The linkage between rural modernization and electrification reached its zenith in the Soviet Union (see, generally, Bailes, 1978). Prior to the Bolshevik Revolution, V.I. Lenin penned the blueprint for the socialist reconstruction of the countryside, *The Agrarian Question* (1976), in 1907. Lenin saw an inevitable shift from a peasant-based agriculture – based on family production units, low-intensity technology and what he saw as 'backward' habits and processes – to a highly mechanized, large-scale production system based on communes as essential to the socialist development of Russia. Central to this vision was the electrification of all of Russia. As James C. Scott notes, Lenin was famous for asserting that 'Communism is Soviet Power plus the Electrification of the whole countryside' (Scott, 1998: 166). For the Soviet Union electricity became the metaphor for centralization, planning, mechanization, the destruction of the Kulak class and the modernization of the entire rural sector. Despite massive investments and numerous attempts to bring electricity to the countryside, 'only one in twenty-five collective farms had electricity by the eve of World War II' (Scott, 1998: 203). Though the Soviet Union aggressively dammed rivers, built power station complexes, felled whole forests and attempted to exploit the country's hydroelectric potential to the fullest – often using forced labour from prisoners in the Gulag system – counterproductive effects could often be observed (Applebaum, 2003).

Electrification, through river basin planning and development, was one of the most significant paradigms and models for rural development. The provision of electricity had become a prominent rural policy throughout the world and in all types of national economic systems. The production and wide-scale distribution of electricity was seen by planners, economists, politicians, bankers and agricultural scientists as holding the potential to revolutionize rural areas and the lives of rural people. It was not the first time – and surely it would not be the last – when 'technological fix' would be advocated to address rural problems.

DEPRESSION AND THE RUN-UP TO WORLD WAR: A POLICY WATERSHED

By the end of the 1940s, the centrality of federal or national governments to rural policy and planning would be complete for nearly all national political cultures. Clearly, the depth, the pervasiveness, and the longevity of the Great Depression, as well as the Second World War, hastened this development. Likewise, this was the period in which the equation that 'agriculture was rural and rural was agriculture' seemed to have been confirmed, despite the actual variety of ways rural people lived and worked. The practical result of this was that the central or federal agricultural ministries became the focal point for rural planning and policy-making and agricultural policy often became the contested landscape for rural and urban interests.

In both the United States and Canada the prosperity of the 1920s very largely bypassed rural areas as a depression gripped the agricultural sector.

At the outset of the Great Depression 44 per cent of the American population lived in rural areas and 22 per cent lived and worked on farms. Farmers suffered from low crop prices, often tied to an urban-biased cheap foods policy, surplus production and chronically low incomes. FDR's administration implemented an array of programmes with a view to increasing the prices that farmers received for their crops and livestock under the broad heading of the Agricultural Adjustment Act of 1933. In 1934, additional support was provided for southern cotton and tobacco farmers through the Bankhead and Kerr Acts respectively. By 1936 these supply-management programmes were partially withdrawn because of a Supreme Court decision which found them to be unconstitutional (U.S. v Butler, 297 U.S. 1, 68, 1936). A substantially different constitutional regime in Canada permitted such supply-management approaches to develop and these have come to typify much of Canada's contemporary agricultural incomes policy.

In 1938 an even more sophisticated programme was developed in the United States in the form of the Agricultural Adjustment Act, which established a complicated system of market quotas and acreage allotments for specific commodities (Cochrane, 1979; Tweeten, 1979). Although farm prices continued to fall, these programmes managed to prevent further drastic declines in agricultural incomes and achieved some degree of stability. The Soil Conservation and Domestic Allotment Act of 1936 was a watershed in American rural planning policy. Its focus upon the conservation of agricultural lands, the reduction of soil and fertility loss, and the need to apply rational and scientific methods to resource use came to be mirrored in the experiences and policies of many other nations' planning systems. In the context of the United States serious federal concern over soil erosion and the need to conserve soil and land fertility preceded the New Deal by several years. While the US Department of Agriculture (USDA) published a number of advisory bulletins and brochures on soil conservation prior to 1928, it was in that year that Hugh Hammond Bennett's *Soil Erosion: A National Menace* was published (Bennett, 1928).

Bennett was a pivotal figure in the establishment of a joint USDA–Department of the Interior effort to address the serious erosion problems which spawned the dust bowls which gripped the Great Plains in the early 1930s. In 1935 the Congress passed the Soil Conservation Act which, among other things, declared soil degradation to be a 'national menace', and crucially the national government began to see agriculture as a 'strategic' issue. Many nations would, in subsequent years, also adopt this same perspective and would come to equate the relative health and well-being of the agricultural sector with national security. 'Soil conservation districts' were formed, which FDR encouraged states to establish through the passage of the necessary enabling legislation (Steiner, 1988). By 1941, 41 states had formed 'soil conservation districts' through which a new set of institutions were introduced into the countryside to provide individual and district-wide soil conservation plans. There can be little doubt that the establishment of soil conservation districts constituted America's first federally inspired rural planning effort. As a consequence of the programme farmers, forest land owners, soil scientists, watershed specialists and planners came together under the programme to form cooperative voluntary organizations to support planning to preserve and enhance rural lands and the communities which depended on these resources for their livelihoods.

With the subsequent passage of the Agricultural Adjustment Act a set of federal incentives was developed with the aim of shifting acreage from soil-depleting crops to soil-conserving ones, like legumes, grasses and other crops. The goal of this programme was two-fold. First, there was a genuine concern over soil erosion and the belief that too many highly erodible acres had been turned under the plough. Second, the programme was seen as a way to reduce commodity surpluses and thereby raise farm incomes. The law created county agricultural adjustment committees which continue to the present day. Combining conservation strategies – usually through land idling and retirement – with income-generating programmes became an established principle which has been carried forward in nearly every major subsequent piece of US federal agricultural legislation. In Canada similar problems also brought a federal government response in the form of the Prairie Farm Rehabilitation Act (PFRA), the focus and emphasis of which was on resource conservation through internal rationalization and farm infrastructure support. PFRA has remained a valuable assistance programme through the region, though its emphasis has shifted to meet other needs (Lapping and Fuller, 1985). As in the case of the United States, the Depression years saw the Canadian federal government become the focal point for public sector rural development activity.

In the United Kingdom the years leading up to the Second World War also saw a renewed interest in rural areas and their problems. With roots in the Victorian era nature movement, British rural planning took on a distinctive 'countryside' or 'Olde England' orientation which reflected, in large part, the long involvement of various groups and organizations, such as the Council for the Preservation of Rural England, the Society for the Promotion of Nature Reserves, the National Trust for Places of Historic Interest and Natural Beauty, the British Empire Naturalists Association, the Society for the Preservation of Birds, and many others. Beyond a strong commitment to preserving the countryside and traditional land ownership and management, access to the countryside and its many amenities was also an important part of the emerging planning regime in the UK. In 1932 the national government passed the landmark Town and Country Planning Act which is, as Paul Cloke has noted, 'the traditional starting point for rural planning' (1983: 75) in Britain. The act permitted local rural governmental authorities to develop planning tools to deal with actual or potential development pressures. Because landowners were given something approaching a 'veto' over any regulations that might be promulgated, the emphasis of the evolving planning system very largely ignored compulsory plan implementation and focused, instead, on the protection of amenities, such as areas of natural beauty, and historic buildings and sites. The effect on rural areas undergoing change was minimal.

On the eve of the Second World War, certain themes in rural planning and policy-making became clear. First, the Depression forced many national and federal governments to review and then to intervene in the agricultural sector. A fear of chronic rural depopulation was often tied to agricultural income policy initiatives. French agriculture, for example, which engaged fully half of the nation's population at the turn of the twentieth century, witnessed a precipitous decline to only a third of all French workers by the end of the Second World War (Hervieu, 1994). Second, agriculture became coterminous with rural, somewhat to the exclusion of other dimensions of rural life. Third, agricultural resource utilization was often the specific focus of public policy. Planning, to the extent that it was institutionalized, tended to address land and resource issues rather than rural socio-economic problems and concerns. Governments were deeply concerned with agricultural and fibre production shortages and vulnerabilities. This was exacerbated during the Second World War. Yet the rudiments of rural planning and policy could be discerned in the years leading up to the war. While in North America little was achieved during the war years *per se* – serious concerns over the fate of rural areas would manifest themselves only in the post-war era – in the United Kingdom a series of important reports were published during the war years – the Barlow Report on the Distribution of the Industrial Population (1940), the Uthwatt Report on Compensation and Betterment (1942) and the Scott Report on Land Utilization in Rural Areas (1942) – which would come to shape the structure, direction and nature of planning in the UK in general and rural areas more specifically (Whitby and Adger, 1993: 71).

POST-WAR RURAL POLICY

The years directly following the Second World War saw several rather profound changes in the way people and governments came to see and understand rural areas and, most especially, agriculture. Though writing essentially about Finland and Scandinavia, Granberg and Peltonen have defined a number of far more pervasive and universal 'sea-changes' in attitudes and policies. Among these was the realization that free markets and *laissez-faire* policy approaches too often failed to guarantee adequate food and fibre production. Second, that the political culture of much of Europe had to be reformed with a focus upon democratizing the countryside given that so many rural people came to support the fascist movement in the pre-war and war years. Perhaps in no place was this more necessary than in Germany, as Baranowski (1995) has so deftly shown. Third, war-time planning and intervention in the marketplace, which had become commonplace, carried over to many sectors following the war, including agriculture. Fourth, food self-sufficiency and income parity became joined as policy imperatives throughout the developed world. These two goals would come to drive agricultural and rural policy for decades to come. In the United Kingdom, as an example, the situation has been aptly characterized in this way:

> Since the Second World War and the shift to an agricultural policy that has emphasized the protection of domestic production and the quest for self-sufficiency, the free market and laissez-faire approach that prevailed previously has often been portrayed as reckless and misconceived. Not only has it been condemned as the cause of extensive rural decline and impoverishment, but also for undermining the country's food security and thereby rendering the UK dangerously exposed to food blockade. (Marsden et al., 1993: 48)

In the United Kingdom four key agricultural production strategies were employed, as embodied in the 1947 Agricultural Act, and stood well into the 1970s when membership in the European Economic Community brought yet another shift in agricultural policy. These included subsidy programmes which guaranteed minimum prices – the deficiency payment schemes – grants for the acquisition of equipment and support for on-the-farm improvements, the establishment of producer-controlled marketing boards to support prices, an extensive system of agricultural research and education, and the imposition of an import protection regime which would secure domestic producers (see, generally, Bowler, 1979, 1986). The practical effect of the 1947 law was to push agricultural production 'at all costs', as Cloke and White (1990: 44) have observed.

Elsewhere similar policies were also put into effect. 'In the Nordic Countries,' for example, 'agriculture was protected from the international market by import restrictions and export subsidies. …' (Granberg and Peltonen, 2001: 298). Finland promulgated an Agricultural Income Act which utilized a very complicated set of calculations in an attempt to peg the incomes of farmers to those of urban industrial workers. This was policy in Denmark, Sweden and Norway as well (Granberg, 2000; Lapping and Forster, 2000). Farmers throughout this region were also aided by a historically rooted and robust farmer's cooperative movement. These, in turn, served as models for farmer's cooperatives elsewhere, as in Ireland, for example (Tovey, 2001).

In North America agricultural policy took similar forms. Canada developed an extensive system of supply-management programmes both to stimulate production and to provide an incomes floor for its farmers. By the 1960s the federal government passed and implemented the Agricultural Rehabilitation and Development Act (ARDA) which 'provided joint federal–provincial funding for soil and water conservation projects and land consolidation schemes designed to increase the productivity of farms' (Weaver and Gunton, 1982: 11). Nowhere was this more important than in the western provinces where grain production defined much of the region's agriculture and rural sector. The Prairie provinces had long been the focus of much of Canada's rural settlement policy. From a population of slightly more than 400,00 in 1901, the region's population grew to just under 2 million by 1921 (Census of the Prairie Provinces, as noted in Jones, 1982: 96). To aid grain producers in the Prairies the federal government supported the development of a transcontinental railroad system with subsidized freight rates to help farmers. This subsidy allowed farmers cheap shipment costs to millers and processing facilities in eastern Canada. To assist farmers in marketing their product throughout the world, the federal government established the Canadian Wheat Board 'to ensure smooth transitions in prices from year to year and to assist in marketing the product' (Cummings, 1989: 61). In these ways Canadian policy aggressively incorporated state intervention in the agricultural economy.

In the United States, the situation has been accurately summed up by Mann and Dickinson when they wrote:

> [f]rom 1938 to the present the particularities of the farm acts have been modified many times to meet varying conditions such as war, differential market conditions for various crops and so on. In general, however, the current government programs incorporated many of the fundamental principles embodied in the New Deal legislation. In particular, government payments are made to those farmers who participate in market quotas, comply with acreage allotments and/or participate in land diversion programs. (Mann and Dickinson, 1980: 306)

Very importantly, as they point out, American agricultural policy during these years led to a situation wherein 'increased productivity has not led to a decline in the need for state support, but on the contrary, as agriculture has become more productive, so state intervention has become more imperative, more extensive, and more expensive. Indeed 'many of these state agricultural policies accentuate and perpetuate overproduction and inequalities in the countryside and hence function to reproduce those very conditions which called forth the need for state support in the first place' (Mann and Dickinson, 1980: 316).

Across the mature capitalist and mixed economies agricultural policy as rural policy reached an almost inevitable end or what Vail et al. have called the 'handwriting on the wall'. 'By the early 1980s', they write, 'the core agricultural policy instruments used for decades in most advanced capitalist nations had created an economically untenable situation. The triple bind of chronic excess production, escalating public farm expenditures, and depressed international food prices was widely attributed to policy measures that inflated domestic farm prices, restricted imports, subsidized exports, and encouraged ceaseless technical and structural "rationalization"' (1994: 1). The 'crisis on the farm' had become a permanent part of the rural landscape.

A SOCIALIST ALTERNATIVE

The socialist revolution which swept through Russia with the conclusion of the Bolshevik Revolution and the ensuing civil war brought profound and far-reaching changes to rural Russia. The new Soviet Union, initially under Lenin and then under Stalin, sought the rapid industrialization of the country by capturing the rural surplus and transferring it to the nation's cities for investment in heavy industries. As Pallot has put it, 'the economic purpose of these transformations was to reorganize agriculture in such a way as to provide for the transfer of resources from the countryside to the towns' (Pallot, 1988: 120). To a very real extent, early Soviet policy had a strongly anti-rural orientation and attempted nothing less than the restructuring of rural life, even to the point of near-destruction.

To accomplish the re-creation of the rural regions of the Soviet Union a process of de-peasantization became 'the order of the day' through a sustained process of the forced collectivization of agriculture and rural labour. Private property was all but abolished and the rural bourgeois or middle class – the Kulak class – was physically removed or eliminated. Soviet policy, especially in the form of the National Economic Plans (NEP) and subsequent five-year plans, was little more than 'a cover for a war against the peasants – some of whom were killed, others deported, and still others locked up in the huge farms under the auspices of the Party – Kolkhozes and Sovkozes', (Furet, 1999: 144). In 1932, for example, a politically inspired and enforced famine was created in the Ukraine to destroy the peasants. Between 5 and 6 million people died in the Ukrainian operations (Furet, 1999: 145–6; see, too, Conquest, 1968).

The major policy tools implemented to bring about this de-peasantization were the mass collectivization of Soviet agriculture, the expropriation of land, the creation of massive state land reserves, the imposition of confiscatory taxes on farmers, and commodity prices well below cost. More specifically, the Soviet model of rural reconstruction relied heavily on the establishment of the *kolkhozes*, or collective farms, which were to be transitional institutions from private farming to the ultimate model of Soviet agrarian collectivization, the *sovkhoze*, or the state farm. However, by 1950 all-Union policy came to favour larger units and smaller cooperative farms were merged into bigger ones. The results of consolidation were felt immediately and these had some profound impacts upon the spatial organization of rural life. As Raun has observed of the Estonian Soviet Socialist Republic, 'from a peak of over 3,000 [*kolkhozes*] at the beginning of 1950, the number of collective farms declined rapidly to 1,137 by the end of 1951 and further to 1,018 by the end of 1952. At the later date they were divided into 934 agricultural *kolkhozes* and 84 fishing *kolkhozes*' (1991: 180). Additional collectivization finally slowed toward the end of the Stalinist era.

Collective farms operated in an environment defined by state controlled central planning mandates, edicts and production goals. Alanen, citing the work of Clarke, has written that 'the Soviet system was basically a variant of a barter economy: exchange, investments in production, etc. were the result of a complicated negotiation system rather than a straightforward command structure. Production goals were nevertheless assigned from above (the Politburo and central planning ministries) down to the bottom, where production plants like *kolkhozes* and *sovkhozes* were located' (2001a: 129).[1]

The Khrushchev era of Soviet leadership brought a number of important changes to farm life and rural policy throughout the Union. Levels of state interference were reduced, and collective farms could now purchase their own machinery, a move which brought about increases in farm productivity. Khrushchev also ordered the dismantling of some of the larger *kolkhozes*, though a number were transformed into state farms, or *sovkhozes*, which were coming to play an ever larger role in all-Union agriculture. With the relaxation in some areas of agricultural policy, small-plot private farming soon emerged and compulsory deliveries from such plots, as well as the high taxes placed upon them, were largely eliminated in 1959. What emerged in the *kolkhozes* was a symbiotic relationship between large-scale farming with small-scale plot agriculture. According to Alanen:

> this symbiotic relationship, which was complementary and economically beneficial to both parties, had originally developed spontaneously during the post-Stalin era. It arose from the inability of the Soviet government and Soviet farms to control the co-operation between private plot farmers and tractor operators, but later *kolkhoz* leaders, and Soviet leaders more widely, came to realise its obvious advantages, and developed the system further with their own goals in mind. (Alanen, 2001: 133; 2001)

The obvious result of all of this, beyond an underperforming agricultural sector, was that 'rural living standards suffered accordingly, agricultural workers became second-class citizens denied the legal rights of town dwellers, and they were poorly remunerated for their work in the collectives' (Pallot, 1988: 120). A more subtle though no less transforming effect was the realignment of territory and rural space throughout the countryside in the USSR. The *kolkhoz*, or the collective farm, was rarely a farm. Rather it was a grouping of farms and villages linked together under a centralized administrative structure. This permitted authorities to begin the consolidation of living space and service units as they sought to achieve certain economies-of-scale in the provision of physical infrastructure and other improvements. Places termed 'non-viable areas' were simply slated to disappear. Adopting something akin to the notion of the 'key settlements' approach (see, generally, Cloke, 1983), which would also find favour in British and American planning circles, Soviet planners sought to consolidate rural institutions, concentrate resources and investments only in community centres that would anchor a number of interconnected settlements, and permit a more 'rational' approach to land utilization and community lay-out. But as Pallot has pointed out, in many regions of the Soviet Union the 'lack of funds meant that village concentration never progressed beyond the planning stage ...' (Pallot, 1988: 126–7). The practical effect and impact of consolidation and the overall poor prospects of those who lived and worked in rural areas was that people, especially the young, left the Soviet countryside in droves. Nove indicates that 'a lack of amenities had led to the out-migration of younger and more skilled peasants' (Nove, 1990: 264). This in turn led to an endemic shortage of agriculture labour which required that soldiers and 'voluntary' brigades of students leave the towns for the countryside when harvests had to be brought in.

Some socialist countries, most especially Poland, Hungary and Bulgaria, experimented with alternatives to the collectivist model of agricultural production. China, however, also sought alternatives to the Soviet model which it adopted after the founding of the People's Republic in 1949. With the 'Great Leap Forward' programme in the late 1950s and early 1960s, the commune system was fully established, private property was all but abolished, agricultural prices were kept artificially low through very tight 'command and control' mechanisms and plans, and grain self-sufficiency was pushed at nearly all costs. The 'cultural revolution' spread havoc across the countryside with the majority of China's farmers operating at a semi-subsistence level. By 1979 and the restoration of some degree of order drastic and significant changes occurred throughout the Chinese rural sector. Decollectivization became the norm with a return to family farming, agricultural prices were allowed to rise, internal trade restrictions on farm produce were relaxed and productivity increased markedly in grain, livestock and horticultural products. Rural incomes also increased, though not nearly as rapidly as incomes in urban and industrial areas (see, generally, Putterman, 1993). Productivity on the farm continued to rise, largely as a consequence of the adoption at the farm-level of new technologies, all the while migration from the countryside to the cities was also increasing. This led to substantial regional disparities between China's urban centres – largely in the coastal provinces and the 'special development zones' of southern and eastern China – and the huge rural hinterlands of interior China. These divisions appear to be growing, and closing the income gap and narrowing the regional disparities continue to be major objectives of Chinese domestic policy (Liu, 2001; Tian, 1999).[1]

It is perhaps one of the great ironies of the Soviet model of the socialist reconstruction of the countryside that it actually adopted as its model, its *sine qua non*, the American system of mass production, the industrialization of agriculture.

THE RURAL PLANNING SYSTEM IN THE MODERN STATE

The years after the Second World War ushered in the 'Cold War' in its many manifestations. As previously noted, agricultural policy and much of what passed for rural policy were often coterminous. In the West – democratic Europe, Australia, New Zealand, Japan and North America – rural planning was often driven by programmes to foster food production to meet domestic needs. Often cloaked in the rhetoric of self-sufficiency and a renewed enthusiasm for the 'rural idyll' after the horror and chaos of world war and genocide, domestic food production at all costs became the hallmark of much of the thinking about rural areas (on the rural idyll, see, generally, Bunce, 1994). Along with this came a strong impulse toward protectionist policies to buttress domestic agriculture. This policy paradigm, what Marsden and his colleagues have called the 'productivist regime', defined agricultural policy specifically and rural policy more generally (Marsden et al., 1993). In much of Europe, this policy objective, which had implications for land-use policy, has remained largely intact even with the accession of nations into the Common Market/European Economic Community and the development of the Common Agricultural Policy (CAP).

Simultaneous with this thrust in agricultural policy was the emergence of various planning regimes which brought greater focus to economic growth, land use and natural resources/ environmental policy. In part a result of war-time mobilization and public sector spending to induce rapid industrialization, this new burst of activity was also derivative of a growing recognition across market-based societies that 'modern industrial society requires public intervention to achieve national goals; assumes that such intervention must touch all fundamental social developments; must be goal-oriented, and effectively coordinated at the center; must be anticipatory rather than characterized by ad hoc solutions and timing dictated by crisis' (Graham, 1976: xii–xiii). In the 1950s, France, for example, established its General Planning Commission (Commissarariat General au Plan) which issued a number of four-year national development plans. The Netherlands created a country-wide econometric model to guide growth and development, while the first social democratic government of newly democratized Japan issued a highly detailed set of national economic projections to guide reconstruction and industrial and spatial development. Even on the international level several new institutions, including the International Monetary Fund and the World Bank, were founded to provide some degree of planned guidance in the realms of trade and fiscal policy. The rural sectors in both developed and developing economies were directly and indirectly effected by this new emphasis on public intervention in markets. It is also worth noting that these periods of interventionist and place-based rural programming have often been followed by a re-emergence of non-interventionist policies, which championed the marketplace and the efficacy of labour and capital mobility. Reviews of national rural planning systems in the United Kingdom and the United States will demonstrate these points.

The full range and nature of the rural planning regime in the UK has been well described and assessed by Gilg (1978). He, amongst others, places particular emphasis upon the importance of the 1947 Town and Countryside Planning Act as the defining statutory authority for the British rural planning system. Cloke concurs with this evaluation and calls the 1947 Act the 'midwife to the organization of radical post-war planning in Britain' (1989: 24).' In essence the act created a system of restrictive controls which sought to achieve national objectives and goals within the framework of a nationally guided but locally operated and enforced system of land-use regulation.

The 1947 Act was subsequently amended a number of times and when one also considers the Local Government Acts of 1972 and 1974 – which altered the nature of local governance throughout the country – the UK had what Gilg has called 'the most comprehensive piece of land-use legislation in the world'. This regime 'gave land-use planners one major weapon – the power through regulation control to permit, impose conditions, or refuse applications for countryside developments, except most of those involved with agriculture and forestry' (1978: 116). Along with the 1947 Act, the UK also passed into law the Distribution of Industry Act of 1945, the New Towns Planning Act of 1946, the National Parks and Access to the Countryside Act of 1949 and the Town Development Act of 1952. The Parks law additionally led to the establishment in 1968 of the Countryside Commission, the successor to the National Parks Commission, which had been an advocate for the national parks system, areas of outstanding natural beauty (AONBs), rural footpaths and a particular conceptualization of the rural UK which had emphasized the leisurely, scenic and recreational values of the countryside. It has also been the source of considerable support for the Countryside Acts of the mid-1960s. Taken together this 'remarkable burst of legislative activity', as Peter Hall has called this period, established much of the contemporary planning regime in the UK (1992: 71).

As the century wore on, however, the needs of rural areas were downplayed as the 'crisis of the inner cities' came to dominate discussion and policy-making. Perhaps the death-knell for regional and rural development policy came with the election of Margaret Thatcher as Prime Minister. Higgins and Savoie report that one of her key ministers quipped that a major achievement of the Thatcher years was 'to bring regional development in Britain to an end and to stop all the nonsense' (Sir Alan Walters, quoted in Higgins and Savoie, 1977: 265). While some rural areas benefited throughout these years, it was the nation's urban areas which received the greatest attention in addition to the larger macro-economic problems confronting the UK.

Perhaps the most far-sighted aspect of the planning system in Britain was the collectivization or nationalization of development rights. This codified the notion that community rights superseded those of the individual land owner.

Additionally, a 'betterment' levy was implemented. Any increase in land values as a result of development was defined as 'the betterment value' and was now subject to a tax. The rationale for this was clear. As Cullingworth has described it, 'all betterment was created by the community, and it was unreal and undesirable (as well as virtually impossible) to distinguish between values created, for example, by particular planning schemes, and those due to other factors such as the general activities of the community or the general level of prosperity' (1985: 173).

In subsequent years a number of changes and amendments to the planning system were made. Indeed, the national planning systems established under the 1947 Act was becoming somewhat fragmented when, in 1974, Scotland assumed responsibility for planning in its territory and Wales achieved similar status and responsibility in 1996 (Alden, 2001). But for the UK, as with much of Western and Southern Europe, the CAP continued to be an important structural reality, largely because it persisted in stimulating agricultural over-production and high surplus food storage costs. The government attempted, in the mid-1980s, to take steps to reduce surpluses by implementing a package of programmes known as ALURE, which stood for Alternative Land Use and Rural Enterprise policies. This allowed for the delineation of environmentally sensitive areas with compensation to the farmer, a wood lots programme to take agricultural land out of production through increased afforestation, and by making it easier to convert farmland into urban uses by easing planning requirements and restraints (MAFF, 1987). Of these programmes it was the designation of environmentally sensitive areas and their removal from agricultural production uses which proved to be the most popular. Whitby and Adger note of it that '[t]he uptake of the agreement offered was remarkably rapid, reflecting some combination of the farmers' enthusiasm for the objectives of the policy, the level of payments offered to those complying, ease of compliance with scheme prescriptions and the individual farmer's view of future levels of price and income' (1993: 69).

In 1985 the government made another change in that it permitted the agricultural ministry to use its budget to support environmental projects not directly linked to production. This was a harbinger of the growing influence and power of emerging interest groups, some of whom were settling in rural areas as part of a modest national trend toward counter-urbanization, along with the well-established rural Tory elite (Cloke, 1985). They tended to see the English countryside not solely as the locus of food and forest products production but also as the generator of prized environmental and leisure amenities. On balance, however, little really changed and the thrust of state policy remained focused on the provision and enhancement of the necessary infrastructure to support agricultural development (Cloke and White, 1990: 57). Writing in 2003, Terry Marsden concluded that

> despite over fifteen years of debate and policy crisis concerning the 'arthritic' nature of productionist support mechanisms within the Common Agricultural Policy (CAP), and the need to shift the emphasis towards a new social and environmental agenda, it is still the case that the main pillar of the CAP remains in this area in terms of funding. Moreover, it still tends to reinforce the logic of agricultural productivist scale economies by rewarding the largest volume producers, as well as 'locking in' many of the less productive producers and those least able to meet the demands such policy-designed 'technological treadmills' require. (2003: 19)

Robinson has concluded that the British experience with rural planning has been 'piecemeal' and 'lacking focus' (1990: 402). Yet Cloke maintains that by adopting a key settlements policy, which concentrated resources into growth centres, Britain generally avoided rural sprawl while successfully providing many of the necessary services for those living in villages and across the countryside (1988: 93). He further notes that rural planning in the UK has been defined by two seemingly contradictory impulses. As he writes: 'almost regardless of which government has been in power, land-use planning has been used to strengthen the economic and political power of development interests'. Yet at the very same time he finds that the planning system also has an 'enduring commitment to long-term conservation' (Cloke, 1989: 43). This seeming paradox or tension perhaps was most apparent in the ongoing battles over development within the country's green belts and came into its clearest focus during the 'appeals-led period' of the Thatcher years when developers made a number of successful appeals to the various Secretaries of State, who have overall operational authority over the British planning system, and who routinely overturned the judgements of local authorities on land use matters to support development at odds with local policies (Tewdwr-Jones, 1996). These tensions are but part of the larger differentiation in the ways that various interest groups, actors and governments understand and conceive of the UK's rural landscape.

'The essential features of the English planning system are … in many ways little changed from those introduced over 50 years ago with the passage of the 1947 Act', as Baker (2000: 163) has observed. And while the Secretary of State for Environment, Transport and the Regions has issued a number of planning policy guidelines (PPGs) and regional planning guidelines (RPGs) over the years, the spatial focus of powers and responsibilities for the rural areas of the UK remain essentially focused at the national and local levels through structural plans at the county level and discretionary plans at the local level.

In the United States the rural landscape was permanently modified by at least one significant Cold War policy imperative. During the Eisenhower administration of the 1950s massive federal support went into the development of a country-wide interstate highway system – known as the National Defense Highway System – which was seen as a means both to decentralize defence production and installations and to provide for the rapid movement of armaments, goods and people across the entire nation. As America's railroad system began to be replaced by trucking as the primary means for the transport of goods, the interstate highway system continued to grow and criss-crossed the country. Fortunate was the rural community with a highway interchange on the interstate system. Many towns and hinterland areas bypassed by the new transportation network simply went into decline. Those places attached to or located near the road network often experienced the reorientation of local land use patterns away from their central business districts or downtowns to the highway interchanges where strip development and sprawl soon came to define the landscape. In subsequent years the rise of 'big box' retailers, epitomized by Wal-Mart, sealed the fate of many downtowns and service centres. This may well have been one of the proverbial 'unintended consequences of policy'. With hindsight it now seems astounding that policy-makers seemed almost unaware that public infrastructure investments, like the interstate highway system, would not have such profound land use and community development implications. The American land-use planning system, such as it was, proved largely ineffective in controlling or containing this transformation (Lapping, 1992: 231).

In time the interstate highway system sought to bypass busy metropolitan areas and a system of circumferential roads was built around them, most at the same scale as the interstates themselves. This brought many hitherto rural places within the 'commuting shed' of major cities and fed counter-urbanization pressures. This led to making communities and land within the rural–urban fringe among the most contested in America (see Furuseth and Lapping, 1999). What Murdoch and his colleagues wrote about the UK is likewise an accurate characterization of the nature of contestation in the United States: these areas 'lie outside the main commuter zones. Here local agricultural, commercial and development interests may be politically dominant and these interests will tend to favour development for local needs. However, these traditional development interests are increasingly opposed by "incomers" who may be middle-class workers or retirees attracted to the area because of its residential environment. Thus the development process is marked by increasing conflict between old and new groups, but with no single interest attaining dominance …' (Murdoch et al., 2003: 13).

Additionally the US federal taxation system provided handsome subsidies for home ownership as opposed to renting and further favoured new construction over renovation and rehabilitation. Taken together these investments and subsidies had the effect of making America a rapidly suburbanizing country, one in which both cities and rural areas were consistently losing population, housing, retail activity and jobs to the suburbs. By 2000 the United States had become, according to the national census of population, a suburban nation.

Unlike the British system, and those operative in most other advanced economies, the American federal or national presence in planning has historically been indirect. Indeed from 1968 to 1976 annual efforts to pass a National Land Use Policy Act failed to garner enough congressional support to become law. Planning authority, then, has traditionally been a local responsibility operating within a national ideological framework that has sought to minimize the role of government intervention and to maximize market-based initiatives and solutions to problems.

Having said this the federal government did provide some of the necessary legal infrastructure and resources to promote rural land-use planning. Again, during the Eisenhower years planning programmes received an enormous boost with the 1954 amendments to the federal 1949 Housing Act. Known as the 'Section 701' programme, the federal government provided matching funds to state and local governments to assist them in establishing and staffing local and regional planning commissions (Graham, 1976: 124, 162). Many of these were located in rural areas and regions.

As the federal government began to disburse more funding to locales through numerous programmes, a 'grants economy' began to emerge wherein local and regional bureaucracies competed with one another for such funding. The federal government refused to take responsibility for vetting which applications ought to receive priority for limited funding. It published 'Circular A-95', also a part of the 'Section 701' system, charging each state with naming an agency that would evaluate and recommend grant applications according to some priority ranking. Indirectly this seemingly bureaucratic rule had the practical effect of pushing a number of states to create or enhance state planning offices with the responsibility to plan, coordinate, review and rank applications, among other functions. Some of these offices, like Vermont's, took the then bold step of establishing rudimentary statewide land use and growth management strategies and plans (Lapping, 1973).

The federal government created much of the legal and policy context within which the states operated. This framework was defined by a set of new environmental laws, the Coastal Zone Management Act of 1972, the Air Quality Act of 1967, the 1972 and 1977 amendments to the Clean Water Act, the Forest and Rangeland Renewal Resources Act of 1974, the National Forest Management Act of 1976 and the Wild and Scenic Rivers Act of 1968 (see, for example, Pyle, 1999). A number of these federal laws concerned themselves with the planning and management of public lands in the federal or national domain. The size of these holdings is especially significant in the American West. In the state of Nevada, for example, fully 90 per cent of all land is federally owned and managed by branches of the US Department of Agriculture and the US Department of the Interior. New management and planning regimes for such lands have had important implications for rural communities dependent upon ranching, mining, forestry and tourism. These programmes were not greeted with universal enthusiasm by all Westerners and a movement against planning and federal ownership emerged known as the 'Sagebrush Rebellion'. It had the tacit support of President Ronald Reagan, himself strongly identified with Western and anti-government interests and the movement to privatize federally owned natural resources. In time this movement morphed into the 'Wise Use' movement which continues to be active across the landscape (Jacobs, 1998). Overall, however, these federal policies and mandates reflected the growing strength of the environmental movement in the United States and the beginning of the 'greening' of rural and wild lands planning and policy (Lapping et al., 1989).

During the 1960s and 1970s important if oftentimes subtle changes within the local planning framework emerged, perhaps to alter forever how Americans understood the land and the countryside. Bosselman and Callies (1972) have called this period 'the quiet revolution in land use control'. Popper saw it as a burst of imaginative 'land-use reform'. This movement, which focused its critique on the limitations of local planning processes, 'succeeded in obtaining the passage of a great deal of innovative land-use legislation embodied in new programs of centralized regulation. Most of these programs are administered at the state level …' (1981: 3). Areas of critical or unique environmental concern, such as the California coast and the San Francisco Bay, New York's Adirondack Mountain region, Massachusetts's Martha's Vineyard island and New Jersey's Pine Barrens and the Hackensack Meadowlands, were among these pioneering or 'quiet revolution' planning innovations. In each case an extra-local planning authority was created to address specific land use issues with an eye toward protecting fragile environmental resources and attempting to balance economic development with emerging environmental priorities. Likewise projects of regional rather than purely local scale, such as the siting of major infrastructure facilities like waste management and disposal facilities, and especially the protection of agricultural lands, also came in for particular attention in many of these new planning systems. Yet if much of the experimentation with new land use planning and guidance systems was to be found at the state level, it was not true that all states were equally involved in the 'quiet revolution'. Among those jurisdictions whose programmes have garnered the most attention are Vermont, Oregon, Florida, California, New York, New Jersey and Maine (Daniels et al., 1989; DeGrove, 1984; Healy and Rosenberg, 1979; Linowes and Allensworth, 1975; Popper, 1984; Rosenbaum, 1976).

No single issue quite defines the nature of rural planning practice in the United States to the degree that the preservation of agricultural land does. Indeed, protecting farmland appears to be a substantial planning effort in many developed nations, as Rachelle Alterman's (1997) recently published comparative study of farmland retention strategies in the United States, Canada, the United Kingdom, the Netherlands, France and Israel reveals.

Those who seek to preserve farmland come to the issue from many different perspectives though in some important ways the problem has also become closely identified with the strong desire 'to save the family farm' in the face of the rapid and widespread consolidation and industrialization of the American food system (Heffernan et al., 1999; Ikerd, 1995). (The family farm has been an important consideration in the European discourse on agriculture as well, see Peterson, 1986). The extent of farm land loss has itself been a matter of considerable debate. The National Agricultural Lands Study of 1981 – carried out jointly by the Pressident's Council on Environmental Quality and the US Department of Agriculture – studied the problem intensively and came to the conclusion that between the years 1967 and 1975 approximately 1.2 million hectares of farmland was taken out of production annually due to retirement, urbanization, infrastructure development, and other factors (Daniels et al., 1989: 164). Immediately the study's findings were attacked, even by economists with the USDA, which helped to sponsor and staff the study commission (Fischel, 1985; Gustafson and Bills, 1984; Simon and Sudman, 1982). Opposition to NALS and the emerging farmland preservation movement contended that increased productivity per acre, due to the adoption of new technologies and better farming techniques, made the loss of land largely irrelevant. In the aggregate, they argued, production was increasing and that, rather than the amount of land in production or the number of farms, was the crucial thing to assess (Bromley, 1991).

It was on the local and regional level, however, that the loss of farmland was perceived to be a substantial issue and it was here that policies were adopted to stem the tide of farmland loss. Literally every state has implemented some programme(s) to respond to the problem. A number of states and many local governments have adopted land use controls on development in the countryside. These measures have included the creation of agricultural districts, the establishment of agricultural zones within more robust zoning and planning systems, the review of public infrastructure investments to determine their growth-inducing potential, incentives for cluster housing, and similar measures. Every state, however, has sought to address the problem of the economics of land conversion, as Alterman notes, or the economics of agriculture itself (1997: 222; Lapping and FitzSimmons, 1982). Thus, all states have adopted some form of differential or preferential land taxation for farmland and most have implemented some form of a right-to-farm law in an effort to protect farmers from the sorts of lawsuits which non-farm rural residents have used to curb legitimate agricultural activities (Lapping and Leutweiler, 1988). Additionally, many states and local jurisdictions have created programmes to transfer or purchase the development rights or obtain conservation easements on agricultural land which lower tax assessments, reduce the costs of operation, provide farmers with capital for investment and retain land in permanent agricultural and open space uses (Pfeffer and Lapping, 1994). A growing number of states are seeking to give further help by providing subsidies for on-farm infrastructure investments to control pollution, direct support for diversification efforts, and a variety of marketing initiatives, including direct-marketing programmes to consumers (the most comprehensive treatments of measures used to protect farming in the United States are to be found in American Farmland Trust, 1997; Daniels and Bowers, 1997).

The farmland preservation problem is really many different problems and varies greatly from place to place. This reflects the diversity of interests within the rural planning mix. Some retention policies are oriented to enhancing the economic viability of agriculture in general. These have tended to find their greatest support among those in the farming community, the agribusiness sector and traditional small town America. Here most of the issues relate to the changing structure of agriculture, chronically low prices, marketing and trade. There is little competition from other potential land uses and so land use controls are seldom seen as necessary. Federal policies have been as important in such places as have local ones.

Others policies seem to be more appropriate for land in rural–urban fringe areas where the preservation of open space and the countryside as an amenity tends to define the planning framework as much as the desire to secure local or regional agricultural economies. The emphasis in such areas has tended to focus on growth controls, preventing rural sprawl, and securing land in permanent agricultural use. Economic viability also is an on-going concern and it is within these communities that support for niche marketing, organic production and direct marketing to consumers, as in farmers' markets and community supported agriculture agreements, are common public policy objectives (Hinrichs, 2000: 16). Land use policy is the overriding planning framework. Others tend to support agricultural land preservation because of a deep concern over the loss of wildlife habitat and the other environmental benefits and services which farming produces. Here land planning is crucial and it is carried out both by local and state governments and, increasingly, also by private land trust organizations which increasingly acquire and manage land for biodiversity. And still other measures reflect the desire among urban consumers for the continuation of a cheap food policy. The larger point is that farmland retention programmes, the single most common rural planning policies implemented across the United States, are derivative of the many different and sometimes conflicting interests that seek to 'drive' rural planning.

CONCLUSIONS

Over the past half-century or so the landscape and political economy of rural planning and policy demonstrate an amazing consistency and lack of imagination in terms of their focus and orientation. In many national and international contexts an emphasis on agriculture as the rural persists. Even in the face of rapidly changing demographic, social, environmental and economic realities, policy frameworks all around the world continue to emphasize agriculture as the key sector for rural regions. As Thomas Rowley (2003), a research scholar with the University of Missouri's Rural Policy Research Institute, has observed of the American situation:

> policymakers continue to believe, or at least act as if they believe, that farming is the sum total of rural life. Therefore, policies aimed at supporting agriculture are touted as 'rural economic and community development' when in truth their effect on the larger economy and community is minimal. Yes they help, but most of the benefit goes to landowners (many of whom neither farm nor even live in the area) and many of the dollars head straight off the land, past the town, and out of the area, never to be seen again.

The situation is little different in the sprawling and highly diverse European Union. As Marsden has observed:

> despite fifteen years of debate and policy crisis concerning the 'arthritic' nature of the productionist support mechanisms within the Common Agricultural Policy (CAP), and the need to shift the emphasis towards new social and environmental agenda, it is still the case that the main pillar of the CAP remains in this area in terms of funding. Moreover, it still tends to reinforce the logic of agricultural productivist scale economics by rewarding the largest volume producers, as well as 'locking in' many of the less productive producers and those least able to meet the demands such policy-designed 'technological treadmills' require. (2003: 19)

Reform of this policy model of rural development – the 'agriculture-as-rural' 'productivist' system – has been slow and halting and remains largely intact. Indeed, farm sector supports, domestic subsidies and import duties in mature economies not only pose significant problems for the rural sectors within the countries where these measures have long been implemented, but have now become a very real impediment for the development of 'third world' nations in that they are a barrier to trade in food commodities, their primary exports. In response to a strongly worded request from several developing nations – India, China and Brazil amongst them – for drastic cuts in such farm supports, the EU Farm Commissioner Frans Fischler declared that the 'third world' nations were 'circling in a different orbit. If they want to do business, they should come back to earth' (Miller and Newman, 2003: A7). In other words, a change in the dominant EU model is highly unlikely.

Change at the far margins is taking place, however, even if the pace is slow, halting and lacking in substantial financial support. The CAP in one form or another has been in place since 1962. Faced with growing demands for a more broadly based, nuanced and sustainable approach to rural development, the EU instituted a 'Second Pillar' of the CAP known as 'Agenda 2000'. The major element of it has been the Rural Development Regulation (RDR) which provides member states with a broad selection of strategies and tools from which each country can choose to meet its own unique and idiosyncratic conditions. To date both France and the UK have chosen comprehensive rural development plans to meet the diverse needs of their rural areas and places. In the case of the UK this has happened simultaneously with some other changes, such as the renaming of the Ministry of Agriculture as the Department for Environment, Food and Rural Development. The approach, which the UK has selected under the terms of 'Agenda 2000', known as the Rural White Paper and the English Rural Development Programme, places a new focus on the revival of rural market towns, rural tourism and development, rural business and enterprise creation and a set of environmental objectives for the agricultural community to achieve as part of its operations. Yet John Bryden argues that while 'Agenda 2000' has promoted some new thinking about a more robust rural development strategy, 'it has turned out to be almost entirely composed of measures for farmers, and little concerned about the rest of the rural population (the majority), enterprises, and diversification. The mid-term review of the CAP (July, 2002) mentions rural development, but the proposals turn out to be focused almost entirely on agri-environmental measures' (2002: 3). Marsden is just as critical of the new rules. As he writes, 'the recent policy reforms under Agenda 2000, in addition to the new rural development regulation, expose a policy framework which will do little to shift the basic philosophy beyond its bias towards the industrial model' (2003: 23).

Even in the face of some new and more localized programmes to support rural areas, such as the LEADER I and II programmes adopted by the European Commission to stimulate innovative solutions to rural problems (Jouen, 2001), renewed attempts by the EU to foster greater regional planning through spatial planning and redistribution and an 'endogenous development agenda' dependent on networking and local initiatives through the European Spatial Development Perspective (ESDP) and similar initiatives (Faludi, 2001), or the initiative of the state of Texas (USA), which recently established an Office of Rural and Community Affairs (ORCA), both to split rural development from agriculture and to more effectively coordinate services and support for rural communities, rural planning and policy around the globe seems to remain fixed in its commitments.

These biases include a focus on the regional over the local, an agricultural-productivist paradigm, a reliance upon scale-economics within a neo-classical framework of analysis, a dependence on the provision of physical infrastructure and mega-projects to generate employment and capital and labour mobility, a fundamental disregard for indigenous peoples living in the rural periphery, and a fragmented and poorly coordinated service delivery environment together with out-dated governance systems. Together these elements consistently expose their own internal contradictions and move us only slightly closer to a model and a reality of genuine rural sustainability (Audriac, 1997; Williams, 2001).

REFERENCES

Alanen, Ilkka (2001a) 'Soviet patrimonialism and peasant resistance during the transition – the case of Estonia', in Leo Granberg, Imre Kovach and Hilary Tovey (eds), *Europe's Green Ring*. Aldershot: Ashgate Publishers.

Alanen, Ilkka (2001b) 'The destruction of Kanepi Kolkhoz', in I. Alanen, J. Nikula, H. Poder and R. Ruutsoo (eds), *Decollectivization, Destruction, and Disillusionment – A Community Study in Southern Estonia*. Aldershot: Ashgate Publishers.

Alden, Jeremy (2001) 'Planning at a national scale: a new planning framework for the UK', in Louis Albrechts, Jeremy Alden and Artur da Rosa Pires (eds), *The Changing Institutional Landscape of Planning*. Aldershot: Ashgate Publishers.

Alterman, Rachelle (1997) 'The challenge of farmland preservation: lessons from a six-nation comparison', *Journal of the American Planning Association*, 63 (2): 220–243.

American Farmland Trust (1997) *Saving American Farmland: What Works*. Washington, DC: AFT.

Applebaum, Anne (2003) *Gulag: A History*. New York: Doubleday.

Audriac, Ivonne (ed.) (1997) *Rural Sustainability in America*. New York: John Wiley and Sons.

Bailes, Kendell (1978) *Technology and Society Under Lenin and Stalin: Origins of the Soviet Technical Intelligentsia*. Princeton, NJ: Princeton University Press.

Baker, Mark (2000) 'The development planning process in six English regions', in David Shaw, Peter Roberts and James Walsh (eds), *Regional Planning and Development in Europe*. Aldershot: Ashgate Publishers.

Baranowski, Shelley (1995) *The Sanctity of Rural Life: Nobility, Protestantism and Nazism in Weimar Prussia*. New York and Oxford: Oxford University Press.

Bennett, Hugh Hammond (1928) *Soil Erosion: A National Menace*. Washington, DC: Government Printing Office.

Berlin, Isaiah (1991) *The Crooked Timber of Humanity: Chapters in the History of Ideas*. New York: Knopf.

Bosselman, Fred and Callies, David (1972) *The Quiet Revolution in Land Use Control*. Washington, DC: President's Council on Environmental Quality, US Printing Office.

Bowers, William L. (1974) *The Country Life Movement in America, 1900–1920*. Port Washington, NY: Kennikat.

Bowler, Ian R. (1979) *Government and Agriculture: A Spatial Perspective*. London: Longman.

Bowler, Ian R. (1986) 'Direct supply control in agriculture: experience in Western Europe and North America', *Journal of Rural Studies*, 2: 19–30.

Bromley, Daniel W. (1991) *Environmental and Economy: Property Rights and Public Policy*. Oxford: Blackwell.

Bryden, John (2002) 'A review of European Union rural policy'. Paper presented at the National Symposium of Rural Policy, University of Nebraska, Lincoln, NB, October.

Bunce, Michael (1994) *The Countryside Idyll: Anglo-American Images of the Landscape*. London: Routledge.

Busch, Lawrence and Lacy, William (1983) *Science, Agriculture, and the Politics of Research*. Boulder, CO: Westview.

Chandler, William (1984) *The Myth of the TVA: Conservation and Development in the Tennessee Valley*. Cambridge, MA: Ballinger Publishers.

Cloke, Paul (1983) *An Introduction to Rural Settlement Planning*. London: Methuen.

Cloke, Paul J. (1985) 'Counter-urbanization: a rural perspective', *Geography*, 12: 13–23.

Cloke, Paul J. (1988) 'Britain', in Paul J. Cloke, *Policies and Plans for Rural People*. London: Unwin Hyman.

Cloke, Paul J. (1989) Land-use planning in rural Britain', in Paul J. Cloke, *Rural Land-Use Planning in Developed Nations*. London: Unwin Hyman.

Cloke, Paul J. and White, Jo (1990) *The Rural State: Limits to Planning in Rural Society*. Oxford: Clarendon Press.

Cochrane, Willard W. (1979) *The Development of American Agriculture: A Historical Analysis.* Minneapolis: University of Minnesota Press.

Conquest, Robert (1968) *The Great Terror.* New York: Macmillan.

Creese, Walter (1990) *TVA's Public Planning: The Vision and the Reality.* Knoxville, TN: University of Tennessee Press.

Cullingworth, J.B. (1985) *Town and Country Planning in Britain*, 9th edn. London: George Allen and Unwin.

Cummings, F. Harry (1989) 'Rural development and planning in Canada: federal and provincial perspectives', *Plan Canada*, 29 (2): 11.

Danbom, David D. (1979) *The Resisted Revolution: Urban America and the Industrialization of Agriculture, 1900–1930.* Ames, IO: Iowa State University Press.

Daniels, Thomas L. and Bowers, Deborah (1997) *Holding Our Ground: Protecting America's Farms and Farmland.* Washington, DC: Island Press.

Daniels, Thomas L., Lapping, Mark B. and Keller, John W. (1989) 'Rural planning in the United States: fragmentation, conflict and slow progress', in P. Cloke (ed.), *Rural Land-Use Planning in Developed Nations.* London: Unwin Hyman.

DeGrove, John (1984) *Land, Growth and Politics.* Chicago: American Planning Association Press.

Dorman, Robert L. (1993) *Revolt of the Provinces: The Regionalist Movement in America, 1920–1945.* Chapel Hill, NC: University of North Carolina Press.

Faludi, Andreas (2001) 'The European spatial development perspective and the changing institutional landscape of planning', in Louis Albrechts, Jeremy Alden and Artur de Rosa Pires (eds), *The Changing Institutional Landscape of Planning.* Aldershot:. Ashgate Publishers. pp. 35–54.

Fischel, William A. (1985) 'The urbanization of agricultural land: a review of NALS', *Land Economics*, 58: 236–259.

Fitzgerald, Deborah (2001) 'Accounting for change: farmers and the modernizing state', in C.M. Stock and R.D. Johnson (eds), *The Countryside in the Age of the Modern State: Political Histories of Rural America.* Ithaca, NY: Cornell University Press.

Friedmann, John and Weaver, Clyde (1980) *Territory and Function: The Evolution of Regional Planning.* Berkeley, CA. University of California Press.

Furet, Francois (1999) *The Passing of an Illusion: The Idea of Communism in the Twentieth Century.* Chicago: University of Chicago Press.

Furuseth, Owen and Lapping, Mark B. (eds) (1999) *The Contested Countryside: The Rural–Urban Fringe in North America.* Aldershot: Ashgate Publishers.

Gilg, Andrew (1978) *Countryside Planning.* London: Routledge.

Graham, Otis L. (1976) *Toward a Planned Society: From Roosevelt to Nixon.* New York: Oxford University Press.

Granberg, Leo (2000) 'Small production and state intervention in agriculture', *Acta Sociologica*, 29 (3): 243–253.

Granberg, Leo and Peltonen, Matti (2001) 'Peasantisation and beyond in Finland and Scandinavia', in Leo Granberg, Imre Kovach and Hilary Tovey (eds), *Europe's Green Ring.* Aldershot: Ashgate Publishers.

Gustafson, Greg and Bills, Nelson (1984) *US Cropland, Urbanization and Landownership Patterns.* Washington, DC: USDA, Agricultural Economics Report No. 520.

Hall, Peter (1992) *Urban and Regional Planning.* London: Routledge.

Hargrove, E. and Conkin, P. (1983) *TVA: Fifty Years of Grass-Roots Bureaucracy.* Urbana, IL: University of Illinois Press.

Healy, Robert and Rosenberg, John (1979) *Land Use and the States*, 2nd edn. Baltimore, MD: Johns Hopkins University Press.

Heathorn, Stephen (2000) 'An English paradise to regain? Ebenezer Howard, the Town and Country Planning Association and English ruralism', *Rural History*, 11 (1): 113–128.

Heffernan, W., Hendrickson, M. and Gronski, R. (1999) *Consolidation in the Food and Agricultural Sector.* Columbia, MO: Department of Rural Sociology, University of Missouri.

Hervieu, B. (1994) *Les Champs du futur.* Paris: Editions Julliard.

Higgins, Benjamin and Savoie, Donald J. (1977) *Regional Development Theories and Their Application.* New Brunswick, NJ: Transaction Books.

Hinrichs, C. Claire (2000) 'Embeddedness and local food systems: notes on two types of direct agricultural markets', *Agriculture and Human Values*, 16: 56–64.

Holmes, R. (1912) 'The passing of the farmer', *Atlantic Monthly*, 110 (4): 523.

Ikerd, John (1995) 'The Industrialization of Agriculture: Why We Should Stop Promoting It'. The Harold Briemeyer Lecture in Agricultural Policy, Columbia, MO: University of Missouri, 16–17 November.

Jacobs, Harvey M. (1998) 'The "wisdom" but uncertain future of the Wise Use Movement', in Harvey M. Jacobs (ed.), *Who Owns America? Social Conflict Over Property Rights.* Madison, WI: University of Wisconsin Press.

Jacobs, Jane (1961) *The Death and Life of Great American Cities.* New York: Vintage Books.

Jones, David C. (1982) 'There is some power about the land – the Western agrarian press and country life ideology', *Journal of Canadian Studies: Revue d'etudes canadiennes*, 17: 3 (Fall).

Jouen, Marjorie (2001) 'The European Rural Model', http://europe.eu.int/comm/archives/leader2/rural-en/biblio/model/a.

Lapping, M. and FitzSimmons, J. (1982) 'Beyond the land issue: the economics and institutional viability of agriculture', *GeoJournal*, 6 (6): 519–524.

Lapping, M. and Leutweiler, N. (1988) 'Agriculture in conflict: right-to-farm laws and the peri-urban milieu for farming', in Willie Lockeretz (ed.), *Sustaining Agriculture Near Cities.* Ankeny, IO: Soil Conservation Society of America.

Lapping, Mark (1973) 'Planning in Vermont: new legal horizons', *Alternatives: Journal of Environmental Studies*, 2: 2.

Lapping, Mark and Forster, Dale (1983) 'Swedish agricultural policy: an integrated approach', *International Regional Science Review*, 7 (3): 293–302.

Lapping, Mark and Fuller, Tony (1985) 'Canadian rural development policy: an interpretation', *Community Development Journal*, 20: 2.

Lapping, Mark B. (1977) 'Radburn: planning the American community', *New Jersey History*, LCV: 2.

Lapping, Mark B. (1992) 'American rural planning, development policy, and the centrality of the federal state', *Rural History: Economy, Society, Culture*, 3 (2): 231.

Lapping, Mark B. (1997) 'A tradition of rural sustainability: the Amish portrayed', in I. Audriac (ed.), *Rural Sustainability in America.* New York: John Wiley.

Lapping, Mark B. (2001) 'Sufficient unto themselves: life and economy among the Shakers in nineteenth century rural Maine', *Maine History*, 40 (2): (Summer).

Lapping, Mark B., Daniels, Thomas L. and Keller, John W. (1989) *Rural Planning and Development in the United States.* New York: Guilford Press.

Lears, Jackson (1981) *No Place of Grace: Antimodernism and the Transformation of American Culture, 1880–1920.* New York. Pantheon.

Lenin, V.I. (1976) *The Agrarian Question and the Critics of Marx*, 2nd rev. edn. Moscow: Progressive Publishers.

Linowes, Robert and Allensworth, Don (1975) *The States and Land Use Control.* New York: Praeger.

Liu, Hui (2001) *Poverty Stricken Areas and Poverty Relief, China's Regional Disparities: Issues and Politics.* New York: Nova Science Publishers.

Mann, Susan A. and Dickinson, James A. (1980) 'State and agriculture in two eras of American capitalism', in Frederick H. Buttel and Howard Newby (eds), *The Rural Sociology of the Advanced Societies: Critical Perspectives.* Montclair, NJ: Allanheld. p. 306.

Marsden, Terry (2003) 'The condition of rural sustainability: issues in the governance of rural space in Europe', in Charalambos Kasimis and George Stathakis (eds), *The Reform of the CAP and Rural Development in Southern Europe.* Aldershot: Ashgate Publishers.

Marsden, Terry, Murdoch, Jonathan, Lowe, Philip, Munton, Richard and Flynn, Andrew (1993) *Constructing the Countryside.* London: UCL Press.

Martin, J.R. (ed.) (1967) *The Economic Impact of the TVA.* Knoxville, TN: University of Tennessee Press.

Miller, Scott and Newman, Matthew (2003) 'EU criticizes proposal for cut in farm supports', *The Wall Street Journal*, 5 September.

MAFF (Ministry of Agriculture, Fisheries and Food) (1987) *Farming and Rural Enterprise.* London: HMSO.

Murdoch, Jonathan, Lowe, Philip, Ward, Neil and Marsden, Terry (2003) *The Differentiated Countryside.* London: Routledge.

Nove, Alec (1990) 'Can the USSR Learn from Hungarian and Chinese Agricultural Experiences?', in K.R. Gray (ed.), *Soviet Agriculture: Comparative Perspectives.* Ames, IO: Iowa State University.

O'Connor, Alice (1992) 'Modernization and the rural poor: some lessons from history', in Cynthia M. Duncan (ed.), *Rural Poverty in America.* New York: Auburn House.

Odum, Howard W. (1934) 'The case for regional-national social planning', *Social Forces*, No. 13.

Pallot, Judith (1988) 'The USSR', in P. Cloke (ed.), *Policies and Plans for Rural People: An International Perspective.* London: Unwin Hyman.

Peterson, Martin (1986) '"The family farm": a review of a central concept in Western European agricultural politics', *Scandinavian Journal of History*, 11 (3): 265–282.

Pfeffer, Max and Lapping, Mark (1994) 'Farmland preservation, developments rights and the theory of the growth machine: the view of planners', *Journal of Rural Studies*, 10 (3): 233–248.

Popper, Frank (1981) *The Politics of Land-Use Reform.* Madison, WI: University of Wisconsin Press.

Putterman, Louis (1993) *Continuity and Change in China's Rural Development: Collective and Reform Eras in Perspective.* Oxford: Oxford University Press.

Pyle, Lisbeth A. (1999) 'The Wild and Scenic Rivers Act: putting policy into practice in West Virginia', in Andrew Gilg et al. (eds), *Progress in Rural Policy and Planning*, Vol. 4. Chichester: John Wiley and Sons.

Raun, Toivo U. (1991) *Estonia and the Estonians.* Stanford, CA: Hoover Institution Press.

Reiner, Thomas (1963) *The Place of the Ideal Community in Urban Planning.* Philadelphia: University of Pennsylvania Press.

Richert, Evan and Lapping, Mark B. (eds) (1998) 'Centennial Symposium on Ebenezer Howard's "Garden Cities for Tomorrow"', *Journal of the American Planning Association*, 64: 2.

Robinson, Guy (1990) *Conflict and Change in the Countryside.* London: Belhaven Press.

Rosenbaum, Nelson (1976) *Land Use and the Legislatures: The Politics of State Innovation.* Washington, DC: The Urban Institute.

Rowley, Thomas D. (2003) 'The rural identity crisis', http://www.rupri, org/

Schmitt, Peter J. (1969) *Back to Nature: The Arcadian Myth in Urban America.* Baltimore, MD: Johns Hopkins University Press.

Scott, James C. (1998) *Seeing Like a State: How Certain Schemes to Improve the Human Condition Have Failed.* New Haven, CT: Yale University Press.

Selznick, Philip (1949) *TVA and the Grass Roots.* Berkeley, CA: University of California Press.

Shi, David E. (1985) *The Simple Life: Plain Living and High Thinking in American Culture.* London and New York: Oxford University Press.

Simon, Julian L. and Sudman, S. (1982) 'How much farmland is being converted to urban use?', *International Regional Science Review*, 7 (3): 257–272.

Singal, Daniel J. (1982) 'Howard Odum and social science in the South', in *The War Within: From Victorian to Modernist Thought in the South, 1919–1945.* Chapel Hill, NC: University of North Carolina Press.

Steiner, Frederick (1988) 'The evolution of federal agricultural land policy in the United States', *Journal of Rural Studies*, 4 (4): 355.

Taylor, Henry C. and Taylor, Anne Dewees (1952) *The Story of Agricultural Economics in the United States, 1840–1931.* Ames, IO: Iowa State University Press.

Tewdwr-Jones, M. (1996) *British Town Planning in Transition: Planning in the 1990s.* London: University College of London Press.

Tian, Xiaowen (1999) 'China's Regional Disparities Since 1978: Main Trends and Determinants'. Singapore: Occasional Paper No. 21. East Asian Institute, National University of Singapore.

Tovey, Hilary (2001) 'The co-operative movement in Ireland: reconstructing civil society', in Hilary Tovey and Michel Blanc (eds), *Food, Nature and Society: Rural Life in Late Modernity.* Aldershot: Ashgate Publishers.

Tweeten, Luther (1979) *Foundations of Farm Policy.* Lincoln: University of Nebraska Press.

Vail, David, Hasund, Knut Per and Drake, Lars (1994) *The Greening of Agricultural Policy in Industrial Societies: Swedish Reforms in Comparative Perspective.* Ithaca, NY: Cornell University Press.

Weaver, Clyde and Gunton, Thomas I. (1982) 'From drought assistance to mega-projects: fifty years of regional theory and policy in Canada', *Canadian Journal of Regional Science*, V (1): 11.

Whitby, Martin and Adger, Neil (1993) 'UK land use and the global commons', in S. Harper (ed.), *The Greening of Rural Policy: International Perspectives.* London: Belhaven Press.

Williams, Jo (2001) 'Achieving local sustainability in rural communities', in Antonia Layard, Simon Davoudl and Susan Batty (eds), *Planning for a Sustainable Future.* London: Spon Press. pp. 235–252.

Part 2

RURAL RESEARCH: KEY THEORETICAL COORDINATES

8

Landscapes of desires?

E. Melanie DuPuis

> What distinguishes the map from the tracing is that it is entirely oriented toward an experimentation in contact with the real. The map does not reproduce an unconscious closed in upon itself; it constructs the unconscious. ... It can be drawn on a wall, conceived of as a work of art, constructed as a political action or as a meditation. (Deleuze and Guatarri, 1987: 12)

In his Prologue to *Nature's Metropolis*, William Cronon tells a story about travelling as a child from the Midwestern countryside into the City of Chicago. He saw that city as 'a cancer on an otherwise beautiful landscape' in which 'I had no wish to linger'. What, he later ponders, made him reject Chicago and embrace the countryside as more 'natural'? His answer is that the stories we tell about the journeys we take in life – country to city, young to old, innocent to corrupt – comprise 'a powerful narrative trajectory, a compelling token of the divided world we inhabit – and yet also reproduce that divided world'. As Cronon and many others have noted, this dualistic thinking is a characteristic of Western thought, and he spends the rest of the 400+ page book showing the 'underlying unity' between Chicago and its countryside, that the country and the city are not 'other' to each other (and othering itself as a mythological project).

Yet, the book never really answers the original question: what is it about Western thought that makes us tell stories in twos: that which we supposedly have (or where we supposedly are) and that which we supposedly want (or supposedly want to go)? Cronon holds that by 'erasing the false boundary' between nature and culture, we can discover commonality and get rid of the Western practice of othering which leads to both environmental destruction and social inequalities. This intellectual project – showing the cultural construction of landscape as a way of erasing nature/culture dualisms – has been picked up by a number of scholars in the last few decades, producing several works, particularly edited volumes (Cloke and Little, 1991; DuPuis and Vandergeest, 1995; Marsden et al., 1993; Nye, 1999). Much of this work follows Cronon, tearing away at the veil of false dualities. Yet, despite these efforts to change the Western 'mind', dualist stories continue to be told. Why?

The recent literature on the nature of Western 'Desire' provides a different answer to Cronon's question. Derived from the psychoanalytic theories of Lacan, Desire has emerged as a major analytical category in postmodernism and cultural studies. The notion of Desire has taken on particular explanatory power in the context of the postmodern critique of Western society. Lacan argued that the human self was always fractured and incomplete, that humans as language-using animals were always seeking to re-unite themselves with the world they were separated from because they had named it. As a result, human social activity could be explained as attempts to make the self whole by possessing or controlling that which lay outside it. This attempt is doomed to failure because the language-using human self is intrinsically unable to fulfil this project of re-unification. 'Desire', therefore, signifies a project of impossible fulfilment, the project of creating the Self through making the dream world the real one. The project remains unfulfilled because it is based in the Self's impossible desire to achieve wholeness by re-uniting with that which it is not. Yet, despite the impossibility of desire, this drive for the fulfilment of wholeness has its effects on the world, effects that can be 'mapped.'

Social theorists politicized Lacan's psychoanalytic theory by situating it – 'historicizing' it – within the social and political context of Western modernity. This historicizing of the Self and its relationship to the world follows broadly in the footsteps of Michel Foucault. Rather than attributing desire to intrinsic human characteristics, a historicized version of desire had a different starting point: namely the Western Self as fractured due to the bifurcated ideologies of modern Western civilization. 'Desire', as a historical construct, thereby became a focus for the critique of modernity. It also became a source for the understanding of the Western drive for power. In post-colonial literature, desire for the other – 'Orientalism' (Said, 1978) – became the motivating factor behind European domination of colonial subjects as a way of creating a European self. In post-structuralist thought, Deleuze and Guatarri (1987) formulated a 'Politics of Desire': a mapping of the effects of human dreaming and the attempts to create the world according to impossible dreams. Originating with classical Western philosophy, the bifurcation between mind/body, nature/culture, self/other, male/female define the Western worldview but also haunt it with unfulfilment, creating the desire to overcome these dualities through power: the control and/or possession of the 'Other'.

Taking Western desire as an intellectual basis creates an intellectual project to re-think landscape that is different from Cronon's. Rather than tearing away the veil of dualisms, the project becomes the analysis of desire's effects. As Deleuze and Guatarri state, the problem becomes 'the map', the effect desire has on reality, rather than 'the tracing', the desire itself. For Cronon, the project is to unveil the idea of pristine nature and show the reality underneath; for Deleuze and Guatarri, the project is to map the effects that the idea of pristine nature have on reality.

Much of the post-modern critique of Western forms of domination and control notes the intimate relationship between domination and the Western scientific drive to control nature, imperialism's drive to control territory and patriarchy's drive to control the 'unruly female' (Merchant, 1989). Because of the insistent Western bifurcation of categories, the other possessed must also be and remain purely that which is not its controlling opposite. To control the other is to both keep it pure and uncontaminated – matter in place, in Mary Douglas's terms – and to erase it through re-unification, since a contaminated other does not bring the fulfilment of unity. In this way, the female must remain purified non-maleness, everything a man is not, although flipping between the civilized (Baker, 1991) and the natural (Merchant, 1989) side of the dichotomy. In tandem, colonial subjects are put in their place by their categorization as female, natural and uncivilized (Said, 1978; Stoler, 1995).

In the study of landscape, nature becomes entangled in the dreams of modernity, a repository of everything civilization is not: pure, uninhabited, unconscious, non-rational, free of inhibitions and intent. In romantic thought, nature becomes the good to civilization's bad while, in scientific thought, nature becomes the realm which culture must control for its own benefit. The actors in the imperial project pursue both of these attitudes simultaneously, for example, practising 'scientific forestry' that creates landscapes meant to fulfil the needs of an urban populace for recreation and leisure through escape from urban settings.

The nature/culture split also becomes the conceptual territory within which Western actors contest the use of landscape. For example, the 'Great Wilderness Debate' (for a thorough collection of the main readings in this debate, see Callicott and Nelson, 1998), that has raged since the 1970s pits the Deep Ecologists, who argue for the preservation of a pristine, people-free nature, against the political ecologists and post-structuralists, who argue that nature is a cultural construct. The unending nature of the wilderness debate exemplifies the way in which the contest over Western dualities becomes entangled in unending, continuous discursive loops. From the perspective of Deleuze and Guatarri, the preservation of wilderness can be seen as the desire to maintain the purity of the non-civilized, non-human landscape as 'the preservation of the world' for the creation of the self.

What becomes clear in the post-structural examination of Western desire is the relationship between the creation and preservation of the 'Other' and the creation and maintenance of identity. Political struggles over identity therefore take place in struggles over the control of the Other, including natural spaces. In the case of the Great Wilderness Debate, the struggle over wilderness can be seen as a struggle over the nature of Western identity, or the struggle over various identities and their relation to various forms of nature. This is particularly evident in

Theodore Roosevelt's writings on wilderness and masculinity. Describing the benefits of Yosemite as a wilderness reserve, he states: 'Every believer in manliness and therefore in manly sport ... should strike hands with the farsighted men who wish to preserve our material resources, in the effort to keep our forest and game beasts ...' In this case, the preservation of wilderness is linked to larger projects of American pioneering settlement and the preservation of the 'character' associated with the early Western white woodsmen, characteristics that seemed to be disappearing in the Western urbanizing populace just about as quickly as old growth forests.

Related to the nature/culture split is the Western bifurcation between the city and the country. Yet, compared to the wilderness debate, the role of Western duality is more complex and more subtle in the contestations over rural landscapes, the 'middle landscape' (L. Marx, 1964), which seems a mixture of human and nature. In the pastoral view of the middle landscape, country is set against both city and wild nature as the balance between the two (Cronon, 1991; Williams, 1973). Yet, the balance of the pastoral is as impossible and problematic a goal as the desire for purity.

As more than one author has noted (Bunce, 1994; L. Marx, 1964; Vandergeest and DuPuis, 1995; Williams, 1973), the rise of the movement to 'save' the countryside came about with the rise of urbanization and a predominantly urban or suburban populace. It also came about with the growth of the critique of urbanism as a degraded form of life. This is particularly evident in the 'countrification' of suburban developments, making every home seem like a small, usually English, manor (Bunce, 1994), and thereby using the countryside as proxy for upward mobility and racial purity. The desire to both preserve and transform the countryside arises with the increasing power of urban elite populations to control their rural counterparts and to escape from the control of less elite urban groups.

As many recent studies have shown, it is the subtlety and contradiction of rural landscapes that make them such a rich resource for examining the ways in which place reflects political struggles over identity. Phillips notes in a historical overview of rural geography that the countryside is being increasingly recognized as 'a racialized, nationalized, aged, sexualized and gendered space' (Phillips, 1998), meaning that researchers have begun to look at the nature of rural landscapes as more than the materialization of an English middle class ideal (or idyll). In this vein, new studies of rural landscape explore the ways in which various groups create themselves in their struggles to create the countryside. In these recent studies, the questions of 'What is rural?' and 'Who is rural?' become intertwined (Mormont, 1990).

Inspired by post-structural theories, such as Hall's 'politics of representation' and Baudrillard's notion of 'simulacrum', in which the imaginary world becomes more socially important than the material world, a number of authors have looked at identity and its relationship to imagined countrysides (Bunce, 1994; Cloke, 1994; Marsden et al., 1993; the articles in Cloke and Little, 1997). In this case, the struggle is not only over how the actual landscape will be shaped but also about its representation. Landscape as representation also has a strong interconnection with the post-structural focus on narrative. How we tell stories about landscape, whether through words, pictures or maps, ends up having an effect on landscape itself, as we try to fit the material world into our ideas of what it should be. The two major Western narratives of landscape parallel the condition of bifurcation and desire: either stories of triumph over the landscape in narratives of progress, science, nationhood and empire or stories of loss, 'The Fall', or stories of decline and the desire for recovery (DuPuis, 2002; Merchant, 1995; Vandergeest, 1995). In rural studies, the narrative of rurality lost is often discussed in terms of Tonnies's notions of *Gemeinschaft* – the community solidarity of the traditional village – replaced by the *Gesellschaft* of purely instrumental relationships in the modern city.

The flip side of this discussion of the rural idyll as social solidarity is often seen in studies of the relation of imagined and real countrysides to marginal identities. In these cases, social elites create the rural idyll either through excluding others from rural communities or by making them invisible. This literature has a long history in California studies, starting with Carey McWilliams's *Factories in the Field* (1971 [1939]) and continuing today with Don Mitchell's *The Lie of the Land* (1996), Matt Garcia's *A World of Its Own* (2001) and, from an urban perspective, Mike Davis's *City of Quartz* (1992). This perspective parallels the contemporary literature in Britain on social exclusion, especially David Sibley's *Landscapes of Exclusion* (1995), Cloke et al. (2001a, 2001b) on the hiddenness of rural homelessness, Valentine's work on rural homosexuality (1997), and new work on rural masculinities (see, for example, the *Rural Sociology* (2000) special issue on the topic). Much of this work extends Mary Douglas's work on pollution as 'matter out of place', although most of this work mysteriously avoids the Interactionist School and Goffman's work on stigma. These studies show how elites create their identity through creating pure, idyllic landscapes devoid of rowdy teenagers, homosexuals, the homeless, or women working the land, in favour of landscapes exhibiting rurality as order, purity, middle-class prosperity, family and masculinity.

The rural idyll, as defined by elites, often paints the countryside as a place of leisure rather than as working nature, places of consumption, not production (see, for example, Woods, 1997). Ironically, in these contests over the meaning of rurality in rural places, farmers themselves can become increasingly 'matter out of place' as the discourse of rurality turns from a productivist to an environmentalist focus (DuPuis, 1995; Halfacree, 1994). Certainly, the open meadows up the road from my house preserve nature in favour of frogs and butterflies but also against Wal-Mart and brussels sprout cultivation.

This newer rural discourse literature moves the arguments about rurality from the production to the consumption sphere, as contests by interest groups over the nature of rurality become contests over the right to consume rural landscapes in particular ways. Desire as a concept, in its simplest form, tends to be used simply as a marker for consumption – that is, the desire to buy. Certainly, the realm of the rural studies literature focusing on the consumption of landscape has been fruitful; however, it does not fully exploit the usefulness of desire as a concept. Desire in its complexity – as the pursuit of impossible dreams – is more than a critique of consumer culture; it is not just another 'culture industry' problem to be unveiled. As sociological thinkers from crisis theorists to queer theorists have shown, capitalist hegemonic control is also an impossible dream; it is never complete. Yet, this dream has major effects. And there are other dreams of the Self that go beyond both production and consumption – imagined nations, motherhoods, masculinities and accumulations. A broader and more complex concept of desire therefore allows for a broadening of the scope of rural landscape studies as well, enabling them to reach greater forms of complexity and understanding.

The rest of this chapter will therefore focus on bringing in a broader literature from both political economy and cultural studies, to see what these works might have to say to rural studies, and what rural studies may have to add to these perspectives. The discussion will cover a number of current topics, which are certainly not exhaustive: imperialism/colonialism/globalization, habitation, sustainability, livelihood/deprivation, nation and justice. Most of the studies discussed below do not specifically take 'landscape' as a research focus. In fact, in many of these studies, what is said about landscape is often the de facto result of other points made. However, they can all be characterized as works that have something to say about desire and its effects on landscapes.

Each topic will examine a few works as exemplars of fruitful directions. As mentioned, the works examined here are not specifically from the rural studies literature. Most of them can be characterized broadly as stemming from the genres of environmental sociology, political ecology, environmental history and the sociology of agriculture. However, each of these sub-disciplines has a lot to say about landscape and desire.

IMPERIALISM/COLONIALISM/ GLOBALIZATION

These topics cover a major portion of the ways in which researchers have chronicled 'Orientalism': the effects of Western desire and its drive to control the Non-western Other. A broad range of studies have examined the desire to control other peoples and other lands, on behalf of the homeland or, in the case of globalization, on behalf of capitalist owners. Of course, plantation agriculture as a 'landscape of desire' figures prominently in the work on landscape and these forms of domination. The relative success of plantation *vs.* smallholders in global commodity systems has significant impacts on rural landscapes (Grossman, 1998; Raynolds et al., 1993, for example). This reflects a struggle over the nature of agriculture in the realm of political economy. While these studies often do not address the issue directly, it is clear that the success of one form of agriculture over another affects the shape of the rural landscape. This brings production back into the formation of landscape and links agrarian and rural studies. Clearly, these struggles also entail different imaginizations, representations and dreams of perfection as well.

The struggle over landscape reflects social struggles not just of class, but also of race. Judith Carney's *Black Rice* describes landscape struggle as reflected in the negotiations between slave and master in ante-bellum South Carolina and colonial Brazil (2001). In this case, African knowledge of rice growing created a system that in many ways enabled them to replicate many of the landscape features of their Rice Coast African homeland. The desire, at least in Carney's narrative, is not necessarily a re-creation of the African 'idyll' but a desire to eat food from home. Plantation owner dependence on slaves for their knowledge of rice growing led to concessions granted to slaves in terms of control over aspects of their lives in this region. Yet, white South Carolinans, in their re-telling of the story of rice, re-created its 'origin story', placing the knowledge of and responsibility for the creation of the rice plantation landscape with the planters themselves. Carney shows the effects of the politics of slavery on the South Carolinan and Brazilian landscapes as well as on the depopulated and war-torn landscapes of Africa during the slaving era.

Struggles over representation take place beyond the Anglo-American landscape. Candace Slater's *Entangled Edens* relates the stories of nature in an area often considered 'wilderness' in the Western mind but very much a middle landscape to Brazilians (2002). In this case, she compares the stories of Spanish conquistadors in search of El Dorado to the stories of the Amazon told by the descendants of escaped slaves and by the gold miners working in the region. Like Deleuze and Guitarri, she shows that the nature/culture divide is a product of the Western mindset, as portrayed in the conquistador's idea of a rich but pliant golden city lying in wait to be conquered and plundered. In contrast, she shows that the stories of escaped slaves and gold miners exemplify a relationship to nature that blurs boundaries between people and nature, and in which each has power to transform the other.

In Donald Worster's *Rivers of Empire* the rural landscape of focus is the Great American desert and the social and political forces that tamed the rivers of the West to water this desert for the purpose of large-scale agriculture (1985). In this case, once again, the play of desire which forms the Western landscape is the seemingly inexorable march of progress, but the story told is the struggle over whether that progress would go to the benefit of smaller family farms or large-scale agriculture owned by urban speculators. Worster's classic introduction, in which he walks along an irrigation ditch and describes the blighted and alienated agro-industrial landscape he sees during his waterside 'idyll,' relates first off who the winners were.

Valerie Kuletz's *The Tainted Desert* (1998) deserves mention under this heading. Kuletz examines the 'landscapes of sacrifice' in the irradiated landscape of the western desert nuclear test zones, a land deemed empty but in fact inhabited by various people, including Native Americans. Her overlapping maps of what this landscape means to Native inhabitants, the scientists of Los Alamos and the military shows that particular landscapes can exemplify multiple, often conflicting desires.

From this viewpoint, the whole realm of new global struggles over landscape can be seen as including struggles over ownership of genetic species, and the conflicting desires of local inhabitants and bioprospectors. As well, the conflicting desires of biotechnology companies and their genetically engineered seeds with the desires of local farmers is a struggle over landscape with major effects. In this case, even the 'nature' of the genetically engineered pollen has an effect as it interacts with the non-engineered varieties in a region. In other words, landscapes can reflect the micro-politics of large-scale struggles for control and domination that are not specifically about what a particular region looks like but which still has enormous effects on a landscape (Marsden, 1992).

HABITATION

As mentioned above, the question of which people get to inhabit a particular landscape is already a major part of the rural studies literature. Beyond the studies listed above, which deal directly with landscape as consumption, one can add to the literature that on the political economy of rural restructuring. From a political economy perspective, researchers have examined the ways in which counter-urbanization and the rise of service class inhabitants has affected the nature of rural landscape in the UK (Urry, 1995, for example), and to some extent in the US and Australia as well. This literature has also included research on how large-scale capitalist restructuring has led to different patterns of rural industrialization and industrial agriculture (Marsden et al., 1993, for example).

However, there are other studies that are not specifically about rural landscapes but which can help explain changes in who lives in the rural landscape. Studies of immigration, particularly when combined with understandings of rural areas' place in global restructuring, can say something about the 'desire to inhabit'. Prominent in this realm are Gouveia's studies of the restructuring of the meat packing industry and the concurrent arrival of Latino immigrants as workers in these new Midwestern meat packing plants (1994). Like Gouveia, Wells's (1996) study of the strawberry growing region of the central California Coast describes the changing landscape of this region in relation to Latinos' desire to inhabit this landscape as workers and in ways that go beyond this role. In the case of Wells's strawberry growing region, certain workers become sharecropping farmers on the foothills of California's Central Coast, leading to significant changes in the landscape of the area. Clearly, the vast literature on industrial hog farming and its effect on Midwestern farm regions describes a struggle over who, and what, will inhabit the landscape.

Who is in the landscape may also be the result of lack of access to work as much as struggles over rural jobs. Fitchen's work on the rural townships of Upstate New York showed that poor people living on welfare benefits often filled in the apartments and trailer parks emptied of people in the now de-industrialized, de-agriculturalized small rural towns (1991). In this case, an area that had been built up in a previous period of rural industrialization experiences serious changes in its inhabitants. Attracted by the inexpensive if run-down housing market, the poor living on benefit cheques were the only people still able to exist in these areas. In many cases, they became the majority population in these small rural towns. Carless and without services, they shopped, phoned and spent leisure time at the remaining local convenience store in the empty storefronts of otherwise empty towns. These are often not the areas of rural amenities prized by counter-urbanizing elites or, if they are, they have not yet been discovered. Once discovery happens, the gentrification and exclusion processes begin again.

LIVELIHOOD/DEPRIVATION

The desire to maintain rural livelihoods is also a project with potential effects on the rural landscape. Generally falling under the sub-discipline of political ecology, the focus of these studies is on the access to resources and the struggle between local users or between local and outside users of resources. These studies, generally covering issues of traditional and peasant agriculture and forest-edge agri-gathering cultures in the South, examine the struggles and claims particular groups have over certain local resources. For example, Rocheleau and Edmunds (1997) look at the multiple, many-dimensional resource claims of groups over particular resources, claims that cannot be represented in two-dimensional maps. These claims, such as that of women over trees, are fluid and constantly under negotiation. Clearly, shifts in negotiational success, as when women lose the right to trees, can have effects on the rural landscape. In such cases, other actors take over or log trees. Therefore, the landscape, representing multiple claims over resources, will mirror the changes in the legitimacy of these claims.

Fairhead and Leach, in *Misreading the African Landscape* (1996), show that narratives of loss, of a continually 'degrading' environment worked for the benefit of national environmental institutions, even though these narratives were untrue, 'Considering the environment as degrading is crucial to the solvency of state environmental institutions even when they do not receive donor support. Making the case for pressing environmental degradation helps to justify state budgetary allocation' (Fairhead and Leach, 1996: 268). This idea of environmental degradation meshes with the idea of the farmer as 'environmental destroyer, and of the need for modernization of resource management and farming techniques' from the modern experts (Fairhead and Leach, 1996: 268). Therefore, romantic narratives of recovery go hand-in-hand with prescriptions of progress as science and expertise become responsible for saving nature. The landscape of desire is one in which both farmers and the land are managed by outside expertise controlling natural and monetary resources.

NATION

Landscapes, real and represented, can also reflect the desires of national identity. Images of the countryside can represent for some the source of national identity. In other cases, a modernized countryside and industrial agriculture can represent national progress. Clearly, the English countryside is a deep source of national identity (Bunce, 1994; Williams, 1973). This is true in other places as well. Vandergeest shows how Thai notions of the village as a timeless repository of national identity overlap with Thai nationalistic ideas of how villages should be re-organized into the realm of rational modernity (Vandergeest, 1995). Peluso and Vandergeest's comparative study of forest law in Indonesia, Malaysia and Thailand (2001) show how national cultures and histories affect the way in which nations manage landscape. Sturgeon (2000) shows how two nations, China and Thailand, and their ways of relating to a particular tribal people inhabiting both sides of a national border, affect the ways in which that tribe utilizes the resources that make up the local landscape.

JUSTICE

While it may seem a surprising inclusion, the environmental justice literature also describes how desire can have effects on a rural landscape. In this case, the desire is less likely to be saving 'the rural' and more likely to be resistance to the involuntary imposition of health risks due to the siting of a hazardous facility in the landscape (Walker, 1995). Yet, once again, this desire can affect the local landscape in terms of what is *not* there, or how the potentially hazardous facility occupies the place. Because of the tendency for the owners of hazardous waste companies to see rural inhabitants – particularly African-American inhabitants of the American rural South – as less resistant to the siting of hazardous facilities, the issue of environmental justice is often a question of racial struggles over what is often a rural landscape. For example, one of the largest and best-known environmental justice protest movements took place in Warren County, North Carolina, a poor and rural county where residents opposed the siting of a large toxic waste landfill (Geiser and Waneck, 1994). In addition, Bullard (1994) describes the landscape of 'Cancer Alley', Louisiana's industrial corridor, as 'paradise lost'. An area previously rich in soil, fish and animal resources, it was overtaken by petrochemical facilities, with drastic environmental consequences. A quote from one resident indicates that the environmental effect of a local toxic waste burning facility was perceived as not only strong odours and continual sickness, but also a deterioration of the local landscape: 'People used to have nice gardens and fruit trees. They lived off their gardens and only had to buy meat. Some of us raised hogs and chickens. But not after [the facility] came in. Our gardens and animals were dying out' (in Bullard, 1994: 56).

SUSTAINABILITY

Sustainability, the desire to organize production in ways that do not deplete resources, is a project that has potential to impact rural landscapes. Sustainability is certainly a landscape of the political imaginary, often represented in the pictorial language of the pastoral (Bunce, 1994; DuPuis, 2002). Exactly what a material effect of the sustainable vision might be is not exactly clear. However, most envision it as far from plantation agriculture and its concomitant social inequities. In fact, the desire for equity and the desire for sustainability can often not be distinguished in visions of the sustainable landscape. It can also overlap with visions of smallholder agriculture, as in American family farms, the efforts to preserve smaller farms in the European Community and the preservation of peasant agriculture worldwide. For example, the writings of the Wageningen School focus on locally based forms of agriculture adapted to particular ecological regions. This work tends to see sustainable agriculture as embedded in local knowledges and social networks (van der Ploeg and Long, 1994).

Urban desire for organic food can also be seen as having potential effects on the landscapes of rural areas. The extent to which organic farms and CSA farms have impacted the nature of rural landscape has not yet been studied. The impact would be most felt in the case of ex-urban farms.

CONCLUSION

As these examples show, it is possible to incorporate desire into the analysis of rural landscapes without abandoning the important conclusions of political economy. This is particularly clear when looking at the question of sustainability. While there is clearly a sustainability narrative of progress (ecological modernization) and a sustainability narrative of loss and recovery, the most important direction in this topic of focus is the work that links sustainability to an overcoming of Western dualities and narratives of landscape. Redclift and Woodgate, for example, argue that we must go beyond the science *vs.* social construction debate because 'we refashion our environments physically *as well as* cognitively' (1997: 62) and how we construct our knowledge of our environment has effects on how we interact with it physically. In other words, the truth uncovered by political economic analysis is something more subtle than 'the real'; it is knowledge.

Yet, what the social constructionist view shows us is that the dualisms of nature/culture, city/country are simply ways the West has known the world so far, and are not the only way to know it. 'Thus, there is no single way in which we, as human beings, relate to external nature. Acceptance of the complex, interactive character of social and environmental change means that simple distinctions between 'social' and 'natural' soon become untenable' (Redclift and Woodgate, 1997: 62). In other words, if we desire sustainable landscapes – landscapes that cannot be defined according to a single, or even a dual, representation or narrative – then we must get rid of Desire, with a capital D, the one based on Western dualisms and their associated narratives of progress or loss, as truth. Instead, we must start with our humble, small-d desires, the ones based in what Donna Haraway calls 'situated knowledges'. Otherwise, sustainability becomes another dream of impossible fulfilment, creating ironies and inequalities in its effects.

REFERENCES

Baker, Paula (1991) *The Moral Frameworks of Public Life: Gender, Politics and the State in Rural New York, 1870–1930.* New York: Oxford University Press.

Bullard, Robert (1994) *Dumping in Dixie: Race, Class and Environmental Quality.* Boulder, CO: Westview Press.

Bunce, Michael (1994) *The Countryside Ideal: Anglo-American Images of Landscape.* London: Routledge.

Callicott, J. Baird and Nelson, Michael P. (1998) *The Great New Wilderness Debate.* Athens, GA: University of Georgia Press.

Carney, Judith (2001) *Black Rice: The African Origins of Rice Cultivation in the Americas.* Cambridge, MA: Harvard University Press.

Cloke, Paul (1994) '(En)cultural Political Culture: A life in the day of a "rural geographer"', in P. Cloke, M. Doel, D. Matless, M. Phillips and N. Thrift (eds), *Writing the Rural: Five Cultural Geographies.* London: Paul Chapman. pp. 149–190.

Cloke, P. and Little, J. (1997) *Contested Countryside Cultures: Otherness, Marginalization and Rurality.* London: Routledge.

Cloke, P., Milbourne, P. and Widdowfield, R. (2001a) 'Homelessness and rurality: Exploring connections in local spaces of rural England', *Sociologia Ruralis*, 41 (4): 438.

Cloke, P., Milbourne P. and Widdowfield, R. (2001b) 'Interconnecting housing, homelessness and rurality: evidence from local authority homelessness officers in England and Wales', *Journal of Rural Studies*, 17 (1): 99–111.

Cronon, William (1991) *Nature's Metropolis: Chicago and the Great West.* New York: W.W. Norton and Co.

Davis, Mike (1992) *City of Quartz: Excavating the Future in Los Angeles.* New York: Vintage Books.

Deleuze, Gilles and Guattari, Felix (1987) *A Thousand Plateaus: Capitalism and Schizophrenia.* Minneapolis MN: University of Minnesota Press.

DuPuis, E. Melanie and Vandergeest, Peter (1995) *Creating the Countryside: The Politics of Rural and Environmental Discourse.* Philadelphia, PA: Temple University Press.

DuPuis, E. Melanie (2002) *Nature's Perfect Food: How Milk Became America's Drink.* New York: NYU Press.

Fairhead, James and Leach, Melissa (1996) *Misreading the African Landscape: Society and Ecology in a Forest-Savanna Mosaic.* Cambridge: Cambridge University Press.

Fitchen, Janet M. (1991) *Endangered Spaces, Enduring Places: Change, Identity and Survival in Rural America.* Boulder, CO: Westview Press.

Garcia, Matt (2001) *A World of Its Own: Race, Labour, and Citrus in the Making of Greater Los Angeles, 1900–1970.* Chapel Hill, NC: University of North Carolina Press.

Geiser, Ken and Waneck, Gerry (1994) 'PCBs and Warren County', in Robert Bullard (ed.), *Unequal Protection: Environmental Justice and Communities of Color.* San Francisco: Sierra Club Books.

Gouveia, Lourdes (1994) 'Global strategies and local linkages: the case of the US meatpacking industry', in Alejandro Bonanno et al. (eds), *From Columbus to ConAgra.* Lawrence, RS: University of Kansas Press.

Grossman, Lawrence S. (1998) *The Political Ecology of Bananas: Contract Farming, Peasants, and Agrarian Change in the Eastern Caribbean.* Chapel Hill, NC: University of North Carolina Press.

Halfacree, K. (1994) 'The importance of "the rural" in the construction of counterurbanization: evidence from England in the 1980s', *Sociologia Ruralis*, 34 (2–3): 164–189.

Kuletz, Valerie (1998) *The Tainted Desert: Environment and Social Ruin in the American West.* New York: Routledge.

Marsden, Terry (1992) 'Exploring a rural sociology of the Fordist transition: incorporating social relations into economic restructuring', *Sociologia Ruralis*, 32: 209–230.

Marsden, Terry, Murdoch, John, Lowe, P., Munton, Richard and Flynn, A. (1993) *Constructing the Countryside.* London: UCL Press.

Marx, Leo (1964) *The Machine in the Garden: Technology and the Pastoral Ideal in America.* London: Oxford University Press.

McWilliams, Carey (1971 [1939]) *Factories in the Field: The Story of Migratory Farm Labor in California.* Santa Barbara: Peregrine Publishers.

Merchant, Carolyn (1989) *Ecological Revolutions: Nature, Gender, and Science in New England.* Chapel Hill: University of North Carolina Press.

Merchant, Carolyn (1995) 'Reinventing Eden: Western culture as a recovery narrative', in William Cronin (ed.), *Uncommon Ground: Toward Reinventing Nature.* New York: Norton. pp. 132–159.

Mitchell, Don (1996) *The Lie of the Land: Migrant Workers on the California Landscape.* Minneapolis, MN: University of Minnesota Press.

Mormont, M. (1990) 'Who is rural? Or, how to be rural. Towards a sociology of the rural', in Terry Marsden, P. Lowe and Sarah Whatmore (eds), *Rural Restructuring: Global Processes and their Responses.* London: David Fulton. pp. 21–45.

Nye, David (1999) *Technologies of Landscape: From Reaping to Recycling.* Amherst, MA: University of Massachusetts Press.

Peluso, Nancy and Vandergeest, Peter (2001) 'Genealogies of the Political forest and customary rights in Indonesia, Malaysia, and Thailand', *Journal of Asian Studies*, 60 (3): 761–812.

Phillips, Martin (1998) 'The restructuring of social imaginations in rural geography', *Journal of Rural Studies*, 14 (2): 121–153.

Ploeg, Jan Douwe van der and Long, Ann (1994) *Born from Within: Practice and Perspectives of Endogenous Rural Development*. Assen, Netherlands: Van Gorcum.

Raynolds, Laura, Myhre, David, McMichael, Philip and Buttel, Frederick (1993) 'The new internationalization of agriculture: a reformulation', *World Development*, 21: 1101–1121.

Redclift, Michael and Woodgate, Graham (1997) 'Sustainability and social construction', in Michael Redclift and Graham Woodgate (eds), *The International Handbook of Environmental Sociology*. Cheltenham: Edward Elgar.

Rocheleau, Dianne and Edmunds, David (1997) 'Women, men and trees: gender, power and property in forest and agrarian landscape', *World Development*, 25: 1351–1371.

Roosevelt, Theodore (1916) 'Wilderness reserves: the Yellowstone Park', in *Outdoor Pastimes of an American Hunter*. New York: Charles Scribner's Sons. pp. 320–353.

Said, Edward (1978) *Orientalism*. New York: Pantheon Books.

Sibley, David (1995) *Geographies of Exclusion: Society and Difference in the West*. New York: Routledge.

Slater, Candace (2002). *Entangled Edens: Visions of the Amazon*. Berkeley, CA: University of California Press.

Stoler, Ann Laura (1995) *Race and the Education of Desire: Foucault's History of Sexuality and the Colonial Order of Things*. Durham, NC: Duke University Press.

Sturgeon, Janet (2000) 'Practices on the periphery: marginality, border powers and land use in China and Thailand'. PhD dissertation, Yale University.

Urry, John (1995) *Consuming Places*. London: Routledge.

Valentine, Gil (1997) 'Making space: lesbian separatist communities in the United States', in P. Cloke and J. Little (eds), *Contested Countryside Cultures: Otherness, Marginalization and Rurality*. pp. 109–122.

Vandergeest, Peter (1995) 'Real villages: national narratives of rural development', in E. Melanie DuPuis and Peter Vandergeest (eds), *Creating the Countryside: The Politics of Rural and Environmental Discourse*. Philadelphia, PA: Temple University Press.

Walker, G. (1995) 'Social mobilization in the city's countryside: rural Toronto fights waste dumps', *Journal of Rural Studies*, 11: 243–254.

Wells, Miriam J. (1996) *Strawberry Fields: Politics, Class, and Work in California Agriculture*. Ithaca, NY: Cornell University Press.

Williams, Raymond (1973) *The Country and the City*. New York: Oxford University Press.

Woods, M. (1997) 'Discourses of power and rurality. Local politics in Somerset in the 20th century', *Political Geography*, 6: 453–478.

Worster, Donald (1985) *Rivers of Empire: Water, Aridity, and the Growth of the American West*. New York: Oxford University Press.

9

Idyllic ruralities

Brian Short

INTRODUCTION: THE RURAL IDYLL – AMBIGUITY AND POWER

A parliamentary Environmental Select Committee, reviewing the UK 1995 Rural White Paper, noted that: 'Rural poverty must be spoken about. Its causes and consequences discussed, perceptions of a rural idyll need to be corrected and hidden pockets of poverty sought out and relieved.'

The rural idyll is an oft-cited but culturally charged phrase which has come to particular prominence within the last 25 years, entering everyday parlance, and which has assumed a shorthand meaning covering a number of interrelated concepts and their contextual relations. It is a phrase often used pejoratively in academic circles, but one that is both ambiguous and powerful. It is, of course, also a phrase which acquires meaning only through the consciousness of, and contradistinction between an assumed 'other' which is un-idyllic – and this chapter will uncover the spatial and temporal correlates of that 'othering', without which the phrase is meaningless (Short, J., 1991: 34).

The origins of the concept have been traced elsewhere, although a full analysis of its constituent parts and their variation over time and space has not been attempted. Raymond Williams, in his well-known contribution to the development of the concept, reveals how, by uncovering successive time slices, demonstrated through literary conventions set within their historical relations, we travel back through British (more specifically English) history to an early-modern sixteenth and seventeenth century concern with a beauty and rural ease, a concern which then appears again in the late fourteenth century work *Piers Plowman*, and then jumps alarmingly back through time and space to a borrowed 'golden age' of Mediterranean Antiquity. And as Eldridge and Eldridge put it: 'Williams offers the image of the escalator. Each time one gets off at the period referred to as the golden age, there will be a contemporary writer to greet us and tell us that such an age has past [*sic*]' (1994: 178). In the long-term overview that follows, we first encounter this archaic and classical phase before moving on to successive representations offered during the medieval, Renaissance and capitalist modern periods, the latter including the climactic years between 1860 and 1930. At the end, we look at later and more recent understandings of the term, together with some of the implications for contemporary rural society and culture.

ARCHAIC AND CLASSICAL DIMENSIONS

In the ninth century BC, Hesiod in his *Works and Days* could already look back to a time of a first Golden Age when:

> the fruitful earth unforced bare them [mortal men] fruit abundantly and without stint. They dwelt in ease and peace upon their lands with many good things, rich in flocks and loved by the blessed gods. (http://sunsite.berkeley.edu/OMACL/Hesiod/works.html)

Hesiod was therefore recommending practical agriculture, social justice and neighbourliness to retrieve such splendour on the fertile plains of Boeotia (Williams, 1973: 24). Later versions, based on third century BC idylls of Theocritus (*c*.310–250 BC) and set now on the Greek islands, on Sicily and around Alexandria, downplayed Hesiod's concern with practical crop production and flock maintenance in favour of a pastoral life within simple communities. Thus Theocritus:

> All rich delight and luxury was there:
> Larks and bright finches singing in the air;
> The brown bees flying round about the well;
> The ring-dove moaning; everywhere the smell
> Of opulent summer and of ripening-tide:
> Pears at our feet and apples at our side
> Rolling in plenteousness; in piles around
> Branches, with damsons burdening to the ground
> Strewn for our feast.
> (Williams, 1973: 26)

The recurrent popular feeling that there is moral value in an agricultural life, and that all progress detracts from some earlier Golden Age can be charted within Roman writings. For example, the Roman Marcus Porcius 'Censorius' Cato (d. 149 BC) blamed Greek influence for his society's increased preoccupation with leisure, and for the large-scale farming methods ('latifundia') that he foresaw would precipitate the decline of ruralism. Although paying scant attention to the totality of Theocritus' work, the pastoral material within his work was taken up by later Roman writers. In his work on intertextuality in Theocritus and Virgil, Richard Thomas demonstrates the transition from Hellenistic traditions to the Virgilian *Eclogues*. Whereas Theocritus had set his idylls in the Sicily of his youth, Virgil (70–19 BC) in the first century BC sets his *Eclogues* or bucolics, lines written by *c*.37 BC, in an imaginary Arcadia (the first to use such a place name), a remote and escapist space. But real rural issues broke through, as Virgil's family was being faced with confiscation and eviction from the family holding to make way for veteran soldiers who were being relocated (Thomas, 1999, 2001). This concern with loss brings tension to the poetry as immediate threats are contrasted with a Golden Age in Arcadia. Whilst preserving the pastoral style of Theocritus, such as the good-natured banter of the shepherds and their love songs, dirges and singing matches, Virgil gave the *Eclogues* an original and more grounded character by introducing real persons and events into the poems, offering what Howard has called 'dialectical temporality' in the interpenetration of 'then' and 'now' as well as 'there' and 'here' (Howard, 2003: 53–4). On the other hand, his slightly later *Georgics* (written *c*.36–29 BC) reflect more realistic rural life, and have been compared to the work of Hesiod. Such practical husbandry advice has become known more generically as 'georgic'. Indeed, George Champion entitled his 1618 volume *The Georgicks of Hesiod*, thereby compounding the confusion between Hesiod the archaic author, and the Roman Virgil, although pastoral and georgic could frequently be combined, since a dalliance or love interest between the farmworkers could sit alongside sterner advice on agricultural matters. By the eighteenth century this would certainly become a common elision within English rural representations.

In all these archaic and classical writers the concept of a blissful Golden Age of ease and comfort, with its spontaneous crops and docile animals, was used to make a contrast with the writers' contemporary environments in which hard work and agricultural knowledge was essential to maintain that fertility of soil that was once divinely given (Glacken, 1967: 132). The concept of antithesis was an important one. Taking the concept one stage further, Varro (116–29 BC) was also to write in his *De Re Rustica*, 'Divina natura dedit agros, ars humana ædificavit urbes' ('Divine Nature gave the fields, human art built the cities'), thereby underlining the contrast even in his contemporary view, between town and country, or between the graces of primitivism and the degeneracy of urban life.

It is important to note that for all these texts, and for others such as the *Beatus Ille* of Horace's Second Epode, where it is a usurer who praises the charms of country life (Beatus ille qui procul negotis … Happy is the man far removed from business affairs), there was either an inherent tension involved, or else the body of work of a writer, such as Theocritus, has not been wholly absorbed. This is important because one element in the Renaissance revival of classical expression was to selectively seize on the pastoral elements in this work, but in a muted, non-confrontational manner. In their incorporation into the Western canon, Williams has it: 'step by step, these living tensions are excised, until there is nothing countervailing, as selected images stand as themselves: not in a living but in an enameled world' (Williams, 1973: 29).

MEDIEVAL IDYLLICISM

Following the collapse of the Roman Empire and the withdrawal of Mediterranean culture from

Western Europe we lose sight of the pastoral tradition for some time. However, it appears again to our view from within the Abbey of St Martin at Tours where Alcuin (735–804), former head of York Cathedral School and who was recognized by Charlemagne as a leading European scholar, wrote several Virgilian-inspired poems. One, possibly by him, runs as a dialogue between winter and spring:

From the high mountains the shepherds come together,
Gathered in the spring light under branching trees,
Come to sing songs, Daphnis, old Palemon,
All making ready to sing the cuckoo's praises.
Thither came Spring, girdled with a garland;
Thither came Winter, with his shaggy hair.
Great strife between them on the cuckoo's singing.
(http://www.parsonsd.co.uk/pastoral.php)

The genre was maintained thereafter but our glimpses are rare. In the thirteenth-century northern French tale of Aucassin and Nicolete, the pastoral idiom is seen again, with happy bantering shepherds as an element within the love story (Sargent-Baur and Cook, 1981). However, the medieval view of the dichotomy between urban and rural life saw the city in a relatively positive light. Rural life was associated with the mundane, with feudal oppression and hardship, and with danger as civilized frontiers were pushed into wilderness areas of wood, heath and marsh, while town life represented freedom and opportunity. Those who could escape from the land to the city for the period of a year and a day might look to become freemen. Craft guilds and neighbourhood associations grew up as urban substitutes for the kin-based ties found in rural areas. Only for the elite did the rural offer leisure in the hunting forests and chases, or in the royal palaces built conveniently within reach of London. For the Christian church, the rural offered the possibilities of asceticism and hard work amid natural beauty. William of Malmesbury's late eleventh or early twelfth century description of the founding of Thorney Abbey in the Fens refers to its image of paradise and a reflection of Heaven itself in its loveliness. Trees, herbs, greenness, a well-nourished soil, fruits and vines were eloquently praised (Glacken, 1967: 313). The creation of the medieval walled monastic gardens also hearkens back to concepts of the Garden of Eden (before the Fall). Works such as *Piers Plowman* by William Langland (*c.*1330–1387) also referred back to simplicity and innocence in its allegorical treatment of social life, and which included the mystery of Christ in the form of a simple shepherd:

In a summer season · when soft was the sun,
I clothed myself in a cloak as I shepherd were,
Habit like a hermit's · unholy in works,
And went wide in the world · wonders to hear.
But on a May morning · on Malvern hills,
A marvel befell me · of fairy, methought.
I was weary with wandering · and went me to rest
Under a broad bank · by a brook's side,
And as I lay and leaned over · and looked into the waters
I fell into a sleep · for it sounded so merry.
(*The Vision of William concerning Piers the Plowman* – the prologue)

Chaucer (d. 1400) also included rural figures among his pilgrims, the poor parson and his brother the ploughman, to represent a simple, Christian lifestyle that the other pilgrims do not. However, the urban–rural duality was insufficient to understand the full range of attitudes to the countryside. Beyond the tamed and blessed farmland was the wilderness, with an uncontrollable nature, and which acquired negative cultural overtones of moral impoverishment, lawlessness and savagery, a place for wandering rather than the stability of the farmer or shepherd. These were the spaces at the Europeans margins of feudal control. But even here the concept is complex, for by the early-modern period the incursions of the medieval period into former wilderness had produced a different image of the wilderness: the wolf and beaver had gone by the thirteenth century, wild boar were only to be found in the hunting forests, and now wilderness was seen with inhabitants who were protectors and teachers of an older pre-Christian environmental wisdom, more in tune with their surroundings than 'cultivated' people. The haunting images of the Green Man in medieval churches (as for example at Southwell Minster or Norwich Cathedral), though pre-dating Christianity, perhaps show such emblematic one-ness with nature, the interwoven foliage and human face being incorporated into environmental story-telling and myth.

THE RENAISSANCE AND THE RURAL IDYLL

The pastoral and georgic traditions were resuscitated more fully within the urban cultures of Renaissance Europe. Imitations of Virgil became an accepted art form by the late fifteenth and early sixteenth century. Politian (1454–94) wrote *Rusticus* describing the year of a Tuscan peasant, and Alamanni (1495–1556) wrote *La Coltivazione* which bears a strong resemblance to the *Georgics* (Williams, 1973: 31). Idealized romantic love was incorporated into the pastoral tradition in the work of Sannazaro (1457–1530), whose *Arcadia* (*c.*1502) represents an early non-dramatic Renaissance pastoral. It appeared in a new edition virtually every other year throughout the sixteenth century.

Within England, Sir Philip Sidney's *Arcadia* (1590, republished in 1593 as *The Countess of Pembroke's Arcadia*), a direct descendant from Sannazaro, underlines the deceit of the rural idyll by the early sixteenth century in that it was written on the site of a former village that had been transformed into an elite parkland landscape. Much of the pastoral by this time was a conventionalized form of allegory or metaphor. The good shepherd of Edmund Spenser's *The Shepherd's Calendar* (1579) for example is allegorical, written in Kent whilst its author was moving into the Earl of Leicester's political faction. It was now understood that such a genre was produced 'under the vaile of homely persons and in rude speeches to insinuate and glaunce at greater matters' (Puttenham, 1589 [1936]: 38) – but such a 'glaunce' was done through layers of accessible meaning that have their own validity. Thus, Pan might represent a Greek god, or Henry VIII, the divine patron of poets, or even Christ. Simple visions of a Golden Age of plenty were by this time once more offered as alternatives to worldly ambition, as idyllic retreats or places of retirement (holiday).

The English versions of these Renaissance works accept many of the conventions of what was by then a genre stretching back over many centuries, at least to Hesiod, and to the eastern Mediterranean. For practical commentators such as John Evelyn (1620–1706) or John Graunt (1620–74), the ill-health of London was obvious compared with the countryside. But the rural idyll was also conceived rather more poetically as one with nymphs, fauns and satyrs of an older world. It was also subject to spatial displacement, and as European settlement expanded overseas Montaigne could write of the inhabitants of North America as having a society resembling a 'Golden Age' (Ashcraft, 1972: 152), thoughts echoed by Sir Walter Ralegh and Michael Drayton. The development of a concern for a lost Golden Age, or age of rural innocence, can be traced through several centuries, each generation conceiving of the concept in its own way. The classical form of poetry which dealt with topographic description was applied to the English landscape by Sidney, which could then be seen as different from its antecedents which were landscapes of the mind beautified by shepherds' songs (Lindenbaum, 1984: 524–41). And subsequent writers expanded on the 'natural' order of landownership which produced order and stability and whose landscaped views were the iconographic representations of that hegemonic power.

The genre could also be stretched to incorporate social values and tributes to actual country houses, as in Ben Jonson's *To Penshurst* and Thomas Carew's *To Saxham*. Christian virtue and the Golden Age of plenty combine to offer the poet and visitor a 'natural' bounty, although one which obscures the labour required to provide the food and drink. Again, the concurrence of such idyllic poetry with the full onset of agrarian capitalism in the seventeenth century is no coincidence. Here was social commentary, made by attaching moral virtue to a vanished and irrecoverable age which could then be contrasted with the ills of the present. Enclosure, for example, was hugely influential at this time in much of rural England, in terms of its landscape, economy, society and also its culture. Enclosure manifested itself physically, but also metaphorically. The enclosed garden, a small encapsulated piece of the rural idyll, could be a metaphor for religious issues or even for the state itself, securely walled against its enemies (Stewart, 1966), and there were many ways in which Elizabethan society used the landscape as a metaphorical tool. The relating of 'Eliza, Queen of shepheardes' within the pastoral form of Edmund Spenser is one example (Montrose, 1980, 1983). Shakespeare has his Gaunt, in *Richard II*, describe England as 'This other Eden, demi-Paradise', with the 'precious stone set in a silver sea', which serves to protect it 'against the envy of less happier lands'.

The use of the Golden Age to highlight the contemporary social, economic or even political problems was well established by this time. We now have the great country houses established in their estates and parklands, based often on the profits from the dissolution or court patronage; the piecemeal enclosure movement gaining strength; the introduction of new plants and communications, experimentation with animal husbandry and the floating of water meadows etc. But whilst Jonson and Carew celebrated the beneficence of their hosts, the landless might also look to a Golden Age to extract another latent theme, that of communal solidarity, which might again provide a future aspiration. Two contrasting themes could therefore be drawn from the concept of the Golden Age: one conservative and unchanging in which the natural order of society was to be preserved, the other a more radical reworking in which the landless might gain land, and become, as in Thomas More's earlier *Utopia* (1516), an owner-occupying small farmer, in a more primitive state.

This primitivism regarded the countryside as arcadian, rather than violent, and innocent as in Spenser's *Faerie Queen* or Shakespeare's setting for *A Midsummer Night's Dream*. In turn the notion of a 'Golden Age' of Tudor England which incorporated Arcadia in a semi-English, semi-mythical idyllic setting, always rural, provided a powerful nostalgia. And the idea of the noble savage culminated in Rousseau's cult of primitivism and a 'new and improved version' of the savage (Symcox, 1972: 223–47).

EIGHTEENTH AND EARLY NINETEENTH CENTURY AGRARIAN CAPITALISM

The eighteenth and nineteenth centuries saw unprecedented urban and industrial growth. In 1750 an estimated 75 per cent of the population worked in agriculture: by 1851 this figure had dropped to just over 21 per cent. The absolute preconditions for a nostalgic regard by many dislocated and proleterianized families for the reassurance of village family ties were thereby enhanced (Bunce, 1994: 9–12). Community and a sense of place were lost to many.

Building on the early-modern preconditions of investment in land in Britain, by the early eighteenth century such visions of rural ease had become more situated in actual English social situations, even if they were still somewhat generalized depictions in what is known as the 'Augustan' style imitating classical writers. But at least the idyll was now to be found in a small estate or farm just outside the reach of what Abraham Cowley called 'this great Hive, the City' (Waller, 1905: 88):

> A farm some twenty miles from town
> Small, tight, salubrious and my own:
> Two maids, that never saw the town,
> A serving man not quite a clown,
> A boy to help to tread the mow,
> And drive, while t'other holds the plough
>
> Matthew Green (1696–1737)

For Williams (1973: 23–47), the simplification of pastoral to a superficial form that lacked the former complexity was associated with the transition from a feudal to a bourgeois culture, and more specifically to an agrarian capitalism. Furthermore, the superficial contrast is between the texts illustrating the pleasantries of rurality when compared with those of urban(e) cosmopolitan and civilized notions of living. Such a starting point could quite easily drift into an essentialist relativism, with 'countryside' and 'town' always seen as in contradistinction and as timeless entities. But the underlying analysis that he posits, and which is taken further by Sales (1983), is that of the highly significant connections between town and country in the form of capitalist economic and social relations. Not just the town in this context of course, since much of the exchange that brought wealth was with colonial or empire possessions, and indeed was formulated on the back of slavery. But the development of such relations gives the concept another time frame, namely that of the development of urban-commercial capitalist relations in Britain. As Sales puts it, 'Capitalism was really quite at home in both the long and short grass of rural England.' Only when the grounded reality of such relations is exposed, and the commodification of the rural appreciated, can we really see the rural idyll for what it actually is. To quote Sales again on pastoral, it is 'the language of the victors' (1983: 17). Or perhaps more precisely it is shorthand for those elements of a discourse which comprise the inter-relationships of the social, political, economic and environmental power struggles within the countryside. Such a discourse, then, used the concept of the neo-pastoral rural idyll as ammunition to demonstrate an argument, to prove a point, to establish a position to defend.

Other such works, alongside an emphasis on descriptions of landscape serenity and beauty, aligned the genre more with the view of the visitor, the tourist, rather than the rural worker or real shepherd. The latter was depicted in works such as François Boucher's *Shepherd piping to a shepherdess* (*c.*1745), and had by now become a traditional figure representing innocence; and those with elite pretensions adopted the figure as a disguise in courtly diversions. Marie Antoinette most famously had 'Le Petit Hameau' stocked with perfumed sheep and goats, and built in 1783 as a Viennese retreat in the rustic half-timbering as seen on Normandy farms, although the interiors were fully equipped with the latest refinements – all this for her enjoyment and relaxation next to the Trianon at Versailles. This new lakeside village had replaced the original village of Trianon, on the orders of Louis XIV in the 1670s, and included a working farm that later supplied the queen with fresh milk, eggs, butter, cream and cheese. Here she played at being a simple shepherdess and entertained close friends and family.

By the eighteenth century, Thomson (1700–48), in *The Seasons* (1726–30), is the first major poet to treat landscape and society as a specifically English issue since the seventeenth century. There were also strong currents of intertextual referencing between the major poets and aesthetic theorists of the time – Thomson, Gilpin, Wordsworth, Coleridge, Cowper and Samuel Johnson. A particular rural landscape view, represented as 'natural', therefore becomes a metaphor for authority over previous delineations of rurality as well as for control over the contingencies of independent livelihood at this time. They were also landscape representations which were employed for a normative argument, and the stepping stone for a discourse which contains both conservative and more radical views. The very landscape itself, as owned by gentlemen in the eighteenth century, could be seen as an image of disinterested aesthetic taste on the one hand, or as resulting from patronage, corruption and self-seeking on the other.

There are other eighteenth century themes observable in rural writing. There is melancholy invoked by the intimate portrait of a country churchyard in Thomas Gray's *Elegy*, a theme of ongoing farewell which continues to pervade the genre. The eighteenth century view of landscape was one of idyllicism, of generalities rather than specifics, and of timelessness. Of interest here also is the increased use of the landscape to reflect moral character, whether in Fielding's *Tom Jones* (1749) or Jane Austen's *Mansfield Park* (1814), and landscape images suffused the output of Wordsworth, as in the *Prelude* (1805), and Coleridge at the turn of the nineteenth century. James has referred to landscape as seen by Wordsworth as 'an active element working on the individual, a "medicine" for the soul suffering from the effects of weariness, doubt and the pressures of an increasingly urbanized society' (James, 1989: 64). And both Barrell (1983) and Payne (1993) have stressed that such eighteenth century paintings, created for the wealthy, were representations of the desirable virtues of the rural poor, rather than observable reality. Jane Austen could write in 1817 of the village of Sanditon, on the Sussex coast, as 'precluded by its size from experiencing any of the evils of civilisation' (Austen, 1987: 140). Books, images and performances were now widespread and contained images of pastoral or georgic convention. Packe, for example, in his *Ancographia* of 1743, describes the countryside of East Kent as an Earthly Paradise, with an 'Inexpressible Consent, Grandeur and Politeness all over the face of the work, that can never be made to appear in the chart, as it does in the grand Book of Nature' (Charlesworth, 1999: 119).

We do not lack empirical underpinning for such a placement. The eighteenth century sees the material transformation of wholesale rural landscapes from farmland to manicured parkland; the moving and creation of villages; the re-planning of much of the lowland farming landscape through enclosure; and the interpenetration of these material changes with an emergent English school of landscape painting which both gave to, and took from, its contextual relations with powerful patrons and their privatized landscapes (Williamson, 2002).

The rural idyll in the late eighteenth and early nineteenth centuries was coloured to a large extent by Romanticism. A typical feature here was the prominence and close relationship of emotion, intuition, aesthetics and morality; another feature was the well-known aversion to modernity and industrialized society. In the representation of nature, these features are reflected in a strong emotional and moral sympathy towards 'natural' beauty, and in the admiration of real rural life and wilderness. In poems such as *Michael: A Pastoral Poem*, Wordsworth harkened back to an earlier eighteenth century lifestyle, intrinsically linked with nature, of the Grasmere Vale Lakeland 'statesman' – the independent small farmer of the region. With typical pastoral as his method, there is the excision of real economic factors and a generalized harking back to an earlier and better age. Melancholy, and a more distanced view of simple rustic figures in tune with their landscapes, are apparent here. But Michael's virtue lay in hard work, not the graceful arts. Both Wordsworth and Coleridge could use the pastoral poem to launch political attacks on those whom they feared were in danger of ruining the landscape – gentlemen and outside capitalists – those with leisure time to cultivate taste and fashionable landscape views. In the work of Thomson and Cowper landscape description was used politically to intervene in arguments about the nature of national government and the corruption of commerce (Fulford, 1996: 5–9; 1998: 59–86). And poets found themselves caught up in the conflicting politics of their patrons, or, as with Wordsworth, exposing the ideologies of 'gentlemanly taste' for landscape in styles that were now revealing highly specific environments, such as the Lake District.

The early nineteenth century images of idyllicism are perhaps most associated with the naturalism of Constable or the abstract romanticism of Turner. The close identification of the former with his native Stour valley has resulted posthumously in the 'Constable country' epithet becoming widely known, and seen by many as the very essence of the rural idyll. By the 1890s Thomas Cook was organizing railway trips to Constable country (Daniels, 1993: 212). But the paintings of these two great artists were in turn rooted in the landscapes of seventeenth century Dutch and French painters, such as Hobbema and Claude. Mainstream artists such as Thomas Sidney Cooper (1803–1902) furnished the Royal Academy with compositions bathed in a golden arcadian, or otherwise limpid, light. Otherwise, interior cottage scenes likewise built on Dutch seventeenth century painters such as Ostade and Teniers, and painters such as Myles Birket Foster (1825–99), developed themes of rustic simplicity and rude health in contrast to the emerging urban squalor. The urban 'other' therefore looked for escapism to a rural, and indeterminately past, lifestyle and landscape.

An ever-present danger when dealing with such representations is to select according to the argument one is making. It is relatively easy to construct an unfolding narrative of a Golden Age viewed through artistic creativity; the Romantic poets of the nineteenth century spun a golden tale of idyllicism, from Wordsworth through to Keats, Shelley, Tennyson or the Brownings. The romantic view was of living beings in their landscapes, and in particular of living beings and landscapes positioned at a distance from industrial society. The increasing concern felt by many nineteenth century literary and artistic individuals for the loss of a bucolic childhood may have become a cliché but it was nevertheless deeply rooted. Of the Kentish childhood of Charles Dickens, for example, it could be written that 'the pastoral images rooted in his personal myth of a rural childhood Golden Age [had] continued to hold a special kind of power over his imagination' (Schwarzbach, 1979: 173). Not only did rural landscapes possess restorative power, but also Victorian moral values. However, it should not be forgotten that, if not equal in their numbers, there were also those who sought to bring the attention of metropolitan classes to the hidden problems of the countryside. For every Dingley Dell there were the lesser regarded thoughts of a pugnacious Cobbett or observant William Howitt. The latter tried hard to reconcile pastoral vision with the reality he saw: 'In many of the southern counties, but nowhere more than in Hampshire, do the cottages realize, in my view, every conception that our poets have given us of them.' Margaret Hale in Mrs Gaskell's *North and South* describes a village where she used to live as 'like a village in a poem – in one of Tennyson's poems' (Keith, 1988: 81–7).

Change was accompanied by nostalgia, regret, even by riot. Those who bestowed patriarchal beneficence in the countryside and expected deference in return might look back wistfully to a Golden Age, but those whose living was threatened by the drainage of marshland, woodland clearance or other 'improvements', and whose common rights vanished with enclosure, could not afford wistfulness and there were local riots in protest at the loss of ancient landscapes and their inherent communal rights. No rural idyll here then, but a growing social gulf between those who consumed landscapes for pleasure and those whose very livelihoods were threatened by the landscape changes. So, the rural idyll did not go completely unchallenged, especially in the multiplicity of attitudes to the countryside during the later eithteenth and nineteenth centuries. We can see real anger at departed rural ways of life, and moral and political authority too, in George Crabbe's *The Village* (1783), itself a rebuttal of Goldsmith's *Deserted Village* (1770), and, for example, in the poetry of John Clare (1793–1864), who saw the enclosure of his village of Helpston as the end of the world he had always known, but whose work was sometimes suppressed because it failed to offer the required pastoral vision (Barrell, 1972; Bate, 2003). For Crabbe:

Fled are those times, when, in harmonious strains,
The rustic poet praised his native plains:
No shepherds now, in smooth alternate verse,
Their country's beauty or their nymph's rehearse ...

I grant indeed that fields and flocks hath charms
For him that grazes or for him that farms;
But when such pleasing scenes I trace
The poor laborious natives of the place,
And see the mid-day sun with fervid ray,
On their bare heads and dewy temples play;
While some, with feebler heads and fainter hearts,
Deplore their fortune, yet sustain their parts –
Then shall I dare these real ills to hide
In tinsel trappings of poetic pride?

(George Crabbe, *The Village*, 1783)

Clearly, the concept of an idyllic rurality is multivalent, composed of interlocking strands which emerge at different times and places. It could be argued convincingly that the concept emerges strongly in England with the commodification of society, with industrialization and urbanization, with the possibility of purchasing positional goods and the greater ease of travel between town and country, and with the identification of nation-states with particular landscapes. Such an emphasis would then place the articulation of the rural idyll, in the British Isles at least, within an emergent eighteenth century capitalism and national consciousness, becoming part of the cultural experience of capitalist society.

THE APOGEE OF THE RURAL IDYLL: 1860–1930

Poor rural housing conditions contributed to rural–urban migration, since despite the squalor of urban slums, the countryside progressively lagged behind the town in amenity. By the mid-nineteenth century, writers continued to present towns as the centres of vice and crime, although many views of the rural population also continued to focus on the moral, intellectual and cultural shortcomings of rural areas. In demographic terms therefore, it was the pull of wider opportunities, combined with the push effect of discontent with conditions at home, which induced the later-Victorian rural exodus (Short, B. 2000: 1271–96). And by the end of the nineteenth century there was a general sense that the close-knit, narrow, rural world of the eighteenth and early nineteenth centuries had vanished for ever. This feeling was undoubtedly emphasized by the gradual realization in the last quarter of the century that the agricultural depression was socially divisive through the loss of rural jobs, curtailment of investment and further deterioration in housing conditions. But even so, it was the squalor and poverty of urban living which became so powerful that a whole 'countryside cultural industry' grew up in response. Escape might be actual in the sense of walking, cycling, golfing or 'weekending' or through the rash of commuter villa-dom with access to middle-class employment; or virtual, through a display of cultural capital in the form of paintings or prints. Paintings such as Foster's *Children Playing* (1886), or the paintings of vernacular Surrey cottages by Helen Allingham (1848–1926) which were exhibited in London in the 1880s and 1890s, supplied the need. These were powerful messages which created an influential set of harmonious images. And when allied to the allegorical artists attempting to demonstrate God's work in nature, such as Linnell (1792–1882), the Victorian sensibilities were further raised. The Pre-Raphaelite Brotherhood from the late 1840s also produced very detailed images of a 'still moment, usually in summer or autumn, heavy with sensations' (James, 1989: 71), as in Ford Madox Brown's *Pretty Baa Lambs* (1851–9).

The importance of tracing the rural idyll in the later nineteenth century therefore lies in part in the influence that was imparted to the growing preservation/conservation movement. Allingham's links with William Morris's 1878 Society for the Protection of Ancient Buildings (SPAB) revealed her genius for the recording of old cottages that were being demolished or renovated at a great rate, and at the same time being replicated in pastiche imitations. Town architects were thereby '[rub]bing out a piece of old England, irrecoverable henceforth by all the genius in the world and the money in the bank … although perhaps on the very next property an architect is building imitation old cottages with lattices!' (Treble, 1989: 54). Elements of nostalgia and romanticism tinged such efforts, allied to a middle-class self-identification and desire to escape from contact with inner-city problems, and incorporation into the conservative values of the elite landowning classes. But that is not to dismiss the very real concerns over the unhealthy state of many industrial cities, the efforts to preserve open spaces for recreation, and a real sense of the destructive power of humankind. Undoubtedly the Commons Preservation Society (1865) and the National Trust (1895) as well as SPAB owed a great deal to late Victorian rural idyllicism. Ruskin and William Morris in turn developed anti-modern ideas in design and craftsmanship and concern for architecture which praised older rural lifestyles. Morris became established in Victorian family circles as an escapist poet of Arcadia. His poetic romance *The Earthly Paradise* begins:

> Forget six counties overhung with smoke,
> Forget the snorting steam and piston stroke,
> Forget the spreading of the hideous town …
> (cited in Marsh, 1982: 13–14)

Real destructive power was horribly all too evident in the First World War. Combined with this was what Paul Fussell has referred to as 'A tradition of imperialist exile from home' (Fussell, 1975: 222–3) as civil servants, planters, military personnel and younger sons remembered the 'Home Counties' with huge and exaggerated affection, as the British Empire spanned ever more remote regions of the globe. The Georgian poets gave rural life huge wartime symbolism, as for example in Rupert Brooke's 'The Soldier' and rural nostalgia was a very powerful incentive to fight, even though patriotism may have stirred but a narrow class view. In many ways, of course, the Great War was the ultimate anti-pastoral, although even the war cemeteries were arranged as English gardens with reference to halcyon pre-war days 'innocent enough to offer hope' (Morris, 1997: 427).

The incorporation of the concept by all shades of political opinion is also marked. The rural idyll, at least in some of its manifestations, is as much an element in the political landscape of the far left as the far right in Britain. Country dancing, for example, became an accepted element in a socialist education and upbringing, and Robert Blatchford's influential socialist text *Merrie England* (1893) became hugely popular. Raphael Samuel's communist family sent him to a progressive school dedicated to nature, crafts, fresh air and folk dancing (Matless, 2000: 81). There were also those who sought a fresh life untainted by urban influences. The 'back to the land' or 'simple life' movements were particularly active between 1880 and 1914, and drew in disparate utopian philosophers, agrarian communes, farm colonies and peasant arts and crafts interest groups to confront the reality of rural existence. A passion for pre-industrial workshop organization and pastoralism gave impetus to organizations such as Ashbee's Guild and School of Handicrafts in the Cotswolds from 1901, or the sculptor and calligraphist Eric Gill's colony at Ditchling below the South Downs from 1907. The aims were generally to improve rural life, not just to escape the towns, but the values could be transposed back to the towns, as with the garden suburb at Hampstead, or at Bourneville. The ideal of a close-knit *gemeinschaft* community, with a nostalgia for a way of life remembered as purer, simpler and closer to nature, with stability, a sense of belonging and an escape from the evils and dangers of the city, underpinned semi-rural suburban developments such as Bedford Park (Bunce, 1994: 158–9), laid out in 1867 at Chiswick with designs by Norman Shaw with its green and inn (even, sometimes, a maypole). In turn the later radical ideas of a fusion between town and country, as advocated in Ebenezer Howard's *Tomorrow: A Peaceful Path to Real Reform* (1898), led to real, if limited, attempts to recreate community at Letchworth Garden City (1903) (Hall and Ward, 1998; Heathorn, 2000: 113–28).

Radical thinking had, in large measure, been aligned with the English rural idyll. By the 1920s there were few villages in favoured parts of the South East without their share of radical intelligentsia, such as the Bloomsbury group which had decamped to Sussex, or the Surrey group of Adrian Boult, G.M. Trevelyan, Clough Williams-Ellis, Edwin Lutyens and others at Shere (Brandon and Short, 1990: 343). Thomas Hardy's evocation of the Wessex Egdon Heath, not typically idyllic countryside but nevertheless viewed as part of a quintessential England, was the subject of work by Hardy's friend Gustav Holst (1874–1934), who aimed to incorporate sounds of the heath into his *Egdon Heath* (1927). He had earlier written *A Somerset Rhapsody* (1907), which, with its rich oboe parts, conveys a full English rural nostalgia and emotional beauty. Holst's contemporary Ralph Vaughan Williams (1872–1958), deriving inspiration from his Gloucestershire countryside, is well known for his fashionable incorporation of folk music, as in his *Pastoral Symphony* (1922), or his incorporation of Housman's poetry in his *On Wenlock Edge* (1909). The collecting of English folk music owed much to Williams, as well as its doyen Cecil Sharp and also to Percy Grainger (1882–1961), or to George Butterworth (1885–1916), whose *Banks of Green Willow* (1913) is a classic languid and poignant musical evocation of the English rural idyll. The more overtly nationalistic music of Edward Elgar (1857–1934) and Frederick Delius (1862–1934) also drew heavily on the English folk traditions. Delius's pastoral miniatures popularized the style of music in such pre-First World War works as *On Hearing the First Cuckoo in Spring* and *A Walk to the Paradise Garden*. These composers joined with Cecil Sharp and others to create a veritable folk revival in the early years of the century (Boyes, 1993; Stradling, 1999: 176–96).

When the work of these composers is added to the writing not only of Hardy, but also that of Kipling, Lawrence and their imitators, as well as the Georgian poets, the period between about 1860 and 1930 must surely be viewed as the high point of the rural idyll as depicted in the aesthetic imagination and creative art, broadly conceived and consumed by a hungry urbanizing populace in Britain. In the 1860s Cadbury's began to use idyllic paintings, such as those of Myles Birkett-Foster, on the covers of their chocolate boxes, to increase their sales – hence the 'chocolate-box' image of common usage. Rural-based fantasy by such writers as Kenneth Grahame or the Powys brothers contributed further to a 'lost world' of Arcadia: 'It is then not only that the real land and its people were falsified: a traditional and surviving rural England was scribbled over and almost hidden from sight by what is really a suburban and half-educated scrawl' (Williams, 1982: 258). Nor was this all, for even Prime Minister Baldwin seized the mood in order to emphasize tradition and stability in his now-famous statement on national identity given at the Annual Dinner of the Royal Society of St George in 1924, in which he referred to 'the tinkle of the hammer on the anvil in the country smithy, the corncrake on a dewy morning ...' and so on. The speech itself is well known, but Baldwin also noted that these rural legacies were sacred and eternal and as something fundamental and innate in the human race. Four years later he was to endorse enthusiastically Mary Webb's *Precious Bane* in the 1928 posthumous edition, thereby creating huge popularity for the 'lyrical intensity ... and fusion of nature and man' in Webb's work in particular, and for rural idealism more generally, as he linked Englishness with the stability of rural life at this uncertain time (Miller, 1995: 89–102; Webb, 1928: 9–11).

This is not to deny the continuing counter-pastoral undercurrent during this period. Within painting, the bucolic renditions which sold in their thousands were also challenged by a counter-myth of rural misery within the late-Victorian social realist movement spearheaded by Luke Fildes (1843–1927), Robert Walker Macbeth (1848–1910) and especially by George Clausen (1852–1944). Writing of the nineteenth century paintings of agricultural life in England, Christina Payne has commented that:

> There were three pervasive myths ... that people were happier in the countryside, that country people are more virtuous, and that country people were more virtuous and happy in the past than they are now. (Payne, 1993: 24)

The several processes of alienation were the subject of works such as Disraeli's *Sybil* (1845) or Charles Kingsley's *Yeast* (1848), contrasting visual beauty of the countryside with the suffering of its residents. More knowledgeable and intimate rural portraits came from Edward Thomas (1878–1917) and Ivor Gurney (1890–1937). Similar themes, set within a Dorset-inspired Wessex landscape, were represented by the very knowledgeable Thomas Hardy, who listed the themes of rural decline as resulting from the move towards mass education, a growing self-consciousness, and economic and technical changes in farming practices and within rural society. In his subtle novels, Hardy posited a tension between landscapes which were used as characters alongside his human actors on the Wessex stage. Of Egdon Heath, 'like man, slighted and enduring', he perceptively wrote:

> Men have oftener suffered from the mockery of a place too smiling for their reason than from the oppression of surroundings oversadly tinged. Haggard Egdon appealed to a subtler and scarcer instinct, to a more recently learnt emotion, than that which responds to the sort of beauty called charming and fair. (Hardy, 1912: 3)

Writing of his own *Jude the Obscure*, Hardy is quoted as saying that the 'grimy' feature of the story shows 'the contrast between the ideal life a man wished to lead, and the squalid real life he was fated to lead' (Draper, 1975: 33), an emblematic statement encompassing much of the reasoning behind the persistence of the rural idyll.

LATER TWENTIETH CENTURY IDYLLS AND MODERNITY

The rural idyll flowered again after the Great War, in inter-war Britain, in country house novels, 'back-to-the-land' movements, and the new capacity for visiting and remaining in countrysides now accessible with the family car (Matless, 1998). For the next few decades increasing numbers of motorists used their cars to explore the countryside. They were actively encouraged to do so by advertising campaigns arranged by the oil companies. For example, during the 1930s the now highly collectable Shell posters by artists such as Paul Nash, Edward McKnight Kauffer and Duncan Grant illustrated an English rural idyll, whilst ironically avoiding the depiction of petrol stations. As advertising insinuates itself into general consciousness through identification of the product with larger social processes, so the identification of Shell's commercial interests with the rural idyll was cannily conceived. Shell's concern was to build on the idea that, as Stuart Laing has written for the later twentieth century, 'somewhere at the far end of the M4 or A12 there are "real" country folk living in the midst of "real" English countryside in – that most elusive of all rustic Utopias – "real communities"' (Laing, 1992: 135; Bernstein, 1992). The 1931 launch of the 'See Britain first on Shell' series includes iconographic scenes from Stonehenge and the New Forest, and in the post-war years hundreds of items of promotional rural material were produced by Shell e.g. *The Shell Guides*, *The Shilling Guides* or the *Shell Nature Book*, or even the long-playing and 45 rpm records *The Shell Records of Nature*. W.G. Hoskins, arch anti-modernist, wrote the *Shell Guide to Rutland* in 1963, when it was already clear that the small county was under threat of extinction.

Thereafter the themes might re-emerge in new media forms, such as radio's rural soap opera *The Archers*, for example, from 1951, in television, including television advertising, or the recent mass market for countryside magazines. But one senses a continual re-working of the essential themes of the rural idyll – what Sales (1983) refers to as the 5 'R's of pastoral – refuge, reflection, rescue, requiem and reconstruction. Rural community remained a signifier of stability and English values, and survived the horrors of the Great War only to re-emerge in the Second World War as symbolic arcadian landscapes to be fought for as depicted by artists such as Frank Newbould in his 1940 '*Your Britain – fight for it now*', or very specifically in C. Henry Warren's 1941 *England is a village*. In his foreword to one of many such anthologies of country writing published in the war, Richard Harman began:

> In an age of destruction there is a re-awakened interest in the things that endure. The hills, fields and rivers of England touch the hearts of all of us because they offer normal living and the natural joys of earth. Life is very much more real and full as we get close to the earth. (1943: 5)

Emerging from the exigencies of the Second World War, the enduring countryside with its farmers who had met the challenge to feed the embattled nation, was generally venerated. The structural realignments during the war were thereafter maintained to nurture a 40-year productivist period as farm wellbeing was allied to rural protectionist policies. But at the same time a growing counter-urbanization brought people and jobs to many countrysides, and with the emergent problems associated with agribusiness, overproduction, pollution and landscape destruction, Britain moved into a post-productivist phase in which the concept of the rural idyll has, in a quite new way, been turned against the farming community by an articulate and powerful new middle class. The rural landscape is no longer seen as resulting from agricultural production, and although cheap food remains important to the urban consumer, the unchanging idyllic countryside is still also required (Burchardt, 2002: 158; Howkins, 2003a, 2003b). Indeed, agriculture is now a minority earner even in the countryside, with more than 20,000 jobs in agriculture being lost between 1998 and 2001 alone (Countryside Agency, 2003: 125), and instead the countryside is to be preserved for consumption, primarily by middle-class residents, many of whom have moved recently from metropolitan backgrounds.

This preservation takes many defensive forms and results in many a strange alliance. David Harvey sees such manifestations which look to aestheticize environments or to emphasize their place-derived qualities as essentially conservative and static – looking to security, stasis and nostalgia in an uncertain, changing and globalizing world. Such a course can, however, lead to pastiche, and the faking of a folksy past (Harvey, 1990: 303). As we have seen, there is nothing very new in this. We obtain these insights through the cultural objects left by those generations, through physical relicts such as model villages or landscaped parks and gardens, and most clearly through landscape painting, photography, postcards and literary sources such as poems, novels and guidebooks. Films such as the 1953 Ealing comedy *The Titfield Thunderbolt* rely on the antithesis of modernity/bureaucracy versus the unchanging village in the story of a community which, despite many mishaps, keeps its own branch line open.

The rural idyll continues to exert a real influence in today's world, essentially as a part of what Alison Light (1991: 16) has termed 'conservatism's shifting appeals'. The power of media and advertising demonstrate the ongoing link between rural epithets and consumerism. The consumer buys (into) the countryside through the link made with products, whether they be cars, duvets, beer, kitchen interiors or other 'heritage' products. And the consumer has been culturally attuned from childhood to make the link between the rural and the 'good' (Houlton and Short, 1995), so the market is primed. The market, too, is seemingly only too ready for a profusion of countryside magazines and an ongoing production of film, radio and television productions extolling the country life – relocating houses, vets, rural vicars (comedic or otherwise), soap operas and series such as *Heartbeat* or *All Creatures Great and Small*. And *Country Living*'s retro-ruralism is clearly highly in demand – with anodyne images of cottage interiors and generalized rusticity. Equally memorable and popular is the televised advertising of a rural paradise selling Hovis bread, using a northern tough, gritty, grainy voiceover but a Dorset town and the music of a central European composer (Dvorak). Clearly the rural idyll remains seriously commercial at the start of the twenty-first century, all the more so since it clearly straddles 'high' and 'popular' culture. Downshifting as one of the processes behind counter-urbanization is particularly relevant here, and is frequently inspired by perceptions of the 'good life' in a more remote countryside.

It must be stressed that this is a relatively recent idea; this is a postindustrial view of the relative merits of the country and the city – and it became strongly commercial in the twentieth century. Furthermore, since the concept is used to sell a product, it can no longer be associated with the distant past or far-off Mediterranean golden sunsets of Claude Lorraine. Currently it is used to sell commercial products of all kinds. This includes houses on the edge of Haywards Heath, Sussex. A contemporary private housing development on the south-western edge of the town thus advertises itself as an 'instant rural idyll', a positional good, with its 'boulevards' of four-bedroomed houses and twin garages, interspersed with the new – and thatched – woodcutter's cottage and almshouses, facing around village greens and wooded streets, with woodland effectively supplying the protection of a gated community and with the commuter station just a short distance away. The local cricket team has been sponsored and protection advertised as offered to the environment and local wildlife. The object for sale and the image have been elided, and the rural idyll is blatantly packaged and commodified.

Rural reality can be very different. Eleven million people, over a fifth of the population of England, currently work and/or live in rural areas. Despite the many different versions of the rural idyll held by many incomers, 25 per cent live in or on the margins of poverty (Cloke and Milbourne, 1992; Cloke et al., 1997). Low wages, under-employment and inadequate housing are everyday realities for large numbers of people. Homelessness, drug abuse, suicides and crime rates are all increasing in the countryside, the Rural Stress Information Network is a fast-growing charity, and rural services are in decline. There are complex issues involved here, of course, and they have been invoked in multi-faceted and somewhat non-selective ways by the Countryside Alliance since the late 1990s in a series of high profile marches. Alongside the perception of the countryside as a pre-industrial idyll, there is also a feeling that it is backward and to be left as soon as possible, especially by the intelligent and aspiring youth. Within contemporary countrysides we therefore have the co-existence of groups referred to as 'pastoralists' and 'modernizers' (Murdoch et al., 2003). But the positive perception of the countryside is so strong that current views continue to elide this symbolic construct with what is essentially a *gemeinschaft* ideal, and with pastoralism and the aesthetic gaze, and thereby to elevate the rural idyll to a disproportionate and distorting status within British culture. When the Tate Gallery, London, mounted an exhibition of Richard Wilson's eighteenth century pastoral views of the countryside in 1982 entitled 'The Landscape of Reaction' with captions demonstrating how Wilson legitimated the mythology of his patrons, critics from the *Daily Telegraph* to the *Guardian* were outraged at the critique of 'this gentle landscape painter' (Daniels, 1993: 2).

The concept of the rural idyll yields a great yet ambiguous power, derived from yoking the apparent beneficence of nature and community in so many different ways. At various times one or the other has been emphasized, and in so doing the concept invokes an ongoing and profound nostalgia which can be harnessed at times of difficulty. This is a fact not lost on politicians, various commercial interests, the media and the advertising industry who understand that different versions of the idyll contain different forms of social relations 'naturalized' within them. And through these agencies, the cultural fantasy, even if only half believed, becomes tangible reality. And so, for those who can afford it, the search for the rural idyll continues. And for those whose circumstances do not allow them to fit within the received and constantly reproduced ideas of the idyll, even as archetypal figures, marginalization also continues. Little and Austin, for example (1996: 101–11), have examined the impact of the concept on the lives of rural women, arguing that it shapes and sustains patriarchal gender relations. The woman of the rural idyll is the wife and mother at the heart of community, not the single professional woman. And the mystification inherent in idyllic concepts of the rural community actually does shape attitudes and behaviour. Counter-urbanization consists in part at least of a middle-class search for belonging, to opt out of the urban social structure into 'country life' into a more authentic, intense and rooted existence. The rural idyll in what Lash and Urry refer to as our 'symbol-saturated society' lures the middle-class resident who in turn thereafter articulates class values based on borrowed pre-industrial ideologies to ensure physical stability in the environment, whilst also bringing different social linkages with them to use in the 'defensive politics' of (their) place (Lash and Urry, 1994: 222, 247; Murdoch et al., 2003: 69–75).

FRAGMENTING THE IDYLL

The rural idyll, then, is a contested term, largely because its ambiguity allows the space for interpretations to be applied as required. Surrounding the phrase is a cultural compound, or as Cloke would have it, 'cultural mêlée' (Cloke, 1994: 179), referring to harmony, permanence, security, inner strength, refreshment and renewal. Somewhere there too are family values, community cohesion, a respect for necessary authority and an emblematic nationhood – all being set within surroundings that are aesthetically pleasing. Lowenthal and Prince (1965: 187–222) see the English ideal landscape as 'compartmented into small scenes furnished with belfried church towers, half-timbered thatched cottages, rutted lanes, rookeried elms, lych gates and stiles'. It should be noted that such a landscape, despite Wordsworth, was that of the south of England, and not necessarily characteristic of many other regions. So can there be a 'national' landscape, and does this equate to the rural idyll within Britain? The concept moves uneasily between, on the one hand, a nostalgic aspiration which, as a positional good, may be purchased in theory, and which is therefore spatially restricted, and, on the other, a national emblem reflecting democracy, liberty and other assumed virtues.

On the whole, the concept of the 'idyll' also has an insecure place among definite literary forms. Its character is vague, often referring to the purely sentimental, and our conception of it is further obscured by the fact that although the noun carries no bucolic idea with it in English, the adjective 'idyllic' has become synonymous with pastoral and the rustic. A dictionary definition thus becomes: 'A poem or prose work describing an idealized rural life, pastoral scenes etc.', although it also includes subsidiary meanings as 'Any simple narrative or descriptive piece in poetry or prose' or 'A charming or picturesque scene or event'. The highly generalized use of the word 'idyll' by the second half of the eighteenth century, both in English and French, was enhanced by the popularity of two works, curiously enough almost identical in date, by two eminent and popular poets. The *Idylles hëroiques* (1858) of Victor de Laprade (1813–83) and the Arthurian romance *Idylls of the King* (1859) of Tennyson (1809–92) enjoyed a success in either country which led to imitations of the title, but possibly by those who had very little idea of its meaning.

The precise subject matter could therefore vary: rural, but what kind of rural? Could it relate to landscape without figures, or did there need to be some semblance of human occupation (even if just as animated props or 'actors on a rustic stage')? As with many definitional projects, the object itself can disappear with greater exposure to a forensic light. Nor must the contingent and contextual issues of different times and spaces be lost in a hurried reductionism. Was there a rural idyll in Scotland? Certainly following Landseer and Scott perhaps, although typically Victorian rural scenes were drawn from lowland, gentle, bounded environments. Scott's most important novels, such as those containing his Rob Roy character, used the idea of the Highland line as dividing wild and lawless countryside from a civilized and cultivated lifestyle (Thorsley, 1972: 293). Even the picturesque was seen by many as a gentle agreement that wild nature could be safely packaged for polite consumption. Was there one in Wales? Again probably following Borrow or the nymphs in Wilson's Italianate *Holt Bridge on the River Dee* (1767). But the timing and form that such idyllicisms took varied, and were very different from those of lowland England, for example, although we must bear in mind that the consumers of the texts were more likely anyway to be English-educated gentlemen. And in England, the predominant space was that of the 'South Country', as depicted by Hilaire Belloc in his *Verses* (1910) or in Edward Thomas's *South Country* (1908), itself a product of an imagined 'other' to the burgeoning urban world of the late nineteenth century, no matter if the rural world too was in crisis (Howkins, 1986: 62–88). And would the concept be relayed differently in different English regions, or is it a national image that is sufficiently flexible to accommodate spatial shifts between regions, whilst still retaining its power? We should recall that one important version of the national landscape was that promulgated by the National Parks movement at mid-century, and that this version was primarily upland and open moorland, quite at variance to the 'south country'. With the contemporary development of devolution and regional awareness within Britain, and the attempt to formulate 'countryside character' landscapes (Countryside Agency, 1999), it may be time once again to examine the spatial distinctiveness of the many different rural idyllicisms (Brace, 2001).

There are other research questions that might be pursued. Did the concept differ in its formulation between Classical Mediterranean slave-based cultures, early-modern capitalism, and through an age of enlightenment to the present day? We have seen that although the concept of a rural idyll is perduring, the precise formulations vary with time and space. The classical Mediterranean version was not that which was received into politer country house society in eighteenth century England, but the extent to which these older versions linger as ghosts within our contemporary rural idylls would again be worthy of further investigation. Part of any such investigation would be the extent to which the greater scientific rationalization of thought ushered in by Bacon's *Novum Organum* in 1620, while paving the way for modern understanding of the interrelationships of the living countryside, nevertheless may have left our sentiments and cultural attitudes largely untouched. Indeed, the unravelling of nature's complexities within children's 'nature studies' has in itself imbued generations of children with the desire not to lose contact with nature; a desire surely reinforced by Western cultural expression in art, music and literature. We might also ask whether the concept is one of the many resistances to modernity, or whether the concept comprises one of the many continuities between 'then' and 'now', resisting altogether any thought of a decisive break between the premodern and the modern? Possibly it simply becomes one further element of a selective and plasticized 'heritage', which can be dipped into as required and endlessly shaped to suit our modern purposes? It is very clear that this cultural theme is also dependent for its power upon the social and economic structures of the society within which it is being deployed. The links between culture and society are thus very clear, and within academic writings concerned with rural studies idyllicism has been a potent, though sometimes unrecognizsed, force. Thus, ideas of the organic (rural) community were powerful within history, sociology and social anthropology (Wright, 1992: 195–217). More recently, studies of the search for 'authentic' rural space and its subsequent defence by class- or gender-based interests have been mounted by anthropologists, sociologists and geographers. The degree to which concepts of the rural idyll are imbedded so deeply within Western (including colonial and postcolonial) consciousness will surely mean that rural studies will return often to examine the many manifestations and implications that arise (Robertson and Richards, 2003).

The wealth and opportunities arising from an urban-industrial Western world have promoted the idea of the rural idyll in both positive and negative ways. Much can be made of the negative reaction to modernity and the city, but a wealthy society has also positively sought out the rural as an important cultural peg on which to hang moral, social and aesthetic values. Raymond Williams saw a spectrum of such representations as having social roots:

> Not just one neutral convention or another, some of them are interested lies, some of them are ways of seeing which are related not to mendacity but to privilege, some of them are much deeper and less conscious limitations of the vision of an inherited or class position, some of them are partial breakthroughs, others are relatively complete insights. (1981: 304–5)

Taking a long-term perspective, as this chapter has done, the rural idyll, as portrayed in creative artistic convention, may be seen to be omnipresent at most, if not all, times in Western urban consciousness, but to emerge most strongly as a discourse – at least as seen through preserved references within cultural production – at particular historical moments of crisis in urban society, for example. Although authors may point to Hellenistic or Alexandrian Greece, Augustan Rome or the eighteenth century English Romanticism as key points of rural idyll 'production', we have seen that the concept can be used as an ongoing point of reference to less complex ways of simpler and more honest endeavour, and keying into fundamental human desires to sustain some harmony with nature and community. And in so doing, the asymmetries of power, whatever the historical *zeitgeist*, moment or place, provide contested representations of this very complex notion.

REFERENCES

Ashcraft, Richard (1972) 'Leviathan triumphant: Thomas Hobbes and the politics of Wild Men', in E. Dudley and M. Novak (eds), *The Wild Man Within: An Image in Western Thought from the Renaissance to Romanticism.* Pittsburgh: University of Pittsburgh Press. pp. 141–81.

Austen, Jane (1987) *Sanditon.* Bolton Abbey: Folio Society edition.

Barrell, John (1972) *The Idea of Landscape and the Sense of Place, 1730–1840: An Approach to the Poetry of John Clare.* Cambridge: Cambridge University Press.

Barrell, John (1983) *The Dark Side of the Landscape: The Rural Poor in English Painting, 1730–1840.* Cambridge: Cambridge University Press.

Bate, J. (2003) *John Clare: A Biography.* London: Picador.

Bernstein, D. (1992) *Shell Poster Book.* London: Hamish Hamilton.

Boyes, G. (1993) *The Imagined Village: Culture, Ideology and the English Folk Revival.* Manchester: Manchester University Press.

Brace, C. (2001) 'Publishing and publishers: towards an historical geography of countryside writing, *c.*1930–1950', *Area*, 33: 287–96.

Brandon, P. and Short, B. (1990) *The South East from AD 1000.* London: Longmans.

Bunce, M. (1994) *The Countryside Ideal: Anglo-American Images of Landscape.* London: Routledge.

Burchardt, J. (2002) *Paradise Lost: Rural Idyll and Social Change since 1800.* London: I.B.Taurus.

Charlesworth, M. (1999) 'Mapping, the body and desire: Christopher Packe's Chorography of Kent', in D. Cosgrove (ed.), *Mappings*. London: Reaction.

Cloke, P. (1994) 'A life in the day of a "rural geographer"', in P. Cloke, M. Doel, D. Matless, M. Phillips and N. Thrift, *Writing the Rural: Five Cultural Geographies*. London: Paul Chapman.

Cloke, P. and Milbourne, P. (1992) 'Deprivation and rural lifestyles in rural Wales II', *Journal of Rural Studies*, 8: 360–74.

Cloke, P., Milbourne, P. and Thomas, C. (1997) 'Living lives in different ways? Deprivation, marginalization and changing lifestyles in rural England', *Transactions of the Institute of British Geographers*, NS 22: 210–30.

Countryside Agency (1999) *Countryside Character* (8 vols). Cheltenham: Countryside Agency.

Countryside Agency (2003) *The State of the Countryside 2003*. Cheltenham: Countryside Agency.

Daniels, S. (1993) *Fields of Vision: Landscape Imagery and National Identity in England and the United States*. Cambridge: Polity Press.

Department of the Environment and Ministry of Agriculture, Fisheries and Food (1995) *Rural England – A Nation Committed to a Living Countryside*. Cm 3016, HMSO (Rural White Paper).

Draper, R.P. (1975) *Hardy: The Tragic Novels*. London: Macmillan.

Eldridge, J. and Eldridge, L. (1994) *Raymond Williams: Making Connections*. London: Routledge.

Fulford, T. (1996) *Landscape, Liberty and Authority: Poetry, Criticism and Politics from Thomson to Wordsworth*. Cambridge: Cambridge University Press.

Fulford, T. (1998) 'Fields of liberty? The politics of Wordsworth's Grasmere', *European Romantic Review*, 9: 59–86.

Fussell, P. (1975) *The Great War and Modern Memory*. Oxford: Oxford University Press.

Glacken, C. (1967) *Traces on the Rhodian Shore*. Berkeley, CA: University of California Press.

Hall, P. and Ward, C. (1998) *Sociable Cities: The Legacy of Ebenezer Howard*. New York: Wiley.

Hardy, Thomas (1912) *Return of the Native*. London: Wessex edn (Holt, Reinart & Winston, 1969).

Harman, R. (ed.) (1943) *Countryside Mood*. London: Blandford Press.

Harvey, David (1990) *The Condition of Postmodernity*. Oxford: Blackwell.

Heathorn, S. (2000) 'An English paradise to regain? Ebenezer Howard, the Town and Country Planning Association and English ruralism', *Rural History*, 11: 113–28.

Houlton, D. and Short, B. (1995) 'Sylvanian Families: the production and consumption of a rural community', *Journal of Rural Studies*, 11: 367–85.

Howard, W. Scott (2003) 'Landscapes of memorialisation', in I. Robertson and P. Richards (eds), *Studying Cultural Landscapes*. London: Arnold. pp. 47–70.

Howkins, A. (1986) 'The discovery of rural England', in R. Colls and P. Dodd (eds), *Englishness: Politics and Culture 1880–1920*. London: Croom Helm.

Howkins, A. (2003a) *The Death of Rural England: A Social History of the Countryside since 1900*. London: Routledge.

Howkins, A. (2003b) 'Qualifying the evidence: perceptions of rural change in Britain in the second half of the twentieth century', in D. Gilbert, D. Matless and B. Short (eds), *Geographies of British Modernity: Space and Society in the Twentieth Century*. Oxford: Blackwell. pp. 97–111.

James, Louis (1989) 'Landscape in nineteenth-century literature', in G. Mingay (ed.), *The Rural Idyll*. London: Routledge. pp. 61–76.

Keith, W.J. (1988) *Regions of the Imagination: The Development of British Rural Fiction*. Toronto: University of Toronto Press.

Laing, S. (1992) 'Images of the rural in popular culture', in B. Short (ed.), *The English Rural Community: Image and Analysis*. Cambridge: Cambridge University Press. pp. 133–51.

Lash, S. and Urry, J. (1994) *Economies of Signs and Space*. London: Sage.

Light, A. (1991) *Forever England: Femininity, Literature and Conservatism between the Wars*. London: Routledge.

Lindenbaum, P. (1984) 'The geography of Sidney's Arcadia', *Philological Quarterly*, 63: 524–31.

Little, J. and Austin, P. (1996) 'Women and the rural idyll', *Journal of Rural Studies*, 12: 101–11.

Lowenthal, D. and Prince, H. (1965) 'English landscape tastes', *Geographical Review*, 55: 187–222.

Marsh, J. (1982) *Back to the Land: The Pastoral Impulse in Victorian England from 1880 to 1914*. London: Quartet Books.

Matless, D. (1998) *Landscape and Englishness*. London: Reaktion.

Matless, D. (2000) 'The predicament of Englishness', *Scottish Geographical Journal*, 116: 79–86.

Miller, S. (1995) 'Urban dreams and rural reality: land and landscape in English culture, 1920–45', *Rural History*, 6: 89–102.

Montrose, L. (1980) 'Eliza, Queen of shepheardes, and the pastoral of power', *English Literary Renaissance*, 10: 153–82.

Montrose, L. (1983) 'Of Gentlemen and shepherds: the politics of Elizabethan pastoral form', *English Literary History*, 50: 415–59.

Morris, M. (1997) 'Gardens "for ever England": landscape, identity and the First World War British cemeteries on the Western Front', *Ecumene*, 4: 410–34.

Murdoch, J., Lowe, P., Ward, N. and Marsden, T. (2003) *The Differentiated Countryside*. London: Routledge.

Payne, Christina (1993) *Toil and Plenty: Images of the Agricultural Landscape in England, 1780–1890*. New Haven, CT: Yale University Press.

Puttenham, George (1589 [1936]) *The Arte of English Poesie* (ed. by Gladys D. Willcock and Alice Walker). Cambridge: Cambridge University Press (reprinted Scolar Press, 1970).

Robertson, I. and Richards, P. (2003) *Studying Cultural Landscapes*. London: Arnold.

Sales, R. (1983) *English Literature in History: 1780–1830 Pastoral and Politics*. London: Hutchinson.

Sargent-Baur, Barbara N. and Cook, Robert F. (1981) *Aucassin et Nicolete: A Critical Bibliography.* London: Grant and Cutler.

Schwarzbach, F.S. (1979) *Dickens and the City.* London: Athlone Press.

Short, B. (2000) 'Rural demography, 1850–1914', in E.J.T. Collins (ed.), *The Agrarian History of England and Wales VII: 1850–1914.* Cambridge: Cambridge University Press. pp. 1232–96.

Short, J.R. (1991) *Imagined Country: Society, Culture and Environment.* London: Routledge.

Stewart, S. (1966) *The Enclosed Garden: The Tradition and the Image in Seventeenth Century Poetry.* Madison, WI: University of Wisconsin Press.

Stradling, R. (1999) 'England's Glory: sensibilities of place in English music, 1900–1950', in A. Leyshon, D. Matless and G. Revill (eds), *The Place of Music.* New York: The Guilford Press.

Symcox, G. (1972) 'The wild man's return: the enclosed vision of Rousseau's Discourses', in E. Dudley and M. Novak (eds), *The Wild Man Within: An Image in Western Thought from the Renaissance to Romanticism.* Pittsburgh: University of Pittsburgh Press. pp. 223–47.

Thomas, Richard F. (1999) *Reading Virgil and His Texts. Studies in Intertextuality.* Ann Arbor, MI: University of Michigan Press.

Thomas, Richard F. (2001) *Virgil and the Augustan Reception.* Cambridge: Cambridge University Press.

Thorsley, P.J. Jr (1972) 'The wild man's revenge', in E. Dudley and M. Novak (eds), *The Wild Man Within: An Image in Western Thought from the Renaissance to Romanticism.* Pittsburgh: University of Pittsburgh Press. pp. 281–307.

Treble, R. (1989) 'The Victorian picture of the country', in G. Mingay (ed.), *The Rural Idyll.* London: Routledge. pp. 50–60.

Waller, A.R. (ed.) (1905) *Poems. Miscellanies, The Mistress, Pindarique Odes, Davideis, Verses Written on Several Occasions Abraham Cowley.* Cambridge: Cambridge University Press (replica reprint 2001 edn, London: Elibron Classics).

Webb, Mary (1928) *Precious Bane*, 2nd edn. London: Jonathan Cape.

Williams, R. (1973) *The Country and the City.* London: Paladin.

Williams, R. (1981) *Politics and Letters: Interviews with New Left Review.* London: Verso.

Williams, R. (1982) *Culture and Society.* London: Hogarth Press.

Williamson, T. (2002) *The Transformation of Rural England: Farming and the Landscape, 1700–1870.* Exeter: University of Exeter Press.

Wright, S. (1992) 'Image and analysis: new directions in community studies', in B. Short (ed.), *The English Rural Community: Image and Analysis.* Cambridge: Cambridge University Press. pp. 195–217.

10

Variations on the rural idyll

David Bell

> Rural life reflects at one and the same time the boundlessness of the imagined landscape and community and the restrictiveness of access to the material and cultural conditions which permit the imagined to be lived out other than in the imagination. (Cloke, 1994: 171)

INTRODUCTION

My aim in this chapter is to take a long, hard look at the rural idyll. I am interested in where it comes from, and in the cultural work that we ask it to do for us. To do this, it is my intention to explore the ways in which the rural idyll is produced and those things its production denies or excludes. Let me begin with some questions: What is the rural idyll? Where can we go to find it? What does it look like? What do we do with it? And what do we want it for? I began pondering these and other questions in the summer of 2001, while participating in my own bit of idyll-catching, on holiday in Catalonia, staying in a converted farmhouse. This setting certainly conforms to the term 'idyll', but also brought into view many of the ambivalences that mark the experience of the rural as idyllic. The farm is isolated, remote; the landscape wild (or at least 'managed wild', since much of it is farmland). The life there is 'simple', rustic; farming is the dominant economic activity (though tourism is rapidly catching up). There are animals (wild and farmed), fine vistas, peace and quiet, traditional foods at the market, workers in the fields. Kids can roam free, climb trees, watch cows. Rural pastimes – walking, bird-spotting, sight-seeing – are abundantly available. This is the rural idyll in its current (Western, or Euro-American?) holiday packaging.[1]

But what has to be hidden, or denied, to make this idyll? And what things threaten to disturb it? While we were there holidaying, for example, there were countless signs of protest at a proposed new road on display. In the protesters' eyes, the road would open up the area to more incomers and thereby 'spoil' the tranquillity and cut through the landscape. Equally evident, however, were banners demanding that the road be built – as an economic improvement to the region. The tourists who seek out the idyll therefore also threaten it – a paradox common to tourist sites (Rojek and Urry, 1997). They threaten it in other ways, too, for example by demanding the kind of idyll they want – for the idyll is imagined in the minds of the tourist as an *entitlement*. The business of living in the region, and of making a living there, can threaten idyll-disturbance, therefore: farm machinery is noisy, some of the agribusiness practices do not sit squarely with the idyllized view of farming (the intensive pig pens, the electric fences, the pesticides), the locals drive their cars too fast. The nearby town has supermarkets full of ready meals and shops selling Nike sportswear. The cows make too much noise – especially after their calves have been taken away from them. The wildlife includes mosquitoes. And as for those bloody tourists …

LOOKING FOR THE RURAL IDYLL

My experience of Catalan country seems to capture, therefore, the *problem* of the rural idyll: that it is a manufactured landscape, the product of a particular moral ordering or act of purification (Sibley, 1995). Purity is always threatened by pollution, of course, and the whole history of rural preservationism tells us about the moral imperatives at work in keeping the rural pure (Lowe, 1989). However, idyll-disturbance always threatens to unsettle our experience, not least by revealing the idyll as manufactured (and therefore inauthentic, which also means un-idyllic). So, let's ask another question: if the idyll doesn't exist in my Catalan holiday, where does it exist? Maybe it only exists in the imagination, as a symbolic landscape, as urbanism's other. Certainly the genealogy of the rural idyll shows it to be an urban construction; the country cannot exist without the city to be its 'not-a'. So the place to find the rural idyll is in the city, since that is where it is made. Idyllization is a symptom of urbanization, then (Williams, 1973). More precisely, as Kathleen Stewart (1996) argues, the idyll (and its attendant otherings) is a product of the bourgeois imaginary that emerged with modern urban-industrial culture, and which sought to produce an ordered social spatialization of margin and centre (see also Shields, 1991; Stallybrass and White, 1986). Here, the rural was (and still is) at once an object of desire (because it is not-modern *in a good way*) and of dread (because it is not-modern *in a bad way*). We shall return to Stewart's argument later.

So, if we are to look for the rural idyll, we have to track it into the bourgeois imaginary. As Michael Bunce (1994) suggests, the 'countryside ideal' exists in our minds, cooked up for us to dream of in popular culture – what he refers to as the 'armchair countryside'. Manifest in diverse cultural forms and practices, we can therefore find the idyll on television, in novels and poems, in shops, even on our plates. Children (and parents) can find it in *Little House on the Prairie*, *Wurzel Gummidge*, *The Secret Garden*, in plastic farm animals and Sylvanian Families. For grown-ups, there's *The Archers*, *The Good Life*, *A Year in Provence*, *Far From the Madding Crowd*, wildlife documentaries, Andy Goldsworthy, Thoreau's *Walden*, the 'new acoustic' folk-rock music, and the classic American western movie. Adverts for a whole host of products and services trade on the positive connotations of the rural. It is visible in trends in interior décor, home furnishings, garden design and clothing ranges. We can eat the idyll for dinner, in foods packed with natural goodness, or in what Michael Pollan (2001) neatly refers to as the 'supermarket pastoral' of organic produce. As Paul Cloke (1994) comments, there are seemingly endless opportunities in the marketplace to buy into the idyll commodified as taste or style. In addition, our continued love of the 'great outdoors' finds ever-new outlets from rambling to abseiling, white-water rafting to pot-holing (and holidays to Catalonia). Politically the idyll can be traced in protectionist bodies such as in the UK the National Trust, the CPRE, the farming lobby and environmental groups. And socio-spatially, it is there in our temporary or permanent migrations to the countryside: second homes for the weekend, *gîtes*, commuter villages, telecottages, retail parks and retirement communities.

If we step back from the specificities of each instance of idyllization listed above for a moment (though we will revisit some of them later), what common themes might we discern that connect them? I think there are three ideal-typical idylls that emerge from this roster, in fact: the pastoral ('farmscapes'), the natural ('wildscapes') and the sporting ('adventurescapes'). The farmscape reflects the agricultural landscape (but artisanal rather than agribusiness). The wildscape is pre-cultural, pre-human, untamed nature – the wilderness. The adventurescape constructs the rural as an adventure playground, drawing on some wilderness motifs but adding a focus on physical endurance and 'limit experiences' (Cloke and Perkins, 1998; Lewis, 2000). These three forms of rural idyll, it seems to me, comprise a mobile combination of the following elements: nature (natural wonders, closeness to nature, etc.), romanticism, authenticity and nostalgia (for simpler ways of life, for example), all stamped onto the land and its inhabitants (plants, animals, people). The exact recipe for rural bliss varies historically and geographically, of course (Bunce, 1994).

The idyll has a very powerful and enduring ideological pull, moreover, and constantly morphs to fit with new times. As G.E. Mingay put it:

> The rural idyll is a changing concept: the countryside at the end of the twentieth century is very different from that of a hundred years ago ... Each generation of country dwellers and observers sees what it wants to see in the land: romantic beauty, nostalgic traces of the rustic past, peace, tranquillity; despoiled landscapes, brutal intrusions of modernization, hurry, noise, pollution. (1989: 6)

In the UK, the recent farming crisis shows both the ideological pull and the morphing of the idyll, with the discourses surrounding BSE, foot and mouth, hunting with dogs, the Countryside Alliance and the future of British farming straining (but ultimately succeeding) to maintain some dyll-ish notion of the centrality of agrarian life to British culture.

This raises yet more questions, of course – and these are, I think, the most important ones. First: What is the rural idyll for? It is first and foremost a symbolic landscape into which is condensed and onto which are projected a whole host of things: identifications, imaginings, ideologies (Darby, 2000). It is, perhaps most importantly, a receptacle for national identity – a symbolic site for shoring up what it means to be English, or Dutch, or whatever (as well as providing regional and local identifications; see Cloke, 1994). More on all of this later. It is also a reminder of the past – usually a golden past now lost in the rush to modernity. The pastoral idyll is the bountiful heartland, the nation's foodstore (at least symbolically); the wild idyll offers a different bounty – the romantic opportunity to commune with nature. It is a restorative resource, a place to go to touch nature (and maybe God), to find peace, to gaze and meditate.

And second: Who is the rural idyll for? If the idyll really is a symbolic resource for shoring up national identity, then that means it should be for everyone, I guess. But, given what I have hinted at earlier about exclusions, about purity and pollution, it should be no surprise to find that the rural idyll is actually an exclusive and exclusionary place (hence past and present calls for land nationalization, right of access, etc.; see Shoard, 1987). The exclusions are, moreover, *symbolic* as well as material. Rural studies has begun to take an interest in those groups excluded from the idyll, or even from the rural – often short-handed as 'rural others' (Cloke and Little, 1997; Milbourne, 1997; Philo, 1992). In these terms, if we think of the city as (albeit not without contestation) the site of multiculturalism and diversity (Young, 1990), then the country is by contrast a monoculture with no space for difference – other than its absolute difference from the urban (though, for an opposing view, see Rankin, 1999):

> [The rural] is a place where gender and ethnic identities can be anchored in 'traditional' ways, far ... from the fragmented, 'mixed up' city. Within the rural domain identities are fixed, making it a white, English, family-orientated, middle-class space ... [T]he rural is extolled for the virtues of peace and quiet, of community and neighbourliness, virtues deemed to be absent from the urban realm. (Murdoch and Marsden, 1994: 232)

Producing this monocultural idyll relies on processes of denial and expulsion, which together produce a category that I will call here the *rural abject* – those people and things dispelled from the idyll, rendered other, cast out (Kristeva, 1982). This means solidifying a set of rural/urban oppositions, where the rural is positively valued and the urban denigrated (an overturning of modernity's prizing of the urban over the rural; see Ching and Creed, 1997). Typical binaries here would be rural = peace/urban = noise, rural = slow/urban = fast, rural = clean/urban = dirty. But it also involves establishing *appropriate* forms of rurality to set alongside inappropriate or debased forms: those that belong in the category I previously named the anti-idyll (Bell, 1997). It is the process of classifying certain rural types into idyll and anti-idyll classes that is, I think, most revealing of the problem of the rural idyll, and I therefore want to spend some more time considering that act of othering here.

PRODUCING RURAL OTHERS

In a previous essay, about horror films set in the countryside, I began to think through the notion of the anti-idyll as the abject underbelly of the rural (Bell, 1997). What that essay attempted to do was trace the ways in which selected mainstream cinematic representations frame ideas about country folk (and city folk) that draw on the cultural resource of the rural idyll, but that also contest it (see also Williamson, 1995). The genre I designated 'hillbilly horror' (aka the rural slasher), which includes movies like *Deliverance* and *Texas Chain Saw Massacre*, stages the urban/rural dichotomy as a site of ambivalence, showing that the production of the rural idyll (as something that urban folk desire) depends on othering forms of rural life that are out of place in such idyllizations. The hillbillies, the rednecks, the rural white trash are thus the abject group here; even though they are country folk, they do not belong in the rural idyll, and so they are depicted as monstrous. As Stewart (1996: 119) puts it, the hillbilly has come to function in the American bourgeois imaginary as a 'tense and contradictory' sign. It is contradictory since it embodies both desire (through notions of authenticity, for example) and dread (figured in notions of backwardness, etc.). In Appalachia, where her ethnography was based, Stewart tracked the sedimentation of this anti-idyllization into the image of the region and its people:

> it became the site of a culture that was irredeemably white, poor, rural, male, racist, illiterate, fundamentalist, inbred, alcoholic, violent, and given to all forms of excess, degradation, and decay. (1996: 119)

Now, I think that the class dimension of this process can also be traced in other moments of rural othering, too. In the UK, for example, the fuss over New Age travellers shows another out-of-place rural group, their economic marginality consigning them to underclass abjection (Hetherington, 2000). Travellers, of course, mobilize their own version of the rural idyll, though this is often dismissed (or othered) as an inauthentic, urban idyllization that fails to appreciate the 'realities' of the countryside (Mosbacher and Anderson, 1999) – showing that the shifting meaning of the rural idyll is constantly being corralled into a dominant or hegemonic version. This hegemonic idyll is so powerful, Cloke (1994) argues, that it renders terms like 'rural poverty' or 'rural deprivation' culturally illegible, since life in the country can never be 'poor' or 'deprived' – though the denial of poverty and deprivation is again labelled as a metropolitan sentimentalization of rural life by some commentators (Mosbacher and Anderson, 1999).

In the case of hillbilly horror, moreover, the hillbillies are carriers of a particular class identity that puts them in a lineage with earlier generations of agricultural labour. They are, in short, *peasants*. However, whereas past peasantries are redignified and incorporated into the rural idyll, as they have been since Victorian times (Marsh, 1982), the hillbilly remains excluded, undignified. The hillbilly is here the bearer of all the negative traits of rural working-class (or underclass) life – small-mindedness, backwardness, atavism, animalism. Now, while these characteristics can in some circumstances be rebranded to fit with the idyll (as authentic, traditional, close-to-nature, etc.), in the case of rural horror they are unrecuperable, excessive. Here the hillbilly is beyond idyllization. As an excluded rural underclass, the white-trash hillbilly occupies a complex position in the cultural politics of rurality, therefore (see also Wray and Newitz, 1997).

What particularly interests me about this idea is the way it suggests the need for rurality to be contained, or domesticated. The rural is threatening here because it is potentially *wild* – as are its people. The islanders in *The Wicker Man* or the mountain men in *Deliverance* exhibit an unconstrained, excessive wildness, perhaps hinting at the fragility of humanity's civilization. They certainly suggest the frailty of the rural idyll, as always threatening to spill over into violence and degradation. In *Deliverance*, for example, city men venture to the country in search of idyllic thrills (riding a soon-to-be-dammed river in Appalachia – an adventurescape), only to find themselves hunted, and to experience the countryside as alien – a place where their urban competences count for nothing. Only by becoming hunters and killers – by becoming the other – can they survive in this anti-idyll:

> At the river's edge they find a jaded and sinister semi civilization (in which degenerate wildness lurks). Safety, reason, and a renaturalized order of things lie in the city, though the foray out into the wilds has left the men themselves marked with Otherness ... Direct encounter is dangerous. (Stewart, 1996: 120)

To maintain the rural idyll as an attractive regenerative place, therefore, requires the containment of these wilding impulses and the construction of an exclusion zone to limit (or prohibit) 'direct encounter'. (Direct encounter is similarly figured as contaminating and ultimately fatal in *The Wicker Man*, a film which plays with paganism on a remote Scottish island.) How is this containment and exclusion achieved? One increasingly prominent way, at least in the UK and the United States, is via the production of enclavic rural idylls or 'fortress villages'. Architecturally and socio-spatially, the process of creating these spaces I shall call *retrofitting rurality*.

RETROFITTING RURALITY[2]

> Rurality can ... be seen as an outcome of processes of class formation as individuals and collectivities attempt to mould rural space into forms which reflect and perpetuate class identity and difference. (Murdoch and Marsden, 1994: 15)

In an excellent conference presentation on his home village, the pseudonymous 'Allswell' in South-West England, Owain Jones (2000) drew attention to the purification processes at work in the production of distinct and exclusive rural communities. Jones's paper explored the imprinting of a new social (and spatial) contract on an existing village, which served to preserve a distinctly nostalgic, middle-class white English version of the rural idyll (see also Murdoch and Marsden, 1994). Its closest relative, in my view, is the enclavic space of the expatriate community, where the hyper-performance of Englishness is pursued as an act of (national) identity maintenance (Edensor, 2002). In the fortress villages of rural England, a hyper-performance of idyllic rusticity keeps 'country life' alive for white-collar commuters, homeworkers and homemakers. Rituals of village life are re-enacted (and reinvented) in order to keep the drama of everyday country folk live – in the case of 'Allswell' the village cricket team occupies centre-stage as an institution for ensuring harmonious social relations. In fact, in Jones's view, 'Allswell' is itself somewhat like a country club – right down to the fact the membership is exclusive, and depends on a system of patronage. Cloke (1994) is thus right to label these middle-class incomers the 'new gentry'.

A second version of retrofitting rurality can be witnessed around England with increasing frequency these days, in the building of new 'executive homes' in small, often gated estates – a building trend evident in parts of England for the past 20 years or more, and with older antecedents (Murdoch and Marsden, 1994). These spaces embody the retrofitting ideal even more clearly than 'Allswell', since they are new-build pastiches of villages, with odd mixes of 'heritage' architectural styles and quaint, rustic names. In their study of rural life in Buckinghamshire, Murdoch and Marsden (1994) detailed the process of planning and designing such new 'concept' settlements, and talked to developers. The houses self-consciously evoke the rural idyll in terms of their styling, borrowing from a mix of past architectural motifs to create a distinct and 'timeless' rural look – to produce 'added-value' housing. One developer Murdoch and Marsden interviewed compared the villages to film sets:

> Our schemes are definitely like film sets. We think that the trend is that people watch a lot of television where colour television creates a tone because the lighting is different, you never see a true colour, you see a technical or bright colour. What we try to do in our house styles is create the same image, that same glamour ... the site looks like a film and the film looks like a site – it's all a bit glitzy. (quoted in Murdoch and Marsden, 1994: 77–78)

Like previous experiments in rural social engineering, including the Victorians' model farms and villages, these settlements consciously seek to create desirable social arrangements through the planning and styling of houses (Havinden, 1981). They are readable as the residence of choice for new traditionalists, that middle-class (or managerialist) group intent on reproducing 'traditional' forms of family and social life as a nostalgic defence against the body-blows of post-industrial consumer capitalism (Leslie, 1993; Probyn, 1990) – as a kind of 'respectable' survivalism or secessionism maybe (Lutticken (2001) draws a line from gated communities to Waco and the Unabomber's Montana shack). Fortress villages are a middle-class pastoral take on the contemporary life-strategy of 'bunkering in' (Kroker and Kroker, 2000), though as Lutticken (2001) says, these estates can appear more like prisons than sanctuaries. Just as the suburbs had previously sought to produce in plan and building form the perfect setting for the bourgeois imaginary (Silverstone, 1997), these new communities clearly represent the architecture of new traditionalism's desires (and dreads) about rural (and urban) living. New traditionalists thus repurpose rural life (imagined as patriarchal nuclear domesticity) as protection against the ravages of urban consumer capitalism:

> They are looking for rural life, which means life in a 'community'. If such a community does not exist, it will be created, as incomers weave together the 'old' and the 'new' into a 'hybrid' rurality, one that seeks to exclude all that these residents have moved away from, i.e. the pernicious effects of urbanism, with its 'fragmented' ways of life, its 'mixed-up' classes and ethnicities, its 'ambivalent' sexualities ... Thus, the new rural communities can be seen as sites for 'anchoring' traditional middle-class identities. (Murdoch and Marsden, 1994: 229)

This line of argument captures some of what is occurring in fortress villages, to be sure. But I think it is also useful to bring in work that has sought to analyse new-middle-class taste cultures, and particularly the roles of notions of the authentic and the exotic in marking distinction in contemporary consumption-based class fractions (May, 1996a, 1996b). It can be argued, I think, that the rural has to be rendered sufficiently authentic and exotic to appeal to these cosmopolitan taste cultures, rather than being seen as provincial (and therefore worthless for taste-marking purposes). This can be read to suggest that rural life has been sufficiently othered that it can be responded to positively, as a kind of 'other within'. I previously made a similar argument in the context of food consumption, arguing that the turn towards traditional indigenous cuisines and ingredients – the renewed popularity of offal, for example – marked a similar transformation in middle-class tastes (Bell and Valentine, 1997). In that sense, we can concur with Aidan Rankin's (1999) otherwise politically dubious assertion that the British countryside is the true site of multiculturalism (though he is deeply suspicious of 'townies' wanting to celebrate 'indigenous' folk cultures). That such a revaluing of the rural is most readily achievable in retrofitted communities (where the rural is grafted onto essentially metropolitan lifestylizations) must be remarked upon here; it echoes some of the comments made by May's Islington respondents, who did not want their exotic to *be too* exotic. This means containing and regulating modes of ruralness (and urbanness) that threaten idyll-disturbance – itself the prime form of anxiety experienced by fortressed ruralities (Loader et al., 2000).

In my neighbouring county of Cheshire, these kinds of housing developments are blossoming, offering an almost theme-park-like form of rural idyll. In the United States, the Disney-designed settlement of Celebration in Florida epitomizes this trend, retrofitting small-town life on reclaimed swampland (Ross, 2000) – and there are certainly no hillbillies in Celebration (Lutticken, 2001). The swamp, as a landscape of rural otherness, is an ironic substratum for the fantasy life of Celebration. The playful, nostalgic pastoralism of gated villages and retrofitted communities like these represents the rural emptied of its otherness, then – the metaphorical draining of the swamp.

Writing about Celebration alongside the Cheshire set, or thinking about the depictions of the rural other in *Deliverance* and *The Wicker Man*, leads me to begin to question an earlier statement about the rural idyll: that it is primarily a resource for national identity. The argument about 'Merrie England' as key symbolic site for constructing and reconstructing Englishness, for example, is well established and relatively uncontested (even if the forms of Englishness allowed access to that symbolic landscape are themselves contested; Matless, 1998). However, given my previous discussions about the new managerialist class (a class associated more centrally with the global than the national or local; see Sklair, 2001), and about the making of the rural idyll in the bourgeois imaginary and in popular culture, I would like to think here about what might at first appear oxymoronic: something I will call the *transnational rural*.

THE TRANSNATIONAL RURAL[3]

Paul Cloke (1994) usefully reminds us of the importance of spatial scale in analyses of the rural idyll. While he is concerned to focus more closely than the national, to consider regional and local refractions of the idyll, my intention here is to pull back, to consider a scale often neglected in rural cultural studies: the global. Marsden et al. (1993) have clearly established that globalization is effecting rural life and rural space in manifold ways. While their focus is primarily on economic and political aspects of global restructuring, they appear mindful of cultural globalization – which will be my main focus here. If we follow the line of argument that processes of globalization impact upon ou experiences of space and place in multiple an diverse ways, what are the implications of global ization for the rural idyll? At first thought, i might be possible to see the rural as straightfor wardly anti-global, as a defensive response to th placelessness of globalization: the rural as *local* The rural idyll is here the imagined homeland – and at least in some political rhetoric, it is a homeland under threat from globalization, and therefore in need of protection and preservation However, Lutticken (2001) reads the multinational corporations' free trade zones as producing another form of secessionism, making them analogous rather than oppositional to fortress villages, therefore (since both mark an active retreat from the nation). Both landscapes, then, can be seen as products of the same thing: global capitalism.

So, if we turn our attention to cultural globalization, I want to pose the following question: Are there also globalized notions of the rural idyll circulating? To take some more-or-less banal examples, what impacts do, for instance, the intimate relationship between the British viewing public and American musical, cinematic and televisual products have on our imaginings of the rural idyll (and anti-idyll)? Similarly, the global tourist industry takes us to ever-more remote rural idylls, as those on our doorstep are no longer idyllic enough (Butler, 1998) – what kinds of idyll do rural tourists consume?

As Urry (2002) notes, the globalization of the tourist gaze – itself subject to intensified mediatization – reprofiles the experiences and expectations that tourists bring to places, and in this sense the transnational rural idyll is a product of that global gaze. The food production business, too, trades globally to keep our larders stocked with nature's bounty (Bonanno et al., 1994). How does the global agro-food business redefine the idyll as a source of endless nourishment? And how do these (and other) sites for the production of the rural idyll intersect and interact? So, while much theoretical attention has focused on globalization and the city, we now need to mirror this with explorations of globalization and the country. Let's look a little more closely at each of my examples.

MEDIA IDYLL

The cross-cultural interplays that result from the entertainment media industries (music, film, television, Web) have been widely cited as producing new hybrid reading positions, while often simultaneously being accused of cultural imperialism (Tomlinson, 1999). My job is not to resolve those never-ending debates here, but to think about the impact of popular TV, music and film on cross-cultural understandings and imaginings of the rural idyll. Remember that the developer interviewed by Murdoch and Marsden (1994) was striving to make his fortress village televisual, recognizing that most people's imaginings of country life come from the mass media (see Phillips et al., 2001). Or, to come at the question another way, think about the codes contained in the western genre (Wright, 1975), and how these have come to give audiences outside the United States a set of symbolic resources for thinking about the country – and, I would argue, their own country, not just America, as witnessed in 2001 through the UK cinema release of the Thai western *Tears of the Black Tiger*. The same argument can be made for those other cinematic treatments of the American idyll (and its attendant anti-idylls), the road movie and the small-town movie (Cohan and Hark, 1997; Levy, 1991). Or, from a different angle, consider the American fixation with English 'heritage cinema's' depiction of idyllic rustic lifestyles (Vincendeau, 2001). Finally, while finishing this chapter I heard (on children's TV show *Blue Peter*) that the children's programme *Postman Pat*, based around a very nostalgized version of English village life, is now shown in 65 countries worldwide. Now, while it might be tempting to read the mass media as homogenizing global culture, a focus on the contexts (and contextualizations) of reception presents a more complex picture, as scholars of globalization are well aware (Tomlinson, 1999). As Phillips et al. (2001) argue, mediated ruralities are subject to complicated, even contested, processes of decoding.

In an insightful essay on the construction of tourist sights and sites, Chris Rojek uses the notion of indexing to refer to the processes by which we build up a set of resources (from the media, popular culture, etc.) that shape our experiences of tourism. While his focus is on the decidedly un-idyllic 'sensation sights' (scenes of celebrity death or atrocity), I think we can usefully borrow his basic idea here. This is his definition of the process:

> The cultural significance of sights engenders representational cultures which increase the accessibility of the sight in everyday life. In theory one might speak of an index of representations; that is, a range of signs, images and symbols which make the sight familiar to us in ordinary culture. The process of indexing refers to a set of visual, textual and symbolic representations of the original object. It is important to recognise that representational culture is not a uniform entity. Rather one might speak of files of representation. A file of representation refers to the medium and conventions associated with signifying a sight. (1997: 53)

This is useful in the context of my earlier discussion of the production of the rural idyll in the bourgeois imaginary, in that it gives us a way to frame the ongoing social and cultural process of idyll-making, and links popular representation to the experience of place. As Rojek says, 'there is much in how our ordinary consciousness of tourist sights functions which suggests that we draw on a collective fund of unconscious symbols, images and allegories' (1997: 54) – a good definition of the role of the bourgeois imaginary in shaping our experience of and desire for the rural idyll, too.

To return, then, to my argument about the entertainment industry's idylls before returning our attention to rural tourism, let's take one more example: the globalizing of country (by which I mean US-based country and western) music. Like the western film, country music can be read for a series of codes. While these might seem unquestionably *American* codes, the appeal of country music outside the United States cannot be wholly accounted for by a kind of cultural fetishism for Americana (though without denying that such a fetishism is at work here). In its mobilizing of codes of 'rustic authenticity', as Barbara Ching (1997: 232) puts it, country music also works cross-culturally, making sense way outside of Dixie or the Grand Ol' Oprey. (We might make similar arguments about the neo-folk music of 'new acoustic' bands in the UK, who draw their inspiration from previous pastoralist performers such as Nick Drake.) The hybridizing of US country music with other forms – a kind of indigenization also familiar to scholars of global culture – contributes to the notion of the transnational rural without suggesting a McDonaldsesque 'McCountry' homogeneity (the kind of criticism usually levelled at country music mega-stars such as Dolly Parton or Shania Twain; Jensen, 1998). In the UK, meanwhile, there has been a long-standing country music subculture, more recently augmented by the surge in popularity of line-dancing as a hybrid dance/fitness/social pastime. Finally, the recent revival in the United States of interest in bluegrass (and its retrofitted relative, 'nu-grass') music, to which established country stars such as Dolly Parton and Patty Loveless are currently turning (or, in their terms, 'returning'), might suggest a revaluing of a previously marginalized cultural product as an 'other within' similar to the revival in offal-eating noted above. (Note that in this bluegrass revival, the hillbilly is heroized, too.) Let's label the versions of rural life portrayed in film, TV and music the media idyll (which has an accompanying set of media anti-idylls, as seen in hillbilly horror).

TOURIST IDYLL

Rural tourism offers us similar opportunities to think about (and visit) the global idyll. The business of giving us holidays in the countryside – whether pastoral or wilderness – demands that forms of idyll are produced as culturally legible for tourist consumption (Duruz, 1999). As I said earlier, we get the idyll we want. From the bucolic joys of the country walk to the adrenaline-exhilarations of white-water rafting, forms of idyllization (and the rendering invisible of otherness) are perhaps more prominent in tourism than anywhere else. As Richard Butler (1998) suggests, moreover, there is a relationship between the intensifying mediatization of the rural, described above, and the production of the tourist idyll. At one level this means that city folk want to holiday in the idyllic places they have seen on screen – Butler cites the tourist impact of movies such as *Field of Dreams* and *The Bridges of Madison County* on their filmic locations and settings. The rural tourism industry increasingly promotes this kind of location-attraction; in the UK, the villages used for filming TV series like *Heartbeat* and *Last of the Summer Wine* are woven into the rural tourist itinerary. Jennifer Craik (1997) similarly explores the synergies between media products and the global tourist gaze on rural life and landscapes, focusing on Beatrix Potter tourism in the English Lake District (which is very popular with Japanese visitors) and tourism to Amish communities in the United States (popularized after Hollywood exposure in the movie *Witness*). More broadly, of course, the tourist idyll has to live up to the expectations concocted from indexing the media idyll.

The production of the tourist idyll also revolves around facilitating a repertoire of tourism practices or performances, including the key activities of sight-seeing and photographing (and increasingly camcordering) in what Rojek and Urry (1997) refer to as 'scenic tourism'. Mark Neumann (1999: 283), for example, calls the Grand Canyon tourist performance the 'wait-ride-circle the parking lot-look-snap the photo-repertoire' (see also Edensor, 2001). Engagements with nature and with the natural landscape (or with its agricultural and/or historical equivalents) are similarly requisite for the rural tourist experience, and outdoor pursuits such as walking, water-sports, picnicking and bird-watching often complete the experience of the tourist idyll. For the more adventurous, there is also a whole 'experience economy' that remakes the rural as the site for, among others:

> house-boating, portaging, mountain-biking, cattle-driving, bob-sledding, tall-ship sailing, tornado-chasing, canyon orienteering, wagon training, seal viewing, ice-berg tracking, racing car driving, hot-air ballooning, rock climbing, spelunking, white water rafting, canoeing, heli-hiking, hut-to-hut hiking, whale kissing, llama trekking, barnstorming, land yachting, historic battle re-enactments, iceboating, polar bearing and dog-sledding. (Thrift, 2000: 49–50)

In terms of the argument about the transnational rural, then, what we can witness here is the emergence of a series of sites and practices that package certain places as tourist idylls. The similarities between images of tourists visiting the Grand Canyon, in Mark Neumann's *On the Rim* (1999), and those of visitors to England's rural tourist sites in John Taylor's *A Dream of England* (1994), for instance, vividily illustrate the globalizing of the tourist performance and gaze (and the staging of that performing and gazing by the tourism industry), while the experience economy commodifies all kinds of rural spaces as sites for adventure (Urry, 2002).

Now, while Edensor (2001) is right to remind us that these tourist performances are culturally (and historically and geographically) specific, and moreover that there are many counterpractices that disrupt these routines, I want to suggest here that the globalization of tourist performances is today linked (in the West) to the production of (and demand for) distinct tourist idylls. Moreover, we must reiterate the role that indexing plays in prefabricating the tourist idyll. The representations we use to produce the tourist idyll are not only from the media, of course; they include other tourist idylls we have visited and, as we shall see below, other elements of popular culture and everyday life. As Craik puts it, 'cultural experiences offered by tourism are consumed in terms of prior knowledge, expectations, fantasies and mythologies *generated in the tourist's origin culture* rather than *by the cultural offerings of the destination*' (1997: 118; emphases in original) – a formulation we can usefully adopt to describe the production of the rural idyll in and out of tourism.

GASTRO-IDYLL

My last example of the production of the transnational rural and the global idyll is food, or more precisely food and farming. In terms of the playing off of idyllic and anti-idyllic versions of the rural, farming gives us lots to work with. All the recent fuss in the UK over farming's (mis)management – BSE, foot and mouth, GM foods – has provided us with a distinct set of anti-idyllic images, which contrast with attempts to idyllize agricultural practices and products, for example through organic farming and 'fair trade' agreements (Goodman and Watts, 1997). Or, to take a mundane example, consider the provision of fresh (or 'fresh') fruit and vegetables in Western food shops. The fetish of freshness – at least an implicit version of what we might label the gastro-idyll – has necessitated a huge restructuring of the agro-food industry, to grow, harvest, package and ship fresh produce year round to our supermarket shelves, deploying resource-guzzling transcontinental 'cool chains' (Friedland, 1994). Like its near-relative, 'healthy' food, the demand for 'fresh' produce signals a desire to eat food closer to nature – or, more accurately, food *presented to us as closer to nature*, since a tinned peach is arguably no more industrial than a 'fresh' peach that has been shipped across continents for us to eat (on 'healthy', 'fresh' and 'natural' foods, see Lupton, 1996). Food standards systems such as *appellations d'origine* further cement a series of gastro-idyllic associations, linking produce to place, people and production techniques, thereby assuring quality (Moran, 1993; though of course, as Watts and Goodman (1997) remind us, 'quality' is something with radically different meanings in the production and consumption of food).

This idyll on a plate, then, testifies to globalization in food production and consumption – though the two processes are often disjunctive. The farming crisis in the UK in the early twenty-first century shows how important the idyllization of farming is symbolically: the reluctance of the majority of consumers to sanction genetically modified (GM) crops, and the concomitant growth in demand for organic food, illustrates how the gastro-idyll exerts a powerful economic as well as cultural pull. What we witness here is, I think, a tension between *the production of idyllic food* and *the idyllic production of food* – which we might embody in terms of UK grocery shopping through the year-round availability of asparagus from Peru (idyllic food – endlessly available, always perfect) and the organic 'veggie box' (idyllic production – locally grown, seasonal, pesticide-free; see Purdue et al., 1997). Organic food mobilizes a form of rural idyll as anti-global, in fact, also drawing in discourses that link it to rurality, nature, health and environmentalism (James, 1993).

In the language of agro-food studies, these products respectively represent the processes of 'precision farming' (Peruvian asparagus) and the 'greening of agriculture' (organic food) (Watts and Goodman, 1997). That precision farming comprises anti-idyllic agribusiness practices is a fact routinely obscured at the site of consumption, not surprisingly (Cook and Crang, 1996). The increasing colonization of organic production, most notably in the United States, is meanwhile cast very negatively as 'polluting' the idyllic production imagined in small family farms with their environmentally friendly (that is, close-to-nature) production regimes, using the same association of agribusiness and anti-idyll (Pollan, 2001; Watts and Goodman, 1997). Elsewhere, however, the 'fad' for organic food is dismissed as yet another deluded urban attempt to dictate country ways; and, moreover, as one that even most urban folk don't want:

> The middle-class people who order boxed organic produce may get depressed after weeks of finding nothing but potatoes and cabbages in their food parcels. But at least they are happy for food to look as if it once grew in the ground or walked on it. Most people aren't. The supermarkets go to great lengths to place meat in absorbent packaging, so that no free blood is visible under the cling film. Fruit and vegetables are washed and waxed so they look bright and clean. The aim is to make food look as if it arrived on earth as a ready meal. The public is more than happy to pay extra for a hamburger that's already in its bun, ready for the microwave. They are not so keen to pay extra for a perfectly dull, organic turnip. (de Lisle, 1999: 91)

Here, though presented in polemic fashion, is the problem for organic food: while food is seen as one of the more palatable ways of deploying cultural capital and exploring otherness (Bell and Valentine, 1997), the food industry has prepackaged for us a set of expectations about food that organic produce does not necessarily deliver. While Leanda de Lisle overstates her case, there has been a reluctance on the part of consumers, certainly in the UK, to take wholeheartedly to organic foods (though its popularity and availability is on the up). As one of the most rudimentary ways in which we seek to connect with nature and the rural, then, the food we eat occupies a symbolic centrality that makes the gastro-idyll a potent and complex sign of our feelings about the countryside – and organic food is perhaps the most potent and most complex site where this is currently played out. I will now attempt to draw the different threads of my argument together, and reiterate some general points about the rural idyll that I have been addressing here.

CONCLUSION

I have attempted to show here that the rural idyll is a product of the bourgeois imaginary, worked up in the processes of urbanization, industrialization and modernization that are still unfolding. The idyll is imagined through the familiar bourgeois impulses of desire and dread (Stallybrass and White, 1986), and is set in opposition to the urban. The rural exists in urban minds, therefore, as a kind of other. Producing the rural idyll demands further forms of othering, too – most notably in terms of producing forms of anti-idyll, based on a set of binaries about who and what belongs in the country and what and who is out of place there. Mechanisms that keep the other excluded include the construction of new-build idylls in the form of gated or enclavic settlements – what we might call fortress villages. These landscapes represent an attempt, I have suggested, to model an imaginary rural idyll for new managerialist (and new traditionalist) home-owners.

Fortress villages can provide us with insights into the exclusions at work in making up the rural idyll, therefore; they can also reveal to us the imaginings of a distinct class fraction – one more often associated with new forms of urban living, however (gentrification, loft-living, urban villages). The aligning of neo-rustic lifestyles with new-service-class taste cultures marks, I think, a distinct articulation of the rural idyll, drawing it into the field of cultural capital (Murdoch and Marsden, 1994). The rural is re-imagined as an 'other within', thus rendering it appealing to cosmopolitan tastes. The presence of new-service-class workers as visitors or inhabitants in idyllized rural settings also lets us explore what I name here the transnational rural – the idea that globalization (a set of processes closely associated with that class) might be producing distinct new types of rural idyll. To interrogate this further, I turned my attention to the production of forms of rural idyll through the entertainment industry ('the media idyll'), through the practices of tourism ('the tourist idyll') and through the agro-food system ('the gastro-idyll'). Borrowing Rojek's (1997) notion of indexing, we can see how food, tourism and the mass media produce particular imaginings of the countryside that we feed into our experiences of it. While this might sometimes lead to disappointment – as the 'real' idyll fails to match our expectations (Rojek labels this 'tourist denial') – my aim here is not to force a separation between 'real' and 'imagined' idylls. Instead, I want to argue finally that the production of the rural idyll, here and now, in early twenty-first century Euro-American cultures, involves a dense, complex and ongoing mediation of images, ideas and experiences (past, present and future). There are plenty of other sites for exploring these ideas. What, for example, is happening in online communities, where visions of 'homesteading' and revisions of the 'global village' can be seen to produce an ambivalent, high-tech virtual idyll (Bell, 2001)? Or, how can we read items of clothing, such as walking boots, as defining certain idyllized ways of being in the rural (Michael, 2000)? The examples I have explored here, therefore, only begin the task of unpacking that process; the many meanings and uses of the rural idyll await further scrutiny.

NOTES

1 So idyllic was the spot, that cameras from Spanish TV news came to capture us on film, for a news feature on rural tourism – a representation that, in its own small way, adds to the media production of this Catalan idyll.

2 The phrase is borrowed from sci-fi folklore, and was used to describe the process of designing the sets for the movie *Blade Runner* – a postmodern fusion, recycling, accreting and juxtaposing past architectural styles.

3 Thanks to Jon Binnie for making this connection, and for prompting me into exploring the transnational rural here. Both he and Tim Edensor also read drafts of the chapter, and provided very useful feedback.

REFERENCES

Bell, David (1997) 'Anti-idyll: rural horror', in Paul Cloke and Jo Little (eds), *Contested Countryside Cultures: Otherness, Marginalisation and Rurality*. London: Routledge. pp. 94–108.

Bell, David (2001) *An Introduction to Cybercultures*. London: Routledge.

Bell, David and Valentine, Gill (1997) *Consuming Geographies: We Are* Where *We Eat*. London: Routledge.

Bonanno, Alessandro, Busch, Lawrence, Friedland, William, Gouveia, Lourdes and Mingione, Enzo (eds) (1994) *From Columbus to ConAgra: The Globalization of Agriculture and Food*. Lawrence, KS: University of Kansas Press.

Bunce, Michael (1994) *The Countryside Ideal: Anglo-American Images of Landscape*. London: Routledge.

Butler, Richard (1998) 'Rural recreation and tourism', in Brian Ilbery (ed.), *The Geography of Rural Change*. Harlow: Longman. pp. 211–232.

Ching, Barbara (1997) 'Acting naturally: cultural distinction and critiques of pure country', in Matt Wray and Annalee Newitz (eds), *White Trash: Race and Class in America*. London: Routledge. pp. 231–248.

Ching, Barbara and Creed, Gerald (eds) (1997) *Knowing Your Place: Rural Identity and Cultural Hierarchy*. London: Routledge.

Cloke, Paul (1994) '(En)culturing political economy: a life in the day of a "rural geographer"', in Paul Cloke, Marcus Doel, David Matless, Martin Phillips and Nigel Thrift, *Writing the Rural: Five Cultural Geographies*. London: Paul Chapman. pp. 149–190.

Cloke, Paul and Little, Jo (eds) (1997) *Contested Countryside Cultures: Otherness, Marginalisation and Rurality*. London: Routledge.

Cloke, Paul and Perkins, H. (1998) '"Cracking the canyon with the Awesome Foursome": representations of adventure tourism in New Zealand', *Environment and Planning D: Society and Space* 16: 185–218.

Cohan, Steven and Hark, Ina Rae (eds) (1997) *The Road Movie Book*. London: Routledge.

Cook, Ian and Crang, Phil (1996) 'The world on a plate: culinary culture, displacement and geographical knowledges', *Journal of Material Culture* 1: 131–153.

Craik, Jennifer (1997) 'The culture of tourism', in Chris Rojek and John Urry (eds), *Touring Cultures: Transformations of Travel and Theory*. London: Routledge. pp. 113–136.

Darby, Wendy (2000) *Landscape and Identity: Geographies of Nation and Class in England*. Oxford: Berg.

Duruz, Jean (1999) '*Cuisine nostalgie*? Tourism's romance with "the rural"', *Communal/Plural* 7: 97–109.

Edensor, Tim (2001) 'Staging tourism: tourists as performers', *Annals of Tourism Research* 27: 322–344.

Edensor, Tim (2002) *National Identities in Popular Culture*. Oxford: Berg.

Friedland, William (1994) 'The new globalization: the case of fresh produce', in Alessandro Bonanno, Lawrence Busch, William Friedland, Lourdes Gouveia and Enzo Mingione (eds), *From Columbus to ConAgra: The Globalization of Agriculture and Food*. Lawrence, KS: University of Kansas Press. pp. 210–231.

Goodman, David and Watts, Michael (eds) (1997) *Globalising Food: Agrarian Questions and Global Restructuring*. London: Routledge.

Havinden, Michael (1981) 'The model village', in G.E. Mingay (ed.), *The Victorian Countryside*. London: Routledge & Kegan Paul. pp. 414–427.

Hetherington, Kevin (2000) *New Age Travellers: Vanloads of Uproarious Humanity*. London: Cassell.

James, Allison (1993) 'Eating green(s): discourses of organic food', in Kay Milton (ed.), *Environmentalism: The View from Anthropology*. London: Routledge. pp. 205–218.

Jensen, Joli (1998) *The Nashville Sound: Authenticity, Commercialization and Country Music*. Nashville, TN: Vanderbilt University Press.

Jones, Owain (2000) 'Is all well in "Allswell"?' Paper presented at the Rural Economy and Society Study Group conference, Exeter, September.

Kristeva, Julia (1982) *Powers of Horror: An Essay in Abjection*. New York: Columbia University Press.

Kroker, Arthur and Kroker, Marilouise (2000) 'Code warriors: bunkering in and dumbing down', in David Bell and Barbara Kennedy (eds), *The Cybercultures Reader*. London: Routledge. pp. 96–103.

Leslie, D.A. (1993) 'Femininity, post-Fordism and the "new traditionalism"', *Environment and Planning D: Society and Space* 11: 689–708.

Levy, Emmanuel (1991) *Small-Town America in Film: The Decline and Fall of Community*. New York: Continuum.

Lewis, Neil (2000) 'The climbing body, nature and the experience of modernity', *Body & Society* 6: 58–80.

de Lisle, Leanda (1999) 'Organic farming: no option for mainstream agriculture?', in Michael Mosbacher and Digby Anderson (eds), *Another Country*. London: Social Affairs Unit. pp. 89–94.

Loader, Ian, Girling, Evi and Sparks, Richard (2000) 'After success? Anxieties of affluence in an English village', in Tim Hope and Richard Sparks (eds), *Crime, Risk and Insecurity*. London: Routledge. pp. 65–82.

Lowe, Philip (1989) 'The rural idyll defended: from preservation to conservation', in G.E. Mingay (ed.), *The Rural Idyll*. London: Routledge. pp. 113–131.

Lupton, Deborah (1996) *Food, the Body and the Self*. London: Sage.

Lutticken, Sven (2001) 'Parklife', *New Left Review* 10: 111–118.

Marsden, Terry, Murdoch, Jonathan, Lowe, Philip, Munton, Richard and Flynn, Andrew (1993) *Constructing the Countryside*. London: UCL Press.

Marsh, Jan (1982) *Back to the Land: The Pastoral Impulse in Victorian England from 1880 to 1914*. London: Quartet.

Matless, David (1998) *Landscape and Englishness*, London: Reaktion.

May, Jon (1996a) 'A little taste of something more exotic: the imaginative geographies of everyday life', *Geography* 81: 57–64.

May, Jon (1996b) 'In search of authenticity off and on the beaten track', *Environment and Planning D: Society and Space* 14: 709–736.

Michael, Mike (2000) 'These boots are made for walking …: mundane technology, the body and human–environment relations. *Body & Society* 6: 107–126.

Milbourne, Paul (ed.) (1997) *Revealing Rural Others: Representation, Power and Identity in the British Countryside*. London: Pinter.

Mingay, G.E. (1989) 'Introduction', in G.E. Mingay (ed.), *The Rural Idyll*. London: Routledge. pp. 1–6.

Moran, Warren (1993) 'Rural space as intellectual property', *Political Geography* 12: 263–277.

Mosbacher, Michael and Anderson, Digby (eds) (1999) *Another Country.* London: Social Affairs Unit.

Murdoch, Jonathan and Marsden, Terry (1994) *Reconstituting Rurality.* London: UCL Press.

Neumann, Mark (1999) *On the Rim: Looking for the Grand Canyon.* Minneapolis, MN: University of Minnesota Press.

Phillips, Martin, Fish, Rob and Agg, Jennifer (2001) 'Putting together ruralities: towards a symbolic analysis of rurality in the British mass media', *Journal of Rural Studies* 17: 1–27.

Philo, Chris (1992) 'Neglected rural geographies: a review', *Journal of Rural Studies* 8: 193–207.

Pollan, Michael (2001) 'Going organic', *Observer Food Monthly*, August: 42–7.

Probyn, Elspeth (1990) 'New traditionalism and post-feminism: TV does the home', *Screen* 31: 147–59.

Purdue, Derrick, Durrschmidt, Jorg, Jowers, Peter and O'Doherty, Richard (1997) 'DIY culture and extended milieux: LETS, veggie boxes and festivals', *Sociological Review* 45: 645–667.

Rankin, Aidan (1999) 'Modern Britain's homogeneity versus the multicultural countryside', in Michael Mosbacher and Digby Anderson (eds), *Another Country.* London: Social Affairs Unit. pp. 187–192.

Rojek, Chris (1997) 'Indexing, dragging and the social construction of tourist sights', in Chris Rojek and John Urry (eds), *Touring Cultures: Transformations of Travel and Theory*. London: Routledge. pp. 52–74.

Rojek, Chris and Urry, John (1997) 'Transformations of travel and theory', in Chris Rojek and John Urry (eds), *Touring Cultures: Transformations of Travel and Theory.* London: Routledge. pp. 1–22.

Ross, Andrew (2000) *Celebration Chronicles.* London: Verso.

Shields, Rob (1991) *Places on the Margin: Alternative Geographies of Modernity.* London: Routledge.

Shoard, Marion (1987) *This Land is Our Land: The Struggle for Britain's Countryside.* London: Paladin.

Sibley, David (1995) *Geographies of Exclusion: Society and Difference in the West.* London: Routledge.

Silverstone, Roger (ed.) (1997) *Visions of Suburbia.* London: Routledge.

Sklair, Leslie (2001) *The Transnational Capitalist Class.* Oxford: Blackwell.

Stallybrass, Peter and White, Allon (1986) *The Politics and Poetics of Transgression.* Ithaca, NY: Cornell University Press.

Stewart, Kathleen (1996) *A Space on the Side of the Road: Cultural Poetics in an 'Other' America.* Princeton, NJ: Princeton University Press.

Taylor, John (1994) *A Dream of England: Landscape, Photography and the Tourist's Imagination.* Manchester: Manchester University Press.

Thrift, Nigel (2000) 'Still life in nearly present time: the object of nature', *Body & Society* 6: 34–57.

Tomlinson, John (1999) *Globalization and Culture.* Cambridge: Polity Press.

Urry, John (2002) *The Tourist Gaze,* 2nd edn. London: Sage.

Vincendeau, Ginette (ed.) (2001) *Film/Literature/Heritage.* London: BFI.

Watts, Michael and Goodman, David (1997) 'Agrarian questions: global appetite, local metabolism: nature, culture, and industry in *fin-de-siècle* agro-food systems', in David Goodman and Michael Watts (eds), *Globalising Food: Agrarian Questions and Global Restructuring.* London: Routledge. pp. 1–32.

Williams, Raymond (1973) *The Country and the City.* London: Chatto & Windus.

Williamson, J.W. (1995) *Hillbillyland: What the Movies Did to the Mountains and What the Mountains Did to the Movies.* Chapel Hill, NC: University of North Carolina Press.

Wray, Matt and Newitz, Annalee (eds) (1997) *White Trash: Race and Class in America.* London: Routledge.

Wright, Will (1975) *Sixguns and Society: A Structural Study of the Western.* Berkeley, CA: University of California Press.

Young, Iris Marion (1990) *Justice and the Politics of Difference.* Princeton, NJ: Princeton University Press.

11

Constructing rural natures

Noel Castree and Bruce Braun

the tactic of … revealing nature to be a 'social construct' has lost much of its initial intellectual potency. (Bartram and Shobrook, 2000: 370)

INTRODUCTION

As our epigram suggests, this is an inauspicious moment to consider the constructedness of rural natures. In relation to those myriad things conventionally labelled as 'natural', rural studies is fast entering a 'post-constructivist' moment.[1] Today the non-human is being granted a constitutive role in rural life within a non-dualistic, anti-essentialist worldview. This moment follows hard on the heels of nature's 'return' to rural studies from the late 1980s and through the 1990s. Over a decade ago, nature gained a long overdue place on rural researchers' agendas by, paradoxically, being de-naturalized. As we shall demonstrate in this chapter, the idea that rural natures were socially constructed inspired 10 years of interesting and important research. Today, though, enthusiasm for the social construction thematic is on the wane. In rural studies, as in several other research fields (like human geography and environmental sociology), this is because the philosophical limits of constructionist arguments have been exposed (see Chapters 12 and 13, this volume). This is a positive development because it has opened the door for approaches to rural natures that do not rest upon ontological separations and the explanatory-normative frameworks that such separations have inspired.

Having, in one sense, pulled the rug from under ourselves, why write about rural studies, nature and social constructionism at all? Our answer is three-fold. First, it is easy to overlook the considerable strengths of constructionist approaches. While post-constructivist frameworks such as actor-network and non-representational theory[2] may now be paradigmatic for many rural researchers, we forget at our peril the considerable intellectual and political resources that constructionism still provides. For instance, while the baroque jargon of some academics might declare that a society–nature dualism never existed, it none the less continues to animate thought and action in myriad everyday sites and situations. 'Wilderness', for instance, is still an immensely important organizing concept in and for rural North America, while the 'rural idyll' still resonates in Britain. Even now, then, the constructionist move of de-naturalizing that which seems self-evidently 'natural' can still be a powerful and productive one. In the second place, we also want to argue that despite its familiarity as a catch-phrase, rural researchers have not always been entirely clear on what 'social constructionism' means. This is because there is no such thing as a generic social constructionist position, only specific *modalities* of social constructionism. Yet few in rural studies have bothered to tease out the differences of degree or kind between these several strands of social constructionist argument. As David Demeritt (2002: 768) notes, 'The '"social construction of nature" is spoken about in such different and often imprecise ways that its [exact] … meaning and implications can be difficult to understand and evaluate'. Finally, there may be good reasons to question whether the 'post-constructivist' moment is really the radical break it seems to be. We argue that it should be considered a *realization* of certain themes within the social constructionist problematic that have, for too long, been confined by unhelpful ontological suppositions.

We hope, then, that this chapter at once clarifies and pushes to the limit what is meant by the 'construction' of nature. The chapter is organized as follows. In the next section we introduce the 'question of nature' in rural studies and revisit the moment when the field experienced its 'constructionist turn'. We do this to remind readers why constructionism proved such an appealing approach for several years from the late 1980s onwards. Following this, we identify several types of constructionist argument. In the third section we then discuss the limits of otherwise different constructionist approaches to rural natures. Despite appearances, this critique does not imply a wholesale abandonment of constructionist positions. Although that *is* one possible implication, we argue that strategic retention of these positions can still be justified. This said, we conclude that social constructionism will soon have very few rural researchers willing to sing its explanatory and normative praises.

RURAL STUDIES, NATURE AND THE 'CONSTRUCTIONIST TURN'

In an oft-cited review of rural studies in the mid-1990s, Paul Cloke (1997: 371) detected 'strong moves to reclaim … nature as a legitimate and provocative component of [a] reconstructed rural studies'. Even if one acknowledges that the conventional idea of rural spaces being somehow more 'natural' than urban ones is misconceived, it is still surprising how marginalized nature was in post-war (and especially post-positivist) rural studies discourse. To the extent that it figured at all it was either a backdrop or a taken-for-granted aspect of rural life rather than a formal subject of analysis. For instance, in a wide-ranging late 1980s' review of rural geography Cloke found no reason to mention nature at all. Yet within a few years nature was on the rural studies agenda (and Cloke's own agenda) in two important ways. On the one side, concerns about environmental degradation and the need for 'sustainability' led some to focus on nature as a biophysical entity with identifiable capacities, limits and (in the case of sentient non-humans) even rights. In the case of several Marxists – like George Henderson (1999: part I) – this focus was inspired more by a curiosity about how 'natural barriers to accumulation' are overcome in farming. But in most cases it has arisen from a concern about environmental degradation and the loss of non-human species. On the other side, several rural researchers took a less realist tack and sought to explore the non-naturalness of nature. In so doing they partook of a wider constructionist turn in the social sciences and humanities regarding 'nature' and collateral concepts such as 'race', sexuality, biology and environment. This turn gathered momentum through the mid-to-late 1980s. Among the first rural studies publications that used the term 'social construction of nature' in a substantive sense were Lowe et al. (1990) and Maunder et al. (1993). Subsequently, numerous empirical and theoretical studies were published that applied, finessed and extended social constructionist ideas.

The positive legacy of this two-pronged engagement with nature lives on today. At present there are several rural researchers still mining these two research veins, while the critique of the society–nature dualism underpinning both has, as noted in the introduction, propelled still other researchers into innovative new domains of inquiry. Though this latter cohort of post-constructivist researchers eschew any literal use of the label 'nature', they none the less remain profoundly concerned with the entities this signifier has traditionally named. As Rouse (2002: 69) explains, 'If the post-constructivist tradition denies that there is any role for "unreconstructed nature" in our understanding …, it is not because we are able to get "outside" of a relatively self-enclosed social world, but because we have never been "inside" one in the first place.'

As a result of these various developments, contemporary rural studies takes nature seriously in several different ways. Though we do not wish to engage in pedantic boundary disputes over where 'rural studies' begins and ends, the following is just a brief list of some of the topic areas where nature currently figures in rural researchers' inquiries: environmental governance; agribusiness; images of rural life; rural wildlife and domesticated animals; commodity chains and food networks; organic farming; post-productivist landscapes; the body; environmental knowledges (lay and expert); and agricultural biotechnology. The theoretical pluralism of rural studies means that in these and other topic areas nature is approached in various (often incompatible) ways. But prior to our current post-constructivist moment, it is fair to say that several theoretical perspectives shared a common commitment to a constructionist worldview. We will review these perspectives in the next section. For now, we simply want to speculate about why constructionism made such a splash in rural studies from the late 1980s until quite recently.

The philosopher of science Jeff Coulter (2001: 82) recently suggested that constructionism is 'more a shibboleth than a coherent domain of … theorizing'. Though many would doubtless agree, it would be remiss to forget that the term 'constructionism' meant something quite definite when it became a watchword in a range of Anglophone academic disciplines more than a decade ago. Typically associated with academics who would class themselves as 'left wing' in some sense or other, the term 'social construction' was first popularized in a famous text by Berger and Luckmann (1967). According to Steven Vogel (1996), though, its deeper origins lie in Georg Lukacs' (1971) seminal critique of fetishism and reification (itself based on a creative reading of Hegel's philosophy). We mention this because it seems to us that the turn to social constructionism in rural studies (as well as a number of other fields) was motivated by a desire to reap the rewards of a double critique: namely, a critique of *misrecognition* and *hypostatization*. Let us explain.

When some rural researchers re-discovered nature in a realist rather than constructionist sense in the early 1990s their actions were entirely predictable. After all, these researchers seemed to be confirming the common-sense idea that rural areas are more profusely natural than towns and cities. What could be more sensible – and important – than examining the ecological limits of agriculture, the state of rural wildlife or the restoration of hedgerows, dry-stone walls or beaver habitats? For constructivists the answer to this question was that de-mystifying nature was more intellectually and politically rewarding. 'Nature', we can hardly forget, has long been a polysemic weasel-word. As William Cronon (1991: 36) rightly insisted, 'We cannot fall into the trap this word has laid for us.' For constructivists, the 'problem' with nature was that its apparent givenness and obviousness was precisely what shielded it from critical scrutiny. To suggest that nature was 'constructed' was thus – given the unproblematized naturalism characteristic of pre-1990s rural studies – radical and liberating. The term construction, rich in its meanings, describes not only a process and an end product but also an act of conceptual construal. Its invocation was a caution against the easy rediscovery of an asocial, biophysical nature that had quietly inhabited rural researchers' ontological worldviews for several decades. It was also a corrective to the 'common-sense' move of equating rural things with natural things.

This had two dimensions. First, the 'constructionist' label suggested that natural things – fields, crops, water-courses, wolves or what-have-you – were often social things in a cunning disguise. The 'rural idyll', for example, could be shown to be a culturally specific and normatively loaded image rather than a faithful depiction of an evergreen countryside (Bunce, 1994; Mingay, 1989). Secondly, this exposure of routine acts of misrecognition was linked to critique of hypostatization. To hypostatize something is to ossify and freeze it when, in fact, the thing so conceived is part of a changeable and dynamic process. To show that apparently natural (and thus 'given') things are, in fact, mutable social constructs was thus very appealing for several rural researchers. It allowed them to contest acts justified in the name of nature. It allowed them to highlight the social relations animating apparently non-social ideas or entities. It enabled them to show that 'nature' was a medium for the expression of power relations. And it offered them the chance to imagine alternative social and ecological arrangements that honestly confronted the non-naturalness not just of ideas of nature but also the phenomena those ideas referred to. In short, the constructionist motif was very empowering for a cohort of 1990s rural researchers and remains so for several of them today. In a sense, 'nature' became a metaphor for these researchers: one which connoted the duplicity of appearances, the hidden hand of power and misplaced concreteness of the visible and perceptible.

Before we end this section, it is worth recalling what has been implicit in the paragraphs above: namely, that constructionist analyses of nature have been an important vehicle for the importation of putatively 'critical' theories into rural studies. We say importation because, as we shall show below, rural researchers have produced little in the way of 'indigenous theory' when it comes to analyses of nature. The specifically 'rural' dimension of the natures being deconstructed and denaturalized has, in the main, had no wider theoretical consequences. The one major exception to this, perhaps, is agro-food studies, where the materiality of rural natures has led to a reworking of concepts imported from Marxian political economy, regulation theory and the like. Notwithstanding the theoretical insignificance of rurality *per se*, the constructionist approach to nature was, during the 1990s, important in building a post-positivist momentum in rural studies that persists to this day. From our perspective this is all to the good. Social constructivism may no longer be a live option in rural studies – a result, no doubt, of the ennui that sets in with all intellectual-political 'turns' within academia. But neither should constructivism be dismissed as a thing of the past – as no more than a curiosity to be dissected and discarded. Instead, it has provided critical leverage for new and productive avenues of inquiry that animate rural studies. We will discuss these new avenues later. But let us first map the landscape of constructionist analyses of nature in rural studies.

PLURAL CONSTRUCTIONS, MULTIPLE NATURES

The term 'social construction' has become almost hackneyed through over-use in a range of disciplines. According to Coulter (2001: 82), 'this feeds the suspicion that it has become virtually devoid of intellectual substance'. Though this suspicion is hardly groundless it is certainly not accurate. As we now try to anatomize constructionism, bear in mind that we are inevitably interposing our own conception of 'construction' in order to place some limits on the material reviewed here. Social constructionist arguments in rural studies (as in other fields) have not always announced themselves as such. For several commentators (like Cloke, 1997), they are the preserve of those on the 'cultural left' of rural studies. But this is surely too narrow an interpretation. Constructionism does not begin-and-end with 'post-prefixed' approaches like Derridean post-structuralism – though this is often assumed to be the case. A more generous interpretation encompasses some political-economy perspectives on nature and actor-oriented studies of the rural, among others. This widening of the constructionist label is not merely a nominal issue. Real intellectual and political differences are obscured when we mis/label some approaches to rural natures 'constructionist' and others not. Finally, we should note that outside the rural studies field there are now several excellent surveys of the constructionist approach to nature (see Demeritt, 1998, 2001, 2002; Hacking, 1999). Demeritt, for example, offers a six-fold typology of constructivist analyses, straddling the distinction between broadly discursive (or representational) and material (or ontological) approaches. However, since Demeritt's and other surveys do not map neatly onto rural researchers' approaches to nature we choose not to use them in any rigid sense here.

We mentioned above that the term construction describes both a process and a product. There have, in our view, been three broad modes of constructionist argument in contemporary rural studies and we can think of each in terms of the process(es) and product(s) they specify. First, there are *material constructionisms* that consider nature as a physical domain. Secondly, there have been *discursive constructionisms* that look at ideas, representations and images of rural nature. Finally, there have been *material-semiotic constructionisms* that encompass materiality and discourse within one analytical frame. Let us take each in turn and offer some illustrative examples.

Post-war rural studies, as we observed earlier, operated with an unproblematized and largely implicit realist conception of nature up until the mid-to-late 1980s. Material constructionism challenged this by suggesting that rural natures, in the bio-physical sense, were not simply 'given'. This argument was made most forcefully by Marxist and neo-Marxist researchers such as Fred Buttel, David Goodman, Jack Kloppenburg and George Henderson. And it was made in relation to those elements of rural nature most subject to conscious material manipulation: namely, crops and cattle. These authors argued that a definite set of social processes, structures and relationships – variously described as 'capitalism', 'industrialism' and 'Fordism', among others – were physically transforming certain rural natures. The 'weaker' version of material constructionism argued that biophysical natures can only be defined *relative to* the socio-economic systems of production in which they are entrained. This meant that the 'naturalness' of a putatively non-social nature was, in fact, contingent and shifting not absolute and stable. Mann and Dickinson's (1978) pioneering work on how capital moves around 'natural barriers to accumulation' was an early example of this weak material constructionism. A stronger version of physical constructionism argued that rural natures were being *materially reconstituted.* For instance, Kloppenburg's (1986) *First the Seed* explained how the 'real subsumption' of biology to capital was possible once the science of hybridization was established in the early twentieth century. This subsumption is an example of what the geographer Neil Smith (1984) called the 'production of nature', a term which suggests that everything from crops to broiler chickens are now mere means to the end of capital accumulation. Given the intensified biochemical alteration of crops and livestock in the current era, Kloppenburg's analysis was, in retrospect, most prescient (see also Prudham, 2003).

This Marxist work illustrates the strengths of material constructionism as an approach to rural nature. First, this research dispelled the 'myth' that rural areas are home to nature by showing that the nature in question was, increasingly, artificial. What's more, this research showed that the social forces remaking nature on farms and in fields extended beyond the rural realm. As Cronon's (1991) classic (though non-Marxist) analysis demonstrated, if what happens in the country is systematically linked to what happens in the city then 'rural nature' cannot simply be *rural* or *natural* any more (see also Brechin, 1999). This theme has, more recently, been the focus of research into agricultural commodity chains, rural industrialization, the regulation of rural production and world food systems (see, for instance, Goodman and Watts, 1997). Secondly, Marxist material constructionisms offered new political, as much as analytical, possibilities. They focused rural researchers' attention onto a key question: namely, who constructs rural natures in what ways, to what ends and with what consequences? Asking this question did two important things simultaneously. On the one hand, it nipped any unreflexive forms of ecocentrist politics in the bud by showing that nature can never be considered separately from specific forms of societal organization. On the other hand, it opened up the possibility of a politics that would seek to remake nature for different socio-ecological ends than those dominant in capitalist societies. In short, by de-reifying rural nature, material constructionisms promised to avoid the Scylla of a naïve naturalism and the Charybdis of a Promethean disregard for nature. This said, one must concede that few Marxist rural researchers have paid much heed to the materiality of constructed nature in its own right and less still to what a political-normative theory of this 'un-natural nature' might look like. In the main, these researchers have paid little attention to the agency of a transformed material nature nor to the evaluative schema one might use to judge the propriety of how this nature is being remade.

The economic and physicalist emphases of Marxist rural research was, in many ways, unusual within the wider rural studies 'turn' to nature from the 1980s onwards. Like much of the social sciences and the humanities, rural studies became swept up in two other turns that informed its new-found interest in nature: namely, the linguistic and the cultural. The latter, of course, were not entirely un-Marxist in the theoretical sense. The works of Gramsci, Stuart Hall and Raymond Williams, for instance, were key influences. But in rural studies, if these figures were appropriated at all, it was rarely in conjunction with an attempt to understand the material construction of nature. George Henderson's (1999) work *California and the Fiction of Capital* remains one of the few exceptions. His analysis of the material remaking of early twentieth century Californian agriculture is accompanied by a sophisticated analysis of the discourses of rural nature that accompanied this physical transformation. Specifically, Henderson analyses such seemingly disparate representational devices as adverts and novels to show how different actors in the Californian landscape made flesh their attempts to exert power and express resistance. The Marxian notion of ideology he works with shows how representations of rural nature serve certain interests that are, in turn, dialectically structured by the economic and ecological context in which they are situated. His point is not that representations *conceal* 'what's really happening' but, rather, that they are material forces in their own right that capture partial truths about their context.

Most rural researchers interested in the discursive construction of nature have, unlike Henderson, focused on discourse alone. Indeed, it is probably fair to say that representations of nature became privileged subjects of analysis in rural studies during the 1990s. This was, no doubt, a legacy of the wider enthusiasm within Anglophone academia for the ideas of Derrida, Foucault, Barthes, Rorty and for those other (usually continental European) theorists preoccupied with the 'power' of language and the performativity of knowledge. Initially, the focus on representing rural natures was counter-intuitive, and that was its appeal. After all, in post-war rural studies as much as in the wider society, representations were usually considered to be more-or-less neutral 'mirrors'. This was most obvious in rural studies' positivist phase but even by the late 1980s there was something novel and arresting about denaturalizing representations of rural nature. These deconstructive manoeuvres took several forms in relation to several aspects of nature. The most general was the argument that the rural itself was a culturally constructed *idea* (not an empirical given) and that its contingent meaning depended upon equally contingent depictions of nature – an argument that went back to Williams's (1973) *The Country and the City*. As Little (1999: 440) put it, while 'nature cannot be conflated with the rural, the countryside does represent one commonly identified spatialization of nature'. Others, more specifically, anatomized particular written, spoken and visual representations of rural nature – not in their own right but as elements within wider taken-for-granted grids of cultural understanding. These kinds of analyses were a key means whereby non-Marxist approaches were able to find a place in rural studies. Feminists, anti-racists and critics of heteronormativity, in particular, were able to show the gender, racial and sexual (not merely class) inflections of many hegemonic understandings of rural natures – be they the 'frontier' lands of the United States, the 'great outdoors' of the Antipodes, or the British fox-hunting countryside. Sibley (2003), among others, has shown how these understandings structure individual and group self-understandings and so become, in a very literal sense, *lived*. This is most evident in research into rural *identities* that shows how conceptions of, and practical engagements with, the natural world define stereotypical self-understandings of manliness, femininity and sexuality (e.g. Brandth, 1995; Bryant, 1999).

However, not all research on rural representation has been indebted to critical post-Marxist (and non-Marxist) theory or to continental linguistics, philosophy and psychoanalysis. For instance, the Wageningen School of rural research is more actor-oriented, looking at how different knowledges of rurality and nature issue from different rural actors because of their location within wider social relations and networks (see, for instance, Haan and Long, 1997). Likewise, a lot of English-speaking research on 'expert' and 'lay' understandings of rural issues is micro-sociological in approach and interested not just in how constructed bodies of 'local knowledge' inform action but in how the legitimacy of those knowledges is challenged and changed (e.g. Lowe and Ward, 1997).

We do not have space here to typologize discursive constructionisms in rural studies. In any case the emphasis of research into representations of rurality and nature has shifted in recent years, as we will see below. Suffice it to say that the discursive constructionisms mentioned above are not all of a piece, except perhaps in that they share some broad strengths that explain why many rural researchers caught discourse fever through the 1990s. First, they gave long overdue recognition that representation 'matters' in both senses of the term (that is, it is important and has tangible effects). Given the conventional view of nature as something given not made, this recognition was especially important. It also allowed researchers to show that specific and contestable depictions of nature typically disavowed their own constructedness. Furthermore, it allowed these researchers to show that representations of nature, in one sense, *create* the 'realities' they purport merely to convey. Thirdly, this demonstrated that there is a politics of representation to be uncovered in which power and resistance are involved (see Maunder et al., 1993). 'Mere representations' of nature were shown to be key sites of struggle that can materially affect people as much as the non-human world. It is worth noting that, on the whole, discursive constructionism in rural studies has functioned as what Demeritt (2002) calls 'philosophical critique' rather than as 'refutatory critique'. The latter aims to expose the *mis*-representations that delude onlookers about the 'realities' of their world. Philosophical critique, by contrast, argues that there is no way to understand those realities that escape 'the looping effect through which conceptual change transforms what has been conceived' (Rouse, 2002: 76).

If these insights now seem rather passé it is because discursive constructionism has become such a central part of the rural studies landscape. But there is a third form of constructionism in rural studies that pushes beyond both the discursive and the material version and which has become prominent in the past few years. It is a material-semiotic form of constructionism that aims to link the discursive and the physical together. We have already mentioned Henderson's work as one example of this. But Henderson, arguably, ultimately separates the discursive and the material, albeit as what David Harvey (1996) would call related 'moments' within a complex but continuous process. Other rural researchers, by contrast, fuse the two. They do so not, as might be supposed, by resorting to motifs like dialectic, interaction, or system. Rather, they aim to remove the representation-reality distinction that underpins these otherwise appealing motifs. Counter-intuitively, they do so by deepening the notion of discursive constructionism so that it describes a seamless process of *materialization*.

An example is Braun's research on rural British Columbia. Writing in the wake of Butler's (1993) adaptation of Derrida to the question of the human body, and of Foucault's work on governmentality, Braun (2000) seeks to avoid any hint of cognitive dualism or 'ontological outsides' of discourse. He shows how geologists like George Dawson 'ordered' and 'enframed' rural British Columbia such that there was never any possibility of a self-evident nature 'beyond discourse' whose raw materiality could be directly apprehended. This is not the same as saying that language physically 'constructs' those things that have conventionally been labelled as both 'rural' and 'natural'. Rather, it is to suggest that understandings of, and interactions with, those things cannot be separated from the discursive practices that make them available to calculation such that there is 'an implosion of the epistemological and the ontological' (2000: 14). This implosion is possible once one realizes (or accepts) that (i) there is no clear-cut outside of language and representation and that (ii) language and representation are practical, world-making devices not passive mirrors of something external to themselves. Braun's 'flat' philosophical worldview suggests that discursive representations are practical tools that help matter to 'materialize' for us both visibly and physically (see also Braun, 2002). It is a worldview that pushes beyond earlier work on representations as 'lenses' that interpose themselves *between* the physical world and people (Braun, 1997).

The contrast with Henderson's work is instructive, marking the difference between material-semiotic constructionisms informed by Marxian theory and those informed by leading French philosophers. Henderson is ultimately a 'materialist' in the sense that the physical environment and the material structure of society together condition the sorts of representations of rural nature that become hegemonic at any moment in history. Braun, by contrast, does not ultimately subordinate the discursive to the material in this way because he sees no ready distinction between the two. This is not to suggest that material-semiotic constructionisms like Braun's trump those expounded separately by rural researchers preoccupied with either discourse or the material environment. At base, one might suggest that the plurality of constructionist positions in rural studies reflects an ongoing uncertainty about two things. The first is what the 'ontological units' of analysis should be in constructionist analysis (that is, where does one draw distinctions so that meaningful 'noun chunks' of reality can be identified). The second is what the causal link between these ontological units is (that is, how does one theorize the relation between a constructing process and a constructed idea or object?).

BEYOND CONSTRUCTIONISM?

For all their strengths, one of the paradoxes of constructionist approaches to nature is that they break down the 'common-sense' separation of society and nature only to (i) reinstate it at another level or else (ii) erase nature altogether. As we saw in the previous section, the 'classic' constructionist move is to denaturalize nature by bringing it – either discursively or biophysically (or both together) – within the realm of the social. But this boundary-transgressing act is only apparently non-dualistic. Depending on the specific modality of constructionism in question, it can reinstate a society–nature divide in several other registers. For instance, claims about the discursive construction of rural natures often imply – by default – that a 'real' nature lies beyond the lattices of language. Where a nature–society dualism is not reinstated it is often (again implicitly) replaced by a monism, whereby what we call nature is almost completely eclipsed. For instance, the argument that many elements of 'nature' in rural settings are biophysically remade 'all the way down' by agri-business gives most of the motive power to capital. Likewise, the non-dualist approach of Braun risks ignoring the agency and affordances of those things that become intelligible to people through their labelling as 'natural'. So what becomes of nature here? We ask this question not to imply that some constructionisms need an antidote of good old-fashioned realism-cum-naturalism. For realism usually partakes of the same society–nature dualism we have just questioned. Rather, we are concerned to know how the presence and agency of what we happen to call 'nature' can be registered without getting stuck on the horns of an irresolvable constructivist–realist dilemma. Equally, we are concerned to know what role those things heretofore captured by abstractions like 'society' or 'social power' play in relation to putatively 'natural' phenomena. These twin concerns are starting to be addressed in rural studies as integral parts of the field's current post-constructivist trajectory.

'Our ontological choices', David Goodman (2001: 182) avers, 'are consequential.' We couldn't agree more. The choices constructionists make (consciously or not) circumscribe as much as they enable. So what is wrong with the dualistic worldview underpinning social constructionist arguments (and their realist counterpoints)? Two problems loom large. First, to say that nature is socially constructed is to grant 'the social' (be it capitalism, culture, discourse or what-have-you) an unwarranted degree of materiality and integrity. While it is possible to 'cut the world at the joints' conceptually, at the ontological level what we call 'the social' cannot be materially disembedded from what we call 'the natural'. Secondly, this suggests that we live (and have always lived) in a world of 'distributed materiality' where the actions of 'social' and 'natural' phenomena upon one another is a *relational achievement* not a case of discrete domains colliding with, controlling or subsuming the other. It thus seems to make little sense to talk about a 'socially constructed nature' just as, conversely, it is unhelpful to talk about an 'asocial nature' that has an independent materiality and ethical value. Arguably, we need a 'post-social' conception of 'rural natures' that is, at the same time, a *post-natural* one too (cf. Knorr Cetina, 2001). In short, for many critics in rural studies and beyond, we need a synthetic, symmetrical appreciation of 'socionatures'.

There are at least four ways in which rural researchers, and other social scientists with interests in nature, have sought to move beyond the social constructionist problematic in recent years. First, some have embraced the actor–network theory (ANT) pioneered and popularized by Bruno Latour and Michel Callon (e.g. Murdoch, 2003). There are arguably two reasons why ANT has appealed to several rural researchers. One is that it provides an elaborate conceptual vocabulary (including neologisms like 'actants' and 'intermediaries') that can be readily put to work in new research contexts. The second reason is that Latour, Callon and other like-minded theorists have a strongly empirical focus to their work. Given rural studies' previous aversion to theory (until the 1980s), the empirical cast of ANT has proven congenial to many in the field because it chimes with their 'grounded' sensibilities (e.g. Eden et al., 2000). Secondly, some rural researchers have been inspired by the ontologies of force and practice expounded by Deleuze, Guattari, Grosz and other relational philosophers. Thrift (2003: 309), for example, has used some of the latter's 'portentous philosophical statements' to consider how both human bodies and non-human phenomena are entangled in country life. Thrift's vision is of rural worlds where numerous assemblages of bodies, matter, extension and action both endure and yet churn in non-essential ways. Thirdly, Thrift's own 'non-representation theory' and a wider precoccupation with 'performance' in cultural studies has inspired others to examine the iterative co-constitution and remaking of bodies, identities, species and landscapes in rural places. This dovetails with a 'dwelling perspective that accents the mundane practices of *engagement* through which people's identities and corporealities are configured and the non-human world is fashioned (Wylie, 2003). This dwelling perspective suggests that human actors learn by doing not simply by knowing, where learning encompasses patterns of thought, bodily habits, physiology, sight, smell and sound. Finally, and moving more into the realm of ecological science, some rural researchers have been inspired by new thinking about how energy and matter pass between people and the non-human world. This new ecological thinking looks at how complexity, disequilibrium and scale-switching all animate human-non-human connections (see Zimmerer, 2000).

Much of this relational thinking is examined elsewhere in this volume; this is why we do not want to discuss it in any detail here. However, we do want to acknowledge that relational approaches offer a promising means of bringing nature 'back in' to rural studies without reverting to a rather rigid realism. Earlier we mentioned those rural analysts with interests in species and habitat loss, conservation policy and environmental pollution. We also observed that what was missing from rural studies' 'constructionist turn' during the 1990s was a due acknowledgement of the materiality of those things we conventionally class as natural things. The relational approaches listed above give analysts the chance to take nature seriously in explanatory and moral terms – be it the fleshy body or the non-human domain. But they all do so, in effect, by doing away with the category of nature altogether in order to explore how different worldly actors with diverse capacities and affordances co-constitute one another.

CONCLUSION

In light of our arguments in the previous section it would not be difficult to conclude with a Latourian turn of phrase. 'We've never been constructivists!' could easily be our refrain. But such a conclusion, aside from glibly calling the work of numerous rural researchers into question, also neglects the positive role constructionist analyses can still play today. There is no neutral ground on which to compare the relative merits of constructionist and post-constructionist approaches. The former denaturalizes rural natures, while the latter questions whether there are discrete 'social' processes that create discernible 'constructions' representationally and/or materially. Given the incommensurability of these two paradigmatic approaches to rural natures, all we can do is ask the pragmatic questions: what kinds of moral-political visions and practical actions are mandated by each approach?; and do we find those visions and actions to be worthy ones? If one chooses to retain the philosophical assumptions underpinning most constructionist positions then the original advantages (detailed in the second section of this chapter) of the denaturalizing argument remain useful today. The exposé of misrecognition and hypostatization still carries force, as does the kind of refutatory critique intrinsic to many social constructionist approaches in rural studies. Also, let us not forget that many researchers in rural studies and beyond are simply unimpressed (or just confused) by the relational thinking of such authors as Deleuze, Guattari, Serres or Latour. Many of these researchers remain staunch realists who have little or no patience for the idea that nature might not, in some significant sense, be natural. And, as John Habgood's (2002) book reminds us, *ideas of nature* still 'perform' in Western societies where their constructedness often goes unnoticed by those deploying them and those feeling their effects. At the same time, even those with a qualified sympathy for post-social approaches to nature – like Castree (2002) and Marsden (2000) – are concerned that they risk throwing out the baby with the bathwater in both explanatory and normative terms. They argue that some things 'matter' more than others and resist the incipient 'levelling' effect of some relational worldviews. At the very least, adherence to a 'strategic constructionism' can still yield cognitive and moral gains in this incipient post-constructionist moment in rural studies.

Constructionism, and its naturalist/realist twin, are like 'vampire figures': they are the intellectual undead that still haunt rural researchers' analyses even as other critics try to take them into new ontological and normative waters. This said, we suspect that the pace of academic innovation will soon mean that those remaining constructionist researchers in rural studies fast dwindle in number. The embers of social constructionism are still aglow in rural studies, but for how much longer? One imagines that the term 'social construction' will suffer the same fate as other buzz-terms of the 1990s, like postmodernism. Whatever their intellectual substance, they will decline in popularity because their familiarity ultimately breeds contempt. But this decline may be more apparent than real. Even if 'strong' constructionisms are unlikely to live on, there is, in reality, only a fine-line separating 'weaker' ones from some of the relational approaches the constructionist problematic paved the way for. Symptomatic of this fact, we predict, is that two issues will preoccupy post-constructivist theorists in rural studies just as they have haunted the constructionist research we have reviewed in this chapter. The first is how to identify and distinguish the 'ontological units' that any and all forms of rural analysis need to be able to do in order to make cognitive and normative claims about the world. The second is how to theorize determination: that is, relative balance of cause and effect between these units in any given context. Even if 'construction' will not be the favoured metaphor of a post-constructivist rural studies, it will still be necessary to make justified ontological and moral choices about what exists, what affects what, what should be valued, and on what ground values can be upheld.

ACKNOWLEDGEMENT

We would like to thank Terry Marsden for constructive comments on an earlier version of this chapter.

NOTES

1 In this chapter we choose not to explore the various meanings and referents of the polysemic word 'nature'. This has been done before, most recently by Habgood (2002). Here we use the term nature to signify the non-human world, though towards the end of the chapter we discuss the human body also.

2 Labels like these inevitably sloganize rather complex philosophical and normative approaches. In simple terms, actor–network theory challenges the categorical and ontological dualism of society–nature and focuses on the co-constitution of 'social' and 'nature' phenomena. Non-representational theory, associated with Nigel Thrift above all others, emphasizes the lived, practised and contingent nature of all existence.

REFERENCES

Bartram, R. and Shobrook, S. (2000) 'Endless/end-less natures', *Annals of the Association of American Geographers* 90 (2): 370–80.

Berger, P. and Luckmann, T. (1967) *The Social Construction of Reality*. London: Penguin Books.

Brandth, B. (1995) 'Rural masculinities in transition', *Journal of Rural Studies* 11 (2): 123–33.

Braun, B. (1997) 'Buried epistemologies: the politics of nature in (post)colonial British Columbia', *Annals of the Association of American Geographers* 87 (1): 3–31.

Braun, B. (2000) 'Producing vertical territory', *Ecumene* 7 (1): 7–46.

Braun, B. (2002) *The Intemperate Rainforest*. Minneapolis, MN: Minnesota University Press.

Brechin, G. (1999) *Imperial San Francisco*. Berkeley, CA: University of California Press.

Bryant, L. (1999) 'The detraditionalisation of occupational identities in farming in South Australia', *Sociologia Ruralis* 39 (3): 236–61.

Bunce, M. (1994) *The Countryside Ideal*. London: Routledge.

Butler, J. (1993) *Bodies that Matter*. New York: Routledge.

Castree, N. (2002) 'False antitheses? Marxism, nature and actor-neworks', *Antipode* 34 (1): 111–46.

Cloke, P. (1997) 'Country backwater to rural village?', *Journal of Rural Studies* 13 (4): 367–75.

Coulter, J. (2001) 'Ian Hacking on constructionism', *Science, Technology and Human Values* 26 (1): 82–6.

Cronon, W. (1991) *Nature's Metropolis*. New York: W.W. Norton.

Demeritt, D. (1998) 'Science, social constructivism and nature', in B. Braun and N. Castree (eds), *Remaking Reality*. London: Routledge. pp. 173–93.

Demeritt, D. (2001) 'Being constructive about nature', in N. Castree and B. Braun (eds), *Social Nature: Theory, Practice and Politics*. Oxford: Blackwell. pp. 22–40.

Demeritt, D. (2002) 'What is the social construction of nature?', *Progress in Human Geography* 2 (3): 255–79.

Eden, S., Tunstall, S. and Tapsell, S. (2000) 'Translating nature', *Society and Space* 18 (3): 257–72.

Goodman, D. (2001) 'Ontology matters', *Sociologia Ruralis* 41 (2): 182–200.

Goodman, D. and Watts, M. (eds) (1997) *Globalizing Food*. London: Routledge.

Haan, H. de and Long, N. (eds) (1997) *Images and Realities of Rural Life*. Assen, Netherlands: Van Gorcum.

Habgood, J. (2002) *The Concept of Nature*. London: Darton, Longman and Todd.

Hacking, I. (1999) *The Social Construction of What?* Cambridge: Cambridge University Press.

Harvey, D. (1996) *Justice, Nature and the Geography of Difference*. Oxford: Blackwell.

Henderson, G. (1999) *California and the Fiction of Capital*. Oxford: Oxford University Press.

Kloppenburg, J. (1986) *First the Seed*. Cambridge: Cambridge University Press.

Knorr Cetina, K. (2001) 'Postsocial relations', in G. Ritzer and B. Smart (eds), *Handbook of Social Theory*. London: Sage. pp. 520–37.

Little, J. (1999) 'Otherness, representation and the cultural construction of rurality', *Progress in Human Geography* 23 (4): 437–42.

Lowe and Ward, N. (1997) 'Field level bureaucrats and the making of new moral discourses in agri-environmental controversies', in D. Goodman and M. Watts (eds), *Globalising Food*. London: Routledge. pp. 256–72.

Lowe, P., Seymore, S., Ward, N. and Ward, P. (eds) (1997) *Moralizing the Environment*. London: UCL Press.

Lukacs, G. (1971) *History and Class Consciousness*. London: New Left Books.

Mann, S. and Dickinson, J. (1978) 'Obstacles to the development of a capitalist agriculture', *Journal of Peasant Studies* 5 (4): 466–81.

Marsden, T. (2000) 'Food matters and matters of food', *Sociologia Ruralis* 40 (1): 20–30.

Marsden, T., Murdoch, J., Munton, R., Lowe, P. and Flynn, A. (1993) *Constructing the Countryside*. Boulder, CO: Westview Press.

Mingay, G. (ed.) (1989) *The Rural Idyll*. London: Routledge.

Murdoch, J. (2003) 'Co-constructing the countryside', in P. Cloke (ed.), *Country Visions*. Harlow: Pearson. pp. 263–80.

Prudham, S. (2003) 'Taming trees: capital, science and nature in Pacific slope tree improvement', *Annals of the Association of American Geographers* 93 (3): 636–56.

Rouse, R. (2002) 'Vampires: social constructivism, realism and other philosophical undead', *History and Theory* 41 (1): 60–78.

Sibley, D. (2003) 'Psychogeographies of rural space and practices of exclusion', in P. Cloke (ed.), *Country Visions*. Harlow: Pearson. pp. 218–30.

Smith, N. (1984) *Uneven Development*. Oxford: Oxford University Press.

Thrift, N. (2003) 'Still life in nearly present time: the object of nature', in P. Cloke (ed.), *Country Visions*. Harlow: Pearson. pp. 308–27.

Vogel, S. (1996) *Against Nature*. Buffalo: State University of New York Press.

Williams, R. (1973) *The Country and the City*. Paladin: London.

Wylie, J. (2003) 'Landscape, performance and dwelling', in P. Cloke (ed.), *Country Visions*. Harlow: Pearson. pp. 136–54.

Zimmerer, K. (2000) 'The reworking of conservation geographies', *Annals of the Association of American Geographers* 90 (2): 356–69.

12

Networking rurality: emergent complexity in the countryside

Jonathan Murdoch

INTRODUCTION

In the view of many commentators, contemporary economic and social life is becoming increasingly dynamic in nature. To take just one example, in *Sociology Beyond Societies* John Urry (2000) argues that current changes in social life are resulting in the displacement of long-standing socio-economic structures by heterogeneous constellations of networks. In Urry's view, such networks have proliferated in the wake of globalization trends which have helped to dissolve endogenous social structures previously encased within strong nation-states. Thus, Urry directs attention to the flows of objects, information and peoples that now configure national and other territories. Writing with Scott Lash, he argues that,

> The movement, the flows of capital, money commodities, labour, information and images across time and space are only comprehensible if 'networks' are taken into account because it is through networks that these subjects and objects are able to gain mobility. Whatever form of institutional governance is dominant, whether markets, hierarchies, the state or corporations, the subjects and objects which are governed must be mobile through networks. (Lash and Urry, 1994: 24)

Network flows emerge from differing points of origin (global, national, regional, local arenas) and carry their cargoes over varied distances. Sociology should therefore direct its attention to the 'heterogeneous, uneven and unpredictable mobilities' that run through given social spaces (Urry, 2000: 38).

It is not only within sociology that this concern for networks has come to the fore. In the political realm there is growing interest in 'policy networks' and the power relations that bind them together. Such networks are now seen as important mechanisms for the formulation and delivery of policies because 'old-style' government – that is, top-down, hierarchical decision-making – is deemed much less effective than 'alliances', 'partnerships' and other collaborative forms (Rhodes, 1997). Policy networks emerge as part of a broader system of 'multi-level governance' (Goodwin, 1998) and allow greater fluidity in political relations, processes and outcomes.

In a similar fashion, studies of new economic formations focus analytical attention upon the 'interdependencies' between firms and other (supporting) institutions in innovative 'hotspots' (e.g. Emilia Romagna, Baden Wurttemberg, the Cambridge sub-region). These interdependencies and other such linkages are assumed to provide benign contexts for effective economic action and development (Amin, 1999). As economic linkages build upon one another in discrete spatial contexts so we witness the emergence of 'clusters' which facilitate both cooperation and competition between firms (Cooke and Morgan, 1993). The combination of these two processes allows trust to be established, new ideas and practices to be disseminated, and modes of innovation to flourish (Storper, 1997).

In their different ways, all these perspectives seem to bear out Castells's (1996) claim that we are now living in some form of 'network society'. They infer that the stable economic and social structures that were progressively built up within European and other nation-states in the early post-war era have given way to much more fluid sets of relations, that is, to flows of people, goods, information, culture and so on, which move around the globe at ever-increasing speeds. A new form of sociality thus emerges as the flows consolidate networks of various kinds. Moreover, the networks not only transport artefacts of various kinds rapidly through space but also serve to construct a variety of differing space-times in line with the precepts of each network alignment. Network complexity and spatial complexity therefore advance simultaneously.

Yet, despite the all-inclusive, 'macro' nature of the perspectives generated by network scholars, what is striking about much of the work conducted in this general area is a neglect of rural space (this is especially noticeable in Castells's work – see for instance, Castells, 1996). This relative silence about rurality implies that networks are mainly to be found in urban locations. If rural areas play any part at all in network configurations it is usually as a neglected realm, an arena where network linkages are weak and underdeveloped. The 'hotspots' of network society are invariably urban in character so that the countryside seems to be lacking processes of network development (the countryside can once again be characterized as 'backward' and 'lacking' – a revised version of the modernization thesis).

The aim of this chapter is to challenge such assumptions by describing the constellations of networks that can now be found in the contemporary countryside. The narrative presented in the following section is divided into three main parts: these deal, first, with the rural polity, secondly, the rural economy and, thirdly, rural society. In each section, the same general approach is taken: the structural nature of the rural is described, the shift to a 'networked' rurality is outlined, and the overall impact of this 'networked' rurality is identified. For the sake of conceptual clarity, the narrative focuses upon a single national rural space – the UK – in order to show how one particular set of national structures is giving way to a much more fluid configuration of socio-economic networks. While it is suggested that the processes identified in this case study area are applicable elsewhere, notably in Europe and the United States, the section presents an analysis of only this one country.

In describing the shift from structure to networks, the chapter also engages with Urry's (2002) claim that the emergence of network society is giving rise to considerable socio-economic complexity as differing networks co-evolve and interact. As we move through the discussion of rural politics, rural economy and rural society we shall see in each case that rural change is becoming more complex in nature as the differing network types coalesce within varied network formations. The chapter goes on to consider these formations in broad spatial terms. It examines a 'regionalization' of rural space which, it is argued, follows the shift from national structure to heterogeneous network. This regionalization process is described using the 'differentiated countryside' typology first introduced in Marsden et al. (1993) and more recently expounded in Murdoch et al. (2003). It is proposed that, in general terms, rural regions are increasingly moving along distinct and diverse development trajectories and that these trajectories are driven by the differing constellations of networks now to be found in the new rural spaces. In short, the chapter argues that an increasingly 'complex' countryside is coming into being.

NETWORK COMPLEXITY IN THE COUNTRYSIDE: THE CASE OF THE UK

The emergence of political networks

For most of its history the British countryside has been characterized by considerable diversity. Differences in climate, soil type and terrain, allied to variations in culture, cuisine, dialect, custom and tradition, ensured that until the mid-nineteenth century forms of life in rural areas were quite heterogeneous in nature (Murdoch, 1996). However, this diversity was undermined, first, by improved transportation links between previously distant rural places and, second, by the shift of industry and thus population into towns following the industrial revolution (Lowe and Buller, 1990). As a result, British rural economies, like such economies elsewhere, became specialized in agricultural production. Moreover, agriculture itself was driven by an increasingly commercial approach, which stimulated technological change and standardization throughout the sector (Wormell, 1978).

From the 1930s onwards, the commercialization and industrialization of agriculture were driven by the state. Before that time, the UK government sustained only a fitful interest in the agricultural industry: although agriculture was seen to have strategic importance during wartime, differing governments were less willing to lend support during times of peace (they preferred instead to concentrate on ensuring the provision of cheap food). With the outbreak of the Second World War, however, a much more robust structure of agricultural support came into being (as part of a new 'food regime' – see Marsden et al., 1993). The war brought state control of farming as the whole sector had to be mobilized to maximize domestic food supply (Murray, 1955). This was imperative in order to diminish dependency on food imports, which were vulnerable to enemy action. As Kirk (1979: 47) puts it, agriculture 'was made into a sort of ward of the state, much like a nationalized industry'.

The political structures that dominated the agricultural sector in the post-war period emerged from this requirement to increase domestic food production rapidly during wartime. The policy led the state into a so-called 'productivist regime', one that entailed that resources be channelled towards increased output and productivity (Ilbery and Bowler, 1998; Wilson, 2001). The term 'productivist regime' refers to the network of institutions – state agencies, farming unions, input suppliers, financial bodies, R&D centres – that were oriented to achieving a sustained growth in food production from domestic resources (Marsden et al., 1993). This regime, which emerged in almost all the major industrial countries in the early post-war period, dominated the agricultural sector in the second half of the twentieth century and provided a coherent structure of political support for rural areas. We shall briefly consider this structure before turning to its dissolution within a networked polity.

At the core of the post-war political structure in the UK was a 'partnership' between the state and the agricultural industry (Self and Storing, 1962). This 'partnership was given form and substance by two representative organizations – the Ministry of Agriculture, Fisheries and Food (MAFF) and the National Farmers' Union (NFU). These two organizations worked closely together in order to orchestrate post-war agricultural policy in line with the precepts of 'productivism'. They evolved a stable and enduring policy-making relationship, one that effectively encompassed the entire national rural space. This structure included the majority of farmers (once it had been 'brought inside' the policy process the NFU could claim to be the exclusive 'voice' of the farming industry, thereby boosting its membership, which reached 200,000 by 1949 – Murdoch, 1988). Likewise, it included an extensive administrative machinery covering the whole rural space (the post-war growth in state intervention in agriculture was paralleled by a growth in the administrative structure of MAFF – Winter, 1996).

The MAFF/NFU 'partnership' was institutionalized within the Annual Price Review, a forum that had been instigated in 1943 and which became the 'font' of national agricultural policy after it was given a statutory basis by the 1947 Agriculture Act. The Review had two main functions: first, the review team were required to consider each year the general economic condition and prospects for the agricultural industry; second, the team would decide what changes were required in the level and distribution of prices and production grants. In order to carry out these two tasks, the Review required the collection and collation of an enormous amount of statistical material, including estimates of aggregate net farm income, its distribution by type and size of farms, the overall level of the subsidy bill in total and the amount to be spent on different commodities and production grants. All these data were used to compile a picture of the so-called 'national farm' (Murdoch and Ward, 1997). Once this picture had been constructed then policies could be put in place which encouraged and coerced farmers to play their part in the national agricultural structure (see Short et al., 2000).

According to Grant (1983), the MAFF/NFU relationship was the 'classic case' of corporatism, with the two main participants agreeing on the priorities for the industry and on how these should be implemented (see also Winter, 1996). The structure of policy-making was firmly hierarchical, with the priorities agreed by MAFF/NFU working their way down to farms so that gradually rural areas came to be reshaped in line with the precepts of productivism. The everyday practices of farmers were increasingly 'disciplined' by the national policy network in which they were enmeshed (Murdoch and Ward, 1997). In effect, agricultural policy determined patterns of development across the whole national agricultural space.

Agricultural policy was able to dominate the rural policy arena in large part because it remained separate from other rural policy processes. However, this separation began to break down in the later years of the twentieth century with the result that the MAFF/NFU policy-making nexus was undermined. Two main sets of changes were important in this regard. First, agricultural policy was undone by its own success: increases in output, production intensity and capital investment revolutionized the agricultural industry; however, they also gave rise to environmental, budgetary, trade and food quality problems (Bowers and Cheshire, 1983; Lowe et al., 1986). These problems combined, leading to concerted pressure for policy reform both from within the policy community and from without (Winter, 1996). Second, other policy relationships emerged in rural areas, notably within European Union (EU) initiatives. For example, under the European Structural Funds, new ways of making decisions about rural development were introduced in which rural local authorities and communities gained a greater role in the design and implementation of development schemes (Ray, 1998; Ward and McNicholas, 1998). Likewise, agri-environmental programmes brought with them new consultative arrangements, involving conservation agencies and groups, at the EU, national and regional levels (Whitby, 1996). All these approaches circumvented the top-down governmental approach associated with the traditional agricultural policy community.

In short, a governmental process of 'rescaling' (MacLeod and Goodwin, 1999) began to take place in the rural policy arena: national policy institutions such as the Annual Review lost their dominance to a much more diffuse network of agencies distributed across various spatial scales (European, national, regional, local) and across sectors (agriculture, economic development, planning, environment) (Edwards et al., 2000). The rural policy framework could now be characterized as a 'multi-level governance' structure of policy delivery (Goodwin, 1998), one held together by a diverse range of political and policy networks. In this new policy structure, networks emerge that encompass a much broader range of social and economic interests than was the case during the period of agricultural productivism. Under the post-war settlement, the state built strong linkages to farmers and their representatives but neglected other rural groups and organizations. In the more recent period, of 'multi-level governance', the state finds itself orchestrating economic and social actions in a diverse range of fields, fields moreover that increasingly intersect. In short, the regulation of rural areas has become both more diffuse and more complex.

The emergence of economic networks

As we have seen above, the scope and nature of national policy has recently begun to shift so that the policy framework associated with the protection of agricultural land and increases in agricultural productivity has begun to give way to more diverse and complex approaches, captured in the term 'multi-level governance'. In this and the following sections we shall argue that this shift has been conditioned not only by changes in the agricultural sector but also by a broader set of socio-economic pressures (Marsden, 1992). Two such pressures are of particular interest: first, the *urban–rural shift of manufacturing and services* and, secondly, *counter-urbanization*. These two interlinked trends have brought new demands upon the countryside and have changed the context in which these demands are assessed. They have also helped to dissolve stable post-war socio-economic structures and have led to the emergence of new networks in the rural arena. We look first at the impact of the urban–rural shift before turning, in the next section, to examine counter-urbanization.

In assessing recent changes in the rural economy, the first point to note is that the productivist regime in agriculture led to a sharp decline in agricultural employment as manual workers were increasingly replaced by mechanized processes. The decline in agricultural employment was mirrored in other primary industries, with fewer and fewer people employed in mining, quarrying and forestry (Murdoch et al., 2003). However, this reduction in primary sector employment was offset in the early post-war years by employment increases in the public services, mainly as a result of growth in the national education and health services. Later, during the 1970s, it became evident that manufacturing and service industries were relocating away from the conurbations and were providing new sources of employment in rural locations (Fothergill and Gudgin, 1982). By the 1990s, as a result of this urban–rural shift in employment, the employment profile of rural England, in aggregate terms at least, looked very much like that for the UK economy as a whole, with around 10 per cent employed in the primary sector, 20 per cent employed in manufacturing and 70 per cent in services (Butt, 1999; North, 1998; PIU, 1999).

In part, this trend was facilitated by changes in the structure of manufacturing industry in an era of 'post-Fordism' so that branch plants could be situated at some distance from company headquarters. Thus, rural areas could benefit from the inward movement of new plants attracted by low rents and wages (Urry, 1984). However, the urban–rural shift was also fuelled by the indigenous growth of small firms in rural areas, with new enterprise creation rates often higher in rural rather than urban locations (North, 1998; North and Smallbone, 2000). According to Keeble and

Tyler (1995: 994), this latter trend reflects a new pattern of 'enterprising behaviour' in which 'rural settlements have been able to attract a relatively high proportion of actual and potential entrepreneurs, largely because of their desirable residential environmental characteristics'. A key component of rural economic success is therefore the preference for rural locations on the part of some of the most dynamic firms and entrepreneurs. Keeble et al. (1992) show that hi-tech businesses have been attracted to accessible rural areas; likewise business and professional services have favoured rural areas in their location decisions, with some of the highest rates of growth evident in accessible rural areas, especially in the south of England (see also Keeble and Nachum, 2002). Thus, while private service employment grew in the conurbations of the UK by 19 per cent between 1981 and 1996, it grew by 49 per cent in the towns and rural areas (Gillespie, 1999: 14; see also Turok and Edge, 1999).

This broad shift in the structure of the rural economy has moved the focus of rural economic inquiry away from approaches that concentrate on the *uniqueness* of rural areas (for example, agricultural economics) to those that concern themselves with *general* economic trends. A number of approaches have recently been developed to analyse the urban–rural shift in these terms. The leading mode of analysis here is political economy, which examines the rural economy through the lens of the restructuring processes that unfold within national and international economic contexts. It therefore considers how rural economic structures conform to national and international economic structures.

Put simply, and rather crudely, political economy considers the 'rural' to be little more than the outcome of the use of space by various 'fractions' of capital. Day et al. summarize this view when they argue that,

> In order to comprehend the most significant dynamic processes in contemporary 'rural' areas it is necessary to focus upon the interactions between various types of economic activity, themselves to be understood as outcomes of complex rounds of investment; it is the particular forms taken by these interactions which will determine the special character of given localities, defined in terms of their overlapping roles in a series of spatial divisions of labour. (1989: 229–30)

In this perspective, the main function of rural areas has, until recently, been to provide opportunities for capital accumulation in agriculture. Thus, the agricultural sector, in partnership with the state, has pushed forward an ambitious programme of investment in an ever-more rationalized agricultural system (Rees, 1984). However, large-scale capital investment has had the effect of releasing resources from the agricultural sector, notably land and labour, thereby opening up the opportunity for alternative uses of rural space by alternative fractions of capital. In short, the restructuring of agriculture has permitted the more general restructuring of the rural economy. This more general restructuring has taken the form of the urban–rural shift of manufacturing and services.

In accounting for the 'extension' of manufacturing and service structures into rural areas the political economy approach puts considerable emphasis on the requirements of capital. Fothergill and Gudgin (1982), for instance, highlight land constraints on business expansion in urban areas and suggest that businesses may be moving to rural areas to take advantage of land resources released from agriculture and other primary industries. Urry (1995: 69) argues that the shift occurs because access to labour power (skills, cost, supply, organization, reliability) is becoming of heightened importance. In this regard, rural areas may gain new locational advantages as the shedding of labour from primary sectors means that there is a cheap, non-unionized, but flexible and technically competent workforce available for hire (Urry, 1984). Thus, in Urry's view, the shift in industry and services takes place because of the labour resources to be found in rural areas.

As these comments indicate, the explanatory strength of the political economy perspective resides in its ability to tie together a whole host of economic factors – including the requirements of capital, the restructuring of industry and changing rural labour markets – within a single, holistic framework (Day et al., 1989). Moreover, it allows rural areas to be considered in exactly the same terms as other spatial zones by focusing upon the underlying structure of the economy and the way this structure underpins particular uses of space. No specifically *rural* mode of analysis is required: urban and rural areas are bound together in *general* processes of economic change in which the structure of the national (or international) economy conditions the structure of the rural economy.

However, while the political economy approach indicates that specific rural resources – land, labour – may prove attractive to capital, it is arguable that the focus on such instrumental uses of rural space misses some key characteristics of the new urban–rural economic formations. In particular, the most dynamic economic areas (whether urban or rural) seem to be comprised of innovative clusters in which 'networks, norms, conventions, trust-based (often face-to-face) interactions and horizontal relations of reciprocity' are to the fore (MacLeod, 2001: 808). It is the generation of these 'quasi-economic' phenomena within specific regional and sub-regional clusters that seems to underpin new economic processes. As Staber observes:

> Networks are seen as an important defining characteristic of industrial districts, binding firms together into a coherent and innovative system of relational contracting, collaborative product development and multiplex organisational alliances. All economic action in industrial districts is said to be embedded in a dense web of network ties among individuals, firms and service organisations. (2001: 537)

In this account, firms, institutions and other economic actors are encouraged to come together in network formations by the existence of 'agglomeration economies'. In general terms, such economies stem from four main characteristics of contemporary economic networks: first, networks congeal around collective resources such as shared infrastructures; secondly, the spatial clustering of industry networks means that local labour markets come to hold specialist skills; thirdly, firms can reduce their costs within spatially proximate inter-firm transactions; and, fourthly, the clustering of economic actors facilitates knowledge transfer, innovation and learning (Malmberg and Maskell, 2002). These characteristics provoke the formation of places where 'mutual knowledge, collaboration and the exchange of information' are facilitated and where 'trust and mutual respect' are fostered (Maillat, 1996: 75). These latter dimensions – trust, respect, collaboration, learning – are not easily captured within the political economy repertoire.

The role played by new industrial network formations in generating economic change in rural areas in the UK has been investigated in recent work by Keeble and Nachum (2002). In their study of clustering processes within the business and professional service sector in South-East England, they show that agglomeration economies span urban and rural locations. In their sample, a significant proportion of rural businesses and professional service firms (57 per cent) work through 'regionalized' clusters, indicating that a large number of rural firms are being incorporated into network formations at the regional level. Evidence gathered in the North-East of England by Laschewski and colleagues (2002) supports this view. In their analysis of local business networks these authors show that efforts are being made by both state institutions and more informal groupings of individual firms to develop stronger cooperative linkages in rural areas. A 'package' of networking activities has been introduced to the area, including social events, circulation of bulletins, information services, workshops and focus groups. All these activities aim at knowledge dissemination and learning.

Interestingly, Laschewski et al. (2002: 383) discover that it is *newcomers* who are most active in the networks, that is, entrepreneurs who have moved into the area for 'quality of life' reasons, that is, to live in a more 'natural' environment. They say: 'Newcomer businesses appear to see greater value in network formation, because of their own lack of contacts and support within a new locality and their experience from being located in other business contexts.' Thus, we might conclude that regional economic networks are emerging in the wake of counter-urbanization-led economic change in the countryside (see following section). In short, economic and social networks mutually reinforce one another.

These findings indicate that the emergence of economic networks in rural areas may be ascribed to some specific characteristics of such areas. First, the very particular physical environment that marks out rural space now serves to attract economic actors that are seeking a good 'quality of life'. Once established in a rural area these actors may engage in 'enterprising behaviour', thereby strengthening both regional clusters and the reach of rural economic networks (Keeble and Nachum, 2002). Secondly, the productive activities that have historically taken place within rural areas can yield industrial structures that interact in very particular ways with the new industrial networks. For instance, Cooke and Morgan (1998) identify how small enterprises, which are often linked to traditional rural industries such as craft production and agriculture, seemingly enable network-like characteristics to be constructed on a local and regional basis (see also Asby and Midmore, 1996; Mosely, 2000). In other words, the rural economy is not simply subject to economic forces unfolding elsewhere (for example, urban locations) but is itself generating complex sets of changes, changes that take rural regions off along a variety of developmental paths (see below). Such complexity is arguably best apprehended within a network, rather than a political economy, perspective so that endogenous and exogenous development processes can be more easily aligned.

The emergence of social networks

The other main shift that has served to undermine post-war structures is an increase in the rural population as a result of urban-to-rural migration movements. For much of the twentieth century, population has shifted away from rural areas in search of employment in cities. The continual shedding of labour from agriculture reinforced a process of urbanization that had been set in train during the Industrial Revolution. However, the economic changes described in the previous section have corresponded with a propensity on the part of more and more households to leave the city in search of a better life in the countryside. This search has been termed 'counter-urbanization' (see Champion, 1989, 2001). Although counter-urbanization is an international trend, Champion (1994: 1504) argues that Britain has been in its 'vanguard': increasing affluence, along with changes in transportation systems, has allowed many people to combine country living with urban employment and more people are now moving away from cities than are moving into them (Boyle and Halfacree, 1998). The countryside is widely viewed as offering a better way of life than that available in urban areas (Countryside Agency, 2002).

The first evidence of counter-urbanization came to light in the 1961 census which showed that although during the preceding ten years the population of metropolitan areas had grown by around 5 per cent, this was matched by growth in nonmetropolitan small towns and rural areas. Between 1961 and 1971 growth in the rural population (5 per cent) outstripped metropolitan growth (3.5 per cent) for the first time. This trend has continued ever since: between 1971 and 1981 metropolitan areas lost over 2 per cent of their population while nonmetropolitan areas increased their share by around 9 per cent; between 1981 and 1991 the population of metropolitan Britain grew by 0.4 per cent, yet the rural areas increased their share of population by over 7 per cent (see Champion, 1994, for a summary of these trends). Further evidence seems to indicate that urban-to-rural migration is continuing. Champion et al. (1998), for example, estimated that for the year 1990/91 there was a net movement of 80,000 people from urban to rural areas and they expected this rate to be maintained throughout the decade. It was more recently calculated that the number of households in England's rural districts would increase by more than one million (that is, 19 per cent) between 1996 and 2016 (Countryside Agency, 2000; King, 2000).

Counter-urbanization has changed the character of rural communities and rural society. The traditional view of rural communities is that they are stable, consensual institutions in which kinship relations bind members into a coherent whole (Bell and Newby, 1972; Day and Fitton, 1975). This holistic view of rural communities has been evident in the many rural community studies conducted in the post-war period, where kinship networks, spatial proximity, close social cooperation and an agricultural orientation were thought to mark out distinctive rural social formations (see Harper, 1989). However, in the wake of counter-urbanization, sociological interest has shifted from the role played by traditional social relations in maintaining the distinctive shape of rural communities towards the way more general social trends play themselves out in the rural context. In its initial phases such work emphasized how rural areas were being drawn into national social structures; but as we shall see, in more recent years sociological accounts have highlighted the importance of social networks in rural communities.

The first attempt to see rural social changes as but one part of more general social trends was made by Pahl (1966). In his study of villages in southern England, Pahl claimed that such communities were best interpreted not in terms of any particular spatial categorization (such as urban or rural), but in terms of *social class*. He showed that the movement of middle-class commuters into the countryside had brought a more complex rural class structure into being. Where a traditional rural working class had previously seen itself only in opposition to a landed class, there was now an ascendant middle class. In other words, the rural class structure was becoming increasingly complex in line with national trends. In Pahl's view, little distinguished rural from urban communities, at least in terms of class structure.

In short, Pahl argued that the movement of middle-class households from urban to rural locations effectively undermined any social distinctions between urban and rural areas. These findings have subsequently been echoed by a number of scholars (e.g. Savage et al., 1992; Thrift, 1989), leading to some profound questions being asked about the contemporary status of 'rurality' (see, for instance, Hoggart, 1989, 1990). However, more recent work by Michael Bell (1994) indicates that rural areas do continue to contain distinctive social forms and ways of life. In describing the social structure of a village in the South of England, Bell shows that middle-class residence in the countryside is part of a search for new forms of belonging. In particular, middle-class residential preferences are strongly linked to the aspiration for a 'country' identity (to be realized through country sports, closeness to nature, village institutions, and so on). In Bell's account, this linkage is made because rural residents believe a country identity somehow lies *outside* normal social relations (that is, typical forms of class belonging). In the view of counter-urbanizers the world of social interests resides primarily in the city, while the countryside operates according to a different, more 'natural' calculus.

In this perspective, counter-urbanization can be seen as an attempt to 'escape' the social through an immersion in 'country life'. Once established in rural locations, counter-urbanizers will seek to consolidate those aspects of the rural that most closely accord with their preconceptions of this spatial area. In particular they will act to safeguard the rural environment through participation in amenity and environmental groups and will put pressure on government (notably through the planning system) to safeguard the countryside's 'natural' and 'timeless' character (Lowe et al., 1977; Murdoch and Marsden, 1994).

This general attempt on the part of middle-class households to dissolve social tensions (such as class conflict) within rural communities can be seen, in Lash and Urry's (1994: 248) terms, as a form of 'aesthetic reflexivity'. Lash and Urry believe social reflexivity has become more pronounced in guiding social actions because of a general 'detraditionalization' of social life. As Urry puts it,

> people's tastes, values and norms are increasingly less determined by 'societal' institutions such as education, family, culture, government, the law and so on. One effect of such stripping away of the centrality of such institutions is that individuals and groups are more able to envisage establishing their 'own' institutions, relatively separate from those of the wider society. (1995: 220)

Urry proposes that the countryside has become a key site for the establishment of new institutions by the mobile middle class. Thus, the movement of middle-class households into the countryside can be seen as an attempt to create new relations, identities and forms of belonging, following the breakdown of traditional social identities.

The significance of the countryside, Lash and Urry (1994: 247) argue, stems from a heightened aesthetic sensibility amongst middle-class groups. This sensibility is oriented to 'old places, crafts, houses, countryside and so on, so that almost everything old is thought to be valuable' (Lash and Urry, 1994: 247). For the reflexive middle class, country places are 'heavy with time' (Lash and Urry, 1994: 250); they bear the markings of the past in their social and physical fabrics, and these markings are used to 'anchor' new middle-class identities.

However, in the view of some commentators the social transformation of the rural population in line with aesthetic or social reflexivity markedly changes the nature of the rural communities that are so valued. For instance, Bauman believes the traditional community cannot survive its reflexive appropriation by the new middle class. He says,

> Since 'community' means shared understanding of the 'natural or 'tacit' kind, it won't survive the moment in which understanding turns self-conscious, and so loud and vociferous ... Community can only be numb or dead. Once it starts to praise its unique valour, wax lyrical about its pristine beauty and stick on pearly fences wordy manifestoes calling its members to appreciate its wonders and telling all the others to admire them or shut up – one can be sure that the community is no more (or not yet, as the case may be). 'Spoken of' community (more exactly: a community speaking of itself) is a contradiction in terms. (2001: 11–12)

In this view, counter-urbanization leads to the replacement of traditional communities by new social institutions and associations. While these institutions and associations might take a variety of forms, in Wittel's view they are likely to be established as networks. He provides the following rationale:

> 'Individualization' presumes a removal of historically prescribed social forms and commitments, a loss of traditional security with respect to rituals, guiding norms and practical knowledge. Instead individuals must actively construct social bonds. They must make decisions and order preferences. [This is] a change from pre-given relationships to choice. Pre-given relationships are not a product of personal decisions; they represent the sociality of communities. In contrast, [individualization] is defined by a higher degree of mobility, by translocal communications, by a high amount of social contacts, and by a subjective management of the network. (2001: 65)

Wittel (2001) describes this 'active construction' of social bonds by reflexive individuals as 'network sociality'. It comprises a social context in which the construction and maintenance of individualized (middle-class) identities depends upon the network linkages in which each individual is enmeshed. Wittel sees network sociality as antithetical to the traditional community:

> The term network sociality can be understood in contrast to 'community'. Community entails stability, coherence, embeddedness and belonging. It involves strong and lasting ties, proximity, and a common history or narrative of the collective. Network sociality stands counterposed to *Gemeinschaft*. It does not represent belonging but integration and disintegration. It is a disembedded intersubjectivity. (2001: 51)

If we take these arguments seriously then counter-urbanization on the part of middle-class households implies that traditional communities in the countryside are increasingly being displaced by networks in which individuals come together around shared interests. The nature of the new associations will be defined by the particular characteristics of these shared interests. As Liepins (2000: 30) puts it: 'people will be simultaneously participating in one "community", as a local network of interaction, whilst also being located in networks and "stretched out communities" of many other kinds' (see also Day, 1998; Silk, 1999; Stacey, 1969). The rural community is therefore fragmented by cross-cutting networks but is recast within individual social networks. Rather than a single 'community' there are now multiple (networked) communit*ies*.

NETWORKS AND SPATIAL COMPLEXITY

The preceding sections use the case study of the UK to describe a broad shift from a national structure of rural regulation, in which rural areas were governed by a hierarchical agricultural regime, to much more fluid processes of political, economic and social change. This shift has been captured using the language of structures and networks: it has been argued that broad national structures have given way to complex constellations of networks in countryside locations. Some of the theoretical resources used to apprehend this increasing fluidity – policy network analysis, economic cluster theory, aesthetic reflexivity – have been outlined and the implications for contemporary understandings of rural change have been discussed.

In this section, the spatial implications of the shift to 'network society' are considered. The general argument here is that the emergence of network configurations in rural areas opens up considerable spatial complexity. In particular, it results in a 'regionalization' of the rural; that is, the national rural space put in place in the post-war period gives way to 'regionalized ruralities' in which differing combinations of networks come into being. These differing network combinations serve to place rural regions on divergent development trajectories. In this section these various development trajectories are described in terms of the 'differentiated countryside' (Marsden et al., 1993), a typology that has been applied to both the UK rural space (Murdoch et al., 2003) and to rural areas in Europe (Hoggart et al., 1996).

The differentiated countryside approach suggests that any understanding of new patterns of spatial differentiation must first consider the combined effects of the urban–rural shift of industry and counter-urbanization and second the spatialization of these combined effects. In essence, it is argued that these two processes serve to accentuate differences between rural locations. These differences can be conceptualized in regional terms as it is argued that changes in the rural economy and changes in rural society coalesce to place discrete regional areas on differing development trajectories.

For instance, when we look at the rural economy it is evident that there are very real variations in the economic structures of discrete rural areas, notwithstanding the generalized character of the urban–rural shift of manufacturing and services. On the one hand, many accessible rural areas have become amongst the most economically advantaged; they are favoured locations for leading-edge industries (notably hi-tech and private service firms) and display higher than average income and GDP levels (Gillespie, 1999). On the other hand, many remote rural areas tend to demonstrate traditional weaknesses such as low wages, low skill levels, economic vulnerability and so on, and these are reflected in continuing low levels of income and GDP per head (see Monk and Hodge, 1995). The dynamic nature of the new rural economy means that the distinctions between these areas are increased.

When we turn to examine counter-urbanization we find the same spatial distinctions emerging. Accessible and environmentally pleasing rural areas tend to attract well-heeled middle-class residents; thus, the levels of prosperity to be found in such areas increase markedly (notably, house prices and average income per capita). However, other more remote and environmentally denuded rural districts (such as ex-mining areas) continue to lose population; they therefore tend to suffer problems of long-term decline, including a large elderly population (the young tend to move out), declining community institutions and scarce employment opportunities. Again, the dynamic nature of contemporary social changes means that such distinctions become ever more deeply entrenched.

In short, the combined impacts of the urban–rural shift in employment and counter-urbanization serve to 'regionalize' the rural (Murdoch et al., 2003). Furthermore, this complex social geography is underpinned by a differential distribution of networks as economically and socially advantaged areas gather together 'thick' socio-economic linkages while peripheral areas manage to construct only 'thin' linkages (see Amin and Thrift, 1994 for explanations of these terms). This distribution of rural networks therefore reflects not only the changed structure of economy and society but also the distinctive patterns of social action that occur in differing rural places.

In order to capture this new, increasingly complex pattern of spatial change, Marsden et al. (1993) introduce the notion of the 'differentiated countryside'. In their view, the countryside can be seen as a series of 'ideal types'. These differing 'types' demarcate the main socio-economic constellations that can now be found in rural areas. The types are as follows:

- The *preserved countryside*, which is evident in accessible rural areas around major cities. It is characterized by economic buoyancy, combined with strong anti-development attitudes. Such attitudes are expressed mainly by middle-class social groups living in the countryside. Due to their large numbers and their skill in exerting political influence such groups are able to dominate local political processes, notably land-use planning (e.g. Murdoch and Marsden, 1994).
- The *contested countryside*, which refers to areas that lie outside the main commuter zones but which comprise attractive living environments. Here local agricultural, commercial and development interests may be politically dominant and they will tend to favour various forms of economic development to meet local needs. However, these traditional groups are now increasingly opposed by retired 'incomers'. Thus the development process is marked by sharp conflicts between old and new groups, with no single interest attaining overall dominance (e.g. Lowe et al., 2001).
- The *paternalistic countryside*, which refers to areas where large private estates and big farms predominate and where the development process is decisively shaped by established landowners. Many of the large estates and farms will be faced with falling incomes (due to problems in the primary sectors) and will therefore be looking to diversify their economic activities. They will seek out new development opportunities and are likely to be able to implement these relatively unhindered. However, these areas are likely to be subject to less development pressure than either of the above two types, partly because the middle class is present in much lower numbers (e.g. Newby et al., 1978).
- The *clientelist countryside*, which can be found in remote rural areas. Here agriculture and its associated political institutions still hold sway. Processes of rural development are largely determined by state agencies, in part, because farming can only be sustained by state subsidy. Thus, local politics is dominated by employment concerns and the welfare of the 'community' (e.g. Munton, 1995).

In these differing areas we find differing combinations of networks (Murdoch et al., 2003). In the 'preserved countryside' we might expect 'thick' clusters of economic networks to be generating rapid economic growth. However, in such areas we also find strong environmental networks comprising local politicians, planners, amenity groups and local residents. Thus, economic change must conform to an overall context of 'preservation' (that is, 'clean' high-tech, rather than dirty low-tech, firms). In the 'contested countryside' we see a more mixed set of development processes. The development networks, which encompass local politicians, small businesses and farm families, articulate local economic concerns while the environmental networks, which comprise counter-urbanizers and members of environmentalist and amenity groups, regard such concerns as posing a threat to valued natural environments. In the 'paternalistic countryside' economic change is mediated by the landed estates, in part because of the inability of preservationist networks to offer any real challenge to traditional and long-standing paternalistic networks. Finally, in the 'clientist countryside' a very traditional set of agricultural and rural development networks tends to dominate economic and political processes. These networks bring state agencies, agricultural groups and rural development interests within a set of 'corporatist' relationships at the local level. Thus, processes of socio-economic change tend to work through traditional socio-political structures.

These four types and their associated networks alert us to the fact that the shift from structures to networks is unevenly expressed in rural areas. While some rural regions are incorporated into dynamic clusters of economic networks and see their rural communities displaced by 'reflexive' social networks, other regions witness a progressive diminution in their economic linkages and a continuing stagnation in community institutions. Thus, we should not assume that any generalized shift in the rural economy, society and polity has taken place; rather, traditional rural institutions still exist but they must now be set alongside the newer socio-economic forms such as 'reflexive' or 'learning' networks. This mixing of 'old' and 'new', which differs from place to place and which gives rise to complex sets of outcomes, underpins contemporary processes of 'regionalization' or 'differentiation' in the countryside.

It follows, then, that any system for managing change in the contemporary countryside must be alert to the precise constitution of the networks in these differentiated spatial arenas. In particular, managerial initiatives should be sensitive to the need for stronger and more coherent systems of coordination and competition in underdeveloped network spaces (Murdoch, 2000). They will also need to work across sectors so that synergies between economic, social and political sectors can be realized. Yet, as the differentiated countryside typology illustrates, in many rural locations the various networks are frequently in conflict with one another, indicating that mechanisms of conflict resolution are also urgently required. Thus, the task facing rural governance institutions is to align differing networks in ways that enhance both the developmental capabilities of any rural area while also preserving its distinctive rural characteristics. It is likely that such an alignment will only be adequately achieved if some system of genuinely multi-level governance is introduced so that the activities of networks of differing sizes and lengths can be coordinated.

CONCLUSION

The theories of political, economic and social change outlined in this chapter chart a shift from solid and limited national trends to diffuse and differentiated patterns of change. Such differentiation tends to be orchestrated by networks that have come to the fore following the dissolution of dominant political, economic and social structures in the post-war period. In the political sphere, the 'policy network' emerges from a crisis in structures of agricultural governance and heralds a more general shift to 'multi-level governance'. In the economic sphere networks appear to lie at the heart of new regional economies based on the generation of innovation, learning and trust mechanisms. In the social sphere the growth of the 'reflexive' middle class promotes 'networked' forms of sociality and these act to undermine traditional community institutions while also giving rise to new valuations of the rural environment and other aspects of rural life.

Yet, despite this seemingly generalized shift in the contemporary countryside, the emergence of 'network society' is spatially uneven. It varies in line with levels of accessibility (or, alternatively, 'peripherality'), with the structure of the local economy, with the state of the local environment, the make up of local society, and so forth. These features combine in different ways in different spatial contexts. We thus witness the emergence of a 'differentiated countryside' in which discrete rural regions develop along quite distinct trajectories of change. These trajectories are determined both by the mixture of networks found in rural locations and the processes of coordination and competition that take place within these network mixtures.

This general conclusion on the shape of 'network society' in the countryside gives rise to a further set of research questions for rural studies. In particular it raises the need to investigate in more detail the relationship between network and countryside location, and the means by which differing mixtures of networks are put in place. This requires an understanding of 'differentiation' processes and the way these processes interact with given network formations (formations that combine both 'endogenous' and 'exogenous' networks, that is, networks that are 'localized' and networks that are 'globalized'). It also requires some in-depth investigations of network formations themselves. We have seen above that networks of differing types can mutually reinforce one another to the benefit of rural regions; also that they can work against one another to the detriment of such regions. The differing ways in which these outcomes are achieved needs to be more closely studied so that enhanced development processes can be put in place. In short, this area of research would aim to work against processes of uneven rural development while recognizing that some amount of differentiation between rural regions is not just inevitable but desirable (in order to allow each region's network strengths to be appreciated and its network weaknesses to be overcome).

A second set of research questions refers to the intermingling of traditional rural characteristics and new network forms. As we have seen above, networks comprise mixtures of long-standing rural features (communities, environments, small firms, etc.) and new socio-economic practices (reflexivity, innovation, learning, etc.). A key research task is thus to identify how the 'old' and the 'new' interact in networks and the shape of the resulting 'hybrids' (again, linked to differing contexts of 'hybridization' that result from processes of differentiation). This leads on to another important issue: the means by which rural assets, resources, ways of life and so on, are mobilized within networks. Under preceding structuralist modes of analysis it was assumed that rural entities would somehow be fixed or stabilized under any new structural regime that might emerge (even if the 'terms of trade' within that regime tended to marginalize the rural in favour of the urban). Within network society, the prevailing social condition is enhanced 'fluidity' so rural entities are now likely to 'flow' up and down networks (and across urban/rural boundaries – see Murdoch and Lowe, 2003). The means by which these flows are engineered, and the likely consequences for rural space, need careful consideration, as does the cultural significance of a more 'fluid' rurality for this notion runs counter to the many widely held beliefs about an 'unchanging' and 'timeless' countryside that still exist in society at large.

REFERENCES

Amin, A. (1999) An institutionalist perspective on regional economic development. *International Journal of Urban and Regional Research*. 23: 365–378.

Amin, A. and Thrift, N. (1994). Living in the global. In A. Amin and N. Thrift (eds), *Globalisation, Institutions and Regional Development.* Oxford: Oxford University Press.

Asby, J. and Midmore, P. (1996) Human capacity building in rural areas: the importance of community development. In P. Midmore and G. Hughes (eds), *Rural Wales: An Economic and Social Perspective.* Aberystwyth: Welsh Institute for Rural Studies.

Bauman, Z. (2001) *Community: Seeking Safety in an Insecure World.* Cambridge: Polity Press.

Bell, C. and Newby, H. (1972) *Community Studies.* London: Allen and Unwin.

Bell, M.M. (1994) *Childerley.* London: University of Chicago Press.

Bowers, J. and Cheshire, P. (1983) *Agriculture, The Countryside and Land Use*. London: Methuen.

Boyle, P. and Halfacree, K. (eds) (1998) *Migration into Rural Areas*. London: Wiley.

Butt, R. (1999) The changing employment geography of rural areas. In M. Breheny (ed.), *The People: Where Will They Work?* London: TCPA.

Castells, M. (1996). *The Rise of the Network Society.* Oxford: Blackwell.

Champion, T. (ed.) (1989) *Counterurbanisation: The Changing Pace and Nature of Population Deconcentration.* Sevenoaks: Edward Arnold.

Champion, T. (1994) Population change and migration in Britain since 1981: evidence for continuing deconcentration. *Environment and Planning A*. 26: 1501–1520.

Champion, T., Atkins, D., Coombes, M. and Fotheringham, S. (1998) *Urban Exodus*. London: Council for the Protection of Rural England.

Cooke, P. and Morgan, K. (1993) The network paradigm: new departures in corporate and regional development. *Environment and Planning C: Society and Space.* 11: 543–564.

Cooke, P. and Morgan, K. (1998) *The Associational Economy.* Oxford: Oxford University Press.

Countryside Agency (2000) *The State of the Countryside.* Cheltenham: Countryside Agency.

Day, G. (1998) A community of communities? Similarity and difference in Welsh rural community studies. *Economic and Social Review*. 29: 233–257.

Day, G. and Fitton, M. (1975) Religion and social status in rural Wales. *Sociological Review*. 23: 867–891.

Day, G. and Murdoch, J. (1993) Locality and community: coming to terms with place. *Sociological Review.* 41: 82–111.

Day, G., Rees, G. and Murdoch, J. (1989) Social change, rural localities and the state: the restructuring of rural Wales. *Journal of Rural Studies*. 5: 227–244.

Edwards, B., Goodwin, M., Pemberton, S. and Woods, M. (2001) Partnership, power and scale in rural governance. *Environment and Planning C: Government and Policy*. 19: 289–310.

Fothergill, S. and Gudgin, D. (1982) *Unequal Growth.* London: Heinemann.

Gillespie, A. (1999) The changing employment geography of Britain. In M. Breheny (ed.), *The People: Where Will They Work?* London: TCPA.

Goodwin, M. (1998) The governance of rural areas: some emerging research issues and agendas. *Journal of Rural Studies*. 14: 5–12.

Grant, W. (1983) The NFU: the classic case of incorporation. In D. Marsh (ed.), *Pressure Politics*. London: Junction Books.

Harper, S. (1989) The British rural community: an overview of perspectives. *Journal of Rural Studies*. 5: 89–105.

Hoggart, K. (1990) Let's do away with the rural. *Journal of Rural Studies*. 6: 245–57.

Hoggart, K. (1998) Rural cannot equal middle class because class does not exist? *Journal of Rural Studies*. 14: 381–6.

Hoggart, K., Buller, H. and Black, I. (1996) *Rural Europe*. London: Arnold.

Ilbery, B. and Bowler, I. (1998) From agricultural productivism to post-productivism. In B. Ilbery (ed.), *The Geography of Rural Change*. London: Longman.

Keeble, D. and Nachum, L. (2002) Why do business service firms cluster? Small consultancies, clustering and decentralisation in London and southern England. *Transactions of the Institute of British Geographers*. NS 27: 67–90.

Keeble, D. and Tyler, P. (1995) Enterprising behaviour and the urban–rural shift. *Urban Studies*. 32: 975–997.

Keeble, D., Tyler, P., Broom, G. and Lewis, J. (1992) *Business Success in the Countryside: The Performance of Rural Enterprise*. London: HMSO.

King, D. (2000) *Projected Household Numbers for Rural Districts of England*. Chelmsford: Countryside Agency.

Kirk, J. (1979) *The Development of Agriculture in Germany and the UK: UK Agricultural Policy 1870–1970*. Ashford: Wye College.

Laschewski, L., Phillipson, J. and Gorton, M. (2002) The facilitation and formalisation of small business networks: evidence from the North East of England. *Environment and Planning C: Government and Policy*. 20: 375–391.

Lash, S. and Urry, J. (1994) *Economies of Signs and Space*. London: Sage.

Liepins, R. (2000) New energies for an old idea: reworking approaches to 'community' in contemporary rural studies. *Journal of Rural Studies*. 16: 23–35.

Lowe, P. and Buller, H. (1990) Overview. In P. Lowe and M. Bodiguel (eds), *Rural Studies in Britain and France*. London: Belhaven.

Lowe, P., Clark, J., Seymour, S. and Ward, N. (1997) *Moralising the Environment*. London: UCL Press.

Lowe, P., Cox, G., O'Riordan, T., MacEwan, M. and Winter, M. (1986) *Countryside Conflicts: The Politics of Farming, Forestry and Conservation*. Aldershot: Gower.

Lowe, P., Murdoch, J. and Norton, A. (2001) *Professionals and Volunteers in the Environmental Process*. Newcastle upon Tyne: Centre for Rural Economy, University of Newcastle upon Tyne.

MacLeod, G. (2001) New regionalism reconsidered: globalisation and the remaking of political economic space. *International Journal of Urban and Regional Research*. 4: 804–829.

MacLeod, G. and Goodwin, M. (1999) Space, scale and state strategy: rethinking urban and regional governance. *Progress in Human Geography*. 23: 503–527.

Maillat, D. (1996) Regional productive systems and innovative milieux. In OECD (ed.), *Networks of Enterprises and Local Development*. Paris: OECD.

Malmberg, A. and Maskell, P. (2002) The elusive concept of localisation economies: towards a knowledge-based theory of spatial clustering. *Environment and Planning A*. 34: 429–449.

Marsden, T. (1992) Exploring a rural sociology for the Fordist transition: incorporating social relations into economic restructuring. *Sociologia Ruralis*. 32: 209–230.

Marsden, T., Murdoch, J., Lowe, P., Munton, R. and Flynn, A. (1993) *Constructing the Countryside*. London: UCL Press.

Monk, S. and Hodge, I. (1995) Labour markets and employment opportunities in rural Britain. *Sociologia Ruralis*. 35: 153–172.

Mosely, M. (2000) Innovation and rural development: some lessons from Britain and Western Europe. *Planning, Practice and Research*. 15: 95–115.

Munton, R. (1995) Regulating rural change: property rights, economy and environment – a case study from Cumbria, UK. *Journal of Rural Studies*. 11: 269–284.

Murdoch, J. (1988) State and Agriculture in Wales. Unpublished PhD thesis, University of Wales, Aberystwyth.

Murdoch, J. (1996) Planning the rural economy. In Allanson, P. and Whitby, M. (eds), *The Rural Economy and the British Countryside*. London: Earthscan.

Murdoch, J. (2000) Networks – a new paradigm of rural development? *Journal of Rural Studies*. 16: 407–419.

Murdoch, J. and Lowe, P. (2003) The preservationist paradox: modernism, environmentalism and the politics of spatial division. *Transactions of the Institute of British Geographers*. 28: 318–32.

Murdoch, J. and Marsden, T. (1994) *Reconstituting Rurality: Class, Community and Power in the Development Process*. London: UCL Press.

Murdoch, J. and Ward, N. (1997) Governmentality and territoriality: the statistical manufacture of Britain's 'national farm'. *Political Geography*. 16: 307–324.

Murdoch, J., Lowe, P., Ward, N. and Marsden, T. (2003) *The Differentiated Countryside*. London: Routledge.

Murray, K. (1955) *Agriculture: A History of the Second World War*. London: HMSO.

Newby, H., Bell, C., Rose, D. and Saunders, P. (1978) *Property, Paternalism and Power*. London: Hutchinson.

North, D. (1998) Rural industrialisation. In B. Ilbery (ed.), *The Geography of Rural Change*. London: Longman.

North, D. and Smallbone, D. (2000) The innovativeness and growth of rural SMEs during the 1990s. *Regional Studies*. 34: 145–157.

Pahl, R. (1966) *Urbs in Rure*. London: LSE.

Pahl, R. (1970) *Readings in Urban Sociology*. Oxford: Pergamon.

PIU (Performance and Innovation Unit) (1999) *Rural Economies*. London: Cabinet Office.

Ray, C. (1998) Territory, structures and interpretation – two case studies of the European Union's LEADER 1 programme. *Journal of Rural Studies*. 14: 79–88.

Rees, G. (1984) Rural regions in national and international economies. In T. Bradley and P. Lowe (eds), *Locality and Rurality*. Norwich: Geo Books.

Rhodes, R. (1997) *Understanding Governance: Policy Networks, Governance, Reflexivity and Accountability*. Buckingham: Open University Press.

Savage, M., Barlow, J., Dickens, P. and Fielding, T. (1992) *Property, Bureaucracy and Culture: Middle-Class Formation in Contemporary Britain*. London: Routledge.

Self, P. and Storing, P. (1962) *The State and the Farmer.* London: Allen and Unwin.

Short, B., Watkins, C., Foot, W. and Kinsman, P. (2000) *The National Farm Survey 1941–1943: State Surveillance and the Countryside in England and Wales in the Second World War.* Wallingford: CAB International.

Shucksmith, M. (2000) Endogenous development, social capital and social inclusion. *Sociologia Ruralis.* 40: 208–218.

Silk, J. (1999) The dynamics of community, place and identity. *Environment and Planning A.* 31: 5–17.

Smith, A. (2000) Policy networks and advocacy coalitions: explaining policy change and stability in UK industrial pollution policy. *Government and Policy.* 18: 95–114.

Staber, U. (2001) The structure of networks in industrial districts. *International Journal of Urban and Regional Research.* 25: 537–552.

Stacey, M. (1969) The myth of community studies. *British Journal of Sociology.* 20: 34–47.

Storper, M. (1997) *The Regional World.* London: The Guilford Press.

Talbot, H. (1997) Rural telematics in England: strategic issues. Centre for Rural Economy Research Report, Department of Agricultural Economics and Food Marketing, University of Newcastle, Newcastle upon Tyne.

Thrift, N. (1989) Images of social change. In C. Hamnett, L. McDowell and P. Sarre (eds), *The Changing Social Structure.* London: Sage.

Turok, I. and Edge, N. (1999) *The Jobs Gap in Britain's Cities: Employment Loss and Labour Market Consequences.* Bristol: Policy Press.

Urry, J. (1984) Capitalist restructuring, recomposition and the regions. In T. Bradley and P. Lowe (eds), *Locality and Rurality.* Norwich: Geo Books.

Urry, J. (1995) *Consuming Places.* London: Routledge.

Urry, J. (2000) *Sociology beyond Societies: Mobilities for the Twenty-first Century.* London: Routledge.

Urry, J. (2002) *Global Complexity.* London: Sage.

Ward, N. and McNicholas, K. (1998) Reconfiguring rural development in the UK: Objective 5b and the new rural governance. *Journal of Rural Studies.* 14: 27–40.

Whitby, M. (ed.) (1996) *The European Environment and CAP Reform: Policies and Prospects for Conservation.* Wallingford: CAB International.

Wilson, G. (2001) From productivism to post-productivism ... and back again? Exploring the (un)changed natural and mental landscapes of European agriculture. *Transactions of the Institute of British Geographers.* NS 26: 77–102.

Winter, M. (1996) *Rural Politics.* London: Routledge.

Wittal, A. (2001) Towards a network sociality. *Theory, Culture and Society.* 18: 51–76.

Wormell, P. (1978) *The Anatomy of Agriculture: A Study of Britain's Greatest Industry.* London: Harper and Row.

13

Non-human rural studies

Owain Jones

INTRODUCTION

Non-human rural studies represent approaches to conceptualizing and studying the rural which work upon the idea that 'the rural' (or any other 'social' phenomenon for that matter), is not formed and practised by human presences, actions and agencies alone. Rather, such formations are woven from the disparate beings, processes and materialities of the world, and the forces that shape them include differing forms of agency which can be variously described as non-human agency, relational agency or collective agency. Another way of putting it is that the world, or parts of it (in this case, 'the rural'), or specific elements of 'the rural', are co-constituted by a wide range of actors working in some form of hybrid, relational arrangement. This means that if (rural) social scientists want to ask questions about rural 'social' formations – what is their nature, how have they come about, how might they (be) change(d)? – they need to take into account the more-than-social world, the non-human presences and processes acting in relation to the social. Such approaches, which question the sharp distinction between the social and non-social (natural/material) world, are gaining prominence in various sectors of the social sciences. Witness for example, the 'more than human geographies' of 'culturenatures' (Whatmore, 2003), Harvey's (1996) 'socio-ecological' formations, and hybridity (Cloke, 2003; Murdoch, 2003; Whatmore, 2002).

These themes draw inspiration from a range of ideas within the social sciences and beyond. For example, the key influences of culturenatures are environmental history, material culture, feminist theory and science studies (Whatmore, 2003: 166). Within science studies, actor-network theory (ANT) has become a prominent means of thinking about the non-human and hybridity. ANT has grown in significance in rural studies (see Murdoch, 2003, and Chapter 12 this volume) and has exerted influence on how non-human rural studies are developing. Therefore much of this chapter deals with ANT, its uses in rural studies, critiques of it and other related concepts such as co-constitution and dwelling.

Murdoch makes the case for the non-human and hybridity in rural studies thus:

> the idea that the countryside is simply a social construction, one that reflects dominant patterns of social relations, cannot adequately account for the 'natural' entities found within its boundaries. There is something beyond the 'social' at work as the countryside displays a material complexity that is not easily reducible to even the most nuanced social categories ... to paraphrase Sarah Whatmore (1999) the countryside is 'more than human'. (2003: 264)

On these grounds he argues that the concept of hybridity, which is as yet 'not in common usage' in rural studies, does have 'the potential to capture the socio-natural complexity of the countryside more easily than traditional modes of representation' (p. 264). Significantly, the suggestion here is that taking non-humans seriously should not be a specialized and discrete form of study, but rather that non-humans are likely to be actively present, and thus deserving of theoretical and empirical attention, in just about any consideration of any rural phenomena. Therefore a general sensitivity to hybridity is needed. To cite Murdoch again,

> In short, while any particular vision of the countryside will continue to focus upon social forms, natural entities or even hybrid objects, it will also need to be aware of the interrelationships that exist between these realms if it is to capture the full range of processes currently running through rural areas. (2003: 280)

If these notions of hybridity – and other related theoretical developments, such as embodiment (Franklin, 2002), and accounts of the partialities and uncertainties of representation – are not taken up, rural social studies will remain firmly fixed in ways that have been so roundly criticized by, amongst others, Latour (1993) and Thrift (2004). In short, their critique of much social science output is that it offers static, partial and flawed representations of 'social' formations (while claiming omnipotence), which can only miss the interconnectedness, the fluidity, the nowness of the world, and all the mass of life process 'beyond' the cognitive, the represented and the strictly social.

The first part of the chapter deals more fully with the ideas sketched out above. The agencies of non-humans are discussed further, along with conceptual pathways through which they can be approached: actor-network theory, hybridity, co-constitution, dwelling and landscape. The second part of the chapter deals more specifically with the presence of animals in rural studies. This is because, first, the study of differing animal presences in the rural has been a growing trend in rural studies, and secondly because differing animal collectives, such as agriculture, hunting and wildlife, can be very significant elements, or actors, within different ruralities. The foundations of 'rural animal studies' are set out, and questions of agency and ethics raised, as these have a particular set of resonances in relation to animals in rural spaces. Finally, as a conclusion, I briefly consider the future of non-human rural studies.

THE AGENCIES OF NON-HUMANS

The essence of non-human rural studies is, then, to pay *particular attention* to the presences of non-human actants in rural formations, not in a way that merely reverses the privilege between human and non-human, but in a way that sees the rural arena in terms of relational, co-constituting processes. The whole point of this approach is the recognition that what happens in the world is not driven by humans alone – a point that becomes obvious in the wake of even the most modest burst of cosmological, geological and/or ecological thinking. But alongside these narrative of grand terraforming agencies, comes the recognition that we share the world with a veritable panoply of things and organisms which are all active players in ordinary ongoing everyday co-constitution of rural space and places. In 'the rural' (be it in the developing or developed world and in differing national, regional or local ruralities) there will be particular combinations, or networks, of things at work to produce it and facets and spaces of it.

These combinations need to be seen in some respects as on the move, as complex, patterned flows of social, economic, ecological and technological changes in the countryside which (re)shape rural space, and rural social formations, economic practices, cultural constructions, and so on. They will be multiple, precarious, and generate internal and external conflicts and tensions. In other respects they will have persisting tendencies which are manifestations of entrenched ideological forces, notably capitalist modes of production and consumption, which are now themselves networked into global systems, material and other forces. More broadly, rural places and territories can themselves be seen as persistences in the context of flows of material, power, money and people. They are constantly being sustained by the human and non-human presences within them, while at the same time, their interconnectedness, which is increasingly of a globalized nature, is implicated in the making and re-making of that sustaining.

To ground these quite abstract ideas I will take two seemingly mundane technological developments in agricultural technology and briefly consider them in non-human agency terms. In many 'advanced' agricultural economies there has been a shift in grass fodder harvesting from hay to silage systems and from small bale to large bale technologies. These transformations, clearly in one sense instigated by human agency, have over time quietly transformed the rural networks they are 'wired' into. Rural landscapes, farm buildings, related technologies, working practices, will shift and adapt as these new presences in the network bed in. Seasonal, casual labour might be less in demand, collective working is replaced by people working in isolation with machines, and new forms of visual iconographies of the rural are produced (big bales being a favourite photographic subject for those generating new glossy, quintessential rural (idyll) images).

This is not to say that these technologies have some kind of inner agency which resembles the common notion of the agency of the liberal, individual human(ist) subject – that is to miss the point of ideas of relational agency. It is to say that these technologies allow new kinds of practices and networks to evolve, and they can contribute to the building of new networks, practices and socio-ecological forms which can go far beyond the initial intentions of the designers, manufacturers, sellers, purchasers and users of the machines. Agency, as Whatmore (1999: 26) suggests, is 'a relational achievement, involving the *creative* presence of organic beings, technological devices and discursive codes' (emphasis added). The word 'creative' emphasizes the space for the individual entity to bring qualities to the achievement, and this needs to be considered in precise, grounded ways, and in ways that allow for notions of differing kinds of agency (Cloke and Jones, 2001, 2004; Jones and Cloke, 2002).

APPROACHING THE NON-HUMAN

How then can we meaningfully represent the non-human in hybrid formations of rurality? A number of approaches have been developed which are relevant to, and are being used and further developed within, rural studies. The foremost of these in terms of use is ANT and this is discussed first, as it is a very clear and increasingly commonly articulated strategy. In relation to ANT the term hybridity is emerging as a broader orientation in terms of the nature of rural spaces and how to approach them. I also discuss co-constitution and dwelling as these too have been deployed to consider the active presences of non-humans.

Actor-network theory

Beyond rural studies, ANT has been deployed in considering the agencies of things, often machines and texts, and how these combine to have creative, relational agencies, in that they make possible, help shape and sustain the complex networks and systems which form lived formations and practices. This approach has become increasingly evident in rural studies, and in studies that, while not operating under the heading 'rural', have been considering what could be called rural phenomena. As an example of the former, Lowe et al. (1997: 197) discuss how 'the story of farm waste pollution … requires that we consider the roles and functions of farmers, environmentalists, Pollution Inspectors, magistrates, cows, fish, snails, slurry, rain, pits, tanks, pumps and various pieces of paper (forms, guidelines, regulations and codes of practice)'. As an example of the latter, Eden et al. (2000) use ANT to consider a river restoration project where the river itself and applied technology are significant actors as well as the human networks involved. These studies show that the non-human devices were enrolled into competing networks and their presences produced effects that were not 'predetermined' but, rather, dependent on the specific co-constitutive relational arrangements they were part of. As these examples show, organisms, natural processes and features, technological systems are some of the most obvious and increasingly studied non-humans in rural formations.

But rurality is dramatically diverse across the world and intensely complex in its formation. As a consequence, the possible range of networks and non-human actants to be considered is vast. In the UK Abram et al. (1998) pay particular attention to the roles of numbers – government statistics – in the formation of rural planning and development policies addressing rural migration pressures and projections. They show that once these numbers are generated and 'pushed' through complex networks of governance, they, in effect, become actors which frame, include and exclude and have 'deep and lasting implications' (1998: 250) for the rural areas they relate to. In Australia Lockie (2004) has brought ideas of collective agency between humans and non-humans to the study of the 'Landcare movement'.

These examples of non-human rural studies not only reflect ANT's 'technical inflection' (Whatmore and Thorne, 2000: 186) but also focus on natural organisms (plants, animals), natural processes (rain) and topographical features (rivers). This inclusion of topographical and organic actors may be particularly appropriate for rural studies. This is not to say the rural is intrinsically more natural than the urban but that the presences of animals and natural features in the rural are very hard to ignore and do play a key part in its material and social construction (Murdoch and Day, 1998). All these beings and things bring something to the party, and thus their roles need to be acknowledged and studied and, as Thrift (1996: 26) states, the promise of ANT is that 'things are given their due'.

However, ANT has been questioned and criticized within the context of rural studies and beyond. Woods (1998: 338) is critical of ANT because it can fail to account for 'the existing social and political [rural] terrains over which networks are constructed'. This relates to wider questionings of ANT's ability to deal with ideas of place (Cloke and Jones, 2001; Thrift, 1999).

I am also concerned that ANT has, somewhat, ironically, an anthropocentric inflection, in that networks in which the natural enrols the social are under-considered in comparison with those where the social enrols the natural. There is the further concern that 'the flattening process [of ANT] leads to an obscuring of differences between different ... "noun chunks" of reality' (Laurier and Philo, 1999: 1014), in other words, between particular things and beings within networks. And lastly there are also questions of a disregard of politics, power relations and victimization in ANT (Bingham and Thrift, 2000; Castree and MacMillan, 2001). This is particularly important with regard to the enrolment of animals (Jones, 2003) and is discussed in the section on rural animal studies.

Hybridity

Both Cloke (2003) and Murdoch (2003) use the term hybridity as a broad approach of which ANT is one part. It captures much of what ANT drives at in that it 'focuses our attention on the zone where the two worlds of nature and society meet' (Murdoch, 2003). The rural is clearly replete with such meetings. Take, for example, an apple orchard. It is an achievement which is as much about nature (apple trees, bees), as it is the social (owners, workers, customers), the cultural (local/regional identity), the economic (food production), the political (food and landscape conservation legislation) and the technological (machines, pesticides), all interwoven through embodied practices (pollination, grafting, pruning, picking, spraying) (Cloke and Jones, 2001).

Franklin (2002) uses the example of the garden as an emblematic space which reflects the profound interpenetrations between the social and natural, and develops sociological and political insights into the condition of late modernity from this. Much of his focus falls upon urban areas, and the extent to which they are penetrated by nature, not least through gardening, but he makes the point that this penetration of Western (sub)urban space by nature can be seen as 're-ruralization' (2002: 158). He throws up the question that hybrid orientated non-human rural studies can start to encroach into the city, following ecological trails, and find, perhaps, the urban-set 'post-ruralities' discussed by Murdoch and Pratt (1993), where 'rurality' is more a question of assemblable, performable, lifestyle and outlook rather than mere geographical location.

At the heart of the notion of hybridity and ANT are questions about the balance between relationality and individuality. Whatmore (2002) accepts, with qualifications, the insights of ANT and of Haraway's notion of cyborgs in her notion of hybrid geography which shares much with the approaches of Murdoch and Cloke. But Whatmore is keen to increase the pressure on the notion of the autonomous individual (human and non-human) and the mind–body, subject–object, self–other divides, in order to build a politics and ethics of affective intercorporality. Critically, Whatmore's view of hybridity is trying to get away from the notion that it entails the joining of two (or more) previously separated entities, and is instead building a view which sees – paraphrasing Latour's (1993) call that 'We have never been Modern' – that we have never been separate.

This view of the world as being intensely hybrid inevitably has theoretical and methodological implications. Importantly, Cloke (2003) and Murdoch (2003) use the term hybridity not only to address the complexity and variation of and within rural formations, but also to call for a theoretical hybridity and plurality which gives rural studies the flexibility to address the variation, complexity and heterogeneity of the everyday (rural) world.

Co-constitution, dwelling and landscape

Notions of the co-constitution and co-construction (Murdoch, 2001) of achievements by human and non-human are clearly very closely related to notions of relational agency and hybridity. What they add, or change, is, in part, a response to some of the criticisms already raised about ANT. First, through these approaches networks are 'placed'. Secondly, and in some tension with Whatmore's notion of hybridity, they offer individual entities in relational processes more space and credence as creative beings, rather than entities whose form is dictated entirely by the network they are in, and this perhaps overcomes some of the problems with ANT highlighted above. Cloke and Perkins (2004) suggest that such an approach offers insights into how places are formed and even performed and give an account of the co-constitution of Kaikoura, in New Zealand through eco-tourism experiences that involve people, animals, technologies and other entities in complex relation.

The notion of dwelling, has, like the approaches outlined above, a concern for the relational in that it sees action in the world as the outcome of the ongoing interplay between bodies, materials and spaces. It has been a source of (not unproblematic) inspiration for human geographers thinking about embodied human relations with place and landscape. Since Ingold's (1993) celebrated reworking of Heidegger's writings on dwelling (see Wylie, 2003) this idea has become central to some efforts to reconfigure human nature relations in such a way that the latter again becomes active. Notably, Macnaghten and Urry (1998) place Ingold's (1993) notions of 'dwelling' and 'taskscape' at the centre of their idea of *Contested Natures*. For those interested in non-human elements of the rural milieu, it should be remembered that Ingold's (1993) explication of dwelling is based around a discussion of the quintessentially rural image of a Bruegel landscape painting (*The Harvesters*). Here a tree, a field of grain, the topography, and people's habitual, embodied, engagements with them, build up specific materialized formations of meaning through practice. Landscapes and 'nature' are thus seen as complex spatial and temporal achievements and the 'relationships with what is taken to be "nature" are embodied, involving a variety of senses and that there are "physical" components of walls, textures, land, plants and so on, which partly constitute such "dwellings"' (Macnaghten and Urry, 1998; 168). So, as with ANT and hybridity, dwelling insists on the need to dissolve the 'category of the social' (Ingold, 1997: 249), but in particular, 'to re-embed (human) relationships within the continuum of organic life' (p. 249), a pressing need in general and particularly in understandings of rural space. Dwelling, then, is a complex performative achievement of heterogeneous actors in relational spatial/temporal settings *which can only be grasped in the practice of their unfolding*. For example, Cloke and Jones (2001) and Jones and Cloke (2002) show how trees in both rural and urban settings play active co-constitutive roles in the productions of dwelt places.

One recent commentary on these uses of dwelling in understanding rural landscapes does suggest that deploying Ingold's notion of dwelling can help (rural) cultural geographies develop an approach which 'reinstates landscape as a sensuous and material milieu', but argues that these ideas of dwelling need to be adjusted and developed to allow a fuller presence of relational material processes in accounts of landscapes, including those which are transitional and fleeting (Wylie, 2003). In his account of ascending Glastonbury Tor, Wylie offers passages of very striking description which bring the specific material configurations of the landscape, and nature and his specific, sensed and embodied engagement with it into play. For example, in the following extract the movement of the approaching paths, and the elevation of the Tor itself, become activated.

> As one climbs up and around the Tor, the landscape uncoils, it begins to encircle as it expands. The sensation of height is considerable, but it is not a height upon whose edge one teeters, rather an expanding volume opens to and enfolds the beholder. (2003: 152)

The non-human, the topography, becomes active, as it surely does in myriad engagements between people and 'rural' landscapes across the world. Cater and Smith state that 'it is important to give due acknowledgement to the role of nature as a key external factor in adventure [tourism]' (2003: 214). In the practices of extreme sports in New Zealand which they study, the landscape becomes a background to the dramatic moment of enactment (bungee-jumping), but clearly it also co-constitutes the experience, from the moment the mountains are first glimpsed in the distance, the steep drive up to the sports centre, the topographies of sheerness, the vistas that wheel as you plunge.

The above ways of approaching the non-human in rural studies are not intended to be exclusive, for that would fly in the face of the methodological plurality and hybridity already called for. They have been reviewed because they are emerging as key theorizations which collectively are trying to 'represent' the creative presences of non-humans in rural formations and which share certain key, characteristics. However, there is another legacy of non-human presence within rural studies, which I feel needs particular attention, namely animals.

ANIMALS IN RURAL STUDIES

The study of animals in rural formations is becoming increasingly visible in a number of contexts. This is, increasingly, being conducted through the theoretical approaches outlined above. But I feel I need to deal with animals in a separate section because, first, there are a lot of animal-focused rural studies which do not fit neatly into the above approaches but, nevertheless, do need representing here. Secondly, animals have very significant and sometimes highly visible and contested roles in rural formations. Thirdly, the nature of animals demands that they be considered in ways that respond to that nature, not least in terms of particular forms of agency and particular demands for ethics, demands which cannot be satisfied if animals become subsumed in a general category of non-human.

Animals have long (probably always) had major presences in the everyday, social, cultural, political, economic and ecological life of many rural places, be it in the form of farm animals, wildlife, hunting animals and prey, sporting and companion animals, and so on. This remains the case today, but, of course, these animal presences and the co-constitutions they bring, will vary dramatically in differing national and local rural spaces. There is now a considerable legacy of what I will term Rural Animal Studies (RAS), which has begun to consider these animal presences. The study of animals within rural space was a minor feature of early twentieth century geography, agricultural geography and, latterly, the distinct form of rural studies which emerged in the 1970s (Cloke, 1989). It generally took the form of agri/economic geography, studying spatial distribution of differing animals as livestock in rural economies and landscapes. RAS continues its interests in animals in terms of their varying presence in food production systems, rural economies and so forth, but it has developed in terms of underpinning theoretical concerns and in terms of considering other forms of animal presences in the countryside. Although this overlaps to a considerable extent with the manifestations and impetuses outlined above, it has also arisen in relation to other contexts and other impulses, notably the wider arena of animal studies and environmentalism, and has begun to confront particularly pressing issues of ethics, agencies and emotions in human–animal interactions in rural contexts.

Animal studies, the cultural turn and environmentalism

The study of animals in rural formations has developed within the wider context of a growing movement of animal studies across the social sciences, notably in sociology and geography. In the former, Arluke and Sanders (1996) argue that the focus of animal research needed to move beyond studying animal presences in traditional societies and begin to theorize and study the symbolic roles animals play in modern industrial society and the staggering range and inconsistency of attitudes and behaviours towards them (p. 3). This, of course, indicates that there has been a focus on the presences of animals in traditional rural societies. Franklin (1999) takes up the challenge analysing the changing presences of animals in modern societies in terms of hunting, pets, food, the zoological gaze, and how these are cross-cut by social difference, and relate to questions of modern and postmodern social formations and conditions. Again, the rural becomes a context but in very differing ways, as the ground for modern agricultural practices. Studies are now emerging which challenge the 'invisibility of animals' in social (rural) space and studies (Tovey, 2003). Similarly, animal geography has emerged markedly since the mid-1990s (Philo and Wilbert, 2000; Wolch and Emel, 1998; Wolch et al., 2003). Broadly, this movement is concerned with how animals are implicated in social formations in all manner of significant, spatialized ways. The range of conceptual approaches and substantive foci is now considerable, but studies addressing farm animals, wildlife, hunting, the symbolism of animals all cross-cut with the differing arena of rural space. Importantly there is now developing a concern for the spatialities of ethics (see Lynn, 1998a, 1998b; Jones, 2003) in human–animal interaction.

In close concord with the development of animal studies, the cultural turn in rural studies opened up considerations of 'otherness' in the rural academic gaze (Cloke, 1997). Studies of childhood, gender, sexuality, ethnicity, poverty, homelessness in what was now understood as socially constructed rural space were becoming key theoretical objectives. This widened the horizons of rural studies and made the idea of studying animal presences seem more pertinent and possible. However, to claim that the cultural turn has fully opened up space for consideration of animals may be premature.

In much of rural cultural geography's engagement with animals, the subjects can appear less as embodied presences and more as symbolic resources enlisted in discourses and politics of the rural. Whatmore suggests the cultural turn intensified 'divisions between the natural and the cultural, championing the "agent" over the "medium" to such an extent that the world is rendered an exclusively human achievement in which "nature" is swallowed up in the hubris of social constructionism' (2003: 165). Work is therefore needed which emphasizes the presence of the animal as a living body, and the complex intimate interactions between people and animals in everyday encounters. Another concern is that the preoccupations of the cultural turn may have focused on animals in rural formations in some ways while neglecting others. As Morris and Evans (2004) suggest 'the cultural turn was silent on agriculture'.

The ongoing, and in many ways positive cultural inflection within rural studies therefore needs to ensure it is not neglectful of major collective *corporeal presences* of animals in rural space, notably animals in agri-economic systems. The relatively sympathetic critiques of the cultural turn by Philo (2000a) and Thrift (2000) may help in this respect, offering a more materialized, embodied and complex notion of cultural processes. And, in this respect, it is significant that a key architect of the cultural turn in rural studies (Philo, 1992) is also a key architect of animal geography (Philo, 1995, 1998; Philo and Wilbert, 2000). Whatmore's (2000, 2003) advocation of animated, hybrid culturenatures, which pays full attention to the active presences and agencies of nature, seems to close the gap between culturally constructed nature/rurality and a lively material nature/rurality in which animals and other non-humans are recognized as actively present.

These developments in (rural) animal studies have also been informed by the growth of environmental concern and related questions about animal rights which have now penetrated deeply into rural studies (Buller and Morris, 2003). Developments in scientific studies of animal natures and behaviours, animal sentience and their mental and emotional capacities (Compassion in World Farming, 2003a, 2003b; Young, 2003) has also begun to add further impetus and challenges to how we see animals in rural spaces, and will help the culturally inflected approaches to recognize animals as being extremely complex evolved entities 'saturated with being' (Whatmore and Thorne, 2000: 186). Building on these foundations, the proliferation of approaches and specific foci in recent RAS emphasizes three themes for (non-human) rural studies: first, an awakening to the idea that animal presences in the countryside have always been more complex, fluid, diverse and contested than the narrow scope of earlier studies had allowed: secondly, a growing sensitivity to further proliferations of animal presences in society driven by, amongst other things, science and technology, novel ecological mixings and the heterogenization of production and consumption systems, which are often embedded or part embedded in rural spaces: thirdly, responses to various destabilizations of modernist assumptions about animals and nature. These three themes are played out in various differing forms of animal presences in rural formations, but for the purposes of this chapter I will focus primarily upon agricultural presences, and also more briefly raise hunting, wildlife and animal culture as other key ways in which animals are studied. It should be added that animal presences in the countryside can take many other forms, such as 'alien' species, companion animals, animals associated with sport, recreation and spirituality. Clearly, the conceptual, political and material boundaries between these categories are messy and contested, and the particular forms of such animal presences will vary markedly between and within national rural cultures. I am painfully aware of the vast global reach and diversity of rural landscapes and I realize that this is necessarily a very partial account. I can only hope that the theoretical ambitions speak to other rural formations not represented here.

Agriculture

A significant form of animal presences in the many rural spaces are animals in agricultural production systems of one kind or another. Sheep, cattle, pigs, poultry, goats and others have long been agricultural staples, the horse and other animals have been and remain means of power, and other animals such as dogs are working animals too. This has ensured that animals have been at the heart of these spaces in material, economic and cultural terms, and that, in very many different ways, they play their part in the co-construction of differing ruralities.

Some of the earliest forms of UK RAS revolved around studying animals in terms of the presence in agricultural and economic systems. Yarwood and Evans show how these early 'livestock geographies' were largely economic or distributional in nature, and, tying into both regional and economic perspectives, 'treated animals as "units of production" ' (2000: 100). This clearly is an important aspect of considering animal presences in rural space, but these approaches generally did not consider animal presences in terms of cultural constructions of the rural, the politics of animals and other questions such as of agency or ethics. Recent developments in agricultural geography have begun to open up approaches where the complex presences of animals may have more visibility. Whatmore points to three main ways this is happening; through opening up questions of nature, food consumption and the body and bodily interactions. The latter include questions about 'the bodies of industrial pig or cow, made larger and leaner by genetic engineering, and hormone supplements' (2000b: 12).

Recent considerations and theorizations of alternative agro-food networks (see Goodman, 2003) offer opportunities to consider the animal and other non-human presences in agricultural and rural formations. Mansfield's analysis of the international surimi seafood industry indeed does do just this by seeking to reveal 'how specific aspects of what we call the "natural world" participate in specific interactions [within network formation]' (2003: 9).

The legacy of earlier rural geographical foci on farm animals has also been continued and developed by Yarwood and Evans. In a series of publications (Evans and Yarwood, 1995, 2000; Yarwood and Evans, 1998, 1999, 2000) they have taken established interests in farm animal distribution and farm animal presence in rural economies, and developed these through research into 'post-productivist' rural space and, in conjunction with new animal geography, have folded these economic perspectives with issues of social construction of the countryside, nature conservation, politics and ideologies of animals, nature and rurality. In their UK geographies of rare breeds (Evans and Yarwood, 2000), they chart how certain types of livestock are implicated in shifts to a post-productivist rural where (tourist) consumption of animals as spectacle has become important within farm diversification, innovative conservation management schemes and local cultural coherences in rural places. They have also mapped the distribution of animal breeds and rare breeds and analysed the reasons for their distribution and fluctuation in numbers. They show that, in the UK, post-Second World War productivist policies reduced the variety of bred types used in farming and that 'by the 1960s, a highly fragmented and uncoordinated patterns of breeds that were low in number had emerged' (2000: 233).

In the UK (and elsewhere) there have been a series of food production crises with animals at their centre. In the 1990s there was a surge in concern about salmonella in chicken eggs. More significant was the bovine spongiform encephalopathy (BSE) outbreak with the resulting export bans on British beef (see Macnaghten and Urry, 1998). In 2001 yet another major crisis hit British agriculture with the outbreak of foot and mouth disease (FMD). The outbreak dominated national news for a number of weeks, bringing images of animals and rural landscapes, communities and farms to the top of television news broadcasts and onto the front pages of the national press.

These crises had a number of significant effects on rural areas, and on the understandings of animal presences within them. First they exposed the sheer presence of farm animals in society. This links to wider discourses about how farm animal numbers have increased on a global scale as meat eating becomes more widespread. 'Already there are twice as many chickens on the planet as humans, plus a billion pigs, 1.3 billion cows and 8.3 billion sheep and goats' (Jowit, 2004). Secondly, they opened up some of the often 'closed' spaces and practices of animal production, such as the 'unnatural' feeding practices at the heart of BSE (see Whatmore, 2002) and the transporting of animals from one part of the country to another for dealer trade and slaughter, as exposed by FMD. Thirdly, they highlighted the role of animals in rural economies, rural regions and links between agriculture and governance, and agriculture and tourism (Bennett et al., 2001, 2002; Scott et al., 2004). Fourthly, they revealed animals' presences in constructions of individual, collective identities, and of places and landscapes. Fifthly, they raised sharp questions about the moral status of farm animals and the complex emotional responses that both producers and consumers may have to their life and to their death (Convery and Bailey, 2002; Wrennell, 2002). Lastly, they highlighted how farm animals are bound up in the construction of 'wild' ecologies, habitats and the countryside more generally. These aspects of animal presences dramatically opened by the FMD outbreak in the UK, will be, in varying forms and combinations, significant within other rural contexts as well.

More broadly, the ongoing crisis in UK agriculture, which incorporates falls in farm incomes, a rise in farm business failures and wider ongoing struggles to reform European Union agricultural policies, has significant implications for animal presences in the countryside. The post-war productionist era, with its relatively settled Fordist systems, has been undone by subsidy, surplus, pressures from globalization and environmental aggravations, and is breaking up as forms of diversification, ranging from local ethically based organic production to high-tech GM-based initiatives, proliferate, and this pattern can be seen repeating, in differing forms across the developed world, while rural areas in the developing world face yet other dynamics that impact on rural spaces and animals within them, not least in relationship to the developed world through forces of globalization.

Beyond the impacts of these crises, other research is beginning to open up the spaces of animal human relationality within agriculture. For example: Holloway (2001) has reported on research into the differences that occur between 'hobby' farmers and 'commercial' farmers in their relationships to their livestock and animal-related agricultural practices; Smith (2002) enters the challenging space of the modern slaughterhouse to expose and question modernity's relationship with farm animals and to ask if their death cries can awake us from our 'ethical apathy'; Wilkie (2002) considers the complex and paradoxical nature relationship between animals and their handlers; and (Sellick, 2004) is beginning to explore the embodied, relational practices of such human–animal interactions.

Hunting

Hunting, as Franklin (1999: 105) points out, has 'enjoyed sustained popularity and growth during the twentieth century'. And it has to be considered as another complex and contested way in which animals are bound up in the construction of material and imagined rural space. This varies between the Western cultures Franklin considers (the United States and UK), but will, inevitably, vary significantly between and within all national cultures where very different sets of hunting-related human–animal interactions will be practised, which are based on dense interactions of ecological, economic, cultural and political/historical trajectories.

In the UK there has been a flurry of studies of hunting as the politics of it has become highly contested and high on the political agenda. These studies are often about the social, cultural and economic dynamics of hunting but do raise questions of animal presences and do, in some instances, try to bring the animals in as actants. Scenes from these activities are part of the quintessential iconography of the 'traditional' countryside. In addition, large swathes of the rural landscape are managed or part managed to satisfy the particular needs of hunting and shooting.

Milbourne (2003) examines questions revolving round the hunting debate, which had fulminated around the possibility of anti-hunting with hounds legislation in the UK (eventually enacted in 2005). Milbourne examines the Countryside Alliance's tactics of insisting that hunting contributes key socio-cultural elements to rural communities, placing this claim in wider contexts of nature–society relations and rural–hunting relations. He calls for sensitivity to the geographies of these relationships within rural areas. Woods (1998, 2000) has also built connections between animal and rural geography, particularly through the politics of hunting and other representations of animals. Here all the powerful visual and written and performed discourses of pro- and anti-hunting movements are depicted as highlighting and destabilizing the imagined and performed animal geographies of the countryside. Woods considers the 'political marginalization' of the animals themselves and how the agency of animals is '(unintended) agency-as-effect'.

Wildlife–nature conservation

Another major presence of animals in rural space is represented by the idea of nature and wildlife. In *The Theft of the Countryside* Marion Shoard (1980) states that in the UK 'for many people … the creatures and plants of our countryside have provided the key to its charms' (p. 183). These presences and efforts to conserve and even enhance them are central to the idea of the rural as a space of nature and even in certain areas, as a space of wildness. Often in tension with the needs of agriculture (but not exclusively) wildlife appreciation and latterly conservation has shaped aspects of countryside legislation and countryside cultures. This has made the countryside a multi-dimensional and contested 'animal space'.

In the very different ruralities of North West America the geographical scales, and the economic, ecological and political mixings are markedly different from that of the UK, as Proctor (1996) shows in his accounts of the debates and disputes about the conservation of the northern spotted owl. But again, what are revealed are complex mixings in which animals are central presences. In any rural space such mixings are likely to occur. In some instances, these may distil into debates and practices of the conservation or reintroduction of 'native' wild animal on one hand and the culling of other 'alien', invasive species on the other (see Whatmore, 2000a on the fate of the Ruddy Duck in the UK). So particular ruralities become defined by particular forms of appropriate and inappropriate animal presences. Buller (2003) has considered the cultural complexities and resonances of 'wild animals' in the European countryside and in particular the increasing incidence of 'big cat' stories which bring a narrative of 'other' nature, danger and wilderness to the otherwise 'tame' English countryside.

The complex spatial formations of wildlife which inevitably, but not exclusively, operate in rural terrains have been studied by Whatmore and Thorne (1998, 2000). To grapple with the political ecologies of such formations (but not exclusively in terms of animals) FitzSimmons considers the possibilities in connecting geography and ecology and suggests that we need to develop 'conscious socio-ecological projects of human liberation and responsibility towards nature' (2004: 44). This, she suggests, is in part to be achieved by going beyond critique and connecting with 'people who are finding ways to share responsibility for the natural world in the city and the country' (p. 45). But, as Lorimer (2004) points out, the burgeoning practices of biodiversity management and environmental governance which are at work in the countryside have received little sustained critical academic attention thus far, and here is yet another whole set of animal (and other non-human) presences in complex relational networks which link ecology, politics, science, in ways that are refashioning certain rural areas.

Animal culture: rural place(s) and politics

Finally I briefly turn to this last heading, as animal presences are bound up with the cultural/political construction of the rural as place. Ridley (1998) argues that 'animals represent one of the chief points of friction between town and country', and this points up the political resonances of animal presences in the countryside. The hunting debate discussed above is but one example of the politics of rurality and animality. Evans and Yarwood (2000) in their study of the British Rare Breeds Trust not only look at the composition of this movement, but also ask questions of 'how animals themselves are used for political ends'. They consider the role of animals in social constructions of rurality and the spatial relationships between people, animals and places. The tendency of many rare domestic breeds to be named after a particular region or place is clearly a sign of, and opportunity for, the mobilization of local and regional politics, economy and identity in which animals play a key role.

This construction of regions can also work through more complex interplays between 'indigenous' wildlife, 'alien' species and farming stock and practices. Matless shows how, in the Norfolk Broads, particular animals – the bittern and the coypu – play their roles in a region where there is 'a tight interweaving of local ecology and economy' (2000: 138) in the production of a regional nature.

Lastly it needs to be stressed that these presences of animals (as so other natures) bring an inherent instability to the processes of politics and identities which links back to the notion of the disruption of human agency central to this chapter. For example, Macnaghten and Urry (1998: 249) conclude *Contested Nature* by discussing the BSE crisis and how animals, and a myriad other actors, were involved in an intensely complex formation of global nature which is 'more or less ungovernable, at least within available practices and discourses'. These are important pointers to the possible density and heterogeneity of rural (political) formations in which animals are active. And heeding Macnaghten and Urry's warning of the ungovernability of complex networks when understood from present (purely social) perspectives, this is precisely why the non-human needs to be taken seriously.

ANIMALS, AGENCY, ETHICS AND THE RURAL

In the opening sections of this chapter the calls for reassessing the distribution of agency throughout social, technological and ecological formations stemmed in part from post-structuralism, science studies and ANT's reconsiderations of the notion of the human subject and the way the world is heterogeneously engineered. In terms of the agency of nature, these have worked alongside developments in ideas of social nature which are now keen to acknowledge that while nature may be socially constructed it is still a vital force. FitzSimmons and Goodman (1998: 194) claim that 'it has been a commonplace in social theory to ignore the specific "agency" and "materiality" of nature'. There is now a concern to consider the agencies of nature and the specific, embodied, agencies of beings of nature (Harvey, 1996: 183) in relational assemblages (Whatmore, 2003). This, inevitably, will entail considerations of not just 'animals' but specific types of animals, and even individual animals in specific relational arrangements.

ANT has been deployed to think about the agency of animals and animals in rural networks and beyond. Philo and Wilbert (2000) in their comprehensive account of human–animal relations and new animal geographies, focus on ANT as 'sophisticated intellectual innovation' which is useful for developing the discussion of the agency of animals, and they conclude that, 'Latour's pioneering experiments open up "a space" for contemplating the agency of non-humans, animals included, and for speculating about them taking their seats in his envisaged parliament of things' (2000: 17).

In rural studies ANT has been deployed to seek out animal presences in networks. Woods (1998, 2000) has followed human and non-human actants in rural networks and to account for the precise outcome of relational interactions at work in both agricultural production and hunting. Whatmore (1997) and Fitzsimmons and Goodman (1998) have traced networks of food production and the role of animals (and plants) as 'quasi-objects, quasi-subjects' (Fitzsimmons and Goodman, 1998: 209) within them. Evans and Yarwood (2000) use an ANT approach for their study of how rare breeds are present in post-productionist rural networks. And, as indicated in the introduction, Lowe et al. (1997) suggest that the story of farm pollution needs to account for the agencies of people, technology, natural processes (rain) and animals (cows, fish, snails) (see also Murdoch, 2003 on BSE).

However, as already shown, there are some questions being raised about some aspects of ANT. As well as the questions highlighted earlier, there are also some grounds for considering it to have some weaknesses or blind spots which may be significant in terms of considering animals in rural studies. Most significantly, it can be viewed as ignoring the 'quite real effectivity of victimisation' (Wise, 1997: 39, cited by Bingham and Thrift, 2000: 299) and bypassing 'questions of unequal power' (Bingham and Thrift, 2000: 299). This, as I have argued (Jones, 2003), seems very important when thinking about how animals are enrolled into networks of production.

It is also important to note that a concern for non-human (animal and plant) agencies has emerged in relation to explorations of bio-philosophies and bio-ethics through ideas which have been developed in environmental philosophy, perhaps most significantly, ecofeminism. For example, Plumwood points out that:

> Once nature is reconceived as capable of agency and intentionality, and human identity is reconceived in less polarised and disembodied ways, the great gulf which Cartesian thought established between the conscious, mindful human sphere and the mindless, clockwork natural one disappears. (1993: 5)

At the same time as this reassessment of agency has been occurring, certain scientists and other people concerned for, and working with, animals, have been demonstrating that animals are much more complex and capable in terms of their mental states of being than Cartesian visions allowed. (See, for example, the recent report on the sentience of farm animals by Compassion in World Farming (2003) and Plumwood (2002)). These reassessments of animal beingness (Gaita, 2002) and animal agencies, not only make demands to take their co-constitutive roles seriously but also reinforce ethical demands as well.

Ethics and dwelling

There are ethical questions to be asked about all the animal presences in the rural which have been highlighted, ranging between the vexed issues of industrial farming, how wilderness and wildlife are managed, and hunting. Tapper observes that 'urban–industrial society, finally, is dependent for animal products on battery – or factory – farming' (1988: 4). The rural has become the space where much of the subjugation of animals on behalf of modern society takes place, and thus it becomes a space of great tension in terms of human–animal relations. This is clearly illustrated by Watts's account of the broiler chicken industry in the United States:

> Broilers are overwhelmingly produced by family farmers in the US, but this turns out to be a deceptive statistic. They are grown by farmers under contract to enormous transnational enterprises – referred to as 'integrators' in the chicken business – who provide the chicks and feed ... Growers are not independent farmers at all: they are little more than underpaid workers ... propertied labourers. (2004: 55)

Murdoch (2003) in his calls for geography to widen 'the circle of our geographical concerns' to include animals builds much of his case around the fate of farm animals – rather than wild animals. He asks in particular that we should 'enlarge our sympathies' to include the almost unimaginable number of animals being exploited within the modern agro-food industry.[1] Clearly this is a question in relation to society as a whole and not just rural society, but inevitably, the density of networks which comprise this use of animals is significant in rural space. The ethical question is also pertinent to society as a whole, but for rural studies these ethical questions can be set in the contexts of agricultural history, and the social, cultural, economic and ecological dynamics of rural areas.

Buller and Morris (2003) explore the development of animal welfare discourses and policies. They point out that there is some movement in this respect, as shown in the progress made by the welfare discourse of the 'Five Freedoms' for domesticated animals, and, in some respects, European Union animal welfare directives stemming from the Amsterdam Treaty. They suggest that this shows that, to an extent, the demarcation of nature/animals as resource, and the human realm as the ethical realm, has been breached.

They rightly point out, however, that while animals are reared for food and for other forms of consumption, there seems to be an inevitable limit to acknowledging their rights as sentient individual beings. Buller and Morris begin to explore academic and welfare policy thinking around this area, posing the question: can production systems be devised that afford domesticated animals some form of 'voice' and some recognition of their independent and animal distinctiveness. The growing discussions about the increase in meat production and consumption (Jowit, 2004) and its implications, and ongoing questions about how meat is produced are critical. So too are other questions about other animal presences and these are considered in the last section.

Dwelling is developing as one means of thinking about animals which may account for their co-constitutive practices in a way which allows for their agency but also their call for an ethical signature (Jones, 2003). All lives are dwelt in so much as this term captures the unfolding of life in the moment, but the moment not as the infinitesimal moving point of present time, but the emplaced, embodied, enspaced, entimed, weight of life, and the patterns and practices that are built out of this. As Whatmore and Hinchcliffe (2003) state, dwelling is about 'the ways in which humans and other animals make themselves at home in the world through a bodily register of ecological conduct'. It is also about relationality and agency – Whatmore and Hinchcliffe (2003) again – about 'more-than-human agents exercising unbidden, improvised and sometimes, disruptive energies'.

The vast divergences of rural animal dwellings, such as those of native wildlife, alien species, pets, recreational sport animals, hunting and hunted animals, farm animals, are driven by economy, culture, technology, ecology, ethics, welfare policies and their own agencies. Dwellings are intrinsically spatial and relational and more or less visible in literal, ethical, economic and cultural terms. Cloke and Jones (2001) and Jones (2003) discuss how dwellings can be either bitter and harsh or benign and fulfilled in relation to the kinds of networks and the way non-human actors are enrolled into their hybrid formations.

THE FUTURE OF NON-HUMAN RURAL STUDIES

The world is continuously on the move. Social studies are always struggling to keep up with it. New arrangements keep extruding out in rhizomic bifurcations as networks form, compete, combine, and divide, fail and reform. Those in rural arenas, as elsewhere, are hybrid assemblages in which the ideological, technological and ecological combine. These can take toxic form, as in food crises, or be more liberational, as in, say, alternative agro-food networks (Goodman, 2002). The consideration of the roles and presences of non-humans will be vital as networks in rural areas proliferate and intensify in density, complexity and variation. As Murdoch (Chapter 12, this volume) suggests, 'rural complexity looks set to increase in both scope and scale'. In other words, an increasing number of non-human actors – 'who [will] come in many and wonderful forms' (Cheney, 1994: 170–1, citing Haraway, 1988), will be busy at relational work, co-constituting multiple ruralities through spatialized practices.

The future of non-human rural studies will rest upon theorizations and research initiatives that seek to highlight the *lived materialities* of these presences within all the flows of human agency and social construction which flow through rural networks and which will continue to receive sustained attention. It will also rest upon addressing the *imaginary entanglements* that collect around such lived materialities and perhaps, particularly in relationship to animal presences, the *ethical* and *emotional* registers of engagement.

Approaching lived materialities inevitably means moving towards *sustained empirical engagements with particular networks and paying attention to the non-human actants in them.* For example, Mansfield's (2003) study of networks of the international seafood industry shows that 'translations mean that nature cannot simply be a resource for human activity, a constraint to be overcome, or a pure figment of social interaction. Rather, distinct aspects of what we call "the natural world" participate in *individual* interactions' (2003: 19). ANT is likely to be a dominant strategy to do this but questions will need to be pursued about the nature of things, places, ethics, politics and power and how these might operate in any rural milieu. Related approaches of co-constitution, and hybridity, will need similar sensibilities.

Notions of dwelling might help leaven ANT approaches with a sensitivity to the emplaced, embodied, processes of co-habitation which is often concurrent to co-constitution particularly when animals and other living actants are involved. Developing insights from ANT, dwelling and more animated versions of social construction of nature, and notions of hybrid geographies (Cloke, 2003; Murdoch, 2003; Whatmore, 2002) may be particularly useful in terms of think the complexities of non-human presences.

This brings me back then to the vital question of the presences of animals in rurality. These vast and diverse populations have been relatively ignored in rural studies thus far. Yarwood and Evans suggest that rural studies has so far paid scant attention to animal and human–animal connections in the much wider senses opened up by animal geographies, and they conclude that 'there is still much work to be done on the place of animals in the countryside' (2000: 99). Such a statement, when coupled with the plea for more engagement with ideas of animal rights and questions of ethics (Murdoch, 2003), and with the questions of agency, new scientific studies of animals, analysis of ecological systems and their conservation (FitzSimmons, 2004), analysis of food networks, seems to open up a vast and diverse range of future trajectories for non-human/animal rural studies. This is reinforced by the likelihood that the intensity and variation of human and animal interactions will continue to proliferate within the general growth of complexity suggested by Murdoch (2003). The increases of ecological mixing driven by deliberate and accidental global flows is one such source of growing complexity. Technological developments, including GM technologies, seem another significant factor here, as GeneWatch state:

> Genetic Modification of animals represents a watershed in our relationship to the natural world and a significant further step towards seeing animals as only commodities to be created for our convenience. (2003: 1)

In imaginary terms, approaches need to be developed which wrestle with the intense complexity of the interplay between the material and how it is imagined, or socially constructed. This is so for the non-human in general, and for the specific presences of animals in rural arenas. For example, Buller (2004) discusses the 'faunistic icons of rural areas' and how, through shifting and mixing material and imaginative elements, animals' presences in the rural are shifting in specific form, in conjunction with wider shifts in nature–society relations. Buller raises the intriguing question 'what happens … when the "wild" creeps back into the domesticated and humanised nature that is the European countryside' (133).

In relation to animal presences, the ethical question must loom large, as Murdoch (2003) points out. But to operationalize this we must pay attention to the emotional dimensions of human–animal interactions. Murdoch (2003) also quotes Hacking (2000) that it is time 'to own up to the messiness of our passions' in our relationship with animals, and Murdoch suggests this means linking 'carefully reasoned theories and descriptions to deeply held sympathies, commitments and affiliations' (2003: 289). This kind of trajectory seems to steer rural studies which are concerned with animals towards the notion of emotional geography (see Anderson and Smith, 2001), which is a call to recognize the role of emotions in the construction of the world, and in interpretations of the world, which continues the movement to accept the *full humanness and complexity of being-in-the world, and in the addressing of that being-in-the-world developed by* post-structuralism and feminism. Such a move clearly links with the ideas of geographers who are thinking about non-cognitive knowledges of the unconscious and the body (Thrift, 2004). Anderson and Smith (2001) argue that 'to neglect the emotions is to exclude a key set of relations through which lives are lived and societies made' (2001: 7). This seems particularly relevant in terms of our relationships with animals (Mabey, 2003) be they farmed, wild, hunting, hunted, 'indigenous' or 'alien'. Recent studies are beginning to conceptualize and investigate the emotional and embodied intimacies, and the complexities, paradoxes and messiness of human–animal interactions (Convery and Bailey, 2002; Holloway, 2001; Sellick, 2004; Smith, 2002; Wilkie, 2002; Wrennall, 2002). Some of this work has emerged in the wake of destabilizations of agricultural practices, but others are more concerned with exploring the ordinary, everyday, presences and experiences of animals in rural space.

Finally, it is very difficult to try to envisage the future of non-human rural studies with also trying to anticipate the direction of the wider field. I have tried not to do this, and such divinations will be present elsewhere in this volume. I will only add here that rural studies should have an eye to what could be called liberational understandings of life in the rural arena across the world. Non-human rural studies can play a part in this by contributing to understandings of hybrid networks which are both liberational and oppressive. Overall, given the diverse range of actants and issues sketched out above, the future of non human rural studies will rest on the kind of methodological pluralism advocated by Cloke (2003) and Murdoch (2003) coupled with creative geographical imaginations that range across spaces and issues and into the specific relational processes they are formed of, and across the discipline boundaries which often confine thinking and research into partial realms (Goodman, 2003; Spencer and Whatmore, 2001).

NOTE

1 Ten billion birds and mammals were raised and killed for food in the United States (Murdoch, 2003). In the UK, Beavis (2002) suggests that '860 million animals are reared each year for food, usually in highly intensive systems'.

REFERENCES

Abram, S., Murdoch, J. and Marsden, T. (1998) 'Planning by numbers', in P. Boyle and K. Halfacree (eds), *Migration into Rural Areas: Theories and Issues*. London: John Wiley. pp. 236–251.

Anderson, K. and Smith, S.J. (2001) 'Editorial: Emotional geographies', *Transactions of the Institute of British Geographers*, NS 26: 7–10.

Arluke, A. and Sanders, C.R. (1996) *Regarding Animals*. Philadelphia, PA: Temple University Press.

Beavis, S. (2002) 'Creature comforts', in *Farming Today?* (Supplement), *Guardian*, Friday 28 June.

Bennett, K., Carroll, T., Lowe, P. and Phillipson, J. (2002) *Coping with Crisis in Cumbria: Consequences of Foot and Mouth Disease*. Newcastle: Centre for Rural Economy.

Bennett, K., Phillipson, J., Lowe, P. and Ward, N. (2001) *The Impact of the Foot and Mouth Crisis on Rural Firms: a Survey of Microbusinesses in North East England*. Newcastle: Centre for Rural Economy.

Bingham, N. and Thrift, N. (2000) 'Some new instructions for travellers: the geography of Bruno Latour and Michel Serres', in M. Crang and N. Thrift (eds), *Thinking Space*. London: Routledge. pp. 281–301.

Buller, H. (2004) 'Where the wild things are: reflexions on the evolving iconography of rural fauna', *Journal of Rural Studies*, 20: 131–141.

Buller, H. and Morris, C. (2003) 'Farm animal welfare: a new repertoire of nature–society relations or modernism re-embedded?', *Sociologia Ruralis*, 43 (3): 1–22.

Castree, N. and MacMillan, T. (2001) 'Dissolving dualism: actor-networks and the reimagination of nature', in N. Castree and B. Braun (eds), *Social Nature: Theory, Practice and Politics*. Oxford: Blackwell. pp. 208–224.

Cater, C. and Smith, L. (2003) 'New country visions: adventurous bodies in rural tourism', in P. Cloke (ed.), *Country Visions*. Harlow: Pearson. pp. 195–217.

Cheney, J. (1994) 'Ecofeminism and the reconstruction of environmental ethics', in K.J. Warren (ed.), *Ecological Feminism*. London: Routledge. pp. 158–178.

Cloke, P. (1989) 'Rural geography and political economy', in R. Peet and N. Thrift (eds), *New Models in Geography, Volume One*. London: Unwin Hyman. pp. 164–197.

Cloke, P. (1997) 'Country backwater to virtual village? Rural studies and the cultural turn', *Journal of Rural Studies*, 13 (4): 367–375.

Cloke, P. (2003) 'Knowing ruralities', in P. Cloke (ed.), *Country Visions*. Harlow: Pearson. pp. 195–217.

Cloke, P. and Jones, O. (2001) 'Dwelling, place, and landscape: an orchard in Somerset', *Environment and Planning A*, 33: 649–666.

Cloke, P. and Perkins, H.C. (2004) 'Cetacean performanc[e] and tourism in Kaikoura New Zealand', *Environmen[t] and Planning D: Society and Space*,

Compassion in World Farming (2003a) *Stop – Look – Listen: Recognising the Sentience of Farm Animals*. Petersfield: Compassion in World Farming Trust.

Compassion in World Farming (2003b) *Understandin[g] Animals: Putting Animal Sentience on the Educationa[l] Agenda*. Conference, 10 May 2003, King's College London.

Convery, I. and Bailey, C. (2002) *Farmers and th[e] Farmed*. Paper given at the 'Farming' session of th[e] 'Emotional Geographies' conference, 23–25 September Lancaster University.

Eden, S., Tunstall, S. and Tapsell, S. (2000) 'Translatin[g] nature: river restoration as nature-culture', *Environmen[t] and Planning D: Society and Space*, 18: 257–273.

Emel, J. and Wolch, J. (1998) 'Witnessing the anima[l] moment', in J. Wolch and J. Emel (eds), *Anima[l] Geographies: Place, Politics and Identity in th[e] Nature–Culture Borderlands*. London: Verso. pp. 1–26.

Evans, N. and Yarwood, R. (1995) Livestock and landscape. *Landscape Research*, 20: 141–146.

Evans, N. and Yarwood, R. (2000) 'The politicisation of livestock: rare breeds and countryside conservation', *Sociologia Ruralis*, 40: 228–248.

Fitzsimmons, M. (2004) 'Engaging ecologies', in P. Cloke, M. Goodwin and P. Crang (eds), *Envisioning Human Geography*. London: Arnold. pp. 30–47.

FitzSimmons, M. and Goodman, D. (1998) 'Incorporating nature: environmental narratives and the reproduction of food', in B. Braun and N. Castree (eds), *Remaking Reality: Nature at the Millennium*. London: Routledge. pp. 194–220.

Franklin, A. (1999) *Animals and Modern Cultures. A Sociology of Human–Animal Relations in Modernity*. London: Sage.

Franklin, A. (2002) *Social Nature*. London: Sage.

Gaita, R. (2003) *The Philosopher's Dog*. London: Routledge.

GeneWatch (2003) 'Genetically modified and cloned animals. All in a good cause? www.genewatch.org.

Goodman, D. (2003) 'The quality "turn" and alternative food practices: reflections and agenda', *Journal of Rural Studies*, 19: 1–7.

Hacking, I. (2000) 'Our fellow animals', *New York Review of Books*, June: 20–26.

Haraway, D. (1988) 'Situated knowledges: the science question in feminism and the privilege of partial perspective', *Feminist Studies*, 14 (3): 575–599.

Harvey, D. (1996) *Justice, Nature, and the Geography of Difference*. Oxford: Blackwell.

Holloway, L. (2001) 'Pets and protein: placing domestic livestock on hobby-farms in England and Wales', *Journal of Rural Studies*, 17: 293–307.

Ingold, T. (1993) 'The temporality of landscape', *World Archaeology*, 25: 152–174.

Ingold, T. (1997) 'Life beyond the edge of nature? Or, the mirage of society', in J. Greenwood (ed.), *The Mark of the Social*. London: Rowman and Littlefield. pp. 231–252.

Jones, O. (2000) 'Inhuman geographies: (un)ethical spaces of human non-human relations', in C. Philo and C. Wilbert (eds), *Animal Geographies: New Geographies of Human–Animal Relations*. London: Routledge. pp. 268–291.

Jones, O. (2003) '"The Restraint of Beasts": rurality, animality, actor network theory and dwelling', in P. Cloke (ed.), *Country Visions*. London: Pearson Education. pp. 450–487.

Jones, O. and Cloke, P. (2002) *Tree Culture: The Place of Trees, and Trees in their Place*. Oxford: Berg.

Jowit, J. (2004) 'Eat less meat and you'll help save the planet', *Observer*, 14 March, p. 10.

Latour, B. (1993) *We Have Never Been Modern*. Hemel Hempstead: Harvester/Wheatsheaf.

Laurier, E. and Philo, C. (1999) '"X" morphising: review essay of Bruno Latour's Aramis or the Love of Technology', *Environment and Planning A*, 31: 1047–1071.

Lockie, S. (2004) 'Collective agency, non-human causality and environmental social movements: a case study of the Australian "Landcare movement"', *Journal of Sociology*, 40 (1): 41–58.

Lorimer, J. (2004) 'UK biodiversity conservation and the co-construction of postnatural landscapes', *Journal of Social and Cultural Geography*, forthcoming.

Lowe, P., Clark, J., Seymour, S. and Ward, N. (1997) *Moralizing the Environment: Countryside Change, Farming and Pollution*. London: University of London Press.

Lynn, W.S. (1998a) 'Animal, ethics and geography', in J. Wolch and J. Emel (eds), *Animal Geographies: Place, Politics and Identity in the Nature–Culture Borderlands*. London: Verso. pp. 280–297.

Lynn, W.S. (1998b) 'Contested moralities: animals and moral value in the Dear/Symanski debate', *Ethics, Place and Environment*, 1 (2): 223–244.

Mabey, R. (2003) 'Nature's voyeurs', *Guardian Review*, 15 March, pp. 4–6.

Macnaghten, P. and Urry, J. (1998) *Contested Natures*. London: Sage.

Mansfield, B. (2003) 'Fish, factory farming, and imitation crab: the nature of quality in the seafood industry', *Journal of Rural Studies*, 19: 9–21.

Matless, D. (2000) 'Versions of animal–human: Broadland, *c.*1945–1970', in C. Philo and C. Wilbert (eds), *Animal Geographies: New Geographies of Human–Animal Relations*, London: Routledge. pp. 115–140.

Merton, R. (1976) *Sociological Ambivalence and Other Essays*. New York: The Free Press.

Milbourne, P. (2003) 'Hunting ruralities: nature, society and culture in "hunt countries" of England and Wales', *Journal of Rural Studies*, 19: 157–171.

Morris, C. and Evans, N. (2004) 'Agricultural turns, geographical turns: retrospect and prospect', *Journal of Rural Studies*, 20: 95–111.

Murdoch, J. (2001) 'Ecologising sociology: actor-network theory, co-construction and the problem of human exemptionalism', *Sociology*, 35: 111–133.

Murdoch, J. (2003) 'Co-constructing the countryside: hybrid networks and the extensive self', in P. Cloke (ed.), *Country Visions*. Harlow: Pearson. pp. 263–282.

Murdoch, J. and Day, G. (1998) 'Middle-class mobility, rural communities and the politics of exclusion', in P. Boyle and K. Halfacree (eds), *Migration into Rural Areas: Theories and Issues*. Chichester: John Wiley & Sons. pp. 186–199.

Murdoch, J. and Pratt, A.C. (1993) 'Rural studies: modernism, postmodernism and the "post-rural"', *Journal of Rural Studies*, 9 (4): 411–428.

Philo, C. (1995) 'Animals, geography, and the city: notes on inclusions and exclusions', *Environment and Planning D: Society and Space*, 13: 655–681.

Philo, C. (1998) 'Animals, geography, and the city: notes on inclusions and exclusions', in J. Wolch and J. Emel (eds), *Animal Geographies: Place, Politics and Identity in the Nature–Culture Borderlands*. London: Verso. pp. 51–71.

Philo, C. (2000) 'More words more worlds: reflections on the "cultural turn" and human geography', in I. Cook, D. Crouch, S. Naylor and J.R. Ryan (eds), *Cultural Turns/Geographical Turns: Perspectives on Cultural Geography*. Harlow: Prentice Hall. pp. 26–53.

Philo, C. and Wilbert, C. (2000) 'Animal spaces, beastly places: an introduction', in C. Philo and C. Wilbert (eds), *Animal Spaces, Beastly Spaces: New Geographies of Human–Animal Relations*. London: Routledge. pp. 1–34.

Plumwood, V. (1993) *Feminism and the Mastery of Nature*. London: Routledge.

Plumwood, V. (2002) *Environmental Culture: The Ecological Crisis of Reason*. London: Routledge.

Proctor, J.D. (1996) 'Whose nature? The contested moral terrain of ancient forests', in W. Cronon (ed.), *Uncommon Ground: Rethinking the Human Place in Nature*. New York: W.W. Norton.

Ridley, J. (1998) 'Animals in the countryside', in A. Barnett and R. Scruton (eds), *Town and Country*. London: Jonathan Cape. pp. 142–152.

Scott, A., Christie, M. and Midmore, P. (2004) 'Impact of the 2001 foot-and-mouth disease outbreak in Britain: implications for rural studies', *Journal of Rural Studies*, 20: 1–14.

Sellick, J. (2004) 'The intimacies of space in human and non-human animal encounters', *Journal of Rural Studies*, forthcoming.

Shoard, M. (1980) *The Theft of the Countryside*. London: Maurice Temple Smith.

Smith, M. (2002) 'The "ethical" space of the abattoir: on the (in)human(e) slaughter of other animals', *Human Ecology Review*, 9 (2): 49–58.

Spencer, T. and Whatmore, S. (2001). 'Editorial: biogeographies: putting life back into the discipline', *Transactions of the Institute of British Geographers*, NS 26: 139–141.

Tapper, R. (1988) 'Animality, humanity, morality, society', in T. Ingold (ed.), *What Is an Animal?* London: Unwin Hyman. pp. 47–62.

Thrift, N. (1996) *Spatial Formations*. London: Sage.

Thrift, N. (1999) 'Steps to an ecology of place', in D. Massey, P. Sarre and J. Allen (eds), *Human Geography Today*. Cambridge: Polity Press. pp. 295–352.

Thrift, N. (2000) 'Afterwords', *Environment and Planning D: Society and Space*, 18 (2): 213–256.

Thrift, N. (2001) 'Still life in nearly present time: the object of nature', in P. Macnaghten and J. Urry (eds), *Bodies of Nature*. London: Sage.

Thrift, N. (2004) 'Still life in nearly present time: the object of nature', in P. Cloke (ed.), *Country Visions*. Harlow: Pearson. pp. 308–328.

Tovey, H. (2003) 'Theorising nature and society in sociology: the invisibility of animals', *Sociologia Ruralis*, 43 (3): 198–215.

Ufkes, F.M. (1998) 'Building a better pig: fat profits in lean meat', in J. Wolch and J. Emel (eds), *Animal Geographies: Place, Politics and Identity in the Nature–Culture Borderlands*. London: Verso. pp. 241–258.

Ward, N. (1999) 'Foxing the nation: the economic (in)significance of hunting with hounds in Britain', *Journal of Rural Studies*, 15 (4): 389–403.

Watts, M.J. (2004) 'Enclosure: a modern spatiality of nature', in P. Cloke, M. Goodwin and P. Crang (eds), *Envisioning Human Geography*. London: Arnold. pp. 48–64.

Whatmore, S. (1993) 'On doing rural research (or breaking the boundaries)', *Environment and Planning A*, 25 (2): 605–607.

Whatmore, S. (1997) 'Dissecting the autonomous self: hybrid cartographies for a relational ethics', *Environment and Planning D: Society and Space*, 15: 37–53.

Whatmore, S. (1999) 'Rethinking the "human" in human geography', in D. Massey, P. Sarre and J. Allen (eds), *Human Geography Today*. Cambridge: Polity Press. pp. 22–41.

Whatmore, S. (2000a) 'Heterogeneous geographies: reimagining the spaces of N/nature', in I. Cook, D. Crouch, S. Naylor and J.R. Ryan (eds), *Cultural Turns/Geographical Turns: Perspectives on Cultural Geography*. Harlow: Prentice Hall. pp. 265–273.

Whatmore, S. (2000b) 'Agricultural geography', in R. Johnston, D. Gregory, G. Pratt and M. Watts (eds), *The Dictionary of Human Geography*, 4th edn. Oxford: Blackwell. pp. 10–13.

Whatmore, S. (2002) *Hybrid Geographies: Natures, Cultures, Spaces*. London: Sage.

Whatmore, S. (2003) 'Culturenatures: Introduction: more than human geographies', in K. Anderson, M. Domash, S. Pile and N. Thrift (eds), *Handbook of Cultural Geography*. London: Sage. pp. 165–167.

Whatmore, S. and Hinchcliffe, S. (2003) 'Living cities: making space for urban nature', *Soundings: Journal of Politics and Culture*, January (This version downloaded PDF from www.open.ac.uk/socialsciences/habitable_cities/habitable_citiessubset/)

Whatmore, S. and Thorne, L. (1998) 'Wild(er)ness reconfiguring the geographies of wildlife', *Transactions of the Institute of British Geographers*, NS 23 (4): 435–454.

Whatmore, S. and Thorne, L. (2000) 'Elephants on the move: spatial formations of wildlife exchange' *Environment and Planning D: Society and Space* 18: 185–203.

Wilkie, R. (2002) *The Emotional Paradox of Livestock Production*. Paper given at the 'Farming' session of the 'Emotional Geographies' conference, 23–25 September Lancaster University.

Wise, J.M. (1997) *Explaining Technology and Social Space*. London: Sage.

Wolch, J. and Emel, J. (eds) (1995) 'Bringing the animals back in', *Environment and Planning D: Society and Space*, 13: 632–636.

Wolch, J. and Emel, J. (eds) (1998) *Animal Geographies: Place, Politics and Identity in the Nature–Culture Borderlands*. London: Verso.

Wolch, J., Emel, J. and Wilbert, C. (2003) 'Reanimating human geography', in K. Anderson, M. Domash, S. Pile and N. Thrift (eds), *Handbook of Cultural Geography*. London: Sage. pp. 184–206.

Woods, M. (1998a) 'Mad cows and hounded deer: political representations of animals in the British countryside', *Environment and Planning A*, 30 (4): 1219–1234.

Woods, M. (1998b) 'Researching rural conflicts: hunting, local politics and actor networks', *Journal of Rural Studies*, 14 (3): 321–340.

Woods, M. (2000) 'Fantastic Mr Fox: representing animals in the hunting debate', in C. Philo and C. Wilbert (eds), *Animal Spaces, Beastly Places*. London: Routledge pp. 182–202.

Wrennall, S. (2002) *Echoes of Emotion; Transformations in the 'Place' of Cumbrian Foot and Mouth Culled Farms*. Paper given at the 'Farming' session of the 'Emotional Geographies' conference, 23–25 September, Lancaster University.

Wylie, J. (2003) 'Landscape, performance and dwelling: a Glastonbury case study', in P. Cloke (ed.), *Country Visions*. Harlow: Pearson. pp. 136–157.

Yarwood, R. and Evans, N. (1998) 'New places for "Old Spots": the changing geographies of domestic livestock animals', *Society and Animals*, 6: 137–166.

Yarwood, R. and Evans, N. (1999) The changing geography of rare livestock breeds in Britain', *Geography*, 84: 80–91.

Yarwood, R. and Evans, N. (2000) 'Taking stock of farm animals and rurality', in C. Philo and C. Wilbert (eds), *Animal Geographies: New Geographies of Human–Animal Relations*. London: Routledge. pp. 98–114.

Young, R. (2003) *The Secret Life of Cows*. Farming Books and Videos.

C Sustainability

14

The road towards sustainable rural development: issues of theory, policy and practice in a European context

Terry Marsden

INTRODUCTION: THE SOCIAL NATURE OF RURAL DEVELOPMENT – FROM DENIAL TO DIVERSITY

The deep crisis of the conventional food and agricultural sector, associated with growing health risks, environmental loss, overproduction of low-quality products and the decline in number of producers and farm workers, is continuing to render the public perception of the agricultural sector within Europe as detached, oversubsidized and unattractive. Even many aspiring politicians and their officials, whether in Brussels, member state governments or regional authorities, regard it as a 'minefield' or 'graveyard' area with little chance of positive political gain or success given the history of policy and market failure, coupled with the slow, labyrinthine operation of the EU's Common Agricultural Policy (CAP) 'reform'.

But despite these macro-conditions, and indeed because of their very origin in the contradictions associated with the agro-industrial model of rural development (see Marsden, 2003), we have to recognize that from a sustainable development perspective – from the point of view of long-term, public needs – the agricultural sector cannot be residualized or left in a state of public denial. From a sustainability perspective, its social, economic and physical role is a central element in achieving a more sustainable society, both for rural and urban publics. Indeed, despite the growing literature by rural researchers on the 'post-productivist' countryside (see Goodman, 2004), it cannot be denied that, from a perspective of sustainability, the socio-ecological sphere of agriculture needs to be re-inserted into broader sustainability debates.

Recent detailed case study research in Europe (van der Ploeg, Chapter 18, this volume; Alonso Mielgo et al., 2001; Knickel, 2001) suggests that there is strong evidence that a new, broader rural development paradigm is beginning to take hold in different parts of rural Europe. This is the case at policy levels, where there are indications that the spread of the bovine spongiform encephalopathy (BSE) crisis to the European continent might have substantially speeded up the undercurrent of rural development policy measures that were prudently initiated with the publication of the *Future of Rural Society* in 1988 and strengthened in subsequent CAP reforms (see CEC, 1988; Potter, 1998). None the less, the paradigm shift is most clearly illustrated by developments in the reality of the European countryside, where increasing numbers of farm families are adopting strategies to strengthen their farm livelihoods alongside fundamentally different lines from the previous modernization approach (Renting and van der Ploeg, 2001).

However, as this chapter shows, this is both a highly competitive and contested process; one that has to compete with the maintenance of an agro-industrial model of agricultural development that continues to devalue and subsume the primary production sector through the adherence and propagation of liberalization and globalization logics, as expressed in the context of World Trade Organization (WTO) talks. Essentially, the political-economic 'position' of agriculture does not face in one direction at present. There are contested dynamics, between a rationalistic agro-industrial model that is largely detached from rural society and uses new technologies (not least genetically modified organisms) and scale economies to produce mass foods for mass markets (see Murdoch et al., 2000), and a variety of so far under-theorized alternative rural development 'counter-movements', which are struggling to create organizationally different types of food supply chains and broader rural development options (Marsden et al., 2000; McMichael, 2000) and otherwise reinforce the multifunctional role of agriculture within rural society. The latter are based upon the encouragement of diversity and specificity, whereas the former are based upon standardization and consumer and corporate retailer-led flexible specialization.

In order to progress more sustainable development, there is a need to consider how the prevailing (and potentially new) conditions are based upon sets of emerging scientific, social-scientific and political-economic relationships. The diverse set of rural development practices that are currently being taken up by rural actors may only be understood and strengthened on the basis of new theoretical frameworks that go decisively beyond the postulates of the previously dominant approach of agricultural modernization and industrialization. The economic mechanisms underlying newly emerging rural development practices – such as the associated reductions in transaction costs and the capture of extra value-added through quality specifications – and the new relations with wider society in which they are embedded are two important points of attention (van der Ploeg et al., 2000). In addition to this, so it seems, the new rural development practices (potentially) embody highly relevant ways for reconstituting nature–society relations within rural development, which may facilitate a re-embedding of farming practices in the local ecology.

In this chapter, I wish to explore the broader role of rural and environmental social sciences within this highly contested and binary developmental terrain. Most of all, it will examine how agroecological and ecological modernization approaches to rural development may be conceptually applicable and useful to the European case, and how this can then be substantiated theoretically and empirically. This may provide a more adequate basis for wholesale policy change in (and public revaluation of) the agro-food sector, as well as further substantiate viable ways for a realignment of agricultural production with wider society – rural and urban alike. This is now recognized by many scholars as essential if sustainable agricultural and rural development is to progress from a position of fanciful definition to real delivery. However, for this to be progressed it is necessary to develop some key empirical and conceptual parameters to strengthen the claims of 'the new rural development paradigm'. We have to recognize that, as Goodman cautions us:

> In its binary formulation, the crisis of industrial agriculture is absolutely central to theorisations and prognoses of a new European rural development paradigm. Yet this crisis, its social and spatial patterns, and the ways in which the 'old' might shape the 'new' is taken as a foregone conclusion, the unexamined assumption is that farmers and other rural actors face an unyielding imperative to embrace the new model … We can accept that the modernisation paradigm does indeed have an uncertain future, but more actor-oriented and behaviourally grounded research is needed to clarify the multi-faceted nature of this 'crisis', whether or not this designation is merited, and if so, its likely evolution over time and space. In this light, a more modest approach to the contours of a successor model and the role of AFFNs might first see their present efflorescence as innovative responses to the ongoing struggle for rural livelihoods. (2004: 11)

The rest of this chapter will address both the question of the vibrant maintenance of the agro-industrial model and the emergence, however uneven, of the new rural development paradigm. It will argue that to fully understand and to progress the latter more conceptual refinement and building is needed. This also needs to take place in the competitive context of both models operating and *co-evolving* over time and space. In the next section, we will see how, indeed, the agro-industrial model has itself been in a process of mutation as the various types of crisis have affected it and, as a response, how it has adjusted its competitive and regulatory features. This is a process of attempting to continually sustain the unsustainable (see Buttel, Chapter 15, this volume). The following sections will review and postulate some of the key features of the distinctive rural development model which is based upon alternative and systemic agroecological and ecological modernization perspectives, incorporating autonomous, distinctive and new socio-technical features. These, in turn, confront, challenge and compete with the agro-industrial model. This is, in a range of perspectives, a true 'battlefield of knowledge' operating over both time and rural space.

THE REALIZATION OF DIFFERENT CONCEPTUALIZATIONS OF SOCIAL NATURE AND RURAL DEVELOPMENT: UNDERSTANDING THE OBSTACLES

Before we explore these issues in more depth, it is necessary to examine how some of the most prominent social-science conceptualizations have tended indeed to avoid social nature and rural development questions and, by default, to socially and politically deny agriculture a central place in their anticipated development logics. Moreover, and interestingly, these social-science approaches have shown no sign of adopting a significant, radical break or autonomy with the agro-industrial model, despite its deepening crisis. Indeed, as the crisis has deepened so have the attempts by these bodies of social science to accommodate many of its regulatory forms.

We can suggest three bodies of socio-political, rural social science that are currently obscuring and constraining a more effective rural development dynamic from taking hold.

The mutable agro-industrial model: the appropriation and subsumption of nature

The agro-industrial model of food production continues to be upheld and reinforced by an alliance of agricultural economists and biological scientists who are still legitimizing the intensification and scale economies imported from American and Australasian modernization theory. This sets a precedent upon reducing the costs and prices of primary products through the continued adoption of technological advancements, corresponding reductions in production costs and continued scale enlargement to reach economies of scale. Large areas of rural space are unable to compete in this 'race to the bottom' scenario and, hence, are forced to rely upon the state for more and more support, which, in turn, acts only to moderate the effects of this treadmill (Marsden, 1998). Under these conditions, rural and agricultural nature continues to be seen as an obstacle to be overcome, something to transform along industrial lines and logics.

The agro-industrial mode is strongly associated with the continued efforts of producers and manufacturers to reduce and/or regularize the importance of nature in the food production process. The disconnection of food production from nature is mediated by two interrelated processes: appropriation – that is, the attempt to replace previously natural production processes by industrial activities – and substitution – or the attempt of industrial capitals to replace natural products in the food system with industrially produced substitutes (Murdoch et al., 2000). While these trends are long-standing, they have taken on a new mutation as the negative social, environmental and economic 'externality effects' have become much more publicized.

The bureaucratic 'hygienic' mode: accommodating the consumer countryside

While the agro-industrial logic is still in a fairly dominant position in global agricultural terms (such as represented in the WTO talks), it has been revised and complemented in some respects by a parallel, more bureaucratic mode of regulation and perspective. This has attempted to 'correct', in highly sophisticated and technocratic ways, the agro-industrial model rather than construct viable alternatives for it. It has done this by accommodating and assuming much of Beck's and Giddens's notions of the 'risk society' (Beck, 1992; Giddens, 1998), by attempting to put into place, in a highly interventionary and bureaucratic fashion, policies that effectively 'police' the food and rural production systems in ways that seem to make them more *hygienic* and environmentally safe production and consumption spaces.

These trends are most noticeably associated with the regulatory effects of the spreading BSE affair, with, for example, national governments and the EU putting into place highly complex 'safeguards' within the regulation of key places in the food supply chains of beef and lamb (see Chapter 1, this volume). These, however, have the effect of ratcheting up the regulatory costs of small producers and food processors and reducing market entry into the most lucrative supply chains (often linked to the main corporate retailers) to those larger producers and processors who can more easily meet the demanded quality criteria that are now set. This further exacerbates the market failure associated with the agro-industrial model by compounding it with higher, hygienic regulatory costs. Somewhat ironically, however, these costs affect disproportionately those producers and processors who may have been marginal to the causes of the health risks in the first place (for example, less intensive producers and small abattoirs sourcing locally). These innovations in supply-chain management and regulation are deemed to be in the public interest, but from a rural development and, indeed, an environmental view, they further constrain the room for manoeuvre of many of the primary producers to realign farming with ecology.

This hygienic mode is also supplemented with what has been termed the post-productivist countryside logic (see Marsden et al., 2001), which has encouraged many of the more recently located ex-urban residential groups to view farming as 'a dirty business', and environmental agencies and pressure groups to more strictly police farmers who may be liable to pollute their cherished consumption spaces (Lowe et al., 1997). Like food safety, environmental regulations result in sharp increases in regulatory costs and associated 'obligatory investment rounds' (Ward, 1993). To some extent, the same mechanism is also associated with animal welfare regulations.

From the point of view of the farm, we can see that the hygienic mode can attack and constrain the producer both vertically, through the food supply chain, and horizontally, through the local revaluation of the countryside environment by ex-urban residential and amenity groups. Of course, all this is done in a spirit of public and state concern for 'the protection of the environment' and for 'cleaning up agriculture'. The irony of this mode of regulatory action and consumer-driven politics is that it is leading to the very opposite of further liberalization. It creates a kind of Foucauldian Panopticon around the agricultural and agro-food sector in that it is increasingly watched, monitored and regulated at arm's length by teams of new agents, such as planners, environmental health officers, retailer category managers and buyers etc.

It is important to realize the consequences for aspects of social nature of this bureaucratic-hygienic mode, and its social science-consumption orientation. It tends to hold particular consequences for distancing farmers, and rural dwellers more generally, from their natures, by fragmenting these into particular and highly regulated components. This is probably best illustrated through the process of *schematization* whereby a panoply of agro-environmental, anti-pollution, local planning, countryside stewardship, animal welfare, anti-overgrazing schemes are introduced as specific palliatives to particular aspects of the inherently multivariate (and systemic) natural crisis surrounding industrial forms of agriculture (see Lowe et al., 1990; OECD, 1989).

This has the effect of 'breaking up' the problem into different technical and bureaucratically convenient palliative packages; one which empowers the regulators and attempts to quell public concern, at least in the short term. It also further constrains farmers in ways that tend often to directly counterpoise the other more productivist policy levers. This makes it all the more difficult for producers to treat what they know as a more environmentally sensitive, holistic approach to 'farming nature', and blocks off many of the more integrated measures that might be taken at farm level to realign farming and nature. It is also at odds with encouraging cultural heritage and diversity, through, for example, threatening several gastronomic and culinary traditions by the hygienic attempts to guarantee and secure 'germ-free' food.

The growth of a profound regulatory burden as a response to the crisis in the industrial mode of agro-food tends, therefore, to strengthen the economic and political power of agro-industrial interests (including the large retailers and manufacturers). Both private and public interest forms of regulation are used (see Marsden et al., 2000) to 'clean up' the industrial system in the ostensible 'public interest'. This has the added consequence of further constraining the real potential for integrated agricultural development as well as providing economic barriers to market entry for many of the smaller producers and processors. For consumers, it also allows the disconnections and distanciations between production and consumption to conveniently continue, with an encouragement that 'safety' comes before, or indeed is a substitute for, sustainability. A growth industry for rural environmental scholars has been found in the evaluation of such hygienic schemata.

The relativist mode: the social construction of nature and the assumption of the rural gaze

Despite the much more all-encompassing character of the social constructivist perspectives on human–nature relations, and their considerable attractiveness to numerous groups of human geographers, sociologists and anthropologists over the past decade (see Redclift and Woodgate, 1997), there are considerable gaps in the approach from the point of view of rural development. Considerable emphasis has been placed upon challenging the more modernist stance that has tended to reduce natural relationships to their social bases. This has led authors, such as

Goodman (1999), to suggest new concepts (such as corporeality, relational ethics and metabolic reciprocities) to at least begin to provide an improved symmetry with (what is perceived as) the modernist social stance (e.g. Bonanno et al., 1994). Much of this has also (somewhat overzealously) co-opted an actor-network theory approach, which has focused upon exploring ways of highlighting human–nature interactions (for a sober reflection of this trend, see Murdoch, 2001).

There is no doubt that this literature has progressed human–nature relations and given greater theoretical and empirical focus to how nature is actively constructed through both social and natural spheres. From a rural development perspective, however, it still provides more questions than answers. It tends to downplay the actual power relations in which actors operate – for example, in food supply chains – and it too readily accepts that nature is both constructed and holds somewhat undefined agency. The BSE crisis has been a classic case of how nature can 'hit back' but it has done so in particular social, political and economic circumstances and in specific ways that have been strongly associated with the described bureaucratic mode.

A major problem with these perspectives has been their failure to generate the very analytical focus many of its advocates espouse. It still needs to begin to provide a real and more substantive micro-analytical account of how social and natural struggles interact in particular places and at particular times. Much of the discussion has, rather, focused upon the broader, aggregated analysis of such areas as alternative social food movements (such as organics) or the role of agro-industrial science in shaping natural interactions.

In short, these perspectives have yet to link with the realities and materialities of agricultural ecologies *in situ*. They have yet to really respond to Norgaard's challenge of exploring the co-evolution of the natural and the social in agroecological relationships (Norgaard, 1994), or test how such relationships may indeed be constructed. *A major question remains: if the materiality of nature is constructed, how and who (which local actors, which experts, which politicians) should be involved in reconstructing it given the damage now exhibited?* So far this question is far too normative for social constructivist rural social science; and there is a sense that it is just too risky a business to go beyond the growing identification of diversity, rather than explore how that diversity is created out of social and political struggles. Indeed, the current state of agricultural nature in Europe is an active and constructed thing; but how can these actant energies be given a more sustainable direction?

REMOVING THE CONCEPTUAL OBSTACLES: THE AGROECOLOGICAL AND ECOLOGICAL MODERNIZATION PERSPECTIVES

Agroecological perspectives

To build upon, rather than completely deny, social constructivist perspectives requires the need to creatively integrate agroecological approaches. Agroecology emerged in the 1980s as an interesting attempt to establish a scientific basis for alternatives to industrialized agriculture, avoiding the resource degrading tendencies with which the latter has become associated (Hecht, 1987). The agroecological framework draws upon different intellectual traditions and disciplines, including peasant studies, ecology and environmentalism, and development theory. While the theory of agroecology was at first mainly developed in the context of agricultural development in the South (Altieri, 1987), in recent years it has been spreading to the United States (Gliessman, 1990) and Europe (Alonso Mielgo et al., 2001).

The agroecological perspective is highly promising in providing a much clearer normative direction to the analysis of human–nature relations in rural development, while also recognizing the very active and symmetrical nature of nature. Sevilla Guzmàn and Woodgate (1999), synthesizing the literature on agroecological approaches as well as pointing to the distinct differences between its framework and the more restrictive 'farming systems' approaches, identify the following key parameters as key elements of the agroecological approach.

Crisis of modernity

Agroecology implies an alternative definition of sustainability from which is generated an ecologically, rather than industrially, oriented discourse. A central concept is 'co-evolution'.

Co-evolution

Unlike 'symmetry', as proposed by social constructivist thought, agroecology refers to the reliant co-development or co-evolution of society and natural factors (Norgaard, 1994). It is recognized that farming systems essentially result out of co-production, the ongoing interaction, mutual transformation and dependency between humans and nature – that is, between the social and the natural. It is through co-production that both the natural and the social are unfolded into particular forms and relations (see also van der Ploeg, 1997). The agro-industrial model has thoroughly changed the nature of co-production and disrupted many interdependencies between the natural and the social, thereby reducing the renewability of both. The question becomes how reversible processes can be put in place to regenerate interdependency over time.

Local farmers' knowledge systems

Local peasant or indigenous knowledge needs to be seen (both in the North and the South) as significantly different from standard scientific knowledge (and its 'hygienic' equivalents – see above) in that they are embedded in local ecology and encoded in culture rather than theoretical and abstract notions. Science is context- independent, yet agriculture is actually defined not just by its bio-physical context, but also with reference to its localized socio-political elements. It is based upon the interdependent accumulation of local, natural and social resources, practices and knowledges. These are not just about maintaining 'old cultures', but they involve the creation of defence systems to enable a constant replenishment of knowledge systems in the face of the dominating 'eco-technocratic discourses' associated with globalization.

Endogenous potential

All agro-social systems have their own endogenous potential; however, it is a matter of how these get articulated and valorized through social and political processes to what extent they are actually effectuated. The social aspects of endogenous potential refers to local knowledge systems, but also to struggles of local groups to resist, propose and actively construct alternatives to industrial modernization and to their capacity to develop social networks to enable these. The ecological dimensions are to be found in the promotion of diversity of agroecological systems and a strengthening and valorization of local ecological specificity.

Collective forms of social action

Social reproduction needs to be realigned (and historically in many instances did coincide) with the ecological management of biological systems (Gliessman, 1990; Toledo, 1990). This involves the construction of collective forms of social action, which attempt – either explicitly or implicitly – to challenge the industrial modes of production, consumption and circulation. The social construction of new patterns of circulation involves the (re-)building of networks that, to varying degrees, are composed of citizens of both the centre and the periphery, the urban and the rural, of producers and consumers. This can occur across and through supply chains as it can through different types of space. Agroecology is thus founded upon new forms of associationalism.

Systemic strategies

The social factors that are at work (for example, ethnic, epistemological, ethical, religious, political, economic, gender-based) cannot be seen in isolation. They must be interrelated within an overall understanding of society. Also, the broad range of bio-physical factors, such as water, soil, solar energy, and plant and animal species, must be conceived according to the ways they interact not only among themselves but also with social factors. This involves an understanding of energy, material, money and knowledge flows generated in processes of production, consumption and circulation within and between systems. Ethical decisions are at the heart of progressing more systemic sustainable systems.

Ecological and cultural diversity

Agroecology aims not only to celebrate cultural and natural diversity but also to progress it and materialize it in new social and natural forms. This means abolishing the notion of a dominant ethnocentric agricultural development model and accepting and promoting diverse pathways of development built upon cultural and physical complexity and richness. This marks agroecology out as distinct from both the globalization logics of the agro-industrial model and its bureaucratic-hygienic counterpart, but also from the diversity 'for its own sake' approaches of the postmodernists and constructivists. It also calls into question the notion of a single, detached price-driven market for foods. Social progress and the progress of agriculture have to be realigned in ways that maintain this diversity.

Sustainable societies

These parameters of agroecology begin to reposition agriculture as a key driver for achieving sustainable societies more generally. Rather than a further segregation of agro-food production from rural society, agroecology proposes to reinforce interlinkages and foster the multifunctionality of farming systems. As Sevilla Guzmàn and Woodgate argue:

> Sustainable societies can only be constructed on the basis of sustainable, locally relevant agricultures ... implying a complete rejection of the homogenizing tendencies of the neo-liberal, global modernization project and the re-direction of co-evolution towards more sustainable ways of living that are based upon the endogenous potential of an infinite diversity of locally relevant agro-ecosystems. (1999: 304)

Europeanizing agroecology: the quest of autonomous ecological modernization

However critical one might be about arguments associated with the rhetoric of sustainability, it is also the case that writers have pointed to the development, albeit fledgling in some cases, of a more ecologically modernizing agenda built upon a diverse theoretical base (see Buttel, 2000; Mol, 2000; Murphy, 2000). This is centrally associated with a European perspective on the development of clean industrial technologies and with the ways in which environmental coalitions and movements begin to affect reluctant state agencies. So far, they have rarely been applied to the rural sphere (see Buttel, 2000; Frouws and Mol, 1999), having been confined to the focus upon 'win-win' solutions in the business context (see Andersen and Massa, 2000). However, the degree to which we might be entering a new phase of modernization which would be much more autonomously 'ecological' is becoming a central theme in rural and agro-food studies (see, for instance, Goodman, 1999); and the rural domain is becoming a central field for exploring the role and meaning of social nature debates (Milbourne, 2003). This could represent a new surge of creative and critical connections between environmental social theory as represented by the broad church of ecological modernization, on the one hand, and a more open and pluralistically engaged rural sociology on the other. It is to this exploration that this chapter wishes to contribute.

A major question here begins: *to what extent are we seeing the arrival of a more autonomous ecologically modernizing process operating in Europe and, as part of this, through rural development trends specifically*? Secondly, *if we believe that this is a viable question, which conceptual development and empirical realities does it suggest with respect to the rural sphere*? With specific reference to rural development, it is suggested here that we are witnessing the development of an ecological modernization process which is significantly different and autonomous in its character from the earlier industrial, twentieth-century modernization process.

As Frouws and Mol theoretically delineate:

> The ecological modernization theory analyses possibilities for a process of 're-embedding' economic practices – in view of their ecological dimension – within the institutions of modernity. This modern 're-embedding process should result in the institutionalization of 'ecology' in the social practices and institutions of production and consumption. The institutionalization of ecological interests in production and consumption processes, and thus the redirection of these basically economic practices into more ecologically sound ones, involves an 'emancipation' or differentiation of ecology. The differentiation of an ecological rationality and an ecological sphere, both becoming *relatively independent from their economic counterparts*, is the logical next step. (1999: 271; my emphasis)

This process is analytically challenging in the sense that at one and the same time it is necessary to distinguish the differentiation of ecological 'spheres' – that is, making analytical space for considering relatively *autonomous ecological spheres* so as to study how ecological actions and practices are becoming steadily institutionalized in the central institutions of modernity, but doing so without completely labelling these as distinct areas in society, to be simply identified empirically. *The process of emancipation from the strictly economic sphere, and the gradual re-embedding of ecology in the institutions of economy, is a central aspect of ecological modernization, creating the spaces for an ecological as well as economic rationality.*

This socio-ecological postulate requires detailed assessment in terms of its relevance to rural development, and it raises a theoretical potentiality for developing a more robust sustainable rural development paradigm (see van der Ploeg et al., 2000; Marsden et al., 2003).

We should, of course, also recognize that this does not deny the maintenance of a conventional economic rationality, working as it may do to limit these new ecological, social, economic and political spaces. However, as some ecological theorists propose, it does question the longevity of what Buttel (2000) refers to as the more established rational frameworks of the 'treadmill of production' and the 'growth machine', which have by no means completely disappeared. Arguably, however, they are in many places, less in their ascendancy and subject to internal and external crisis tendencies (such as food scares, legitimization and ethical concerns, pollution incidents and long-term health problems, and not least the increasing and uneven global regulation of greenhouse gases), as well as deep structural tendencies which are seen as increasingly contradictory. Despite these internalized problems in the governmentality of a strictly economic rationality, it does not seem to reduce the possibility of strong reactive ('backlash') politics from taking ground back from the ecologically modernizing agenda. This is, for instance, one interpretation of the US Bush Administration and its arguments against the signing of the Kyoto agreement and the renewed faith in the domestic exploitation of oil reserves (Watts, 2002). Nevertheless, social-ecological debates and discourses are gaining ground in different guises and through different types of social, political and economic practices. While these may confront the former structural changes, associated with globalization of corporate capital, for instance, they are by no means simply marginal or subject to the marginalization effects of corporate states, firms and their social and political logics. Rather, and from a rural perspective, they may suggest a new centrality for many of the features of rural life that the earlier industrially based modernization process tended to marginalize: for instance, aspects of agroecological development as part of rural development (see Jokinen, 2000; Rannikko, 1999), the development of decentralized and more sustainable rural communities as a central part of settlement hierarchies, and more mobile and ICT-based sharing of experiences in rural and urban areas (Andersson, 2003). Indeed, we might suggest that one central element of ecological modernization is the very redefinition of the spatial and social balances between the more mobile urban and rural living experiences and frameworks; and it also concerns the re-alignment, more specifically, between nature, quality, region and local producers and consumers for a more ecological rural resource base.

Some may see these notions as going too far. Moreover, we have to recognize that there are significant distinguishing features between what Christoff (1996) and Toke (2002) depict as forms of 'weak' and 'strong' forms of ecological modernization, suggesting a caution about both the direction and pace of 'autonomous ecologism'. Nevertheless, we need to analytically explore the modalities between the economic and ecological rationalities, on the one hand, and the uneven development of 'weak and 'strong' ecological modernization tendencies, on the other. In addition, this challenges us to provide ecological modernization with a more robust theoretical basis; one that deepens the political-sociological perspective of ecological modernization such that the way forward would at least include not just:

> empirical debates over the potentials and limits of environmental engineering and industrial ecology, but rather [would] deepen the links to political-sociological literatures which will suggest new research problems and hypotheses. (Buttel, 2000: 64)

We see here then a significant theoretical and empirical challenge both for rural sociology, on the one hand, and the broader field of environmental social science on the other. Both could benefit from at least engaging with the emergence of ecological modernization theory, if only to address in a restricted sense the very longevity and resilience of the twentieth century 'growth machine' and 'treadmill of production'. More optimistically, such a deeper engagement could begin to set the coordinates for mapping Frouws and Mol's (1999) more ecologically embedded and conceptually autonomous model of ecological modernization.

Rural development becomes in this context a potentially rich sphere to assess these ecologizing tendencies, and to test the contingencies and frameworks involved in rebuilding a more viable and robust rural development perspective which at least begins to suggest how such ecological modernization notions might be more effectively progressed. Indeed, just as with the industrial mode of modernization, a new ecological paradigm also needs a viable, critical and normatively engaging social science. This is particularly the case with ecological modernization given, as we shall see below, the spatially variable and context-dependent ways in which it actually expresses itself. What seems clear is that there is a lack of coherence in ecological modernization, one which can be partially or contradictorily adopted by national, regional and local governments, and one which may need particular confluences of strategic and local interests and actors to operate in new and innovative ways.

Looking at the recent rural sociological, and particularly the environmental social science literature, one begins to see this tendency being reported. One further key question is how far and fast will it travel, and how do we best equip ourselves as scholars to have a growing relevance in understanding and mediating its required and contested knowledges?

There are some significant questions here. We should perhaps now recognize that the scholarly challenge has begun to change; for it is no longer sufficient to just critically examine the problems of such ageing regimes. We need to be reconstructing as well as de-constructing models and frameworks which suggest *how things could work in different and more socio-ecological ways over space and time*. We need to visualize and articulate how actions and cases at one level can build up into significant projects of change more broadly.

By doing this we can begin also to integrate some of the over-drawn dichotomies and debates between political economy and actor strategies and networks, social constructivism and realism, globalization and localism, economistically determined productionism and culturally confined consumptionism. We need more engaging conceptual and theoretical formulations which help and guide us to not only make sense of the 'new ruralities' which confront us, but also allow a more *reconstructivist* as well as *critical interpretive* role in both rural and environmental social science.

Exploring and taking forward ecological modernization debates may, therefore, be one significant way of meeting the growing need within rural studies for improved theoretical engagement. As Frouws and Mol conclude in their analysis of the ecological modernization of Dutch agriculture:

> Environmental sociology seems, in this respect, indeed capable of being a 'formative power' in the development of rural sociology. Its contribution is especially valuable to clarify the all-embracing impact of the environmental question on the technological and institutional reconstruction of agriculture, and to address its social, political and economic implications ... This brief exploration also revealed, between the lines, the socio-political contestability and indeterminate outcome of ecological modernization as a political program for agricultural change in the Netherlands. (1999: 286)

CONCLUSIONS: INTEGRATING AGROECOLOGY AND ECOLOGICAL MODERNIZATION IN RURAL DEVELOPMENT – CREATING AUTONOMOUS RATHER THAN DEPENDENT DEVELOPMENT

No one would suggest that these principles of agroecology or ecological modernization can be straightforwardly superimposed upon any one agriculture or any one agrarian space in a European context. And with regard to agroecology, their resonance has, thus far, been mainly applied and explored in a southern regional and rural context, while ecological modernization has been the preserve of a distinctively European agenda. Nevertheless, *these principles need to be used as important assessments of how far current developments in alternative rural development practices in the European context represent sustained and, moreover, autonomous moves towards a more agroecological system.* Do the current multifarious struggles adopted by farmers and rural actors across Europe in the face of the crisis of the agro-industrial model add up to any form of agroecological break with the past? Do they engender some of the features of the agroecological model? Do they represent viable pathways to sustainable agricultural and rural development and, indeed, deserve more state encouragement? Or, alternatively, do we see them as merely partial compromises with the mutating agro-industrial model, as Goodman (2004) argues, struggling to create new markets that may disconnect and fragment the social and the natural in new, more pervasive ways?

In beginning to progress answers to these questions it is necessary to distil the conceptual implications that can be drawn from a series of empirical projects recently conducted in Europe. These begin to witness the competitive development of the *rural development paradigm* as an autonomous rather than dependent process; one that is attempting to carve out its own paradigm of thinking and working which is distinct from the prevailing agro-industrial model and its attendant scientific and economic rationalities. As outlined in the introductory sections to this chapter, this is made all the more difficult by the mutable tendencies of the agro-industrial model, not least in its attempts to occupy the food 'quality' agenda and introduce new 'clean' technologies which some would regard as a branch of (weak) ecological modernization. Nevertheless, recent research indicates such autonomous tendencies with regard to the social and technical reorganization of agro-food, and its re-linking and re-localizing with local rural development. We can identify several key features.

Associationalism

Alternative food supply chains are based upon new networks and clusters of activity which cross-cut conventional supply chain relationships. These can involve a variety of face-to-face relationships, proximate or extended at-a-distance relationships between producers, processors and consumers (see Knickel, 2001; Renting et al., 2003).

De-bureaucratization: working by ecological and social conventions as opposed to technical and imposed regulation

Working in alternative quality food niches and networks creates in some cases opportunities for actors to work outside the conventional *hygienic* mode of regulation. This means that actors and producers are forced to develop working relationships based upon new social and ecological conventions, and, even more importantly, sustain these over time and space in the context of the dominant commercial regulatory mode.

Clustered cooperation and networking

Opportunities are thus created for clustered cooperation, like the collective purchasing of farm inputs, monitoring and evaluating quality conventions, or sharing marketing and processing facilities. In some instances regional state agencies financially support these ventures (Marsden et al., 2003).

Retro-innovation

A central feature of the new rural development paradigm concerns the appearance and evolution of a distinct form of innovation and knowledge system: retro-innovation. This can be defined as developing expertise that combines elements and practices from the past (often chronologically before the modernization paradigm) and the present, and configures these elements for new and future purposes (see Stuiver and Marsden, under review). Such examples include the cluster of innovations associated with the development of environmental cooperatives in the Netherlands (Renting and van der Ploeg, 2001; Wiskerke, 2003). Here 'old' forms of silage making, feed production and manure application tend to ecologically restore soils as well as maintain production levels. Another example concerns the redevelopment of nineteenth century recipes for meats based upon the re-invention of traditional breeding, curing and processing techniques in the new organic food networks spreading across rural mid-Wales.

Ecological entrepreneurship

Whilst many of the principles of ecological modernization (as outlined above) tend to assume a new, and to some extent leading role for institutions and state agencies, it is now being discovered through empirical work that a significant aspect of the growth and importance of alternative networks in agro-food relates to a distinctive form of 'ecological entrepreneurship', whereby key actors begin to construct and sustain networks, retro-innovate and relate to the new emerging consumer markets. These new breeds of entrepreneur are able to mix alternative ecological strategies with new market-based developments. These actors are an important element in the process of ecological modernization, and indeed may resist the intervention of the state and institutions in the development of these new economic ventures (see Marsden and Smith, 2005).

Spatial capture of locale

These distinctive features associated with the new rural development paradigm begin to establish, in variable combination, what we might call new 'spatial havens', new agrarian and rural spaces that provide a freedom to develop these socio-technical niches without the burden of conventional regulations and controls. Here, new social and environmental landscapes begin to take shape; they are free in the sense that they are released from the 'grip' of the conventional modernization project. Here new labour and community practices can begin to take hold and, indeed, be experimented. Under these conditions, it is the actual degree of distance from the agro-industrial regulatory system that producers and processors can create, which then influences the degree of success of these new initiatives. To enable their development, therefore, such networks not only need to create their own internal mechanisms (that is, associationalism, network-building, retro-innovation and ecological entrepreneurship), but also need external, institutional support to assure and defend their social and spatial boundaries – boundaries that can sustain the benefits of exit and 'lock-out' from the prevailing agro-industrial and state-supported landscapes which surround them. Hence, the battleground of knowledge which typifies the process of competition between the 'new' and 'old' rural development paradigms holds a distinct and highly competitive social and agrarian geography – a geography that is indeed a physical battlefield in itself, and one which then helps to shape either the success or failure of the assemblage of social and technical constructions which constitute the new rural development paradigm. This is, therefore, as Goodman (2004) alludes, a highly spatially contingent process.

A central part of the assessment of the degree to which a more agroecological or ecological modernizing rural development paradigm is indeed emerging should be, therefore, the theorizing of rural development practices in Europe with reference to detailed case studies and wider empirical research relating to their impacts. We also need a better understanding of the extent to which these practices may gain momentum in ways that significantly – both socially and spatially – challenge the agro-industrial, hygienic logics outlined earlier. It is only by doing this that scholars will be able to assess how aspects of social nature become more effectively embedded in rural development.

The most recent *hygienic mode* adopted by much of European food policy has tended to break or at least fracture the environmental question into specific boxes, making it more difficult to make the above holistic connections. The new parameters outlined in conclusion here represent real ecologically modernizing quests by rural actors to recreate those holistic connections. Past state intervention, even that which pertained to 'the environment', has tended to discourage the bringing together of many of the connections necessary for agroecology/ecological modernization to flourish. This has represented a pervasive form of *constructed marginalization* of the latter over much of the past 20 years in the European context. As a result, it has been all the more difficult for key actors to harness the spatial, natural, regional and knowledge-based resources necessary to progress real rural development. This has led to contradictions and to rural development being seen as a contradictory set of practices that only score partially on the Sevilla Guzmàn and Woodgate (1999) or Frouws and Mol (1999) criteria and principles. However, there are significant exceptions and clusters.

What we witness are actors in rural development struggling to form new associations and linkages at the same time as European state policies are tending to regulate nature through highly bureaucratic and legalistic means. Under these conditions, we can witness different, highly context-dependent struggles to create alternative rural development practices. None of these fit ideally with all of the parameters of the agroecological or ecological modernizing paradigm, but some make significant inroads and adopt at least some of the principles. All of them demonstrate new rural development practices that attempt to define and build upon nature in different and innovative ways. All are reincorporating nature back into rural development and agro-food developments. This means establishing new associations, links with the locality and region, and new types of food supply chains.

The aim here has been to bring together two strands of rural and environmental social science (agroecology and ecological modernization) in ways that deepen our critical understanding of how nature is becoming reconstituted through the pursuit of rural development practices. The ascendancy and conceptual development of both of these perspectives is needed in the quest to understand the new rural development paradigm. This is one important method of building upon the critical understanding of the agro-industrial, hygienic and social constructivist approaches, all of which have, in their own ways, tended to deny and fragment the social and natural embeddedness and autonomy of alternative rural development. If sustainable rural development is to have a chance in rural Europe, we will need a more robust theoretical and empirical base from which to progress the citation of interesting examples to the reconstruction of a new round of real rural and sustainable *modernization*.

REFERENCES

Alonso Mielgo, A.M., Sevilla Guzmàn, E., Jiménez, N.M. and Guzmàn Casado, G. (2001) Rural development and ecological management of endogenous resources: the case of mountain olive groves in Los Pedroches *comarca* (Spain), *Journal of Environmental Policy and Planning* 3, pp. 163–175.

Altieri A. (1987) *Agroecology. The Scientific Basis of Alternative Agriculture*. Boulder, CO: Westview Press.

Andersen, M.S. and Massa, I. (2000) Ecological modernization: origins, dilemmas and future direction, *Journal of Environmental Policy and Planning* 2 (4), pp. 337–345.

Andersson, K. (2003) Regional development and Structural Fund measures in two Finnish regions. In K. Andersson, E. Eklund, L. Granberg, T. Marsden (eds), *Rural Development as Policy and Practice. The European Umbrella and the Finnish, British and Norwegian Contexts*. SSKH Skrifter No. 16. Research Institute, Swedish School of Social Science, University of Helsinki, Helsinki.

Beck, U. (1992) *Risk Society: Towards a New Modernity*. London: Sage.

Bonanno, A., Busch, L., Friedland, W.H., Gouveia, L. and Minzone, E. (eds) (1994) *From Columbus to ConAgra: The Globalization of Agriculture and Food*. Lawrence, KS: University of Kansas Press.

Buttel, F. (2000) Ecological modernisation as social theory, *Geoforum* 31, pp. 57–65.

CEC (1988) *The Future of Rural Society*. Commission of the European Community: Brussels.

Christoff, P. (1996) Ecological modernisation, ecological modernities. *Environmental Politics* 5 (3), pp. 476–500.

Frouws, J. and Mol, A. (1999) Ecological modernisation theory and agricultural reform. In H. de Haan and N. Long (eds), *Images and Realities of Rural Life:*

Wageningen Perspectives on Rural Transformations. Assen, Netherlands: Van Gorcum. pp. 269–286.

Giddens, A. (1998) *The Third Way: The Renewal of Social Democracy*. Cambridge: Polity Press.

Gliessman, S.R. (1990) *Agroecology: Researching the Ecological Basis for Sustainable Agriculture*. New York: Springer Verlag.

Goodman, D. (1999) Agro-food studies in the 'age of ecology': nature, corporeality, bio-politics, *Sociologia Ruralis* 39, pp. 17–38.

Goodman, D. (2004) Rural Europe redux? Reflections on alternative agro-food networks and paradigm change, *Sociologia Ruralis* 44 (1), pp. 3–16.

Hecht, S. (1987) The evolution of agroecological thought. In M.A. Altieri (ed.), *Agroecology: The Scientific Basis of Alternative Agriculture*. London: IT Publications. pp. 1–20.

Jokinen, P. (2000) Europeanisation and ecological modernisation: agri-environmental policy and practices in Finland, *Environmental Politics* 9 (1), pp. 138–167.

Knickel, K. (2001) The marketing of Rhöngold milk: an example of the reconfiguration of natural relations with agricultural production and consumption. *Journal of Environmental Policy and Planning* 3 (2), pp. 123–136.

Lowe, P., Clark, J., Seymour, S. and Ward, N. (1997) *Moralising the Environment. Countryside Change, Farming and Pollution*. London: UCL Press.

Lowe, P., Marsden, T.K. and Whatmore, S. (eds) (1990) *Technological Change and the Rural Environment*. London: David Fulton.

Marsden, T. (1998) New rural territories: regulating the differentiated rural spaces, *Journal of Rural Studies* 14, pp. 107–117.

Marsden, T.K. (2003) *The Condition of Rural Sustainability*. Assen, Netherlands: Van Gorcum.

Marsden, T., Milbourne, P., Kitchen, L. and Bishop, K. (2003) Communities in nature: the construction and understanding of forest natures, *Sociologia Ruralis* 43 (3), pp. 238–256.

Marsden, T.K., Flynn, A. and Harrison, M. (2000) *Consuming Interests. The Social Provision of Foods*. London: UCL Press.

Marsden, T.K., Renting, H., Banks, J. and van der Ploeg, J.D. (2001) The road towards sustainable agricultural and rural development, *Journal of Environmental Policy and Planning* 3 (2), pp. 75–83.

Marsden, T.K. and Smith, E. (2005) Ecological entrepreneurship: sustainable development in local communities through quality food production and local branding, *Geoforum* 36, pp. 440–451.

McMichael, P. (2000) The power of food, *Agriculture and Human Values* 17, pp. 21–33.

Milbourne, P. (2003) Nature–society–rurality: making critical connections, *Sociologia Ruralis* 43 (3), pp. 193–196.

Mol, A. (2000) The environmental movement in an era of ecological modernisation, *Geoforum* 31, pp. 45–56.

Murdoch, J. (2001) Ecologising sociology: actor-network theory, co-construction and the problem of human exemptionalism, *Sociology* 35, pp. 111–133.

Murdoch, J., Marsden, T. and Banks, J. (2000) Quality, nature and embeddedness: some theoretical considerations in the context of the food sector, *Economic Geograph* 76, pp. 107–126.

Murphy, J. (2000) Ecological modernisation, *Geoforu* 31, pp. 1–8.

Norgaard, R. (1994) *Development Betrayed*. Londo Routledge.

OECD (Organisation for Economic Co-operation ar Development) (1989) *Agricultural and Environment Policies. Opportunities for Integration*. Paris: OECD.

Ploeg, J.D. van der (1997) On rurality, rural developme and rural sociology. In H. de Haan and N. Long (eds *Images and Realities of Rural Life: Wageningen Perspe tives on Endogenous Rural Development*. Asse Netherlands: Van Gorcum. pp. 39–73.

Ploeg, J.D., van der, Renting, H., Brunori, G., Knickel K Mannion, J., Marsden, T., de Roest, K., Sevil Guzmán, E. and Ventura, F. (2000) Rural developmen from practices and policies towards theory, *Sociologi Ruralis* 40, pp. 391–408.

Potter, C. (1998) *Against the Grain: Agri-environment Reform in the United States and the European Unio* Wallingford: CAB International.

Rannikko, P. (1999) Combining social and ecological sus tainability in the Nordic forest periphery, *Sociologi Ruralis* 39, pp. 394–410.

Redclift, M.R. and Woodgate, G.R. (1997) Sustainability an social construction. In M.R. Redclift and G.R. Woodgat (eds), *The International Handbook of Environment Sociology*. Cheltenham: Edward Elgar.

Renting, H. and Ploeg, J.D. van der (2000) Reconnectin nature, farming and society: environmental coopera tives in the Netherlands as institutional arrangemen for creating coherence, *Journal of Environmental Polic and Planning* 3 (2), pp. 85–101.

Renting, H., Marsden, T.K. and Banks, J. (2003 Understanding alternative food networks: exploring th role of short food supply chains in rural developmen *Environment and Planning A* 35 (3), pp. 393–411.

Sevilla Guzmàn, E. and Woodgate, G. (1999) *Fro Farming Systems Research to Agroecology. Technica and Social Systems Approaches for Sustainable Rura Development*. European Commission, Report 45/98.

Stuiver, M. and Marsden, T.K. (forthcoming) The promise o retro-innovation for rural development: theorising beyon the modernisation paradigm. Wageningen: Department o Social Sciences, Wageningen University. Mimeo.

Toke, D. (2002) Ecological modernisation and GM food *Environmental Politics* 11 (3), pp. 145–163.

Toledo, V. 1990. The ecological rationality of peasant production. In M. Altiesi and S. Hecht (eds), *Agroecology and Small Farm Development*. Boca Raton, FL: CRC Press. pp. 53–60.

Ward, N. (1993) The agricultural treadmill and the rural environment in the post-productivist era, *Sociologia Ruralis* 33, pp. 348–364.

Watts, M. (2002) Green capitalism, green governmentality, *American Behavioural Scientist* 45, pp. 1313–1317.

Wiskerke, J.C.S. (2003) On promising niches and constraining sociotechnical regimes: the case of Dutch wheat and bread, *Environment and Planning A* 35 (3), pp. 429–448.

15

Sustaining the unsustainable: agro-food systems and environment in the modern world

*Frederick H. Buttel**

INTRODUCTION

For more than two decades scholars, policy-makers and citizens groups have employed and debated the notion of agricultural sustainability. There is general agreement that sustainability can be defined, in part, as environmental or ecological soundness of the production system or agro-food commodity chain. There is also general agreement that while ecological soundness is the core dimension of sustainability, economic viability and 'justice' or 'equity' are also significant dimensions (Allen, 1993).

Sustainability is employed variously as a critique and sometimes as a defence of prevailing agricultural practices and institutions. On one hand, some observers insist that modern agriculture tends to fall far short of a satisfactory level of sustainability; not only is agriculture seen as compromising environmental quality, but in addition agriculture is seen to be socio-economically unsustainable (in that net income and returns to equity capital continue to decline) and unjust (in that 'family-type' production is being marginalized while industrial-scale production is in ascendance, and in that agriculture is increasingly based on poorly paid immigrant or minority labourers and involves a lack of attention to the 'rights' of animals). For these groups, agriculture's lack of sustainability is employed as a critique of the agro-food system. Other groups, however, particularly mainstream farm organizations, commodity groups, agricultural researchers and agricultural ministry officials, consider most farming operations to be sustainable in the sense that they are largely based on renewable natural resources and provide food and fibre efficiently and cheaply. These groups may have concerns about the economic sustainability of agriculture, but these concerns are not specific to the viability of household production units. These groups seldom see agriculture as being unsustainable from the vantage point of social justice.

The fact that there can be such disparate views about agricultural sustainability demonstrates that sustainability is a contested notion. Some observers believe that because sustainability is vague and includes heterogeneous, if not conflicting and incommensurable, indicators, the notion should be jettisoned. My view, however, is that agricultural sustainability remains a useful notion – partly on account of the fact that it is not so static or formulaic that it cannot be debated and contested. Sustainability is also useful as an intellectual notion and policy framework because there is perhaps no other concept that better reflects a desirable vision for agro-food systems in the future. Sustainability will serve quite adequately as a measuring stick and roadmap for the future if two points are stressed in its definition. First, most views of sustainability explicitly or implicitly include the criterion that sustainable systems are those that are capable of being continued *indefinitely or for some reasonably long period of time*. Generally, this notion of indefiniteness has a primary, if not an exclusively, ecological referent; thus, it is presumed

*After a long, courageous battle with cancer, Frederick H. Buttel passed away in January 2005, before this Handbook went into production. We are honored to present here one of the last publications of a most prolific and influential author in the field of rural studies.

that unsustainable systems are those that will at some future point be rendered dysfunctional because of ecological problems, risks, or calamity. I would suggest that such an imagery – that 'unsustainable systems' must inevitably undergo progressive environmental degradation such that they collapse and no longer 'function' – is highly problematic.[1] I will suggest below that contemporary unsustainable agro-food systems have an enormous amount of staying power and a capacity to maintain their dynamics despite socio-environmental costs and risks and despite the depletion of the 'natural capital' of agroecosystems. Thus, what we need to understand social scientifically is how it can be the case that unsustainable agriculture is capable of being sustained for lengthy periods, if not indefinitely. Second, sustainability is not so much an end-point as it is a process; the notion of sustainability reminds us that there will always be new ways that agro-food systems can be rendered more ecologically sound, more economically viable and more socially just.

In this chapter I will provide an overview of the issue of agricultural sustainability. My focus will be on the sources and antecedents of agricultural unsustainability. While I will devote some attention to mechanisms by which agriculture's lack of sustainability can be addressed, the overall thrust of my argument is that there is a very strong dynamic toward 'sustaining the unsustainable' in global agro-food systems today. My emphasis will be primarily on the United States, where this dynamic of sustaining the unsustainable is particularly powerful, but material from other developed as well as developing countries will be provided. I will also limit myself to the principal environmental impacts of agriculture. I will thus de-emphasize the environmental impacts of agriculture where these are minor compared to other sources (for example, agriculture's contribution to climate change) and will also not consider how global environmental changes will affect the future of agriculture (but see the brief, but admirable treatment of these topics by Tansey and Worsley, 1995).

A VERY BRIEF HISTORY OF AGRICULTURAL SUSTAINABILITY

The notion of agricultural sustainability emerged in roughly the late 1970s as an outcome of the first major decade of struggle over the status and future of the agricultural environment. Prior to the 1970s, most mainstream environmental groups in North America were generally preoccupied with two types of issues: *metropolitan environmental issues* such as industrial and municipal pollution, water pollution, energy consumption and so on, and *wildland preservation issues* such as wilderness protection and endangered species. At this time North American environmental groups devoted very little attention to problems of the agricultural environment.[2]

Growing interest in the agricultural environment had four major sources. First, in the early 1960s, Rachel Carson's (1962) *Silent Spring* called attention to the ecological problems associated with insecticides and other pesticides. Note, however, that Carson's exposé on pesticides did not pertain mainly to agroecosystems and agricultural workers. Instead, 'silent spring' pertained to dead birds (due to pesticides accumulating up the food chain and interfering with avian reproduction) rather than to degradation of the agricultural environment, reflecting the fact that Carson's emphasis (and the emphasis of the American environmental movement at the time) was on wildlife, wildlands and protected areas. Second, in the early 1970s it was becoming apparent that agricultural run-off and nonpoint pollution accounted for a similar share of water pollution as did manufacturing industry and municipalities. A number of environmental groups thus came to recognize that continued progress toward clean air and water could not be made without dealing directly with the ecological problems of agriculture (Pimentel and Pimentel, 1996). Third, Jim Hightower's (1973) scathing analysis of the US public research system called attention to the fact that the new technologies it produces tended to marginalize family-size farms and to be harmful to the agricultural environment. Fourth, the 1973–74 OPEC oil embargo and energy crisis created considerable concern that modern mechanized and chemical-dependent agricultures were vulnerable to future energy shortages (Pimentel and Pimentel, 1996).

Most of the Western countries experienced a similar series of popularized revelations about the environmental costs of agriculture. While these revelations about agriculture's environmental problems were relatively new, the tendency in Europe is that they were extensions of more longstanding conservation concerns such as wildlife, countryside access, and so on. None the less, by the middle of the 1970s there was growing attention being paid to the environmental consequences and implications of agriculture. A good many scholars, activists and citizens came to see 1970s' agriculture as being ecologically destructive and vulnerable (Lockeretz, 1982). The 1970s thus witnessed a good many declarations that agriculture was in the throes of incipient ecological crisis (Merrill, 1976). Authors such as Allaby and Allen (1974) and Oehlaf (1978) wrote persuasively that modern agriculture was rapidly approaching ecological collapse and the energy crisis was likely to push agriculture into a state of persistent crisis.

Interestingly, the sustainability notion was originally a response to the growing realization during the late 1970s and early 1980s that modern agriculture was *not*, in fact, immediately vulnerable to ecological shortcomings. For example, in the early 1980s the energy crisis blew over, the relative costs of petrochemical inputs returned to their earlier parameters, many of the more toxic pesticides began to be phased out, irrigation water was in tighter supply but still available, and so on. Defenders of modern agriculture hastened to point out that there was no extant ecological contradiction of agriculture – to which environmental critics increasingly responded by saying that, *in the long term, agriculture was still 'unsustainable'*.

It is well known that the concept of sustainability would be propelled to further visibility through its being appropriated by the World Commission on Environment and Development in its renowned work, *Our Common Future* (WCED, 1987). Sustainability became the shorthand for 'sustainable development' – a compromise notion that aimed to appeal to both environmental NGOs and developing country groups, which had clashed over earlier environmental movement icons such as the 'population bomb' and 'limits to growth'. The notion of sustainability is now applied to virtually all realms of economy, environment and society at all imaginable scales – from individual lifestyles to the global carbon economy. Even so, sustainability and sustainable development continue to be thought of substantially in terms of the rural renewable-resource sectors such as agriculture, forestry and fisheries.

Now, 15 years past 'Brundtland' – the common shorthand for *Our Common Future*, referring to Gro Harlem Brundtland, the WCED chair and the book's senior author – it is increasingly apparent that the current unsustainable condition of agriculture is by no means novel. That is to say, most economic sectors and most national economies can be characterized in terms of the sustainability of unsustainability. Agriculture, however, has some particularities, such as its fundamental tie to land, the fact that this fundamental input is immobile in space and cannot be manufactured, and the enormous ecological variability of agroecosystems and agricultural biota. These particularities are reflected in the fact that the antecedents of the unsustainability of agriculture are generally quite specific to this sector, and that the social processes that continue to render sustainable agriculture's unsustainability are also distinctive.

THE UNSUSTAINABILITY OF MODERN AGRICULTURE

There are two common views about the unsustainability of modern agriculture that are often brought forward in discussions of agricultural sustainability in ways that fail to illuminate the causes of these problems or the steps that need to be taken to deal with them. The first is that ecological attitudes or land ethics (or the lack thereof) are major factors in agricultural resource management. Research employing a range of methodologies has repeatedly found that farmers' environmental (or other) attitudes and values bear little relation to their practices and conservation behaviours (see the summary in Buttel et al., 1990: ch. 4). The second such view is that the main structural change in agriculture over the past century or so, and accordingly the main structural root of the lack of ecological sustainability, is the trend toward larger (and fewer) farms. There is growing evidence, however, that while social structural processes (rather than farm operator attitudes) are the principal antecedents of the unsustainability of agriculture, the scale of farm operations *per se* is not the most fundamental factor. Farm size is, to be sure, associated with or related to these other structural processes, but farm size or scale *per se* tends not to be the direct cause of unsustainability.

In this section of the chapter I will outline the three major direct causes of the unsustainability of agriculture: specialization/monoculture and the creation of spatial homogeneity; the intensification of production through expanded use of external chemical, energy and irrigation inputs; and the concentration of livestock in space and the spatial separation of crop and livestock production. In the section that follows I will focus on the social and political processes that lead to the persistence of the low degree of sustainability of modern agriculture.

Before discussing in detail the three overarching direct causes of the tendency of agriculture to be unsustainable, it is useful to note some of the social forces that have led to these tendencies. Specialization, intensification through external inputs and concentration of livestock production have occurred for a variety of interrelated reasons. Much of the impetus for these changes was technological. Public and private agricultural research and development (R&D), for example, led to major advances in mechanization. Specialization – accompanied by monoculture, diminished biodiversity and spatial homogeneity – was a strategy used by farmers to spread the fixed costs of larger and more expensive machines over more acres. Indeed, apropos the earlier comments about farm size and scale, the imperative to specialization led some farms to become larger as the fixed costs of investments were spread over larger acres; in the main, however, farmers specialized by deleting crop and livestock enterprises and by using fewer lines of machinery on the same number of acres/hectares. Further, as I will note shortly, agricultural R&D in the agro-chemical sphere led to the expanded availability of chemical fertilizers and biocides which made large-scale monoculture biologically and socio-economically feasible (Goodman et al., 1987; Goodman and Redclift, 1991); public research also contributed enormously to knowledge about the use and management of these external inputs. Public and private agricultural R&D also resulted in major developments in large-scale confinement technology for hog, poultry, dairy and beef production.

The major socio-ecological changes in agriculture were by no means simply technological phenomena. Ongoing changes in the political economy of agriculture induced the development of new technologies and made it logical for individual agriculturalists to use them. For example, the development of commodity programmes during the Great Depression and shortly after the Second World War subsidized the cultivation of particular commodities (while bypassing most), making large-scale cultivation of supported commodities the optimal strategy for obtaining state support (Goodman and Redclift, 1991). Growing agricultural productivity in the post-Second World War period resulted in a series of state strategies to dispose of surplus agro-food commodities – initially, through foreign aid programmes (Friedmann and McMichael, 1989), and ultimately through 'free-trade' policies (McMichael, 2000). The expansion of the world market in basic agricultural commodities intensified the competition in agriculture, and increasingly dictated that individual farmers aggressively optimize their agricultural operations. In the main, they did so through specialization, monoculture and widespread use of external chemical, petroleum and irrigation inputs. Structural change in agriculture was also driven more generally by the class and labour market dynamics of industrial capitalism. Market forces contributed to much of the social differentiation in agriculture that occurred in the twentieth century and the early twenty-first (Lobao and Meyer, 2001).

The discussion that follows will primarily stress the ecological dimension of sustainability. It should be stressed, however, that each of the three major socio-ecological drivers of the twentieth century trend toward the unsustainability of agriculture has had economic social sustainability dimensions as well as an ecological dimension. This overarching process, which is conveyed best by Cochrane's (1993) notion of the treadmill of technology, not only rendered agriculture ecologically unstable (and thus economically and socially unsustainable). In addition, the condition of twentieth century agriculture, as depicted so vividly by Cochrane's use of the treadmill metaphor, also created the socio-economic conditions suitable for the rapid development and deployment of the technologies of unsustainability.

As noted earlier, market forces within twentieth century capitalism propelled the social differentiation of agriculture, resulting in increasing socio-economic disparities among cultivators and producers. Rising wages due to the growth of trade unions and labour parties, for example, made hired agricultural labour more expensive, and made mechanization – and thus specialization and so on – more logical or imperative. Industrial development attracted farm people and farm youth to industrial and other nonfarm labour markets. Agriculturalists in need of hired labour increasingly turned to ethnic and racial minority group members, many of whom were non-citizens and who were paid very low wages. The spread of the logic of 'flexible' industrial production into the sphere of food and fibre commodities has contributed to sustainability problems by encouraging more livestock and vegetable production to be undertaken through contractual, 'integrated' relations with huge processing and manufacturing firms that are increasingly engaging in cut-throat national and global competition (Bonanno et al., 1994).

Specialization, monoculture and spatial homogeneity in agriculture

At the turn of the twentieth century, more than 75 per cent of American farms raised cattle, milked cows, raised hogs, or raised chickens, and more than 60 per cent of farms raised corn, raised hay, or raised vegetables. By 1997, only 6 per cent had milk cows, only 6 per cent raised hogs, only

5 per cent raised chickens, 23 per cent raised corn and 3 per cent raised vegetables (www.usda.gov/nass/pubs/trends/timecapsule.htm; accessed 24 April 2003). These data underscore the extraordinary trend away from mixed farming and spatial heterogeneity of agriculture and toward specialized agriculture and spatial homogeneity. These trends, in fact, largely occurred over the very brief time period, from the end of the Second World War through the 1960s, in most of the industrial countries. In ecological terms, the specialization of agriculture at the farm or enterprise level has translated into monoculture, the diminution if not cessation of crop rotations, and decreased biodiversity of agro-ecosystems (Brookfield, 2001; Jordan, 2002). As Gliessman (1998) has noted, monoculture and spatial homogeneity of agriculture are the single most important causes of a lack of sustainability.[3] Monoculture and homogeneity undermine the biological integrity of soils and make soils more vulnerable to soil erosion and nutrient run-off. Monoculture and homogeneity also obstruct the biological pathways that close biogeochemical cycles in agricultural systems, rendering them more 'leaky' and more vulnerable to off-site transport of soil, nutrients and chemicals (Jackson, 2002; Tansey and Worsley, 1995: ch. 2; Vandermeer, 1990).

Intensification through use of external inputs

Without the external ecosystemic subsidies of external chemical, energy and irrigation inputs, large-scale monoculture would be infeasible. Expanded use of chemicals and energy is the direct complement of monoculture, and has been driven by essentially the same set of factors. Monoculture enables intensification of agriculture – and thus its apparent high productivity – but also disrupts the major natural cycles that make life possible. Monoculture interrupts the natural cycles that sustain soil fertility – particularly the nitrogen and mineral cycles – and thus leads to a dependence on artificial forms of fertilization. Monoculture also involves the simplification of the species structures of ecosystems and thereby interrupts the biological pathways that enable the regulation of pest and pathogen populations. Accordingly, monoculture increases the dependence of farmers on externally supplied biocides such as insecticides, herbicides and fungicides. In addition, irrigation water and the energy required to deliver irrigation water are external inputs into many agricultural systems, particularly those in semi-arid and arid zones.

Chemical fertilizers, biocides and irrigation are associated with significant sustainability shortcomings (Pimentel and Pimentel, 1996). Fertilizers and biocides are major threats to quality of water (both surface and subsurface) and soil (National Research Council, 1993). Fertilizers and pesticides enter surface and subsurface waters through run-off and infiltration. The use of these chemicals in combination with monoculture tends to lead to soil erosion unless there is aggressive management (such as reduced tillage, buffer strips alongside surface waters) to prevent it. Irrigation is leading to widespread land destruction through salinization (National Research Council, 1993). In addition, the more that nitrogen fertilizer is used and other forms of nitrogen continue to be 'overproduced', the natural denitrification processes of the world's ecosystems are becoming unable to cope, with the excess nitrogen running off into waterways (Tansey and Worsley, 1995: ch. 2).

Concentration of livestock in space, and the spatial separation of crop and livestock production

Earlier some basic data on the distribution of livestock across space in US agriculture were discussed. Since roughly 1945, US animal agriculture, and animal agriculture in most other industrial countries, has been radically transformed. The predominant tendency is that contemporary livestock production occurs on a small percentage of American farms, generally at a relatively large scale. The average herd size for dairy farms is now over 110 cows, and the average hog farm produces over 2,000 hogs per year. Further, US agricultural regions have become highly specialized in the livestock they produce. Hog, dairy and poultry production are also concentrated in a handful of America's 50 states. The only partial exception is that of cattle raising; cattle raising occurs in all of the American agricultural regions, though it is also the case that huge cattle feedlots are concentrated in a few states in the Great Plains. None the less, the specialization of livestock production at the enterprise level combined with regional specialization of livestock production have led to an extraordinary degree of spatial concentration of livestock production. That is, livestock are now produced on a handful of the total number of farms, and their production occurs largely under large-scale confinement conditions on a small proportion of the country's agricultural land surface (McBride, 1997).

There are two highly negative sustainability concomitants of change in the structure of livestock production (Congressional Research Service, 1998; National Research Council, 1993:

ch. 11). The first is that the concentration of livestock in space has made animal manures increasingly a point (as opposed to a diffuse or nonpoint) pollution problem. Large-scale confinement production tends to involve large shares of the nutrients not being returned to the land, and in the process becoming sources of water and air pollution.

The second highly negative implication of change in the structure of livestock production has to do with the spatial separation of crop and livestock production. Due to this profound spatial separation, animal manures are generally becoming unavailable for (or are used only very inefficiently in) nutrient provision in cropping systems. The concentration of livestock in space, and the spatial separation of crop and livestock production, has transformed animal manures from being valuable sources of nutrients to being a waste disposal problem (often euphemistically referred to as a 'waste management challenge'). The general lack of animal manure as a soil amendment has in turn had negative implications for soil quality, soil erosion and dependence on external sources of nutrients.

REPRODUCING THE UNSUSTAINABILITY OF AGRICULTURE

As noted earlier, the unsustainability of agriculture is not accompanied by demonstrable trends toward the social or ecological demise of mainstream agricultures. What are the forces and processes that enable the social and ecological reproduction of the unsustainability of agriculture? In this section of the chapter I will briefly discuss the five most fundamental factors that account for this seeming contradiction.

Agricultural technoscience

In very substantial measure, the public and private agricultural research systems comprise a fundamental mechanism for the social and ecological reproduction of unsustainability in agriculture. To be sure, very few public or private agricultural researchers would see their roles in this way. In the main, they view their roles as responding to farmers' technical needs (either technical needs that are currently manifest or those that are expected to emerge in the future) through providing either public goods or privately appropriable input commodities. In substantial measure, though, one can see that farmers' current and future technical needs tend to revolve around adapting to the social and ecological costs of agriculture's unsustainability. Research on solutions to pest, pathogen, weed control, soil fertility, animal disease problems and animal growth promoting agents are mostly related to the imperative to increase productivity in a context of monoculture, intensification and livestock confinement. In recent years the public and private research establishments have invested heavily in research on reduced tillage practices and precision (or 'site-specific') agriculture in order to patch up the problems caused by large-scale monocultural production.

When I use the imagery of agricultural technoscience 'patching up' problems, I have in mind two closely related but specific hypotheses or propositions. First, the (public as well as private) agricultural R&D process becomes highly oriented to solving the immediately experienced technical problems of large-scale, capital-intensive producers while confining the search for these solutions to those that are consistent with currently existing production systems. Second is the notion that these technical solutions superficially solve or control these problems without addressing their root causes in ecological (or social) unsustainability. Commercial genetic engineering research on agricultural crops has been particularly targeted to solving the technical problems of large-scale monoculture (for example, corn borer and corn rootworm infestations in continuous corn,[4] the difficulties of weed control in monocultural soybean production). Likewise, in the United States the vast bulk of research in the applied animal sciences is devoted to solving the technical problems of large-scale, vertically integrated CAFO[5] production. In recent years there have also been enormous public and private investments in enhancing the capacity of huge integrated CAFO livestock producers to control the water and air pollution that is inherent in large-scale confinement livestock production (Buttel, 2003).[6]

The exoneration of agricultural producers from environmental and other regulations

Historically farmers have tended to be exempted from much of the regulatory oversight that nonfarm industries are obliged to comply with. Farmers have been exonerated from regulation for three different reasons. First, particularly in political economies such as that of the United States that have had a strong yeoman or family farming tradition, farmers have tended to be exempted from statutes such as clean air and water regulation because of their being perceived as struggling small business people who ought not to be 'burdened' with compliance with legislation that was aimed at manufacturing industry and municipalities. Thus, in the US context, federal Clean Air Act and Clean Water Act legislation has involved exemptions for farmers. Under these Acts, regulatory authority was delegated to the states, and then enforced only unevenly.

The second reason that farmers have tended to be exonerated from environmental and other regulations is that most farm production units are small businesses, and the bureaucratic costs of ensuring farmer compliance with regulations were considered to be prohibitive. There are many-fold more farms in the United States than there are manufacturing establishments, and it was considered an ineffective use of regulatory resources to ensure compliance among small (relative to industry) agricultural production units. Third, much of the environmental impact of agriculture has tended to be of a 'nonpoint' rather than 'point' character. Nonpoint pollution such as nutrient run-off is difficult to measure, and regulatory benchmarks are difficult to establish and to do so equitably. Regulation of nonpoint pollution has historically tended to be far less than that of point pollution sources. Indeed, to the degree that there has been a trend toward more strict environmental regulation of agriculture, this is due, in part, to the fact that increasingly larger, more industrial and highly specialized monocultural or CAFO producers are becoming point sources of water and air pollution. Thus, in the United States, the Environmental Protection Agency has decided to implement environmental regulation of concentrated animal feeding operations (so-called CAFOs) even though there is fierce opposition to such regulation among most livestock producers.

Inexpensiveness of the inputs that override unsustainability

In substantial measure the unsustainability of agriculture is masked by the fact that the three principal technologies of intensification – chemical fertilizers, plant protection chemicals and irrigation – remain inexpensive in relative terms. Ahearn's (1998) research on input use and productivity in American agriculture shows that input usage in the early 1990s has continued its long historical trends of expansion of chemical use, decline in the use of land and rapid decline in the use of labour. The overarching reason for these trends is that of relative price shifts, particularly the declining relative prices of chemicals and the rising relative prices of labour. The principal role of the three intensification technologies is to compensate for the interruptions of natural cycles caused by monoculture, spatial homogeneity, the decline of biodiversity, the spatial concentration of livestock production and the spatial separation of crop and livestock production. Inexpensiveness of the intensification inputs enables the ecological and social costs of the lack of sustainability to continue to be masked or compensated for.

Distanciation of producers and nonfarming citizens from the costs of agricultural production

Neither agricultural producers nor nonfarming citizens tend to be spatially proximate to the sustainability problems caused by modern agriculture. One of the core features of agricultural unsustainability is that farm producers are able to externalize many of the costs of intensive, specialized monocultural production through air and water transport of pollutants. In addition, most nonfarming citizens do not directly experience the externalization of the costs of unsustainability.

Distanciation (also referred to as 'distance' or 'distancing') is the degree to which there is 'severing of ecological and social feedback as decision points along [a commodity] chain are increasingly separated along the dimensions of geography, culture, agency, and power. … [T]he concept of distancing highlights the increasingly isolated character of consumption choices as decision makers at individual nodes are cut off from a contextualized understanding of the ramifications of their choices, both upstream and downstream' (Princen et al., 2002: 16). Developments in the means of transport, new production practices (vertical integration, such as subcontracting and outsourcing), and new patterns in the world trading system have given rise to a tendency for social groups to be spatially and socially removed from the areas in which their production and consumption activities interact with the biophysical environment. As Princen (2002: ch. 5) has noted, geography or space is the principal, but not the only, dimension of distanciation. Geographic distanciation refers to the degree to which the sites of decision-making about extraction and production and consumption are far removed spatially. All things being equal, to the degree to which there is high spatial distanciation of extraction and production-consumption, primary resource producers have less awareness of and knowledge about the conditions under which these materials become extracted, valuated, transported and converted into industrial goods. Geographic distanciation thus tends to be accompanied by differences in bargaining power between extractive producers and industrial producers. Geographic distanciation also tends to be associated with lack of ecological feedback on important natural resource management issues at the site of extraction; extractive producers tend to be unaware of the social and environmental conditions of industrial transformation, while industrial firms tend to be unaware (or face less of an imperative to become aware) of the social and environmental conditions of extraction.

Most nonfarming citizens live in large urban places with water pollution abatement facilities in place to ensure that farm pollutants do not contaminate household and commercial water sources. Particularly in the United States, in which there is a profound spatial separation of agriculture from human settlements (which are increasingly in suburbs on the perimeters of major metropolitan areas), nonfarming citizens seldom encounter the homogeneous agricultural landscapes, the salinization of farm land and the growth of CAFOs in agricultural producing regions. As a result, neither farmers nor nonfarming citizens tend to bear the full brunt of the costs of unsustainability. The costs of unsustainability are concentrated among the riverways that drain the great monocultural crop producing regions, and among the hypoxia regions of the seas (especially the Gulf coasts of Louisiana and Texas) adjacent to these waterways (see Buttel, 2003). Thus, only a small handful of citizens – for the example, the nonmetropolitan households who face polluted groundwater – experience these costs directly. The distanciation of the majority of both producers and nonfarming citizens from the immediate effects of unsustainability leads to the tendency that there is little outrage among either the farming or nonfarming communities that these problems occur, and little sentiment to directly address these problems.

Global neo-liberalization and offshore veto of environmental regulation

The late twentieth century and the early twenty-first century have been an era of global neo-liberalization. Global neo-liberalization refers to the temporal coincidence of 'globalization' (the international integration of agricultural and food product markets and growth in the volume of world agricultural trade), 'trade liberalization' ('freer' trade or reduction of state subsidies of agriculture), and the diminution of the size and scope of the state sector (at least with respect to social welfare). Nominally, global neo-liberalization of agriculture involves major changes in national agricultural policies such as the reduction of national 'market-distorting' subsidies of agriculture, the reduction of trade barriers on agricultural goods, the subordination of national food regulatory standard-setting to globally 'harmonized' standards, and especially the removal of import barriers and nontariff barriers to trade. Perhaps most fundamentally, though, global neo-liberalization is a set of pro-corporate policies that are aimed at creating profitable opportunities for private firms in the agro-food and other sectors with the idea that stimulating private investment and encouraging mobility of investment capital will yield social benefits.

Since the ratification of NAFTA in 1994 and WTO in 1995, there has been a definite trend toward increasingly synchronized global markets in most agricultural commodities. The covariation of the implementation of these trade liberalization pacts with the downward trend in farm commodity prices suggests a causal relationship of some sort. Ray et al. (2003) have presented simulations and arguments suggesting that WTO has led to downward pressure on global farm product prices for a complex set of reasons: the 1996 USA Farm Bill's dismantling of the policy machinery for managing supply and supporting prices (and the continuation of these measures in the 2002 Farm Bill), the ratcheting up of decoupled emergency farm subsidies in the United States, the EU's persistent agro-export subsidies and the decline of agricultural tariffs. An equally important impact of these trade pacts is that they make it more difficult to advance alternative policies. Most fundamentally, global neo-liberalization of agriculture involves the actual (or, in the case of the United States, a self-imposed) offshore veto over agricultural policy alternatives that derives from the commitment to trade liberalization agreements such as WTO (and NAFTA). Nelson (2002), for example, points out how US government obligations under the WTO – and perhaps even more fundamentally, the fact that government non-compliance with the WTO agricultural agreement would undermine US leadership in seeking to deepen trade liberalization – might serve to rule out many alternatives for domestic support of and investment in agriculture. The establishment of increasingly hypercompetitive conditions in world agriculture that have led to progressively lower price ceilings for most commodities also serves as an implicit veto over national policy alternatives. As such, they represent a kind of offshore veto of prospective national policies that might seek to de-emphasize productivity increase as the major goal for agricultural policy-making and to stress the ecological, economic and social sustainability of agriculture.

It should be stressed, however, that as much as global neo-liberalization of agriculture is a palpable trend, many of the components of global neo-liberalization are proceeding in highly discontinuous and contentious ways. In practice, NAFTA – and especially WTO – has been far less efficacious in 'liberalizing' agriculture than many proponents and opponents of these policies accept. In the current uni-polar world, in which the USA dominates economically, militarily and politically, the United States tends to get its way to a considerable degree. But while all of the US administrations since the 1980s have favoured trade liberalization and the current administration does so with particular vigour, the current administration has permitted – sometimes even encouraged – protectionism of various sorts (for example, massive emergency payments to farmers in the late 1990s, huge subsidies to producers of cotton, import restrictions on sugar, and many of the farm support provisions of the 2002 Farm Bill). Also, there is still sufficient support for farmers (not to mention steel and textile manufacturers) in the US Congress that it has continued to vote for huge farm subsidies, and seems scarcely more prepared than 50 years ago to let American farms be annihilated through unfettered global market forces. Nearly every world nation wants to be part of the World Trade Organization, but most would prefer that WTO's agricultural rules be substantially changed (with some preferring a scaling back of liberalization while others desire even more liberalization). Yet because almost every nation reasons that it cannot survive economically without being in the world trading system, there does not seem to be a truly viable policy alternative or a viable coalition in opposition to current trade liberalization agreements and policies.

At time of writing there is a level of shared, though differentiated anti-WTO militancy among the Group of 21 developing countries (which derailed the Cancun WTO Ministerial in 2003) and the anti-globalization movement. On one hand, the Group of 21 countries and the anti-globalization movement share some common views about the shortcomings of the WTO; in particular, both groups are strongly opposed to US and EU agricultural subsidies. On the other hand, the two groups have very different aims. The anti-globalization movement, which includes a number of radical sustainable agriculture groups, is generally committed to a fundamental roll-back of the Uruguay Round Agreement on Agriculture (URAA), if not the entire WTO regime. The Group of 21 (G21) countries, by contrast, are essentially seeking implementation of the spirit of the URAA – that their trade access to US and EU markets be increased by phasing out most industrial-country subsidies and protections of agriculture – in order to augment the G21's role in the world trading system. It is not clear that the two groups comprise a potent coalition for addressing neo-liberalization of agricultural policy in a manner that would be consistent with ecological and socio-economic sustainability (see Ray et al., 2003, for an exploration of the national and international policy provisions necessary to do so).

The fact that the G21 countries and the bulk of the developing world are more interested in pursuing the implementation rather than the dismantling of the URAA and WTO says a great deal about both USA hegemony and how USA dominance of the world food regime and the world economy now shapes how developing countries see their interests. The aftermath of 9/11, and especially the Iraq war, have led to recognition of American hyperpower and to the notion that it has become a 'rogue superpower' both militarily and in terms of international relations. It is useful to recognize, however, that American unilateralism in international agreements has a much longer history (see Anderson, 2000) and is not merely a readily reversible policy choice by the Bush Administration. It should also be recognized that the United States's hyperpower stature conveys to it enormous powers to shape global agro-food systems purely through its domestic policies, which have led to the generalization to most of the rest of the world of a set of forces that have dramatically affected food economies. By increasingly setting the global competitive standard, USA power and hegemony have contributed to the global generalization of organizational models of food provision (vertical integration, multiple sourcing, outsourcing, distanciation, raw materials cheapening and industrialization). Neo-liberalization policies are increasingly advanced more through bilateral deals between developing countries and the United States than through WTO and other multilateral regimes. These changes induced through US dominance of the world food economy have served to 'lock in' the developing world as a whole, including the G21, to commitment to the intensive, vertically integrated agricultural model.

VISIONS OF THE ROAD TO SUSTAINABLE AGRICULTURE

It is commonplace for scholars to refer to 'the sustainable agriculture movement' even though for some time it has been apparent that there are many different types of agro-food sustainability movements and strategies (Buttel, 1997). As much as sustainability is a fairly tangible handle on how to understand agriculture and its problems, this has not led to a clear consensus on what sustainability of agriculture means or on what changes must be made to ensure a more sustainable agriculture in the future. There are, in fact, a number of competing visions for how to achieve sustainability in agriculture. I will explore a number of the key ideas that have been developed for achieving sustainable agriculture, and each is the focal point for activism. These major strategies examined here include agro-food localism, contestation of public and private research priorities, increased environmental regulation of agriculture, eco-taxation and related agroenvironmental incentives, and multifunctionality policies. In this section of the chapter I will interrogate each of these visions of the road to sustainable agriculture in terms of whether each is sufficiently robust to address the powerful forces leading to and socio-ecologically reproducing the unsustainability of agriculture.[7] I will discuss multifunctionality at particular length, since multifunctionality brings forth several particularly promising ideas for research and policy-making that deserve greater attention in the future.

Localism and foodshed strategies

One of the most longstanding focal points for activists and citizens interested in enhancing the sustainability of agriculture has been to promote the localization of the food system. This strategy is based on two compelling observations. First, unsustainable agriculture involves 'long', distanciated commodity chains in which nonfarming citizens or 'consumers' have little idea of who produced their food, where it was produced, or the conditions under which it was produced. The 'foodshed' approach to sustainability is based on imagining that such a hypothetical food distribution catchment area is somewhat analogous to a watershed. Foodshed researchers (e.g., Kloppenburg et al., 1996) see the foodshed notion as a framework for understanding the social and spatial character of the food system – particularly its corporate domination and the way that the separation of producers and consumers impoverishes people and communities all along the food chain – in order to be able to change it. The second observation is that while mainstream agriculture tends to be unsustainable, there is a certain quarter of agriculture – its organic or sustainable producers – who are already actively engaged in making agriculture more sustainable. The basic assumption is that the way to make agriculture more sustainable is to build on the efforts of producers who are already making a commitment to agricultural sustainability. The link between the two observations is that efforts to localize food systems are one of the best ways of enhancing opportunities for sustainable producers while simultaneously building links between nonfarming citizens/consumers and sustainable growers.

Localism/foodshed strategies have the advantage that they build on ongoing practices of sustainable farmers and consumers and that the framework is an attractive one for combining research and activism. In the USA it is arguably the case that from an activism and social movement standpoint localism is the predominant wing of the sustainable agriculture movement ensemble. The shortcoming of localism, however, is that it makes few inroads into countering the powerful drivers and reproduction mechanisms that are leading to the sustainability of unsustainability. Localism/foodshed strategies are clearly worthwhile, but their ultimate import may well be limited to creating circumscribed networks of producer–consumer linkages within selected bioregions (DeLind, 2003).[8]

Redirection of agricultural research priorities

It has been noted several times already in this chapter that there is a definite technoscientific dimension of the unsustainability of agriculture. Particular technological forms, such as synthetically compounded fertilizers, chemical biocides, animal confinement technology, and so on, have contributed substantially to agriculture's lack of sustainability. Further, we observed earlier that contemporary agricultural research institutions, both public and private, essentially function to patch up the technical problems experienced most directly and profoundly by the very farmers – large monocultural crop producers and large, industrial-type confinement livestock producers – who are at the leading edge of agriculture's unsustainability. Further, for many, sustainable agriculture or organic agriculture is, first and foremost, a set of technologies. Thus, any realistic vision for sustainable agriculture must address the role of technoscience in sustainability and unsustainability.

For these reasons, many sustainable agriculture advocates have focused their efforts on a critique of mainstream agro-food technoscience, and on advocacy of an alternative research agenda. Supporters of sustainable agriculture have long held that the public agricultural research system's priorities revolve around supporting unsustainable practices. And when public researchers did begin to devote some attention to environmental or sustainability research, they tended to focus on decreasing the externalities of large-scale, monocultural or CAFO producers, rather than focusing on the research priorities of greatest interest to family-scale producers and organic farmers. In recent years the sustainable-agriculture critique of mainstream agro-food technoscience has placed particular emphasis on the negative features of molecular biology research in general and genetically modified (GM) crops and foods in particular (see, for example, Kloppenburg and Burrows, 2001).

It is clearly of the utmost importance for the future of sustainable agriculture that public – and eventually private – R&D priorities be restructured to permit more emphasis on production techniques that involve more biodiversity and that are based on realizing the 'integration efficiencies' (Harwood, 1985) of diverse, locally adapted cropping systems. This said, it is also unlikely to be the case that more appropriate technology alone can be an effective counterforce to the power drivers that are leading to the reinforcement of the unsustainability of agriculture. It could be argued, in fact, that were the institutional conditions, such as national and international policies, to be more propitious for sustainable agriculture, the sustainable-type technology that currently exists, along with that which would be induced in short order, would comprise a satisfactory technological basis for a much more sustainable agriculture.

Environmental regulation of agriculture

Agriculture has traditionally been a 'frontier sector' (Princen, 2002; Princen et al., 2002) in that its peripheral spatial position and its connections to the yeomanry/peasantry enabled producers to externalize environmental and other societal costs on other peoples and regions. This still remains the case to a significant degree in North America. In some of the industrial countries, however, agriculture has increasingly found itself on the radar screen of environmental regulators. None the less, it is generally recognized that if environmental regulatory authorities were to apply to agriculture the same standards and penalties employed in regulation of industrial pollutants, the penalties for noncompliance with regulations would render monocultural and CAFO production very expensive. As an example, there is evidence that CAFO production of hogs is associated with two types of air pollutants – fine particulates (which are associated with lung diseases) and sulphur dioxide (a neurological agent) – which are difficult to abate. Enforcement of air quality regulation with the same aggressiveness that is evident with respect to diesel engine exhaust or municipal air quality would conceivably render CAFO hog production increasingly infeasible (Buttel, 2003).

It seems likely that agriculture will be increasingly expected to comply with general environmental regulations; this trend seems especially likely to the degree that agriculture production units become larger businesses which become increasingly important as point sources of air and water pollution. There are, however, several potential concerns about the regulatory-driven strategy for sustainable agriculture. One concern is that the early twenty-first century is, if anything, an era of the dismantling of 'command-and-control' regulatory practices. There is now under way a global shift toward new types of collaborative and voluntary environmental policy strategies (Jordan and Wurzel, 2003; Mol et al., 2000). Thus, the strict, inflexible regulatory practices that would subject agriculture to the same standards as nonfarm industry are increasingly tending to be phased out in favor of alternative regulatory forms, and these more 'collaborative' and decentralized regulatory forms may be particularly unsuitable for environmental protection in agriculture. Second, unsustainability is by no means confined to large-scale, CAFO-like production units. Stricter implementation of environmental regulation in agriculture would disadvantage many household-scale producers. Third, larger producers, in fact, may be best positioned to make the investments required to comply with stricter environmental regulations. Fourth, agricultural sustainability is not coterminous with a relative absence of air and water pollution. A truly sustainable agriculture means much more than relatively pollution-free production practices and facilities; sustainable agriculture not only minimizes on-site and off-site pollution, but also involves a high degree of 'resource efficiency', a minimization of dependence on finite supplies of nonrenewable resources, and de-distanciation of agro-food systems. The regulatory-driven route to sustainable agriculture alone is unlikely to lead to this more encompassing vision of agricultural sustainability.

Eco-taxation and other forms of market incentives

As noted earlier, one of the contemporary trends in environmental policy-making consists of experiments with alternative environmental policies that are more collaborative or involve market incentives. Incentives are by no means new in the realm of agriculture. Most industrial countries have employed incentives to encourage soil conservation and improved waste management practices, and it is likely that incentives will continue to be employed to enhance the environmental performance of agriculture.

Perhaps the most intriguing type of market incentive programme, however, would be one that shifts the costs of the incentives from the state to the agricultural producer. It is generally agreed that eco-taxation in agriculture could play a major role in inducing a shift toward sustainable agriculture. Significant taxes on the use of particular agro-industrial inputs, such as nitrogen and phosphorous fertilizers and synthetic biocides, could play a major role in reducing the ability of cheap inputs to continue masking the ecological and social costs of modern agriculture.

To some degree there are already significant green taxes in agriculture in some European countries (for example, 15 per cent taxes on agricultural chemicals in Norway). In the United States and most industrial countries, however, green taxes have not yet been implemented. In fact, there are a number of 'negative green taxes' in many American states because agricultural chemicals are often exempted state sales taxes; 'negative green taxes' basically amount to a tax subsidy of the use of these chemicals. Many US advocates of the eco-taxation approach are now pushing for 'environmental tax shifting', which involves repealing tax exemptions on pesticides and fertilizer that subsidize or encourage their use while reducing other taxes (for example, on net farm income) in order to make the shift 'revenue equal' – a crucial consideration in the current political environment of obsession with 'no new taxes' and preoccupation with the efficiency of government programmes. At the present time, however, the tendency for politicians to be skittish about implementing new major sources of tax revenue seems to make it unlikely that eco-taxation will play a major role in agriculture. At some future point, however, eco-taxation on agricultural chemicals represents a promising approach for neutralizing some of the major drivers of the unsustainability of agriculture.

Multifunctionality

Multifunctionality consists of two interrelated notions: the first is that, in addition to production of food and fibre (and other marketable goods), agriculture has a number of *other, mostly non-commodity, outputs*. Agriculture's non-commodity outputs include environmental protection, flood control, ecosystem services, maintenance of landscape or habitat, rural development, maintenance of agricultural heritage or culture, and so on. Non-commodity outputs, because they do not have a market price, tend to be underproduced. Most of the non-commodity outputs of agriculture are also unrecognized. Many supporters of sustainable agriculture argue that agricultural researchers and professionals need to reconceptualize agriculture in terms of its being a set of goods and services produced within, and other activities that occur in the context of, *multifunctional agricultural landscapes* (see the essays in Bland and Buttel, 1994). A second meaning of multifunctionality is that of a *type of agricultural policy*: a policy that involves investments in and payments for the non-commodity functions of agriculture to ensure that they are provided at optimal levels (see OECD, 2001).

For several decades now there has been a growing role played by groups that advocate that agricultural commodity and farm support programmes include provisions that contribute to environmental quality rather than subsidize overproduction and unsustainability (Harvey, 1998). For example, the 2002 US Farm Bill debate, particularly in the Senate, focused not only on how much to invest in conservation programmes; in addition, there was a significant degree of discussion about whether and how there should be a prominent role for what might be called 'green payments' to farmers. Prior to the late 1990s, however, there was not much talk of 'green payments' as such; most of the conservation content of Farm Bills affected whether and how land is farmed.

The notion of green payments – also often referred to as agri-environmental payments – has emerged for several reasons. First, the notion of green payments implies that if we are going to pay or compensate farmers, we should ask for something in return. Thus, in return for subsidies, parliaments and the public ought to expect farmers to do something – essentially to protect the environment, conserve natural resources, provide 'ecosystem services' and so on – in return. This appeals to many groups who feel that farm programmes ought to have 'strings' or accountability. The green payments notion owes its existence as a terminology in part to the new policy discourses that have emerged from the Uruguay Round Agreement on Agriculture (the World Trade Organization's current agricultural agreement) and subsequent global policy debate. Green payments have been advocated for their ability to combine conservation and provision of ecosystem services with farm income support, and for their ability to deliver more benefits more efficiently and to more people.

What would or should green payments pay for? Among the possible 'green things' that we can pay farmers to do would be the following. Some 'green' goods and services are already being provided by farmers, and require no changes in their current management practices. Farmers in certain locations enhance the environment simply by being farmers (for example, maintaining open space for tourism). Other 'green goods' require a change in cropping patterns or management practices; conservation policy can create other environmental goods by encouraging farmers to change their management practices (for example, farmers could improve water quality if they lowered soil erosion rates through adoption of conservation tillage and use of buffer strips along waterways).

With so many arguments in favour of a conservation payments programme, why not immediately adopt one? There are two apparent reasons. The first is that a green payments programme would be difficult to design and administer. For example, a number of political and pragmatic implementation issues would need to be addressed. What is the 'value' of various environmental goods? What emphasis should be placed on conservation versus income support? Should farmers who have previously adopted environmentally friendly practices be given green payments? Should green payments programmes be based on performance or practices? How should agencies monitor and verify that conservation contracts are being complied with? What agencies will have primary enforcement responsibility? Will the practices and environmental goods targeted differ regionally? How should bidding for contracts be carried out? Should there be payment limitations for green payments?

In addition to these admittedly formidable implementation issues, there are broader political issues at stake. One is that around the world today a comprehensive programme of 'green' agricultural payments is known as 'multifunctionality', and multifunctionality is recognized as one component of the European bargaining position in the World Trade Organization negotiations. We live in an era in which the US government sees Europe as insignificant and troublesome, and tends instinctively to see European ideas and proposals as being irrelevant. The basic policy goal of supporters of multifunctionality is to expand the definition of 'green box' policies in the World Trade Organization agriculture agreement to include those that attain multifunctional objectives. European national government advocates of multifunctionality suggest that multifunctionality payments are far preferable to trade barriers, and express willingness to trade off trade barriers for multifunctionality. Multifunctionality is now endorsed by a number of world nations, including much of Europe plus Japan and South Korea. But multifunctionality is rejected by the United States, Canada, Australia and other agricultural exporting powers because these countries prioritize international market expansion over environmental quality and protection of domestic producers. Developing countries tend to be suspicious of multifunctionality because it seems as if it is nothing more than a sophisticated justification for rich countries to continue to subsidize their farmers. Thus, the second reason why multifunctionality as a specific type of policy will encounter formidable obstacles is that this policy is seen as another example of European protectionism.

It seems exceedingly unlikely at this point that the current ('Doha') Round of the WTO negotiations will result in agreement on a multifunctionality agenda. As noted earlier, the possibility looms of an impasse on the Doha Round Agricultural Agreement, and the scramble for a formula that will be a compromise position among the key actors to the Doha Round could very well lead to a multifunctionality-type approach. In addition, Article 20 of the Uruguay Round Agricultural Agreement (the current agriculture agreement of the WTO), which permits what are called 'non-trade concerns' to be included in the next WTO round, represents a 'foot in the door' that is now being approached aggressively by several European countries, particularly the Norwegians (http://odin.dep.no/ld/mf/).

Even if the Doha Round makes no provision for national multifunctionality policies toward agriculture, the fact remains that the stylized type of farm programme envisioned when the WTO was being negotiated in the early 1990s (essentially no domestic production subsidies or export subsidies) has proven to be politically infeasible *as a package* for most countries, especially those that are not major agro-exporting powers. The provisions of the United States's 2002 Farm Bill may be WTO-compliant in a narrow technical sense, but they clearly violate the spirit of the World Trade Organization's agriculture agreement. For the foreseeable future, farm programmes that do not have farm income maintenance as a major objective are not realistic; when income in the agricultural sector is low, payments of some sort must compensate for the difference. The objective – *or at least the stated objective* – of keeping people on the land is also a political necessity. Multifunctionality provides a defensible framework for rationalizing what are likely to be continuing subsidies of farmers and farm sectors.

CONCLUSION

The overarching story of this chapter has been two-fold. First, contemporary global agricultures are largely unsustainable. In the main, though, the unsustainability of agriculture can be 'sustained', or reproduced, for considerable periods of time by way of forces such as agricultural-technoscientific problem-solving, vacuums of environmental regulatory jurisdiction, and distanciation. For the short to medium term, the various actors in agro-food systems will tend not to experience unsustainability as ecological collapse or disaster. Instead, unsustainability involves progressive deductions from the natural capital or 'cultivated capital' of agriculture.[9]

The second message of this chapter is that the sustainability and unsustainability of agriculture involve powerful global and national forces that generate widespread, though often contradictory and inconsistent, social responses. Our brief discussion of some possible routes to sustainable agriculture confirms what many readers have accepted intuitively for some time: there is no straightforward road to a more sustainable agriculture. Each of the five key movements and activist strategies – agro-food localism, contestation of public and private research priorities, increased environmental regulation of agriculture, eco-taxation and related agroenvironmental incentives, and multifunctionality policies – for achieving sustainability has its advantages and attractions, but none is a certain path to sustainability.

In this chapter I have drawn disproportionately on the US experience and US evidence, but I have striven to discuss sustainability issues in a general manner that is relevant to other industrial countries. Diversity among the industrial countries is arguably most relevant to assessing the roads to agricultural sustainability. Indeed, my assessment of the viability of the five major approaches is based to a considerable degree on the American context. I would hasten to stress, however, that assessment of these alternative roads to agricultural sustainability cannot be done meaningfully without taking into account the structural and cultural specificities of the diverse nations and regions in the advanced industrial world. Thus, for example, the multifunctionality, reform of technoscientific research priorities and eco-taxation roads to sustainable agriculture seem to be much more likely to be implemented in Europe than in North America. By contrast, localism and environmental regulation of agriculture seem to be the more likely routes to sustainable agriculture in North America. Further, the European countries themselves exhibit considerable diversity in the forces propelling, reproducing and contesting the unsustainability of agriculture.

One of the striking areas of difference between the United States and most of Europe lies in the degree to which the political economy of agriculture and food is 'regulated' through the auspices of off-farm supply chains, particularly food retail. Marsden's admirable treatments of rural and agro-food sustainability in Europe in general and the UK in particular (Drummond and Marsden, 1999; Marsden, 2003) are heavily based on the primacy of privately related supply chains in the agro-food system. In the US agro-food system, by contrast, control is exercised much closer to the production end of the supply chain (for example, by the American Farm Bureau Federation, agricultural commodity groups, the agricultural input industry, and the public and private agricultural research establishments); the retail sector is not as concentrated as in Europe and has yet to become the dominant player in the agro-food system.

In general, the prospects for substantial movement toward sustainable agriculture seem far greater in most of Europe than in North America. In most European countries there is far less spatial separation between countryside (or rural landscape) and agriculture than is the case in the United States and Canada. North America's agricultural heartland, stretching from the Corn

Belt through the Northern and Southern Great Plains, is an agricultural grain desert that attracts few tourists. Most of these US regions have low population densities and few metropolitan centres and population concentrations. As a result, the political pressures from nonfarming citizens for a more sustainable agriculture tend to be less in the United States than in most of Europe. Even so, the sustainable agriculture movement in the United States and Canada has considerable strength and vitality, and there are rapidly growing and tightly knit networks of sustainable producers and consumers (Thrupp, 1996). These North American agricultural sustainability movements, however, have a strong urban flavour, and are increasingly centred around urban-based community-supported agriculture networks, organic food cooperatives and farmers' markets. These observations suggest that there will be many different paths to a more sustainable agriculture.

NOTES

1 The notion that unsustainability and ecosystem perturbation tend not to result in ecosystem collapse is typically referred to as the 'new ecology' position. Botkin (1990) is generally recognized as one of the pioneers in the development of this non-equilibrium approach to ecological theory.

2 It should be stressed, however, that European environmental groups have had a more longstanding concern with the agricultural environment. This seems to be mainly because high population density and lengthy settlement history have caused there to be very little 'wildland.' Agricultural zones and tourism or recreation zones thus overlap to a far greater degree than in North America.

3 Spatial homogeneity (or what Pretty, 2002 refers to as 'moonscapes') is the large-scale (bioregional scale) expression of monoculture, which is a circumstance in which one crop is grown extensively at a field or farm level.

4 Corn rootworm is an especially interesting case in point. It is generally recognized that until the establishment of widespread corn monoculture in the US Midwest after the Second World War, corn rootworms were not major pests (Sutter, 1999). Monsanto's most recent Bt corn product, which involves genetically engineering corn to produce a toxin that kills rootworms, is a prototypical 'patch-up' technology in that it solves an immediate technical problem caused by large-scale monoculture but does so in an unsustainable manner that creates as many or more problems than it solves. Thus, for example, mainstream entomologists believe that Bt rootworm corn will be very highly subject to rootworm resistance (Gray, 2000). Transgenic technology is almost certain to lead to larger-scale corn and soybean monocultures in the American heartland. The likely failure of transgenic rootworm hybrids will intensify the corn rootworm problem and require more desperate – and, in all likelihood, highly unsustainable measures – such as new Bt hybrids and increased use of insecticides.

5 CAFO is the acronym for 'concentrated animal feeding operation'. CAFO was a term originally used by American regulatory agencies as a benign synonym for 'animal factory' or 'factory farm', but the term is now used by opponents of large-scale confinement production in recognition and critique of the desire of state regulatory agencies not to offend mainstream agricultural interests.

6 This is not to suggest, of course, that if agriculture were more sustainable there would be no need for productivity-increasing or 'maintenance' research. But maintenance research for sustainable agriculture would mainly stress the enhancement of the natural resource efficiency of agriculture and maintenance of the integration efficiencies of agriculture.

7 It should be noted that I will not treat the sustainable agriculture and organic farming movements as a unique vision or strategy. These movements are clearly of tremendous importance (Buttel, 1996). But it is also the case that sustainable agriculture social movement organizations and their adherents vary a great deal in terms of their preferred strategies for addressing sustainability questions in agriculture. It should also be noted that this overview of routes to sustainability is by no means comprehensive in that there are a number of strategies (for example, expansion of the conservation titles of the Farm Bill, eco-labelling, and consumer/consumption-driven strategies) that are not treated here for reasons of space. For useful treatments of consumer/consumption-driven approaches, see DuPuis (2002) and Goodman and DuPuis (2002).

8 The international agricultural development expression of localism is typically that of self-reliance (see Pretty, 1995).

9 In this regard I find Webb's (2002) notion of 'cultivated capital' as a modification of the notion of natural capital to agriculture to be a very useful formulation. Webb notes that while the notion of natural capital has some heuristic use in understanding agricultural sustainability, it needs to be recognized that agriculture tends to be a site of creation of 'cultivated capital'; accordingly, cultivators, if they tend the land sustainably, can increase the capital stock within agroecosystems. By contrast, the notion of natural capital as employed by ecological economists and others has a static character based on models of non-renewable resources.

REFERENCES

Ahearn, M. (1998) 'Ag productivity continues healthy growth', *Agricultural Outlook*, May: 30–33.

Allaby, M. and Allen, F. (1974) *Robots Behind the Plow*. Emmaus, PA: Rodale Press.

Allen, P. (1993) 'Connecting the social and the ecological in sustainable agriculture', in P. Allen (ed.), *Food For the*

Future: Conditions and Contradictions of Sustainability. New York: Wiley. pp. 1–16.

Anderson, K. (2000) 'The Ottawa Convention banning landmines, the role of international non-governmental organizations, and the idea of international civil society', *European Journal of International Law*, 11: 91–120.

Bland, W. L. and Buttel, F. H. (eds) (2004) *New Directions in Agroecology Research and Education*. Madison, WI: Center for Integrated Agricultural Systems, University of Wisconsin.

Bonanno, A., Busch, L., Friedland, W. H., Gouveia, L. and Mingione, E. (eds) (1994) *From Columbus to ConAgra*. Lawrence, KS: University Press of Kansas.

Botkin, D. (1990) *Discordant Harmonies*. New York: Oxford University Press.

Brookfield, H. (2001) *Exploring Agrodiversity*. New York: Columbia University Press.

Buttel, F. H. (1997) 'Some observations on agro-food change and the future of agricultural sustainability movements', in D. Goodman and M. Watts (eds), *Globalising Food: Agrarian Questions and Global Restructuring*. London: Routledge. pp. 344–365.

Buttel, F. H. (2003) 'Internalizing the societal costs of agricultural production', *Plant Physiology*, 133: 1656–1665.

Buttel, F. H., Larson, O. F. and Gillespie G. W. Jr (1990) *The Sociology of Agriculture*. Westport, CT: Greenwood Press.

Carson, R. (1962) *Silent Spring*. Greenwich, CT: Fawcett.

Cochrane, W. W. (1993) *The Development of American Agriculture*, 2nd edn. Minneapolis, MN: University of Minnesota Press.

Congressional Research Service (1998) *Animal Waste and the Environment*. CRS Report 98–451. Washington, DC: Congressional Research Service.

DeLind, L. (2003) 'Considerably more than vegetables, a lot less than community: the dilemma of community-supported agriculture', in J. Adams (ed.), *Fighting for the Farm: Rural America Transformed*. Philadelphia, PA: University of Pennsylvania Press. pp. 192–206.

Drummund, I. and Marsden, T. (1999) *The Condition of Sustainability*. London: Routledge.

DuPuis, E. M. (2002) *Nature's Perfect Food*. New York: New York University Press.

Friedmann, H. and McMichael, P. (1989) 'Agriculture and the state system: the rise and fall of national agricultures, 1870 to the present', *Sociologia Ruralis*, 29: 93–117.

Gliessman, S.R. (1998) *Agroecology: Ecological Processes in Sustainable Agriculture*. Chelsea, MI: Sleeping Bear Press. p. 357.

Goodman, D. and DuPuis, E. M. (2002) 'Knowing food and growing food: beyond the production-consumption debate in the sociology of agriculture', *Sociologia Ruralis*, 42: 6–23.

Goodman, D. and Redclift, M. R. (1991) *Refashioning Food*. London: Routledge.

Goodman, D., Sorj, B. and Wilkinson, J. (1987) *From Farming to Biotechnology*. Oxford: Blackwell.

Gray, M. E. (2000) 'Prescriptive use of transgenic hybrids for corn rootworms: an ominous cloud on the horizon?', in *Proceedings of the Illinois Crop Protection Technology Conference*. Urbana–Champaign, IL: University of Illinois Extension. pp. 97–103.

Harvey, G. (1998) *The Killing of the Countryside*. London: Vintage.

Harwood, R. (1985) 'The integration efficiencies of cropping systems', in T. C. Edens et al. (eds), *Sustainable Agriculture and Integrated Farming Systems*. East Lansing, MI: Michigan State University Press. pp. 64–75.

Hightower, J. (1973) *Hard Tomatoes, Hard Times*. Cambridge, MA: Schenkman.

Jackson, D. (2002) 'Restoring prairie processes to farmlands', in D. L. Jackson and L. L. Jackson (eds), *The Farm as Natural Habitat*. Washington, DC: Island Press. pp. 137–154.

Jordan, A. and Wurzel, R. (eds) (2003) *New Instruments of Environmental Governance: National Experiences and Prospects*. London: Frank Cass.

Jordan, J. R. (2002) 'Sustaining productivity with biodiversity', in D. L. Jackson and L. L. Jackson (eds), *The Farm as Natural Habitat*. Washington, DC: Island Press. pp. 155–168.

Kloppenburg, J. Jr and Burrows, B. (2001) 'Biotechnology to the rescue? ten reasons why biotechnology is incompatible with sustainable agriculture', in B. Tokar (ed.), *Redesigning Life*? London: Zed. pp. 103–110.

Kloppenburg, J. Jr, Hendrickson, J. and Stevenson, G. W. (1996) 'Coming into the foodshed', *Agriculture and Human Values*, 13: 33–42.

Lobao, L. and Meyer, K. (2001) 'The great agricultural transition: crisis, change, and social consequences of twentieth century US farming', *Annual Review of Sociology*, 27: 103–124.

Lockeretz, W. (ed.) (1982) *Environmentally Sound Agriculture*. New York: Praeger.

Marsden, T. (2003) *The Condition of Rural Sustainability*. Assen, Netherlands: Van Gorcum.

McBride, W. D. (1997) *Change in U.S. Livestock Production, 1969–92*. Agricultural Economic Report No. 754. Washington, DC: Economic Research Service, US Department of Agriculture.

McMichael, P. (2000) *Development and Social Change*, 2nd edn. Thousand Oaks, CA: Pine Forge Press.

Merrill, R. (ed.) (1976) *Radical Agriculture*. New York: New York University Press.

Mol, A. P. J., Lauber, V. and Liefferink, E. (eds) (2000) *The Voluntary Approach to Environmental Policy*. Oxford: Oxford University Press.

National Research Council (NRC) (1993) *Soil and Water Quality*. Washington, DC: National Academy Press.

Nelson, Fred (2002) Aligning US Farm Policy with World Trade Commitments. *USDA, Agricultural Outlook*, Jan-Feb: 12–16.

Oelhaf, R. (1978) *Organic Agriculture*. Montclair, NJ: Allanheld, Osmun and Co.

OECD (Organization for Economic Cooperation and Development) (2001) *Multifunctionality: Towards an Analytical Framework*. Paris: OECD.

Pimentel, D. and Pimentel, M. (1996) *Food, Energy, and Society*. Niwot, CO: University Press of Colorado.

Pretty, J. J. (1995) *Regenerating Agriculture*. Washington, DC: John Henry Press.

Pretty, J. J. (2002) *Agri-Culture*. London: Earthscan.

Princen, T. (2002) 'Distancing: consumption and the severing of feedback', in T. Princen et al. (eds), *Confronting Consumption*. Cambridge, MA: MIT Press. pp. 103–131.

Princen, T., Maniates, M. and Conca, K. (2002) 'Confronting consumption', in T. Princen et al. (eds), *Confronting Consumption*. Cambridge, MA: MIT Press. pp. 1–20.

Sutter, G. R. (1999) 'Western corn rootworm', in K. L. Steffey et al. (eds), *Handbook of Corn Insects*. Lanham, MD: Entomological Society of America. pp. 64–65.

Ray, D., de la Torre Ugarte, D. and Tiller, K. (2003) *Rethinking US Agricultural Policy: Changing Course to Secure Farmer Livelihoods Worldwide*. Knoxville, TN: Agricultural Policy Analysis Center, University of Tennessee.

Tansey, G. and Worsley, T. (1995) *The Food System*. London: Earthscan.

Thrupp. L. A. (1996) *New Partnerships for Sustainable Agriculture*. Washington, DC: World Resources Institute.

Vandermeer, J. H. (1990) 'Intercropping', in C. R. Carroll, J. H. Vandermeer and P. M. Rosset (eds), *Agroecology*. New York: McGraw–Hill. pp. 481–516.

Webb, P. (2002) 'Cultivated capital: agriculture, food systems and sustainable development'. Discussion Paper No. 15. Bedford, MA: Food Policy and Applied Nutrition Program, Tufts University.

World Commission on Environment and Development (WCED) (1987) *Our Common Future*. Oxford: Oxford University Press.

16

Social forestry: exploring the social contexts of forests and forestry in rural areas

Paul Milbourne, Lawrence Kitchen and Kieron Stanley

INTRODUCTION

> The challenge lies in the understanding we have achieved of the potential contribution which forestry can make to development and rising welfare. For contrary to what many outsiders believe, forestry is not, in its essence, about trees. It is about people. It is about trees only so far as they serve the needs of people. (Westoby, 1987: 92)

This statement produced by Jack Westoby, a prominent figure within the Forest Department of the United Nations Food and Agriculture Organization, indicates a set of broader social and welfare concerns associated with forestry and forests within advanced capitalist countries. While his assessment that forestry does *not* involve trees may be a little extreme, it is clear that Westoby is attempting to move beyond dominant policy constructions of forestry as the scientific management of groups of trees and of forests as merely a land resource towards a more socially relevant form of forestry. A key purpose of this chapter is to extend the ideas proposed by Westoby to social science approaches to forests and forestry. We suggest that there are three main reasons for undertaking such a task.

First, it is clear from reviewing the academic literature on forests and forestry that the 'socialization' of forestry discussed by Westoby has not really fed into dominant social science approaches to the subject. Indeed, it remains the case that studies of forestry have largely shunned the types of critical theoretical perspective that have been adapted by rural researchers over recent years. Forestry research has remained largely preoccupied with providing detailed descriptions of shifting policy contexts and land resource issues and conflicts bound up with forestry and forests. This is not to say that such concerns are not valuable in their own right but we want to argue that there is a need to embed these concerns within broader and more critical theoretical literatures so that we may gain more meaningful understandings of the roles played by forestry and forests within (rural) society.

A second reason that we consider it necessary to develop a broader social agenda for studies of forests and forestry relates to recent developments that have been made within human geography and sociology concerning 'social nature'. While the term social nature is itself bound up with multiple and contradictory meanings within the social sciences, at its core is a recognition that nature needs to be understood in relation to broader sets of social, cultural, political and economic processes and structures. In this chapter we want to extend this idea of social nature to social forestry, which we propose is concerned with the connections between forestry/forests and broader economic, socio-cultural and governance processes that shape rural areas.

Third, it is clear from the Westoby quotation that there have been efforts made by key actors within the forestry sector to re-construct forestry and the role of forests within society. In fact, recent years have witnessed attempts by the private forestry sector to respond to the challenges of environmentalism, while public sector agencies have sought to develop new environmental and social agendas for forestry. In relation to these latter attempts, the terms multi-purpose or social forestry have been introduced to indicate shifts from productivist to post-productivist modes of forestry. Unfortunately, relatively little attempt has been made by rural researchers to critique these claimed shifts within the context of theoretical approaches developed within the social nature literatures. A key concern of this chapter is to provide such a critique, by drawing on regulationist and governance approaches.

Our aim within this chapter is to provide a critical account of the broader social contexts of forestry in advanced countries. We do this by making connections between theoretical literatures on social natures and recent studies that have provided new perspectives on forest/forestry–society relations. The chapter is structured around three main sections, within which we develop linkages between social/cultural nature and forestry literatures. In the first, we focus on the political economy, regulation and governance of forests. The second section considers the social construction and consumption of forests and the third discusses the ways that forests and forestry have been drawn into debates on social justice and rural development. While each of these sections makes connections between nature and forestry literatures, these connections are developed further in the conclusion, where we propose new agendas for research on forestry and forests within rural studies.

THE PRODUCTION, REGULATION AND GOVERNANCE OF FORESTS

Important attempts have been made over the past couple of decades to import critical theories of political economy, regulation and governance into accounts of nature–society relations. For more than two decades social scientists have considered how nature is drawn into systems of capitalism and, in particular, the historical political-economic processes through which capital has come to appropriate and, in many cases, exploit different natural resources (soils, minerals, water, and so on) (see Smith, 1984; FitzSimmons, 1989; Braun and Castree, 1998; Castree and Braun, 2001). According to Katz (1998), within such accounts nature is constructed as an '"open frontier" for capitalism in the sense of an absolute arena of economic expansion'. Through these processes of appropriation and exploitation, it is argued, nature became transformed into 'second' or 'capitalist' nature, defined in commodity terms 'within the specific logics of capitalist production and competition accumulation' (Castree and Braun, 1998; see also Harvey, 1996). In Escobar's (1999) view, this form of 'capitalist nature' came to represent a hegemonic regime, with dominant ideas of 'the production of nature' destroying any distinction between first and second forms of nature.

By the 1970s it has been claimed that dominant attitudes towards nature in advanced countries started to change as a result of a variety of factors, including the de-colonization of 'developing' countries, the rise of environmental pressure groups and the oil crisis of 1973 (see Katz, 1998). These factors led to the creation of new meanings of nature and shifts in the capital–nature relationship. However, Katz claims that this capital–nature relationship remained essentially the same, based on the exploitation of the latter by the former. Challenges from environmentalism, for example, were able to be effectively incorporated into remarkably similar accumulation strategies. Such a regulationist account of capitalism's responses to this 'crisis of nature' has been taken further by others. For example, Bridge and McManus (2000) have utilized ideas of regulation theory to make sense of key shifts within particular natural resource industries. As is discussed elsewhere within this book (Goodwin), regulationist approaches are concerned to understand the shifting modes of regulation utilized to overcome periodic obstacles associated with systems of capitalism. Within regulation theory, capitalism is viewed as composed of a regime and system of accumulation – the modes of economic production and the links between processes of production and consumption – and a mode of social regulation – regulatory structures, procedures and institutions that enable the continuation of the regime of accumulation (see Goodwin and Painter, 1993; Jessop, 1990; Peck and Tickell, 1992). From the regulationist perspective, the accumulation system is inherently contradictory and so needs to be stabilized around particular regimes of accumulation through the social mode of regulation (Goodwin et al., 1995).

Bridge and McManus (2000) argue that regulationist approaches are able to improve our 'understanding of the challenge posed to capitalism by environmentalism and how particular fractions of capital have embraced ecological modernization in light of this challenge' (2000: 13). They call for new regulationist accounts of nature through the development of 'intermediate concepts' of regulation that are sensitive to different spatial, temporal and sectoral contexts (see also Castree, 2002; Marsden, 2001). Elsewhere within rural and food studies it is possible to see such meso-level approaches to regulation. Goodwin et al. (1995), for example, explore economic, political and socio-cultural change in four areas of rural Wales through the regulationist lens, while regulation theory has been employed within studies of the agricultural and food industries (for example, Drummond and Marsden, 1996; Kenney et al., 1991; Marsden, 2001; Marsden et al., 1986). In relation to forestry, Bridge and McManus (2000) provide a detailed regulationist account of the forestry industry in British Columbia, Canada. What their study highlights is that the industrial forestry sector has overcome recent challenges to its operations from environmentalists by altering its mode of social regulation, that is, through the incorporation of discourses of sustainability into forestry, without any major consequences for the accumulation system (see also McManus, 2002).

Regulation theory has also been used to explore the ways in which forestry and mining economic systems come together within the South Wales Valleys in the UK (Milbourne et al., 2005). The Valleys forest was established in the 1920s as one of the first Forestry Commission plantations in Britain and was expanded over much of the twentieth century. It developed around an industrial model of forestry, involving extensive land acquisitions from estate owners, the clearance of farm tenants and buildings, relatively little consultation with local communities, and the mass planting of coniferous trees to boundary edges. The development of forestry in this area also connects with coal mining, as the early forests supplied timber for pit props, provided employment for ex-miners and was able to clean up local landscapes that had been scarred by more than a century of mining activities. More recently, though, it is argued that forestry in this area has witnessed a series of 'crises' in its system of accumulation as the collapse of the local mining sector, declining employment opportunities within the local state forestry sector and the global fall in timber prices have removed the economic rationale for timber production. Such 'accumulation crises' have led to the promotion of new forms of multi-purpose or social forestry based around changes to the social mode of regulation, which have sought to include a broader range of policy actors within the governance of the forest, promote consultation with local communities living in the forest and develop new social meanings for the forest. These attempts to re-construct forestry, though, have themselves encountered a series of obstacles, as particular sets of historical and place-based nature–society relations formed around ideas of industrial nature have led to the development of considerable social distance between forestry, forests and local communities.

More generally, regulation theory can be used to interpret wider shifts in modes of forestry that have been identified within advanced capitalist countries. Over the past few years, it has been generally acknowledged within the academic and policy literatures on forestry that the state forestry sector in various countries has witnessed a transition from productivist to post-productivist systems. Linked to debates on shifts within agricultural production (see Evans et al., 2002; Wilson, 2001), it is claimed that traditional modes of industrial forestry, based around the intensive production of timber, have been replaced by multifunctional forms of forestry that place a greater emphasis on the broader economic, social and environmental functions of trees, woodlands and forests (see Mather, 2001). Unfortunately, relatively little critical scrutiny has been given to such claimed shifts in the mode of forestry. Many of the claims made about the wider socio-economic and environmental benefits of forestry have come from those located within the forestry sector. Social scientists have tended to accept these claims without any real critical assessment having been undertaken. We consider that further regulationist accounts of forestry may indicate whether the emergence of multifunctional or social forms of forestry represents a shift away from previous systems of accumulation or merely a change to the mode of social regulation of forestry.

Related to these regulationist accounts of shifting modes of forestry has been a growing body of literature that has emerged over recent years that has related theories of governmentality and governance to rural spaces, natural environments and forests. Developed by Foucault (1991), the governmentality approach involves a critique of the art of government – including the visibilities, technicalities, rationalities and identities that are bound up with governing (see also Bryant, 2002). As Murdoch and Ward state, this approach is concerned:

> to understand how the state becomes a *centre*, or more accurately, an ensemble of centres, that can shape, guide, channel, direct and control events and persons distant from it in both space and time. Governmentality refers to the way that this centre, or ensemble of centres, 'problematizes' life within its borders and seeks to act in response to the resulting 'problematizations'. In short, governmentality refers to the methods employed as the state both represents and intervenes in the domains it seeks to govern, and how territorial integration is thereby achieved. (1997: 308)

While Murdoch and Ward (1997) use the idea of governmentality to critique the UK government's use of statistics on agriculture, a recent paper by Stanley et al. (2005) has made linkages between governmentality and forestry in the UK. The paper charts key policy changes within the state forestry sector and highlights the visible governance of forestry and forests:

> forests themselves are of course especially visible … the felling of forests, the marking of paths and the sign posting of boundaries are other clear examples of how forests are visibly governed by the Forestry Commission. Yet this visibility is not limited to the forest spaces to be governed, but in the visible means in which they are governed; that is by maps, statistics and so on that delineate a field to be governed. These are then, not just forest spaces but their *representations* that *enact* governable spaces. Decisions are therefore seemingly made based upon representative instruments such as maps and statistics at a distance from the forest spaces. (2005: 3)

Attention has also been given to important shifts in governance systems from those based around structures of government to those involving a wider range of governmental and non-governmental actors. According to Stoker, the term governance has come to indicate 'the development of governing styles in which boundaries between and within public and private sectors have become blurred' (1996: 2). While there is general acceptance of the key contours of 'new' systems of governance – including the complexity of these systems, the increased significance of partnership forms of working, and efforts to devolve power down through policy networks – there has also been a great deal of discussion about the role of hierarchy and power within these systems. For some, governance needs to be viewed as a complete break from previous systems of government, involving 'governing without government' (Rhodes, 1996) or 'governing at a distance' (Murdoch and Ward, 1997), while others, such as Jessop (1997), argue that new governance systems operate within the 'shadow of hierarchy', pointing to the continued/increased influence of the central state and uneven sets of power relations within these systems.

Over the past few years, ideas of governance have been applied by rural researchers to examine the ways in which rural spaces are governed. Attention has been given to attempts by central government to devolve rural policy to the regional and local scales (MacKinnon, 2002; Murdoch, 1997), to the increasing complexity of local governance systems (Edwards et al., 2001; Milbourne, 1998), and to the continued role of government within these systems (Cloke et al., 2000). There has also been some work on the shifting governance of forestry in rural areas, although this has been largely focused on state forestry in the UK and the United States. Efforts have been made to consider the implications for governance of the broader shifts from productivist to post-productivist modes of forestry discussed earlier in the chapter. It is claimed that a key component of post-productivism is a new system of forestry governance; one that involves a broader range of policy actors and the devolution of decision-making to the local community level (see Mather, 2001).

Work has focused on the vertical and horizontal networks within which the forestry sector is located. In relation to the first of these, researchers have focused on the connections between the different scales of environmental and forestry governance, from the global (for example, environmental treaties), through the regional (European Union directives) to the national and local scales. It is clear from work in this area that these linkages are rather complex, with overarching supra-national environmental and forestry directives leading to a variety of national and local outcomes. For example, Kitchen et al. (2002) highlight how national policy shifts towards social forestry in the UK have been interpreted in a variety of ways at the local level, with a number of local forestry personnel actively resisting the development of new (national) modes of forestry governance.

More generally, greater participation in decision-making by non-state bodies alongside state agencies has spread throughout the forestry sector, whether land ownership is under state or private control. The blurring between public and private sectors that Stoker (1996) identifies as a key aspect of governance has increasingly become the locus of forestry practices, complicated by divisions between forest communities and wider consumers of forest spaces (such as visitors, businesses and conservationists). However, it is clear from studies in various countries that while modes of forestry governance have shifted dramatically over recent years, the state continues to play a dominant role in governing forest spaces. In the UK, for example, the government-funded Forestry Commission has been active in promoting new forms of forest governance, developing new forms of partnership working and funding research on new modes of forestry governance. It is also the case that the Forestry Commission has drawn in a broader range of policy agencies in order to resist attempts to limit its scope of operations. In the early 1990s, for example, political attempts to privatize the state forestry sector led to the Forestry Commission entering into a discursive coalition (after Hajer, 1995) with several non-state agencies to promote a broader social forestry agenda. As Winter comments:

> the Forestry Commission in a new alliance with environmental groups mounted a campaign to persuade government of the need to retain the state forests within the public sector for reasons of maintaining public access and implementing many of the new multiple objectives policies that had now been accepted. (1996: 299)

Other research on the shifting governance of forestry has focused on the formation of new horizontal networks, involving the development of new governing networks of policy actors and community groups in particular places. For example, attention has been given to community forestry in the US (Baker and Kusel, 2003; Burns, 2001; Gray et al., 2001), the development of community partnerships in forestry management in the UK (Brown, 2002; Kitchen et al., 2004a; Slee and Snowdon, 1996, 1999; Snowdon and Slee, 1998), and community forest governance in Canada (Beckley, 2000; Parkins, 2002). In addition, research by Buchy and Race (2001) has explored community involvement in new systems of forestry governance in Australia. Much of this work has been concerned with debates surrounding endogenous development or 'bottom-up' strategies that seek to involve local actors and their knowledges, and how these concerns are integrated into forest policy, governed by national forest strategies. While such strategies have been seen by some as a potential area for individual empowerment from state structures (Herbert-Cheshire, 2000), others are more sceptical. MacKinnon (2002), for example, argues that the 'underlying shift towards community action and local involvement is mediated and implemented by local and regional agencies'. In relation to the development of community-based forestry, it is far from clear whether what is being witnessed is a genuine attempt to empower local groups to participate in decisions on the governance of forests or a more cynical effort on the part of forestry policy actors to spin a rhetoric of 'bottom-up' governance that is bound up with broader policy constructions of social forestry. As Macnaghten et al. (1998) comment, in the UK context, not all groups remain convinced of the shift towards new modes of forestry governance:

> professional rural specialists and recreation providers appear to have a somewhat higher opinion of the Forestry Commission than do many members of the public. The former understand the Commission to be a *leader* in the rural recreational field, as part of its commitment to multiple-purpose forestry, whereas many of the latter continue to see it loosely as a commercially-driven timber-producing organisation, albeit with recreational and environmental fringes. (1998: 3)

Similarly, Buchy and Race's (2001) study of community forestry schemes in Australia highlights the continuation of a top-down approach and the inadequate reflection of local inputs in forestry policy, planning and management decisions. Effective community involvement in forestry management, Buchy and Race argue, requires the design of effective participatory processes, rather than mere consultation. Writing about British forestry policy, Tsouvalis also points to the persistence of traditional forms of power relations that privilege the centre within new forms of community forestry governance. Discussing the development of urban and community forests and the National Forest, she suggests that:

> although community participation is often hailed as a necessary prerequisite for the achievement of these new meaningful, composite formations, such communities of interest, as the FC [Forestry Commission] calls them, are rarely involved in the policy making process. This remains a primarily top-down procedure. (2000: 203)

It is also the case that the development of new modes of forestry governance has been bound up with a great deal of local specificity. In a more general discussion of the shifting scales of environmental governance, Lowe and Murdoch (2003) suggest that: 'the implementation of environmental policies must be set within local economic, political, and social conditions as these conditions ensure considerable variation in modes of environmental governance' (2003: 761). A recent paper by Milbourne et al. (2004) has pointed to the significance of local structures and processes in mediating the imposition of new forms of forestry governance in particular places. Using case studies of two forests in the UK – one in southern Wales, the other in the English Midlands – they highlight how the natural features of these forests (for example, the types and sizes of trees, and the densities of the forests), the socio-cultural, economic and political contexts of local spaces, the local assemblages of policy actors, and previous modes of local forestry governance all impact on the development of new governing systems for these forest spaces. In the South Wales forest, which is characterized by densely planted coniferous plantations, previously productivist modes of forestry and high levels of social disadvantage within local communities, attempts to include local people within new devolved governance structures have encountered considerable local animosity. By contrast, the more recent development of the National Forest has been based around post-productivist forms of forestry – characterized by economic, environmental and social objectives –

which has led to more inclusive forms of forestry governance at the community level.

CONSTRUCTING AND CONSUMING FORESTS

It is now widely accepted within social scientific literatures on nature that understandings of nature need to move away from 'productivist vision[s] of human society' (Whatmore and Boucher, 1993) to capture the processes through which nature is socially constructed and consumed by different groups and individuals within society. Important here has been a desire to see nature through society and culture, and to understand the ways that people make sense of nature. An impressive body of literature has emerged that has explored the multiple, shifting and contested meanings that are awarded to nature and has developed new conceptualizations of the increasingly blurred boundaries of nature and culture (see, for example, Braun and Castree, 1998; Castree and Braun, 2001; Soper, 1995; Whatmore, 2002).

Within rural studies, work has examined important linkages between discourses of nature and rurality, as the countryside has been constructed as 'the spatialization of nature' (Cloke et al., 1996a) and the 'natural' components of rural areas have been protected through a suite of land-based regulatory mechanisms (see Macnaghten and Urry, 1998). Furthermore, it is apparent from national surveys as well as local rural social studies, that people attach a great deal of importance to the natural attributes of rural spaces (see Bell, 1994; Jones, 1995). It is equally clear that nature has also emerged as a significant social and cultural battleground in many rural areas over recent years, as different groups draw on different ideas of nature to promote their own constructions of rurality (Milbourne, 2003; Woods, 1998).

These social and cultural approaches to nature are evident within recent studies of forests conducted in different countries. Schama (1995) discusses how trees, woodlands and forests have been drawn into the historical constructions of particular national identities. For example, he shows how the Bialowieza forest on the borders of Lithuania and Poland has been a source of Lithuanian, Polish and Jewish identities over time. In particular, the forest was a focus of Lithuanian–Polish resistance against the occupations by Russian, Prussian and German forces in the nineteenth and twentieth centuries. In Germany, he suggests that forests have had a strong influence on the national psyche, beginning with the defeat of the Romans in the Germanic forest. In the nineteenth century, German artists, writers and philosophers portrayed the forest as a dark and dangerous but virtuous Germanic realm against an over-civilized Napoleonic France. More recently, Schama claims that Hitler's Third Reich drew on the dark, mythical, cultural roots imbued in the Germanic forests.

A contrasting historical construction of forests emerges in England, based on the myth of a sylvan, greenwood idyll. Schama sees the creation of this English identity myth as an essentially Saxon reaction to Norman despotism, which culminates in the popular myth of Robin Hood; here outlaws lived free from oppression in a state of benevolent anarchy in the forest. Schama argues that this greenwood fantasy was useful in forging an English identity and that by Tudor times it was inscribed in the national culture at all levels. A later manifestation of this forest identity was the symbolic 'hearts of oak' used in the building of the English navy, particularly in the wars against France. A fourth case study used by Schama is that of the United States. He argues that the discovery of the giant redwoods in California in 1852 constituted a botanical expression of America's heroic nationalism. Their immense scale reflected perfectly the potential power of the new nation. Moreover, through their age and antiquity America as a nation acquired a previously unrecognized gravitas and a naturalistic religiosity.

A body of literature is emerging in the UK on the social and cultural constructions of woods and forests in different spatial contexts. Tsouvalis (2000), for example, considers the multiple, material, socio-cultural, historic and symbolic relations connected with forest nature in Britain's state forests. In particular, she focuses on the creation and contestation of Royal Forests, the British identification with ancient, native, broadleaf woodland, reactions to the planting of 'alien' conifers and the re-invention of state forestry in Britain. Other work, by Henwood and Pidgeon (1998) on forest identity in North Wales, points to strong connections between forests and local and national identities. They suggest that people place considerable value on trees, woods and forests within their local environments. Furthermore, local forests in their study areas tend to be discussed in relation to national forms of Welsh identity. Commercial forestry, it is claimed, serves to consolidate feelings of Welsh national identity, with the initial planting of these forests resisted as a further imposition on Wales by the British state. More recently, though, these forests have come to be accepted as a key constituent of the Welsh landscape, with attempts to harvest these forests also becoming bound up with issues of national identity.

Research in the UK has focused on the ways that forests are constructed and consumed by those groups that make use of forest spaces. Burgess (1995, 1996) explores people's perceptions of community woodlands in England. Involving different groups – including young people, minority ethnic groups, women with young children – and work undertaken within the woodlands with these groups, the research highlights how experiences of woodlands differ according to gender, age and ethnicity. Similar findings emerge from a study by Macnaghten and Urry (2000) of the bodily engagements of different social groups with particular woods and forests. While their work reveals a common sense of closeness to nature amongst research participants, it also points to differences in the ways in which these woodlands and forests are constructed by their participants (again linked to ethnicity, gender and age). For example, they claim that minority ethnic participants viewed the woods mainly as opportunity spaces for recreation, younger people desired to use woods in physically active, but organized, ways, while a widespread fear of wooded spaces was evident among female discussants.

Other research has explored constructions and consumptions of forests from the perspectives of local communities living within these spaces. A study of the English National Forest by Cloke et al. (1996a, 1996b) was focused on three communities based within it. Not only did this study highlight how understandings of the forest vary according to different social groups but it also reveals the importance of local context in influencing nature–society relations. For example, attitudes to the forest and forestry agencies exhibited marked variations between their ex-mining, agricultural and sub-urban communities, with these differences bound up with local socio-cultural, political and economic structures, as well as the types of local nature that surrounded these communities.

Our own research has also provided community-focused accounts of forests (see Bishop et al., 2002; Kitchen et al., 2002, 2004b; Marsden et al., 2003; Milbourne et al., 2004, 2005). Based around qualitative research in ten communities located in four forests – South Wales Valleys forest, National Forest, Great North forest and Central Scotland forest – our work points to the complexity of understandings of and meanings attached to forests in different socio-natural contexts. In particular, it indicates the significance of the forest within the place-based identities of these areas. However, it is clear that these forests are awarded different sets of meanings by different communities and different groups within communities. For example, the National Forest was generally constructed by local community groups as making a positive contribution to the development of new forms of place-based identity within its area. By contrast, while communities within the South Wales Valleys forest pointed to the cultural significance of their local forest, it was perceived in more negative terms as something that was visually impressive but 'just there'. Across the four study forests, it is also apparent that particular community groups make sense of their local forests in different ways. For older groups, the established coniferous forests were sometimes seen as having taken away the more natural landscapes of previous decades. The forests were often constructed as places of fear and crime by women and particularly women with young children, while for many young people in these communities the forest represented a space of escape, within which it was possible to undertake activities away from the gaze of adults and authority.

What also emerges from this (and other) research is the importance of seeing forests through local communities; that is, of making connections between understandings of local forests and the broader socio-cultural, political and economic contexts of these areas. The selected study areas are characterized by mining histories and high levels of socio-economic disadvantage. Some have also been associated with industrial forms of forestry. Within these areas, then, dominant attitudes have formed that see local nature in economic terms, as a resource that has been exploited by capital and the state. In South Wales, where an industrial forest was imposed on to the local landscape, this viewpoint also extends, as we have seen, to forest nature. High levels of poverty and social exclusion in these areas have led to the formation of dominant attitudes towards local forests that are less concerned with the aesthetics of nature and more with the extension of social problems into local forest spaces (for example, the use of illegal drugs and the dumping of stolen cars in these forests). Furthermore, in the South Wales Valleys forest, historical mining cultures permeate the contemporary forest landscape, with the location of a mining museum and mining artefacts (such as a fly-wheel and ventilation shafts) within the forest, and the use of past local cultures and politics to create new community-based identities for the present-day forest (see Milbourne et al., 2005).

While the work reviewed thus far in this section has been concerned with woodlands and forests, it is also worth pointing to a recent study by Jones and Cloke (2002) that provides an interesting discussion of the cultural significance of trees in particular spaces and places. Working around ideas of actor-networks and hybridity, Jones and Cloke explore the cultural meanings attached to trees by individuals in different local contexts – including orchards and cemeteries. In doing this, they highlight the ways in which trees interact with and act upon physical and cultural milieu to (re)produce meanings in time and space. Their study also points to the important role played by trees in shaping meanings associated with particular places and spaces.

FORESTRY, DEVELOPMENT AND SOCIAL JUSTICE IN RURAL AREAS

The third of the substantive themes relating to social forms of forestry that we want to discuss within this chapter concerns the increasing connections that can be identified between forestry, development, area-based regeneration and social justice issues in rural areas. In relation to the linkages between forestry and rural development, a growing body of literature can be identified that has considered the broader economic roles of multi-purpose forestry in rural areas. While attention has long been directed to tourism in forests and, in particular, the establishment of visitor centres, camp sites and other facilities for visitors, more recently researchers have focused on the broader development potential of forestry and forest spaces.

Within the European Union research has examined the connections between social forestry and rural development within 11 countries (Elands and Wiersum, 2003). In the introduction to their report, Elands and Wiersum make the following statement about the potential roles of forestry within broader processes of rural development:

> Rural development concerns the strengthening of the liveability in [*sic*] rural areas by means of improving and/or restructuring the rural economy and by improving rural identity. Forestry can contribute towards rural development by either contributing improved or innovative production processes or by providing an ecological infrastructure for an attractive rural landscape. (2003: i)

In an earlier paper on the same subject, the authors identify five rural development discourses that emerge as policy increasingly shifts towards multi-purpose forestry and a recognition of its ecological, amenity and social benefits (Elands and Wiersum, 2001). The first they term the agri-ruralist discourse, which emphasizes the continued significance of agriculture to rural development and the improved integration of forestry within agricultural practices and economies. Second, they identify a hedonist discourse within which rural areas are constructed as recreational spaces for urban populations and forests play a role in enhancing the amenity, recreational and landscape attributes of the rural landscape. Third is a utilitarian rural development discourse that seeks to optimize forestry's contribution to the rural economy. A community stability discourse represents their fourth discourse, with forestry seen as a vehicle for supporting and regenerating rural communities. Finally, they propose a nature conservation discourse within which forestry is constructed as possessing an intrinsic ecological, even wilderness, value.

Elands and Wiersum (2003) point to the increasing political prominence of forestry in rural development, in particular to the European Union's Forest Strategy, which seeks specifically to strengthen forestry's role in European rural development. This pan-European research also identifies considerable regional variation in forestry and rural contexts, and the roles of forestry within rural development across the European Union (see, for example, Franklin et al., 2004). Consequently, they suggest:

> it is not possible to develop one uniform approach to multifunctional forestry for rural development. Rather, depending on local conditions, there is a need to develop region-specific approaches to the optimization of the multiple role of forests and forestry within rural development. (Elands and Wiersum, 2003: i)

In the UK, researchers have focused on the establishment of community forests in rural spaces surrounding major urban centres and the National Forest in the English Midlands as examples of multi-purpose forests that include a development remit. Attention has been given to the juxtaposition of forest spaces, declining local economies and disadvantaged local communities within these new projects of social forestry. In addition to increasing the level of tree cover within these areas, a key objective of these projects is to use the forest to bring about environmental, economic and social forms of regeneration. Referring to the National Forest Project,

Cloke et al. (1996b) suggest that the intention was to transform the 'wasteland' associated with post-mining landscapes into a 'wonderland' space characterized by an improved natural environment, increased employment within the green economy and sustainable local living. A more recent paper on the National Forest confirms this broader economic and social vision of this forest project (Milbourne et al., 2004), with a development officer for the National Forest quoted as stating that 'the forest is providing a sort of framework and context I suppose [for] a new way of living and the opportunities in terms of new jobs. We are keen to play a part through the forest creation sort of helping create a better environment in which people can live' (2004: 18).

Alongside these concerns with the developmental and regenerative roles of forestry in particular rural areas, critical attention has also been given to the connections between forestry, forests and social justice. Important here has been the adaptation of theories of environmental justice to the study of forest–society relations. Ideas of environmental justice stem from two main sources. First, the 1992 global Earth Summit in Rio – with its focus on sustainable development and the Agenda 21 report – has produced a growth in social science writings on the socio-natural environment, with serious attempts made to recognize the social and environmental elements of sustainable development and 'the needs and rights of the poor and disadvantaged' (Cahill and Fitzpatrick, 2001). As Low and Gleeson argue, 'sustainable development without environmental justice is an empty formula' (1998: 14).

Second, a more developed and, arguably, more critical literature on environmental justice has emerged in the United States. For Mutz et al. (2002), environmental justice brings together the powerful forces of environmentalism and civil rights. Two dimensions of the relationship between social (in)justice and the environment are emphasized: first, that poverty is associated with environmental degradation; and second, that 'relative wealth is regarded as determining access to what we might call environmental goods (or, indeed, the imposition of environmental bads)' (Dobson, 1998). It has been this second dimension that has received most attention in relation to advanced capitalist countries.

A key argument within the environmental justice literature is that degraded physical environments tend also to be marginalized social environments populated by working-class, poor and minority ethnic groups. Numerous case studies have explored the processes through which environmental risks and hazards linked to 'nature industries' have come to be associated with marginalized communities, and the ways in which these communities deal with these environmental impositions (see, for example, Scandrett et al., 2000). The primary concerns of the environmental justice movement are the distributive equity of environmental goods and bads, particularly as they are constituted in relative wealth, ethnicity, race and class. From a political science perspective, Schlosberg (1999) argues that the environmental justice movement has the potential to engender more critical pluralist forms of democracy and discursive participation in decision-making processes at local levels (Dryzek, 1992). For forested communities this implies input to decisions concerning property rights, natural resources and land use.

In the United States, a couple of studies have highlighted the ways in which environmental injustice has become embedded in the social order of forest governance. Romm (2002) discusses social and environmental injustice bound up with forestry in California. He argues that the state's forest policies continue to reflect the (racist) ideas of early twentieth century wildlife and conservation pioneers and the industrial logging and timber interests. Consequently, forest policy has come to represent the ideas of a small powerful elite of white people, with 'people of colour' – who constitute the majority of the manual forestry workforce – excluded from opportunities in what Romm sees as a 'white/forest reserve'. A further study by Carey (2002) considers the environmental justice issues associated with sustainable forests in New Mexico. He points to conflicts between Hispanic communities, the state Forest Service and environmental groups over the meanings of sustainable forestry. Hispanic communities claim that the forest resource should not be over-exploited and call for an appreciation of natural values in a working landscape. The state Forest Service promotes policies towards industrial forestry 'in the context of mainstream Anglo-American culture', while the environmental lobby seeks to return the forests back to a wilderness. However, Carey suggests that these different viewpoints cannot be divorced from the unequal power relations between these three groups, with both the Forest Service and the environmental lobby rejecting calls for collaboration with local communities in order to maintain their strong power bases.

In the UK, recent research has pointed to the operations of open-cast mining in state forests, with local landscapes continuing to be altered by the needs of extractive capital and working-class communities being largely powerless to resist (Cloke et al., 1996b; Kitchen et al., 2004a, 2004b). Furthermore, many people within such communities have come to accept the unilateral actions of forestry agencies in shaping their local environments, whether in terms of previous planting programmes that were imposed on their local areas or the clear-felling of forested landscapes. The connections between forestry, forests and social justice, though, are clearly more complicated than those indicated so far. Kitchen et al.'s (2004b) study of the connections between community forests and social exclusion in six ex-mining communities in England and Scotland points to examples of social inclusion as well as exclusion. Their research demonstrates the potential of social forestry projects to contribute to more general programmes of anti-poverty and social inclusionary work through the improvement of, and increased involvement of people in their local environments. However, Kitchen et al. also suggest that forests need to be seen as exclusionary spaces in those areas where industrial forestry prevails and governance structures do not include the interests of local communities. This would appear to be the situation in the South Wales Valleys forest (see Bishop et al., 2002).

CHALLENGING RURAL RESEARCH ON FORESTS AND FORESTRY

Within this chapter we have sought to develop more challenging agendas for rural research on forests and forestry – more challenging in the sense that these new agendas may be seen as threatening the dominance of conventional approaches to these subjects and also that the task of developing new theoretically informed accounts of forestry and forests will no doubt be a demanding one. Compared with other areas of rural studies, it is clear that there remains a great deal of work to be done by researchers before it can be claimed that we possess any real critical appreciation of the position of forestry and forests *vis-à-vis* rural economic, socio-cultural and governance structures. The subject has remained locked into forestry policy discourses with researchers seemingly reluctant to acknowledge, let alone engage with the types of theoretical approach that have enlivened key components of rural studies over recent decades. While this closeness to policy is also a feature of agricultural studies, the similarity would appear to stop here as researchers of agriculture have been much more open to engage with theoretical perspectives developed elsewhere.

The chapter has set out what we consider to be significant themes for the study of forests and forestry in rural areas. A key intention has been to make connections between critical theoretical literatures on nature/natural environments and forestry and, in so doing, introduce the term 'social forestry' into social science accounts of forestry. While we do not claim to be the first to use this term – as it first emerged within the forestry policy discourse – we do want to award it different sets of meanings. For us, the social forestry approach emerges from academic approaches to social nature that have sought to re-engage the natural and social realms rather than from the policy realm. It signifies an acknowledgement of the broader contexts of forests and forestry within (rural) society. Social forestry research is therefore concerned to provide critical evaluations of the connections between forestry/forests and those economic, socio-cultural and political processes and structures that are influential in shaping rural areas.

The new agendas for social forestry research outlined in this chapter have the potential to re-invigorate a subject that has, for a long time, been a marginal concern for rural researchers. With the continued significance of woodlands and forests (in rural land use terms) in many advanced countries over the past few decades, changing systems of forestry and the development of sustainability discourses, it would seem that there are a number of interesting avenues of research that allow connections to be made with wider theoretical debates within rural studies. In this chapter we have discussed three broad themes bound up with social forestry that we feel are worthy of further investigation by those interested in researching forestry and forests in rural areas. These themes engage with processes of production, regulation and governance; social construction and consumption; and development and social justice. It is clear, though, that relatively little work has explored forestry within these broader and more critical contexts, and that the work that has been undertaken has generally focused on a handful of advanced countries – most notably the UK, US, Canada and Australia.[1]

There remains a need to develop the social forestry research themes discussed in this chapter in more critical, comprehensive and in-depth ways. To bring this chapter to a close, we propose five areas of research on social forestry which, if developed, should ensure a more prominent position for forestry research within rural studies in future years. First, it is clear that a broader range of case studies in different countries would help to sensitize theoretical accounts of social forestry to different spatial and governance contexts. Second, there remains a need to explore further the relationships between forestry and other 'nature industries' in rural areas, including the agricultural, mining, water and energy sectors. Third, further attention needs to be directed towards the claimed transition from productivist to post-productivist modes of forestry. While there is little doubt that the forestry sector has sought to extend its interests into the social sphere over the past couple of decades, it is far from clear whether new forms of social forestry have been accompanied by new accumulation and governance systems, or merely represent shifts within the mode of social regulation. Fourth, more research is needed on forest cultures and, in particular, the processes through which meanings of forests are developed at national, regional and local scales. At the local level this will involve an increased emphasis being placed on making sense of community and individual understandings and experiences of forests. Finally, it should be recognized that the distinctions between the three themes raised in the chapter are, in many ways, blurred and forestry researchers need to make more meaningful connections between the varied processes of production, regulation, consumption and justice that are bound up with forests and forestry in rural spaces.

NOTE

1 We recognize that a larger volume of academic work has engaged with these themes in advancing countries.

REFERENCES

Baker, M. and Kusel, J. (2003) *Community Forestry in the United States: Learning from the Past, Crafting the Future*. Washington, DC: Island Press.

Beckley, T.M. (2000) 'Sustainability for whom? Social indicators for forest-dependent communities in Canada', SFM Project Report 2000–34. Edmonton, Alberta: Northern Forestry Centre.

Bell, M.M. (1994) *Childerley: Nature and Morality in a Country Village*. Chicago and London: University of Chicago Press.

Bishop, K., Kitchen., L., Marsden, T. and Milbourne, P. (2002) 'Forestry, community and land in the South Wales Valleys'. Papers in Environmental Planning Research No. 25. Cardiff: Department of City & Regional Planning, Cardiff University.

Braun, B. and Castree, N. (eds) (1998) *Remaking Reality: Nature at the Millennium*. London: Routledge.

Bridge, G. and McManus, P. (2000) 'Sticks and stones: environmental narratives and discursive regulation in the forestry and mining sectors', *Antipode*, 32: 10–47.

Brown, L. (2002) 'Laggan Community Forest Project'. Unpublished Report: Louise Brown Research.

Bryant, R.L. (2002) 'Non-governmental organizations and governmentality: 'consuming' biodiversity and indigenous people in the Philippines', *Political Studies*, 50: 268–292.

Buchy, M. and Race, D. (2001) 'The twists and turns of community participation in natural resource management in Australia: what is missing?', *Journal of Environmental Planning and Management*, 44 (3): 293–308.

Burgess, J. (1995) *Growing in Confidence: Understanding People's Perceptions of Urban Fringe*. Woodlands, London: Countryside Commission.

Burgess, J. (1996) 'Focusing on fear: the use of focus groups in a project for the Community Forest Unit', *Area*, 28 (2): 130–135.

Burns, S. (2001) 'A civic conversation about public lands: developing community governance', in G.J. Gray, M.J. Enzer and J. Kusel (eds), *Understanding Community-Based Forest Ecosystem Management*. New York: Food Products Press. pp. 271–290.

Cahill, M. and Fitzpatrick, T. (2001) 'Environmental issues and social welfare', *Social Policy and Administration*, 35 (5): 469–471.

Carey, H.H. (2002) 'Forest management and environmental justice in northern New Mexico', in K.M. Mutz, G.C. Bryner and D.S. Kenney (eds), *Justice and Natural Resources: Concepts, Strategies and Applications*. Washington, DC: Island Press. pp. 209–223.

Castree, N. (2002) 'False antitheses: Marxism, nature and actor-networks', *Antipode*, 34: 111–146.

Castree, N. and Braun, B. (1998) 'The construction of nature and the nature of construction', in N. Castree and B. Braun (eds), *Remaking Reality: Nature at the Millenium*. London: Routledge.

Castree, N. and Braun, B. (eds) (2001) *Social Nature: Theory, Practice, and Politics*. Malden, MA: Blackwell.

Cloke, P., Milbourne, P. and Thomas, C. (1996a) 'The English National Forest: local reactions to plans for renegotiated nature–society relations in the countryside', *Transactions of the Institute of British Geographers*, NS 21: 552–571.

Cloke, P., Milbourne, P. and Thomas, C. (1996b) 'From wasteland to wonderland: opencast mining, regeneration and the English National Forest', *Geoforum*, 27 (2): 159–174.

Cloke, P., Milbourne, P. and Widdowfield, R. (2000) 'Partnership and policy networks in rural local governance', *Public Administration*, 78 (1): 111–133.

Dobson, A. (1998) *Justice and Environmental Sustainability: Conceptions of Environmental Sustainability and Justice*. Oxford: Oxford University Press.

Drummond, I. and Marsden, T. (1996) 'Regulating sustainable development', *Global Environmental Change*, 5 (1): 51–63.

Dryzek, J.S. (1992) 'Ecology and discursive democracy: beyond liberal democracy and the administrative state', *Capitalism, Nature, Socialism*, 3 (2): 18–42.

Edwards, B., Goodwin, M., Pemberton, S. and Woods, M. (2001) 'Partnerships, power and scale in rural governance',

Environment and Planning C: Government and Policy, 19: 289–310.

Elands, B.H.M. and Wiersum, K.F. (2001) 'Forestry and rural development in Europe: an exploration of socio-political discourses', *Forestry Policy and Economics*, 3: 5–16.

Elands, B.H.M. and Wiersum, K.F. (2003) *Nature Forest and Society: Forestry and Rural Development in Europe*. Wageningen: Wageningen University.

Escobar, A. (1999) 'After nature: steps to an antiessentialist political ecology', *Current Anthropology*, 40 (1): 1–30.

Evans, N., Morris, C. and Winter, M. (2002) 'Conceptualizing agriculture: a critique of post-productivism as the new orthodoxy', *Progress in Human Geography*, 26 (3): 313–332.

FitzSimmons, M. (1989) 'The matter of nature', *Antipode*, 21: 106–120.

Foucault, M. (1991) 'Governmentality', in G. Burchell, C. Gordon and P. Miller (eds), *The Foucault Effect*. London: Harvester Wheatsheaf. pp. 87–104.

Franklin, A., Bishop, K., Marsden, T. and Milbourne, P. (2004) 'Forests and forest land in the context of broader land use planning'. Papers in Environmental Planning Research No. 29, School of City and Regional Planning, Cardiff University.

Goodwin, M. and Painter, J. (1993) 'Local governance, the crisis of Fordism and uneven development'. Paper presented to the 9th Urban Change and Conflict Conference, University of Sheffield.

Goodwin, M., Cloke, P. and Milbourne, P. (1995) 'Regulation theory and rural research: theorising contemporary rural change', *Environment and Planning A*, 27: 1245–1260.

Gray, G.J., Enzer, M.J. and Kusel, J. (eds) (2001) *Understanding Community-Based Forest Ecosystem Management*. New York: Food Products Press.

Hajer, M. (1995) *The Politics of Environmental Discourse: Ecological Modernisation and the Policy Process*. Oxford: Oxford University Press.

Harvey, D. (1996) *Justice, Nature and the Geography of Difference*. Oxford: Blackwell.

Henwood, K. and Pidgeon, N. (1998) 'The Place of Forestry in Modern Welsh Culture and Life'. Report to the Forestry Commission, Land Use Planning and Forestry Authority Wales. University of Wales Bangor.

Herbert-Cheshire, L. (2000) 'Contemporary strategies for rural community development in Australia: a governmentality perspective', *Journal of Rural Studies*, 16 (2): 203–215.

Jessop, B. (1990) 'Regulation theories in retrospect and prospect', *Economy and Society*, 19: 153–216.

Jessop, B. (1997) 'The regulation approach', *Journal of Political Philosophy*, 5 (3): 287–326.

Jones, O. (1995) 'Lay discourses of the rural: developments and implications for rural studies', *Journal of Rural Studies*, 11: 35–49.

Jones, O. and Cloke, P. (2002) *Tree Cultures: The Place of Trees and Trees in Their Place*. Oxford and New York: Berg.

Katz, C. (1998) 'Whose nature, whose culture? Private productions of space and the "preservation" of nature', in B. Braun and N. Castree (eds), *Remaking Reality: Nature at the Millennium*. London: Routledge. pp. 46–63.

Kenney, M., Loboa, L., Curry, J. and Goe, R. (1991) 'Agriculture in US Fordism: the integration of the productive consumer', in W. Friedland, L. Busch, F.H. Buttel and A.P. Rudy (eds), *Towards a New Political Economy of Agriculture*. Boulder, CO: Westview Press. pp. 174–184.

Kitchen, L., Marsden, T. and Milbourne, P. (2004a) 'The management of Brechfa Forest: an investigation into the potential and support for community involvement'. Papers in Environmental Planning Research No. 27. Cardiff: School of City & Regional Planning, Cardiff University.

Kitchen, L., Marsden, T. and Milbourne, P. (2004b) 'Social forestry in the post-industrial countryside: making connections between social inclusion and the environment'. Papers in Environmental Planning Research No. 28. Cardiff: School of City & Regional Planning, Cardiff University.

Kitchen, L., Milbourne, P., Marsden, T. and Bishop, K. (2002) 'Forestry and environmental democracy: the problematic case of the South Wales Valleys', *Journal of Environmental Policy and Planning*, 4 (2): 139–155.

Low, N. and Gleeson, B. (1998) *Justice, Society and Nature: An Exploration of Political Ecology*. London: Routledge.

Lowe, P. and Murdoch, J. (2003) 'Mediating the "national" and the "local" in the environmental policy process: a case study of the CPRE', *Environment and Planning C*, 21: 761–778.

MacKinnon, D. (2002) 'Rural governance and local involvement: assessing state-community relations in the Scottish Highlands', *Journal of Rural Studies*, 18 (3): 307–324.

Macnaghten, P. and Urry, J. (1998) *Contested Natures*. London: Sage.

Macnaghten, P. and Urry, J. (2000) 'Bodies in the woods', *Body and Society*, 6 (3–4): 166–182.

Macnaghten, P., Grove-White, R., Weldon, S. and Waterton, C. (1998) 'Woodland Sensibilities: Recreational Uses of Woods and Forests in Contemporary Britain'. Report by the Centre for the Study of Environmental Change, Lancaster University for the Forestry Commission.

Marsden, T. (2001) 'Food matters and the matter of food', *Sociologia Ruralis*, 40: 20–29.

Marsden, T., Milbourne, P., Kitchen, L. and Bishop, K. (2003) 'Communities in nature: the construction and understanding of forest natures', *Sociologia Ruralis*, 43 (3): 238–256.

Marsden, T., Munton, R., Whatmore, S. and Little, J. (1986) 'Towards a political economy of capitalist agriculture: a British perspective', *International Journal of Urban and Regional Research*, 10: 498–521.

Mather, A.S. (2001) 'Forests of consumption: postproductivism, postmaterialism, and the postindustrial forest', *Environment and Planning C: Government and Policy*, 19: 249–268.

McManus, P. (2002) 'The potential and limits of progressive neopluralism: a comparative study of forest politics in Coastal British Columbia and South East New South Wales during the 1990s', *Environment and Planning A*, 34: 845–865.

Milbourne, P. (1998) 'Local responses to central state social housing restructuring in rural areas', *Journal of Rural Studies*, 14 (2): 167–184.

Milbourne, P. (2003) 'Hunting ruralities: nature, society and culture in "hunt countries" of England and Wales', *Journal of Rural Studies*, 19 (2): 157–171.

Milbourne, P., Marsden, T. and Kitchen, L. (2005) 'The complexities of social nature: forestry, mining and community in the South Wales Valleys'. Copy available from P. Milbourne, School of City & Regional Planning, Cardiff University.

Milbourne, P., Marsden, T., Kitchen, L. and Stanley, K. (2004) 'The complexities of social forestry in rural Britain'. Paper presented to the Social Forestry Working Group at the XI World Congress of Rural Sociology, Trondheim, 25–30 July.

Murdoch, J. (1997) 'The shifting territory of government: some insights from the Rural White Paper', *Area*, 29 (2): 109–118.

Murdoch, J. and Ward, N. (1997) 'Governmentality and territoriality. The statistical manufacture of Britain's national "farm"', *Political Geography*, 15 (4): 307–324.

Mutz, K.M., Bryner, G.C. and Kenney, D.S. (eds) (2002) *Justice and Natural Resources: Concepts, Strategies and Applications*. Washington, DC: Island Press.

Parkins, J. (2002) 'Forest management and advisory groups in Alberta: an empirical critique of an emergent public sphere', *Canadian Journal of Sociology*, 27 (2): 163–184.

Peck, J. and Tickell, A. (1992) 'Local modes of regulation? Regulation theory, Thatcherism and uneven development', *Geoforum*, 23 (3): 347–364.

Rhodes, R.A.W. (1996) 'The new governance: governing without government', *Political Studies*, 44: 652–657.

Romm, J. (2002) 'The coincidental order of environmental injustice', in K.M. Mutz, G.C. Bryner and D.S. Kenney (eds), *Justice and Natural Resources: Concepts, Strategies and Applications*. Washington, DC: Island Press. pp. 117–137.

Scandrett, E., Dunion, K. and McBride, G. (2000) 'The campaign for social justice in Scotland', *Local Environment*, 5 (4): 467–474.

Schama, S. (1995) *Landscape and Memory*. London: HarperCollins.

Schlosberg, D. (1999) *Environmental Justice and the New Pluralism: The Challenge of Difference for Environmentalism*. Oxford: Oxford University Press.

Slee, R.W. and Snowdon, P. (1996) 'An economic appraisal of rural development forestry in Scotland', *Scottish Agricultural Economics Review*, 9: 9–19.

Slee, R.W. and Snowdon, P. (1999) 'Rural development forestry in the United Kingdom', *Forestry*, 72 (3): 272–284.

Smith, N. (1984) *Uneven Development: Nature, Capital and the Production of Space*. Oxford: Blackwell.

Snowdon, P. and Slee, R.W. (1998) 'An appraisal of community based action in forest management', *Scottish Forestry*, 52 (3/4): 146–156.

Soper, K. (1995) *What is Nature?* Blackwell: Oxford.

Stanley, K., Marsden, T. and Milbourne, P. (2005) 'Governance, rurality and nature: exploring emerging discourses of social forestry in Britain', *Environment and Planning C*, 25.

Stoker, G. (1996) 'Governance as theory: five propositions'. Mimeo. Department of Government, University of Strathclyde.

Tsouvalis, J. (2000) *A Critical History of Britain's State Forests*. Oxford: Oxford University Press.

United Nations Environmental Programme (2002) *Global Environmental Outlook 3*. London: Earthscan.

Westoby, J. (1987) *The Purpose of Forests: Follies of Development*. Oxford: Blackwell.

Whatmore, S. (2002) *Hybrid Geographies: Natures Cultures Spaces*. London: Sage.

Whatmore, S. and Boucher, S. (1993) 'Bargaining with nature: the discourse and practice of "environmental planning gain"', *Transactions of the Institute of British Geographers*, 18: 166–178.

Wilson, G.A. (2001) 'From productivism to post-productivism ... and back again? Exploring the (un)changed natural and mental landscapes of European agriculture', *Transactions of the Institute of British Geographers*, 26 (1): 77–102.

Winter, M. (1996) *Rural Politics: Policies for Agriculture, Forestry and the Environment*. London: Routledge.

Woods, M. (1998) 'Mad cows and hounded deer: political representations of animals in the British countryside', *Environment and Planning A*, 30 (7): 1141–1330.

17

Commodification: re-resourcing rural areas

Harvey C. Perkins

Where profit is, loss is hidden nearby. (Anonymous)

INTRODUCTION

Capitalism's propensity for constant change means that rural landscapes and the experiences of people who live in them, or who visit them for various purposes, are continually changing. These changes relate to the organization of land, labour, capital and technology and are integrally connected with economic and social crises (Smith, 2000). Capital's response to these crises is constantly to seek new ways to accumulate. Consequently, established rural environments, productive processes and social arrangements are continually being modified or abandoned and replaced, sometimes in the same location, but often elsewhere, with new built forms and new environments for production and consumption. Commodification is an integral part of these processes and social arrangements and therefore underpins the establishment of new rural geographies and ensembles of rural production and consumption which may be understood as re-resourced rural areas. This chapter interprets a number of recent developments in re-resourced rural areas as they relate to commodification processes in countries of the industrialized world.

Immediately problematic in this argument is how one is to define 'rural' or as it is sometimes known, the 'countryside' (Cloke, 1993; Dickens, 2000; DuPuis and Vandergeest, 1996). Early discussions of the rural rested on definitions focusing on the activities that take place in it and the rural was thus equated with agriculture, horticulture, forestry and nature conservation, their landscapes and associated rural economic functions. Contrasted with the city, the rural was remote, on the periphery and sparsely populated. These ideas are still important in popular Western interpretations of rurality but where economic and social restructuring have occurred a greater sense of dynamism has been introduced into our understanding of such spaces and associated social settings (Hoggart and Paniagua, 2001).

This understanding requires consideration of political-economic *and* socio-cultural perspectives on the rural. As exemplified by the opening paragraph, political economists argue that the meaning of rural space is linked to wider dynamics of national and international political-economic activity, and that production and consumption in rural areas are configured in varying ways from society to society. This is because rural areas are influenced by economic regulation and governance in general, and the strengths and weaknesses of the 'regional and national capitalisms' of which they are a part. They are also embedded in social systems of production distinctive to particular societies and regions (Peck, 2000). In general political-economic terms though, changes in the countryside are often responses to signals from afar; and because rural residents are connected to networks having their origins in, or which are closely linked to, urban settings, in many respects rural people are culturally urban. Global influences on rural areas and their residents have resulted from de-industrialization, free trade, corporatization, privatization and re-regulation, all of which have led to ongoing and significant rural transformations (Le Heron and Pawson, 1996: 7–19). Mormont (1990) noted that these transformations mean that rural space, society and economy are therefore not autonomous or discrete (Hoggart, 1990). In addition, rural areas comprise many new and specialized forms of land use, which exist to varying degrees independently of the action of rural populations, and are home for, or are visited by diverse peoples (Pahl, 1968).

In this context, the rate and nature of rural change are influenced strongly by the local, regional and global regulatory regimes in place at any one time, affecting opportunities for entrepreneurial activity and opening up new spaces for economic activity and associated commodification. Some of these regulations impinge directly on how land may be used and range from local and national land use, resource and environmental management and planning instruments, through to international agreements such as the United Nations Framework Convention on Climate Change and its associated Kyoto Protocol. Typically, such arrangements are put into effect by public regulatory bodies whose jurisdiction is mandated by national legislation and who regulate land use, and water and air quality through local and regional plans. These regulations are often a site for considerable contest and conflict over appropriate uses of land, the values that should be applied to environmental management, and the interests it should serve (Memon and Perkins, 2000). Rural environmental regulation exists alongside, and is sometimes intimately connected with, other regulatory forms, and here too, local, national and international influences are at work. Direct industry regulation, the organization of trading in agricultural products and the indirect regulation of rural activity through such things as agricultural subsidies, and active government involvement in rural cultural, social and economic policy-making, may be developed and managed regionally or nationally, but they are closely tied to international agreements such as the General Agreement on Tariffs and Trade and the European Union's Common Agricultural Policy. The effects of these regulations work themselves out in myriad ways and there is now a very considerable literature on the restructuring effects of regulation, de-regulation and re-regulation on rural areas and associated production and environmental management in a wide variety of national and regional settings (for example, Argent, 2002; Fold, 2000; Gilg, 1992; Griggs, 2002; Harberl et al., 2003; Krausmann et al., 2003; Le Heron and Pawson, 1996; McCarthy et al., 2003; Ray, 1998; Smailes, 2002). Regulatory change can have widespread effects. A new regulatory regime in one region can have quite dramatic effects on rural regions and opportunities for investment and commodification in other parts of the world. Farmers, free trade and agricultural organizations in Australasia, for example, have welcomed recent announcements from European Union agriculture ministers that subsidies paid to European farmers will be reduced (Robson, 2003). They hope the reduced subsidies will mean lower production levels in Europe and greater opportunities for sales of New Zealand and Australian produce in European markets. Such an outcome will have significant environmental and social effects on the agriculturally productive regions of New Zealand and Australia.

Advocates of socio-cultural perspectives on the rural, on the other hand, have interpreted rurality in social constructivist terms and have emphasized agency, negotiation and the changing meaning of rural spaces. Rather than focus only on rural space, writers using this perspective have highlighted the important processes by which spaces become places (Massey, 1995; Perkins, 1989). This, in part, involves consideration of agency relating to families, social networks, communities and associations as they interact together in and around rural space. Also important are the idealized visions of the rural held by many people and which are today produced and reproduced using a variety of media such as television and film. Rurality in general terms becomes therefore the repository of the good things we have lost as a result of urbanization. This nostalgia for a rural past is manifested in many ways, most commonly as anti-urbanism and an idealization of the small town or village in North America, particularly the US (Meinig, 1979) and as the rural idyll in the UK, with its emphasis on the countryside as a green and pleasant land which is safe, clean, healthy and enjoyable (Aitchison et al., 2000; Newby, 1979; Short, 1991). In the UK this idealization is picturesque (that is, aesthetic) though recently Phillips, et al. (2001) argued that British television programmes with a rural theme represent both idyllic views of rurality and normative middle-class values about appropriate social behaviour in rural society. In North America the farm, agrarian ideology and small town and village existence are interpreted as a desirable way of life rather than as a picturesque ideal (Bunce, 1994). These sentiments form the basis of imagined geographies which inevitably are based on fictional accounts, or at least partially fictional and sanitized accounts, both of which seek to hide 'inconvenient' aspects of rural land use and the social realities of past and contemporary rural poverty and powerlessness (Hughes, 1992; Short, 1991). Imagined geographies are important elements used in attempts to sell rural places and are the stimulus for various forms of rural consumption and social conflict.

The rural, and interpretations of rurality, are therefore complex and underpinned by material and symbolic factors. These factors are no better exemplified than in the re-resourcing of rural areas and the creation of a bewildering array of continually developing commodities in the form of products, processes, activities and technologies operating at a variety of physical scales and being controlled on a continuum stretching from local operators through to transnational corporations. Without any hope of creating a comprehensive list of such commodities it is important for illustrative purposes to at least attempt a brief, general listing. They include first the expansion of already *well-established* agricultural and horticultural products and activities often underpinned by new technological developments, investment strategies and market opportunities, and operated at different scales from those of the past. Second, are *new* agricultural and horticultural products and activities, again underpinned by technological innovation and diversified markets, but also influenced by changes associated with lifestyle, health and fashion. Included here are niche foods and beverages, organic products, animal fibres and speciality timbers. Third, are peri-urban smallholdings, known popularly as lifestyle blocks or hobby farms. This facet of counter-urbanization has been accompanied in other settings by the ongoing growth of villages and country towns and the movement of former urban dwellers into peri-urban residential settings. Questions relating to changing consumption patterns, idealizations of the rural and anti-urbanism are never far from the surface in these cases. Fourth, is the use of rural landscapes by regional film commissions and place promoters as sites and settings for making both short and feature films. Last in this list of new rural commodities are those associated with recreation and tourism. This category covers a myriad activities ranging from small-scale leisure-related consumption opportunities available to urban residents and tourists in peri-urban areas to the establishment of significant built environments in places distant from cities, often incorporating elements of physical recreation in wilderness settings, the commercialization of cultural performance, guided encounters with nature or the more familiar trio of sun, relaxation and associated pleasurable activities. These leisure- and tourism-related products and activities are often heavily dependent on new recreational and transport technologies, and are linked closely with new consumption patterns, including the rise in tourism participation, increased opportunities for personal mobility, particularly globally and increased numbers of people creating their identities around leisure participation and the purchase of associated products.

These new forms of commodity have in some cases reproduced well-established rural spaces and in others produced new ones. The new rural spaces comprise new resource bases and changed landscapes, and new meanings, practices and imaginations of rural areas. To understand such changes it is essential to interrogate contemporary analyses of commodification and consumption.

COMMODIFICATION AND CONSUMPTION

A useful approach to interpreting commodification is to take Best's (1989) perspective as a starting point. This involves what he calls the 'society of the commodity' and its later transformation into first the 'society of the spectacle' and then the 'society of the simulacrum' (see also Harvey, 1989a; Lash and Urry, 1987, 1994). In the society of the commodity, commodification represents an inversion of exchange value over use value. Objects become commodities when they take on an exchange value over and above their use values and are able to be traded. Marx was not concerned with the commodity *per se* but rather its 'fetishization in capitalist conditions of production and exchange, its magnification to a point where it subsumes and mystifies underlying [and dominating] relations of production' (Best, 1989: 26).

The transformation of the society of the commodity into the society of the spectacle resulted from the changed empirical conditions of late capitalist society and economy (Debord, 1983). Commodification in these terms is a product of a way of living where individuals consume a world made by others rather than producing their own. Spectacle is a complex notion which refers at the simplest level to mass media society but more deeply to the vast institutional and technical apparatus of late capitalism which obscures the experience of continuing alienation. The spectacle therefore pacifies and depoliticizes and takes the form of a permanent 'opium war' (Debord, 1983: 14 quoted in Best, 1989: 29). The narcotics in this case are commodified forms of leisure and entertainment which, while they seem to meet people's needs and satisfy them, are in fact a new form of deprivation which take us from Marx's *being into having* to Debord's *having into appearing*. Here, image takes precedence over material objects and 'the universalization of the commodity form is to be seen as the reduction of reality to appearance, its subsumption to commodity form, its subsequent *commodification*' (Best, 1989: 32).

A further transformation in commodity-form can be discerned from the work of Baudrillard (especially 1983a, 1983b), whose analyses of postmodern society pointed to exchanges carried out in the arena of signs, images and information. In such a society, he argued, commodification involves the absorption of the object into the image, thereby allowing exchange to take place in semiotic form. Commodities reflect sign-exchange values. The focus here for producers and consumers is the conspicuous nature of social meaning, and the commodity will often involve abstract signifiers that can be unrelated to the reality of the commodified place, practice or object. As the commodity is eclipsed by the sign, so it implodes into its imagery, and is characterized by simulacrum in which 'previous distinctions between illusion and reality, signifier and signified, subject and object, collapse, and there is no longer any social or real world of which to speak' (Best, 1989: 24).

The three-stage transformation suggested by Best, from commodity through spectacle to simulacrum, offers a framework to interpret the commodification and consequent re-resourcing of rural areas. The society of the commodity, with its inversion of exchange value over use value, is a very familiar idea in the understanding of rural commodity production in which new food and fibre products are constantly being developed and marketed to replace those that have become unprofitable for a variety of reasons. These new commodities may be developed as part of a diversification of a range of products coming from a particular region to cater for new niche markets, or to take advantage of new or improved production and distribution technologies. Equally the idea of spectacle, particularly in the context of rural leisure and tourism, is now a widely accepted facet of production and consumption processes in rural areas. In these terms, recreation and tourism involve the visual consumption of signs or spectacles which are produced as sites, and are 'transformed into aestheticized spaces of entertainment and pleasure' (Meethan, 1996: 324). Recreation and tourism in the society of the simulacrum is also now an idea that has gained acceptability in the research community (Dann, 1998). It implies a step further into hyper-reality and has as its focus the consumption of representative commodities which have no form or basis in reality (Lash and Urry, 1987, 1994). These hyper-real commodities are sights and sites which are:

> representations of representations or signs of signs. Hence there is a division between natural and semi-natural attractions on the one hand, and artificial ones on the other. The difference lies not in natural versus man-made, but in whether the attraction serves some original function before it becomes touristic. (Pretes, 1995: 4).

Ideas such as these relate to postmodern interpretations of consumption and its reshaping of economy and culture (Dann, 1998; Meethan, 1996; Pretes, 1995). In these terms consumption relates to sign-values; social interaction often occurs around objects of desire. Consumers seek out new commodities signifying sign-exchanges relating to novelty, desirability, the attainment of cultural capital and the importance of intense thrill and spectacle. Consumption is therefore closely linked to advertising which can offer these things to consumers. As Gottdiener put it:

> If anything the pervasive power of advertising has heightened the extent to which commodities of all types are fetishized and made to symbolize attributes that are craved. Advertisers market everything from shoes to toothpaste and underarm deodorant to cars as providing the lucky purchaser with special powers that they would not otherwise have. Products are fetishized because they are bought in the belief that they can enhance the purchasers' abilities to attain their desires for sex, success, notoriety, uniqueness, identity or a sense of self, privileged social status, and personal power. Through the all-pervasive power of advertising in today's society, what Marx only glimpsed one hundred years ago is now an ordinary fact of daily life – people see themselves and others through the possession of commodities. Goods are the tools that signal to others who we want them to think we are and who we want to be. (2000: 4)

Years of advertising have reinforced the fetishization of commodities. The reach of material objects has been extended by commodification to encompass most aspects of human life in industrialized countries. 'There is no want or need that does not already have its correlate in some object manufactured for profit. Consumer society is fetishization writ large' (Gottdiener, 2000: 9). Advertising is therefore closely related to conspicuous consumption (Veblen, 1970 [1899]) in which commodities are sign values, convey social meaning and form the basis of status hierarchies based on social distinctions. Ritzer's (1996) discussion of the 'McDonaldization' and 'McDisneyization' of society in late modernity is also relevant here. Consumption sites are increasingly places where people seek out experiences that are predictable, efficient, calculable and controlled. Such sites are sold and managed using Disney's techniques of branding, marketing, pricing, ancillary product sales, safety and staff performance. Commodification, underpinned by consumption processes and related advertising, consequently have dramatic effects on people and places. Much of the discussion of spectacle and simulacrum has therefore been linked with questions about the transformation of local communities and cultures as they interact with new capitalist stages.

An alternative approach can be taken to the interpretation of commodification. This occurs particularly among anthropologists (e.g., Steymeist, 1996), who use the term 'commoditization' and who have argued that the approach to commodification outlined above lacks a sufficient sense of agency. They take a stronger cultural perspective and write in ways that challenge the notion that commodification is a process that works itself out in the same way in all places as part of capitalism's ever-spreading embrace. In short, they doubt the usefulness of those universalizing elements of the neo-Marxist interpretation of capitalism and consequently also its significant emphasis on inversion, abstraction, false consciousness and domination. Rather, they believe that commodification is a process that is negotiated by actors in particular places to meet particular situations and requirements and it therefore differs in form and content from place to place. In this view, most places are interpreted as being more or less affected by the capitalist market economy and the challenge for social scientists is to interpret the great variability in impact of market influences on those places and the experiences to be had there. Further, social scientists also need to interpret how places are socially constructed and reconstructed in an ongoing and emergent fashion by a purposeful set of actors either in conflict or cooperation with each other. This also demands comparative analysis where the unique and general aspects of commodification in particular places can be highlighted. Methodologically, this requires a situational approach grounded in the history of places and events (Steymeist, 1996).

This alternative view of the commodification of places and custom is particularly useful when discussed in the context of Massey's (1995) account of the contested nature of place meaning and identity. She noted that:

> The description, definition and identification of a place is thus always inevitably an intervention not only into geography but also, at least implicitly, into the (re)telling of the historical constitution of the present. It is another move in the continuing struggle over the delineation and characterization of space-time … it may be useful to think of places, not as areas on maps, but as constantly shifting articulations of social relations through time; and to think of particular attempts to characterize them as attempts to define, and claim coherence and a particular meaning for, specific envelopes of space-time … the identity of places, indeed the very identification of places *as* particular places, is always in that sense temporary, uncertain, and in process. (1995: 188–190)

Commodification may be seen in these terms as one of a number of processes in the creation of place that must be investigated in specific time–space locations at the intersection of the global and local. Increased commercialization, or new forms of commercialization of a place and its people, will not therefore destroy them in the sense of making place and culture meaningless; rather this commercialization will take the form of a new importation around which local and global actors will compete and/or cooperate in the ongoing and emergent construction of the meaning of the place.

Ideas such as these are useful aids in the interpretation of recent, sometimes dramatic, changes to rural areas. They can best be illustrated by first emphasizing exemplars of particular classes of commodity and then highlighting, using regional cases, how they have worked themselves out and influenced particular localities.

NEW RURAL COMMODITIES

Established agricultural and horticultural products

This class of commodity, representative primarily of Best's society of the commodity, includes a vast array of familiar agricultural and horticultural food and fibre staples. In the context of a discussion about the re-resourcing of rural areas such commodities are often being grown in new, perhaps expanded ways, sometimes in new sites often influenced by new technological developments, modes of economic and environmental regulation, investment strategies, market opportunities and operated at different scales from those of the past. Their production patterns and management vary considerably from country to country. Importantly, these commodities result from processes of 'agro-industrialization' in which the constraints of nature have been lessened over time by the application of technological and organizational innovations. Each commodity can be thought of as being the product of a 'farm to table' agro-commodity chain having five stages: scientific and technical inputs to production, the growing of the products, processing, distribution and consumption (Page, 2000). The implication of this, when thought about in terms of commodification processes, is that the production of agricultural and horticultural staples is always in a state of change as the factors above come to bear on them and as actors in agro-commodity chains negotiate issues of chain integration (Le Heron and Roche, 1999).

In this context, dairying is a good example of rural commodification involving a well-established agricultural industry currently experiencing change. Milk production, which in many countries was once largely a family affair involving small farms and herds, is now also a site for very significant transnational corporate investment and international trade concern (Le Heron and Roche, 1999). Large herd dairying (several hundreds if not thousands of cows per unit) requires very large farms, extensive capital investment, intensive pasture management often involving irrigation, and is associated with significant environmental problems associated with excrement disposal, fertilizer run-off and related water quality issues. Where less profitable small-scale dairy farms or other pastoral farming types have been replaced with large herd dairy farms significant landscape change has occurred with the removal of hedgerows and shelter systems to accommodate mobile irrigation units. This has both aesthetic and animal welfare implications that are sometimes not addressed adequately. Large herd dairying also introduces a new demographic profile to rural areas. Large dairy farms require a range of staff, some with pasture management expertise and others who specialize in working with the animal side of the business. This influx of workers often increases the number of families and children in rural areas, thus boosting school rolls and diversifying local community life.

Because of its profitability, dairy commodity production at the local level and its consequent connections to local landscape, environment and community are integrally tied into global investment flows and matters of international diplomatic debate and action. Dairying has, for example, become an issue in the World Trade Organization's attempts to pursue less regulated trade in agricultural products. Recently, the United States and New Zealand governments jointly brought a case to the World Trade Organization's appellate body alleging that Canada had provided illegal subsidies to its dairy industry. The dispute relates to a dairy export scheme the Canadians devised to replace a scheme already ruled as being illegal by the World Trade Organization in 1999. The New Zealand Minister of Agriculture noted that 'Canada's illegal export subsidies cost New Zealand $NZ80 million a year because of their depressing effect on world dairy prices. "That's more than $NZ5,500 per year for each New Zealand dairy farmer," he said' (New Zealand Press Association, 2001: 9).

New agricultural and horticultural products

This class of commodity, representative of Best's society of the commodity but to an extent also relating to the society of the spectacle and to Gottdiener's discussion of consumption, concerns economic changes emphasizing the diversification of rural commodity production to serve largely urban markets with niche products catering particularly for well-resourced consumers. The production of such commodities is again underpinned by technological innovation, but more particularly is influenced by changes associated with consumer lifestyle, health and fashion. Included here are high value niche foods and beverages, animal fibres and speciality timbers. The key to high returns for such products revolves around branding and advertising strategies which combine desirable images of often exotic places and novel consumer goods, promising one or a combination of quality, social status, novel or stylish experience, the attainment of cultural capital, and better health.

The list of these products is significant and includes such diverse categories as boutique vineyards and wines; nuts; speciality animal, bird and fish meats; edible fungi; flowers; dairy products; fruit and berries; and animal and plant fibres. It also includes, in some settings, attempts by producers to add value to their products by establishing themed touristic sites associated with their production. Good examples include such 'commodity parks' as 'the Big Pineapple', 'the Big Strawberry' and 'Nut World' on the Sunshine and Gold Coasts of Queensland in Australia (though see Mansvelt, 1999 for an example of such a commodity park associated with an agricultural staple, milk).

While scale of production in these new rural commodities may not be as extensive when compared with conventional rural commodities, corporate interests and global financing are well represented. Some of this production is also dependent on sophisticated transport arrangements allowing rapid shipping of perishable goods, such as cut flowers and some forms of fruit, across the world. Underpinning this type of commodity production is a relentless search by grower entrepreneurs for new crops to commodify. It may involve re-thinking the potential for commodification of plants and animals long known about but until now not thought to be profitable. It may also involve looking for new areas in which to grow 'conventional' plant crops and animals but in new ways or for distant markets willing to pay high prices for products delivered 'out of season'.

Such production can have significant landscape effects as land use changes. In locations where such niche production becomes concentrated, significant community change may also result. The putative positive elements of these effects have recently encouraged policy-makers in Europe to see niche markets and regional speciality food products as having the potential to help develop peripheral, that is, lagging, regions, although there is yet little empirical evidence to support this idea (Ilbery and Kneafsey, 1999).

A particularly good example of a category of new agricultural and horticultural products displaying many of the characteristics discussed above is organically grown food and fibre products. Organic rural production is presently in a state of flux. With the advent of chemically based and scientifically rational farming in the early twentieth century, organic production became marginalized, and with some exceptions by the 1960s, was largely practised for subsistence purposes. The growth of environmental consciousness in the late twentieth century helped consolidate interest in the production and consumption of organic foods, primarily for their human and environmental health-giving effects. This was the foundation for small-scale, commodified organic production for niche markets in nearby urban centres, and in some cases, for global distribution (Coombes and Campbell, 1998).

It is clear from the literature that the culture of organic production varies from country to country depending on the nature of environmental concerns, dominant agricultural and horticultural forms, and political and regulatory arrangements associated with land use (Kaltoft, 2001). Considerable differences are evident between European activity that is relatively well developed (rising in the 1990s to 2 per cent of total European production and in Austria and Sweden incorporating 10 per cent of farmers and agricultural land) (Michelson, 2001; Page, 2000) and organic production in countries such as the United States, Canada, Australia and New Zealand. Those countries display greater variation in sophistication of organic production techniques, markets and farmer participation, and have very limited production levels when compared with those from conventional agriculture. Despite these international differences it is possible to make some generalizations about the social and environmental organization of much organic production. In many places it is characterized by small family or cooperative enterprises, earning relatively low returns, working in peri-urban sites of small size surrounded by conventional agriculture and horticulture. The exceptions to this are somewhat larger operations, comprising 'new' organic farmers, perhaps former conventional producers, operating organic dairy farms, vineyards, market gardens and orchards, with higher levels of capitalization but still run as family or small cooperatively owned production units. Considerable tension often exists between organic producers and their conventional farming neighbours over questions of landscape management, particularly weed control and care of hedgerows and shelterbelts. These tensions revolve around questions of landscape aesthetics and the meaning of 'messy', herbicide- and pesticide-free organic landscapes when compared with the landscapes of scientifically rational horticulture and agriculture. Tensions also arise over questions about the 'appropriate' use of rural land and arguments about the degree to which organic agriculture, which does without artificial fertilizers, realizes the productive potential of the land (Egoz et al., 2001).

In the 1990s, advocates and practitioners of organic agriculture and horticulture were confronted with a further commodification of organic production. Realizing the potential for profit, particularly in European and North American markets, multinational corporate interests have recently begun to invest in larger-scale organic production in, for example, California (Buck et al., 1997), New Zealand (Campbell and Liepins, 2001; Egoz, 2002) and Ireland (Tovey, 1997). This runs counter to the small-scale, cooperative ethos of earlier modes of organic production and is creating considerable heartache for some small-scale organic farmers. Organic agriculture and horticulture under this new regime of commodification looks much like conventional rural commodity production. This is difficult to accept for 'traditional' organic producers who interpret organic activity as an alternative to the agricultural *and* social mainstream. Despite these concerns Campbell and Liepins (2001) argue, from a New Zealand perspective, that corporate interests have not undermined the production protocols mandated by organic certification agencies.

Ironically, and typical of capital's quest for accumulation, the recent developments in the commodification of organic production and its connection with global capital flows and markets, comes at a time when corporations and governments are also actively investigating the potential to profit further from the application of genetic engineering in food and fibre production. Stark contradictions such as this are an inevitable part of rural commodification, and as they work themselves out the meaning of the rural will change accordingly.

Counter-urbanization

Recent manifestations of counter-urbanization, the movement of formerly urban residents into rural areas, to small rural settlements, smallholdings or stand-alone housing, perhaps sited on large rural lots, relates to Best's societies of the commodity and spectacle and requires paying attention to debates about consumption and changing place meaning. The examples of counter-urbanization being discussed here are largely occurring in peri-urban areas, within commuting distance of significant urban centres. The exception is the recent movement of people often far from urban centres, aided by employment-related computer and telecommunications technology. These are people who once resided in major urban centres, perhaps working in the print media, advertising, publishing and aspects of public relations, who have gone to live in, and work from, remote, semi-remote or village residential settings and make a living via the computer and telephone.

Longstanding expressions of counter-urbanization have more often involved the movement of people to reside in rural settlements, villages and towns, while still depending for employment on opportunities in larger urban settings. Much research in this area has been on rural gentrification, which still continues in many places (see, for example, Birdsall (2001) for a North Carolinian example and Aitchison et al. (2000) for a British discussion) and the remodelling of old rural buildings – churches, barns, etc. – for residential use. Dahms and McComb (1999), synthesizing British, US, Australian and Canadian research, and focusing specifically on south Georgian Bay on the outer edge of Toronto's urban field, suggest also that 'amenity areas' up to 200 km from metropolitan centres continue to be particularly attractive to tourists, retirees and businesses established to cater for these populations.

Smallholdings are a relatively recent form of counter-urbanization in some countries, particularly New Zealand, Australia, Canada and the United States. These take various forms, ranging from large gardens (up to 2 hectares) to relatively significant blocks of land used for small-scale but productive purposes (up to 20 hectares). The environmental effect of these smallholdings is to increase the diversity of peri-urban landscapes. This diversity includes the visual impact of plant crops and animals not commonly found in large-scale farming or horticultural operations. The choice of plant crops and animals made by smallholders is often influenced as much by fashion and sentiment as by economic considerations and residents and visitors to the countryside, when travelling through concentrations of smallholdings, are confronted by new and unusual sights. Longstanding rural residents, often engaged in conventional agriculture or horticulture, sometimes have strong negative views about smallholdings and smallholders. These are based on perceptions that smallholdings represent an attack on agrarian values, specifically that they represent wasteful use of valuable productive land.

Cognizant of the potential for profit in meeting the demand for smallholdings, peri-urban land owners have sold their larger farms for subdivision to well-capitalized developers, who subdivide land and market it in ways that attract financially well-resourced newcomers. This process has, in some settings, been actively facilitated by local government, who benefit from increased tax bases. A recent example of such entrepreneurial development in New Zealand has been the establishment of rural residential subdivisions in which houses are built amongst tree or other plant crops. Depending on climate, soil type and market opportunities these may grow, for example, grapes, olives, nuts, apples, citrus fruit or oak trees (inoculated with edible fungi) or timber trees. Typically, the entrepreneur purchases and subdivides the land, builds infrastructure such as roads, water and sewerage reticulation and community recreational facilities, plants the crop and shelter systems and then sells housing sites on moderately sized blocks on which purchasers build a house of their chosen design. Arrangements for crop maintenance on the housing sites vary, with some purchasers entering into a contractual arrangement with a management company which manages the crop over the entire subdivision, while others buy lots to tend separately from their neighbours.

In many cases the form of counter-urbanization being discussed here rests on a desire for new rural residents to engage in conspicuous consumption, particularly when displaying wealth by building large architecturally designed housing and accompanying gardens, tennis courts and swimming pools, or restoring or modernizing older well-appointed houses once owned by rural plutocrats or other notables. It also rests on anti-urban and rural idyllic sentiment so that rural smallholdings are marketed using various forms of lifestyle advertising of the 'live the dream' sort. Connections may be established here with the representation of niche foods because consumers of rural smallholdings are also again presented with desirable images promising a combination of quality, social status, novel or stylish experience, the attainment of cultural capital, and better health. Not mentioned in the advertisements are such issues of smallholding ownership as significant commuting times to work or recreation in the city, and for those without the financial resources to employ staff or contractors to look after their rural idyll, considerable physical and time-consuming work.

Many smallholders and other ex-urban rural residents are unused to the realities of agricultural production. It comes as a shock that what they thought of as a green and pleasant land is actually a working environment that is often noisy, smelly, dirty, brutal and violent (Short, 1991). Some ex-urbanites attempt to sanitize their new surroundings, by, for example, requesting local government to tightly regulate what they see as noxious activity. This brings them into direct conflict with agriculturalists, horticulturalists and foresters whose capacity to make a living is threatened. In some jurisdictions local government has passed 'right to farm' by-laws to protect producers' interests. These conflicts centre squarely around contested meanings of the rural and the power the combatants have to consolidate their construction of rurality (Massey, 1995).

Quite apart from the changes wrought in the landscape by peri-urban smallholdings a number of other effects are also noticeable. The arrival in the countryside of wealthy ex-urbanites and the consequent inflationary effect on land and property prices has the effect of driving out some former residents of modest means, particularly if they are not landowners (Aitchison et al., 2000; Newby, 1977). This raises another point that should not be overlooked. Not all migrants to rural areas are wealthy. In many countries, low-income urban residents migrate to peripheral rural towns and other settings to lower the costs of housing and other services and to find employment. These and other low-income rural residents, depending on their skills, are often able to find work servicing smallholders who are often very dependent on their new neighbours for support (Hansen, 1995). Smallholding owners also enable the survival of rural services such as schools and rural retail activity. In some regions smallholdings have become very important economically and companies that were established mainly to suit conventional farmers are now offering advisory and other services to smallholders (*The Press*, 2001: 16). While these positive spin-offs are often welcomed, they come at a cost to rural dwellers engaged in conventional rural production.

A further and final commodification effect of counter-urbanization is that ex-urbanites also introduce new cultural and economic forms to rural areas. These range from a plethora of equipment retailing and servicing outlets focused on the needs of smallholders to up-market retailing and cafés serving the relatively sophisticated recreational needs of people who are fundamentally urban but who live in the country. These retailing outlets also attract urbanites to the country for recreational visits and these activities and the development of tourism sights and sites in rural areas are at present a major force for rural change.

Recreation and tourism

It is in the context of rural commodification associated with recreation and tourism (Watson and Kopachevsky, 1994) that all three of Best's characterizations of society come into play with one another. Commodified rural recreation and tourism cover myriad activities ranging from small-scale leisure related consumption opportunities available to local urban residents and tourists in peri-urban areas, to the establishment of significant built environments in places distant from cities, often catering for international visitors (Crouch, 1999).

In discussing the processes of commodification in recreation and tourism, Britton (1991) emphasized that discrete and categorized landscapes are set aside as recreational spaces to satisfy the functional necessity for recreation as an important element in the reconstitution of human capital. These recreational spaces are overlain by elements of the tourist production system and are recognizable because of their markers, which may include brochures, guide books, on-site plaques, reproductions (photographs, art prints, souvenirs), educational material, television travel programmes, reviews in lifestyle magazines, or the incorporation of tourism sights into cultural symbols or national icons. Markers therefore comprise any information or representation that labels a sight as a sight (Britton, 1991: 463).

Markers and the tourist attractions which they represent are produced in three ways: first by co-opting existing cultural attractions or places of interest for touristic purposes; second, by creating new, purpose-built attractions; and third, by engaging in touristic rehabilitation of socially and economically depressed regions. These attractions (their specific sites and their surrounding regions) are commodified first as the tourist industry works to connect its products, and thereby enhance their meaning, to already existing, often attractive public goods (for example, rural landscapes); and second by assimilating originally non-touristic ventures into the tourist system (for example, retail outlets) thereby imparting touristic meanings to specific places and sites. In Britton's (1991) view the primary mechanism for imparting these meanings (apart from the effect of propinquity) is advertising, packaging and market positioning, that is, by selling places; and in the process by suggesting that the tourist will get more than in reality the place can offer (Gold and Ward, 1994; Kearns and Philo, 1993).

Such place marketing or promotion is nicely illustrated by Hughes's (1992) study of the Scottish Tourist Board's selling of the Highlands for tourism purposes. He argued that:

> It has become conventional to speak of 'selling places' [and that] places can be sold and tourists can be treated as place consumers ... That places *can* be sold slips, by elision, into places *will* be sold. The final challenge to the commodification of place is seen as the shaping of a pricing mechanism. (1992: 39)

Interpreting promotional texts, Hughes suggested that the Scottish Tourist Board has created a sanitized and picturesque mythical Highlands quite unlike the real poverty-stricken and oppressed Highlands of history. Tourists are essentially sold stories of this place which blur the boundaries between history and fiction.

> The purpose behind this management of Scottish myths is overtly commercial and has resulted in the commodification of images of place ... Thus places are being constructed in the image of tourism, both socially and physically ... The philosophical shift, embodied in this commercially-motivated representation of place, is from historical representation to fictional depiction. (1992: 39)

In similar cases, regional film commissions and related organizations intent on boosting regional economic fortunes attempt to attract film-makers to rural areas. Their objective is in part to stimulate economic growth directly by building a film industry; but in conjunction with these developments regional tourism organizations also attempt to sell rural places portrayed in films to tourists. *Braveheart* (the Scottish Tourist Board) and *The Lord of the Rings* (Tourism New Zealand) are very good recent examples (Preston, 2000).

New recreational rural commodities are often based on locations not previously commercialized but important for recreation. Alternatively, these locations may not have been recreational sites at all. As the commodification process progresses, new sites are drawn into the commercial embrace and are given new meaning. Such sites may include elements of the built and natural environment (Pawson and Swaffield, 1998). While some sites involve new commercial uses of privately owned areas, others involve the privatization or at least commercialization of publicly owned areas such as national parks and similar places reserved for nature conservation. Because of the relative rarity of such conservation areas they are sought after by recreationists who are prepared to pay for using them. This strategy is particularly attractive to governments of a neo-liberal persuasion because user fees are an opportunity, other than through direct taxation, to raise funds. Effectively, in cases such as this, taxpayer-funded public areas become playgrounds largely for those who can afford to pay. As the commodification process in national parks progresses such areas experience considerable management pressures over who should use parks and what activities should be conducted in them (Robson, 2001). Such situations are often the subject of intense policy debate.

Many new forms of rural recreation and tourism are commodifications of well-established activities that have been practised by rural local residents as part of their everyday recreation. Under changed economic circumstances, necessitating an entrepreneurial re-evaluation of the potential profitability of those activities, what was once local recreation becomes a commodity sold to visitors to the region. Good examples include opportunities for boating, rafting, cycling, fishing, horse riding and walking. In their newly commodified forms they may have new technology added to them which makes them attractive to visitors. Jet boating, rafting and kayaking on white water, four-wheel drive vehicle adventures, mountain biking, some eco-tourism activities and walking using sophisticated back country lodges are all examples of technologically based commodified rural recreational activities catering for local and international visitors on a fee-paying basis (Cloke and Perkins, 1998, 2002). The search for new forms of tourism commodity has meant that the meaning of some rural spaces has been re-made to the extent that completely new ways of thinking about and managing them are established, taking the form of rural place-myths (Cloke and Perkins, 1998).

These changes mentioned above have occurred in conjunction with the rise of post-mass tourism or post-tourism and the search for niched tourism experiences focused on the consumption of signs and symbols (Cloke and Perkins, 1998, 2002; Perkins and Thorns, 2001; Urry, 1990/2002, 1992, 1995). Mitchell (1998), in her study of heritage shopping villages, argues that it is important to take cognizance of the roles of individual entrepreneurs and their interactions with local government in particular localities in order to gain a better understanding of the intensity and range of the commodification processes at work (see also Harvey, 1989b; Kneafsey, 2001). She situates her discussion in debates about the idealization of the countryside and reactions to the chaotic nature of urbanization and nostalgia for a simpler rural life. Consistent with these anti-urban views, in the years after the Second World War, wealthy and well-educated urban residents began to use peri-urban regions for recreational purposes which in turn created recreational amenity landscapes and reproduced the mythology of the countryside ideal. Recognizing the potential for profit, in recent times entrepreneurs in conjunction with local governments have attempted to satisfy consumers' desire for this imagined countryside. Their investments have re-created pre-industrial village landscapes and reproduced pre-industrial commodities. The first of these contributes to visual representation of the ideal, the latter gives the ideal its material form (Mitchell, 1998: 275). Both depend on, and reinforce, a strong sense of nostalgia among consumers.

Such efforts therefore require the restoration or reconstruction of vernacular buildings or streetscapes upon which tourists can gaze and perform. Such landscapes must be unified and present a visual experience consistent with the imagined landscape, one devoid of 'jarring elements' that would detract from the visual simulacrum (Dorst, 1989: 29, quoted in Mitchell, 1998: 275). Simply visualizing the countryside is not all there is. Visitors also want to buy products, obvious reproductions that symbolize the 'idealized [pre-industrial] mode of production' (Dorst, 1989: 64, quoted in Mitchell, 1998: 275). The consumer purchases both an object (for example, a quilt), and the ideal of the object (quiltness) with all its attendant associations (Dorst, 1989: 64). This conjunction of visual appeal and efficient retailing is of course reminiscent of Ritzer's (1996) McDonaldization thesis.

Mitchell also discusses how the idea of creative destruction may be applied to rural commodification. It has five stages: early commodification, advanced commodification, pre-destruction, advanced destruction and post-destruction. As postmodern consumers in search of nostalgia return to rural roots, they engage in initially limited opportunities for the consumption of rural tradition thus providing entrepreneurs with the opportunity to profit. This allows them to reinvest, thus extending the degree of commodification of rural areas. In time, if left unchecked by regulatory agencies like local government, further commodification begins to undermine the features that initially attracted consumers. Where that occurs, and as a result of still further commodification, consumption levels begin to drop and the area begins to decline as levels of congestion, noise and other ills of overcrowding drive visitors away. Finally, as opportunities to accumulate level off, entrepreneurs seek new landscapes in which to invest and the old area is abandoned perhaps later becoming a site for re-investment of another sort.

Less dependent on peri-urban sites, but also sometimes associated with them (Schöllmann et al., 2000), are rural recreation and tourism of an adventurous nature which have, for example, become popular in Europe, Canada, Australia and New Zealand. Cloke and Perkins (1998, 2002) used the example of the commodification of adventure tourism in New Zealand to discuss these issues. There, many well-established local outdoor recreational activities have become commodified for touristic purposes as part of a mix of activities provided to international tourists visiting the country. This commodification has had significant place-related impacts. The representations of adventure tourism have altered the place-meanings of parts of rural New Zealand. Places previously represented as paradisal and having historically important scenery have been attributed new meanings associated with fresh, youthful thrills, and the conquering and admiration of nature. Adventure tourism has played its part in the production of a 'designer rurality' which is 'placed' alongside other rural forms reflecting pastoralism, alpine splendour, and the like. As adventurous tourism becomes commodified, then, so do the places in which that tourism is sited. Place-change involves important social and cultural changes as well as physical impacts (Morris, 1995: 180–181). The rise of adventure tourism attractions is intimately associated with the translation of places, and physically challenging and exciting recreational activities, into commodities. The emergence of new tourist attractions such as these represents part of a continuing progression of the social relations of capitalism dictated by profit motives and revolving around the nexus of exchange (Watson and Kopachevsky, 1994).

CONCLUSION

The central argument of this chapter has been that commodification is an integral part of the re-resourcing of rural areas. It works itself out in myriad ways across the globe as capital seeks to accumulate and interacts with national and international regulatory arrangements and local production and consumption practices. In this process particular interconnected and overlapping *forms* of rural commodity are maintained, adapted and created, and so therefore, are rural landscapes, productive processes, technologies, social arrangements, activities and practices. Consequently, the meaning of the rural is also continually changing for residents, visitors and those who view it from afar.

Most obvious among these rural commodity forms are a wide range of *products*, and I have illustrated these by discussing examples of well-established and new agricultural and horticultural commodities; a diverse range of rural settlement types associated with counter-urbanization; short and feature films incorporating rural landscapes made with the support of regional film commissions and place promoters; and a plethora of recreation and tourism products and activities. These products are integrally linked to commodified forms of *production*, some of which are well established, but these exist alongside many new ways of doing things. Rural areas are therefore sites in which old and new production practices and technologies are applied, developed and interact with each other. Some of the products and production processes discussed above are closely linked to commodity forms which may best be discussed using the terms *attraction* and *experience*. The sale of new and 'boutique' foods and beverages, often at the point of production; the diversification in patterns of counter-urbanization; and providing a significant array of commercial rural recreation and tourism opportunities are based on the re-making of the rural as a set of places that are attractive to those with money to spend on consumption goods and fashionable experiences. It follows, therefore, that *land* and *lifestyles* are centrally important commodity forms arising from the process of rural commodification. Particular types of rural lifestyle are available for purchase by those who can afford to do so. Land, perhaps the most basic of rural commodities, and the lifestyles of the people who live on it, or who visit it irregularly, are also subjected to a variety of material and symbolic forces as it is marketed, exchanged, subdivided, regulated, landscaped, ploughed, fertilized, planted, built on and fought over. The changing meaning of the rural and the ways people make a living in rural areas is intimately tied up with the ways these forces work themselves out.

This raises an interesting question about social agency in rural commodification. Many of the studies noted in this chapter speak directly to that question, and combine elements of neo-Marxist and situational approaches in the study of significant individuals, groups and private and public sector agencies in the commodification process. This focus on entrepreneurialism is important and there is considerable scope for more work in this area, particularly in the links between regulatory change and commodification processes.

Also alluded to in the literature, but to a far lesser extent, are studies of resistance to commodification in rural areas. More work needs to be done in this area as well. This is where Massey's (1995) work is instructive. If one accepts that commodification is less about simply destroying culture, duping consumers and making places meaningless (Cohen, 1988), and more about change based on new importations to places around which local and global actors compete and/or cooperate in the ongoing and emergent construction of meaning, then there is considerably more scope to interpret interactions between agents of commodification and those who resist it. There is scope to focus on how people choose to engage in commodification processes, how they evaluate their options, and how social and cultural values, and sentiment, influence decision-making processes. Questions of constraints on agency are also important, as discussed briefly by Mitchell (1998) when she alluded to the possibilities, albeit challenging, of communities taking greater control through planning processes of the early stages of the creative destruction process to ensure that advanced stages do not develop and 'kill the goose that laid the golden egg'. Similarly, researchers such as Johnston and Edwards (1994) in the context of a study of the commodification of Himalayan mountaineering highlight how the commercialization of cultural performance and associated tourist places is often contested. This underscores the possibility that commodification may be limited by local actors seeking to ascribe new or different meanings to cultural performances and places (Meethan, 1996; Steymeist, 1996). It also raises questions about whether such commodified performances are the 'selling' or 'presentation' of culture (Steymeist, 1996). Game (1991), using an urban example, discusses an ongoing contest in a suburb of Sydney, Australia, known as Bondi Beach. She outlines the battles between those who would 'revitalize' Bondi and make it an international tourist attraction, and local residents who, while they are not unified in their desires for the beach area, want to resist this extra degree of commercialization and keep Bondi for Australians. Studies such as these emphasize the potentially active, reflexive roles of rural residents, both recent and established, and visitors to those areas. They also help focus attention on the ways these actors deal with exhortations to consume. Goss (1993: 664) was mindful of this when he suggested that 'consumers should not be conceived of as merely passive receivers of cultural meanings produced by advertisers'.

ACKNOWLEDGEMENTS

Thanks to David Fisher and Suzanne Vallance, Environment, Society and Design Division, Lincoln University; Judi Miller, School of Education, University of Canterbury; Richard Stevens, Horticulture Management Group, Lincoln University; David Thorns, School of Sociology and Anthropology, University of Canterbury; Ann Winstanley, ESR Ltd, Christchurch; and Stephen S. Birdsall of the Department of Geography, University of North Carolina–Chapel Hill for helpful comments on an earlier draft of this chapter.

REFERENCES

Aitchison, C., MacLeod, N. E. and Shaw, S. J. (2000) *Leisure and Tourism Landscapes: Social and Cultural Geographies*. London and New York: Routledge.

Argent, N. (2002) 'From pillar to post? In search of the post-productivist countryside in Australia', *Australian Geographer*, 33 (1): 97–114.

Baudrillard, J. (1983a) *Simulations*. New York: Semiotext(e).

Baudrillard, J. (1983b) *In the Shadow of the Silent Majority*. New York: Semiotext(e).

Best, S. (1989) 'The commodification of reality and the reality of commodification: Jean Baudrillard and post-modernism', *Current Perspectives in Social Theory*, 19: 23–51.

Birdsall, S. S. (2001) 'Tobacco farmers and landscape change in North Carolina's old belt region', *Southeastern Geographer*, XXXXI (1): 65–73.

Britton, S. (1991) 'Tourism, capital, and place: towards a critical geography of tourism', *Environment and Planning D: Society and Space*, 9: 451–478.

Buck, D., Getz, C. and Gutham, J. (1997) 'From farm to table: the organic vegetable commodity chain of Northern California', *Sociologia Ruralis*, 37: 3–20.

Bunce, M. (1994) *The Countryside Ideal: Anglo-American Images of Landscape*. London: Routledge.

Campbell, H. and Liepins, R. (2001) 'Naming organics: understanding organic standards in New Zealand as a discursive field', *Sociologia Ruralis*, 41 (1): 21–39.

Cloke, P. (1993) 'The countryside as commodity: new rural spaces for leisure', in S. Glyptis (ed.), *Leisure and the Environment: Essays in Honour of Professor J. A. Patmore*. London: Belhaven Press.

Cloke, P. and Perkins, H. C. (1998) 'Cracking the canyon with the awesome foursome: representations of adventure tourism in New Zealand', *Environment and Planning D: Society and Space*, 16: 185–218.

Cloke, P. and Perkins, H. C. (2002) 'Commodification and adventure in New Zealand tourism', *Current Issues in Tourism*, 5 (6): 521–549.

Cohen, E. (1988) 'Authenticity and commodification in tourism', *Annals of Tourism Research*, 15: 371–386.

Coombes, B. and Campbell, H. (1998) 'Dependent reproduction of alternative modes of agriculture: organic farming in New Zealand', *Sociologia Ruralis*, 38 (2): 127–145.

Crouch, D. (ed.) (1999) *Leisure/Tourism Geographies: Practices and Geographical Knowledge*. London and New York: Routledge.

Dahms, F. and McComb, J. (1999) 'Counterurbanization, interaction functional change in a rural amenity area – a Canadian example', *Journal of Rural Studies*, 15 (2): 129–146.

Dann, G. (1998) 'The pomo promo of tourism', *Tourism, Culture and Communication*, 1: 1–16.

Debord, G. (1983) *Society of the Spectacle*. Detroit: Red and Black.

Dickens, P. (2000) 'Society, space and the biotic level: an urban and rural sociology for the new millennium', *Sociology*, 34 (1): 147–164.

Dorst, J. D. (1989) *The Written Suburb*. Philadelphia, PA: University of Pennsylvania Press.

DuPuis, E. M. and Vandergeest, P. (eds) (1996) *Creating the Countryside: The Politics of Rural and*

Environmental Discourse. Philadelphia, PA: Temple University Press.

Egoz, S. (2002) 'The landscapes of organic agriculture: a New Zealand case study'. Unpublished PhD thesis, Lincoln University, Canterbury, New Zealand.

Egoz, S., Bowring, J. and Perkins, H. C. (2001) 'Tastes in tension: form, function, and meaning in New Zealand's farmed landscapes', *Landscape and Urban Planning* (Special Issue: Bridging Human and Natural Sciences in Landscape Research), 57: 177–196.

Fold, N. (2000) 'Oiling the palms: restructuring of settlement schemes in Malaysia and the new international trade regulations', *World Development*, 28 (3): 473–486.

Game, A. (1991) *Undoing the Social: Towards a Deconstructive Sociology.* Toronto: University of Toronto Press.

Gilg, A. (ed.) (1992) *Restructuring the Countryside: Environmental Policy in Practice.* Aldershot: Avebury.

Gold, J. and Ward, S. (eds) (1994) *Place Promotion: The Use of Publicity and Marketing to Sell Towns and Regions.* Chichester: John Wiley.

Goss, J. (1993) 'Placing the market and marketing place: tourist advertising of the Hawaiian Islands, 1972–92', *Environment and Planning D: Society and Space*, 11: 663–688.

Gottdiener, M. (ed.) (2000) *New Forms of Consumption: Consumers, Culture and Commodification.* Lanham, MD: Rowman and Littlefield.

Griggs, P. (2002) 'Changing rural spaces: deregulation and the decline of tobacco farming in the Mareeba–Dimbulah Irrigation Area, Far North Queensland', *Australian Geographer*, 33 (1): 43–61.

Hansen, E. C. (1995) 'The great Bambi war: Tocquevillians versus Keynesians in an Upstate New York county', in J. Schneider and R. Rapp (eds), *Articulating Hidden Histories: Exploring the Influence of Eric R. Wolf.* Berkeley, CA: University of California Press.

Harberl, H., Erba, K-H., Krausmann, F., Adensam, H. and Schulz, N. B. (2003) 'Land use change and socio-economic metabolism in Austria. Part II: land use scenarios for 2020', *Land Use Policy*, 20 (1): 41–49.

Harvey, D. (1989a) *The Conditions of Postmodernity.* Oxford: Oxford University Press.

Harvey, D. (1989b) 'From managerialism to entrepreneurialism: the transformation in urban governance in late capitalism', *Geografiska Annaler*, 71B (1): 3–17.

Hoggart, K. (1990) 'Let's do away with the rural', *Journal of Rural Studies*, 6: 245–257.

Hoggart, K. and Paniagua, A. (2001) 'What rural restructuring?', *Journal of Rural Studies*, 17: 41–62.

Hughes, G. (1992) 'Tourism and the geographical imagination', *Leisure Studies*, 11: 31–42.

Ilbery, B. and Kneafsey, M. (1999) 'Niche markets and regional speciality food products in Europe: towards a research agenda', *Environment and Planning A*, 31: 2207–2222.

Johnston, B. R. and Edwards, T. (1994) 'The commodification of mountaineering', *Annals of Tourism Research*, 21 (3): 459–478.

Kaltoft, P. (2001) 'Organic farming in late modernity: at the frontier of modernity or opposing modernity?', *Sociologia Ruralis*, 41 (1): 146–158.

Kearns, G. and Philo, C. (1993) *Selling Places.* Oxford: Pergamon.

Kneafsey, M. (2001) 'Rural cultural economy: tourism and social relations', *Annals of Tourism Research*, 28 (3): 762–783.

Krausmann, F., Haberl, H., Schulz, N. B., Erb, K-H., Darge, E. and Gaube, V. (2003) 'Land use change and socio-economic metabolism in Austria. Part I: Driving forces for land use change, 1950–1995', *Land Use Policy*, 20 (1): 21–39.

Lash, S. and Urry, J. (1987) *The End of Organized Capitalism.* Cambridge: Polity Press.

Lash, S. and Urry, J. (1994) *Economies of Signs and Space.* London: Sage.

Le Heron, R. and Pawson, E. (1996) *Changing Places: New Zealand in the Nineties.* Auckland: Longman Paul.

Le Heron, R. and Roche, M. (1999) 'Rapid reregulation, agricultural restructuring and the reimaging of agriculture in New Zealand', *Rural Sociology*, 64 (2): 203–218.

Mansvelt, J. (1999) 'Consuming spaces', in R. Le Heron, L. Murphy, P. Forer and M. Goldstone (eds), *Explorations in Human Geography: Encountering Place.* Auckland: Oxford University Press.

Massey, D. (1995) 'Places and their pasts', *History Workshop Journal*, 39: 182–192.

McCarthy, S., Matthews, A. and Riordan, B. (2003) 'Economic determinants of private afforestation in the Republic of Ireland', *Land Use Policy*, 20 (1): 51–59.

Meethan, K. (1996) 'Consuming (in) the civilized city', *Annals of Tourism Research*, 23 (2): 322–340.

Meinig, D. W. (1979) 'Symbolic landscapes: some idealizations of American communities', in D. W. Meinig (ed.), *The Interpretation of Ordinary Landscapes: Geographical Essays.* New York: Oxford University Press.

Memon, P. A. and Perkins, H. C. (eds) (2000) *Environmental Planning and Management in New Zealand.* Palmerston North: Dunmore Press.

Michelson, J. (2001) 'Recent development and political acceptance of organic farming in Europe', *Sociologia Ruralis*, 41 (1): 3–20.

Mitchell, C. J. A. (1998) 'Entrepreneurialism, commodification and creative destruction: a model of post-modern community development', *Journal of Rural Studies*, 14 (3): 273–286.

Mormont, M. (1990) 'Who is rural? or How to be rural: towards a sociology of the rural', in T. Marsden, P. Lowe and S. Whatmore (eds), *Rural Restructuring.* London: David Fulton.

Morris, M. (1995) 'Life as a tourist object in Australia', in M. Lanfant, J. Allcock and E. Burner (eds), *International Tourism: Identity and Change.* London: Sage. pp. 96–109.

New Zealand Press Association (2001) 'NZ dairy outcry sat on', *The Press* (Christchurch, New Zealand), 5 December.

Newby, H. (1977) *The Deferential Worker.* London: Allen Lane Penguin Books.

Newby, H. (1979) *Green and Pleasant Land: Social Change in Rural Britain*. Harmondsworth: Penguin Books.

Page, B. (2000) 'Agriculture', in E. Sheppard and T. J. Barnes (eds), *A Companion to Economic Geography*. Oxford: Blackwell. pp. 242–256.

Pahl, R. E. (1968) 'The rural–urban continuum', in R. E. Pahl (ed.), *Readings in Urban Sociology*. Oxford: Pergamon Press. pp. 263–305.

Pawson, E. and Swaffield, S. (1998) 'Landscapes of leisure and tourism', in H. C. Perkins and G. Cushman (eds), *Time Out? Leisure, Recreation and Tourism in New Zealand and Australia*. Auckland: Addison–Wesley Longman. pp. 254–270.

Peck, J. (2000) 'Places of work', in E. Sheppard and T. J. Barnes (eds), *A Companion to Economic Geography*. Oxford: Blackwell. pp. 131–148.

Perkins, H. C. (1989) 'The country in the town: the role of real estate developers in the construction of the meaning of place', *Journal of Rural Studies*, 5 (1): 61–74.

Perkins, H. C. and Cushman, G. (eds) (1998) *Time Out? Leisure, Recreation and Tourism in New Zealand and Australia*. Auckland: Addison–Wesley Longman.

Perkins, H. C. and Thorns, D. C. (2001) 'Gazing or performing? Reflections on Urry's tourist gaze in the context of contemporary experience in the Antipodes', *International Sociology*, 16 (2): 185–204.

Phillips, M., Fish, R. and Agg, J. (2001) 'Putting together ruralities: toward a symbolic analysis of rurality in the British mass media', *Journal of Rural Studies*, 17: 1–27.

Preston, J. (2000) 'Film and place promotion'. Unpublished Master of Applied Science thesis, Lincoln University, Canterbury, New Zealand.

Pretes, M. (1995) 'Postmodern tourism: the Santa Claus industry', *Annals of Tourism Research*, 22 (1): 1–15.

Ray, C. (1998) 'Culture, intellectual property and territorial rural development', *Sociologia Ruralis*, 38 (1): 3–20.

Ritzer, G. (1996) *The McDonaldization of Society*, rev. edn. Thousand Oaks, CA: Pine Forge Press.

Robson, S. (2001) 'Parking problems', *The Press* (Christchurch, New Zealand), 13 October, p. 4.

Robson, S. (2003) 'EU farm subsidies reduced', *The Press* (Christchurch, New Zealand), 27 June, p. B1; 'NZ set to reap benefits from EU reforms', *The Press* (Christchurch, New Zealand), 28 June, p. A3.

Schöllmann, A., Perkins, H. C. and Moore, K. (2000) 'Intersecting global and local influences in urban place promotion: the case of Christchurch, New Zealand', *Environment and Planning A*, 32: 55–76.

Short, J. R. (1991) *Imagined Country: Environment, Culture and Society*. London: Routledge.

Smailes, P. J. (2002) 'From rural dilution to multifunctional countryside: some pointers to the future from South Australia', *Australian Geographer*, 33 (1): 79–95.

Smith, N. (2000) 'Marxist geography', in R. J. Johnston et al. (eds), *The Dictionary of Human Geography*, 4th edn. Oxford: Blackwell. pp. 485–492.

Steymeist, D. H. (1996) 'Transformation of Vilavilairevo in tourism', *Annals of Tourism Research*, 28 (1): 1–18.

The Press (2001) 'Assistance available for small blockholders', *Anon* (Christchurch, New Zealand), 11 October, p. 16.

Tovey, H. (1997) 'Food environmentalism and rural sociology: on the organic farming movement in Ireland', *Sociologia Ruralis*, 37 (1): 21–37.

Urry, J. (1990/2002) *The Tourist Gaze*. London: Sage.

Urry, J. (1992) 'The tourist gaze "revisited"', *American Behavioural Scientist*, 36: 172–186.

Urry, J. (1995) *Consuming Places*. London: Routledge.

Veblen, T. (1970 [1899]) *The Theory of the Leisure Class: An Economic Study of Institutions*. London: Allen and Unwin.

Watson, G. L. and Kopachevsky, J. P. (1994) 'Interpretations of tourism as commodity', *Annals of Tourism Research*, 24 (4): 643–660.

18

Agricultural production in crisis

Jan Douwe van der Ploeg

INTRODUCTION

In this chapter I will argue that world agriculture is involved in a persistent and many-faceted crisis. This crisis is not just externally induced; it stems also from 'internal' relations, that increasingly interact with, and enforce the impact of, outside pressures exerted on agriculture. The underlying interdependencies are such that policy interventions aiming at the alleviation of, at least some of, the expressions of the crisis, often trigger a deepening of other aspects within the same or in other time–space locations. The discussion of this complex and persistent crisis will be used to develop a re-actualized definition of the old but never disappearing 'agrarian question'. I will pay special attention to an often unnoticed aspect, that is that the use of natural resources is subjected to an undeniable process of degradation – the negative effects of which also extend to the labour process and the quality of food.

Subsequently, the main developmental tendencies of today's world agriculture will be discussed. It will be argued that they contain the seeds of the 'coming crisis', that will characterize world agriculture during the twenty-first century. Counter tendencies to this coming crisis will be analysed in order to reveal alternative trajectories and a newly emerging paradigm.

THE CURRENT CRISIS: ITS IMMEDIATE EXPRESSIONS

The current crisis in agricultural production is characterized, at the global level, by a multitude of often highly contrasting, but strongly interrelated expressions. In many developing countries agricultural production offers insufficient employment opportunities, food and dignity to those who are in desperate need to have a place to work and live. Instead, nearly everywhere, the dynamics of the main agricultural systems imply an ongoing expulsion of agricultural labourers and a marginalization of small farmers. Even the instruments designed to alleviate these problems, such as land reform, often become a vehicle for further marginalization (Thiesenhuisen, 1995; van der Ploeg, 1998). Income levels are low, whilst those who are self-employed periodically face prolonged periods of deprivation (Griffin, 1981).

Specific 'pockets' apart, levels of production are mostly in a state of chronic stagnation. And in those (limited) time–space locations where abundant and, at first sight, continued growth occurs, this expansion is rarely sustainable and often wreaks considerable ecological damage.

Notwithstanding the many differences between developing and industrialized countries (Europe, North America, Australia and New Zealand), the latter face similar trends and problems in their agricultural sectors. There is an ongoing rural exodus (even in periods of extended urban unemployment; Saraceno, 1996), income levels are sometimes painfully low[1], which often induces negative effects in the wider rural economy. Agriculture continues to be a source of considerable ecological damage, whilst the quality of the food produced is increasingly contested (see Bussi, 2002, for a well-documented overview). At the same time regulatory schemes intended to mediate these problems often do not achieve the intended results and are sometimes counterproductive (Frouws, 1993, Frouws et al., 1996).

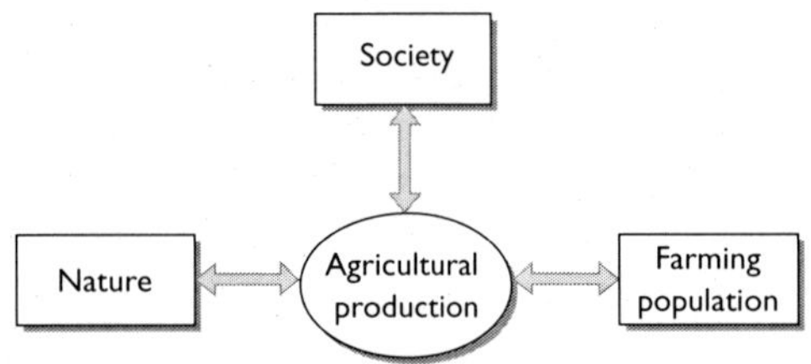

Figure 18.1 The embedness of farming

For many decades, the disquiet resulting from this situation has been dampened by repeated policy proposals and programmes aiming at further scale enlargement and more 'entrepreneurial behaviour' on the part of the farmers. It was thought that the combination of these two changes would solve most problems once and for all. However, it is now increasingly recognized that such 'ways forward' are part-and-parcel of the 'squeeze on agriculture' (Owen, 1966). Instead of securing adequate income levels and stable prospects, accelerated scale-enlargement and entrepreneurial farming strategies are now feeding and pushing the 'treadmill' (Ward, 1993; Marsden, 1998). At the same time, state regulation is blocking a full unfolding of alternatives that are, time and again, being rediscovered or newly created at grass-roots level.

Within this general pattern, there is a notable social and spatial division of labour, both in the developing and the industrialized world. Over the past five decades clear 'growth poles' emerged, whilst the surrounding areas were increasingly restructured into 'hinterlands'. At global level there are now clear, albeit manifold and complex, relations between centres and peripheral areas. These relations have evolved around a new worldwide regime that favours (a) an increased specialization of many Third World countries in the production of cheap commodities and raw material for Western agriculture, (b) the control over the associated commodity flows by a few agribusiness groups (Bonnano et al., 1994) and (c) a nearly 'structural' dependency of Third World countries on the West for the import of a growing proportion of domestically consumed food products. Together these relations compose, as it were, a Gordian knot (de Janvry et al., 2002) – nearly impossible to unravel and to redress, whilst simultaneously being an ugly expression of dependency, underdevelopment, exploitation, subordination and exclusion. This framework entails considerable fluidity as production centres for fruits and vegetables, meat, etc., are constantly moving over the globe, whilst other products, such as fats, oils and sugar, are increasingly replaced by substitutes (Ruivenkamp, 1989; Cook, 1994; Llambi, 1994). Yet, its basic features remain remarkably stable (Arce and Marsden, 1993). In the West people face a range of unintended consequences, partly related to the emergence of animal diseases (some of which threaten public health) and partly expressed in a growing distrust in the 'food industry'[2] (Ritzer, 1993; Schlosser, 2002).

UNDERLYING PATTERNS

Agricultural production represents an activity that is aligned, both materially and symbolically, with nature, society and the interests and prospects of those directly involved in agriculture and food production (see Figure 18.1).

Agriculture is built on the use of local and regional eco-systems, whilst its main objects of labour represent 'living nature' (Sevilla Guzman and Gonzalez, 1990; Toledo, 1992; van der Ploeg, 2003). Hence agriculture, as a productive activity, needs not only to be geared to the specificities of these eco-systems and the implied natural resources – it needs also to reproduce them continuously.

At the same time, agricultural production is expected to meet a range of changing and expanding societal needs (Delors, 1994; Cork Declaration, 1996; Depoele, 1996; Fischler, 1996, 1998; Countryside Council, 1997; Groupe de Bruges, 1998; Scottish Office, 1998). This required alignment presumes an ongoing and active process of coordination. Failure to achieve such coordination is likely, sooner or later, to have serious natural or social repercussions. This may occur through massive ecological degradation, serious stress and diseases in plant and animal systems and increasing natural counter productivity. Additionally it may give rise to socio-political movements that reject current agricultural practices and demand more or less radical measures to improve environmental conditions, animal welfare, food quality and to protect landscapes and natural values.

Finally there are the interests and prospects of those directly involved (or wanting to be involved) in agricultural production. This axis represents the 'aspirations for emancipation' of farmers, rural labourers, landless people, etc. Even if broad societal needs are being met and agricultural production is well aligned with the need to reproduce nature, a considerable agrarian question[3] remains if those directly involved in agriculture are living in misery, or lacking any prospect for further improvement of their 'rural condition'.[4]

Taken together these interrelations compose a three-dimensional space, in which agricultural development processes can be located (see Figure 18.1). This same space also allows for an assessment of the coordination mechanisms through which agricultural production needs to be embedded in order to create the required coherence *vis-à-vis* nature, society and primary producers.

For many centuries, coordination was internalized within, and secured by, *regional farming styles* (Dumont, 1970; Hofstee, 1985).[5] Regional agriculture was attuned, by necessity, to the available (and slowly evolving) eco-systems. Regional markets, sometimes accompanied by international markets, regulated the outflow of products and their prices. These markets functioned as 'outlets' for farming – yet they did not function as the organizing principle that determined the structure and development of agricultural production. Sometimes painful socio-political struggles (and/or for example, migratory processes) were necessary to achieve at least minimal levels of alignment with the interests and prospects for those involved in primary production.

With the enlargement of markets, new expressions of agricultural crises emerged, which in their turn (especially following the 1880 and the 1930 crises) led to the creation of agricultural policies that included different types of interventions into these markets. Even in those countries (particularly England and the Netherlands) where a strictly liberal policy of non-intervention dominated, agriculture was strongly defended (and hence 'protected') through the development of an extensive structure for applied research, extension, schooling and training, as well as through 'quality-oriented policies' that aimed at preventing opportunism and fraud. The markets *as such* proved to be inadequate coordination mechanisms (Koning, 1982; de Hoogh and Silvis, 1988). The disequilibrium between a dispersed supply on the one hand, and an often highly concentrated demand (agro-industrial groups, trading companies) on the other, as well as the unpredictability of the weather, harvests and animal health, made intervention indispensable. Agrarian policy is, therefore, a crucial, albeit continuously contested, institution: it is needed to provide the coordination that cannot be achieved solely through the market. In this respect it is telling that farmers too have made many attempts to coordinate their relationships with markets (for example through the creation of cooperatives and collective auctions). Some of these attempts fell by the wayside, some have been very successful, whilst others have 'degraded' over time into agencies that are indistinguishable from other market institutions.

Within the thus emerging and basically national matrices, regional farming styles evolved in such a way that agriculture developed, at regional, national and supranational levels, into a richly chequered whole, one in which each regional style probably represented the most optimal solution within the framework presented in Figure 18.1. The resultant differences in farming styles were captured by the English farmer Robertson Scott who noted:

> Every year that I live in the country, and every year that I know more of what the people who work the land of the United Kingdom are doing, I realise more fully the profound agricultural truth underlying the remark of a skilled Dutch farmer to an English landowner ... 'If you should come to Holland to farm, you would imitate me, but if I were to go to farm in England I would imitate you.' (1912: ix)

During the second half of the twentieth century, the once organic relationship between eco-systems and regional farming styles, became increasingly unravelled. Agricultural production was disconnected, in an often far-reaching way, from the eco-systems. Resources that were once derived from nature (and reproduced and improved in, and through, the process of agricultural production) were increasingly replaced by new artefacts produced and commercialized by agro-industry. As a consequence the degree of commoditization increased considerably (Long et al., 1986; van der Ploeg, 1990; Bernstein and Woodhouse, 2000) whilst several new dependency relations emerged (analysed by Benvenuti, 1989, as technological-administrative relations, through which the nature of the production process became increasingly prescribed and governed). On the other side of the equation relations between agricultural production and society increasingly came under the control of agro-industry, wholesalers and, more recently, of the large retail chains that nowadays control considerable shares of food supply to the consumers (Wrigley and Lowe, 1996; Marsden et al., 2000).

Due to these changes, agricultural production became increasingly embedded in, dependent upon and consequently (re-)structured by a complex whole of interacting market-agencies that, to a considerable degree, prescribe and control the agricultural process of production. This control, on both the input and the output side of the farm, increasingly intertwines with state control exerted through different regulatory schemes associated with agricultural policy. Together these forms of prescription, control and sanctioning make for a new, now dominant, coordination mechanism which I will refer to, following Rip and Kemp, as a *socio-technical regime*, that is:

> the grammar or rule set comprised in the coherent complex of scientific knowledge, engineering practices, production process technologies, product characteristics, skills and procedures, ways of handling relevant artefacts and persons, ways of defining problems – all of them embedded in institutions and infrastructures. (1998: 340)

Agricultural production is increasingly regulated by such socio-technical regimes. It is important to note that these socio-technical regimes are selective. They define what is to be done, how agricultural production is to be organized and to be developed. Therefore, regimes also delineate 'irrational' and 'non-valid' practices and patterns. Equally, regimes define and create resources as well as defining networks and demarcating 'knowledge' and 'ignorance' (Wiskerke, 2002). Regimes are, to echo Law (1994), 'patterning both the social and the material'. At the same time, though, the thus regulated agricultural production has increasingly come to be at odds with nature, society and the prospects and interests of those directly involved. The expressions of this mismatch are far-reaching and multi-dimensional. Some are well known and have been widely discussed (as, for example, those summarized in the first section of this chapter), other effects are less visible and sometimes even hardly explored.[6]

Referring back to Figure 18.1, the impact of the currently reigning socio-technical regimes might be summarized as (a) the institutionalization of unsustainability (Marsden, 2003), (b) the emergence of a chronic distrust of consumers in the quality and safety of food and (c) an accelerated marginalization and exclusion of increasing numbers of primary producers.

THE CHANGING NATURE OF THE AGRARIAN QUESTION

Historically speaking, the gravitational centre of the agrarian question has always resided in the axis linking agricultural production and the farming population (see Figure 18.1). The (re-) organization of production, and sometimes processing and marketing, has often run counter to the interests and prospects of the involved actors. Time and again these (re-)organizations have led to intolerable social conditions which triggered socio-political struggles through which access to and control over the land was, and still is being, claimed. An excellent analytical overview is contained in Paige, 1975; see also Huizer, 1973 and Wolf, 1969). Examples abound from around the world: the *peones* and *golondrinas* in the endless *haciendas* of Latin America (Pearse, 1975), the *precaristas* in Central America, the *mezzadri* in Italy, the *Krestjanin* in Russia, the *hierboeren* in the North of the Netherlands, the Diggers in England and the crofters of Scotland – time and again they highlight this element of the agrarian question.

Currently though, the agrarian question refers to and summarizes a far more generalized disarticulation. The once organic relations between agriculture, society and nature have been deeply disrupted and in a seemingly irreversible way, whilst the third axis (between agriculture and farmers) has been converted into a complex and contradictory constellation that implies the inclusion of some and the exclusion of many. Through this process a range of new internal contradictions has been introduced. From an analytical point of view, this multiple disarticulation is due to the rise and dominance of the currently reigning socio-technical regimes. Through the 'mechanics' (technical prescriptions, legal requirements, artefacts as plants and animals, trading procedures, expertise, etc.) of these regimes, the processes of agricultural production become removed from the once self-evident parameters entailed in nature, society and the interests and prospects of the involved actors. Instead they are re-oriented, directly and indirectly, towards new 'system requirements' that reflect, above all else, the logic of financial capital embodied in processing and trading companies as well as the need for extended state control.[7]

At the global level, the initially often contrasting regimes are increasingly being brought in line. This occurs in a number of ways: through the dismantling of national agricultural policies and associated institutions (whilst simultaneously creating global coordination mechanisms as WTO, EuropGap etc.); through increased and centralized control over genetic resources by a limited number of international companies; and through a world-wide submission (if not replacement) of local knowledge systems by new, scientifically grounded expert-systems. Thus, globalization occurs *not* through the internationalized flows of commodities, ideas and people,[8] *but* through the subordination and consequent reorganization of local and regional farming systems to just one grammar, that is, the one entailed in, and imposed by, the increasingly interlocking socio-technical regimes (Goodman and Watts, 1997; McMichael, 1994). As a consequence, for the first time in history, the agrarian question is becoming a universal, instead of a localized problem.

DEGRADATION AND COUNTER-PRODUCTIVITY

On the ideological level, large-scale, intensive and specialized agriculture is nearly always presented as highly efficient and indeed as the only possible response to current hunger as well as to the future need to feed a rapidly growing population. However, a growing number of empirical studies worryingly suggest that the opposite is true. Whilst on the one hand the development of productive forces allows for continuously growing yields,[9] the mechanics of the reigning regimes introduce, on the other, backward tendencies that exclude a proper unfolding and use of the new productive potential. To this latter phenomenon we will refer, following Ullrich (1979), who, in his turn built on Illich (1978), as counter-productivity, a concept that embraces a range of regressions that negatively affect levels of productivity[10] and, especially, the income levels of those directly dependent on primary production. Counter-productivity not only has negative effects as far as the nexus between agricultural production and society is concerned (this applies especially in Third World countries), it also is having negative effects on the interests and prospects of primary producers and (indirectly) also on nature.

Van den Dries (2002) presents a detailed comparison of 'traditional' local agriculture and newly emerging, highly modernized types of farming in Barroso (Northern Portugal). Local farming styles are part and parcel of long-established 'farmers' managed irrigation systems', whilst the construction of new, state-controlled irrigation systems is associated with the emergence of a new, large-scale and intensive, type of farming. Although water was and remains the limiting (that is, the most scarce) productive resource, it is notable that water is used far less efficiently within the new irrigation and production practices than in the 'traditional' ones. Gross income per cubic metre of irrigation water has decreased, throughout the modernization process, from 660 escudos/m^3 to 210 esc/m^3. The fall in net income has been even sharper: from 500 to 110 esc/m^3.

A similar phenomenon has been detected by Ventura (1995), who compared 'traditional' Chianina-breeding in Italy with the newly emerging, large-scale feedlot type of meat production. Ventura focuses on a comparison of energy use and shows that the 'energy efficiency of small-scale, artisan farms developed around the "closed cycle" is higher than that of industrial farms and this is true in both absolute and relative terms. On small-scale traditional farms the production of one kilogram of beef requires 8800 Kcalories and on industrial farms more than 10,000 Kcalories.' In addition, 'artisan farms use less non-renewable energy (54 per cent compared to 62 per cent)' (1995: 229–230). The same type of counter-productivity can be noted in terms of nitrogen use. Whilst in the 1950s the total N-efficiency of North West European dairy farming averaged at 60 per cent in the 1990s it had decreased to a worrying average 18 per cent. That is, from every 100 kilograms of nitrogen introduced into the production system, only 18 were used effectively (Bruchem et al., 1999).

Building on empirical work of Commandeur (1998, 2003), van der Ploeg (1998) shows that in industrialized pig breeding an increasing number of sows is needed in order to produce the same number of piglets. 'In artisanal styles, grounded on natural cycles, 12 sows are needed to produce 1000 piglets, whilst in industrialized styles in which the economics of breeding are optimized, nearly double this number of sows is required, that is 22, in order to produce the same number of piglets'(1998: 19). The same applies to dairy husbandry where longevity has been reduced, on the average, to just a little more than two lactation periods. That is, animals are increasingly converted to 'throw away products'. Both their life span and their production during life are reduced (see also Adam, 1998). What increases, though, is the production per animal per year. In combination with an accelerated replacement rate, this makes for the highest return on total investments.

In many parts of Central and Latin America, where land is the scarcest resource, there is currently a massive process of *ganaderizacion* (that is, the conversion of mixed types of peasant production into large-scale cattle breeding oriented at international markets). In a recent empirical study, Gerritsen (2002) shows that this particular type of modernization implies a reduction of production per hectare from 3,777 pesos/ha to just 487 pesos/ha, whilst the decrease in income and employment per hectare is even more accentuated.[11] In North West Europe, where not land but total production is directly or indirectly the limiting factor,[12] a range of studies has shown that the amount of employment and total income generated by different styles of farming sharply decreases with the shift from local low external input types of farming to the large-scale,

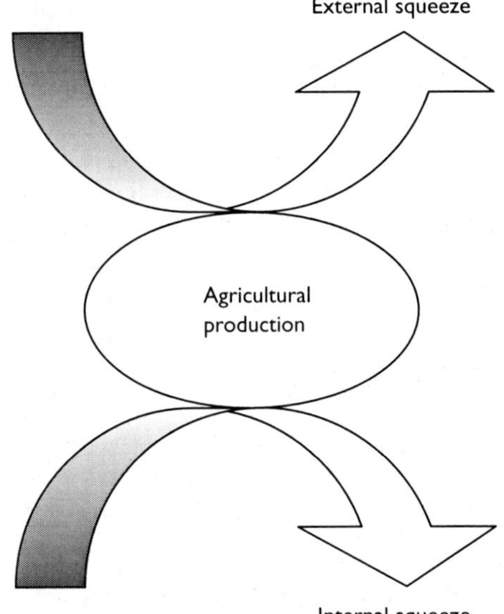

Figure 18.2 ***The double squeeze on agriculture***

industrial type of farming. Case studies of Friesian dairy farming indicate that under a given level of production, such a shift implies a decrease in income as a percentage of total value of production from 42 to 29 per cent (van der Ploeg, 2003: 261). Put in another way, the style of farming economically realizes a surplus of 12.25 euro for every 100 kg of milk produced, compared to 7.55 euro/100 kg in intensified styles of dairy farming (van der Ploeg, 2000: 504). If the total quota for the province of Friesland were produced under an economical farming style (that is, on relatively small-scale farms, where the use of external inputs is limited) 6,295 farms could be maintained. Under intensive styles (large-scale, highly modernized), the same amount of milk can be produced on 4,500 farms (1997 data). The former choice does not imply a 'poverty sharing' type of solution. The remarkable feature of these calculations is that under intensive farming styles the total provincial agrarian income would be some 210 million euro, whereas under economical farming styles it would be 290 million euro (Antuma et al., 1993; see also van der Ploeg, 2003: 373 and 308 for a scenario approach based on these differences).

At first sight, indications of counter-productivity such as those referred to above might appear as mere anachronisms or irrational deviations from the march of progress. Taken together with many other examples, they emerge, however, as typical of the effects produced by the reigning socio-technical regimes and the associated development trajectories. Once agricultural production is 'entrapped' in the rigidities of a particular regime and dependent upon a particular development path, specific forms of counter-productivity are the price to be paid. Once a high technology cowshed and milking parlour have been constructed, fixed costs per unit of space have increased and require a higher milk yield. This in turn implies particular selection and culling decisions, concentrate levels, etc., that imply a decrease in longevity. And so on and so forth. Equally, while a change from peasant agriculture to large-scale, extensive cattle breeding may run counter to societal needs and to the conservation of biodiversity, the need for optimal returns on invested capital makes them unavoidable, 'logical' decisions.

THE DOUBLE SQUEEZE ON AGRICULTURE

At a more general level, the dynamics of the current socio-technical regimes in and around agriculture can be understood as resulting from two intertwined processes of appropriation and redistribution. At the level of agribusiness groups, large trading companies and retail chains there is ruthless competition at the expense of prices paid to farmers. Thus, through the commodity relations in which agricultural production is embedded there is an ongoing appropriation of value; a redistribution between agro-industrial groups and those directly involved in primary production that evidently favours the former. Secondly, within primary production itself, there is a parallel process of redistribution and appropriation. Spurred by high cost levels and continuous pressures on output prices, farmers tend to expand the size and scale of their enterprises. As a consequence, agricultural production is increasingly concentrated in a limited number of farms, which show *higher* cost levels (due to the expansion of the farm and due to the often associated change-over towards new, expensive technologies that allow for an extended scale of operations) and, consequently, *reduced* margins (per unit of end product, per animal, per hectare, etc.). Only the enlarged scale makes for an increased entrepreneurial income.

The first squeeze (from agribusiness on agriculture) is thus followed *and nourished* by the second one, which is internal to agriculture and which implies a redistribution of value (and associated benefits such as, for example, productive employment, income and prospects for the future) towards a small growth pole within the agricultural sector. Together these two squeezes tend to strongly decrease the available value added (and income) derived from agriculture and thereby continually reinforce each other, locking farmers into a race to the bottom.

Thus, at the level of the farm enterprise, this ongoing pressure for scale enlargement becomes one of the most central criteria for farm development – at least for large and quickly expanding farms. In consequence, agriculture is increasingly distanciated from the co-production of man and living nature which formerly characterized it. Under these circumstances, different forms of counter productivity might emerge and may even be welcomed as long as they are instrumental to further scale increases and further appropriation.

THE COMING CRISIS: THE FRAGILITY OF AGRICULTURAL PRODUCTION IN THE TWENTY-FIRST CENTURY

Agricultural production is increasingly encapsulated in and patterned by the prevailing socio-technical regimes, although, as I will point out later, farmers are also continuously trying to 'escape' from these regimes. Before discussing these 'counter-movements', I will discuss the main developmental trends within the dominant regimes.

First, new technological developments in processing, transportation and distribution (but also in the ability to transform complete eco-systems: see Halweil, 2000, 2003) allow for an ever-widening disconnection between 'raw material' and end products, between the *loci* of primary agricultural production and those of processing and consumption. These growing 'gaps' are not only spatial but also temporal.[13] 'Freshness', for instance, can be maintained almost indefinitely.[14] An exemplary illustration is the *latte fresco blu* (blue fresh milk) project of the Parmalat group in Italy. By first separating the liquid and solid elements of fresh milk (with new techniques for micro filtration), then treating these separated elements individually and later recombining them, 'blue milk' is created. In order to protect national dairy farming and the associated cooperatives, Italian legislation defines 'fresh milk' very precisely as milk that is transformed within 48 hours of milking and consumed within the following three days. However, an exception to this definition was negotiated for 'blue milk', making it possible to market it as 'fresh' even when the period between the first operations and the final 'remaking' of the milk is extended for several months. Thus, massive imports from, for example, Eastern Europe (where milk price is just half that in Italy) would become possible, marketing it in Italy as *latte fresco blu*. The losers in this change would be Italian dairy farmers, who would lose a considerable part of their markets, whereas the Parmalat group would have benefited enormously, capturing a far higher proportion of the added value than before.[15] However, after considerable protests by farmers, political parties and green movements, the legal exemption was strongly modified.

The combination of new technologies, political power to adapt (if needed) existing laws and world-wide flows of agricultural commodities, allows for a phenomenon that was first analysed by Feder (1977) as 'strawberry imperialism'. This process might also be described as the outsourcing of agricultural production. Agro-industrial groups increasingly re-allocate the origin of commodity flows from one source to another, which allows them to impose their own conditions on production, delivery, quality, etc., putting the concerned producers in a position of strong dependency and insecurity. Over-exploitation of natural resources and the introduction of cheap, often illegal labour are important consequences. Through this pattern of outsourcing the agrarian crisis is being reproduced on a world-wide scale.

Secondly, such 'squeezes of the first order' are increasingly associated with the creation of mega-farms, that concentrate into one enterprise the production that was previously distributed over tens or even hundreds of family farms. Once mainly limited to specific areas of the United States (i.e. California), these mega-farms are now emerging everywhere and in all sectors. Mega-farms *cannot* be understood as family farms reproduced at an enlarged scale. They enter, instead, directly in competition with existing family farms by appropriating, concentrating and recombining (using often new technologies and/or cheap labour) the available (and limited) resources, such as, for example, land, water, quota, 'environmental space' and marketing possibilities. Within agriculture they emerge as new *latifundia*, converting family farms into the new *minifundia*. This results, as argued before, in a considerable reduction of the total wealth produced in the countryside.

Thirdly, a new danger pops up that, at first sight is hardly believable, but which, in the newly emerging constellation, will be a persistent feature. That is, that the security of food supply will be increasingly threatened. The very combination of outsourcing and mega-farms will make for a fragility in food production. Whilst peasant and family farms continue to produce food even under unfavourable conditions, the mega-farms (which are already more vulnerable due to the reduced margin per unit of end product) will prove to be 'runaway industries' (as described by Barros Nock, 1997 and Gonzalez Chavez, 1994). When production ceases to be profitable, they will de-activate production, move capital to other sectors and/or re-orient their resources to other destinations, leaving a huge gap in the food supply.

Fourthly, we might anticipate, in the decades to come, an important reversal of the rural exodus. Over the past five decades or so, many people have moved, for amongst others reasons as a result of the 'modernization' of agriculture, from the countryside to the towns. However, experience of life in the townships, *favelas, barriades* and *pueblos jovenes* has shown that the development of industry and services can only absorb a fraction of the labour force into formal job arrangements. For too many people their destiny is, as the bitter Brazilian expression goes, one of being *lixo umano* (human litter) (Athias, 1999). It is, therefore, not remarkable that the millions of people actually involved in land occupations and the creation of new settlements in Brazil mainly have *urban* backgrounds (Cabello Norder, 2004). They return to the countryside to make a new and better living in agriculture. Such re-peasantization (which occurs also elsewhere, albeit less visibly and less massively) might very well become a major feature throughout the Third World. This implies that outsourcing and the creation of mega-farms (and the associated appropriation of land, water, space, credit, marketing possibilities) will be increasingly contested in and through socio-political struggles that are literally a matter of life and death.

For a completely different set of reasons, contestation will also be a keyword in considerable parts of the 'Western world', at least in Europe. Since the *repeuplement de la campagne* (as first described by Kayser, 1995) is now a generalized phenomenon all over Europe, a reconstruction of countrysides, landscapes and rural communities into harbours for mega-farms surrounded by idle, 'superfluous' areas will not easily be accomplished.[16]

NEWLY EMERGING RESPONSES

Whilst a minority of farmers has internalized, and is *de facto* propelling the 'race to the bottom', others have been actively constructing a range of responses to the current crisis in agricultural production – some of which might be built upon to provide more widespread solutions to address the 'coming crisis'.

A very widespread response to the squeeze on agriculture has been an increased engagement in pluri-activity. Although part-time farming has been, throughout agrarian history, a permanent, albeit highly variable, phenomenon, pluri-activity definitely cannot be considered as a 'remnant of the past'. It is a widespread and quickly growing feature of today's agriculture: 'An ... attribute of peasant households of growing significance in contemporary developing countries is their multi-activity character' (Ellis, 1993: 4).[17] This also holds true in the more developed countries. In the Netherlands, where farming has been highly 'modernized' over the past five decades,[18] 70 per cent of the so-called professional farms have either the man, the woman, or both, earning a part of the family income through nonfarm activities. On the average Dutch dairy farm, 33 per cent of the available family income is derived from pluri-activity (LEI, 2002), and on the average arable farm more than 50 per cent of the family income is derived from other activities (Wiskerke, 1997). The continuity of many farms, as well as intra-generational succession, would be unthinkable without these additional incomes. Many cases from the Netherlands reveal a phenomenon that is also described in empirical studies from developing countries. That is, that farm investments rely to a considerable degree, if not exclusively, on the earnings derived from pluri-activity. In this way dependency relations with the banking circuit (and the associated financial and transaction costs) are avoided. The farm enterprise is literally 'grounded' on own resources instead of outside capital, thereby enlarging the autonomy of the farmer to pursue farming styles other than those that focus on achieving maximum returns to capital.

Compared to data from the 1960s, 1970s and 1980s, the relevance of pluri-activity (both as a percentage of involved farmers and in terms of income contribution) has risen considerably in recent years (Bryden et al., 1992; de Vries, 1995). It is important to recognize that pluri-activity is no longer (if it ever has been so) the prelude to a

Table 18.1 ***Farmers' definition of the 'beautiful, well-organized farm'***

Features	Irrelevant (%)	Important (%)	Decisive (%)
Produce high quality	14	60	26
To be able to get the work done	1	76	22
No undue stress for the family	5	74	21
Produce large quantity with low costs	13	64	21
Proper balance of own and bank capital	12	70	18
Good technical results	23	60	17
Low external costs	15	69	16
Ambitious investments	15	66	9
Be larger and more modern than others	74	23	3
Dedicate the least time possible	74	25	1

definite closure of the farm. Instead it is (or has become) a crucial condition for the very continuation of the farm (or more generally, the *domus*). Pluri-activity also represents an expression of the emancipation aspirations of many rural women, as well of many men, who want to develop further a small farm that is not reliant on the formal banking circuits (Bock and de Rooij, 2002). And finally, pluri-activity is no longer associated with poverty (as it definitely used to be). It is, on the contrary, an expression of new levels and forms of *wealth*, as has been convincingly argued by Kinsella et al. (2000) and Wilson et al. (2002).

A second important line of defence results from the style of farming economically (van der Ploeg, 2000). Farming economically is basically a strategy aiming to contain monetary costs associated with investments, loans and expenditure on external inputs. Farming economically, therefore, can be equated with 'low external input agriculture' as it is known in Third World discourses (Reijntjes et al., 1992). Central to this farming style is the mobilization, use, development and reproduction of *internal resources* as opposed to resources produced and commercialized by agribusiness groups. In the heyday of the modernization era (1960–1990) the strategy of farming economically and the subsequent creation of a low-external-input agriculture enabled many farms to remain viable without entering into the logic of modernization. Today, farming economically seems to have become the dominant style, at least within North West Europe, a phenomenon that has largely gone unnoticed. It provides farming families with a way of countering the increasingly threatening situation of limited quotas, decreasing prices, high costs associated with land and quota, and the obligation to farm in a more environmentally sound way (Kinsella et al., 2002): 'In following such [a] strategy, ... farmers may be less vulnerable to the ups and downs of the market than ... big, capital-intensive farms' (Djurfeldt and Gooch, 2002). Thus, farming economically is, for many farming families, a viable alternative to scale increases. In the style of farming economically, historically created resources (fertile and productive grasslands, a specific herd, skills, experiences, savings, etc.) do not only allow for a relatively autonomous farming practice in the present – these concrete (i.e. non-exchangeable) resources also carry promises for the future.[19]

Together, farming economically and pluri-activity have been conceptualized as central pillars of the 'regrounding' process – that is, basing the practice of farming on self-owned and self-controlled resources (van der Ploeg et al., 2002c). The main effect of regrounding is a certain 'distanciation' of the agricultural process of production *vis-à-vis* the markets for factors of production and non-factor inputs. In this way farmers avoid a high degree of commoditization, the subsequent submission of both production and development to the 'logic of the market' and the rigidities of prevailing socio-technical regimes. In the following section I will indicate how the process of regrounding might be considerably expanded.

A third pillar centres basically around fine-tuning. In particular farming styles a gradual, but ongoing intensification (rising yields) and an associated increase in the surplus per hectare and/or per cow, is an important ordering principle. Quantity and quality of labour are decisive driving forces for such ongoing intensification, whilst craftsmanship is a central mechanism. The process of production is, consequently, ordered in a characteristic way (Leeuwis, 1993; Wiskerke, 1997; van der Ploeg, 2003), in which 'quality', 'sufficient labour to get the work done' and 'absence of stress' are important guiding principles. The relevance of this mode of ordering *vis-à-vis* the current crisis resides in its capacity to avoid expressions of counter-productivity in

Table 18.2 ***Strategies to deal with price decreases***

Strategic responses	Irrelevant (%)	Important (%)	Decisive (%)
Strive for higher yields	9	65	24
Be more careful with investments	13	64	21
Enlarge flexibility of the farm	4	73	19
Farm more economically	17	65	16
Develop more branches (diversify)	24	59	15
Anticipate markets more quickly	29	55	12
Anticipate policy more quickly	32	64	7
Enlarge the farm	57	32	6

which other practices are ensnared. At the same time the skills intrinsic in this farming style allow, time and again, for additional increases in both yields and income levels per object of labour.[20]

It seems almost inevitable that the coming crisis will translate into, more or less, generalized price decreases. However, when reflecting upon the consequences of such decreases, it is crucial to focus on their *differential* impact. In this respect, the historically developed repertoire of farmers' responses to adverse market conditions (especially fine-tuning, farming economically and the engagement in additional economic activities) is an important element. This repertoire is not 'buried in the past', rather it is continuously being revitalized. In a recent survey Dutch farmers ($n = 765$), were asked to define 'the beautiful and well-organised farm';[21] Table 18.1 summarizes some of the most important outcomes.

Although carefully ordered relations with the markets clearly form part of the image of the 'beautiful farm', they are not decisive (Bagnasco, 1988; van der Ploeg, 1990). It is the *internal* organization of the farm enterprise (a good balance between supply of family labour and work to be done, between own savings and loans and high quality and good technical results, which depend on the craftsmanship contained in the farming family) that enables a farm to confront the markets. This is clearly reflected in the responses to another question: 'is the beautiful and well-organized farm able to deal with price decreases and with considerable market fluctuations?' An astonishing 91 per cent of farmers said yes.

This self-confidence (which evidently cannot be used as a simple legitimation for changes in the reigning market regime) is mirrored in another range of responses that concern the strategies to counter price decreases, which are summarized in Table 18.2.

Again, the contrast is considerable. Whilst further enlargement is, within agro-political discourse, often represented as the 'model-type' (if not the only valid) response, in practice most farmers think that, as a strategy, such a response is irrelevant. They prefer a combination of strategies that include fine-tuning and farming economically.

The same research project (see van der Ploeg, 2003, for a general overview) indicated that it is the highly indebted, large scale and specialized farm enterprises (that is, the 'vanguard farms', or as French and Italian colleagues would say *les grands intensifs* or *le aziende di punta*), that are most vulnerable to these anticipated price decreases. Tracy (1997) also identifies the vulnerability of this particular group: 'price cuts involved in a move to a market-oriented agriculture bear most heavily … on those who have invested heavily in modernizing their buildings and equipment, and who as a result have little flexibility in changing their patterns of production'.

It is somewhat ironic to note that those farmers who have frequently been labelled as the 'most efficient ones' now emerge as those least able to adapt to changing market conditions. It is telling that it is also this group of farmers who make up a majority of the 20 per cent of farmers who receive 80 per cent of CAP spending, as the Commission is well aware (Buckwell, 1997). While it will be painful, this kind of *subventionierte Unvernunft* (as Priebe, 1985, called it), cannot be continued indefinitely.

A fourth and more recent set of responses is often summarized under the heading of new rural development activities and practices. Through the development of such activities farmers aim at the integration of new (non-agricultural)[22] economic activities within the farm (Broekhuizen et al., 1997; IATP, 1998; DVL, 1998; Coldiretti, 1999; Stassart and van Engelen, 1999; Joannides et al., 2001; SARE, 2001; Scettri, 2001). Such activities might embrace agro-tourism, the remunerated management of landscape and natural values, energy-production, the supply of care facilities, etc. (see, respectively, Banks, 2002;

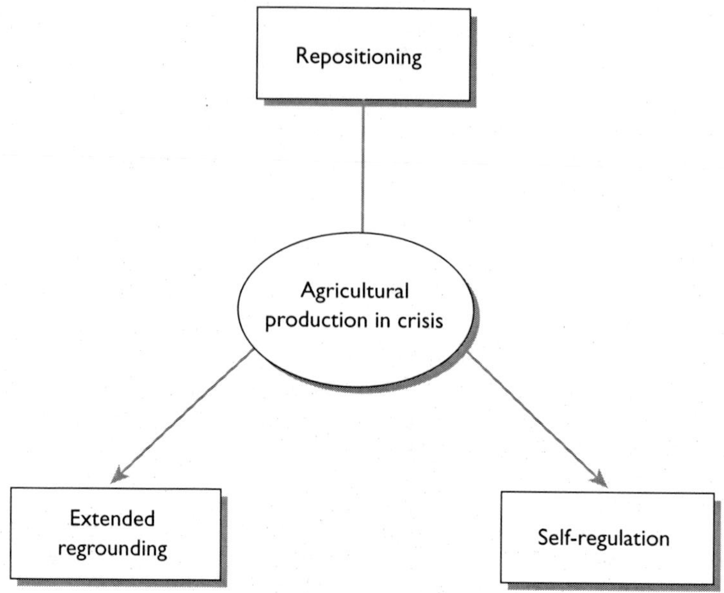

Figure 18.3 ***Ways out of the current agricultural crisis***

Knickel, 2002; Oostindie, 2002). Alongside this 'broadening' type of activities there is another set of activities, currently often grouped together under the heading of 'deepening'. Deepening refers to all those activities that aim at a substantial increase of the value added per unit of end product. Typical expressions are the production of high quality products and regional specialities, organic farming and direct marketing (see respectively de Roest, 2000; Roep, 2002; Knickel and Hof, 2002).

A recent, comparative European research programme[23] (van der Ploeg, Roep et al., 2002) showed that there is not only a broad, 'grass-roots' involvement of European farmers in broadening and/or deepening, but also that such activities generate an additional net value-added of nearly 8 billion euro. By comparison, the total net value of Dutch agriculture, a 'giant' within Europe, was, at the time, 6.8 billion euro.[24] If this extra income component is seen as 'the tip' of new RD practices, there is a strong underlying foundation. The same research programme showed that farming economically and pluri-activity generated a further 36.7 billion euro 'additional farm family income'. In combination with the earnings derived from broadening and deepening activities, this implies that (on 1998 data) 50.5 per cent of total 'farming family incomes in Europe are derived from these old and new 'defence mechanisms'. A more recent European-wide survey confirms this finding and indicates as well that the associated income levels will probably keep growing considerably (Oostindie et al., 2002).

PATTERNS UNDERLYING NEW ALTERNATIVES

Throughout the sometimes bewildering variety, complexity and fragility of the newly emerging responses to the crisis in agricultural production a new paradigm is emerging that reintegrates the relations between agriculture, nature, society and the prospects of the primary producers. Repositioning, extended regrounding and self-regulation are the key words of this newly emerging paradigm that can be discerned in the multitude of current practices (see also Figure 18.3). Taken together, these three processes imply a re-introduction of cultural, ecological and social capital into the process of agricultural production. They reduce dependency upon external financial capital and give a new importance to, the previously neglected, social, ecological and cultural capital as assets for ordering and coordinating production. Thus, the extension of regrounding, repositioning and self-regulation comprise the major axis of a newly emerging paradigm (Ikerd, 2000a, 2000b, 2000c; van der Ploeg et al., 2000a).

Repositioning implies a redefinition and reconstruction of the interrelations between agricultural production and the (output) markets. It represents a shift from financial capital (embodied in food processing industries and retail chains) towards new inter-relations between food production and consumption (Miele, 2001). Repositioning entails, in the first place, *diversification*, in which agricultural production is oriented towards a wider array of markets, including those for so-called 'green services'. Secondly, it entails the *re-internalization* into the farm of tasks and activities that formerly have been externalized to outside institutions, particularly agribusiness groups. Re-internalization might involve the transformation of products into end-products and/or the direct commercialization of (part of) the production. This relates to a third element, the elaboration of *new institutional patterns for commercialization*.[25] This last element is currently being translated at higher levels of aggregation that make for the re-emergence of *le terroir* (see, for example, Brunori and Rossi, 2000 on 'wine routes', Roep, 2000, 2002 on regional specialities). Together diversification, re-internalization and the engagement in new institutional patterns build upon, and simultaneously strengthen, polyvalence (Goodman and Wilkinson, 1993), 'the living historical legacy of heterogenous practices and organizational forms [that] creates the capacity for polyvalent responses' (Goodman, 1999).

In a wider sense, repositioning also has broader effects on agriculture's relations with society at large. By delivering a range of good, fresh, safe, tasty and high quality products, through new and controllable circuits, food production once again becomes a vehicle for generating and maintaining trust. More generally, repositioning builds on, relates with and sometimes feeds and strengthens important changes in the consumption of food and 'rural services' (Marsden et al., 2000; Atkins and Bowler, 2001; Renting et al., 2003).

Repositioning builds on the practices of deepening and broadening, whilst at the same time helping them unfold further and mutually integrating them in order to create synergy. In short, cultural capital, that is the capacity to engage with, and be recognized by 'others', in order to reach higher levels of trust, is strategic for further repositioning.

At the level of agricultural production, repositioning translates back into (and is in turn supported by) the development of a mixed resource base as well as by the multiple use of single resources (Wiskerke, 2001). That is, repositioning is not external to agricultural production (nor to the farm enterprise), but is reflected *and* grounded in an associated restructuring of agricultural production. The highly specialized farm (embedded in a technological support structure that delivers many services and artefacts) is replaced by a 'mixed' (or 'integrated') farm that disposes of all required resources and elements in order to function in a relatively autonomous way (Rabbinge et al., 1997). Required machinery is available, seed, feed, fodder, young animals, but also the needed water, the required energy, etc., are produced in the farm itself (or within a network of cooperating farms).

Extended regrounding refers, in this context, to the development of an integrated and autonomous resource base and its multiple use. The available resources are, as it were, 're-configured'. That is, extended regrounding is about the re-introduction of nature into the agricultural process of production. In short, it is about agriculture being 'regrounded', literally, on ecological capital (Guzman Casado et al., 2000). This extension of regrounding implies several steps. These include, the development of *ecological structures* at micro- and meso levels (that is, within the farm and at local/regional level; Visser, 2000). Such an ecological structure might be an existing (and probably extended) pattern of hedgerows, ponds, small pieces of land in between meadows laying fallow, all of which contain specific natural values. But it might also be created anew, in a goal-oriented way (Primdahl, 1999). Next comes the *revitalization of food webs*, that is, the intricate 'web' of micro-organisms, worms, herbivores, parasitoids, insects, moles and birds – that provide 'a network of consumer-resource interactions among a group of organisms, populations or aggregate trophic units' (Smeding, 2001: 84). Such food webs might improve and sustain productive capacity, for example through enhancing the nitrogen delivery capacity of the subsoil, but may also sustain a range of 'higher order natural values', such as birds of prey. In his brilliant discussion on food webs Smeding argues that 'one important solution … for agriculture in the industrialized countries could be the development of farming systems that are economically based on utilization of biodiversity and that also harbour conservation worthy species' (2001: 131; see also Almekinders et al., 1995; Altieri, 1999). Solid food webs, embedded in adequate ecological structures, might also considerably strengthen the resilience of plant–animal production systems, and reduce the levels of 'stress' which pose a major problem in today's agricultural production systems. In synthesis, through extended regrounding agricultural production is (once again) based on local ecology. The relation with nature is restored in so far as agriculture is re-based on *ecological capital*, whilst it simultaneously (re)produces a wider array of specific natural values. The production of 'green services' (landscape, natural values, a healthy environment, clean water, etc.) and the production of agricultural commodities are no longer separated (or at best 'positioned alongside each other'), but are mutually reinforcing, being the one a condition for the other and vice versa (Gerritsen, 2002).

Extended regrounding as outlined here might be understood as a further extension of the defence mechanisms discussed above. But whilst pluri-activity and farming economically are to be seen basically as an endeavour to disconnect agricultural production from *financial capital* and the circuits controlled by it, extended regrounding aims at its perfection by firmly re-introducing *ecological capital* as bedrock for agricultural production.

Both extended regrounding and repositioning might result in a range of mutually re-enforcing economic benefits. The more agriculture is grounded on food webs, the more variable costs (especially the ones related to fertilization and crop protection) might be reduced. In this way a further unfolding of the style of farming economically is possible, whilst levels of productivity are maintained (for an exemplary case see Brussaard et al., 2003). An increased resilience and a reduced stress in the plant and animal systems might translate into less harvest losses, fewer diseases and hence in reduced costs for herbicides, pesticides, veterinary bills and medicines. Equally, there might be a positive effect on longevity, which again translates into increased benefits and reduced costs.

Once the required balance has been created, the multiple use of resources will result in additional cost reductions, which, according to available empirical studies (Milone and Ventura, 2000; van der Ploeg et al., 2003) might be considerable. This potential has been forcefully outlined by Saccomandi (1998), who, building on the work of Scherer (1975) and Panzer and Willing (1981), made clear that *economies of scope* will occur in multiproduct firms that allow 'for better utilisation of both inputs and outputs'. Given two products, q1 and q2, it applies that:

$$C(q1, q2) < C(q1,0) + C(0,q2) \text{ for } q1 > 0 \text{ and } q2 > 0$$

(Saccomandi, 1998: 123)[26]

Beyond this type of cost reduction, the use of an 'integrated resource base' in farming implies yet another series of potential cost reductions. Manure is 'fed' into crop production, which is partly used for feed and fodder for the herd, and so on. That is, monetary expenses and, especially, the dependency on economic circuits associated with the purchase of fertilizer and concentrates are reduced (Bos, 2002). This in turn will imply a considerable reduction in transaction costs. This same reduction often entails the possibility of further fine-tuning the process of production.

Re-internalizing once externalized activities and tasks might also imply that the value added per unit of end-product is increased, as current examples of high quality products (de Roest and Menghi, 2000), regional specialities (Roep, 2000), organic production (Miele, 2001) and direct marketing (Schuite, 2000) have shown in a convincing way. Equally, the creation of new, well-functioning short circuits for marketing might considerably reduce transaction costs (Saccomandi, 1998; see also Knickel and Hof, 2002 and Ventura, 2001), whilst the presence of a strongly diversified range of outputs (that is, operating on *different* markets) will reduce market-induced risks (Huysman, 1978).

By taking together all these potential cost reductions, gains in efficiency and price increases, it might be argued that extended regrounding and repositioning contain and represent an important line of defence against the (double) squeeze on agriculture (again, see Figure 18.2).

The dual approaches of regrounding and repositioning help meet a number of interests and aspirations of those directly involved in agricultural production. First, they play an important role in helping defend and improve income (and often also employment) levels. They also create better prospects for the medium and long term. But there is more than this at stake. The management of an integrated and multiply used resource base, and the simultaneous management of all the associated activities, require new professional identities and capabilities that are light years away from the 'degraded' work (Braverman, 1974; Lacroix, 1981) that typifies agricultural production today. The overview of, insights into and control over the (widely enlarged) range of activities (and associated 'growth factors') is increased considerably, which translates into an increased governance over the complex whole (contrasting considerably with the current levels of 'external prescription and sanctioning' as described by Benvenuti, 1989 and Frouws, 1993). *Self-regulation* is increased (at an immediate level), which in its turn (especially when more positive effects at higher levels of aggregation are desired) requires an extended domain of self-regulation: more room for manoeuvre is needed. It is through this (immediate and extended) self-regulation that restructured (that is, regrounded and repositioned) agriculture is made possible *and* brought into line with the interests and prospects (with the emancipation aspirations) of those involved.

An increased self-regulation, understood as re-introducing social capital into agricultural production (whilst simultaneously distancing it from the prevailing socio-technical regime), might also render, in itself, considerable economic gains, precisely because it offers the potential to overcome, or avoid the many different forms and expressions of counter-productivity that result from the present day intertwining of production and regime. The outcomes might be substantial. In a recent study Swagemakers (2002) shows that, through a range of small, interconnected 'novelties' (each aiming at the correction of specific forms of counter-productivity), income levels were increased by an average 20 per cent on arable farms and by more than 50 per cent in the case of dairy farms (2002: 161, 169; see also Osti, 1991).[27]

World wide there are many, highly interesting expressions of the struggle for self-regulation. In this respect I refer to: the impressive struggles of peasants in Chiapas and Guatemala to regain control over their livelihoods and to restructure their farming practices (Toledo, 2000; van der Vaeren, 2000); the *Movimento dos Sem Terra* in Brazil (Cabello Norder, 2004); the emergence of environmental cooperatives in the Netherlands (Renting and van der Ploeg, 2001); and the further unfolding of organic agriculture (Miele, 2001; Alonso Miego et al., 2000). All these examples show an intricate combination of repositioning, extended regrounding and self-regulation. The social struggle for self-regulation is not without meaning, it is not aiming for an abstract, futile claim for the non-interference of 'others'. It is, instead, inspired and informed by, and simultaneously pointing towards, radically improved farming practices.

Whether or not such a new way out will materialize will depend on many factors and, probably, on unpredictable crises (for example, war, BSE). A central obstacle in the coming struggles will be, without doubt, the now dominating socio-technical regimes. Currently defended, expanded and strengthened in order to contain the worst expressions of food *in*security, of environmental degradation, etc., they effectively block the required repositioning, extended regrounding and self-regulation. Their capacity effectively to deliver their stated objectives (enhanced food security, environmental protection, etc.), will probably prove to be their Achille's heel. When citizens and consumers in general lose trust and realign (both symbolically and materially) with those farmers engaged in the struggle for self-regulation, far-reaching and massive transitions might well be possible.

NOTES

1 A recent study in the Netherlands (de Hoog and Vinkers, 2000) revealed that more than 40 per cent of Dutch farm families have an agricultural income below the social minimum. When additional incomes, derived from pluri-activity, are taken into account, it is still 25 per cent of all farming families having a total family income below the legally defined social minimum. In Italy the MPAF (2003) analysis refers to similar findings.

2 Here we might refer to antibiotics, dioxine and chemical residuals of pesticides in meat, to salmonella in eggs and poultry, to aflatoxine in milk and dairy products, and more generally to BSE and the (potential) dangers of GMOs (Ecologist, 2003).

3 In this respect it is telling that for Mariategui (1928), 'el problema de la tierra' and 'el problema del indio' represented the two sides of one and the same coin.

4 In this respect, the body of literature that focuses on 'rural livelihoods' composes an important extension of previous studies on the agrarian question. See Chambers and Conway (1992), Farrington et al. (1999) and Niehoff and Price (2001).

5 Underlying the work of both authors (and the larger traditions they belong to) is a focus on culture. This is especially explicit in Hofstee's work. Local peasant culture, then, emerges as the main coordination mechanism. In this respect his work anticipates the later work of Scott (1976) on the 'moral economy'.

6 Further on I will pay special attention to these effects, especially when it comes to the 'counter-productivity' of the newly emerging, industrialized farming systems.

7 Both requirements presuppose a considerable degree of standardization to allow for planning and control. Precisely at that point state interests and financial capital coincide and interact with modern science (Christis, 1985).

8 The oldest available farm accounts (from the sixteenth century) clearly show that already at that time European farming was integrated in, indeed, global circulation of many commodities (Slicher van Bath, 1958; for a summary in English see van der Ploeg, 2003: ch. 2). On the other hand, in *today's* agriculture some 80 per cent of food is consumed within the same area in which it has been produced. The simultaneous co-existence of the global and the local, in the past as well as in the present, is striking.

9 This is not only due, as is often claimed, to scientific breakthroughs, but as much to labour grounded intensification, labour investments in terraces, irrigation, land improvement, etc.

10 Special reference might be made here to the negative effects of high degrees of commoditization and market dependency upon levels of intensity (see Long et al., 1986; van der Ploeg, 1990, 1991; Zuiderwijk, 1998).

11 This mirrors, as it were, the contradictions that were already existing for a long time at the interface between *latifundia* and *minifundia* (Feder, 1973, and especially his synthesis of the well-known CIDA studies). What is new, though, is that these contradictions now emerge *within* the peasantry sector.

12 Directly where quota are applied (as in dairy farming, sugar beet production, intensive husbandry) and indirectly through environmental and spatial limitations.

13 From the 1970s onwards technologies for defraction and recombination have been used and perfected. However, the scale on which they are applied nowadays makes for far-reaching shifts.

14 The production of 'freshness' can, however, also fail spectacularly, as was shown by the Ahold retail chain in the Netherlands, that offered 'fresh' rabbit during Christmas time in 2002. In the kitchens these rabbits proved to be rotten, stinking, and in short – uneatable. Afterwards it turned out that they had been bought more than a year before in China and kept, since then, in refrigeration.

15 It is telling, though, that often only small transport flows are needed to cause similar price effects and value shifts. The shipment, a few years ago, of relatively small quantities of so called 'chip potatoes' from the United States to Europe, for instance, provoked a drastic and long fall of potato prices, which translated, through all the price interdependencies, into a general setback in European arable farming.

16 This is reflected in the development of political views that regard the role of agriculture and countryside in modern societies. See especially the debate on the so-called European Model of Agriculture (COPA, 1998).

17 Illustrative for the conceptual confusion about what the peasantry is supposed to be is Kearny (1997), who basically argues that multi- or pluri-activity implies that peasants cease to be peasants.

18 Within the modernization paradigm, 'real' farm entrepreneurs are highly specialized and dedicate their time exclusively to farming. Pluri-activity is understood as a hindrance for modern farming and is seen as a remnant of the past.

19 This also explains why the massive destruction of cattle, that followed the 2001 outbreak of foot and mouth disease, was so extremely painful for and strongly contested by the concerned farmers (van der Ziel, 2003).

20 In more precise terms: in and through fine-tuning, the technical efficiency is raised, allowing for an operation at (constantly moving) frontier functions (Yotopoulos, 1974; see van der Ploeg, 2003 for an application to current agriculture).

21 This 'folk-concept', which is used much in everyday language and communication in the countryside, functions as an important guiding image in the organization and development of farms. It is a highly differentiated notion; in different farming styles it will be operationalized in different ways. It will be clear that this farmers' concept does not necessarily coincide with the notions that prevail within science and policy-making.

22 That is, 'non-agricultural' when related to classic definitions of agriculture. It could be argued nowadays that the management of, for example, landscape and natural values, can be seen, *intrinsically*, as an agricultural activity. That is, with the progress of rural development, all kinds of social definitions are 'moving'.

23 Covering seven countries representing some 84 per cent of the total agricultural production of the EU.

24 Evidently, there are considerable regional variations here. In Spain the additional net value-added obtained through deepening and broadening was 3.4 per cent of total agricultural NVA, whilst in Germany it was 17.8 per cent.

25 In this respect Whatmore and Stassart (2001) rightly refer to the 'importance of entanglements of consumer and producer knowledge practices to the assemblage and sustainability of new institutional forms which endeavour to transact the spaces between production and consumption'.

26 In which C refers to cost-level.

27 Such effects might occur even more if growth and development in such farms no longer follow the sclerotized pattern of spurred scale-enlargement, but rather the tracks of further repositioning and regrounding.

REFERENCES

Adam, B. (1998) *Timescapes of Modernity: The Environment and Invisible Hazards*. London: Routledge.

Almekinders, C.J.M., Fresco, L.O. and Struik, P.C. (1995) The need to study and manage variation in agro-ecosystems. *Netherlands Journal of Agricultural Science*, 43: 127–142.

Alonso Miego, A., Jimenez Romera, M. and Guzman Casado, G. (2002) The production of organic olive oil: the OLIPE co-operative in the Pedroches region. In J.D. van der Ploeg, A. Long and J. Banks (eds), *Living Countrysides: Rural Development Processes in Europe: The State of the Art*. Doetinchem: Elsevier. pp. 120–127.

Altieri, M.A. (1999) The ecological role of biodiversity in agroecosystems. *Agriculture, Ecosystems and Environment*, 74: 19–32.

Antuma, S.J., Berentsen, P.B.M. and Giessen, G.W.J. (1993) Friese melkveehouderij, waarheen? Een verkenning van de Friese melkveehouderij in 2005; modelberekeningen voor diverse bedrijfsstijlen onder uiteenlopende scenario's. Wageningen: Vakgroep Agrarische Bedrijfseconomie, Wageningen Agricultural University.

Arce, A. and Marsden, T. (1993) Social construction of international food: a new research agenda. *Economic Geography*, 69 (3): 293–311.

Athias, G. (1999) MST transforma excluidos urbanos em militantes: fazenda em Porto Feliz foi ocupada por desempregados e sem-teto. *O Estado de Sao Paolo,* Sao Paolo, segunda–feira, 15 March.

Atkins, P.J. and Bowler, I.R. (2001) *Food in Society: Economy, Culture, Geography*. London: Arnold.

Bagnasco, A. (1988) *La Costruzione sociale del mercato, studi sullo sviluppo di piccola impresa in Italia.* Bologna: Il Mulino.

Banks, J. (2002) Tir Cymen: a whole farm agri-environmental scheme in Wales. In J.D. van der Ploeg, A. Long and J. Banks, *Living Countrysides: Rural Development Processes in Europe: The State of the Art.* Doetinchem: Elsevier. pp. 33–43.

Barros Nock, M. (1997) Small farmers in the global economy: the case of the fruit and vegetable business in Mexico. PhD thesis. The Hague: ISS.

Benvenuti, B. (1989) *Produttore agricolo e potere: modernizzazione delle relazioni sociali ed economiche e fattori determinanti dell' imprenditorialita agricola.* Rome: CNR/IPRA.

Bernstein, H. and Woodhouse, Ph. (2000) Whose environments, whose livelihoods? In Woodhouse, P., Bernstein, H. and Hulme, D. (eds), *African Enclosures? The Social Dynamics of Wetlands in Drylands.* Oxford: James Currey. pp. 195–214.

Bock, B.B. and de Rooij, S.J.G. (2000) Social exclusion of smallholders and women smallholders in Dutch agriculture. Final national report for the EU project: Causes and mechanisms of social exclusion of women smallholders. Wageningen: Wageningen University and Research Centre.

Bonnano, A. et al. (eds) (1994) *From Columbus to ConAgra: The Globalization of Agriculture and Food.* Lawrence, KS: University of Kansas Press.

Bos, J. (2002) *Comparing Specialised and Mixed Farming Systems in the Clay Areas of the Netherlands under Future Policy Scenarios: An Optimisation Approach.* Wageningen: Wageningen University and Research Centre.

Braverman, H. (1974) *Labour and Monopoly Capital: The Degradation of Work in the 20th Century.* New York: Monthly Review Press.

Broekhuizen, R. van, Klep, L., Oostindie, H. and van der Ploeg, J.D. (eds) (1997) *Renewing the Countryside: An Atlas with Two Hundred Examples from Dutch Rural Society.* Misset: Doetinchem.

Bruchem, J. van, Schiere, H. and van Keulen, H. (1999) Dairy farming in the Netherlands in transition towards more efficient nutrient use. *Livestock Production Sciences*, 61: 145–153.

Brunori, G. and Rossi, A. (2000) Synergy and coherence through collective action: some insights from wine routes in Tuscany. *Sociologia Ruralis*, 40 (4): 409–423.

Brussaard, L., Rossingh, W. and Wiskerke, H. (2003) Special Issue of *NJAS, Wageningen Journal of Life Sciences*, 51.

Bryden, J.M, Bell, C., Gilliatt, I., Hawkins, E. and MacKinnon, N. (1992) *Farm Household Adjustment in Western Europe, 1987–1991.* Final report on the research programme on farm structures and pluriactivity, Volume 1 and 2, ATR/92/14. Brussels: European Commission.

Buckwell, A. (1997) Towards a common agricultural and rural policy for Europe. *European Economy*, No. 5. Brussells: EC, Directorate-General for Economic and Financial Affairs.

Bussi, E. (2002) *Agricoltura e alimentazione: impegni, risorse e regole per lo sviluppo.* Relazione al Convegno dell'Istituto Cervi, Reggio Emilia.

Cabello Norder, L.A. (2004) *Políticas de Asentamento e Localidade: os desafios da reconstituçao do trabalho rural no Brasil.* Wageningen: Wageningen University.

Chambers, R. and Conway, G. (1992) *Sustainable Rural livelihoods: Practical Concepts for the 21st Century.* University of Sussex. Institute of Development Studies Discussion Paper, No. 296, February.

Christis, J. (1985) Technologiekritiek: een confrontatie tussen Ullrich en Habermas, *Krisis*, 20: 30–51.

Coldiretti (Movimento Giovanile) (1999) *Nuova impresa, idee ed evoluzione dei giovani agricoltori in Italia.* Rome: Edizione Tellus.

Commandeur, M. (1998) *Gesloten bedrijven: verscheidenheid in de zeugenhouderij.* Wageningen: Circle for Rural European Studies, Wageningen University.

Commandeur, M. (2003) *Styles of pig farming: A Techno-sociological Inquiry of Processes and Constructions in Twente and the Achterhoek.* Wageningen: Wageningen University and Research Centre.

Cook, I. (1994) New fruits and vanity: symbolic production in the global food economy. In A. Bonanno, L. Bush, W. Friedland, L. Gouveia and E. Mingione (eds), *From Columbus to ConAgra: The Globalisation of Agriculture and Food.* Lawerence, KS: University of Kansas Press, pp. 232–248.

COPA/COCEGA (1998) *The European Model of Agriculture: The Way Ahead.* Pr(98)12F2, 2 April 1998, Brussels.

Cork Declaration (1996) A living countryside. Conclusions of the European Conference on Rural Development: Rural Europe – Future Perspectives. Cork, Ireland, 7–9 November.

Countryside Council (Raad voor het Landelijk Gebied) (1997) *Ten Points for the Future. Advice on the Policy Agenda for the Rural Area in the Twenty-first Century.* Amersfoort: RLG Publication, 97/2a.

Delors, J. (1994) *En Quete d'Europe: les carrefours de la science et de la culture.* Rennes: Editions Apogee, Collection Politique Européenne.

Depoele, L. van (1996) European rural development policy. In W. Heijman, H. Hetsen and J. Frouws (eds), *Rural Reconstruction in a Market Economy.* Mansholt Studies 5. Wageningen: Wageningen Agricultural University. pp. 7–14.

Djurfeldt, J. and Gooch, P. (2002) Farm crisis, mobility and structural change in Swedish agriculture, 1992–2000. *Acta Sociologica*, 45 (2).

Dumont, R. (1970) *Types of Rural Economy: Studies in World Agriculture*. London.

DVL (ed.) (1998) *Verzeichnis der Regional-Initiativen: 230 Beispiele zur nachhaltigen Entwicklung*. Ansbach: Deutscher Verband für Landschaftspflege.

Dries, A. van den (2002) *The Art of Irrigation: The Development, Stagnation and Redesign of Farmer-managed Irrigation Systems in Northern Portugal*. Wageningen: Circle for Rural European Studies, Wageningen University.

Ecologist, The (2003) GM will remove consumer choice. August.

Ellis, F. (1993) *Peasant Economics: Farm Households and Agrarian Development*, 2nd edn. Wye Studies in Agricultural and Rural Development. Cambridge: Cambridge University Press.

Farrington, J., Carney, D., Ashley, C. and Turton, C. (1999) Sustainable livelihoods in practice: early applications of concepts in rural areas. In ODI, *Natural Resource Perspectives*, No. 42, June.

Feder, E. (1973) *Gewalt und Ausbeutung, Latinamerikas Landwirtschaft*. Hamburg: Hoffman and Campe.

Feder, E. (1977) *Strawberry Imperialism*. The Hague: Research Report Series, Institute of Social Studies.

Fischler, F. (1996) Europe and its rural areas in the year 2000: integrated rural development as a challenge for policy making. Opening speech presented at the European Conference on Rural Development: Rural Europe–Future Perspectives. Cork, Ireland, 7–9 November.

Fischler, F. (1998) Food and the environment: agriculture's contribution to a sustainable society. In WUR, *Compendium van een driedaagse confrontatie tussen wetenschap, samenleving en cultuur*, Wageningen, 16–18 April. Wageningen: Wageningen Agricultural University.

Frouws, J. (1993) Mest en Macht: een politiek-sociologische studie naar belangenbehartiging en beleidsvorming inzake de mestproblematiek in Nederland vanaf 1970. *Studies van Landbouw en Platteland*, No. 11. Wageningen: Wageningen Agricultural University.

Frouws, J. et al. (1996) Naar de geest of naar de letter: een onderzoek naar knellende regelgeving in de agrarische sector. *Studies van Landbouw en Platteland*, No. 19. Wageningen: Wageningen Agricultural University.

Gerritsen, P.R.W. (2002) *Diversity at Stake: A Farmers' Perspective on Biodiversity and Conservation in Western Mexico*. Wageningen: Circle for Rural European Studies, Wageningen University.

Gonzalez Chavez, H. (1994) El empresario agricola en el jugoso negocio de las frutas y hortalizas de México. PhD thesis. Wageningen: Wageningen Agricultural University.

Goodman, D. (1999) Agro-food studies in the 'Age of Ecology': nature, corporeality, bio-politics. *Sociologia Ruralis*, 39 (1): 17–38.

Goodman, D. and Wilkinson, J. (1993) Towards a polyvalent agro-food system. Unpublished paper.

Goodman, D. and Watts, M. (1997) *Globalising Food: Agrarian Questions and Global Restructuring*. London: Routledge.

Griffin, K. (1981) *Land Concentration and Rural Poverty*. Hong Kong: Macmillan Press.

Groupe de Bruges (1998) Landbouw in Europa: over de noodzaak van een ommekeer. Amsterdam: Balie.

Guzman Casado, G.I. et al. (2000) *Introduccion a la agroecologia como desarrollo rural sostenible*. Madrid: Ediciones Mundi-Prensa.

Halweil, B. (2000) Where have all the farmers gone? *World Watch*, Sept/Oct: 12–28.

Halweil, B. (2003) Home grown. The case for local food in a global market. Worldwatch Paper 163. Washington, DC: Worldwatch Institute.

Hofstee, E.W. (1985) Groningen van grasland naar bouwland, 1750–1930. Wageningen: PUDOC.

Hoog, K. de and Vinkers, J. (2000) De beleving van armoede in agrarische gezinsbedrijven. Wageningen: Wageningen University and Research Centre.

Hoogh, J. de and Silvis, H. (1988) EG-landbouwpolitiek van binnen en van buiten. Wageningen: PUDOC.

Huysman, B. (1978) Uncertainty and decision making in small farmer agriculture. Wageningen Agricultural University.

Huizer, G. (1973) *Peasant Rebellion in Latin America*. Harmondsworth: Pelican Latin American Library.

IATP (1998) *Marketing Sustainable Agriculture: Case Studies and Analysis from Europe*. Minneapolis, MN: Institute for Agriculture and Trade Policy.

Ikerd, J. (2000a) Sustainable agriculture: a positive alternative to industrial agriculture. <http://www.ssu.msissouri.edu/faculty/jikerd/papers/ks-hrtld.htm>.

Ikerd, J. (2000b) Sustainable farming and rural community development. University of Missouri <http://www.ssu.msissourr.edu/faculty/jikerd/papers/ND-NFCD.html>.

Ikerd, J. (2000c) Sustainable agriculture as a rural economic strategy. University of Missouri <http://www.ssu.msissouri.edu/faculty/jikerd/papers/sa-cdst.hrtld.html>.

Illich, I. (1978) Fortschrittsmythen. Reinbek: Suhrkamp.

Janvry, A. de, Murgai, R. and Sadoulet, E. (2002) Rural development and rural policy. In B. Gardner and G. Rausser (eds), *Handbook of Agricultural Economics*, Vol. 2A. Netherlands: Elsevier Science.

Joannides, J., Bergan, S., Ritchie, M., Waterhouse B. and Ukaga, O. (2001) *Renewing the Countryside*. Minneapolis, MN: Minnesota Institute for Agriculture and Trade Policy.

Kayser, Bernard (1995) The future of the countryside. In J.D. van der Ploeg and G. van Dijk (eds), *Beyond Modernization: The Impact of Endogenous Rural Development*. Assen: Van Gorcum. pp. 179–190.

Kearny, M. (1997) *Reconceptualizing the Peasantry: Anthropology in Global Perspective*. Boulder, CO: Westview Press.

Kinsella, J., Wilson, S., De Jong, F. and Renting, H. (2000) Pluriactivity as a livelihood strategy in Irish farm households and its role in rural development. *Sociologia Ruralis*, 40 (4): 481–496.

Kinsella, J., Bogue, P., Mannion, J. and Wilson, S. (2002) Cost reduction for small-scale dairy farms in County Clare. In J.D. van der Ploeg, A. Long and J. Banks (eds), *Living Countrysides: Rural Development Processes in Europe: The State of the Art*. Doetinchem: Elsevier. pp. 149–159.

Knickel, K. (2002) Energy crops in Mecklenburg–Vorpommern: the rural development potential of crop diversification and processing in Germany. In J.D. van der Ploeg, A. Long and J. Banks (eds), *Living Countrysides: Rural Development Processes in Europe: The State of the Art*. Doetinchem: Elsevier. pp. 49–61.

Knickel, K. and Hof, S. (2002) Direct retailing in Germany: farmers markets in Frankfurt. In J.D. van der Ploeg, A. Long and J. Banks (eds), *Living Countrysides: Rural Development Processes in Europe: The State of the Art*. Doetinchem: Elsevier. pp. 103–113.

Koning, N.B.J. (1982) Agrarische gezinsbedrijven en industrieel kapitalisme. *Tijdschrift voor Politieke Economie*, 6 (1): 35–66.

Lacroix, A. (1981) Transformations du procèes de travail agricole, incidences de l'industrialisation sur les conditions de travail paysannes. Grenoble: INRA.

Law, J. (1994) *Organizing Modernity*. Oxford: Blackwell.

Leeuwis, C. (1993) Of computers, myths and modelling: the social construction of diversity, knowledge, information and communication technologies in Dutch horticulture and agricultural extension. Wageningen: Wageningen Agricultural University.

LEI (Landbouw Economisch Instituut) (2002) *Landbouw-Economisch Bericht 2002*. The Hague: LEI.

Llambi, L. (1994) Comparative advantages and disadvantages in Latin American nontraditional fruit and vegetable exports. In P. McMichael (ed.), *The Global Restructuring of Agro-Food Systems*. Ithaca, NY: Cornell University Press. pp. 190–213.

Long, N. et al. (1986) The commoditization debate: labour process, strategy and social network. Papers of the Departments of Sociology, No. 17. Wageningen: Wageningen Agricultural University.

Mariategui, J.C. (1928) *Siete ensayos de interpretacion de la realidad Peruana*. Lima: Biblioteca Amauta.

Marsden, T. (1998) Agriculture beyond the treadmill? Issues for policy, theory and research practice. *Progress in Human Geography*, 22 (2): 265–275.

Marsden, T. (2003) *The Condition of Rural Sustainability*. Assen: Royal van Gorcum.

Marsden, T., Banks, J. and Bristow, G. (2000) Food supply chains approaches: exploring their role in rural development. *Sociologia Ruralis*, 40 (4): 424–438.

Marsden, T., Flynn, A. and Harrison, M. (2000) *Consuming Interests. The Social Provision of Food*. London: UCL Press.

McMichael, P. (ed.) (1994) *The Global Restructuring of Agro-Food Systems*. Ithaca, NY: Cornell University Press.

Miele, M. (2001) Creating sustainability: the social construction of the market for organic products. Wageningen: Wageningen Univeristy and Research Centre.

Milone, P. and Ventura, F. (2000) Theory and practice of multi-product farms: farm butcheries in Umbria. *Sociologia Ruralis*, 40 (4): 452–465.

MPAF (Ministero delle Politiche Agricole e Forestali) (2003) La poverta in agricoltura: una mappa del rischio e del disagio rurale in Italia. Rome: Eurispes.

Niehoff, A. and Price, L. (2001) Rural livelihood systems: a conceptual framework. UPWARD Series on Rural Livelihoods No. 1. Wageningen Agricultural University.

Oostindie, H. (2002) The integration of care activities on farms. In J.D. van der Ploeg, A. Long and J. Banks (eds), *Living Countrysides: Rural Development Processes in Europe: The state of the Art*. Doetinchem: Elsevier. pp. 63–71.

Oostindie, H., Ploeg, J.D. van der and Renting, H. (2002) Farmers' experiences with and views on rural development practices and processes: outcomes of a transnational European survey. In J.D. van der Ploeg, A. Long and J. Banks (eds), *Living Countrysides: Rural Development Processes in Europe: The State of the Art*. Doetinchem: Elsevier. pp. 213–231.

Osti, Giovanni (1991) Gli innovatori della periferia, la figura sociale dell'innovatore nell'agricoltura di montagna. Trento: Reverdito Edizioni.

Owen, W.F. (1966) The double developmental squeeze on agriculture. *American Economic Review*, LVI: 43–67.

Paige, J. (1975) *Agrarian Revolution: Social Movements and Export Agriculture in the Underdeveloped World*. New York: The Free Press.

Panzar, J. and Willing, R. (1981) Economies of scope. *American Economic Review*, 71 (2): 268–272.

Pearse, A. (1975) *The Latin American Peasant*. Library of Peasant Studies, No. 1. London: Frank Cass.

Ploeg, J.D. van der (1990) *Labour, Markets and Agricultural Production*. Westview Special Studies in Agriculture, Science and Policy. Boulder, CO: Westview Press.

Ploeg, J.D. van der (1991) Autarky and Technical Change in Rice Production in Guinea Bissau: on the importance of commoditisation and decommoditisation as interrelated processes. In M. Hasell and D. Hunt. *Rural Households in Emerging Societies: Technology and Change in Sub-Saharan Africa*. Oxford: Berg.

Ploeg, J.D. van der (1998) Land Reform: Unfinished past and open future. Lecture on the occasion of the 80th anniversary of Wageningen Agricultural University.

Ploeg, J.D. van der (2000) Revitalizing agriculture: farming economically as starting ground for rural development. *Sociologia Ruralis*, 40 (4): 497–511.

Ploeg, J.D. van der (2003) *The Virtual Farmer*. Assen: Royal Van Gorcum.

Ploeg, J.D. van der, Roep, D., Renting, H., Banks, J., Alonso Mielgo, A., Gorman, M., Knickel, K., Schaefer, B. and Ventura, F. (2002) The socio-economic impact of rural development processes within Europe. In J.D. van der Ploeg, A. Long and J. Banks (eds), *Living Countrysides: Rural Development Processes in Europe: The State of the Art*. Doetinchem: Elsevier. pp. 179–191.

Ploeg, J.D. van der, Long, A. and Banks, J. (2000c) Rural development: the state of the art. In J.D. van der Ploeg, A. Long and J. Banks (eds), *Living Countrysides: Rural Development Processes in Europe: The State of the Art*. Doetinchem: Elsevier. pp. 8–17.

Ploeg, J.D. van der, Verhoeven, F., Oostindie, H. and Groot, J. (2003) Wat smyt it op: een verkennende analyse van bedrijfseconomische en landbouwkundige gegevens van VEL & VANLA bedrijven. <http://www.velvanla.nl>

Priebe, H. (1985) *Die subventionierte Unvernunft, Landwirtschaft and Naturhaushalt*. Berlin: Siedler Verlag.

Primdahl, J. (1999) Agricultural landscapes as production and living places: on the owner's versus producer's decision making and some implications for planning. *Landscape and Urban Planning*, 46 (1–3): 143–150.

Rabbinge, R., Lantinga, E., Goewie, E. and Oomen, G. (1997) *De renaissance van het gemengde bedrijf: een weg naar duurzaamheid*. Wageningen: Wageningen Agricultural University.

Reijntjes, C., Haverkort, B. and Waters-Bay, A. (1992) *Farming for the Future: An Introduction to Low External Input and Sustainable Agriculture*. Leusden, London: ILEIA/MacMillan.

Renting, H. and van der Ploeg, J.D. (2001) Reconnecting nature, farming and society: environmental co-operatives in the Netherlands as institutional arrangements for creating coherence. *Journal of Environmental Policy & Planning*, 3: 85–101.

Renting, H., Marsden, T. and Banks, J. (2003) Understanding alternative food networks. *Environment and Planning* A, 35: 393–411.

Rip, A. and Kemp, R. (1998) Technological change. In S. Rayner and E.L. Malone (eds), *Human Choice and Climate Change*. Vol. 2. Columbus, OH: Battelle Press. pp. 327–399.

Ritzer, G. (1993) *The McDonaldization of Society: An Investigation Into the Changing Character of Contemporary Social Life*. Thousand Oaks, CA: Pine Forge Press.

Robertson Scott, J.W. (1912) *A Free Farmer in a Free State. A Study of Rural Life and Industry and Agricultural Politics in an Agricultural Country*. London: Heinemann.

Roep, D. (2000) Vernieuwend werken; sporen van vermogen en onvermogen (een socio-materiele studie over verniewuing in de landbouw uitgewerkt voor de westelijke veenweidegebieden). *Studies van Landbouw en Platteland*, No. 28. Wageningen: Circle for Rural European Studies, Wageningen University.

Roep, D. (2002) Value of quality and region: the Waddengroup Foundation. In J.D. van der Ploeg, A. Long and J. Banks (eds), *Living Countrysides: Rural Development Processes in Europe: The State of the Art*. Doetinchem: Elsevier. pp. 87–99.

Roest, K. de (2000) *The Production of Parmigiano–Reggiano Cheese: The Force of an Artisanal System in an Industrialised World*. Assen: Van Gorcum.

Roest, K. de and Menghi, A. (2000) Reconsidering 'traditional' food: the case of Parmigiano–Reggiano cheese. *Sociologia Ruralis*, 40 (4): 439–451.

Ruivekamp, G. (1989) *De invoering van biotechnologie in de agro-industriele produktieketen. De overgang naar een nieuwe arbeidsorganisatie*. Utrecht: Jan van Arkel.

Saccomandi, V. (1998) *Agricultural Market Economics: A Neo-institutional Analysis of the Exchange, Circulation and Distribution of Agricultural Products*. Assen: Van Gorcum.

Saraceno, Elena (1996) Jobs, equal opportunities and entrepreneurship in rural areas. Paper presented at the European Conference on Rural Development: Rural Europe – Future Perspectives. Cork, Ireland, 7–9 November.

SARE (2001) The New American Farmer: Profiles of Agricultural Innovation. USDA Sustainable Agriculture Research and Education (SARE) program. http://www.sare.org/newfarmer/toc.htm. Published by the USDA SARE program.

Scettri, R. (ed.) (2001) *Novità in campagna: innovatori agricoli nel sud Italia*. Rome: ACLI Terra/IREF.

Scherer, F. (1975) *The Economics of Multiplant Operation*. Cambridge, MA: Harvard University Press.

Schlosser, E. (2002) *Fast Food Nation. The Dark Side of the All-American Meal*. Boston, MA: Houghton Mifflin.

Schuite, H. (2000) *Pioneers in Agriculture: A Study on Direct Sales and on Farm Transformation in the Province of Gelderland*. Wageningen: Rural Sociology Group, Wageningen University.

Scott, J.C. (1976) *The Moral Economy of the Peasant*. New Haven, CT: Yale University Press.

Scottish Office, Land Reform Policy Group (1998) *Identifying the Problems and Identifying the Solutions*. Edinburgh: Scottish Office.

Sevilla Guzmàn, E. and Gonzalez, M. (1990) Ecosociologia: elementos teoricos para el analisis de la coevolucion social y ecologica en la agricultura. In *Revista Espanola de Investigaciones Sociologicas*. 52: 7–45.

Slicher van Bath, Bernard (1958) Een Fries landbouwbedrijf in de tweede helft van de zestiende eeuw. In *Agronomische Bijdragen, vierde deel*. Wageningen: Veenman & Zonen.

Smeding, F.W. (2001) *Steps Towards Food Web Management on Farms*. Wageningen: Wageningen University.

Stassart, P. and van Engelen, G. (eds) (1999) *Van de grond tot in je mond. 101 pistes voor een kwaliteitsvoeding*. Vredeseilanden-Coopibo and Fondation Universitaire Luxembourgeoise.

Swagemakers, P. (2002) Verschil Maken: novelty-productie en de contouren van een streekcooperatie. *Studies van Landbouw en Platteland*, No. 33. Wageningen: Wageningen University and Research Centre.

Thiesenhuisen, W.C. (1995) *Broken Promises: Agrarian Reform and the Latin American Campesino*. Boulder, CO: Westview Press.

Toledo, V. (1992) La racionalidad ecologica de la produccion campesina. In E. Sevilla Guzman, and M. Gonzalez de Molina (eds), *Ecologia, campesinado e historia*. Madrid: Las Ediciones de la Piqueta. pp. 197–218.

Toledo, V. (2000) *La paz en Chiapas, ecologia, luchas indigenas y modernidad alternativa*. Mexico, DF: Ediciones Quinto Sol.

Tracy, M. (1997) *Agricultural Policy in the European Union*. Agricultural Policy Studies, esp. ch. 8 (also available on http://ourworld.compuserve.com/homepages/APS_BELGIUM/issues.htm).

Ullrich, O. (1979) *Weltniveau*. Berlin: Rotbuch Verlag.

Vaeren, P. van der (2000) Perdidos en la Selva, un estudio del proceso de re-arraigo y de desarrollo de la Comunidad Cooperativa Union Maya Itza, formada por campesinos guatemaltecos, antiguos refugiados, reasentados en el Departamento de El Peten, Guatemala. Amsterdam: Thela Publishers.

Ventura, F. (1995) Styles of beef cattle breeding and resource use efficiency in Umbria. In J.D. van der Ploeg and G. van Dijk (eds), *Beyond Modernization: The Impact of Endogenous Rural Development*. Assen: Van Gorcum.

Ventura, F. (2001) *Organizzarsi per Sopravvivere: un analsisi neo-istituzionale dello sviluppo endogeno nell'agricolura Umbra*. Wageningen: Wageningen University.

Visser, A.J. (2000) Prototyping on farm nature management, a synthesis of landscape ecology, development policies and farm-specific possibilities. *Aspects of Applied Biology*, 58: 299–304.

Vries, W. de (1995) Pluri-activiteit in de Nederlandse landbouw. *Studies van Landbouw en Platteland*, No. 17. Wageningen: Wageningen Agricultural University.

Ward, N. (1993) The agricultural treadmill and the rural environment in the post-productivist era. *Sociologia Ruralis*, 33 (3/4): 348–364.

Whatmore, S. and Stassart, P. (2001) Metabolizing risk: the assemblage of alternative meat networks in Belgium. Workshop on International Perspectives on Alternative Agro-Food Networks: Quality, Embeddedness, Bio-Politics, University of California, Santa Cruz.

Wilson, S., Mannion, J. and Kinsella, J. (2002) The contribution of part-time farming to living countrysides in Ireland. In J.D. van der Ploeg, A. Long and J. Banks (eds), *Living Countrysides: Rural Development Processes in Europe: The State of the Art*. Doetinchem: Elsevier. pp. 163–175.

Wiskerke, J.S.C. (1997) *Zeeuwse akkerbouw tussen verandering en continuiteit: een sociologische studie naar diversiteit in landbouwbeoefening, technologieontwikkeling en plattelandsvernieuwing*. Wagening: Circle for Rural European Studies, Wageningen Agricultural University.

Wiskerke, H. (2001) Rural development and multifunctional agriculture. Topics for a new socio-economic research agenda. *Tijdschrift voor Sociaalwetenschappelijk onderzoek van de landbouw*, 16 (2): 144–119.

Wiskerke, J.S.C. (2003) On promising niches and constraining sociotechnical regimes: the case of Dutch wheat and bread. *Environment and Planning A* 35: 429–448.

Wolf, E. (1969) *Peasant Wars of the Twentieth Century*. New York: Harper and Row.

Wrigley, N. and Lowe, M.S. (eds) (1996) *Retailing, Consumption and Capital. Towards a New Retail Geography*. Harlow: Longman.

Yotopoulos, P.A. (1974) Rationality, efficiency and organizational behaviour through the production function: darkly. *Food Research Institute Studies*, XIII (3): 263–274.

Ziel, T. van der (2003) *Verzet en verlagen, de constructie van nieuwe ruraliteiten rond de mkz-crisis en de trek naar het platteland*. Wageningen: Wageningen University.

Zuiderwijk, A. (1998) *Farming Gently, Farming Fast: Migration, Incorporation and Agricultural Change in the Mandara Mountains of Northern Cameroon*. Leiden: CLM.

19

Neo-endogenous rural development in the EU

Christopher Ray

INTRODUCTION

This chapter looks at certain aspects of the sociology of rural development. By rural development, we will be referring both to *activity* that occurs in rural areas in pursuit of socio-economic vibrancy and to the domain of *intervention*, that is, the rural policies of national (and sub- and supra-national) administrations and of NGOs with an interest in rural areas. We will not, however, be concerned with the 'rurality' of rural areas *per se* but with the socio-economic regeneration of *territories* (which also happen to have certain rural characteristics). The theoretical fundamentals involved are shared with the domain of urban regeneration: the common objective being to devise an approach to regeneration that can tackle the differential socio-cultural and economic trajectories within countries as well as on the transnational level.

The fundamental idea has been called *neo-endogenous* (rural) development (ignoring its rather unfortunate acronym: NERD). The endogenous part refers to the animation of development along a bottom-up trajectory: that is, when the search for development resources and mechanisms focuses onto the local territorial level. Generally, a synonym for endogenous would be participative. The 'neo' part, whilst not challenging the integrity of bottom-up dynamics, identifies the roles played by various manifestations of the *extralocal*. Actors in the politico-administrative system (through the national up to the European level) as well as in other localities are all seen as part of the extralocal environment of rural development and as potentially recruitable by localities in support of their regeneration strategies.

The other point to mention here is that the politico-economic framework for this chapter is that of the EU and its component countries and regions. Despite the prominence of the participative approach in Third World development studies/practice, the challenge for neo-endogenous rural development is to devise a coherent theory and modus operandi for the contemporary conditions of the EU.

From the policy direction, the neo-endogenous approach is based on the idea that socio-economic well-being (of the presently disadvantaged rural economy) can best be brought about by restructuring public intervention away from individual sectors in favour of a mosaic of local/regional territories. It is an alternative to the practice of central authorities of designing interventions which deal with sectors of social and economic life in isolation from each other and/or which assume that socio-economic problems can be solved by standard measures, regardless of location or culture. According to this viewpoint, vulnerable or ailing territories need not resign themselves to being victims of broad, exogenous, political and economic forces; potentially, localities can effect change in their favour. Central to the approach is that a local area has, or must acquire, the capacity to assume some responsibility for bringing about its own socio-economic development.

In terms of (rural) development, the neo-endogenous approach has two other primary characteristics. First, economic and other development activity is reoriented to maximize the retention of benefits within the local territory by valorizing and exploiting local resources – physical and human. Second, development is contextualized by focusing on the needs, capacities and perspectives of local people; the development model emphasizes the principle and process of 'local participation' in the design and implementation of action and through the adoption of cultural, environmental and 'community' values within a development intervention. The rhetoric offers the prospect of local areas assuming greater influence over their futures by reorienting development around local resources and by setting up structures to sustain the local development momentum following an initial official intervention.

The ongoing project that is the sociology of neo-endogenous rural development is steadily accumulating empirical research results (individual studies of local initiatives; comparative case studies such as the PRIDE study of local partnerships – Westholm et al., 1999; policy evaluations and good-practice guides generated by organizations such as the LEADER Observatory – http://www.rural-europe.aeidl.be). Yet there is also a place for theorizing of a speculative kind. By its very nature, the neo-endogenous approach is exploratory and evolving and social scientists are actors in this dynamic, offering conceptualizations as potentially useful tools for local practitioners and policy officials. This empirical–speculative duality is also the basis for the present chapter. As for its structure, the next section introduces the theoretical principles which are informing the sociological study of neo-endogenous development. This is followed by a discussion of the politico-administrative circumstances which led to the emergence of, and which continue to sustain, the neo-endogenous approach in official interventions.

The rest of the chapter is given over to an exposition of its primary thesis: the way in which neo-endogenous development operates simultaneously on three planes: intra-territorial, vertical (politico-administrative context) and inter-territorial.

THEORETICAL PRINCIPLES

The comprehensive study of neo-endogenous development draws on a number of general and interrelated conceptual approaches: social economy, economic coordination and multi-level governance – which will be introduced here – and the sociology of forms of capital (whose introduction appears in a section entitled 'the intra-territorial plane').

Social economy

According to Sayer and Walker (1992), proponents of the social economy idea argue that modern economies can only be fully understood if due attention is paid to the social context. For them, economics is an inherently humanistic endeavour, the study of which introduces issues of 'co-operation, democracy and meaningful labour [that are] the hallmarks of a just order' (1992: 1). For authors such as Sayer and Walker, the social economy represents a normative approach to the design of (rural) development intervention. Simultaneously, it is a call, on the one hand, for an alternative to the 'triumphant individualism' which is often used as a characterization of the contemporary capitalist society and, on the other, to recognize, more explicitly, that the functioning of real-world economic life actually depends on significant levels and forms of social 'tissue' and regulation.

According to such authors, the pursuit of the social economy idea requires attention to be focused on the division of labour and, thereby, on the lived experience of participants:

> the differences built into human geography are, in large measure, those of the division of labour ... [this] works its magic on the consciousness of participants in differentiated societies. (Sayer and Walker, 1992: 5)

As we shall see below, the neo-endogenous approach to rural development is a response to the idea of social economy. It is humanistic not only through the inclusion of local participation in territorial development activity but also in looking for potential resources that are embodied in local people. Neo-endogeneity is also, by definition, about local collective activity and about defining development broadly in terms of the quality of life. Rural development is thus about economic activity that is, at least partially, embedded in local territories. Neo-endogenous development activity can be seen as an attempt to reorientate the division of labour.

However, the term social economy has also taken on a more specific meaning. In seeking to place development and the economy firmly in a context that is, at once, social and local (territorial), some advocates such as Archibugi (2000) chose to emphasize the domain of voluntary associativeness. The actors in this domain comprise cooperatives and mutuals (which operate in the market), and associations (which do not). Being, inherently, expressions of local expectations, requirements and resources, such entities are thought to embody the essentials of social economy. They do this not only by operating according to the principles of solidarity, participation and civic purpose but also by either not trading to make a profit or by using profits for (local) social purposes. Also known as civil society, this domain has entered into the intervention rhetoric of governments (see, for example, the 2000/2004 Rural White Papers in the UK). The EU also subscribes to this use of social economy, being aware, according to Archibugi:

> of the need to pay particular attention, in the context of its initiatives for economic development, employment and cohesion in the Union, to the potential of these economic actors which cannot be referred to the traditional profit-orientated enterprise, or to public bodies, and therefore form a third sector of the economic system. (2000: 207)

Civil society is part of the neo-endogenous approach but it has to be regarded as only one among a range of potential actors. The approach is concerned with the construction of modes of production, definitions of development and of official interventions that may be more propitious for (economically and socio-culturally) ailing and vulnerable territories as well as for wider society. Yet, neo-endogeneity does not have to entail opting out of modern capitalism and, whilst it elevates the role of these organizations in the pursuit and definition of development, it by no means confines itself to such ideas. Indeed, the recruitment of this sector is, in some situations, seen by local actors instrumentally: as a means to engage more fully in globalizing capitalist society (whether ultimately individualistic or not).

Economic coordination

In 'The Great Transformation', written in the 1920s, Polanyi set out a manifesto (McRobie, 1994). Alarmed at the negative effects of modern global capitalism – the individualization of society, environmental and cultural degradation – he called for action mainly on the part of national governments to recreate the conditions of solidarity so as to protect the social, cultural and economic well-being of communities and nations. Today, the dynamics of, and interrelationship between, Polanyi's forms of economic integration – reciprocity, redistribution and market exchange (and their associated wider patterns of social organization, respectively duality, centricity and atomistic individualism) – continue to pervade the theorization, policy-making and practice of rural socio-economic development. This is particularly so in the case of the neo-endogenous approach.

Studies of neo-endogenous development are concerned, in the first instance, with how these relations manifest themselves at the local and regional level across Europe: that is, whether the distribution of the relations contributes to an understanding of differential socio-cultural and economic vibrancy. More importantly, however, is the investigation of whether, and how, these factors can be manipulated so as to create the conditions for territorial development.

The reciprocity/duality form clearly has utility in explorations of the pursuit of intra-territorial social vibrancy but, increasingly, it is being used – juxtaposed with the market exchange form – in speculative models of territorial economic organization (below). This manifests in contemporary interest in the concepts of social capital and territorial solidarity (local patriotism).

The appearance of the territorial approach in public interventions is also, if implicitly, concerned with redistribution. Yet, the redistribution involved is of a less direct type than that usually understood by the term. It is redistribution in which the potential beneficiaries are charged with the responsibility of being partners in the process; only by organizing themselves so as to be more able to take advantage of an intervention will they be selected to enter into the process (Kovách, 2000; Ray, 2001).

As for market exchange (and atomistic individualism), this is also an explicit component of neo-endogenous development. As will be seen, territorial initiatives are rarely, if ever, concerned with disengagement from capitalist society. Rather, they have as their objective the cultivation of a mode of production that would respond to their presently disadvantaged situation while also couching the definition and methods of economic regeneration in a social and cultural framework.

The experience of neo-endogenous development across rural Europe has also shown that the operationalization of the mix of economic integration/social organization triad occurs along the three interconnecting planes: intra-territorial, vertical politico-administrative and inter-territorial (interlocal). The emerging awareness of the latter is bringing to the analytical framework yet another conceptual approach consisting of general systems theory/network analysis. Although at the speculative stage, this is the idea that, in neo-endogenous development, although territories are being credited with an enhanced sense of agency, there may be causal factors that emerge out of the growing transnational connectivity between localities which partly compromise the strategic actions of individual territorial actors (see section entitled 'the inter-territorial plane').

EMERGENCE AND DRIVERS OF NEO-ENDOGENOUS DEVELOPMENT ACTIVITY

Although at a discursive level it is still possible to talk of '*the* countryside' and of generic 'rural development', the reality is that increasing amounts of the socio-economic life of non-urban space are being organized on the basis of *territories*. Rural development space is taking on the appearance of a mosaic of territorial entities with various relations to a country's politico-administrative system; some are embedded within a Local Authority whilst others are more independent of the public sector. They vary in scale from areas with perhaps as many as 100,000 people to the very local level (single settlements and their catchment areas: the level of community development). Recently in the UK, the territorialization of (disadvantaged) rural space has manifested as Health Action Zones and Education Action Zones (both probably of a temporary nature) as well as a further round of Enterprise/ Employment designations. These initiatives join designations of a much longer vintage, particularly those concerning nature and landscape. This chapter, however, will be concerned solely with rural development territories, not least because they may turn out to be long-lasting structural elements of a new rural economy.

Territories defined according to a rural development rationale emerge along two, usually intersecting, trajectories. One trajectory is that of official interventions – by governments and their agents and the institutions of the EU. Particularly prominent on the pan-European stage has been the LEADER Community Initiative launched by the Directorate-General for Agriculture in 1991.[1] (The Objective 5b programme, although also a manifestation of the same politico-administrative initiative as LEADER, was of a rather different nature from what is being discussed in this chapter.) Within the member countries of the EU can be found other examples of the territorial approach to rural development, including '*contrats de pays*' in France, Local Partnership Companies in Ireland, PRODER in Spain and POMO in Finland (Buller, 2000; Westholm et al., 1999). The territorialization approach also emerges endogenously, as local autonomous initiatives. These are particularly evident in collaborative actions involving food producers or the tourism sector but also include local regeneration initiatives of a more comprehensive nature (see Midmore et al., 1994 for a study of one of many examples). Yet even apparently endogenous initiatives, sooner or later, engage with the extra-local level as they seek funds and other forms of assistance. Moreover, as Buller notes in relation to France, rural development territories are, in one sense, created by projects, whether as official interventions or bottom-up initiatives.

Thus, these rural development territories have multiple functions: they are (often newly created) units in which government, European and NGO policies are implemented (a further channel for public and other funds); they are geographical clusters of potential collective strategic activity by component (socio-cultural, economic and politico-administrative) actors; many are the domain of new organizations – local partnership-committees – which function as interlocutors between a locality and its politico-economic environment; and they provide rationales for reviving or inventing local cultural identity (and as invitations to local actors to subscribe to/exploit that identity).

More generally, rural development territories are definable as much by their relationship to the extralocal level as by the concept of rurality. It is not about local areas pursuing development autonomously of the wider politico-economic environment. The notion of pure endogenous development in which change is animated solely by local actors independent of assistance from external agents is useful but only as a heuristic device: what Weber would have called an 'ideal type'. What follows below is a response to calls by commentators such as Lowe et al. (1995) who argue for the theorization of rural development to go beyond endogenous and exogenous models by focusing analysis onto the dynamic interactions between local areas, their component actors and the political, economic and natural environments in which they unavoidably exist: explicitly and/or implicitly. Neo-endogenous development retains a bottom-up core in that local territories and actors are understood as having the potential for (mediated) agency, yet understands that extralocal factors, inevitably and crucially, impact on – and are exploitable by – the local level.

But before embarking on an exploration of the components of the neo-endogenous idea, we should consider briefly some of the main factors that have led to the emergence, and which influence the evolution, of the approach. By identifying these factors, we will indicate the components of a framework for the ongoing critical analysis of manifestations of the approach.

A major, and very tangible, factor is that of the EU, and in particular the modes of intervention introduced by the Future of Rural Society policy (CEC, 1988), and on a broader scale the dynamics of the EU as a political and cultural space. FORS signalled a renewed attempt by the EU to tackle the problem of socio-economic divergence across rural Europe, this time by targeting interventions onto territories of particular needs (the 'standard problems'). As an exercise in geographical fine-tuning, the interventions that resulted from FORS referred to notions of policy efficiency as well as to the politico-economic imperative of the EU (the pursuit of a pan-European level economic playing field and of popular support for the EU). Yet the territorial approach also entailed thinking about the necessity of broadening the scope of rural development, from *sectors* to the dynamics of coherent *territories*. One driver of neo-endogenous interventions was, therefore, the tentative moves to challenge the primacy of agricultural production in rural policy and, thus, to widen the client group from farmers to include all rural actors. European funds were to be channelled into specific regions (Objective 1, 5b and subsequently 6) designated through a process of negotiation between national governments (and their regional components) and the European Commission. These regions then provided the geographical framework for the EU's radical experiment in territorial/neo-endogenous rural development: the LEADER Initiative. LEADER, although relatively insignificant in terms of its overall level of funding (see Ray, 1998), was both an intensification of the territorial approach (focusing onto local territories of less than – and, in some cases, much less than – 100,000 people) and more ambitious in terms of redefining the actors and process of rural development intervention. Moreover, the initiative was clearly, if implicitly, part of the European politico-economic project; decentralization of policy implementation to local action groups was accompanied by the encouragement (later strengthened to a requirement) to participate in pan-European networks. The latter, in particular, can be seen as an attempt on the part of the EU to cultivate a transnational dynamic by which local territories would supplement their localist perspective with a truly European identity.

This resonates with another factor behind territorialization: the politico-administrative dynamic towards decentralized government and the associated idea of the 'managerial state' which characterize the approaches of administrations at national and European levels. The politico-administrative rationale for neo-endogenous development, on the one hand, can be a matter of efficiency: targeting funding on those localities where there is most need or apparent capacity to respond to the opportunity. On the other hand, it is a manifestation of state managerialism whereby the state enters into arrangements with (partnerships of) local actors contracted to deliver policy objectives, justified by the expectation that effective policy implementation requires interventions to be responsive to the local context and to allow for the active participation of local actors from the private and voluntary, as well as public, sectors. In this sense, neo-endogenous development is, essentially, a manifestation of the contemporary fashion in what is fast becoming mainstream European politics.

Alongside these 'top-down' explanations of the rise of the neo-endogenous approach are those reflecting various bottom-up dynamics. Bottom-up regionalism in the form of political and economic mobilization by the 'cultural regions' of Europe has been under way since the 1960s, with a further boost to activity in the 1980s (Jones and Keating, 1995). Rationales based on the inequities of socio-economic trajectories and the 'internal colony' thesis have been joined more recently by a rhetoric made up of elements of participative democracy, Green ideology and the value of local/cultural identity in a pluralistic Europe. This 'organic' social movement – together with the lobbying activities of environmental NGOs – worked to bring into the realm of rural policy the idea of local culture as both a stock of resources available for economic exploitation and an end of development in its own right. Thus, the notion of rural Europe as a mixture of primary production activity and generic rurality was supplemented by rural Europe as the domain of indigenous cultural identity (in contrast to the inherently more cosmopolitan culture of the urban domain). The role of neo-endogenous rural development intervention as a mechanism for Europeanization was thus able to respond to (or exploit) the regionalist dynamic, or more precisely, a *rural*–regional dynamic. In addition to the decentralization process occurring within nation-states, territorial cultural actors began to look to Europe and the European Commission for support and official funding, this time through rural policy.

The relationship between local cultural identity and economic development leads into another factor in the rise of neo-endogeneity – what Lash and Urry (1994) have called 'economies of signs and space'. This is the idea that a new mode of production-consumption is emerging as a logic of advanced consumer society. The forces of globalization have given rise to a 'reflexive human subjectivity' (p. 5) in which the individual – as consumer or citizen – has acquired an enhanced capacity for agency. Consumer/global capitalism has led production increasingly to replace material and labour value with 'design value'. The trajectory is towards production and consumption being based less on utility and more on the symbolism of the good or service; it is the production and consumption of *signs*. The rationale of the capitalist dynamic and that of the actions of individuals are driven (insofar as they are consumers) by a (greater or lesser) consciousness of the aesthetics of consumption and therefore of production ('aesthetic reflexivity'). Capitalism is increasingly driven by the valorization and accumulation of this subjectivity. One feature of this phenomenon is the valorization of local places, for example in '*produits typiques*' and 'cultural/green tourism', representing a melding of indigenous culture, synthetic stereotypical symbolism (for Lash and Urry's allegorical consumers) and generic rurality values ('localness', 'environmental quality', 'close community' etc.; see, for example, Tregear, 1998). Thus, in devising the neo-endogenous approach to rural development exponents (top-down and bottom-up) were able to make reference to the signals emanating out of contemporary capitalism.

Finally, in this section, we should mention the category of locality-based professionals in influencing policy implementation on the ground. Lowe et al. (1997) demonstrated the way in which Pollution Inspectors from the English National Rivers Authority effectively mediated between farmers and policy designers. As such, the decisions taken by this category of actor were formed through the personal stance that individuals adopted between opposing discourses. They developed coping strategies in the face of ideological and practical pressures. Lowe et al. categorized these individuals as field-level bureaucrats (after the work of Lipsky on street-level bureaucrats). A study of the EU's LEADER initiative (Ray, 1999) found that this category of actor (the co-ordinator of the local action group) comprised *reflexive* individuals who sought, however subtly, to influence the nature of the intervention in their locality. Generally, they brought to the implementation of LEADER already-formed beliefs about the efficacy of participatory action. Their causal influence was also a function of their Europhile attitudes (either pre-existing or acquired as a function of their exposure to LEADER) which enabled them to make associations between the principles of the bottom-up/decentralization approach and seeing the local initiative as part of a pan-European dynamic. It also, incidentally, enabled many such actors to emphasize socio-economic inclusion for all members of a *territory* above that of a single (that is agricultural) sector.

THE THREE PLANES OF NEO-ENDOGENOUS DEVELOPMENT

We can now turn to an examination of the three planes of neo-endogenous development. Two planes – the intra-territorial and the politico-administrative context – have been the foci of much research activity to date. The other – inter-territorial – is a relatively new line of research.

The intra-territorial plane

A TERRITORIAL MODE OF PRODUCTION The territorialization of the rural economy is based on the notion that a new mode of production may be emerging. This new mode is still capitalistic in that individual economic units pursue optimal profitability which, in turn, drives the process of capital accumulation. Equally, principles of self-interest and competition are present. The essence of the new mode of production, however, is that economic activity is refracted through a collective, territorial logic in which the accumulation process requires the territorial integration of capital other than in its purely financial form.

For present purposes, we are interested in the forms of capital as cultural/symbolic, educational and social. Although Bourdieu (1973, 1986) produced the definitive analysis of the interrelationship between these forms, his analysis led him to focus on how social structure results in unequal access to (and therefore material benefit from) these forms of capital. The idea of neo-endogenous development, however, is an attempt to recast some of Bourdieu's ideas into a more optimistic model of territorial rural development.

Cultural capital can be thought of as territorial intellectual property or place-specific factors of production. Social capital describes cooperative/trustful relationships between actors. Theoretically, the presence of social capital reduces transaction costs and enables participants to reap the benefits (socio-psychological and economic) of cooperative activity. Educational capital incorporates the concept of human capital and the cultivation of personal life chances. Equally as important, however, is the dynamic relationship between the accumulation of educational capital and that of social and cultural capital.

The new mode of production appears to require both individual and collective reflexive action to build a pool of territorial common resources. The various actors of a territory pursue their separate goals but also engage in voluntary, cooperative activity, directly and indirectly, to accumulate the forms of capital at the territorial level. Thus, the type of social relations that guide the operation of the new mode would seem to be a mixture of market exchange and voluntary reciprocity.

A territorial mode of production, therefore, would be based on the discovery, or creation, of place-specific resources and a strategic process of capital accumulation. Indeed, it is the logic of accumulation that dictates the nature of production and provides the rationale for voluntary collective activity. Moreover, capital – social, cultural/symbolic and educational – is primarily embodied in people as individuals and so it is the dynamic relationship between individuals, social groups and territories that is of crucial interest in the mode of production. The accumulation of forms of capital may occur within and between all realms: private business, domestic, educational, public administration and voluntary associations.

A territorial mode of production also distinguishes itself by investing the relationship between producer and consumer with an element of symbolic exchange. To those of a Marxist persuasion, industrial capitalism tends to create alienation. As producers, humans recognize their input less and less in the product, over the fate of which the worker has no control. The intrinsic satisfaction that might formerly have been possible in artisan production or production for one's own use is greatly diminished. Similarly, from the consumer side, social relations with the producer tend to be dominated by market exchange. Commodity fetishism – whereby a direct (even emotional) relationship between producer and consumer is impoverished – is the result. The territorial mode of production appears to adopt this analysis by deliberately and strategically reducing alienation and fetishism. The former is addressed by the elevated status of local people as embodiments of territorial capital accumulation whilst the defetishization of territorial products and services involves the cultivation of their symbolic component so that they come to be identified with the specific territory of origin. Consumption and production become imbued with the culture of the producing territory: territories of people with a politico-cultural identity. Thus, culture becomes commoditized but in the form of territorial (common) intellectual property. Yet this is a particular type of commoditization which serves as one of the mechanisms in the accumulation of each of the territorial forms of capital.

These ideas are being developed with reference to general systems theory (Ray, 2001). This requires local actors to be thought of as the component organisms of a (open boundary) system and each territorial economy is itself conceptualized as an organism in a wider system (a pan-European system of rural territorial economies). Following the ideas of Luhmann (1995), social systems can be thought of as systems of communication, an important characteristic of which is autopoieses, defined as the capacity of organisms to monitor, regulate and adapt to the overall conditions within the system and between the system and its environment. Autopoieses occurs through the management of information by organisms about each other and, thereby, about the system as an entity. In a territorial mode of production, autopoieses would operate as a function of the logic of capital accumulation. Component actors would modify their production activity so as to be consistent with the collective, territorial needs. In this way, food production and distribution would become embedded in the territorial logic; the mode of production would act as social (including ecological) control. Local actors – separately and mutually – would act so as to avoid system shocks such as food scares and environmental degradation. Having subscribed voluntarily to a territorial logic, producers would assume an aspect of self-regulation. In other words, a particular 'world of production' would come into being, that is, 'sets of practices involving conventions and the shared ways of understanding they entail' (Hudson, 2001: 27).

TERRITORIAL STRATEGIC ACTIVITY Having come into being by whatever means, territorial initiatives have two basic modes of strategic action, encapsulated in the term culture economy (Ray, 2001). One mode involves the construction and projection of a (new) territorial identity to the outside, that is, the incorporation of cultural resources into a territorial identity in order to promote the territory as an entity. This relates to new territorial development initiatives in which, either using an existing organization (Local Authority, development agency, etc.) or through a new cooperative structure, a territory seeks to establish and promote its identity. In order to pursue its external strategic objectives (such as securing public funding, establishing a strong presence within national and regional policy-making or participating in networks of local initiatives), a local development group portrays itself as being founded on a territory that is coherent and distinctive, and which would be more effective in the pursuit of local needs than other politico-administrative, territorial units.

In the other mode, the emphasis is still on territorial strategies but here the new territorial initiative is engaged in promoting itself internally: to the communities, businesses, associations and official bodies of the local area. Central to the idea of neo-endogenous development is the reinvigoration of new economic opportunities, innovation and a socio-cultural vibrancy that counter economic vulnerability and traditional forces for emigration. The territorial identity invites local capital and entrepreneurship to commit themselves to the culture-territory by presenting common territorial strategic images which businesses and other bodies can exploit. Once the territory has been reconstructed as a coherent entity, with a utility, it can function as a catalyst for local cooperative action and to generate a sense of culture-territorial solidarity in people and enterprises.

SOCIAL CAPITAL Much academic work has been done on the social form of capital, particularly on the local/regional level. The term social capital focuses on the nature of interpersonal and intergroup relationships and how these drive or hinder collective activity. Putnam (1993) defines social capital as features of social organization such as networks, norms and social trust that facilitate coordination and cooperation for mutual benefit. In a very general sense, the term refers to the resources and socio-economic dynamic (including resistance to change) that result from, and recreate, social ties. It is social in that it concerns trust which leads to associativeness and describes therefore both a certain type of behaviour and a set of ethical principles. It is capital because, as Bourdieu explains, it accumulates over time as a function of being invested.

The notion of trustful relationships has been applied to development studies in two basic ways. It has been presented as the foundation of territorial economies, enabling them to be more effective entities of wealth creation and international trade. The other way has been to emphasize trustful relationships on the local scale as promoting a vibrant form of society that is nurturing, humanistic and politically active. Flora (1998) suggests that this reflects an underlying difference in assumptions: social capital can describe either an innate capacity within human nature (the embeddedness perspective) or the means that can be adopted in pursuit of an end (the rational choice perspective).

Putnam, and Gramsci (1981), are regarded as seminal writers on the concept of civil society and trust. Osti (2000) characterizes the Gramscian approach as placing an emphasis on voluntary associations as a tool to mediate the effect on localities of internationally powerful forces (transnational corporations). In the context of the declining role of the nation-state, Gramsci argues that the vacuum must be filled by voluntary associations that can participate in the delivery of social services and assume the mantle of protest and advocacy for social and political groups. The cultivation of spaces of social capital will result in local, popular 'cognitive mobilization' leading, in turn, to political and civic engagement by individuals. The Putnam approach, on the other hand, emphasizes the capacity of voluntary associations to nurture a 'good society'. Associations are also seen as representing the values of democracy, equity and inclusiveness, a view that reflects Putnam's interest in the internal form of associations. Participation in voluntary associations, Putnam argues, encourages people to acquire and reproduce the ethical values of trust and generalized reciprocity (generalized exchange of rights and obligations). This occurs as an outcome of the non-hierarchical structure of associations, the repeated displays of trust and communal altruism, and the frequent and face-to-face interactions that associations allow and require. By contrast, Fukuyama (1995), writing on the role of trust in the trajectory of whole civilizations and contemporary global regions emphasizes that a high stock of social capital works by reducing transaction costs and thereby 'frees' capitalist enterprises from the constraints of kinship or individualism, promoting economic growth, competitiveness and, nowadays, flexible production.

To date, however, it is unclear whether social capital can be a driving force for territorial economic development. Neither is it clear whether it can be consciously cultivated to recreate the particular type of society with which it is associated. Yet, the term retains an importance in the working hypothesis of neo-endogenous development. This, and the fact that that the concept has begun to infiltrate the rhetoric of official rural policy, require the undertaking of further critical research.

PARTNERSHIPS AND LOCAL DECISION-MAKING BODIES The animation *of* territorial development requires organization *by* the territory. The present consensus is that neo-endogenous development is best animated and sustained by a partnership made up of representatives of the private business, public and voluntary sectors of the territory.

Partnership working – collaborative arrangements between public bodies, or between the public, private and voluntary sectors – is increasingly being used as a mechanism to introduce and manage neo-endogenous development. The partners pool their resources in the pursuit of a common policy objective, in this case the socio-economic regeneration of a territory. In theory, the partners cultivate consensual strategies and thereby integrate their separate responsibilities or contributions. The agentic potential lies in the requirement to focus development activity and project financing onto the potential, innate resources of the territory (Edwards et al., 1999; Westholm et al., 1999).

The experience of partnerships for socio-economic development, however, has not always been unproblematic. This is largely because, if attempts to operationalize neo-endogenous development focus primarily on the restructuring of the responsibilities and powers of 'institutional' actors (including those representing interests in the voluntary and private business sectors), analysis has to contend with development being a *discourse*. First, there is the interpretation of neo-endogenous development as *effective intervention*. In this, extralocal actors redesign their modus operandi so as more effectively to achieve their objectives; more effective, that is, than an explicitly exogenous, top-down imposed approach. The effectiveness rationale may, however, be driven by an imperative of financial stringency, it may be ideologically driven as in the political restructuring of states, as they move away from the welfare state model, or it may be part of a technical solution to wider politico-economic goals, as in its use by the EU to reduce socio-economic divergence between regions and thereby promote the European Single Market and political integration. Second, the approach may be advocated as a *legitimacy-seeking* device. Organizations, whether regional, national or international, may seek to enhance the legitimacy of their agendas for change by arguing that endogenous development and the pursuit of their interest agenda are mutually compatible. Thus, for example, environmental NGOs seek to insert their extralocal objectives into local development initiatives and thereby benefit from the legitimacy acquired from being ideologically and practically associated with action based on local, popular participation. Edwards et al. (1999) provide a complementary analysis of how the creation of a partnership is accompanied by one of a set of official discourses which influence the nature of subsequent territorial agency. In their analysis of England and Wales, they identify two main discourses that are implicated in new partnerships: the one stresses the function of partnerships to enhance local *competitiveness* and *innovation*, while the other places greater emphasis on local popular *participation* in social and economic development activity. Each discourse, according to Edwards et al., influences the type of membership that is recruited into a partnership and, thereby, the ethos that prevails over local interpretations of 'development'. A competition/innovation discourse tends to favour the role of the private business sector whereas participation indicates a shift in emphasis towards community (voluntary) organizations.

Partnership-working can also be problematical where the boundary of the territory for which a partnership has been given responsibility does not conform to, or even transgresses, existing politico-administrative boundaries (Ray, 1998) and, although from a neo-endogenous development logic, this is justifiable and even necessary, it often raises issues of democratic legitimacy (accountability) in the minds of civil servants (from the local to national level) and elected representatives. Rarely is there a direct link of accountability, in the sense of representative democracy, between a partnership and local residents. As a result, local authority members of a partnership can feel compelled to draw back from the partnership area as the primary unit of action and instead to confine their influence to only those territorial parts for which they have an electoral remit. Tensions can also emerge between the partnership and the wider politico-administrative system with which it has to engage (for purposes of legitimacy, monitoring and financial support); in some cases, bureaucrats in national and regional administrations feel threatened or confused by radical new bottom-up initiatives (for example, Lehto and Rannikko, 1999).

Territories in politico-administrative context

THE POLICY PROCESS On the second plane of neo-endogenous development, territorial initiatives are conceptualized as being at one end of the policy transfer chain. They represent a new category of recipient for the redistribution function of the EU and the state (Kovách, 2000). The work being undertaken in the field of political science, for example, into policy transfer across national boundaries (for example, Dolowitz and Marsh, 2000) provides a useful analytical framework.

When the state/EU introduce a neo-endogenous development intervention (and/or a local initiative seeks support from government), what is being transferred through the system? Clearly, public funds flow along the channels of the system to these localities: from the centre (the Commission) and from national, regional and municipal government. Kovách also identifies the devolution of *power* to localities, that is, power to design and implement development activity at and by the local level. The objective of the centre is that the new approach will be more effective, and less costly, than orthodox rural development interventions in bringing about socio-economic vibrancy. Another type of content being transferred is that of ideas relating to the techniques and understanding of neo-endogenous development to officials within the politico-administrative system, local animators and potential local beneficiaries. The system, in other words, has a capacity to reform reactionary (dysfunctional) components by demonstrating the way that the rest of the system is working.

Policy transfer analysts are also concerned with the degree of transfer that occurs over time and with identifying causal factors. This concerns the stability of whatever is moving along the channels. If total stability exists, then ideas and experiences will be *copied* from one local politico-economic context to another in unadulterated form – a straight transfer process. Alternatively, the process might work by *inspiration*: the source of the process causing creative reactions in the receiving nodes/territories. Evidence from anthropology and sociology suggests that, in keeping with the logic of neo-endogenous development, local actors will mediate incoming ideas and interventions (see above, p. 283). The reflexive practitioner is a category of professional who functions as both a catalyst for local action and a mediator between the local territorial unit and the extralocal politico-administrative environment. The more successful operators seem to possess a high degree of what Goleman (1995) calls emotional intelligence. Their catalytic effect occurs through the interpersonal contagion of emotional intelligence, encouraging people to participate in the local development dynamic, offering, for example, ideas of how individuals and groups might initiate local social change. Further research on territorial socio-economic development might explore further the role of the reflexive practitioner.

EUROPEANIZATION Analysis of the neo-endogenous policy process has also begun to isolate a Europeanization effect. This term describes two different, yet interacting phenomenon. First, it refers to a plane along which actions are increasingly being organized. These actions include the public interventions of the European Commission in pursuit of the politico-economic goals of the EU and take the form of programmes and regulations (essentially a top-down trajectory). Yet, as is the case in many politico-administrative systems at the national level, the ethos and modus operandi of the Commission are in a perpetual state of reinvention. The actions also include those of manifold socio-economic and politico-administrative actors throughout the EU (as well as the pre-accession CEECs) who variously construct European-level strategies in pursuit of their own goals (and which almost inevitably involve engagement with the interventions of the Commission): local authorities, private sector businesses and NGOs.

Secondly, Europeanization refers to complex processes of cultural identity construction. On the one hand, interventions, as well as bottom-up actions, are raising the prospect of a pan-EU culture or, if not, then at least a new mosaic of transnational shared meanings and resources. Information exchanges, collaborative working, marketing networks, etc. seem, in various ways, to be constructing a pan-EU awareness. Yet, at the same time, the opposite trajectory can be observed in the revival/creation of localized cultural identities throughout Europe. Political regionalism has been joined by local socio-economic development initiatives based upon the valorization of local cultural resources. The latter takes the form both of a bottom-up social movement and of specific public interventions.

Europeanization is an interesting concept because it describes the broader, ideological agenda driving the adoption by the European Commission of decentralized approaches to rural socio-economic development. Yet, the term also describes the sometimes strategic, sometimes unconscious, embryonic change in policy focus of local initiatives, politico-administrative bureaucrats and non-governmental organizations towards the European Commission and its interventions and away from the nation-state.

EVALUATION Finally, on the second plane of neo-endogenous development is to be found the policy evaluation issue. The period since the early 1980s has seen a growth in public sector managerialism. In this, the centre (the state, or the EU as supra-state) devolves some responsibility for intervention design and implementation downwards while simultaneously formalizing its tools of control.

The emerging ethos brings with it a somewhat more complex conceptualization of the policy-intervention process (for example, Palfrey and Thomas, 1996). Decentralization does not absolve the state/supra-state from its obligations to pursue efficiency (value for money) or to ensure that interventions reach the intended beneficiaries. Yet, at the same time, decentralization recruits new actors (social, economic, political) into the process and redefines the modus operandi of officials. Management is imposed by reducing the expected impact of an intervention to a list of quantified outputs. Subsequently, the performance of agents contracted to implement a given intervention – as well as the understanding of the intervention's significance – are evaluated according to these quantified, immediately observable outputs.

The ethos of the managerial state presents particular challenges in the realm of neo-endogenous socio-economic development (Ray, 2000). In this approach, the pursuit of socio-economic vibrancy is re-organized around local or regional territories. Component socio-cultural and economic actors from the private, public and voluntary sectors are invited to reflect on the strategic choices available for the development of their area and to participate in the design and implementation of development projects. Thus, the neo-endogenous approach tends to switch the emphasis away from tangible outputs achievable in the relatively short term by raising the importance of processes such as awareness-raising, confidence-building, the cultivation of local participative society and the widening of the meaning of the term development to include a socio-cultural and environmental dimension.

Whenever the state or the EU adopts a neo-endogenous approach, however, the process emphasis is inevitably subsumed in, and compromised by, the evaluation imperative of managerialism with its basis of quantified, observable outputs. The tension between these two logics remains to be resolved.

The inter-territorial plane

The refocusing of rural development away from sectors and towards territories represents, simultaneously, a process of atomization and of increasingly complex connectivity between places. For the former, enterprises, local authorities, development agencies and the voluntary sector are variously involved in the creation of territorial, neo-endogenous initiatives that are both opportunistic of circumstances and responsive to the imperative of the emerging governance of Western society. Local and extralocal forces interact in order to create rural development entities each with their own identities and a sense that they can, or must, find some way to manage their own resources in the pursuit of the local collective well-being. However, local development initiatives are also, and inevitably, tied in to all manner of policy and public funding frameworks, to capitalist relationships on regional, national, European and even global scales, and to global social movements such as environmentalism.

Gradually, rural Europe is being restructured into a dynamic mosaic of territorial development initiatives between which new relationships are constantly being created. These interrelationships are, on the one hand, encouraged by the European Commission, not only to support the neo-endogenous approach but also because this serves, indirectly, wider EU politico-economic integration objectives. On the other hand, territorial initiatives and their component socio-cultural and economic actors are variously taking a strategic approach to the cultivation of inter-territorial linkages. In some cases, they are soft linkages – involving seminars and study visits, whilst, in others, they have a harder nature, taking the form of cooperation projects. Time scales vary from the short term and one-off to those intended to develop over the longer term. Some linkages are diadic whilst others involve several participant territories. Geographically, linkages are being cultivated on the intra-regional, intra-national and transnational levels.

From its inception, the EU's LEADER intervention included incentives for participating territories to create such transnational linkages. In LEADER II, between 1997 and early 2000, some 290 transnational cooperation projects (TNCs) involving nearly 1000 interlocal connections emerged, as well as a series of thematic seminars open to all local group representatives. The evolution of this (partial) pan-European connectivity of territorial initiatives was a function of the strategies of at least three types of actor. The European Commission, through the LEADER Observatory organization, exerted a causal influence in designing the opportunity into LEADER and in retaining the power to decide which TNC project proposals were to be financed. The local territories themselves also had a clear influence. The evolution of linkages was not to be dictated by the centre but rather by local territories taking the initiative to create each mini-network and to manage them once established. Finally, administrations at national and regional level played a role in network development, the support (practical and financial) provided to embryonic TNCs varying greatly across the EU.

What is the significance of these linkages in terms of the neo-endogenous approach to rural socio-economic development? To answer this, we need to consider the nature of flows along the linkages (Ray, 2001). One category of flow detected could loosely be called culture. It includes projects promoting *linguistic contact* between participating areas such as the injection of multilingualism into a project by imparting to tourism operators a basic competence in the languages of the other areas participating in the joint project and the production of multilingual marketing leaflets. Other projects entail a flow of *cultural memory* by reinvigorating historical trade links.

Another category concerns the flow of goods and consumers. Many projects anticipated the exchange of *products* between participating areas. Usually under a common marketing logo, each territory agrees to promote and sell the products of its partners alongside its own. Thus each local producer gains access to wider markets without having to bear alone the marketing costs of such a venture, while each 'host' area, on the other hand, benefits from having its local range of produce supplemented by other complementary, yet exotic, products. Other projects involve the exchange of *customers* in the form of tourists. For example, areas create a joint marketing initiative in which a similar type of holiday (for example, outdoor pursuits) is offered in a number of contrasting or complementary locations, the aim being to attract customers to one area and then 'pass them around' other participating areas during a single season or over successive years.

A third category emphasizes flows of awareness-raising. Projects, in addition to their explicit objective, can generate a feed-back response. Both the material results and the glamour of international contact can raise awareness among actors in a participating locality of the benefits of their participation in, or financial and political support for a project and the neo-endogenous approach in general. Such projects were used tactically by coordinators faced with an unsupportive local institutional environment or by sceptical local businesses and voluntary organizations. They also serve to raise the visibility of an area and thereby attract the interest of regional and national NGOs in a territorial development initiative.

The fourth category is that of regulation and methods. Many inter-territorial projects involve cooperation to explore practical solutions to production, marketing or environmental conservation problems common to the participants. The objective is jointly to devise a set of common methods which each participant could use or subsequently adapt to the local context. Although generally of an exploratory and preliminary nature, such projects might represent a template for a future collaborative culture, in which ways of structuring and controlling the rural development system are devised by the territorial components themselves. Projects that create standards of production and service are, potentially, manifestations of group quasi-regulation in which territories are voluntary participants.

In terms of the wider significance of these inter-territorial linkages, a number of observations can be made. First, the linkages help to assist the process of cultural capital accumulation. A local initiative may find the economic value of its cultural resources enhanced through membership of a wider project; for example, a linkage based on a shared ecosystem identity will generate capital to supplement the local territorial resources of participants. Similarly, quality assurance schemes and marques, by creating territorial professional standards, increase the earning potential of each component cultural capital. Each territory assesses the trade-off between trying, on the one hand, to establish unique intellectual property and, on the other, being able to exploit a wider resource.

The capitalist economic model to which these ideas appear to belong has much in common with the 'network organization' discussed in Fukuyama (1995). Whether within a territorial initiative or between economic actors in different territories, the principles are the same. Economic entities form various cooperative trading arrangements, which will bring benefits to each partner while retaining their economic and legal autonomy. The arrangements may be ad hoc or of a longer-term nature. Cooperations can take the form of horizontal or vertical development of a producer network. Cooperation is justified on a utilitarian basis; cooperation will occur if the partners believe that it will serve their separate needs. In the economic domain, the rationale may be seen in a reduction in operating/transaction costs and/or a stronger presence in the marketplace. A local economy may acquire economic power – in the marketplace and in political lobbying – by being a member of a number of overlapping trading networks, each with its own purpose and sets of common resources. This would be to imagine a rural Europe consisting of autonomous territorial entities and of a multitude of overlapping cooperative arrangements at intra-regional, intra-national and transnational levels.

Apart from the economic domain, another result of inter-territorial linkages is an intensification of the interaction that is at the heart of the neo-endogenous theory of rural development, that is, between the local and the extralocal levels. The flows of people, experiences, products and languages noted above result in parades of cultural signs in each participating territory. Sometimes, if the common identity constructed proves to be particularly effective, this may lead towards cultural mixing; each territory (insofar as it is a participant in the project) becomes a mirror image of the others so that the primary development entity is not the component territories but, rather, the cooperative networks. In other cases, however, cooperation may serve not only to create a shared development dynamic but also to intensify the specificity of each local culture through repeated acts of juxtaposition of different territorial examples.

Finally, linkages have the potential to create new territorialized sectoral groupings of political power across the EU, creating sufficient critical mass to lobby collectively for changes in EU policy. Other clusters of lobbying power might emerge on the basis of 'minority' cultures and reformations of socio-economic groups, such as farmers.

However, the history of capitalism is one of shifting geographical inequality and knowing this requires us to speculate on the potential fate of local rural economies as they become drawn into the new politico-economic space of Europe. Could transnational cooperation as a central concept in the new rural development model also contain within its logic the potential to cultivate a new form of politico-economic divergence across Europe?

Parallels can be drawn here with Taylor's (1995, 2000) research on global cities in the modern world system. The political economy of global cities in the era of contemporary globalization provides us with three important notions. First, commercial vibrancy is a function not solely, or even primarily, of local enterprise creation but of the collective ability of a city to create connectivity with other nodes in networks so as to be able to profit from net inflows. Second, this results in a hierarchy of city-nodes. Third, the logic of this international connectivity is that the opportunities for a city's development are not restricted to within its regional or national boundaries. Global cities do not (or no longer) emerge as a function of 'central place theory'; they can be centres for distant global regions (for example, London's relationship to the African and Middle East regions).

Taylor's ideas are useful for this study of rural development. The territorial entities emerging through the neo-endogenous approach may themselves find opportunities in the cultivation of transnational linkages. This is not to undermine the crucial importance of focusing on internal factors and processes in order to create and then sustain a territory as a development entity. Rather, after a certain stage, further development will require strategic participation in international (pan-European) networks. The potential of inter-territorial linkages is that they indicate a path along which the apparently mutually exclusive territorial strategic actions might be accommodated and expanded upon. These scenarios, concerning the nature of territories being launched into a 'rural development market', are: competitive territoriality (in which territorial entities compete with each other to create and maintain niche markets and to secure development finance from public, private and NGO sources); cooperative territoriality (groups of territories forming collaborative actions); and solidarity (creating clusters of collective rural development power to lobby for extralocal protection through, for example, legislation in favour of territorial intellectual property rights). Adopting Taylor's analysis would suggest that, inevitably, a hierarchy would emerge as some territories and collaborations position themselves more successfully than others. As territories cultivate linkages with other areas, some may situate themselves as centres of a number of overlapping networks so that they emerge, through their connectivity, as the rural equivalent of Taylor's global cities, albeit on a European, rural scale.

NOTE

1 Officially announced in 1988.

REFERENCES

Archibugi, F. (2000) *The Associative Economy*. Basingstoke: Macmillan.

Bourdieu, P. (1973) Cultural reproduction and social reproduction. In R. Brown (ed.), *Knowledge, Education and Cultural Change*. London: Tavistock.

Bourdieu, P. (1986) The forms of capital. In J. G. Richardson (ed.), *Handbook of Theory and Research for the Sociology of Education*. New York: Greenwood Press.

Buller, H. (2000) Recreating rural territories: LEADER in France. *Sociologia Ruralis*, 40 (2): 190–199.

CEC (Commission of the European Communities) (1988) *Future of Rural Society* (4/88). Brussels: CEC.

Dolowitz, D. and Marsh, D. (2000) Learning from abroad: the role of policy transfer in contemporary policy making. *Governance*, 13 (1): 5–24.

Edwards, W., Goodwin, M., Pemberton, S. and Woods, M. (1999) Scale, territory and rurality and the government of governance in rural development. Paper presented at the Congress of the European Society for Rural Sociology, Lund, Sweden.

Flora, J. (1998) Social capital and communities of place. *Rural Sociology*, 63 (4): 481–506.

Fukuyama, F. (1995) *Trust: The Social Virtues and Creation of Prosperity*. New York: The Free Press.

Goleman, D. (1995) *Emotional Intelligence*. New York: Bantam Books.

Gramsci, A. (1981) *Modelli etici, diritto e transformazioni sociali*. Rome: Laterza.

Hudson, R. (2001) *Producing Places*. New York: Guilford Press.

Jones, B. and Keating, M. (1995) *The European Union and the Regions*. Oxford: Clarendon.

Kovách, I. (2000) LEADER, a new social order and the Central and East European Countries. *Sociologia Ruralis*, 40 (2): 181–189.

Lash, S. and Urry, J. (1994) *Economies of Signs and Space*. London: Sage.

Lehto, E. and Rannikko, P. (1999) Implementation of the EU LEADER II programme and struggle for local power in Finland. Paper presented at the Congress of the European Society for Rural Sociology, Lund, Sweden.

Lowe, P., Clark, J., Seymour, S. and Ward, N. (1997) *Moralizing the Environment: Countryside Change, Farming and Pollution*. London: UCL Press.

Lowe, P., Murdoch, J. and Ward, N. (1995) Beyond models of endogenous and exogenous development. In J.D. van der Ploeg and G. van Dijk (eds), *Beyond Modernization*. Assen, Netherlands: van Gorcum.

Luhmann, N. (1995) *Social Systems*. Stanford, CA: Stanford University Press.

McRobie, K. (1994) *Humanity, Society and Commitment: on Polanyi*. Montréal: Black Rose Books.

Midmore, P., Ray, C. and Tregear, A. (1994) *The South Pembrokeshire LEADER Project: An Evaluation*. Department of Agricultural Sciences, The University of Wales, Aberystwyth.

Osti, G. (2000) LEADER and partnerships: the case of Italy. *Sociologia Ruralis*, 40 (2): 172–180.

Palfrey, C. and Thomas, P. (1996) Ethical issues in policy evaluation. *Policy and Politics*, 24 (3): 345–362.

Putnam, R. (1993) *Making Democracy Work*. Princeton, NJ: Princeton University Press.

Ray, C. (1998) *New Places and Space for Rural Development in the European Union: An Analysis of the UK LEADER I Programme*. Working paper 34, Centre for Rural Economy, University of Newcastle upon Tyne.

Ray, C. (1999) *The Reflexive Practitioner and the Policy Process*. Working paper 40, Centre for Rural Economy, University of Newcastle upon Tyne.

Ray, C. (2000) Endogenous socio-economic development in the European Union: issues of evaluation. *Journal of Rural Studies*, 16: 447–458.

Ray, C. (2001) *Culture Economies*. Newcastle upon Tyne: CRE Press.

Sayer, A. and Walker, R. (1992) *The New Social Economy*. Oxford: Blackwell.

Taylor, P. (1995) Beyond containers: internationality, interstateness, interterritoriality. *Progress in Human Geography*, 19: 1–15.

Taylor, P. (2000) World cities and territorial states under conditions of contemporary globalisation. *Political Geography*, 19 (1): 5–32.

Tregear, A. (1998) *Artisan Producers in the UK Food System*. Agricultural Economics Society conference, University of Reading.

Westholm, E., Moseley, M. and Stenlås, N. (1999) *Local Partnerships and Rural Development in Europe*. Cheltenham: Countryside and Community Research Unit, Gloucestershire University.

E Power

20

Global capital and the transformation of rural communities

Thomas A. Lyson

INTRODUCTION

Over the past 50 years, neo-classical, free-market capitalism has vanquished all challengers as *the* development paradigm. The collapse of communism in the former Soviet Union and Eastern Europe coupled with China's turn down the capitalist road has left the door open for the unfettered spread of capitalism around the world. Today, traditional communities and local economies are being woven into global circuits of mass production and consumption. As more and more aspects of community life are commodified, local residents are transformed from citizens who historically played an active role in the civic life of their towns and village into consumers whose main goal in life is to keep the global engine of accumulation running. In the West, life is increasingly lived at work and the shopping mall. Home is a place to park the car, watch television and sleep.

It is the large national and multinational corporations that have come to set the development agenda. In a scenario of corporate-led economic development, communities become places where production and consumption are concentrated. More to the point, most of what gets produced and consumed in a global economy, where and when it is produced, and by whom, is decided in corporate boardrooms in London, New York, Tokyo and other financial centres around the world. As Wendell Berry (1996: 409–410) noted: 'The ideal of the modern corporation is to be anywhere (in terms of its own advantage) and nowhere (in terms of local accountability). The message to country people, in other words, is Don't expect favors from your enemies.'

Rural communities today have fewer and fewer degrees of freedom to chart their own economic development paths. The self-sufficient, industrially diverse communities of the past are being forced into a corporately orchestrated global development game. Global capital flows to places that offer the highest return on investment and rural communities are forced to amass arsenals of business incentives in hopes of attracting jobs. Much like a high stakes poker game, one community's business incentives are bid against another community's incentives in an effort to win a new employer, a new shopping mall, or some other corporate-directed enterprise. In this game, of course, there are no real winners. Obviously, communities that invest in incentives of one sort or another and fail to stimulate growth are losers. So too are localities that get carried away in their efforts to attract new businesses and sweeten the pot too much. These communities often find that they have bartered away their ability to improve the lives of their most disadvantaged residents. Rural communities and small towns, because they have less to offer prospective employers, are clearly placed in a structurally disadvantaged position *vis-à-vis* larger urban places. Ironically, many rural communities have become trapped at the bottom of a system that they unwittingly helped to create (Fasenfest, 1993; Lyson, 1989).

The extent to which communities compete against one another to attract footloose firms has been well documented (Cobb, 1982, 1984; Lyson, 1989). When communities compete, corporations win. Here is how the game is played. Recently in the United States an upstate New York community was in line to acquire an aluminium manufacturing plant that was looking to leave a nearby state. The New York Industrial Agency, along with the State Commerce Department and local development officials, helped to find a suitable site for the plant and were in the process of arranging industrial revenue bond financing when the owner of the plant received a call from the Governor of South Carolina. According to the owner, 'He [the Governor] said that he heard that I was trying to locate a plant and he wondered if he could fly up in his personal jet, pick me up, and fly me down to South Carolina to show me what they had to offer' (McKeating, 1975: 378).

What South Carolina had to offer was a building that was 30 years newer than the one in New York at one-fifth the cost. South Carolina also offered a 10-year moratorium on most taxes. This reduced the company's tax bill to about 3 per cent of the taxes it would pay in New York. To top it all off, South Carolina offered to pay the cost of moving the company to the state, and the cost of training all of the workers hired. Not surprisingly, the company moved to South Carolina (McKeating, 1975).

In the sections below I examine how rural communities have been transformed by global capital. I show how the so-called 'free-market' neo-classical paradigm in economics has resulted in unequal economic development. Following this, I introduce a body of theoretical and empirical literature anchored to the concept of civic community that provides an opening to challenge corporate-led global development. While global capitalism is currently the master development paradigm around the world, there is accumulating evidence that the environmental and social dislocations associated with it are not sustainable. An alternative development paradigm organized around local problem-solving and civic engagement is emerging in many localities to challenge the free-market/neo-classical model of global capitalism. Framed as a double movement (Polanyi, 1944), a civically organized, problem-solving model of development recasts the corporate community as a civic community. Corporate capitalism that is concerned with economic efficiency and productivity and with business growth and profits succumbs to community capitalism that is concerned with social and economic equity and household and community welfare. To understand how such transformations are possible, we must first turn to rural community life before global capitalism and the corporate community became the *sine qua non* of life today.

RURAL COMMUNITY AS IDEAL TYPE

Less than 150 years ago most rural communities in North America and Europe articulated closely with, and were defined by, agriculture and other extractive industries. All family members, including husbands, wives and children, contributed their labour to the economic maintenance and survival of the household. While there was a well-established division of labour along gender and age lines in many rural households, there was not a well-articulated and formalized occupational structure within most rural communities. In this social and economic context, the household, the community and the economy were tightly bound up with one another. The local economy was not something that could be isolated from society. Rather the economy was embedded in the social relations of the household and the rural community.

Rural communities served as trade and service centres for local populations. They also served as places that nurtured participation in civic and social affairs in an area and as such could be viewed as nodes that anchored people to place. And, as most commentators have noted, schools, churches and other civic institutions were key factors in solidifying and defining community boundaries (Loomis and Beegle, 1957).

Some manufacturing of durable and non-durable goods took place in rural communities, although it was the burgeoning cities that experienced the onslaught of industrialization. Nevertheless, many rural communities had metal-working enterprises, wood-working shops and related activities. In the United States, for instance, the Census of 1870 shows that in the three most rural northern New England states, Vermont, New Hampshire and Maine, there were 12,162 manufacturing establishments. On average, these places employed less than 10 workers.

Sawmills, blacksmith shops, flour and grist mills, wagon-making enterprises and leather-related industries, such as saddle/harness shops and shoe factories, predominated.

Much of this economic activity was organized around small, skilled, artisan shops. Artisan shops are places that employ a small handful of workers and do not use water or steam power in the production process, but rather rely on hand- or foot-powered machinery. In Europe also, rural artisans played a critical role in the social and economic fabric of local communities. As Farcy (1984: 235) notes, during the later part of the nineteenth century, artisans 'were relatively well-integrated into rural society, not only on account of who their clients were, but also on account of their own work on the land'.

A set of distinct occupational titles that reflected a rigid and formalized division of labour was of little use in most rural areas. In fact, the US Census acknowledged the ambiguity in attempting to classify rural nonfarm workers into existing occupational schemes in the 1870 Census:

> As communities advance in industrial character, functions become separated, and distinct occupations become recognized … [however] [i]n many of the communities of the land it is difficult to draw distinctions much finer than those between the agricultural, the mining, the mechanical, and the commercial pursuit of professions. Indeed, even this is not practicable, since it is a matter of notoriety that … the occupations of carpenter and farmer, or blacksmith and farmer, or farmer and fisherman, are frequently united in one person. In large and more prosperous communities a clear separation between such incongruous occupations takes place; yet still, the carpenter, for instance, in nine out of ten counties in the United States, performs a half a dozen functions which, in cities, are recognized as belonging to distinct trades. (United States Census, 1872: 804)

The idea of 'economic embeddedness' is clearly important for understanding how goods and services were produced and rural communities were organized in the 1800s, and it has considerable value in helping understand the relocalization of some production activities in the developed world today. We know from a small, but growing, body of research in sociology and allied disciplines that there are many different ways rural people 'make a living' and provide for their material needs today. Working for wages in a job and buying goods and services in the marketplace are the ways most individuals typically think about the contemporary economy. Indeed, from the perspective of neo-classical economics, the modern economy is one in which families and workers engage almost entirely in formal market transactions bereft of any social or cultural meaning. Beyond the marketplace of the economists, however, lies an economic terrain rich in substance and meaning. Households and communities provide the context in which economic transactions transpire. The 'market' in neo-classical terms, is but one of many venues for 'economic' activity (cf. Eikeland and Lie, 1999; Fuller, 1990).

THE EMERGENCE OF MODERN ECONOMIC FORMS

With few exceptions, prior to the 1800s, economy and society were woven of the same cloth. Agricultural production and manufacturing were organized along very similar social lines. This was the era of proto-industrialization and small-scale family farming. Labour in both manufacturing and agriculture was relatively undifferentiated and there were few specialized work roles. The broad range of labour skills held by one individual and the relative smallness of the production enterprises has been labelled 'craft-production' by many contemporary observers (Piore and Sabel, 1984).

As a system of economic production, craft-based manufacturing and agricultural enterprises produced a diverse array of goods for local markets. We would call this 'customized production' today. The working landscapes of most North American and European nations then, consisted of identifiable conglomerations of economic activities that met local needs. Regional and national markets, such as they existed in the early part of the nineteenth century, were small relative to the aggregate demand of the local markets.

An economic revolution, which began in England in the 1700s, spread through Europe and North America in the mid-1800s with the advent of mass production techniques in manufacturing and the concentration of manufacturing in large urban centres. The system of craft production that dominated the economic landscape for centuries began to give way to relatively large-scale ensembles of production activities organized in one central location. While it has been assumed that advances in technology were the driving force behind mass production, recent historical scholarship has begun to show that the rise of the factory system was not due to superior forms of technical efficiency but rather to a capitalist philosophy of 'so many hands, so much money' (Robinson and Briggs, 1991). That is, the amount of profit was tied directly and almost exclusively to the amount of labour employed. For example, the best-known way for a bicycle manufacturer to increase its profits would be to add more bicycle-makers to the factory. In strictly economic terms, early mass production increased gross profits, but did not necessarily raise the rate of profit.

Over time, of course, technological improvements in manufacturing processes emerged. Water, steam and later electrical power supplanted human labour in production. Manufacturing output became standardized and routinized. At the same time, efficient transportation networks opened up regional and national markets. For the first time in history, mass markets articulated with mass production. Workers in factories that adopted mass production techniques became increasingly differentiated along task lines as capital in the form of machinery was substituted for labour in many industries. In the United States, the culmination of this transformation from craft production to mass production was most evident in the assembly lines of the Ford Motor Company. In fact, the system of mass production manufacturing that is organized around assembly line forms of social organization have taken on the name 'Fordism' in the contemporary economic organization literature (see Lobao, 1990 for a discussion).[1]

FILTERING DOWN

During the first half of the twentieth century, manufacturing employment increased greatly. Most of the growth in industrial jobs took place in and around cities. It was not until the last part of the twentieth century that manufacturing enterprises began moving out of the city (or bypassing it entirely) in favour of more rural locations (Marsden et al., 1990). Wilbur Thompson (1965) was one of the first economists to identify a 'filtering down' process whereby urban firms relocated to rural areas. According to Thompson, the life cycle of many manufacturing industries consists of three phases. The first phase is characterized by an emphasis on research and development. Science and engineering play a critical role in this phase. Since workers with these skills are typically found in cities, firms in the first phase are found predominantly in urban areas. The second phase of the filtering down process is characterized by product growth and market development. Management skills are key in this phase. As firms grow, their initial expansion takes place in cities.

By the third, or mature phase, of development production technology has become standardized and goods are manufactured in long production runs. In the third phase, labour becomes the key variable cost in the production process and firms now seek out low cost production areas. It is during this phase that plants relocate to rural communities. Not only are there typically abundant supplies of cheap labour in rural areas, but rural communities are often eager to subsidize relocation from urban locations through an abundance of incentives such as tax breaks, training programmes and financing schemes (Cantwell and Iammarino, 2000; Falk and Lyson, 1988).

Of course, the filtering down process does not stop in the rural communities of the more advanced industrial nations, but continues trickling on down to the developing world. The economic landscape of rural areas and small towns in the United States is dotted with the vacant factories of low wage industries that abandoned these places in favour of even lower wage areas of the world. Consider the following income chain. Connecticut, a mostly urban state in the Northeastern United States has the highest per capita income in the US. Incomes in Connecticut are twice as high as they are in Mississippi, a mostly rural state in the American South. Mississippi has the lowest per capita income in the US; however, Mississippi's income is twice that of Puerto Rico, a Commonwealth of the US and historically a favourite off-shore location for US and European manufacturing plants. But, Puerto Rico's wages are twice those found in the nearby Dominican Republic. And income in the Dominican Republic is twice that of its neighbour, Haiti. While no one industry or firm proceeds through the filtering down process in a linear fashion, there is abundant evidence that the economic benefits associated with the product cycle accrue to the urban places in the advanced industrial countries, while rural areas everywhere are more or less disadvantaged.

GLOBAL CAPITALISM AND COMMUNITY WELL-BEING AFTER THE SECOND WORLD WAR

In the decades after the Second World War, it was widely assumed that global capital would benefit rural and urban communities alike. During this era, social scientists (and especially economists) consistently demonstrated in their research and writing that larger-scale, capital-intensive, industrial enterprises not only were 'good' for the economic health of a country as a whole (Galbraith, 1967; Kerr, 1960), but also enhanced the social and economic well-being of workers, families and communities (Averitt, 1968; Hodson, 1983;

Lobao, 1990; Stolzenberg, 1978; Tigges, 1987). A conceptual framework to explain the positive effects of large firms on individuals and communities was provided by labour market segmentation theorists (Beck et al., 1978; Doeringer and Piore, 1971; Goldthorpe, 1984) who showed that the lowest paid, least desirable jobs were most often found in smaller, labour-intensive, peripheral firms (Harrison, 1994: 20–21). In contrast, firms within the core, because they are, by definition, larger, more productive, more capital-intensive and associated with national or multinational corporations, have been able to pay their workers higher wages than firms in the periphery. Falk and Lyson (1988), Lobao (1990) and others revealed that communities in which the economy is dominated by core sector enterprises fared much better on virtually every measure of socio-economic well-being than communities affiliated with the secondary labour market or the periphery.

The current restructuring of the global economy toward increased corporate integration is premised on the assumption that core firms (that is, large national and multinational corporations) will be the primary engines of change and development (Barber, 1995; Harrison, 1994; McMichael, 1996b). According to this perspective, over the long run, rising productivity should translate into higher wages and presumably more prosperous communities (Thurow, 1996), even though over the short run some workers and communities may fare less well than others.

In a system tending toward global accumulation and regulation, the nation-state's role in directing economic development and in protecting the welfare of workers and communities has been weakened (McMichael, 1996a). In the United States and the UK, for example, the deindustrialization of large segments of the manufacturing economy in the 1970s and 1980s showed that the state did little to prevent large multinational corporations, those frequently identified as core sector enterprises, from succumbing to competition from lower cost competitors in other parts of the world (Bluestone and Harrison, 1982). The lessons for local communities were clear. As Tolbert et al. note for the United States:

> History suggests that large corporations rarely, if ever, make good neighbors. From the coal mining communities of Appalachia (Caudill, 1962) ... to the automobile and steel cities of the Midwest (Bluestone and Harrison, 1982), and even to the so-called 'high-tech' enclaves in the Northeast (US Congress, 1995), the story has been the same. The social and economic fate of the community is integrally tied to the competitive position of the corporation in the global economy. Over the long term, the vitality of all globally oriented industries and the communities that are dependent on them will be challenged. (1998: 402–403)

Since at least the 1980s, the task of sheltering workers and communities from the disruptions of the marketplace has increasingly devolved from the nation-state to local communities (Grant, 1995; Herbert-Cheshire, 2000; Mander and Goldsmith, 1996; Mohan, 2000). This devolution has sparked a re-examination of the 'bigger is better' model as the favoured blueprint for economic development. A small, but growing, body of theory and research primarily in Europe has focused attention on small firms, regional trade associations, industrial districts and local entrepreneurs as potentially important, though often neglected, agents of development.

CIVIC COMMUNITY AND BALANCED SOCIO-ECONOMIC DEVELOPMENT

Piore and Sabel (1984) set forth a set of precepts by which advanced industrial societies organized around smaller-scale, flexibly specialized production enterprises can contribute to both economic growth and individual and community welfare. Bagnasco and Sabel (1995), Fukuyama (1995), Perrow (1993), and others (cf. Pyke and Segenberger, 1992) have further illuminated the conceptual foundation for an economy in which smaller-scale, flexibly organized, municipally supported units of economic production can serve as a significant source of goods and services in advanced industrial societies. And, over the past 15 years, a small body of empirical research has demonstrated that economies organized around smaller-scale, locally controlled economic enterprises are associated with a more balanced economic life and high levels of social welfare (Lyson and Tolbert, 1996; Piore and Sabel, 1984).

The theoretical and conceptual underpinnings of a more localized economic system were laid down by social scientists interested in European industrial districts (Perrow, 1993; Sabel, 1992; Zeitlin, 1989). Zeitlin (1989: 370), for example, has noted that smaller-scale, locally oriented production and distribution systems 'require a broad set of infrastructural institutions and services to coordinate relationships among economic actors' and compensate for the inefficiencies of a fragmented system of production. Relatedly, Sabel (1992) noted that the success and survival of locally based economic systems is directly tied to the collective efforts of the community to which they belong. Similarly, Perrow (1993: 298) states that 'Small organizations are linked together by a sense of community of fate, rather than a link based on employees sharing the goals of the owners and top executives of a big organization.'

In the United States, the work of Walter Goldschmidt (1946/1978) and C. Wright Mills and Melville Ulmer (1946/1970) have illustrated the benefits of smaller-scale, locally oriented enterprises. These studies were commissioned by the US Senate at the end of the Second World War to answer the question: Does 'big business' enhance or dampen community welfare and well-being? Both Goldschmidt and Mills and Ulmer showed that communities in which the economic base consisted of many small, locally owned firms manifested higher levels of social, economic and political welfare than communities where the economic base was dominated by a few, large, absentee-owned firms.

Goldschmidt studied rural/agricultural communities in the Central Valley of California. The economy of one rural community, Dinuba, was supported by relatively small, family-operated farms. The other rural community, Arvin, was anchored to large, corporate-run enterprises. According to Goldschmidt (1978: 420), these communities were 'selected for their divergence in scale of farm operations'. However, they were also very similar in 'most fundamental economic and geographic factors, particularly richness of potential resources, agricultural production, relationship to other communities, and the more general techniques and institutional patterns of production'. Using a broad array of data collection and analysis techniques, Goldschmidt concluded that 'the community surrounded by large-scale farm operations offered the poorer social environment according to every test made'.

Mills and Ulmer (1946/1970), on the other hand, looked at manufacturing-based communities in Michigan, New York and New Hampshire and were particularly interested in evaluating the 'effects of big and small business on city life'. In the foreword to their report, Senator James E. Murray, Chairman of the Special Committee that commissioned the study, noted that 'for the first time objective scientific data show that communities in which small businesses predominate have a higher level of civic welfare than comparable communities dominated by big business' (cited in Mills and Ulmer, 1946: v).

> In particular, Mills and Ulmer (1970: 124) found that: 1) The small business communities provided for their residents a considerably more balanced economic life than did big business cities; 2) The general level of civic welfare was appreciably higher in the small business cities; 3) The differences between city life in big- and small-business cities were in the cases studied due largely to differences in industrial organization – that is, specifically to the dominance of big business on the one hand and the prevalence of small business on the other.

To explain differences in civic welfare, Mills and Ulmer turned to two intervening variables. First, they believed that the level of 'civic spirit' or civic engagement in a community was directly related to levels of socio-economic welfare. According to Mills and Ulmer (1970: 125), 'Civic spirit may be said to exist in a city where there is widespread participation in civic affairs on the part of those able to benefit a community by voluntary management of civic enterprises. These enterprises may consist of attempts to improve the parks, obtain better schools, make the streets broader, etc.' Simply put, communities with high levels of civic spirit manifested higher levels of well-being and welfare.

Second, and more importantly, Mills and Ulmer identified the *economically independent middle class* as the driving force behind civic engagement. Not only was the economically independent middle class more prevalent in communities not dominated by big business, but it was this group of economic actors that 'usually took the lead in voluntary management of civic enterprises' (Mills and Ulmer, 1970: 125).

Mills and Ulmer offered several reasons why the economically independent middle class plays a key role in organizing and managing civic life:

> For one thing, he [*sic*] usually has some time and money available with which to interest himself in these matters. He is, on average, fairly well educated. His work in conducting a small business trains him for initiative and responsibility. He is thrown into constant contact with the administrative and political figures of the city ... Furthermore, the small businessman often stands to benefit personally as a result of civic improvement. (1970: 141)

The perspectives set forth by Goldschmidt's (1946/1978) study of rural communities in California and Mills and Ulmer's (1946/1970) research on urban communities in Michigan, New York and New Hampshire fits within a renewed interest in civil society, civic community and civic engagement that is challenging the assumption that a more globally integrated and corporately managed economy is the 'best' and perhaps 'only' development path that will lead to enhanced social and economic welfare for workers and communities (Barber, 1995; Putnam, 1993; Tolbert et al., 1998). The civic community perspective posits that small to medium-size

Table 20.1 ***Types of rural communities***

Corporate community	Civic community
Neo-classical economics	*Problem-solving*
Modernization	Sustainability
Globalization	(Re)localization
Production model	*Development model*
Concerned with economic efficiency and productivity	Concerned with economic and social equity
Emphasis on business growth and profits	Emphasis on household and community welfare
Global mass production and mass consumption	Local craft production and consumption
Articulated model	*Disarticulated model*
Large vertically or horizontally integrated multinational corporations competing in a global market	Smaller, locally controlled enterprises organized into industrial districts, regional trade associations, producer cooperatives
Ideal form is large firm	Ideal form is small firm
Corporate middle class	*Independent middle class*
Positions in corporate hierarchies (e.g., professional, managerial, administrative occupations)	Independent middle class composed of small-business owners, farmers, self-employed professional workers
Political processes	*Political processes*
Not communism	Democracy
Motors for development	*Motors for development*
Human capital	Civic engagement
Free markets	Associations
Individual actions	Social movements

production enterprises can serve as the foundation of modern industrial economies. A civic community is one in which residents are bound to place by a plethora of local institutions and organizations (Irwin et al., 1997). Business enterprises are embedded in institutional and organizational networks (Bagnasco and Sabel, 1995; Piore and Sabel, 1984). And, the community, not the corporation, is the source of personal identity, the topic of social discourse and the foundation for social cohesion (Barber, 1995).

TWO MODELS OF DEVELOPMENT: CORPORATE COMMUNITY AND CIVIC COMMUNITY

While a comprehensive theory of the civic community as it relates to social and economic development is still being constructed, the contours of such a theory can be discerned. Table 20.1 compares the corporate community, a construct anchored to the dominant neo-liberal, neo-classical/market-based approach to development, to the civic community. Each of these constructs should be viewed as an ideal type. As such, they offer alternative development pathways for rural communities as they deal with the challenges of global capitalism.

Neo-classical economics vs. problem-solving

While corporate community and civic community reference bodies of social science theory, when they are applied to issues of social and economic development, they do so under the rubric of neo-classical economics and problem-solving. The more applied versions of these theoretical approaches are often couched as modernization and sustainability. The motor of development in modernization is the market economy. Modernization processes take root and are most successful in those societies that have free markets. According to the neo-classical/modernization scenario, economic globalization is the ultimate and preferred outcome of development.

While neo-classical economics is unquestionably the dominant theoretical paradigm for most economists, many sociologists also (and sometimes unwittingly) anchor their work in this perspective. Some of the variants to the neo-classical model including the 'new structuralism' (Wharton, 1994) and the 'new institutionalism' (Britton and Nee, 1997), have found adherents within the sociological community. Dual labour market theory (Saint-Paul, 1996) and dual economy (Beck et al., 1978) are also derivative from the neo-classical model. All of these approaches assume the primacy of the market as the principle organizing mechanism in modern societies, though each offers a specific critique of, or amendment to, how efficient markets should work.

In contrast to modernization, advocates of sustainability/sustainable development shed the strait-jacket of economic determinism and look for an explanation of development that is driven by social processes other than economics. Both Alexis de Tocqueville (1836) and Karl Polanyi (1944) have provided starting points for inquiries into civic community.

Tocqueville shows that the norms and values of civic community are embedded in distinctive social structures and practices. In particular, Tocqueville points to civic associations as problem-solving cornerstones of the civic community. Writing from this perspective, Robert Putnam (1993: 90) notes 'a dense network of secondary associations both embodies and contributes to effective social collaboration'. Esman and Uphoff (1984: 40) also relate civic community to development when they report 'A vigorous network of membership organizations is essential to any serious effort to overcome mass poverty under conditions that are likely to prevail in most developing countries.'

Polanyi, on the other hand, offers a perspective in which the 'economy' is seen as a problem-solving mechanism to meet the material needs of a society through a process of interaction between humans and their environment. As a leading Polanyi scholar, Fred Block (1990: 39) notes, 'The pursuit of human livelihood was structured by kinship, by religion, and by other cultural practices that had very little to do with the economizing of scarce resources. This means that models of formal economics in which individuals maximize economic utilities through competitive behavior cannot easily be applied to such societies.'

In many ways, civic community is the antithesis of free-market, neo-classical economics. Rather than pursue 'rational' self-interest and assume that everyone else will do the same, 'citizens in a civic community, though not selfless saints, regard the public domain as more than a battleground for pursuing personal interest' (Putnam, 1993: 88).

Production model vs. development model

The corporate community model is at heart a 'production' model. The analytic emphasis is on economic efficiency and productivity. Low cost production is not only the 'guiding' principle, it is the 'only' principle. The civic community perspective can best be described as a 'development' model. Economic efficiency is but one yardstick by which to measure success or failure. Equity issues within the community are given equal weight. Decisions are not made solely on economic grounds, but on social grounds as well.

Articulated model vs. disarticulated model

The civic community approach is oriented toward local social and economic systems while the corporate community approach is directed toward economic globalization. The desired outcome for neo-classical economics is a global (mass) market articulating with standardized, low cost, mass production. Sustainable development, on the other hand, rests on production and consumption maintaining at least some linkages to the local community.

In the corporate model, the ideal form of production is the large firm. Large firms are able to capture 'economies of scale' and hence produce goods more cheaply than smaller, and presumably less efficient firms. From the neo-classical perspective, large producers link with large wholesalers, large wholesalers link with large retailers, and large retailers serve the mass market. Large multinational corporations are the driving engines in the development scenario.

The civic community perspective advocates smaller, well-integrated firms cooperating with each other to meet the needs of consumers in local (and global markets). The ideal economic form is the 'industrial district' (Piore and Sabel, 1984). Firms share information and combine forces to market their products. The state supports this economic venture by ensuring that all firms have access to the same resources (information, labour, infrastructure, etc.).

Corporate middle class vs. independent middle class

From the neo-classical/modernization perspective, a worker's social class position is part and parcel of the corporate hierarchy. As the corporation goes, so go the employment prospects of the individual. Not surprisingly, in an economy dominated by large corporations an individual's engagement with the civic affairs of the local community is tempered by his or her allegiance to the corporation.

The economically independent middle class is rooted in the local community. As Mills and Ulmer (1946/1970) showed, the independent middle class is more likely to participate in civic affairs and concern itself with finding solutions to local social problems. What is 'good' for the socio-economic health and well-being of the local community is integrally tied to the welfare of the small business community.

Political processes

Civic communities flourish in a democratic environment. Community problem-solving around social, economic and environmental issues requires that all citizens have a say in community life. Indeed, citizen participation in voluntary organizations and associations is a cornerstone of a civic community.

The free-market neo-classical system of corporate community development, on the other hand, does not necessarily benefit from democracy. Benjamin Barber (1995: 15) recently noted 'Capitalism requires consumers with access to markets; such conditions may or may not be fostered by democracy.'

Motors for development

From the corporate community/modernization perspective, a person's human capital is translated into better labour force outcomes in a free market. Human capital can take many forms but it is typically manifested in an investment in education, work experience and the like, that a person has acquired. Individual rational actions operating through free markets are at the core of the neo-classical model of economic development. It is assumed that the 'market' is the institution that is best able to efficiently allocate human resources.

The motors for development from the civic community perspective are civic engagement and social movements. Civic communities are best seen as problem-solving places in which residents come together in formal and informal associations to address common social problems. Communities that have rich associational and organizational structures nurture social capital and civic engagement and are best able to meet the social and economic needs of all of their residents. Instead of individual rational actors being the foundation of the community, groups of individuals organized into social movements are core to the civic community approach.

CONCLUSION

If small business, the economically independent middle class and civic engagement offer more positive social and economic outcomes for households and rural and urban communities, why then did most of the West virtually abandon this development path in favour of one in which large corporations assumed the dominant role? More to the point, why did the social sciences during the 1950s and 1960s uncritically accept the underlying tenets of the neo-classical/modernization paradigm?

Part of the answer may have to do with the emergence and growth of big business as the primary engine for military production during the Second World War. Once established, the military industrial complex became a model for large-scale, corporate industrial organization in general. The consequences of an economy increasingly organized around large-scale economic enterprises for workers and communities were unknown.

While at least two empirical studies in the United States (Goldschmidt, 1946/1978; Mills and Ulmer, 1946/1970) and social commentary in Europe (Jouvenel, 1957; Roepke, 1948) affirmed the social and economic benefits of small business on community life and the deleterious effects of big business, little, if anything, was done to stem the trend toward corporate concentration and economic globalization. Indeed, the decades after the Second World War saw the foundation being laid for the multinational capitalism that is omnipresent today. According to Bluestone and Harrison:

> In the years following World War II, a host of public policies promoted and facilitated the centralization and concentration of control over private capital. Especially in the form of tax breaks to business, these political and legal 'incentives' were often publicly justified as potential 'job creation' devices. But whatever the official rationale, the 'de facto' outcome of government policy was to promote and protect concentrated economic power ... throughout the world. (1982: 126)

One consequence of the concentration of economic power was that 'post-war economic growth ... promised to generate the material basis for ... raising workers' standard of living ... This wealth in turn made it possible for government to legitimate the new order … by greatly expanding the "social wage": that amalgam of benefits, worker protections, and legal rights that acts to generally increase the social security of the working class' (Bluestone and Harrison, 1982: 132).

But as Bluestone and Harrison and others (cf. Bowles et al., 1983) correctly noted, as global competition heightened and corporate profits fell in the 1970s, the willingness of capital to honour the social contract with labour and the ability of Western economies to afford large and growing social safety nets would come to an end (Bluestone and Harrison, 1982: 139).

The implications of the civic community perspective for local economic policy and community development programmes are clear. Communities dominated by one or more large national or multinational firms are vulnerable to greater inequality, lower levels of welfare and increased rates of social disruption than localities where the economy is more diversified. In the current era of economic globalization and political devolution, an effective economic development strategy should be geared toward fostering an economically independent middle class everywhere. Policies to promote and strengthen regional trade associations, local industrial districts, producer cooperatives and other forms of locally based entrepreneurship should be part and parcel of a comprehensive community-based economic development strategy.

Social scientists would be well-served to revisit the core assumptions that underlie their understanding of both socio-economic attainment processes and approaches to regional and community development. The hegemony of neoclassical economics among the social sciences has dampened attempts to develop other frameworks for understanding community and economic development processes. The problem-solving perspective outlined above offers a pragmatic alternative to the reductionist, efficiency-based models that are currently propelling global capitalism. The re-emergence of an economy organized around locally coordinated, smaller-scale, technologically sophisticated and globally competitive enterprises is both theoretically and practically possible.

If rural and urban communities are going to regain at least some control over their destinies, they will need a social science blueprint that incorporates the dimensions of civic community. Development is a process that should rest on democratic problem-solving, not corporate profits.

ACKNOWLEDGEMENTS

Support for this research was provided, in part, by funds from the Cornell University Agricultural Experiment Station in conjunction with USDA/CSREES regional research project NE-1012.

NOTE

1 Interestingly, Fordism as an organizing paradigm was not easily transferred to Europe. For a discussion see Tolliday and Rabasco Espariz (1995) and Dassbach (1994).

REFERENCES

Averitt, Richard T. (1968) *The Dual Economy: The Dynamics of American Industry Structure*. New York: W.W. Norton.

Bagnasco, Arnaldo and Sabel, Charles F. (1995) *Small and Medium Size Enterprises*. London: Pinter.

Barber, Benjamin R. (1995) *Jihad vs. McWorld*. New York: Times Books.

Beck, E.M., Horan, Patrick M. and Tolbert, Charles M. (1978) 'Stratification in a dual economy: a sectoral model of earnings determination', *American Sociological Review*, 43: 704–20.

Berry, Wendell (1996) 'Conserving communities', in Jerry Mander and Edward Goldsmith (eds), *The Case Against the Global Economy*. San Francisco: Sierra Club Books. pp. 407–417.

Block, Fred (1990) *Postindustrial Possibilities*. Berkeley, CA: University of California Press.

Bluestone, Barry and Harrison, Bennett (1982) *The Deindustrialization of America*. New York: Basic Books.

Bowles, Samuel, Gordon, David M. and Weiskopf, Thomas E. (1983) *Beyond the Wasteland*. Garden City, NY: Anchor Press.

Britton, Mary C. and Nee, Victor (1997) *New Institutionalism in Sociology*. New York: Russell Sage Foundation.

Cantwell, John and Iammarino, Simona (2000) 'Multinational corporations and the location of technological innovation in the UK regions', *Regional Studies*, 34: 317–332.

Caudill, Harry M. (1962) *Night Comes to the Cumberlands*. Boston, MA: Little, Brown.

Cobb, James C. (1982) *The Selling of the South*. Baton Rouge, LA: Louisiana State University Press.

Cobb, James C. (1984) *Industrialization and Southern Society: 1877–1984*. Lexington, KY: University of Kentucky Press.

Dassbach, Carl H.A. (1994) 'The social organization of production, competitive advantage and foreign investment: American automobile companies in the 1920s and Japanese automobile companies in the 1980s', *Review of International Political Economy*, 1: 489–517.

Doeringer, Peter B. and Piore, Michael J. (1971) *Internal Labour Markets and Manpower Analysis*. Lexington, MA: Heath.

Eikeland, Sveinung and Lie, Ivar (1999) 'Pluriactivity in rural Norway', *Journal of Rural Studies*, 15: 405–415.

Esman, Milton J. and Uphoff, Norman T. (1984) *Local Organizations: Intermediaries in Rural Development*. Ithaca, NY: Cornell University Press.

Falk, William W. and Lyson, Thomas A. (1988) *High Tech, Low Tech, No Tech: Recent Occupational and Industrial Changes in the South*. Albany, NY: SUNY–Albany Press.

Farcy, Jean Claude (1984) 'Rural artisans in the Beauce during the nineteenth century', in G. Crossick (ed.), *Shopkeepers and Master Artisans in Nineteenth Century Europe*. London: Methuen. pp. 219–238.

Fasenfest, David (1993) 'Cui Bono?', in C. Craypo and B. Nissen (eds), *Grand Designs*. Ithaca, NY: ILR Press. pp. 119–137.

Fukuyama, Francis (1995) *Trust*. New York: The Free Press.

Fuller, Anthony M. (1990) 'From part-time farming to pluriactivity', *Journal of Rural Studies*, 6: 361–373.

Galbraith, John Kenneth (1967) *The New Industrial State*. New York: Signet.

Goldschmidt, W. (1946/1978) *Small Business and the Community*. Report of the Smaller War Plants Corporation to the Special Committee to Study Problems of American Small Business. Washington, DC: US Government Printing Office. Reprinted as W. Goldschmidt (1978) *As Your Sow*. Montclair, NJ: Allanheld, Osmun.

Goldthorpe, John H. (1984) 'The end of convergence: corporatist and dualist tendencies in modern western societies', in J.H. Goldthorpe (ed.), *Order and Conflict in Contemporary Capitalism*. Oxford: Clarendon Press. pp. 315–343.

Grant, Don Sherman II (1995) 'The political economy of business failures across the American states, 1970–1985: the impact of Reagan's new federalism', *American Sociological Review*, 60: 851–73.

Harrison, Bennett (1994) *Lean and Mean*. New York: Basic Books.

Herbert-Cheshire, Lynda (2000) 'Contemporary strategies for rural community development in Australia: a governmentality perspective', *Journal of Rural Studies*, 16: 203–215.

Hodson, Randy (1983) *Workers' Earnings and Corporate Economic Structure*. New York: Academic Press.

Irwin, Michael, Tolbert, Charles and Lyson, Thomas (1997) 'How to build strong towns', *American Demographics*, 19 (2): 42–47.

Jouvenel, Bertrand de (1957) *Sovereignty: An Inquiry into the Political Good* (trans. J.F. Huntington). Chicago: University of Chicago Press.

Kerr, Clark (1960) *Industrialism and Industrial Man*. Cambridge, MA: Harvard University Press.

Lobao, Linda (1990) *Locality and Inequality*. Albany, NY: SUNY–Albany Press.

Loomis, Charles P. and Beegle, J. Allan (1957) *Rural Sociology*. Englewood Cliffs, NJ: Prentice Hall.

Lyson, Thomas A. (1989) *Two Sides to the Sunbelt: The Growing Divergence Between the Rural and Urban South*. New York: Praeger.

Lyson, Thomas A. and Tolbert, Charles (1996) 'Small manufacturing and nonmetropolitan socioeconomic well-being', *Environment and Planning A*, 28: 1779–1794.

Mander, Jerry and Goldsmith, Edward (1996) *The Case Against the Global Economy*. San Francisco: Sierra Club Books.

Marsden, Terry, Lowe, Philip and Whatmore, Sarah (1990) 'Introduction: questions of rurality', in T. Marsden, P. Lowe and S. Whatmore (eds), *Rural Restructuring*. London: David Fulton. pp. 1–20.

McKeating, Michael P. (1975) 'New York losing the race for new industry', *The Empire State Report*, 1 (October): 378.

McMichael, Philip (1996a) 'Globalization: myths and realities', *Rural Sociology*, 61: 25–55.

McMichael, Philip (1996b) *Development and Social Change*. Thousand Oaks, CA: Pine Forge Press.

Mills, C. Wright and Ulmer, Melville (1946/1970) *Small Business and Civic Welfare*. Report of the Smaller War Plants Corporation to the Special Committee to Study Problems of American Small Business. Document 135. US Senate, 79th Congress, 2nd session, 13 February. Washington, DC: US Government Printing Office. Reprinted in Michael Aiken and Paul E. Mott (eds) (1970) *The Structure of Community Power*. New York: Random House. pp. 124–154.

Mohan, Giles (2000) 'Dislocating globalization', *Geography*, 85: 121–133.

Perrow, Charles (1993) 'Small firm networks', in R. Swedberg (ed.), *Explorations in Economic Sociology*. New York: Russell Sage Foundation. pp. 377–402.

Piore, Michael J. and Sabel, Charles F. (1984) *The Second Industrial Divide*. New York: Basic Books.

Polanyi, Karl (1944) *The Great Transformation*. New York: Farrar and Rinehart.

Putnam, Robert M. (1993) *Making Democracy Work*. Princeton, NJ: Princeton University Press.

Pyke, Frank and Segenberger, Werner (1992) *Industrial Districts and Local Economic Regeneration*. Geneva: International Labour Office.

Robinson, Robert V. and Briggs, Carl M. (1991) 'The rise of factories in nineteenth-century Indianapolis', *American Journal of Sociology*, 97: 622–656.

Roepke, Wilhelm (1948) *Civitas Humana: A Humane Order of Society*. London: W. Hodge.

Sabel, C.F. (1992) 'Studied trust: building new forms of cooperation in a volatile economy', in R. Swedberg (ed.), *Explorations in Economic Sociology*. New York: Russell Sage Foundation. pp. 104–144.

Saint-Paul, Gilles (1996) *Dual Labour Markets*. Cambridge, MA: MIT Press.

Stolzenberg, Ross M. (1978) 'Bringing the boss back in: employer size, schooling and socioeconomic achievement', *American Sociological Review*, 43: 813–828.

Thompson, Wilbur (1965) *A Preface to Urban Economics*. Baltimore, MD: Johns Hopkins Press.

Thurow, Lester G. (1996) *The Future of Capitalism.* New York: W. Morrow.

Tigges, Leann M. (1987) *Changing Fortunes.* New York: Praeger.

Tocqueville, Alexis de (1836) *Democracy in America.* London: Saunders and Otley.

Tolbert, Charles M., Lyson, Thomas A. and Irwin, Michael (1998) 'Local capitalism, civic engagement, and socioeconomic welfare', *Social Forces*, 77: 401–427.

Tolliday, Steven and Rabasco Espariz, Ma. Esther (1995) 'The transfer of Fordism: the first phase of the spread and adaptation of Ford's methods in Europe, 1911– 1939', *Sociologia del Trabajo*, 25: 133–161.

United States Census (1872) *Ninth Census – Volume III. The Statistics of the Wealth and Industry of the United States.* Washington, DC: US Government Printing Office.

US Congress (1995) *Corporate Restructuring and Downsizing.* Hearings before the Committee on the Budget, House of Representatives, 104th Congress, first session, 23 March. Washington, DC: US Government Printing Office.

Wharton, Amy (1994) 'Assessing the new structuralism', *Current Perspectives in Social Theory*, Supplement, 1: 191–213.

Zeitlin, J. (1989) 'Introduction', *Economy and Society*, 18: 367–373.

21

Regulating rurality? Rural studies and the regulation approach

Mark Goodwin

INTRODUCTION

This chapter examines the use that rural studies has made of the conceptual framework provided by regulation theory. This use has become quite widespread over the past decade, both in research on agricultural change and on wider forms of rural transformation. Even when such a framework is not used explicitly, it is often drawn on as a broad 'political-economic backdrop' to these accounts of rural change. Yet confusion still surrounds the application of regulationist ideas within rural studies. Concepts derived from a regulationist framework are often applied inappropriately, and it is sometimes unclear whether authors who use the term 'rural regulation' are referring to regulationist theory or to a narrower understanding of regulation in a legal or administrative sense (what Clark (1992) terms 'real' regulation). One of the aims of this chapter is to clear up some of the misunderstandings that currently surround the use of regulation theory within rural research. Another is to question what a focus on the rural might mean for the further development and application of the regulation theory. Peck has noted that the regulation approach:

> works typically ... to identify institutionally and geographically distinctive modes of economic development, seeking to hold together an appreciation of the generic features of capitalism (such as the appropriation of nature and human labour, its surplus generating dynamics, its crisis-proneness) with an understanding of its specific (institutional) forms in time and space. (2000: 63)

Work on both the changing nature of rural economies and on the institutional structures of rural governance can contribute to the development of this framework, by helping to specify the way that such generic features of capitalism take particular forms in particular rural spaces.

The term regulation approach is used here deliberately. Regulation theory has been described as 'one of the main theoretical industries' of recent years (Thrift, 1994: 366) but despite this, or perhaps because of it, no single theoretical direction has emerged. Instead we find a wide range of studies that have seized upon the ideas of the regulation school and 'developed them in many different directions. We must speak of an approach rather than a theory. What has gained acceptance is not a body of fully refined concepts but a research programme' (Aglietta, 1998: 41–42). What I also seek to do in this chapter is to look at some of the key aspects of this 'research programme', and at how these have been, and might be, deployed within rural studies. Initially the chapter sets out to clarify some of the key parameters of the regulation framework, before moving on to examine the use made of regulationist concepts within rural studies. Two distinct areas of research can be identified in this respect – those concerned with the rural economy, and those examining wider aspects of rural change. After looking at both of these uses, the chapter concludes by discussing the ways in which future rural research might be informed by continuing developments within the regulation approach.

THE REGULATION APPROACH

The regulation approach has its origins in work undertaken by a group of French economists in the mid-1970s. Put simply, these academics were concerned to analyse the 'regulation' (perhaps better translated as regularization or normalization) of the economy in its broadest sense, beginning from the insight that continued capital accumulation depends on a series of social, cultural and political supports. The approach therefore 'aims to study the changing combinations of economic *and extra-economic* institutions and practices which help to secure, if only temporarily and always in specific economic spaces, a certain stability and predictability in accumulation – despite the fundamental contradictions and conflicts generated by the very dynamic of capital itself' (Jessop, 1997a: 288). As a method of analysis, then, regulation theory starts from the premise that the reproduction of capitalist social relations is not guaranteed by the abstract relations that are the defining features of the capitalist mode of production. Rather, both crises in the accumulation process and phases of expanded production (when these occur) are the products of more concrete institutional structures, political and social processes and cultural discourses.

From this original insight, developed some twenty-five years ago, has come an approach to political economy which now embraces several distinct 'schools' and covers at least three 'generations' (see Jessop, 1990, 1997a, 1997b). As might be expected, the output from this approach is rich and complex – Boyer and Saillard's handbook of the 'Parisian School' alone ran to 54 chapters and almost 600 pages (1995), while Bob Jessop's recent edited overview of the regulation approach as a whole comprises five volumes extending over 2832 pages (2001a, 2001b, 2001c, 2001d, 2001e). The body of this work has considerably extended and deepened some of the original regulationist's economic concerns, to encompass issues of space, government and governance, and discourse and identity (Jessop, 1997b). But before we look at these new developments, I want to spend a short while clarifying some of the chief concepts of the original Parisian School.

The theoretical agenda of this school was originally developed as a critique of neo-classical economics. In contrast to the latter's stress on the continued reproduction of capitalism through the 'laws' of supply, demand and exchange, the early regulationists sought to explore the ways in which economic relations were always socially embedded and socially regularized. As a counterpoint to the abstract and atomistic behaviour of 'rational economic man', the regulation approach stressed the ways in which norms of production and consumption were socially and culturally produced. This in turn led to an emphasis on the variability of capitalist accumulation – on the ways in which 'modes of economic calculation and norms of economic conduct' (Jessop, 1997a) differ, both historically and geographically. Indeed, according to Boyer (1990: 27), the variability of economic and social dynamics, in both time and space is 'the central question' for regulation theory. As we shall see below, this explicit stress on uneven development and variability runs counter to the way regulation theory has been viewed by some authors within rural studies, where critiques have been put forward on the basis that it cannot account for the diversity of rural economies.

The emphasis on the social context of economic forms – such as the wage relation or the price mechanism – cannot but help to foreground the social and political struggles which are continually fought over them. Again in contrast to the rather disembodied (and disembedded) emphasis of neo-classical economics, the regulation approach laid a stress on human agency and social conflict. As Jessop explains:

> Thus, to the mechanistic accounts of *reproduction* offered by neo-classical economics and structural marxism, Aglietta and his colleagues counterposed the dynamic concept of *régulation*. This emphasizes the historically contingent economic and extra-economic mechanisms, which lead economic agents to act in specific circumstances [involving] a wide range of institutional factors and social forces ... whilst far from forgetful about the essentially anarchic role of exchange relations (market forces) in mediating capitalist reproduction, the regulation approach stresses the complementary functions of other mechanisms (institutions, norms, conventions, networks, procedures, and modes of calculation) in structuring, facilitating and guiding (in short, 'regulating') accumulation. (1997b: 506)

With this background in mind we can now seek to clarify four distinctive concepts developed by the early Parisian regulationists (see Jessop, 1997a for further details). An *industrial paradigm* refers to the dominant technical and social division of labour. Examples might be mass production, or flexible accumulation. An *accumulation regime* refers to a complementary pattern of production and consumption which is reproducible over a long period. The type of post-war Keynesian demand management experienced in parts of Western Europe would be one such example. A *mode of regulation* (MOR) is an ensemble of rules, norms, conventions, patterns of conduct, social networks, organizational forms and institutions which can help to stabilize an accumulation regime. In covering economic and extra-economic processes it is usually analysed in terms of five dimensions: the wage relation; the enterprise form; the nature of money; the state; and international regimes. Lastly, when these three complement each other sufficiently over a long enough period to secure a long wave of economic expansion and social stability, the resulting complex is referred to as a *model of development*. The best-known example of such a model is Aglietta's identification of American Fordism. Drawing on these concepts, the Parisian School was able to explore the trajectory of post-war economic growth in Europe and North America, link this to earlier stages of capitalism, analyse the Fordist crisis which began in the 1970s, and speculate about how this crisis might be resolved (Jessop, 1997a: 291).

Subsequent work within the regulation approach has extended these concerns into new theoretical and empirical areas. In particular, a new generation of research began to develop a regulationist view of the state (Esser and Hirsch, 1989; Jessop, 1993); explore how space and scale are implicated in mechanisms of regulation and governance (Goodwin and Painter, 1996; Peck and Tickell, 1992, 1995); analyse how non-class movements and identities impact upon the mode of regulation (Bakshi et al., 1995; Steinmetz, 1994); and explore the discursive constitution of economic and political space (Jenson, 1990, 1995). Critiques of regulation theory in rural studies have tended to ignore this newer work, preferring instead to base their discussion on a somewhat vulgar caricature of very early Parisian work. Indeed, the general point noted by Boyer and Staillard that, 'many advances have passed unnoticed in so far as some appreciations of regulation theory still bear on old and superseded founding works' seems particularly apt in relation to rural research (1995: 15, cited in Jessop, 1997b: 513). Page, for instance, claimed that 'regulation theory is unsuited to the study of agriculture because it takes all industries to be variations on a basic theme drawn from the specific histories and geographies of one or two sectors' (1997: 136), and suffers from 'a reductionist tendency to collapse all industrial development into the binary opposites of Fordism and post-Fordism' (1997: 150; see also Goodman and Watts, 1994). This is a significant misreading of the regulationist approach, which should not be equated to studies of Fordism and post-Fordism.[1] Moreover, neither is it true that those using regulation theory to study rural change have collapsed their analysis into the 'binary opposites of Fordism and post-Fordism'. Indeed, as we shall see in the next section, there are many accounts which stress variability and difference, and over a decade ago Cloke and Goodwin explicitly warned against 'any attempt to fit the complexities of rural restructuring into grand conceptual containers such as Fordism and post-Fordism' (1992: 327). What we actually find are a very complex set of debates drawing on regulation theory to understand the rural economy, and it is to these that we now turn.

REGULATION AND THE RURAL ECONOMY

One major problem with the use of the regulation approach in rural studies has been confusion over the very term 'regulation' itself, and this needs to be addressed at the outset of this section. This is partly a relatively simple issue of recognizing the difference between words and concepts – a warning which is especially important when applied to the term 'regulation' (see Jessop, 1995a: 308). However, this warning has not always been well heeded in rural research, where the concepts (or rather the words) have been applied rather indiscriminately in recent years. When used within the regulation approach, the concept 'regulation' does not mean rule-making, or legal regulation. However, this has been an accepted use in rural research, especially in the areas of agricultural price support and market deregulation. Instead, regulation should be understood in a far broader sense, to refer to the regulation of the economy as a whole.

This confusion partly stems from the polysemy of the English term 'regulation'. In French there is a distinction between *règlementation* and *régulation* – *règlementation* refers to regulation in the sense of rule-making (in this case by the state), while *régulation* refers to regulation in the regulation theorists' sense of contingently emerging regulatory effects operating at the level of the economy as a whole. In English, however, one word is used to convey both meanings. In some of the rural studies literature it is not always possible to tell which of the two meanings is being referred to when the word 'regulation' is used.

Buttel, for instance, identifies a whole tradition of agrarian political economy, which he variously labels as 'agri-food political-sociological neo-regulationist studies', 'agri-food system regulationism' and 'agri-food regulationism' (2001: 172, 174). He claims the central problematic of this work is 'how state practices and rules governing food systems are changing, and how these changes in state practices shape agri-food system changes', which would seem much more concerned with *règlementation* than *régulation*. However, he also notes how such work is concerned with the way that changes in state activity 'are altered as a response to structural trends or crises in the food system', which is more akin to *régulation* (2001: 174). The matter is doubly complicated because the legal regulation (and deregulation) of, say, agricultural markets and commodity prices, or land rights, *may* be interpreted as playing a role within the regulation of the economy as a whole. In other words *règlementation* can contribute to *régulation*.

The emphasis on the word 'may' is intentional – for the administrative and legal ordering of rural processes can be analysed as regulatory practices in their own right, and do not have to be linked in any way to a regulation approach. For instance, Bell and Lowe (2000), Marsden (1995) and Munton (1995) all discuss legal elements of rural regulation, and use some variant of 'regulating rurality' or 'regulated rurality' in the titles of their papers, without conceptually situating these within a regulation approach. On the other hand Banks and Marsden (1997) do use a regulationist framework to situate changes to the legal and administrative regulation of the UK dairy industry, and in doing so warn against reducing the concepts of the regulation approach to a binary identification of Fordism and post-Fordism.

There are many other researchers who have also explicitly drawn on a regulationist conceptual framework to inform their work on the rural economy. This has not been straightforward – as we have already hinted, their inquiries do not necessarily map neatly onto regulationist concerns. This makes it more incumbent on researchers to specify precisely which of the regulationists' concepts they are using, and to set out the manner in which they are using them. In terms of an industrial paradigm, neither mass production nor flexible accumulation are concepts that can be unambiguously applied to the countryside of Western Europe and North America – and still less to other parts of the world. There may be some scope for looking at the introduction of Fordist branch plants into rural areas in the 1960s, or at the increasing mechanization and standardization of agriculture and food production. In terms of the latter, Kim and Curry (1993) claimed that chicken broiler production was to become more Fordist in the 1980s not less so. They drew on regulationist concepts to argue that as a mass production industry, overwhelmingly located in low cost areas (especially low labour costs), it was typical of the type of production which underpinned 'peripheral Fordism' (see Lipietz, 1987). There may be room within this type of work to examine the recent growth of information technology and 'knowledge industries' in the countryside, but it remains the case that the dominant technical and social division of labour in the countryside has differed from that found in urban and suburban areas for much of the post-war period.

The concept of regime of accumulation, deployed to examine the complementary pattern of production, consumption and reproduction, has been drawn on in two main ways within rural research. First, some authors have examined those technological developments in agriculture which increasingly linked post-war agricultural production to an economy based on mass production and mass consumption, especially via a shift to an increased 'everyday' consumption of both animal products and of high value-added manufactured foods (see Friedmann, 1993; Kenney et al., 1989, 1991; Sauer, 1990). According to Kenney et al. such a shift firmly integrated 'farmers into Fordist circuits of capital and commodities', as both producers and consumers (1991: 185). They go on to chart how Fordism as a model of development 'both made possible and called forth a fundamental revolution in the food delivery system, marked by the spread of supermarkets and the fast food chain'. Under such a system, 'the form in which food was delivered to consumers resembled that of other consumer markets' (1989: 135). They make the general conceptual point that a regulation approach provides the foundation for an understanding of agricultural transformation by allowing a simultaneous examination of the ways in which farms and farmers were integrated into wider circuits of production and consumption; of the effects on farming of external economic processes such as technological development and the opening of new markets; of the role of the state in promoting agricultural restructuring; and of the political, social and cultural responses of farmers to such restructuring. Likewise, Sauer used the concepts of the regulation school to analyse the links between Fordism as a model of development in the Federal Republic of Germany, and the economic trajectory of the family farm. He concludes that not only did industrial change affect agriculture, but also that agriculture played a significant role in Fordist industrialization – by supplying a reservoir of labour, by providing foodstuffs and by acting as a specialized market for industrial goods and services (1990: 263). He also uses the work of Lipietz to analyse how the post-war validity and stability of market guarantees for agricultural produce helped to constitute a 'social mould', within which farmers operated.

Secondly, some researchers have applied the concept of regime itself to rural change, in identifying successive 'moments' of agricultural production. Both Kenney et al. and Sauer draw on regulationist ideas to view the crisis of agriculture in the United States and the FRG in the 1980s as one engendered by the broader crisis of Fordism itself, and discuss how agriculture might be affected by a wider transition away from Fordist sets of social relations. Marsden also notes how the 'retreat of Fordism has led to the production of new, more varied markets and consumption practices' in rural areas, 'associated with increased levels of mobility of capital, labour and most importantly consumers' (1992: 214). Conceptually, he argues that it is in this context of a Fordist transition that researchers should examine 'the social and regulatory structures and practices which underlie new developments in the countryside' (p. 214).

Friedmann and McMichael have perhaps adapted the notion of regime of accumulation most formally, to identify successive 'food regimes' (1989; see also McMichael, 1995, 1996). Using Wallerstein's world-systems approach as well as a regulationist framework, they examine how national and international policies have developed over time to produce 'global food regimes' that govern the structure of both food production and consumption. They identify three such regimes, all based on different types of commodity chains and production systems which had to be politically constructed, coordinated and maintained across the borders of the constituent countries of the regime. The initial regime was anchored by the economic and political role of British imperialism, within a global trading system based on the import of raw materials and industrial foods from the colonies. This broke down into instability between the wars, to be replaced by a post-Second World War regime centred on the global disposal of over-produced food from the United States and other OECD countries, helped by foreign food aid. They identify a third regime emerging in the 1980s and 1990s, based around international trade agreements and the globalization of neo-liberal financial 'adjustments'. Le Heron and Roche (1996) also map out three such regimes, but they do not anchor their analysis in the study of global political systems. Instead, they tend to foreground agricultural production itself, to identify an initial regime operating from the 1870s to the 1930s, based on growing urban mass markets; a second regime lasting from the 1930s until the 1970s centred on grain-fed livestock and meat-based, durable industrial foodstuffs; and an emergent third regime based on freshness and naturalness.

There have been critiques of both these broad attempts to apply the concept of regime of accumulation to agriculture (Goodman and Redclift, 1991; Goodman and Watts, 1994). In particular, Page (1996) and Page and Walker (1991) have criticized the work of Kenney et al. for its historical inaccuracies, arguing instead that the complex of agro-industries which dominated the American mid-West in the late nineteenth century provided the 'antecedent innovations' (1991: 308) for the Ford assembly line itself. In their view, far from being influenced by industrial Fordism, they make the claim that the farming communities of the Midwest amounted to 'Fordism before Ford', and state that the integration of farmers into mass markets as producers and consumers did not derive from Fordism but was responsible for propelling nineteenth-century industrialization on a path which resulted in Fordism. Agricultural industries were thus 'not a derivative of Fordism, but contributors to it' (Page, 1996: 379). However, to add historical detail would seem to deepen the regulationist approach rather than overturn it – after all Page and Walker are still working with broad regulationist notions of Fordism as a model of development and mode of regulation. They may disagree with the supposed binary 'collapse' (Page, 1996: 379) of all industries into the Fordist/post-Fordist divide, correctly pointing out that agriculture was never Fordist and that 'forcing farming into either mould only obscures its complexity and diversity' (Page, 1996: 380). But as we have noted above, the regulation approach does not actually do this – it works at a much broader scale than that of individual sectors and the criticism that the regulation approach reduces all post-war history to a binary transition between Fordism and post-Fordism would be better aimed at its 'superficial reception' (Jessop, 1997b) rather than at the approach itself. Indeed, recent generations of regulation theory have sought precisely the same kinds of connections that Page and Walker draw between the economy, society and culture of particular regions. For instance, Peck and Tickell (1995) and Goodwin et al. (1993) both use regulation theory to provide accounts of the development of particular cities and regions, and show in the process that the regulation approach is not incompatible with nuanced historical accounts that stress diversity and complexity.

More recent regulationist work on the rural economy has also sought to highlight the transformation between regimes, by highlighting the notions of crisis and restructuring. The stress here is on periods of change and breakdown, rather than on the make-up and constitution of relatively stable periods of growth. In particular, Drummond et al. have drawn on what they call 'new wave' regulation theory, which they argue allows for 'a more localized understanding of periods of stability, crisis and change in capitalism, making more room for localized and national strategies as part of a response to global level crisis and restructuring' (2000: 113–4). They use the regulationist-inspired work of Peck and Tickell (1992, 1994a, 1994b, 1995) and Moulaert and Swyngedouw (1989) to investigate the possibility of different forms of agricultural crises operating at different scales within different countries. By drawing on regulation theory conceptually they identified a number of critical issues for analysis – the way any period of restructuring involves winners as well as losers, especially when a phase of expanded consumption is unlikely; the disjuncture between the established mode of regulation – within which farmers tend to define strategies and horizons of action – and any emergent regime of accumulation; the manner in which change is resisted; and the differential form of experimentation during drawn-out periods of crises. They then empirically examine these issues through a study of contemporary agricultural crises in Australia, New Zealand and Britain. In many ways this paper represents an exemplary use of regulation theory, and shows the value of employing a regulationist approach to analyse the rural economy.

The concepts are carefully delineated and used to set up a number of research issues, which are then investigated empirically. Regulation theory is not used to make any grand claims, and the concepts are not confused with regulation as a legal or administrative practice (see also Drummond and Marsden, 1999, for further empirical work within a regulationist framework, this time on the sugar industry in Barbados and Australia).

RURAL REGULATION BEYOND THE ECONOMY

Newer regulationist work has also sought to move beyond the economy to examine the political, cultural and social constitution of particular modes of regulation. Cloke and Goodwin (1992) and Goodwin et al. (1995) have claimed that, of all the regulationist concepts, 'mode of regulation' is perhaps the one which is best suited for use within rural research. As we noted earlier, when used within the regulationist approach, the concept of the mode of regulation (MOR) refers to the multiple social, cultural and institutional supports which come together to sustain and promote economic growth. In other words analysis of the MOR is concerned with specifying the social and institutional context within which expanded economic production can occur. For instance, Fordism as an MOR at a national scale in the UK involved a labour relations system based on collective bargaining, the separation of ownership from control in large monopolistic enterprises, 'national' money and consumer credit, mass marketing and a Keynesian welfare state (Jessop, 1992). The state operated demand management policies in the fiscal sphere and underwrote a minimum level of working-class consumption to help complete the 'virtuous circle' between production and consumption. However, these core features of Fordism as an MOR were expressed in concrete form differently and unevenly, both within the UK, and between it and other countries. As Boyer has written, 'the plurality of forms of economic and political *régulation* is not the exception but the rule' (2000: 279; emphasis in the original).

Thus, contrary to claims made by authors such as Page (1996), Page and Walker (1991) and Goodman and Watts (1994), we do not have to be looking for an overarching and undifferentiated movement from one regime to another in order to use regulation theory, and we are not disqualified from using it because the particular aspect of rurality which we are studying does not seem to fit some ideal-typical model of a Fordist mode of regulation. Indeed, in many ways we would expect it not to fit, as the processes and practices that combine to make up any particular mode of regulation will only do so in specific historic and geographic contexts. As an example we can look at the variable experiences of suburbanization in the United States and Europe in the immediate post-war period, which has implications for how we interpret the development of 'pressurized' rural areas. Florida and Feldman (1988) and Florida and Jonas (1991) have analysed the role of housing in US Fordism. They have drawn on the regulation approach to examine how the particular trajectory of US Fordism helped to shape the distinctive pattern of suburbanization (which would have had severe implications for those rural areas surrounding major cities), and how that development in turn underpinned the rise in post-war economic production and accumulation by helping to stimulate and enhance consumer demand through privatized mass consumption. Florida and Feldman go on to point out that the particular form of privatized US suburbanization was 'the product of unique historical conditions' (1988: 188) and as such was only one of a range of potential outcomes, all socially determined, that were possible within the basic parameters set by Fordism. In this case, the specific form of (sub)urbanization which emerged in the United States 'was ultimately shaped by contextual circumstances peculiar to the US political economy, including the unique position the US held in the developing world-economy of capitalism in the post-war period' (Florida and Jonas, 1991: 352). Other contributory factors included the particular social and political resolutions of the 'New Deal' conflicts in the inter-war period, which led to a unique and very restricted 'class accord' between capital and labour, as well as the early growth of assembly-line production and a very fine-grained division of labour.

Elsewhere in the world, post-war urban and regional development was shaped by very different constellations of political, economic and cultural processes. In much of Europe, for instance, consumer demand was stimulated by public housing and mass transport, rather than the privatized modes which emerged in the United States. This more 'social-democratic solution' (Florida and Feldman, 1988: 198) to the contradictions of capitalist development, led to a different form of urbanization, where public and co-operative housing was more prevalent, and where state intervention was more established across a range of services. In the UK, for instance, the social democratic compromise which politically and socially underpinned Fordism included an acceptance of both technocratic land-use planning (at the scale of town *and* country) and high levels of state intervention in the built environment, leading to a range of measures designed to tackle the pre-war problems of urban sprawl and deprivation (Goodwin et al., 1993). Again this type of development held major, but different, implications for rural areas, especially those adjacent to urban centres, or those that were designated with National Park status (see Rees and Lambert, 1985, for an overview of post-war urban and regional development in the UK).

In practice, then, because regulation is a continuous, contested and highly variable set of processes there will always be a 'variety of institutional structures, political practices, regulatory mechanisms, and social norms' (Goodwin et al., 1995) within any one MOR. Using this conceptualization, the rural researcher is able to analyse the differential constitution and effects of particular MORs as they operate across rural areas. Certain rural areas, for instance, have been more affected by the recent crisis in agriculture than others, whereas some have become more heavily used as the new 'industrial' spaces of either high-tech manufacturing or knowledge-driven service economies. Still more have been increasingly commodified, and act as spaces of consumption within tourism, leisure and heritage industries. Varying strategies of regulation will condense around the particular contradictions and conflicts which emerge from these social and economic changes. These will be further differentiated because subnational agencies are often the very medium through which these regulatory practices are interpreted and delivered. Hence, the policies of rural development boards, development commissions, local and regional governments and other sub-national state agencies will affect the particular form of the MOR within specific rural areas, and although the broad parameters of regulation will be set by the national state, we will in practice find a plurality of regulatory strategies operating both within and between different rural areas (on the 'Differentiated Countryside' see Murdoch et al., 2003).

Based on this type of conceptualization, Goodwin et al. (1995) set out to empirically explore the particular types of regulation to be found in rural Wales. This work pointed, for instance, to a continuing series of economic, social and political changes which were having a wide impact on the lives of rural residents in Wales. A quarter of those in rural Wales were living in, or on the margins of, poverty, with two-fifths surviving on extremely low incomes. Agriculture was in decline as a source of employment, and instead the rural economy was based around service sector employment. The study also found a significant number of residents suffering from housing, health and mobility problems. As well as helping to dispel the persistent image of idyllic rural society, studies such as this are able to link rural change to broader sets of social and economic shifts. Yet the reordering of society produced in rural Wales was not the same as that experienced elsewhere – Goodwin et al. found a rather unstable society, based around low incomes, seasonal, temporary and part-time employment and high levels of poverty and deprivation, in contrast to more favoured rural areas within England. This raises several questions about the ways in which different rural spaces are regulated differently within the same overall regime of accumulation. Rural Wales is not governed by the same norms, customs and social practices as those which are prevalent in the gentrified commuter enclaves of parts of rural England, and its economy is set on a totally different trajectory.

In contrast to this focus on economic, social and cultural change, MacKinnon (2000, 2001) has examined some of the institutional components of the MOR in his work on rural Scotland. He uses the work of the 'British-school or third-generation regulation approach, associated with a focus on the role of the state and the geography of regulation' (2001: 824) in order to explore how regional state agencies act as institutional channels to mediate and filter the effect of wider regulatory mechanisms. Empirically he shows how a shift at the national level towards neo-liberal forms of regulation was mediated in the Scottish Highlands by pre-existing practices and institutional structures operating at lower spatial scales. He concludes that local and regional agencies are able to 'mould (national) regulatory tendencies selectively to local conditions' but 'the precise nature of this two-way filtering process in terms of which pre-existing practices survive and which regulatory mechanism are transformed will be shaped by institutionalized norms and conventions which are themselves the product of ongoing interaction between state agencies and their regional spaces' (2001: 830). Crucially, MacKinnon notes that regulationist ideas will need to be linked to other concepts to fully inform research on government, governance and state institutions. Here he draws on the work of Painter and Goodwin, who argue that while a regulationist approach would lead to the conclusion that;

> the local state and local governance cannot be fully understood outside their roles (positive and negative) in the ebb and flow of regulation ... neither can they be fully understood within them. The institutions and practices of local government have their own histories and patterns of development. Explaining their changing character thus requires a theory of governance, a theory of the state and empirical ... research, as well as a theory of their impact on (economic) regulation. (1995: 347)

Thus, MacKinnon draws on the work of both Offe on institutional filters and Foucault on governmentality to supply theoretical agendas on state theory and governance respectively, in order to augment his regulationist understanding of rural institutions operating as part of an MOR.

This point is critical for two reasons. First it reminds us that the aims of the regulation approach are actually quite modest:

> It is not a complete theory of social and economic restructuring, nor ... does it contain a substantive account of the path of development of particular economic spaces. Rather ... it is a method or an analytical approach, which allows an assessment to be made of the effectiveness of regulation in different places and at different times ... It helps to explain why the character of capitalism varies over time and space, while its essential features endure. (Goodwin and Painter, 1996: 640)

In terms of rural research, it can be used to explore the varied nature of regulation in different rural areas, and the differential contribution made by divergent rural spaces both to continued capitalist accumulation and to periods of crisis and restructuring. Thus, regulation theory cannot be expected to provide a complete explanation of the changing character of the rural state, or culture or social relations – these are aspects of its explanatory tools and not the object of its explanation. There is a danger in seeking to over-extend its conceptual reach, which is simply centred around an 'overall analysis of accumulation as a socially embedded, socially regulated process' (Jessop, 1997b: 526).

Secondly, we can take from this the need to continually develop the regulation approach, partly through deepening and applying regulationist concepts, but also through employing conceptual frameworks and expertise located outside mainstream economic analysis. It is to this that we now turn in the conclusion, by discussing some of the potential ways in which the regulation approach might be applied to rural studies as part of a future research agenda. If done properly this will not just be a one-way process. The regulation approach may well shed considerable light on a series of rural problematics, but rural research can in turn help to develop and refine the original concerns of the regulation school, by providing new empirical evidence of the varied and uneven nature of capitalist change.

CONCLUSION: TOWARDS A REGULATIONIST RURAL RESEARCH AGENDA

As we have seen above, there is now a substantial body of work within rural studies which draws on concepts and frameworks provided by the regulationist approach. What is now needed is to extend this work into areas of inquiry that can benefit both rural research and the regulationist agenda. In many ways we can achieve this by building on what already exists rather than searching for completely new avenues of research. From MacKinnon's work on rural Scotland, for instance, we can take the idea that regulation is 'a *process* constituted through a set of material and discursive practices' (2001: 827; emphasis in the original). He uses the ideas of Painter and Goodwin, who have sought to rework regulation theory as a method of analysis by stressing the 'ebb and flow of regulatory processes through time and across space' (1995: 827; see also Goodwin and Painter, 1997). As part of this reworking, Painter and Goodwin criticize the reliance of some regulationist work on the concept of mode of regulation, observing that

> the term 'mode' is often understood as implying a completed system, rather than one in the process of formation ... the notion of modes of regulation overemphasizes the functionality, stability and coherence of regulatory relations and underemphasizes change, conflict and development during their period of operation ... Most of the time ... regulation is neither perfect nor wholly absent. Rather, it is more or less effective, depending on the mix and interaction of the various factors involved. (1995: 340–341)

Viewed in this way, the notion of regulation as a process raises all kinds of research questions concerning the ways in which the materialization of local practices gives concrete expression to broader regulatory tendencies. In terms of rural research, a regulation approach that treats regulation as process is able to deal rather more subtly with the complexity and diversity of rural areas, and with their continued variance from some imagined Fordist or post-Fordist 'norm'.

This leads us on to thinking about the generation of regulation (or, conversely, of processes that undermine regulation) as organized in and through key *sites* and spaces. These should become the focus of future research that attempts to identify the nature of such sites, and of the ways in which they operate to support or impair broader regulatory processes. MacKinnon (2001) identifies both local and regional enterprise companies as two key sites of regulation in the Highlands of Scotland, and he charts how they have been used by central government – with variable success – to transmit a neo-liberal economic agenda to the communities of rural Scotland. But most rural research that uses regulationist concepts remains pitched at a fairly general level, and more work is needed that unpacks the mechanics and practices of regulation at a local level. These sites need not necessarily be limited to state agencies or other institutions of governance – they may well be found in a host of social, cultural and economic settings, both within and beyond the countryside.

If we accept that each set of social practices which go to make up the interactions that can be identified as contributing to (or undermining) regulation has its own key sites, then it follows that these interactions take place across space. Hence, the spatiality of regulation is integral to its effectiveness, or the lack of it, as the processes of regulation are constituted geographically. Work remains to be done on how different scales of regulation mesh together within rural areas – in particular how top-down processes driven by national and international agencies, such as the WTO, EU or GATT, interact with local and regional concerns. This in turn could give rise to more comparative research – both between rural areas within the same country, and between rural areas in different nations – on the different sets and scales of regulatory practices and their economic and social effects. There is considerable room, for instance, to explore the different reactions of various European countries to the restructuring of the EU price support mechanisms, and also to examine the different responses of the US and EU agricultural industries to increasing globalization.

From the work of Drummond et al. (2000) we can take the idea that issues of crisis and restructuring are ripe for investigation within a regulationist framework. However, there remains considerable room for studying the political construction of such crises, and for examining the ways in which notions of rural crises are politically articulated. The research of Drummond et al. is important for beginning to specify different stages of agricultural crisis (which shift from the conjunctural to the structural), but it is less forthcoming on the political moments that actually help to constitute these different stages. Indeed, because crises are never purely economic, but are always politically mediated (see Hay, 1996), we require more work on how economic failure and instability in the countryside become translated into different forms and levels of crisis. In other words, there is a space for research that is able to specify the economic and the extra-economic constitution of different forms of crisis and their attempted resolution (see Jones and Ward, 2002 on the interpretation of urban policy as an ongoing attempt at crisis management). In some ways, such specification would mirror the recent work of Hoggart and Paniagua (2001a), who have used the regulation approach to interrogate the meaning and understanding of notions of 'rural restructuring'. They argue that the idea of restructuring has been inadequately specified within the rural literature, and that the use of the term has been over-extended and exaggerated. Instead they use a 'theoretical vision drawn from regulation theory' (2001a: 43) to set out the broad parameters within which such restructuring can be analysed and understood. To delineate these parameters they draw on 'the regulation theory base that Cloke and Goodwin (1992) outline' (2001a: 46), which assesses rural transformations along the three dimensions of economic change, socio-cultural recomposition and state reformulation. In a separate paper (Hoggart and Paniagua, 2001b), they then apply this framework to analysing each element of these transformations in rural Spain. There would certainly seem a future role for a regulation approach in helping to set out and delineate the broad contours of rural change, and assess them empirically against more abstract concepts such as 'restructuring' or 'crisis'.

Drummond et al. point to another potential area of fruitful research when they note that 'an important aspect of a structural crisis is the way that established norms, practices and ideologies underpinning farming are destabilised or delegitimized' (2000: 119). They point out how ideologies that constitute normal and legitimate farming practices at one moment are gradually called into question as crisis deepens. Indeed they become part of the crisis, as established meanings and ways of life are increasingly challenged. They use the example of the farm debt crisis in New Zealand agriculture to argue how 'central ideological concepts in the farm/finance relationship like "equity", "credit worthiness" and who constituted a "good farmer" were completely undermined'. In the end, much of the ensuing conflict between farmers and their banks was one of 'meaning as well as of finance' (2000: 119). This points to the central role of culture and meaning within the establishment of coherent regulatory practices. Once these meanings are undermined and called into question, then the effectiveness of such practices is also damaged in regulatory terms. There is scope here to build on this work, to see how different components of regulation are discursively constructed and negotiated within the countryside, and explore how different sets of knowledges, experiences and expertise are used to promote or disturb new and existing modes of regulation.

Analysis of the state was recognized as a weakness in first-generation regulation theory (Jessop, 1997b; Jones, 1997), but this has been remedied by newer work which has attempted to rethink the relation between the state and economy (see especially Jessop, 1993, 1995b). Indeed, Hoggart and Paniagua claim that one of the attractions of using a regulationist approach within rural studies 'is the importance it attaches to the state' (2001a: 46). However, as we noted above, regulationist ideas will need to be linked to other concepts to fully inform research on government, governance and state institutions – but there is tremendous scope for undertaking this task. We will have to avoid the dangers of eclecticism, and of attempting to combine fundamentally incompatible theoretical frameworks. In this regard, Jessop has argued that regulation theory is commensurate with neo-Gramscian strategic relational state theory, as each is grounded in an inclusive or integral understanding of economy and state respectively (1997c: 53–54). Goodwin and Pemberton (forthcoming) have adopted this approach in an analysis of the changing nature of rural politics in West Wales, and Cloke and Goodwin (1992) drew on the Gramscian concepts of hegemonic and historic blocs to set out a framework for analysing rural political change. It remains the case, however, that we 'still lack a solid empirical base from which to derive clear judgements about change in state action in rural areas' (Hoggart and Paniagua, 2001a: 53).

Another area of future rural research that might be undertaken within a regulationist perspective is that on consumption. It has been one of the central tenets of the regulation approach, from the Parisian School onwards, that modes of development can best be stabilized when there is a 'virtuous circle' of economic growth based around an integrated cycle of production and consumption. Despite this, studies of the consumption element of such a circle have been far less in evidence than those focusing on production. Marsden (1999) has recently pointed to the growth of the 'consumption countryside'. He notes how rural space now plays a 'key role in the political economy of the modern consumerist state', and how rural places become areas where 'highly positional consumption can be practised and expressed' (1999: 507). In earlier work he highlighted the importance of consumption to any broad shift in rural regulation, when he stated that it is, 'the opportunities rural areas, their resources and people offer for both urban and rural consumers which herald a new stage for rural areas in the Fordist transition' (1992: 215).

But despite this call, there have been precious few studies that have given any empirical analysis of the consumption countryside, and its role in the Fordist transition, from a regulationist perspective. If rural areas are becoming increasingly important as the 'variable repositories of consumption relations' (Marsden, 1999: 507), this seems an area where future work could make a real contribution, not just to our understanding of the countryside, but to the development of the regulation approach as a whole, by filling in some of the missing elements within the 'virtuous circle'.

In conclusion, we have seen how the regulationist approach has inspired a wide range of productive research on rural areas. In doing so, it has opened up new and significant avenues of inquiry for rural studies, and will continue to do so. Far from closing down research by forcing it into a binary strait-jacket, the regulation approach allows us to locate and conceptualize rural change within those wider processes which are attempting to 'regulate' the continuing crises and contradictions of capitalism. By focusing on the socially embedded and socially regulated nature of accumulation, the regulation approach offers an analysis of the connections between rural economy, society and polity – and of the way that these connections help to drive the distinctiveness, and the diversity, of rural change. But critically, by introducing new empirical evidence on the nature of economic, political, social and cultural change in rural areas, researchers can contribute to the development of the regulation approach, as it seeks to 'hold together' (Peck, 2000) an appreciation of the generic features of capitalism with an understanding of how these are manifested differentially over particular spaces at particular times.

NOTE

1 It should be noted that not all studies of Fordism are regulationist, and not every regulationist study is concerned with the supposed transition from Fordism to post-Fordism. Indeed, most regulationist accounts would claim that we are still in a period of crisis after the collapse of the Fordist model of development – the identification of a 'shift' to post-Fordism can largely be traced to debates in organizational management and economic geography on industrial restructuring and flexible accumulation (for critiques see Amin and Robins, 1990; Gertler, 1988; Sayer, 1989). The concerns of these debates – which did often focus on the labour process and industrial organization at individual firm or sector level – were much narrower than those of the regulation approach.

REFERENCES

Aglietta, M. (1998) 'Capitalism at the turn of the century: regulation theory and the challenge of social change', *New Left Review*, 232: 41–90.

Amin, A. and Robins, K. (1990) 'The re-emergence of regional economies? The mythical geography of flexible accumulation', *Environment and Planning D, Society and Space*, 8: 7–34.

Bakshi, P., Goodwin, M., Painter, J. and Southern, A. (1995) 'Gender race and class in the local welfare state', *Environment and Planning A*, 27: 1539–1554.

Banks, J. and Marsden, T. (1997) 'Regulating the UK dairy industry: the changing nature of competitive space', *Sociologia Ruralis*, 37: 382–404.

Bell, M. and Lowe, P. (2000) 'Regulated freedoms: the market and the state, agriculture and the environment', *Journal of Rural Studies*, 16: 285–294.

Boyer, R. (1990) *The Regulation School: A Critical Introduction*. New York: Columbia University Press.

Boyer, R. (2000) 'The political in the era of globalization and finance: focus on some regulation school research', *International Journal of Urban and Regional Research*, 24: 274–322.

Boyer, R. and Saillard, Y. (eds) (1995) *Théorie de la régulation. L'État des savoirs*. Paris: La Découverte.

Buttel, F. (2001) 'Some reflections on late twentieth century agrarian political economy', *Sociologia Ruralis*, 41: 165–181.

Clark, G. (1992) '"Real" regulation: the administrative state', *Environment and Planning A*, 24: 615–627.

Cloke, P. and Goodwin, M. (1992) 'Conceptualizing countryside change: from post-Fordism to rural structured coherence', *Transactions of the Institute of British Geographers*, 17: 321–336.

Drummond, I. and Marsden, T. (1999) *The Condition of Sustainability*. London: Routledge.

Drummond, I., Campbell, H., Lawrence, G. and Symes, D. (2000) 'Contingent or structural crisis in British agriculture?', *Sociologia Ruralis*, 40 (1): 111–127.

Esser, J. and Hirsch, J. (1989) 'The crisis of fordism and the dimensions of a 'postfordist' regional and urban structure', *International Journal of Urban and Regional Research*, 13: 417–437.

Florida, R. and Feldman, M. (1988) 'Housing in US Fordism', *International Journal of Urban and Regional Research*, 12: 187–210.

Florida, R. and Jonas, A. (1991) 'US urban policy: the postwar state and capitalist regulation', *Antipode*, 23: 349–384.

Friedmann, H. (1993) 'The political economy of food: a global crisis', *New Left Review*, 197: 29–57.

Friedmann, H. and McMichael, P. (1989) 'Agriculture and the state system', *Sociologia Ruralis*, 29 (2): 93–117.

Gertler, M. (1988) 'The limits to flexibility: comments on the post-Fordist vision of production and its geography', *Transactions of the Institute of British Geographers*, 13: 419–432.

Goodman, D. and Redclift, M. (eds) (1991) *The International Farm Crisis*. London: Macmillan.

Goodman, D. and Watts, M. (1994) 'Reconfiguring the rural or fording the divide', *Journal of Peasant Studies*, 22 (1): 1–49.

Goodwin, M. and Painter, J. (1996) 'Local governance, the crisis of Fordism and the changing geographies of regulation', *Transactions of the Institute of British Geographers*, 21: 635–648.

Goodwin, M. and Painter, J. (1997) 'Concrete research, urban regimes and regulation theory', in M. Lauria (ed.), *Reconstructing Urban Regime Theory*. Thousand Oaks, CA: Sage. pp. 13–29.

Goodwin, M. and Pemberton, S. (forthcoming) 'Rural politics, state projects and local political strategies: accumulation and hegemony in the countryside', *Journal of Rural Studies.*

Goodwin, M., Cloke, P. and Milbourne, P. (1995) 'Regulation theory and rural research: theorising contemporary rural change', *Environment and Planning A*, 27: 1245–1260.

Goodwin, M., Duncan, S. and Halford, S. (1993) 'Regulation theory, the local state and the transition of urban politics', *Environment and Planning D: Society and Space*, 11: 67–88.

Hay, C. (1996) *Restating Social and Political Change.* Buckingham: Open University Press.

Hoggart, K. and Paniagua, A. (2001a) 'What rural restructuring?', *Journal of Rural Studies*, 17: 41–62.

Hoggart, K. and Paniagua, A. (2001b) 'The restructuring of rural Spain?', *Journal of Rural Studies*, 17: 63–80.

Jenson, J. (1990) 'Representations in crisis: the roots of Canada's permeable Fordism', *Canadian Journal of Political Science*, 24 (3): 653–683.

Jenson, J. (1995) 'Mapping, naming and remembering: globalisation at the end of the twentieth century', *Review of International Political Economy*, 2 (1): 96–116.

Jessop, B. (1990) 'Regulation theories in retrospect and prospect', *Economy and Society*, 19: 153–216.

Jessop, B. (1992) 'Fordism and post-Fordism: a critical reformulation', in M. Storper and A. Scott (eds), *Pathways to Industrialisation and Regional Development.* London: Routledge.

Jessop, B. (1993) 'Towards a Schumpeterian workfare state? Preliminary remarks on post-Fordist political economy', *Studies in Political Economy*, 40: 7–40.

Jessop, B. (1995a) 'The regulation approach, governance and post-Fordism: alternative perspectives on economic and political change?', *Economy and Society*, 24 (3): 307–333.

Jessop, B. (1995b) 'Towards a Schumpeterian workfare regime in Britain? Reflections on regulation, governance and the welfare state', *Environment and Planning A*, 27: 1613–1627.

Jessop, B. (1997a) 'Survey article: the regulation approach', *Journal of Political Philosophy*, 5: 287–326.

Jessop, B. (1997b) 'Twenty years of the (Parisian) regulation approach: the paradox of success and failure at home and abroad', *New Political Economy*, 2: 503–526.

Jessop, B. (1997c) 'A neo-Gramscian approach to the regulation of urban regimes: accumulation strategies, hegemonic projects and governance', in M. Luria (ed.), *Reconstructing Urban Regime Theory.* Thousand Oaks, CA: Sage. pp. 51–74.

Jessop, B. (ed.) (2001a) *Regulation Theory and the Crisis of Capitalism, Vol. 1: The Parisian Regulation School.* Cheltenham: Edward Elgar.

Jessop, B. (ed.) (2001b) *Regulation Theory and the Crisis of Capitalism, Vol. 2: European and American Perspectives on Regulation.* Cheltenham: Edward Elgar.

Jessop, B. (ed.) (2001c) *Regulation Theory and the Crisis of Capitalism, Vol. 3: Regulationist Perspectives on Fordism and Post-Fordism.* Cheltenham: Edward Elgar.

Jessop, B. (ed.) (2001d) *Regulation Theory and the Crisis of Capitalism, Vol. 4: Country Studies.* Cheltenham: Edward Elgar.

Jessop, B. (ed.) (2001e) *Regulation Theory and the Crisis of Capitalism, Vol. 5: Developments and Extensions.* Cheltenham: Edward Elgar.

Jones, M. (1997) 'Spatial selectivity of the state; the regulationist enigma and local struggles over economic governance', *Environment and Planning A*, 29: 831–864.

Jones, M. and Ward, K. (2002) 'Excavating the logic of British urban policy: neo-liberalism as the "crisis of crisis management"' *Antipode*, 34 (3): 473–494.

Kenney, M., Lobao, L., Curry, J. and Goe, R. (1989) 'Midwestern agriculture in US Fordism. From New Deal to economic restructuring', *Sociologia Ruralis*, 29: 131–148.

Kenney, M., Lobao, L., Curry, J. and Goe, R. (1991) 'Agriculture in US Fordism: the integration of the productive consumer', in W. Friedland, L. Busch, F. Buttel and A. Rudy (eds), *Towards a New Political Economy of Agriculture*. Westview Press: Boulder, CO. pp. 173–188.

Kim, C-K. and Curry, J. (1993) 'Fordism, flexible specialisation and agro-industrial restructuring', *Sociologia Ruralis*, 33: 61–80.

Le Heron, R. and Roche, M. (1996) 'Globalisation, sustainability and apple orcharding, Hawkes Bay, New Zealand', *Economic Geography*, 72: 376–397.

Lipietz, A. (1987) *Mirages and Miracles.* London: Verso Press.

MacKinnon, D. (2000) 'Managerialism, governmentality and the state: a neo-Foucauldian approach to local economic governance', *Political Geography*, 19: 293–314.

MacKinnon, D. (2001) 'Regulating regional spaces: state agencies and the production of governance in the Scottish Highlands', *Environment and Planning A*, 33: 823–844.

Marsden, T. (1992) 'Exploring a rural sociology for the Fordist transition; incorporating social relations into economic restructuring', *Sociologia Ruralis*, 32: 209–231.

Marsden, T. (1995) 'Beyond agriculture? Regulating the new rural spaces', *Journal of Rural Studies*, 11 (3): 285–296.

Marsden, T. (1999) 'Rural futures: the consumption countryside and its regulation', *Sociologia Ruralis*, 39: 507–520.

McMichael, P. (ed.) (1995) *The Global Restructuring of Agro-food Systems*. Ithaca, NY: Cornell University Press.

McMichael, P. (1996) 'Globalisation: myths and realities', *Rural Sociology*, 61 (1), 25–55.

Moulaert, F. and Swyngedouw, E. (1989) 'A regulation approach to the geography of flexible production systems', *Environment and Planning D: Society and Space*, 7: 327–345.

Munton, R. (1995) 'Regulating rural change: property rights, economy and environment – a case study from Cumbria, UK', *Journal of Rural Studies*, 11 (3): 269–284.

Murdoch, J., Lowe, P., Marsden, T. and Ward, N. (2003) *The Differentiated Countryside*. London: Routledge.

Page, B. (1996) 'Across the great divide: agriculture and industrial geography', *Economic Geography*, 72: 376–397.

Page, B. (1997) 'Restructuring pork production, remaking rural Iowa', in D. Goodman and M. Watts (eds), *Globalising Food*. London: Routledge.

Page, B. and Walker, R. (1991) 'From settlement to Fordism: the agro-industrial revolution in the American midwest', *Economic Geography*, 67: 281–315.

Painter, J. and Goodwin, M. (1995) 'Local governance and concrete research: investigating the uneven development of regulation', *Economy and Society*, 24: 334–356.

Peck, J. (2000) 'Doing regulation', in G. Clark, M. Feldman and M. Gertler (eds), *The Oxford Handbook of Economic Geography*. Oxford: Oxford University Press.

Peck, J. and Tickell, A. (1992) 'Local modes of social regulation? Regulation theory, Thatcherism and uneven development', *Geoforum*, 23 (3): 347–363.

Peck, J. and Tickell, A. (1994a) 'Searching for a new institutional fix: the after-Fordist crisis and the global–local disorder', in A. Amin (ed.), *Post-Fordism: A Reader*. Oxford: Basil Blackwell.

Peck, J. and Tickell, A. (1994b) 'Jungle law breaks out: neo-liberalism and global–local disorder', *Area*, 26 (4): 317–326.

Peck, J. and Tickell, A. (1995) 'The social regulation of uneven development: "regulatory deficit", England's South East and the collapse of Thatcherism', *Environment and Planning A*, 27: 15–40.

Rees, G. and Lambert, J. (1985) *Cities in Crisis*. London: Edward Arnold.

Sauer, M. (1990) 'Fordist modernisation of German agriculture and the future of family farms', *Sociologia Ruralis*, 30: 260–279.

Sayer, A. (1989) 'Post-Fordism in question', *International Journal of Urban and Regional Research*, 13: 666–695.

Steinmetz, G. (1994) 'Regulation theory, post-marxism and new social movements', *Comparative Studies in Society and History*, 36 (1): 176–212.

Thrift, N. (1994) 'Globalisation, regulation, urbanisation: the case of the Netherlands, *Urban Studies*, 31: 365–380.

22

The state and rural polity

Alessandro Bonanno

INTRODUCTION

The objective of this chapter is to illustrate the current debate on the role of the state in contemporary capitalism. In particular, attention will be paid to those studies that investigate state actions in relation to the evolution of agro-food, the environment and alternative and emerging uses of rural space. While the vast majority of works on the state pertain to the study of political systems and settings, analyses of the role that the state played in the evolution of rural areas have been qualitatively important. This chapter will review them along with general theories of the state. The chapter opens with a socio-historical account of the evolution of the characteristics and actions of the state in mature capitalism. The emergence and crisis of the Fordist state are analysed in the opening section. This is followed by a review of 'Fordist' theories of the state, those state theories that were developed in reference to the post-Second World War interventionist state. The third section of the chapter briefly reviews salient changes that characterized the emerging globalization era and the issues concerning the nation-state and the establishment of transnational state forms. The chapter continues with a review of recent transnational theories of the state. In this section emphasis is given to theories that directly relate to agro-food and the use of rural space. The presentation of these theories is followed by a brief illustration of two recent examples of change in rural areas that are relevant to understand the significance of the theories presented before. The final section illustrates some areas of interest for future research and intellectual endeavours.

THE STATE UNDER FORDISM

Under capitalism, the development of rural space has always been affected by the action of the public sector, that is, the state. Prior to the 1930s, however, the state played a much more limited role to the point that it is rare to find nineteenth-century and early twentieth-century studies of agriculture and rural life which focus primarily on the state. Classical theorists such as Marx, Weber, Chayanov and Kautzsky, to name just a few, analysed the rural focusing on a number of structural and cultural variables while paying limited attention to the public sphere. This situation drastically changed in the twentieth century and particularly in the decades following the Second World War. From that time onward, it has been almost impossible to conceptualize the rural without considering the role played by the state.

Much of this change is connected to the post-war implementation of Fordist socio-economic strategies. They contemplated a much larger, interventionist state that successfully sustained steady growth, balanced mass production and mass consumption, allowed very high levels of production and productivity, and pacified labour by steadily increasing wages, job security, opportunity for advancement and expanding welfare (Harvey, 1990; Lipietz, 1992). 'High Fordism' is the term used to signify the consolidated Fordist capitalism of the post-Second World War era (Antonio and Bonanno, 1996). The United States was the main innovator and leader of the early phase of Fordist transition. It began the post-war era with its productive system intact and as the world's hegemonic military and economic superpower. Geopolitical splits with communist bloc and procommunist or unaligned postcolonial nations prevented global consolidation of capitalism and animated the United States's 'permanent war economy'. US 'military Keynsianism' or 'guided capitalism' lacked the comprehensive redistribution, social programmes and planning of post-war European corporatism or social democracy, but it did increase substantially state expenditures, state regulation of the economy, and social welfare and public goods. While most major capitalist societies rebuilt from the war, US firms dominated its huge home market and much of the world market in the 1950s and 1960s. Regardless of major service sector growth, manufacture led the post-war US economic expansion. Federally subsidized creation of suburbs (single family homes and a highway system) and related growth in automobile production and of the standard middle-class consumer package (for example, autos and home appliances) forged a new mass consumer society. Employing breakthroughs in mainframe computer and other information processing technologies, managers of the primary sector's vertically integrated, corporate firms rationalized, centralized and automated production and administration. Massive expansion of higher education facilitated a new wave of managerial professionalization, expanded the ranks of highly trained technical, financial and legal specialists, and extended greatly basic research for the military and corporate sectors. Innovations and growth of mass media and mass entertainment, especially TV, and massive expansion of the retail sector (for example, shopping centres and chain stores) revolutionized marketing. Greatly enhanced private sector means of regulating mass consumption and Keynesian management of aggregate demand (for example, through state spending, monetary regulation, economic 'fine-tuning') fostered High Fordism's unparalleled coordination of production and consumption, low unemployment, steady accumulation and high rates of profit.

During the post-war '*capital–labour accord*', union membership peaked, but unions generally cooperated with management, trading aspirations for stakeholder rights in capital and shared control of the labour process for higher wages and benefits and stable employment. Increased affluence, social security and educational opportunity, especially for white males, upgraded standards of life and enhanced democratic legitimacy for middle-class and unionized, blue-collar workers. Policy intellectuals and the popular press proclaimed the United States to be a new type of hyper-modernized, middle-class democracy that averted classical capitalist crises and class conflicts and effectively ended debates over basic ideology (Hodgson, 1978: 67–98). Class, racial and gender divides all remained sharp, corporate hierarchies were streamlined, and labour market segmentation hardened. Yet many critics and advocates, alike, held that High Fordism worked so well for the enlarged and politically decisive middle strata that alternatives were hard to imagine. Even New Left charismatic leader Herbert Marcuse asserted that the United States was already a 'society without opposition'. He held that the structure of 'total administration' and technical 'coordination' retains 'stupefying' and 'exhausting' work, but it delivers so effectively and widely the entertainment and consumer goods that 'indoctrinate and manipulate', that it destroys critical sensibilities and desire for liberation and integrates the working class into the system (Marcuse, 1964).

In this context, agriculture and rural space still appeared significantly different from their rapidly growing urban and industrial counterparts. Yet, its development was shaped by the same mechanisms that generated a relatively stable and sustained system of capital accumulation and institutions which supported it. High Fordism enhanced inclusion of marginalized people, raised the social wage substantially, and, in the social democracies, sharply increased labour participation. Equal opportunity was advanced, though the lowest strata benefited little and sharp inequalities between the primary and secondary sectors, production workers and professional employees, and races, ethnic groups and genders, were primary facets of the pattern of rationalization and bureaucratization steadily controlled by the state. State intervention stabilized agricultural production, regulated commodity and labour markets and sponsored research that quickly translated into enhanced agricultural and food production and productivity. The state also regulated rural out-migration and the development of alternative uses of rural space. Substituting for economically declining farming, the period's expansion of rural industrialization, new residential and recreational areas, rural tourism and new environmental projects represented examples of state planned and regulated Fordist transformation of rural space. Under post-war High Fordism, rural development often meant subscription to state managed plans and the transformation of rural actors into 'clients' of an interventionist state (Offe, 1985). State-sponsored socially and economically oriented intervention was paralleled by the growth of highly bureaucratized and controlling institutions designed to administer planned growth.

FORDIST THEORIES OF THE STATE

The post-war augmented intervention of the Fordist state in the economy and society prompted the development of an intense debate on the role that the political apparatus played in society. Dwelling on classical observations on the nature of the state in capitalism, the concept of the state was equated with its historical form of the *nation-state*. In essence, it was recognized that under capitalism political authority evolved from its city and regional forms (that is, proto-capitalism city-states and regional states) to a nation-based institution. Indeed, the creation of broader capitalist markets and nations was considered parallel to the development of nationally unified political institutions: the nation-states of capitalism (that is, Braudel, 1982, 1984). Additionally, this nation-centred vision of the state was maintained to refer to colonialist and/or imperialist periods of mature capitalism. Colonialism first and imperialism later were viewed as expansions of the sphere of influence of powerful nation-states over less developed colonial states (Carnoy, 1984).

In this context, debates on the state saw it as (1) an institution which is the direct instrument of the capitalist class (Domhoff, 1967, 1979; Miliband, 1969, 1970; O'Connor, 1973, 1986); (2) an institution endowed with relative autonomy from capitalist domination (Block, 1977, 1980; Offe and Ronge, 1979; Poulantzas, 1978); (3) an institution possessing almost total autonomy (Levine, 1987, 1988; Skocpol, 1979, 1985); and (4) an institution exhibiting a historical mix of instrumentalist and autonomous roles (Campbell and Lindberg, 1990; Friedland, 1983, 1991; Hooks, 1990; Jenkins and Brents, 1989; Prechel, 1990).

The instrumentalist view of the state

The instrumentalist view argues that the state in capitalism is either 'an instrument for promoting the common interests of the ruling (capitalist) class' (Offe and Ronge, 1979: 346) or 'a committee of the ruling class directly manipulated by the members of this class' (Carnoy, 1984: 214). Instrumentalists take seriously Marx's contention that the modern state is but the 'exclusive political sway' of the ruling class (Marx and Engels, 1955: 11) and have sought to demonstrate how the capitalist class politically intervenes to assure its own interests. The instrumentalists emphasize the direct control that the ruling class exerts in all fundamental aspects of society, including the economy and polity.

There are two types of instrumentalist theory. The camp of Miliband (1969, 1970) and Domhoff (1967, 1979) argues that the state bureaucrats are part of the ruling class and are effectively bound to the ruling class through common socialization and educational backgrounds. This view argues that it is possible that members of other classes may enter the ruling capitalist class, but it is the latter that controls the state apparatus. Another instrumentalist view is advanced by O'Connor (1973, 1986, 2001). 'State monopoly capital theory' asserts that monopolistic-corporate fractions of the capitalist class exercise direct control over the state. This approach maintains that the control that the monopolistic-corporate fractions exercise over the economy translates into control of the state. According to this approach, the role that the state plays in the development of rural space has always been guided by the ruling class interests of maximizing profit, controlling subordinate groups and managing unwanted consequences of capitalism. Rural policy, in this view, is the direct expression of class policy and unless radically opposed will lead to further concentration of power and marginalization of weaker segments of society.

The relative-autonomy view of the state

The relative-autonomy approach asserts the partial independence of the state (superstructure) from the economy (structure) (Block, 1977, 1980; Offe and Ronge, 1979; Poulantzas, 1978). These authors draw from the works of Marx (Marx and Engels, 1964), Gramsci (1971, 1975), Habermas (1975), Horkheimer (1972), Lukács (1971) and Marcuse (1964) to emphasize the roles that ideology and legitimation processes play in the development of capitalism. In this context, the state is viewed as an entity which reproduces class relations not because it is directly controlled by the capitalist class or fractions of that class, but because the state officialdom is interested in reproducing the social relations that support the dominant socio-economic system (Offe and Ronge, 1979).

Proponents of this approach suggest that the independence of the state from the capitalist class derives from its ability to mediate the short-term interests of class fractions and, simultaneously, to ensure the continuation of capitalism as a dominant mode of production. At the same time, however, the state can only favour the interests of capital to a certain degree because of its need to legitimize its actions. Legitimization involves maintaining a consensus of political strategies which are conducive to the maintenance of the status quo (Bonanno, 2000).

Offe (1985) contends that the inability of capitalism to maintain adequate levels of economic growth generates crises of accumulation that increasingly require the intervention of the state. Such intervention tends to penetrate new and expanding spheres of action outside the traditional normative competence of the state (Block, 1980). Accordingly, the problem of implementing legitimation increases with the recurrent crises of capital accumulation. Accumulation and legitimation, then, remain the conditions under which modern capitalism can expand. However, legitimation contradicts accumulation as resources are withdrawn from the social actors in charge of accumulation (capitalist class) to be utilized by the social actors in charge of public administration (the officialdom of the state). Changes in favour of one or the other groups of actors could trigger a crisis. A reduction in the economic solvency of the state often engenders its inability to regulate the socio-economic sphere and to overcome present elements of crisis. An increase in transfer from the economic elites to the state would signify a reduction of the potential for accumulation and a drainage of capital which would be detrimental for both groups in question. It follows that the state is called upon to mediate the contradiction between accumulation and legitimation, a task which the state is only partially able to manage.

According to this view, state intervention in the rural is the outcome of the state officialdom's desire to foster capital accumulation but also to do it in manners that do not jeopardize social stability. Programmes for the support of small less productive farms, economically depressed rural areas, rural development, and rural social services are all viewed as necessary steps to legitimate state action which ultimately supports those very forces that cause socio-economic instability and marginalization. It follows that a relative autonomous state allows democratic forms of participation in decision-making processes, albeit in the context of a class-dominated state and society. The state, in fact, cannot transcend capital accumulation. Therefore, measures that promote alternative uses of the rural space cannot transcend profit generation.

The autonomy or 'state-centred' view of the state

State-centred theorists use a Weberian framework to contend that the state has its own interests and agenda apart from the ruling capitalist class, but that the degree of autonomy of the polity can vary in significance (Hooks, 1990; Levine, 1987, 1988; Skocpol, 1979, 1985; Skocpol and Amenta, 1986; Skocpol and Finegold, 1982). This approach emphasizes the politicized workings of the state in developing and implementing new policies. Skocpol and Finegold (1982) used the case of agricultural policies of the New Deal to assert that the state acts autonomously by implementing policies contrary to the wishes of the capitalist class. For these authors, the development of American agriculture in the decades preceding and immediately following the Second World War was strictly the outcome of a political project rather than the result of the working of economic-centred forces. Similarly, Prechel (1990) uses the steel industry in the Second World War to show that the state chose production targets and forced expansion of production against the wishes of the steel manufacturers, supporting the idea of the state's autonomy. Accordingly, the state autonomy view indicates that state action is determined by the exclusive interplay of political forces. In this respect, the use and development of rural space is affected by those actors who control the polity and does not necessarily follow the interests of economic-based groups.

The mixed approach to the state

The aforementioned theories have generally been employed in exclusive terms. An analysis explicitly rejecting the separation between the instrumentalist and relative autonomy positions is provided by Friedland (1983, 1991). Centring his work on agriculture and food, Friedland assumes that the role of the state in society is not given, but rather depends upon specific historical circumstances or the autonomous posture of the state in society. Employing the cases of various agricultural commodities, he demonstrates that the state is simultaneously called upon to organize various interests of dominant capitalist elites and to mediate between capitalist's interests and opposing interests emerging from other classes (an example of relative autonomy theory). However, he further demonstrates that in specific instances the state also operates as an instrument of the capitalists, as the latter directly and effectively control the action of the former (an example of instrumentalist theory). Empirically, he concludes, neither theory is sufficient to describe the complex patterns of state involvement in society. Paradoxically, each theory becomes correct under different historical circumstances. This set of conclusions begs the question of a closer analysis of the genesis and development of rural policies as they can be affected by a number of actors inside but also outside the polity sphere.

THE POST-(NEO) FORDIST STATE UNDER GLOBALIZATION

By the mid-1970s in the United States and Western Europe, labour unrest, race riots, campus disturbances, increased inflation, signs of slower economic growth, falling corporate profits, the rise of anti-Western movements (for example, Islamic fundamentalism), worldwide challenges to Western modernization models, and new political alignments (for example, OPEC) undermined Fordism. Critics declared that the High Fordist state was in a 'legitimacy crisis' whereby it could no longer coordinate effectively and deliver on its promises, especially to mesh demands for social benefits, public goods and regulation with the need for economic growth (e.g., Habermas, 1975; O'Connor, 1973). In this climate, the corporate right mobilized politically and culturally against the social side of state-centred intervention (Akard, 1992). Marked by the Reagan and Thatcher electoral victories, ascendant 'neo-conservatives' revived free-market economics, and fused it with cultural conservatism that blamed High Fordist strategies as the cause of economic and socio-cultural ruin. In agriculture and rural affairs, High Fordism was identified with overproduction, commodity undervaluation, farm crises, pollution, but also the influx of immigrants which destabilized community life and social well-being. Rural space appeared ungovernable and the state lacked the necessary tools to successfully overcome the crisis.

The end of Fordism opened up a period of transition in which Fordist institutions and practices were attacked and changed significantly. A most influential (post)-Fordist work, David Harvey's *The Condition of Postmodernity* (1990), portrayed a new system of restructured firms and labour markets, weakened labour unions and working-class political parties, and diminished welfare rights and the role of the state. This English geographer's views stressed long-wave shifts of international capitalism, but he emphasized especially the spatial reconfiguration of capitalism on a global basis. Although speaking of 'postmodernity', he held that these structural changes are simply the most recent phase of a technologically based long-term process of 'time–space compression' that makes the world's farthest reaches increasingly accessible to each other. Under capitalism, he argued, time–space compression is accelerated in great bursts following major profit squeezes and market crises, during which capitalists seek technical innovations (for example, steamships and railway trains, radio and electric power, TV and computer) that could accelerate capital's 'turnover time' and speed realization of profit and reinvestment. In his analysis the time–space compression mandates a reorganization of the position occupied by the state. Harvey and a number of other contemporary observers stressed that the nation-centred form of the state has been transformed by the global expansion of social relations. It is, therefore, increasingly difficult to equate the concept of the state with the nation. For the first time, the notion of the state began to be discussed in terms of the transnational form of the state, that is, a political sector whose dimensions and jurisdiction transcended the limits of the nation.

Written at the very moment that the Soviet bloc was disintegrating, Francis Fukuyama's (1989, 1992) 'end of history' thesis proclaimed a new age of unchallenged global neo-liberalism. He believed that, in the wake of collapsed communism and failed post-war liberalism (both of which reflected misguided economic egalitarianism and social engineering), 'liberal democracy' (that is, free-market capitalism) was the *only* option for any nation aspiring to be modern. He speculated that we may now be 'at a point where we cannot imagine a world substantially different from our own, in which there is no apparent or obvious way in which the future will represent a fundamental improvement over our current order ...' (1992: 51). During the mid- and later 1990s, the US stock-market boom generated a 'new optimism' and more celebratory views of the new global capitalism appeared in a genre of books aimed at high-tech and dot.com entrepreneurs, investors, managers, symbolic analysts and other successful professionals. Often draped in postmodern attire, this type of post-Fordism is reminiscent of 1960s' claims about a 'postindustrial society', 'end of ideology' and post-Marxian and post-Weberian capitalism free of the rigid bureaucracies and class contradictions of earlier capitalist regimes and so drastically altered that an entirely new socio-economic world appears on the horizon. In this view, the interventionist state of the previous decades is seen as a 'mistake' that must be carefully avoided. Effective rural policies became equated with the significant reduction of state intervention and the reliance of free trade agreements that promised to allow highly effective rural producers (both in agricultural and non-agricultural sectors) to reap the benefits of their enhanced productivity and quality of production.

Thomas L. Friedman's best-selling *The Lexus and the Olive Tree* (2000) is, perhaps, the most comprehensive (albeit not the most extreme) expression of the optimistic view of globalization. It is the exemplar of what has been termed the 'Washington consensus' – the idea that the US neo-liberal or *laissez-faire* model should guide globalization. Friedman argues that the United States is perfectly situated for global competition and for its role as globalization's 'ultimate benign hegemon and reluctant enforcer'. Referring to 'Americanization–globalization', Friedman unabashedly resurrects the post-war idea of the United States as the world's *lead society* (2000: 367–383). He admits that globalization can be disruptive and elicit an occasional dangerous 'backlash', and is less cautious than Fukuyama in his embrace of neo-liberalism. However, he contends that an 'integrated', 'cosmopolitan' 'globalization system', composed of global free markets, new technologies and organizations, and rational investors, has replaced the 'divided', 'frozen', 'Cold War System'. The new order, he claims, represents a global revolution – a *democratization* of technology, finance and information that has created a 'fast world', or non-hierarchical, open, state intervention-free, dynamic 'web' with the 'Internet as its backbone' (2000: 8–9, 44–72, 200). Exalting the virtues of neo-liberal globalization and believing that modernization is driven by global free markets and investor choices, Friedman claims that the 'Electronic Herd' of e-trading individual investors and multinational companies grow the economy so effectively that all nations aspiring to be modern must converge toward the US model. Refusal to don the 'Golden Straightjacket' (that is, globalization) condemns a nation to marginality (2000: 104–111).

Opposing the above-mentioned positions, globalization critics argue that the neo-liberal celebration of the market and rollback of the political sphere limits genuine citizenship to the middle class and above and reduces democracy to a 'stockholders republic'. Neo-liberalism liberates markets and their participants from regulation by representatives of the overall citizenry and by the institutional and socio-cultural worlds in which they are embedded. Although recognizing that nurturance of responsible citizenship and effective political mediation is no easy matter, critics contend that genuine democratization requires much wider participation than consumers and investors and a much broader purview than the individual self-interest that rules in market choices. Overall, critics contend that market-driven globalization has two basic problems: the erosion of democracy within nation-states and the absence of an international civil society and political sphere to mediate transnational finance, commerce and organizations (Barber, 1996: 20).

TRANSNATIONAL THEORIES OF THE STATE: AGRICULTURE AND FOOD, THE ENVIRONMENT AND NEW USES OF RURAL SPACE

The end of Fordism opened up a new phase in the debate on the role of the state which was now centred on scholarship on the emergence of post- and/or neo-Fordist social arrangements under globalization. Studies pertaining to agriculture and food, the environment and new uses of rural space in particular have been very sensitive to the changes that the end of Fordism entailed for the rural. Perhaps more than others, this group of scholars produced copious and novel contributions to the definition of the role of the state in the transnational era (e.g., Bonanno et al., 1994; Friedland, 1983, 1991; Friedmann and McMichael, 1989; Heffernan, 1990; Heffernan and Constance, 1994; Marsden, 1994; McMichael, 1996).

The transnational state and class control

Four general positions exemplify the various understandings of the role of the state in the global post-Fordist era. The first position interprets the transnationalization of the state in *terms of class control*. This position is exemplified by the work of William H. Friedland (1991, 1994, 1995). Friedland maintains that the food policy is a critical national concern because 'unless there is a constant flow of food in abundance and relatively cheaply to the urbanized and industrialized populations of the advanced capitalist countries, domestic unrest will increase' (Friedland, 1991: 51). Therefore, agriculture and food issues are fertile ground for research on the transformation of the nation-state and the emergence of different forms of transnational or supranational states.

While there was a global economy before the 1960s, economies were strongly tied to their nation-states. The defeat of the United States in the Vietnam War marked the end of the dominance of the nation-state as a political-economic form. About the same time, Friedland continues, the world system changed from one based on political-economic interests dominated by national bourgeoisies to a transnational system dominated by transnational corporations (TNCs). TNCs that were once attached to their host countries (as multinational corporations or MNCs) have become 'less concerned with specific national interests, national markets, and/or internal organization which is nationally-based and more concerned with a global orientation' (Friedland, 1991: 52–53).

For Friedland, since the TNCs' global orientation can only be partially controlled by any one nation-state, 'it follows that some new form of state form must emerge to function as the "executive committee" of a new transnational bourgeoisie' (1991: 53). Nation-states can only control TNCs within their own territorial jurisdictions. Therefore, TNCs are currently unregulable by any existing state form. Though the TNCs are unregulable, they would prefer to minimize uncertainty regarding regulation issues. For Friedland, it is this need to reduce uncertainty that necessitates the creation of the transnational state. Historically, he continues, the state has emerged to minimize uncertainties in the accumulation of capital and to create a climate of business confidence. This process must be continued in the new transnational scenario if accumulation is to continue without unbearable contradictions. More specifically, given the continuation of conflict among various fractions of capitals (particularly between domestic capital and transnational capital) and between the interests of capital in general and subordinate social groups (environmental movements, consumer movements, ethnic and minority groups, organized labour, etc.), the organizational, mediative and legitimizing roles must be performed in the new global scenario. Following Hechter and Brustein's (1980) historical analysis of the emergence of the capitalist state, Friedland argues that the emergence of the transnational state must be linked to the issue of control of social opposition. In fact, it was through the action of controlling opposition that the nation-state emerged in the early phase of the expansion of capitalism. The primitive nation-state was created at the behest of feudal lords who were struggling with the burghers and wanted protection for their property, rights and prestige. 'This state, in turn, would be captured by the bourgeoisie in the revolutionary struggles that marked the transition to modern capitalist states in Great Britain and France' (Friedland, 1991: 55). The question of the state, then, is recast in terms of what opposition is developing at the global level which, in turn, will shape the terms of the emergence of a transnational state. Friedland predicts that it will be the opposition of national capitals to international capitals that will characterize the social control function of the transnational state.

Friedland's theory is useful to understand the effects that the emergence of anti-global groups and movements has on the shaping of transnational forms of the state. Particularly important are the implications that the demands of anti-global groups create for the governing of, and policies concerning, agro-food, the use of the environment and rural space. Anti-global forces challenge the productivistic and global nature of agricultural production and rural space uses. Particularly under attack are the environmental, safety and health consequences of globally generated and genetically engineered food products as well as calls for further deregulation of related economic activities. Following Friedland, it can be argued that the nature of future policies concerning the rural will be determined in the context of a struggle that sees a coalition of spatially and politically heterogeneous forces battling transnational capitalist groups and their political allies.

The transnationalization of the nation-state

The second position maintains that *the nation-state has already been transnationalized.* This posture is exemplified by the work of Philip McMichael and his associates (McMichael and Myhre, 1991; McMichael, 1996, 2000). McMichael argues that the transnationalization of the economic sphere has transformed the nation-state into a transnational state that now performs global regulation. This process is based on the integration of the nation-state into capital circuits which are increasingly transnational. First, the nation-state is faced with diminishing control of the activities of transnational financial capital (TFC) structures. For McMichael, TFC is the 'anchor' of a new globally constructed regime of accumulation. TFC is replacing national wage relations protected by welfare states with a global wage system organized by global capital. This decreased control affects the existence of the nation-state by generating a shift of power within the nation-state in favour of finance ministries and to the detriment of programme-oriented ministries. In other words, there is an effective uncoupling of nation-state from banking capital and control of international monetary relations. Second, the establishment of a global agro-food system has diminished the state capacity to control the composition of local food and agricultural production. The latter is increasingly extroverted and oriented toward the production of inputs for livestock feeds and processed foods for affluent markets. The net result of this situation is that the state is no longer a 'political mediator' between global capital and national bourgeoisie and the working class. Rather, it has assumed the role of facilitator of the requirements of global capital and the imposition of global wage relations. McMichael contends, however, that the disorganization of the nation-state leaves some niches open for local resistance. For example, it may be possible that small farmers, local capitalists and labourers might mobilize to create local and sub-national agro-food systems that fill the spaces left open by a food system organizing globally.

According to this theory, rural policy is affected by the disengagement of the 'social state' and the development of global circuits of production and consumption. Because of the growth of these circuits, rural areas are increasingly dependent on the actions of global capital and placed in competition with other areas scattered around the world. This limited autonomy further translates into additional incentives to compete globally and to allow global forces to take advantage of the availability of local natural and human resources. In the context of a dominant free market ideology, this situation is depicted as one of the primary conditions for socio-economic growth. However, it is often the cause of increased social and economic marginalization and decay. A promising alternative consists of uses of rural space that rest outside global circuits and, therefore, provide shelter from global exploitation.

The dialectic of the nation-state–transnational state relationship

The third position consists of authors who underscore the dialectic nature of the relationship between national and transnational state forms and the importance that the nation-state still retains in terms of processes of capital accumulation and social legitimation. Terry Marsden (Marsden, 1992, 1997a, 1997b; Marsden et al., 1996) argues that regulation has been a constant feature of capitalism and that contemporary global capitalism is regulated through new, different mechanisms than in the past. This change has been triggered by political and economic projects of deregulation which diminished the nation-state's centrality and powers. Deregulation, however, implies, he continues, forms of re-regulation as capitalism requires basic forms of management and control of social relations. Accordingly, the deregulation of the nation-state-based system entails processes of re-regulation which are centred on transnational state forms. The two forms of the state are dialectically related and are the outcomes of processes of creation and re-creation of social relations. Following these arguments, he maintains that the transition away from old nation-state-based forms of regulation is a process that concomitantly takes place at the macro and micro levels. New transnational social relations are mediated through local actors and the ensuing social relations which are expressions of local interpretations of reality. Simultaneously, it is erroneous to assume that external (global) social relations have no impact at the local level. The two dimensions must be read as two intrinsically interrelated sides of the same process.

Koc (1991, 1994) contends that the nation-state system originally brought together different regions and localities into one nation governed by one code of law. More recently, the nation-state has played the role of 'universal regulator' through the introduction of standardization (regulatory norms in production, distribution and consumption of goods) and the generation of global mechanisms of regulation and coordination of trade such as the United Nations, the GATT, the World Bank and the IMF. The global accumulation crisis and the rise of TNCs to power engendered problems for this system as the reorganization of global production promoted by the TNCs created limits to the sovereignty of nation-states. However, although the globalization of the economy has limited the ability of the nation-state to perform its historical roles, Koc maintains that the nation-state still plays fundamental roles which are unresolved transnationally. First, at the legitimative level the nation-state has historically been an agent which has homogenized and controlled ethnically, religiously and politically diverse groups. These groups have been brought together through processes of legitimation which have culminated in the establishment of the 'nation' and the ideological and normative dimensions associated with it. Second, and at the level of accumulation of capital, the nation-state has been able to organize and maintain conditions amenable to capital accumulation within its territory by controlling labour and protecting capital. Under globalization, the ability of the nation-state to perform these actions has been somewhat eroded and it has not been replaced by any of the emerging transnational state forms. In this respect, capital accumulation still largely depends on the action of nation-states.

The re-regulation of social relations is the key element in understanding present conditions and future developments of rural areas and policies. While re-regulation has been sponsored by transnational actors, the ways in which it is actually carried out directly involve local social groups. Moreover, processes of social control demand the existence of forms of local power (the state) that are culturally mediated and embedded in local ways and traditions. The weakening of these processes of social control could trigger locally based struggles that often find their strength in indigenous culture and themes. It follows that the contradictory nature of emerging state forms is embodied in the inability of the transnational state to fully control the local and in the concomitant inability of the nation-state to exercise adequate control over global processes.

Contradictory convergence

The fourth position focuses on the concept of contradictory convergence and the resulting relationship between TNCs, the state and subordinate classes under globalization (Bonanno, 1993, 2000; Bonanno and Constance, 1996). Bonanno argues that the globalization of production makes the traditional theories of the nation-state inadequate for understanding the role of the state in a global economy. He contends that both TNCs and subordinate classes have a crucial yet divergent interest in the development of a transnational state. TNCs need a political mediator and/or facilitator at the transnational level to provide a business climate for capital accumulation and mediate legitimation demands from subordinate classes and between capitalist class fractions. Subordinate classes need a transnational state to address their demands concerning health and safety for workers, consumer protection, environmental regulation, food security and other pertinent issues. In the case of subordinate groups, the transnationalization of the economy has weakened their position as well as the ability of the nation-state to satisfy their requests. It follows that subordinate classes also need a regulatory/mediating political entity that is able to address their demands. According to Bonanno, global economic restructuring – accomplished primarily through shifting of production across national borders, reliance on low wage labour and concentration of capital – has severely limited the nation-state's actions aimed at protecting the social and economic gains obtained by subordinate classes in previous decades. He contends that the ongoing attempts to create supranational state forms such as the EU, NAFTA and MERCOSUR will only partially resolve the inability of the nation-states to regulate and coordinate capital accumulation since these are not global entities but rather regional or 'common interest' organizations. He concludes that the present 'transnational-state vacuum' represents both a danger and an opportunity for progressive actors.

According to this theory, the emergence of new forms of the state is the contested terrain where the future of rural regions will be decided. Current conditions point to the creation of much less interventionist forms of the state that weaken popular participation and enhance the power of economic and financial elites. Simultaneously, the issue of social legitimation is unresolved as opposition to transnational forms of regulation is mounting. A host of nation-states have embraced pro-free-market ideologies and practices that favoured the concentration of wealth and power and created social instability. This situation engendered opposition that manifests itself primarily through the emergence of anti-global coalitions. The broad struggle between these two groups will shape rural regions and policies.

INSTANCES OF RURAL CHANGE AND THEORIES OF THE STATE

A brief illustration of two key examples of rural change in the global era can further elucidate the meanings of the above-mentioned theories of the state. The first example refers to the establishments of CAFOs (confined animal feeding operations) in the state of Texas, in the United States (see Bonanno and Constance, 2001). In the context of a nation (the US) that has adopted an overt neo-liberal posture, the state of Texas is one of the most explicit cases of regional governments who embraced an even stronger form of market liberalization and decreased state intervention. By eliminating and/or reducing regulations in key areas such as environmental rules and labour use, promoting economic and tax incentives for firms, and creating an overall attractive business climate, the state of Texas has sought to attract corporate investments. Rural areas have been particularly affected by these state policies as they have become the destination of a variety of corporate investments. In particular, corporate hog and poultry production activities have relocated to Texan rural areas taking advantage of weak zoning regulations, cheap natural resources and friendly local political and social environments. Indeed, the proliferations of CAFOs was originally viewed by residents and the state officialdom alike as a vector for the promotion of local socio-economic growth.

The establishment of CAFOs generated two different outcomes. The first was increased profitability for food producing corporations, while the second was overt environmental degradation and decreasing community well-being. Community-based actions against the establishment of CAFOs emerged and local groups requested state intervention. Local community groups' objectives were to halt the proliferation of CAFOs, and to introduce stronger rules governing their implantation. This hard-fought battle, however, did not produce the positive results for local communities. While originally members of the state officialdom seemed sympathetic to local groups' requests, in the end they not only did not stop CAFOs proliferation but avoided addressing the issue of their regulation.

The second case refers to a price fixing scandal involving Archer Daniels Midland (ADM), one of the world largest food transnational corporations (see Bonanno et al., 2000). This case refers to the production of lysine, an amino acid that if added to agricultural waste can transform it into feed. ADM officials understood lysine's tremendous commercial potentials and sought its genetically engineered production. The original plan was to control the market by out-competing other companies through dumping, an illegal operation. While this strategy achieved the original goal and ADM quickly became the largest world producer of lysine, it also turned out to be highly expensive and economically unprofitable. The large volume of lysine that flooded the market depressed its price keeping it well below profitability levels. ADM decided to correct this situation through price fixing, another illegal activity. Consorting with other large world producers, ADM set up a cartel which established production quotas and the commodity price. When a dissatisfied executive reported these illegal activities to the authorities, ADM was charged with price fixing and a host of other criminal charges. Employing its economic clout and mobilizing a powerful legal team, ADM was able to avoid serious punishment. Indeed, the company received a light fine and managed to have the entire case sealed, that is, no one could access the trial records. Conversely, the manager who alerted authority was found guilty of receiving a bribe and sentenced to a lengthy prison term.

The class control and transnationalization of the nation-state theories would interpret these cases as examples of the transnational capitalist class's control of the state and the decreased ability of the state itself to address demands from subordinate classes. In the first case, corporate food producers used the local state to advance their agenda and to limit opposition. In the second case, a powerful corporation was able to control the judiciary branch of the state to its advantage. The dialectic and the contradictory convergence theories, however, can also be effectively employed to stress the fact that the transnational class control of the state engenders resistance that emerges at the local level. Additionally, while this resistance was effectively controlled in both instances, the inability of the state to address legitimation issues points to the question of the social sustainability of emerging forms of the state. The point is that while these theories do represent important contributions to the understanding of the evolution of the state in the global era, the complexity of changes in the rural space demand additional investigation and analysis.

CONCLUSION: FUTURE DIRECTIONS

The increased globalization of social relations has fundamentally altered the terms of the debate on the role and action of the state. Accordingly, increased attention will be placed on the effects that enhanced global relations will have on 'forms' of the state. A first area of interest consists of the effects that globalization will have on the regulatory capacity of the nation-state. There is substantial consensus that the nation-state will maintain most of its powers for the near future. However, the extent to which the nation-state will be able to exercise these powers and what actors will primarily affect state actions remain issues to be explored. This situation is particularly relevant because the nation-state has been the locus where subordinate groups have been able to express their needs, even in situations in which they contrasted those of the ruling classes. The crisis of the nation-state, therefore, has been equated with a crisis of participatory democracy. The extent to which democratic spaces will be available under globalization will be one of the most important issues addressed by students of the state.

A second and related area of interest will be the emergence of transnational forms of the state. Propelled to the forefront of many scientific agendas by recent events such as the creation and expanded role of the WTO, the establishment of transnational state forms is proposed in terms of both emancipatory and repressive outcomes. Will the expansion of the transnational state translate into a safe haven for corporate interests in which key concerns, such as those for safe food, the environment and agricultural production, will be compromised in the name of enhanced free trade? Or will it create the conditions for much expended locally directed actions freed from the constraints of an overwhelming nation-state?

A third area of interest refers to the evolution and consequences of state actions directed at the establishment of more advanced forms of free-market-based economic and social relations. The adoption of neo-liberal postures has been a feature of a significant number of nation-states. In these cases, the state itself has been instrumental in 'opening' the areas that were once regulated by state action. In effect, the state itself has been actively involved in diminishing its own powers and spheres of influence. The extent to which this contradictory posture can and will be maintained is an area that requires further investigation.

Finally, the crisis of the nation-state and the undetermined nature of transnational state forms will attract more attention on local and/or regional forms of the state. The opening of spaces for these sub-national state forms and their close proximity to local actors beg further analytical probing. Can they represent alternatives to more established state forms? Can they address the issues of democratic participation and free governance? Can they be forums for discussions on the future of rural areas and agro-food activities? These are questions that will certainly be on the agendas of those whose interests include the state, rural space, and agriculture and food.

REFERENCES

Akard, Patrick J. (1992) 'Corporate mobilization and political power: the transformation of US economic policy in the 1970s', *American Sociological Review*, 57: 597–615.

Antonio, Robert J. and Bonanno, Alessandro (1996) 'Post-Fordism in the United States: the poverty of market-centered democracy', *Current Perspectives in Social Theory*, 16: 3–32.

Barber, Benjamin R. (1996) *Jihad vs. McWorld: How Globalism and Tribalism are Reshaping the World*. New York: Ballantine Books.

Block, Fred (1977) 'The ruling class does not rule', *Socialist Revolution*, 7 (3): 6–28.

Block, Fred (1980) 'Beyond relative autonomy: state managers as historical subjects', in R. Miliband and J. Seville (eds), *Socialist Register*. London: Merlin Press. pp. 227–240.

Bonanno, Alessandro (1993) 'The agro-food sector and the transnational state: the case of the EC', *Political Geography*, 12 (4): 341–360.

Bonanno, Alessandro (2000) 'The crisis of representation: the limits of liberal democracy in the global era', *Journal of Rural Studies*, 16: 305–323.

Bonanno, Alessandro and Constance, Douglas (1996) *Caught in the Net: The Global Tuna Industry, Environmentalism, and the State*. Lawrence, KS: University Press of Kansas.

Bonanno, Alessandro and Constance, Douglas (2001) 'Corporate strategies in the global era: mega-hog farms in the Texas Panhandle region', *International Journal of Sociology of Agriculture and Food*, 9 (1): 5–28.

Bonanno, Alessandro, Constance, Douglas and Lorenz, Ether (2000) 'Powers and limits of transnational corporations: the case of ADM', *Rural Sociology*, 65 (3): 440–460.

Bonanno, Alessandro, Friedland, William H., LLambí Luis, Marsden, Terry, Belo Moreira, Manuel and Schaffer, Robert (1994) 'Global post-Fordism and concepts of the state', *International Journal of Sociology of Agriculture and Food*, 4: 11–29.

Braudel, Fernand (1982) *Civilization and Capitalism 15th–18th Century, Vol. 2: The Wheels of Commerce*. London: Collins.

Braudel, Fernand. (1984) *Civilization and Capitalism 15th–18th Century, Vol. 3: The Perspective of the World*. London: Collins.

Campbell, John L. and Lindberg, Leon N. (1990) 'Property rights and the organization of economic activity by the state', *American Sociological Review*, 55 (5): 634–647.

Carnoy, Martin (1984) *The State and Political Theory*. Princeton, NJ: Princeton University Press.

Domhoff, William (1967) *Who Rules America?* Englewood Cliffs, NJ: Prentice Hall.

Domhoff, William (1979) *The Powers That Be*. New York: Vintage Books.

Friedland, William H. (1983) 'State formation and reformation in California grapes'. Paper presented at the conference: The Political Economy of Agriculture, Ann Arbor, University of Michigan, August.

Friedland, William H. (1991) 'The transnationalization of agricultural production: palimpsest of the transnational state', *International Journal of Sociology of Agriculture and Food*, 1: 48–58.

Friedland, William H. (1994) 'The new globalization: the case of fresh produce', in Alessandro Bonanno, Lawrence Busch, William Friedland, Lourdes Gouveia and Enzo Mingione (eds), *From Columbus to ConAgra: The Globalization and Agriculture and Food*. Lawrence, KS: University Press of Kansas. pp. 210–231 .

Friedland, William H. (1995) 'Globalization, Fordism–Postfordism, agricultural exceptionalism: the need for conceptual clarity'. Paper presented at the workshop The Political Economy of the Agro-Food System in Advanced Industrial Countries, University of California, Berkeley, September.

Friedman, Thomas L. (2000) *The Lexus and the Olive Tree*, rev. edn. New York: Anchor Books.

Friedmann, Harriet and McMichael, Philip (1989) 'Agriculture and the state system', *Sociologia Ruralis*, 29 (2): 93–117.

Fukuyama, Francis (1989) 'The end of history?', *National Interest*, 16: 3–18.

Fukuyama, Francis (1992) *The End of History and the Last Man*. Harmondsworth: Penguin Books.

Gramsci, Antonio (1971) *Selection from Prison Notebooks*. New York: International Publishers.

Gramsci, Antonio (1975) *Quaderni del Carcere: Il Risorgimento*. Rome: Editori Riuniti.

Habermas, Jürgen (1975) *Legitimation Crisis* (trans. Thomas McCarthy). Boston, MA: Beacon Press.

Harvey, David (1990) *The Condition of Postmodernity*. Oxford: Blackwell.

Hechter, Michael and Brustein, William (1980) 'Regional modes of production patterns and state formation in Western Europe', *American Sociological Review*, 85 (5): 1061–94.

Heffernan, William D. (1990) 'The transnationalization of the poultry industry'. Paper Presented at the Twelfth World Congress of Sociology, Madrid, Spain, August.

Heffernan, William D. and Constance, Douglas H. (1994) 'Transnational corporations and the globalization of the food system', in Alessandro Bonanno, Lawrence Busch, William Friedland, Lourdes Gouveia and Enzo Mingione (eds), *From Columbus to ConAgra: The Globalization of Agriculture and Food*. Lawrence, KS: University Press of Kansas. pp. 29–51.

Hodgson, Godfrey (1978) *America In Our Time*. New York: Vintage Books.

Hooks, Gregory (1990) 'From an autonomous to a captured state agency', *American Sociological Review*, 55 (1): 29–43.

Horkheimer, Max (1972) *The Eclipse of Reason*. New York: Seabury Press.

Jenkins, Craig and Brents, Barbara (1989) 'Social protest, hegemonic competition, and social reform', *American Sociological Review*, 54 (6): 891–909.

Koc, Mustafa (1991) 'Globalization, compartmentalization and the New World Order'. Paper presented at the International Conference 'Globalization of the Agriculture and Food Order', Columbia, MO, University of Missouri.

Koc, Mustafa (1994) 'Globalization as a discourse', in Alessandro Bonanno, Lawrence Busch, William Friedland, Lourdes Gouveia and Enzo Mingione (eds), *From Columbus to ConAgra: The Globalization of Agriculture and Food*. Lawrence, KS: University Press of Kansas. pp. 265–280.

Levine, Rhonda (1987) 'Bringing class back in: state theory and theories of the state', in Rhonda Levine and Jerry Lembcke (eds), *Recapturing Marxism*. New York: Praeger. pp. 10–28.

Levine, Rhonda (1988) 'Theoretical developments of the question of state autonomy: the case of New Deal industrial policies'. Paper presented at the Annual Meeting of the American Sociological Association, Atlanta, GA, August.

Lipietz, Alan (1992) *Towards a New Economic Order: Post-Fordism, Ecology, and Democracy*. New York: Oxford University Press.

Lukács, Georg (1971) *History and Class Consciousness*. Cambridge, MA: MIT Press.

Marcuse, Herbert (1964) *One Dimensional Man: Studies in the Ideology of Advanced Industrial Society*. Boston, MA: Beacon Press.

Marsden, Terry (1992) 'Exploring a rural sociology for the Fordist transition', *Sociologia Ruralis*, XXXII (2/3): 93–112.

Marsden, Terry (1994) 'Globalization, the state and the environment: exploring the limits and options of the state activity', *International Journal of Sociology of Agriculture and Food*, 4: 139–157.

Marsden, Terry (1997a) 'Creating space for food: the distinctiveness of recent agrarian developments', in David Goodman and Michael J. Watts (eds), *Globalizing Food: Agrarian Question and Global Restructuring*. London: Routledge. pp. 169–191.

Marsden, Terry (1997b) 'Reshaping environments: agriculture and water interactions and the creation of vulnerability', *Transactions of the Institute of the British Geographers*, NS, 22 (3): 321–337.

Marsden, Terry, Cavalcanti, Salete Josefa Barbosa and Irmao, Jose Ferreira (1996) 'Globalization, regionalization and quality: the socio-economic reconstruction of food in the San Francisco Valley, Brazil', *International Journal of Sociology of Agriculture and Food*, 5: 85–114.

Marx, Karl and Engels, Frederick (1955) *The Communist Manifesto*. New York: Russell & Russell.

Marx, Karl and Engels, Frederick (1964) *Capital*. New York: Vintage Books.

McMichael, Philip (1996) 'Globalization: myths and realities', *Rural Sociology*, 61 (1): 25–55.

McMichael, Philip (2000) 'The power of food', *Agriculture and Human Values*, 17: 21–33.

McMichael, Philip and Myhre. David (1991) 'Global regulation vs. the nation-state: agro-food systems and the new politics of capital', *Capital & Class*, 43 (2): 86–106.

Miliband, Ralph (1969) *The State in Capitalist Societies*. London: Weidenfeld & Nicolson.

Miliband, Ralph (1970) 'The capitalist state: reply to Nicos Poulantzas', *New Left Review*, 59: 83–92.

O'Connor, James (1973) *The Fiscal Crisis of the State*. New York: St Martin's Press.

O'Connor, James (1986) *Accumulation Crisis*. New York: Blackwell.

O'Connor, James (2001) 'Introduction to 2001 edition of The Fiscal Crisis of the State', *Capitalism, Nature, Socialism*, 12 (1): 99–114.

Offe, Claus (1985) *Disorganized Capitalism*. Cambridge, MA: MIT Press.

Offe, Claus and Ronge, Volker (1979) 'Theses on the theory of the state', in J.W. Frieberg (ed.), *Critical Sociology*. pp. 345–356.

Poulantzas, Nicos (1978) *State, Power, Socialism*. London: New Left Books.

Prechel, Harland (1990) 'Steel and the state: industry politics and business policy formation', *American Sociological Review*, 55 (5): 648–668.

Skocpol, Theda (1979) *States and Social Revolutions*. Cambridge: Cambridge University Press.

Skocpol, Theda (1985) 'Bringing the state back in: strategies of analysis in current research', in Peter Evens, Dietrich Rueschemeyer and Theda Sckocpol (eds), *Bringing the State Back In*. Cambridge: Cambridge University Press. pp. 23–46.

Skocpol, Theda and Amenta, Edwin (1986) 'States and social policies', *Annual Review of Sociology*, 12: 131–157.

Skocpol, Theda and Finegold, Kenneth (1982) 'State capacity and economic intervention in the early New Deal', *Political Science Quarterly*, 97: 255–278.

23

The rural household as a consumption site[1]

Sonya Salamon

INTRODUCTION

This chapter describes the transformation of rural society from a production orientation to a one oriented toward consumption, albeit consumption embedded in a productive landscape. Suburbanization of the USA and Europe is driving the transformation with the result that suburban tastes and practices are fundamentally accounting for national tastes and life styles. Farmers have become a minority in rural places both in the US and Europe, with the dual processes of the agricultural transition and suburbanization. The transformation is ironically evident in the altered consumption patterns of farm households, who of course are still producers. Suburbanization has stimulated a desire by farm households to look and to act like everyone else – because they relish the national consumerism way of life. Farmers, however, simultaneously represent authentic rural people to newcomers in rural places. Suburbanites view farm life nostalgically, expressed in a desire to recreate an imagined evocation of farm homes from the early twentieth century. Their consumption of farming resembles the tourist fascination worldwide with old people and places, rather than the reality of contemporary life. Thus, I delineate a curious paradox of suburbanites seeking to reproduce a synthetic or gentrified version of farm life – 'Country Cute' as one suburbanite describes her decorating scheme – while denigrating actual farmers. Contemporary farmsteads, with their metal equipment sheds and brick bungalows, lack the charm of Country Cute and represent obstacles to the suburban domestication of a productive way of life. In fact, where farmers and suburbanites share the landscape on the urban fringe, and farmland has become an amenity for urban newcomers, the working farmer daily faces a contested rural space. I show how changing gender roles and other pivotal factors altered consumption patterns among farm households in ways that makes them resemble their suburban neighbors more than their direct ancestors. Their capitulation, I argue, provokes household consumption that crafts a farmstead that is a synthetic evocation of a rich heritage.

Rural society, through much of the twentieth century, was understood as fundamentally shaped by land tenure (Newby, 1980). Access to and control of land explained social status. Household consumption patterns were firmly linked to the productive activities of the agrarian system. Because farming communities were relatively stable populations, other factors in addition to land, such as citizenship and family reputation contributed to the status of a rural household (Adams, 1994; Hatch, 1992). As a consequence of the agricultural transition in industrialized societies, farmers constitute a small minority of rural populations, and farming no longer is the engine driving the rural economy. Farm households, however, remain dependent on access to land for their livelihood. But factors other than land tenure, and occupations other than farming now dominate the rural social hierarchy and shape rural consumption (Lobao and Meyer, 2001). Understanding these fundamental societal changes, particularly population movements associated with the agricultural and rural transformations, is critical to explaining how consumption patterns have shifted among rural households.

The last decades of the twentieth century witnessed a rebound or renaissance of rural populations in the USA and Europe that brought urban newcomers in large numbers to rural places, where people had outmigrated for much of the previous century (Johnson and Beale, 1998). These newcomers provoked social and economic changes that corresponded with the decline in agriculture's importance to rural economies. Newcomers came to formerly agrarian communities, settling in adjacent often upscale subdivisions, or in scattered sites across the countryside, both as full-time residents or as occasional second-home occupants. These new populations are described for England by Bell (1994) for the US by Herbers (1986), and ethnographically in the US by Fitchen in the Northeast (1991) and Salamon in the Midwest (2003). Coupled with the newcomer influx to rural places was an emergence of a mass media that reflected and was driven by the suburbanization process. Ritzer analyzes how consumption was marketed by centering on domesticity of the home, its grounds, as well as clothing and leisure activities (2000).

Eventually the suburbanization or rural-gentrification processes, and the population dominance of non-farmers in rural places brought non-agrarian, consumption factors to underlie the rural social structure, despite its contextualization in a productive landscape. Furthermore, suburban standards for the good life are based on appropriated rural symbols from farmers – decor of home and garden or cultivation of recreational animals (horses for example). In turn farm households, though authentic rural people, have appropriated the conspicuous, consumer aesthetics of their imitators – for that is what is marketed as the good life. Suburbanized households sentimentalize rural culture and life as a rustic idyll. Harris (2000) describes typical suburban aesthetics as 'salvaging' the material culture of the past by making rural life clean, neat, quaint, and romantic. History, he argues looks more abundant and cluttered due to copious objects accumulated by the insatiable shopping in rural flea markets of today's consumers. In this eclectic, contradictory aesthetic: 'Rusty cowbells stand side-by-side in our kitchens with streamlined, chrome-plated Cuisinarts (Harris, 2000: xiii).' Utilitarian kitchen utensils, such as butter churns or milk cans are turned into decorative lamps or stoneware crocks that are displayed as objets d'art in suburban livingrooms. Sanitized old barn-siding on a wall of my dean's office evokes the family-farming heritage of a college that teaches industrialized agriculture. Mimicry of the suburban aesthetic, means that farmers acquire domestic articles that are a shallow imitation of their authentic rural origins.

This rural transformation is occurring in many places. Dubbink (1984) describes how consuming, suburbanized life styles emerged in the rural California landscape to conflict with traditional customs of hunting and forestry, and Lawrence-Zúñiga (1999) describes similar processes in rural Portugal. Ultimately, as Harris points out regarding consumption of rurality, 'Quaintness rides roughshod over authenticity (2000: 34),' by creating a simulation of the past.

> A Chicago exuburban resident of Kane County, 40 miles west of the city undergoing rapid suburbanization, prowls the St. Charles flea market for farm collectables. In particular, she searches for an antique milk can that will help realize her home-decoration theme of Country Cute. Farmers are relevant to this suburbanite's simulation of rurality as the source and meaning for this iconic farm artifact.

FROM PRODUCTION TO CONSUMPTION IN POSTAGRARIAN RURAL SOCIETIES

Farmland wastefully developed on the rural/urban fringe was described over fifty years ago as an irresistible tide (Firey, 1946). Despite farming giving this rapidly changing region its agrarian character, residential and commercial development quickly result in farmers becoming a minority in the postagrarian fringe (Beale, 1989; Fitchen, 1991). By the 1990s the farm-family way of life was similarly transformed, resembling suburban life more than the life of farm-forebears, even a generation ago. Farm women and many farm men commute to urban centers for work, for the fringe benefits in particular that help support their continued farming. Farm children, no longer excused from school in accordance with the demands of busy seasons, now participate in the same extra-curricular activities as do their suburban peers. Time-honored farm-family practices – the shared family noon or evening meal – are less frequently experienced. Farm parents like suburban parents execute intricate daily schedules chauffeuring children to and from team sports, clubs, music lessons and play dates. To assure that their children are competitive, academically and socially farm youth have become over-scheduled like suburban youth. Doherty (1997, 2000) makes the case that the over-scheduling that farm families imitate actually interferes with important aspects of family life. Investment in education and extra-curricular activities emerged as a consumption priorities because farm youth no longer are expected nor encouraged to farm. Parents often consider their operations too small to absorb children, in the context of the ongoing concentration of farms, that Lobao and Meyer (2001) skillfully describe.

Land takes on new meaning, after the transformation from agrarian-production to a postagrarian-consumption, as rural places become destinations for newcomer-residents and/or tourists, according to Urry (1995). Farmland in the postagrarian era is an amenity that provides the aesthetic experience of being close to nature or a view of open space for those who live in the countryside but are unconnected with productive activities (Dubbink, 1984; Halle, 1993). That is, the rural countryside particularly on the rural/urban fringe, is consumed as a landscape by all except those few farmers still engaged in production. Farm fields are a backdrop for the windows of neighboring subdivisions or used for recreational or 'entertainment farming' activities such as pick-your-own produce, horseback riding, petting zoos or cornfield mazes. Some surburbanites living on the fringe choose to commute long distances for work, convinced that their families benefit from the countryside's restorative nature. For these newcomers the preference is for a middle-landscape, that despite development remains rustic and picturesque because farmers keep the scenery tended and neat (Urry, 1995; Pollan, 1998). A new homeowner in exurban Chicago, for example, bemoans losing productive symbols along the road he travels, since his arrival three years ago: 'They should have kept the old farm building there to remember how the farm was.' Similarly a professional employed in the suburbs commented positively about rapid development: 'It would be nice to have some country left … This might sound corny, but I like watching the farmers plant in the spring and watching the corn grow.'

Farmers are useful for managing the scenery but an inconvenience to newcomer-suburban households when the narrow roads must be shared during their daily commute to work. A young farmer in exurban Chicago, who can see a new subdivision from his kitchen window, describes enduring frequent hostile interactions with newcomers during busy seasons when he moves machinery on what were formerly farm roads: 'People get so mad because I'm only going 20 mph. If you took their middle finger away people would be lost.' Another farmer commented similarly about farming on the fringe: 'They hate the farm machines on the roads because they have to slow down.' Moving his equipment at night is one Illinois farmer's solution to contested country roads.

A retired farmer in his 80s, the owner with his farmer-sons of over 2000 acres on the rural/urban fringe west of Chicago, comments insightfully on how emerging meanings for farmland bring into conflict farmers, who retain a productive meaning with suburbanites who are consumers of the farmland as an amenity.

> People come out here and they're 'Open space, open space.' I say, 'Look, your open space is my corn field. Let's try and keep that in perspective.' To us it's a way of making a living. The only thing that keeps it open is the fact that we make money by farming it. And we own it. But open space to a city person means, 'Oh look, there's nothing here.'

These changes from a productive to a consumption-driven countryside cause the various groups inhabiting transformed rural spaces to contest the spaces in roads, public spaces, churches and schools. Newcomers and oldtimers may inhabit the same rural space but live in different worlds, according to how they define their meaning of those things constructing every day life (Appadurai, 1986). Farmers and oldtimers remember a different meaning for community, for example, prior to the transformation of an exurban area 40 miles west of Chicago. Explained a retired farm woman:

> The cars go right behind you, on your bumper. Like they can't wait to push you out of the way. … One lady sat in her car and she goes like this [throws her arms in the air]…. So I just pulled off the road and let her go by. We never had that before when it was rural – we waved at everybody because everybody knew us (Salamon, McGuire and Basic, nd).

When land undergirded the productive rural social structure those families controlling the greatest land holdings, whether working it themselves or hiring others to work their land, generally wielded more community power (Hatch, 1979, 1992; Newby, 1980; Salamon, 1992). Economic survival, farm goals, and size of land holdings in agrarian communities worldwide typically defined household consumption patterns (Bennett, 1982; Netting, 1984). The gender hierarchy of farm families underscored the household consumption hierarchy. Prior to World War II, the household division of labor gave women responsibility for activities on the immediate farmstead – chickens and eggs, garden, and dairying – that supplied resources for household consumption, and the surplus was marketed. Men were responsible for producing the bulk of what was sold off the farm, that financed production expenses and land purchases (Barlett, 1989; Fink, 1986). Farm production, under masculine control, took priority over the feminine controlled household production. Men, therefore, dictated how family financial and labor resources were committed, because family financial decision-making and gender hierarchy privileged the farm over household demands (Adams, 1988; Barlett, 1993; Gasson and Errington, 1993; Kohl, 1976; Wilk, 1989a). Thus, the farm 'came first' in priority over domestic spending. For example, a German-American farm woman did not use the money she earned during the Great Depression for household goods. 'I had some money from my egg route and I wanted to buy some kitchen curtain material and [my husband] said, "No, we had to save it for land." I was looking at today and he was looking at tomorrow,' she recalled (Salamon, 1992: 129). Farm homes reflected these priorities by being furnished with utilitarian, modest possessions that supported family labor and production priorities. Life took place in the kitchen and through the back door – the front door or formal parlor were rarely used (Adams, 1994).

After World War II agriculture became more industrialized and the division of labor within the farm household also changed (Adams, 1988, 1994). For example, what was formerly domestic production controlled by women, such as poultry and eggs, shifted to control by men when transformed into larger-scale commercial production (Fink, 1986). A postwar financial strategy for smaller farm-operators, who desired higher levels of household consumption, was to opt for part-time farming combined with an off-farm job by the male and/or the woman (Barlett, 1989, 1993). Traditionally, those utilizing off-farm employment strategies were relegated a lower community social standing than those whose land resources permitted farming full-time. After the US farm crisis of the 1980s, however, greater numbers of farm women took full-time jobs off-farm because the medical and other fringe benefits were crucial to the farm-household economy. By the early 1990s, for example, Illinois farm women, whether from a household adopting sustainable farming systems or maintaining a conventional farming system, were almost uniformly commuting to jobs in nearby towns or cities (Salamon et al., 1997). When women's off-farm wages and benefits became critical to the farm's financial sustainability, female priorities gained the potential to alter the household's consumption patterns (Lobao and Meyer, 2001). Increasingly, farm households wanted the middle-class urban or suburban standard of living featured in consumer magazines that included a comfortable, stylish home, an attractive garden, annual vacations, and shared leisure. In general, these priorities involved more abundant consumption than what typified farm households even a generation earlier (Barlett, 1993; Löfgren, 1999; Wilk, 2000).

Another consequence of the industrialization of agriculture and changing social relations in the wider society was a lessening of kin-embeddedness of farm households. This kinship system transformation in part was triggered by the emergence of the nuclear farm-family household to replace a residence pattern that incorporated two generations of an extended family together or next door on the same farmstead, like the peasant stem family (Adams, 1994; Creed, 2000). When household membership is nuclear, how its resources are accumulated or spent is less subject to monitoring or criticism by the parental generation (Salamon, 1992). Erosion of kin-embeddedness was also fostered by farm women in large numbers entering the non-farm labor market, as well as the concentration of farms that spurred the outmigration of farm youth to non-farm occupations (Lobao and Meyer, 2001). Mothers-in-law or the husband's parents no longer could observe household spending. Furthermore, parents were grateful to have one child and his or her family willing to farm family land, and earn less than siblings who left to work in cities. Under these circumstances, the older generation was mindful of offending by too much criticism of the successor's life style.

Farm women entering the labor force was also facilitated by the decline of farm-family size to that of urban family size. Smaller families eventually caused a decline in the size of kin-networks

(Fuguitt et al., 1989). Smaller kin-networks decreased some of the responsibilities and obligations incurred by farm households, or typically the farm woman. As the household 'kin-worker', farm women working off the farm no longer had either the time or perhaps the inclination to maximize kin embeddedness through elaborate daily and holiday rituals based on the consumption of handmade items (Di Leonardo, 1987). These world-wide changes in what had been the distinctive residence, kinship, and labor organization of farm households have meant that family goals for being a good child or a good parent shifted toward urban or suburban lifestyle practices or goals (Wilk,1989a). Collier (1986) identifies a shift in gender conceptions in a single generation in a rural Andalusian village. Women went from emulating the Virgin Mary with modest demeanor to emulating stylish advertisements on Spanish television, as education rather than land became more important to determining their children's future. Decreased kin-group priorities, highlights the privatization but more importantly the suburbanization trend among farm households. With a family's land holdings accounting less for community status, and the divergence from utilitarian, agrarian household practices, domestic consumption rose in importance among farm families, especially for material possessions for the home (Miller, 1995; Ritzer, 2001). Other factors also influenced this farm household tranformation.

After World War II, US farm women in particular, were encouraged to devalue the farming way of life according to Adams's description of change in a Southern Illinois county (1988, 1994). For example, educational programs of the US Department of Agriculture, Cooperative Extension Service aimed at farm women, communicated a message that their homes and family clothing should strive to emulate urban or suburban styles. That is, farm life was essentially disparaged as 'hick' or lower class, and suburban or urban middle-class ways of life were extolled as better, more cosmopolitan. Similarly, negative depictions of a backwardness, shown on television situation comedies or movies (e.g. *Deliverance* in the USA or *Cold Comfort Farm* in the UK), are rural stereotypes firmly fixed in the urban consciousness, despite clear evidence to the contrary. As recently as the 1980s University of Illinois students from suburban Chicago, assumed Illinois farmers did not have indoor plumbing, although they regularly drove back and forth to campus through the countryside without ever passing an outhouse. Farm people, however, even those with advanced college degrees, absorbed the powerful negative media images and were determined 'not to look like farmers' in their homes, dress, or behaviors when venturing off the farm. Farm households disposed of family antique furniture, dishes, implements and homes, and decorated, remodeled, or built homes according to styles featured in national home and clothing magazines. Thus, rural household implements, dress, and other aspects of life were disparaged and consumption, based on urban and suburban trends, were effectively promoted to rural people.

Once these fundamental cultural changes took place, the rationale was explicit for farm-households to shift some financial resources previously allocated for productive uses to elevate levels of domestic consumption. New consumption priorities emerged across rural communities. The rural home like the suburban home became the '... locus of power, control and self-expression' and the land surrounding the home became an extension of interior decoration and symbolic of the homeowners power over the environment, as it is for suburbanites (Sheehy, 1998: 119).' Illinois farm women reported consulting decorators for advice about upscaling their homes. With consumption shaping social structures, new rural hierarchies replaced the historic land-based class differences in the USA, and cross-culturally (Baudrillard, 1998; Hearn and Roseneil, 1999; Ritzer, 2001). For example, the transformation from utilitarian production-aesthetic to that of a consumer-aesthetic is reflected in a analogous reordering of domestic priorities that accompanied the economic and social transformation of rural Spain noted above (Collier, 1986). Similarly, in rural Portugal the appropriation of urban and suburban style homes by rural people signifies a profound shift toward domesticity and a suburbanized lifestyle and priorities (Lawrence- Zúñiga,1999). Farmers have an inherent disadvantage for competing with homes and possession in the new consumption driven rural social system. Unlike the more fluid capital of suburbanites, farmers have financial resources tied-up in land and equipment, so have less available for instant investment in consumption. In rural Spain small landowners whose family members did not migrate to cities are now among the poorest and most overworked households (Collier, 1997).

The acceptance of suburbanization aesthetics and priorities in the countryside is tied to the longing of farmers and other native rural households for the respectability of looking like everyone else (that is not like rural or farmers), in their appearance, their daily lives and homes. Newcomers, in contrast, want to imitate farmers, albeit a simulation of nostalgic farm life, as seen below.

CONSUMPTION THEORY AND NEW RURAL HOUSEHOLDS

Anthropologists, archaeologists in particular, have long been attuned to what material culture tells us about social structure, social relations and consumption. According to Appadurai (1986) material culture, or the things people acquire or make, should be read as symbols through which social status is exhibited and maintained. In addition material culture reflects a moral economy that structures everyday experiences. People can manipulate things/possessions to represent changes in their status or identity, relative to those around them.

Historians examining culture change and the evolution of taste in the USA focus on the process by which a popular culture emerged in the twentieth century that was, '... less regional, less ethnic, and less rooted in a particular sense of place (Kammen, 1999: 52).' The flourishing of popular culture coincided with the rise of consumer culture and shaped its content. The consumer culture that rural people (both newcomers and oldtimers) now share reflects a homogenization effect, in which rural culture is more generic – similar in an entire region – where previously it was authentically attached to a specific locale. Culture became dislocated from place via homogenization into a popular rural culture, and disembedded from the social relations that originally were the source of its unique cultural elements (Giddens, 1990). Nostalgia for the past – for a rural life now lost or that never existed – provides a fertile ground for the advertising images that create consumption desires, according to theorists (Harris 2000; Ritzer, 2001). Advertisers use generic rural images to market products to urban consumers by a 'commodification of the rural myth,' for the good life, in an analysis of USA advertising by Goldman and Dickens (1983). European's attraction to American popular culture and to the frontier apparently led urban Americans, the arbiters of taste, to realize that rural material culture was of value. During the late twentieth century important museums opened Americana wings that displayed rural quilts, pottery, woodworking, furniture, architecture and even tools as high art. The connection between Americanism and commercial culture emerged to make rural material culture attractive and even a validation of taste in personal possessions (Kammen, 1999). But the rural material culture with cache was nineteenth century or early twentieth century rather than the present. For instance, old barns are the subject of numerous books, but farmers today if they use a barn have a metal one lacking the rustic charm of older wooden structures (see for example Babcock, 1996).

How farming material culture was transformed from utilitarian to decorative uses reflects style changes in the wider society. Kammen (1999) argues that the establishment of rural artifacts as art, reflects the democratization of culture in which elitism declined regarding what material things were considered tasteful. Consumption of material culture in the form of old farm implements, tools, kitchen utensils, and textiles constitutes a reordering in the wider society of what is art or high culture. Using rural material culture, validated as art, permits a reimagining of regional and national identities and an affirmation of cultural pluralism, in a non-threatening way. Mullin (2001) argues an opposite process to that proposed by Kammen. She describes how Southwest Indian utilitarian objects such as pots, blankets, and jewelry became defined as high art in the early twentieth century due to the advocacy of elite East Coast women searching for alternatives to European art. They succeeded in creating an aesthetic and establishing a market for Indian pots, rugs, and jewelry. American Indians are of course the most rural of USA minorities (Snipp,1989). To collect rural 'antiques' as art merges the utilitarian and the creative in objects of material culture having integrity, due to their authentic origins. Furthermore, because the older, rural objects are humble and even mass produced art, they are affordable symbols of taste. The pursuit of these objects allows the middle-class shopper to indulge in the consuming society's favorite leisure pursuit of spending money, albeit economically. Of course, the Southwest Indian market has a low end, but the small group of Santa Fe women successfully established consumption at the high end, and sustained that market.

Beginning with the mass-cultural mail order catalogues from Sears and Roebuck in the early twentieth century, USA rural people were integrated with national popular culture. After the relatively late electrification of the rural countryside and development of better roads, rural people were more thoroughly exposed to the mass media of radio, newspapers, films, and eventually television and the Internet. Locally-owned small-town businesses that previously supplied rural consumers with goods and services gave way in the late twentieth century to Wal-Marts. Regional shopping malls now provide rural places with the same variety of shopping options that urban people have. Rural people, by the 1960s and 1970s, were drawn into national processes that intertwined leisure time, shopping, and consuming (Ritzer 2001). But these consumption sources were of course locale free. Consuming at malls or over the internet no longer reinforces local communities as when shopping was confined to a town's main street. Small-town main streets withered and died with the restructuring of agriculture and the advent of regional shopping centers. If a main street had charm and a town a good location near a highway, downtown could be reborn as a tourist destination. When antique shops and boutiques catering to urbanites replace main street shops in suburbanized towns, shopping there fulfills the desire of people to buy authentic collectibles from the agrarian past, and thereby the consumption of rurality is fostered. Rural people in contrast are shopping at malls and WalMart (Salamon, 2003).

Rural tools, domestic utensils, and fabrics for the suburban household possess integrity because they embody regional character, cultural distinctiveness, that merges art with the utilitarian (Mullin, 2001). Harris (2000) explains how the quaintness aesthetic restores a sense of time to consumption, provides an instant personal history, that is lost with modern appliances and furnishings and household mobility. A nineteenth century milk can or butter churn despite being mass-produced, represent popular culture of a nostalgic or sentimentalism about rural life – symbols of a life somehow more authentic than the present. By buying an 'antique' object pursued in rural flea markets, suburbanites seek self-expression through reproduction of a rustic ideal. The search recreates the past as a form of recreation. Rural farmsteads in the past, of course, were not neat, mowed, decorative, or rustic, but dirty, smelly, scruffy, and utilitarian when production was paramount for the household (Adams, 1994).

Rural households observed the emerging market for their artifacts and began to value the 'junk' stored in their attic or the barn. Family-farming relics once relegated to the attic, such as milk cans, butter churns, or pie-safes were unearthed, refinished and displayed proudly in the livingroom as antiques. Several Illinois brothers still farming rue the day they trashed all of a grandmother's unused glass canning jars by throwing them out the attic window, because '... now those jars are worth $6 apiece.' Unlike a farm family's authentic heritage, however, finding quaint, cosy things as a folk archeologist – necessarily remains consumption of an imitated life that is a sanitized version of the original (Harris, 2000; Ritzer, 2001). Rural quaintness allows suburbanites to compensate for their lack of roots or personal history in a place, through purchasing an instant, unstained heritage.

As the suburbanization process proceeded in the countryside, rural gradually became defined as a life style, not a place (Riley, 2002). Baudrillard explains such a shift from a unique place to a generic style as a 'simulation' of rurality – whether demonstrated by upscale overalls or a scarecrow decorating a front lawn (Baudrillard, 1983; Ritzer, 2001). In the USA young women in particular choose to wear denim overalls, evoking the utilitarian dress of farming past. Overalls are actually a popular variation of the ubiquitous blue jeans. Suburban and urban high school and university students wear overalls, thinking their dress is like a farmer and that they thereby share a rustic identity. They could not be further from the truth. Farmers would not be caught dead in overalls – because to farmers overalls make them resemble the popular misconception of their being hicks. Only elderly farmers over 70 or 80 years old wear overalls, out of force of habit, to work in the fields, and they are long retired.

Suburban households in general aim with home decoration to emulate domesticity, as well as bring prestige to the household (Miller, 1995). The suburban good life conceptualized by the style guru Martha Stewart, is a home decorated with selected objects that display tasteful consumption. Thus, if Country Cute is the decorative schema it follows that front door wreathes or applique flags or garden ornaments are changed with the seasons to unite the outdoors with the indoors. These decorative elements celebrate ownership by manipulation and modification of the home and the grounds and reflect distinction of personal taste, irony or display (Sheehy, 1998). Rural decorative touches are homage to the home as the ultimate consumer good, both a shrine to Arcadia and a symbol of wealth (Wilk, 1989b). The home and front garden are identity markers in the absence of the other markers that rural people formerly used to judge rural households (Chevalier, 1998).

Competitive community relations are now localized through decorative simulations of homes as upscale farmsteads (Baudrillard, 1998). Rural and farming as a life style are imitated with old wooden farm wagons displayed in front yards as a planter for flowers, or whimsical plastic ducks (with kerchief) beside a flower bed. Country *kitsch* in the form of decorative painted wooden cutouts, windmills, baskets, or small statues, replace authentic productive and functional arrangements of buildings and equipment. Similarly, functional rural vegetable gardens are replaced by ornamental flower gardens. Quaint decorative touches make rural life for suburbanites seem much cosier than the present, and therefore less threatening or competitive. Quaintness is a primary aesthetic of modernity for these reasons, according to Harris (2000). A highly mobile suburban population has made recreation of the rural past among the most popular decorating styles of the USA, albeit sanitized rurality. Perhaps to provide some authenticity to where they live suburbanites attempt to recreate what was eradicated to build the subdivision, just as developers name subdivisions Rolling Meadows or Countryside Manor for what was destroyed (Kuntzler, 1993). Of course, the rural life evoked by suburban residents through objects tries to evoke the nineteenth century, not the high tech twenty-first century. Suburbanites pretend they are living a rural life by using Country Cute as a decorative style. For them the unreal has become reality– because they have simulated some visual traits of farm life (Ritzer, 2001). Although the newest suburbs are situated in the midst of the remaining productive landscape, in the postagrarian countryside farmers, their old or new possessions, as well as their land are eventually consumed and thereby symbolically transformed.

Rural society and the rural landscape were reinvented by suburbanite newcomers. Farm households were coerced or were incorporated willingly into the new social hierarchy, based on consumption. Rural places have become a sort of rural theme park – where the real, in this case original rural households like farmers are remaking themselves, their homes, and their lives so that it seems as if life 'imitates the imitation' (Huxtable, 1997:63)' (Ritzer, 2001). In the process what farm households chose to consume underwent a fundamental alteration, that reflects how they now define themselves (Csikszentmihalyi and Rochberg-Halton, 1981). Many farmers, as authentic rural people like other members of a consumer-driven society have bought into the unreal, decorative simulation of rurality, created by suburbanite newcomers and the media (Goldman and Dickens, 1983). In contrast, the real farm life being simulated had a specific, unique identity built on strong social networks, stability, a sense of community, and family reputation that contrasts with the loose networks, mobility, lack of community attachment, and history-less populations that typify suburban places (Salamon, 2003).

RURAL HOUSEHOLDS AND THE CULTURE OF CONSUMPTION

Like the developers who destroy meadows, forests, and streams by building suburban subdivisions – and then name the development after what was destroyed as we saw above, suburbanite newcomers simulate a vanished rural way of life their coming has shattered. As suburbanites' consumption patterns come to dominate the countryside, formerly rural communities lose many characteristics of a real place. Most crucial, is that the transformed community is divorced from place because the new places are not authentic, but a generic rural, a pseudo memory or a nostalgic experience sugar-coated as a result of being merchandised (Riley, 2002). A major challenge for rural researchers is to understand why farmers, as authentic rural people, buy into the consumption styles of simulated rurality. First, what is the evidence for this transformation of farm-household consumption?

Farmstead aesthetics, when dedicated to the utilitarian needs of the farming production system incorporate front doors rarely used, home interiors tailored for efficiency, and functional landscaping – trees for wind breaks and a vegetable garden to supply the kitchen. Family life takes place in the kitchen (Salamon, 1992). In the suburbanized landscape aesthetics of farmsteads and subdivision homes diverge from the past. In 1992 I interviewed several dozen farm couples in their homes all over Illinois. Only the Country Cute concept explained living rooms decorated with arrangements of long-eared rabbit stuffed dolls dressed in nineteenth century clothes and posed among tiny toy early American chairs, toy kitchens, or school desks. In addition, I saw farmstead flower gardens decorated with wooden cutout ducks or a little girl bending over bed so that her panties showed, but few vegetable gardens. Fancy rush baskets (purchased at home-shopping parties) were arranged as casual country, with commercial arrangements of dried flowers or fruit. These home decorations were devised by the farm women, the men quickly assured me. Their disavowment occurred when I was ushered past the decorated area to the corner of the house defined as male where the office, or a billiard table was located. Kitchens were also decorated with Country Cute decorative items, but women no longer baked, canned, or even froze vegetables and fruits. Families eat out often. One household favored the local Chinese takeout restaurant for which they placed orders over their cellphones while driving their children from various after-school classes. Farmsteads were spiffed up with landscaping – perhaps in an attempt to disguise them as productive staging areas. Livestock, whose smells and noise are not quaint, have largely been abandoned, at least in the Midwest. Like subdivisions named for meadows and forests, farms and these gentrified farmsteads testify to an agrarian life transformed.

The triumph of Country Cute among farmers represents a shift in gender roles and priorities, and in particular a feminization of the home and farm that resembles suburban consumption patterns. This feminization of most 1990s farmsteads contrasts with the utilitarian, efficient, masculine, productively organized place that working farmstead represented when I began to study farmers in the mid-1970s (Ritzer, 2001). Perhaps it was all beginning when in the late 1970s every farm we visited sported a new microwave oven as a Christmas present. Even the masculine-aimed pesticide advertisements represented gender in new ways by the 1990s. Many now claim feminine qualities of nurture and care to show that they are environmentally more sensitive (Kroma, 2002). Clearly showing the feminine side of farming is expected to sell pesticides.

As shown previously, due to the restructuring of agriculture, urban sprawl, and the preference for small-town life in Europe and the USA, farming people are more likely now, than in the past, to live by neighbors with whom they neither share work, history, nor culture (Herbers, 1986; Olson and Lyson,1999; Salamon, 2003). As a consequence for original rural residents, place-based relationships have declined as social control processes, and consumption-based criteria have correspondingly grown in importance for social control (Giddens, 1990). For example, farmers prior to these changes criticized those considered too extravagant and viewed frugality positively. These cultural practices maintained modest farmsteads and relegated expensive purchases such as a boat to a shed, hidden from prying neighbors (Salamon, 1992). When in the early 1980s farm crisis Midwest farmers leveraged their equity to buy land at inflated prices, fiscally conservative farmers termed those who crashed financially, bad managers, deserving to lose land or the farm. Furthermore, a moral judgement was attached to the loss. Not only were agrarian standards for status and reputation altered after these events, a sense of community with shared norms and trust was also diminished (Dudley, 2000). When newcomers settled in the same rural places these transformation process were only enhanced. Loss of trust, of course, is associated with the decline of place-based relationships and consumer society (Giddens, 1990; Ritzer, 2001). Among oldtimers, behavior was customarily placed in the context of a family reputation produced by multiple generations in the community. If you know your neighbors mainly through their conspicuous consumption outside the home, trust is difficult to generate.

The advent of farmers now living in close proximity to suburbanites altered historic rural community productive norms, and replaced them with norms based on consumption. The local status hierarchy therefore fundamentally became shaped by the consumption gauges of beauty, display, and domesticity (Miller, 1995). If farmers wanted to compete in the transformed society, their way of life – their home, their possessions, and their children rearing– had to be recast according to suburban priorities. Farmers no longer 'wanted to look like farmers,' either on the farm or in the community.

When upscale consumerism transforms a rural community those of modest means are forced to find housing elsewhere. Consumption takes a particular profile when these families are clustered as they are in one Illinois village with shabby housing stock that is affordable to rural, working-poor people (Salamon, 2003; Salamon and Tornatore, 1994). Newcomers to this village are not urbanites seeking cheap housing; most were born in rural places. These lower-income newcomers say they prefer small-towns as having the advantages of low-cost housing, a safe environment, and white neighbors, to stabilize their lives. Prior to their move to this tiny village of 200, most families moved often (mean 5.7 times) from one rural town to another. Owning even a rundown home thus improved the position of the previously mobile newcomers. Poor newcomers were attracted by the surplus of affordable housing. A retired farmer, commenting on the village transformation, stated, 'There's been a drastic change in the last 20 years. Poor bought property.' Once six lower-income households were established during the 1970s, members of their networks followed in a chain migration linking these households through kinship or friendship. Newcomers' working-poor incomes or educational levels, however, were not that different from those of oldtimers. But these characteristics were all the two groups share.

Differing perspectives about the responsibilities of community membership divide oldtimers and newcomers. The Old Guard, as the newcomers referred to them, regardless of how they earned a living share attitudes about the meaning of property and suspicion of those whose meaning for property differs from theirs. Property upkeep represents ownership pride and place-identification with the town to oldtimers. For the Old Guard, good citizenship is defined as maintaining painted and repaired exteriors, mowed lawns, and a neat appearance of homes. A retired widow expressed the basic resentment about newcomers, 'They don't keep their places up like they should. To keep a town going, you need people who are interested in the town. They aren't.'

After newcomer families moved in, the village appearance changed. Although the decayed housing stock was not created by the newcomers, homes that became rundown under the ownership of the elderly Old Guard remained dilapidated after acquisition by the working poor, whose small incomes were stretched to make the home purchase. Furthermore, a disorderly appearance of the homes, especially along the main road, developed under newcomer occupation. Objects not normally found around village yards, such as inoperative cars, toys, and garbage, now present a daily provocation. The Old Guard value a home for status enhancement and claim village space for the utilitarian aim of maintaining higher property values. Newcomers attach a meaning to their homes as a workplace used for survival; they claim village space for economic productivity. Although the Old Guard might use a home for a side business such as lawn-mower or watch repair, neatness of a home and business is proof of a good craftsman. Spare parts would not be stored on the front lawn, according to Old Guard norms, and cars should be kept in good working condition. Vehicles beyond the requisite one per household adult are termed 'junk' by the Old Guard. Four newcomer families had at one time a combined total of more than 30 nonfunctioning automobiles parked in yards, along curbs, and filling a vacant lot.

Another source of friction concerned responsible garbage disposal. Newcomers do not have regular trash service. Rather than pay for weekly pickups, newcomers accumulate trash for weeks before they make a 'run to the dump.' Irritation with the resulting piles of garbage was expressed by a former village board member: 'A couple families throw garbage, including old diapers, out on the lawn and leave it.' Garbage, like old cars and weedy grounds, is an expression of the newcomers' *habitus* or their taken-for-granted concrete, informal practices of home, neighboring, and family that reflect social background (Bourdieu, 1984).

Perhaps the Old Guard unreasonably expects middle-class property practices from newcomers, who like the inner-city poor, are successors to deteriorated housing and a village with fewer opportunities for employment or social mobility than in the past. Thus, the struggle between village status groups is motivated by nonmaterialistic issues of cultural ideas, respect, and dignity (Hatch, 1979). This transformed village points up that relative attitudes and expectations of newcomers and Old Guard toward property – personal and public – are of paramount importance in postagrarian communities that become purely residential. Property appearance and values, are of course, suburban priorities.

Because the rural poor are pushed out of upscaled suburbanized towns into affordable ones like this shabby village, (or a trailer park), given the poverty of its newcomers and their *habitus* it will become the persistent rural slum of the regional commuting zone (Duncan, 1999; Falk and Lyson, 1993). The village's public spaces are highly contested due to conflicting consumption practices, and despite newcomers being a minority. These space disputes highlight how in postagrarian communities the appearance of public spaces takes priority over the social connections supported there. Suburbanization has far reaching effects among diverse types of rural households; poor rural people must live somewhere.

A paradox exists in the consumption trends of rural households, particularly farm households. In a society driven by consumption, farms remain a site of production. Farmers remain important symbols of what is authentic and positive about America despite the majority of the US residing in suburbs (Palen, 1995). For this reason, advertising employs farmers symbolically for their core values, in the mass marketing of consumer goods (Goldman and Dickens, 1983). Farmers, however, are ashamed to look like farmers due to their also paradoxically being packaged as lower class, hicks, uncouth or simple minded. Television programs such as *Hee Haw, Green Acres, Beverly Hillbillies, Northern Exposure*, *All Creatures Great and Small,* or movies such as *Deliverance, Tobacco Road,* or *Sweet Home Alabama, Cold Comfort Farm* portray rural people (particularly rural white Southerners or 'white trash') with mean-spirited stereotypes. In a politically-correct time it is permissible to mock rural, white folk as country bumpkins who dress in overalls, use lower-class speech, and live in rough, cluttered unsophisticated homes and nosey communities. Yet other movies such as *A Thousand Acres*, *The River,* and *Country*, or even the twenty-five years of annual Farm Aid benefit rock concerts are media sympathetic to farmers, especially the small family operated farm. By buying into the idea of not looking like a farmer (the negative stereotype they have consumed) farmers opt for consuming the simulated rural life (Baudrillard, 1998; Ritzer, 2001). That is, by consuming the commodified symbols of rural life used by suburbanites to simulate their lifestyle, albeit in the nineteenth not twenty-first century, authentic rural people are losing their distinctive identity along with their storied way of life.

FUTURE QUESTIONS FOR STUDY OF RURAL HOUSEHOLD CONSUMPTION

Do all rural people identify with the decorative simulation of nostalgic rural charm (Harris, 2000)? Certain categories of farmers, for example those adopting sustainable farming systems, resist the lure of the consuming society. In a controlled comparison study of Illinois farm households, in the same community, adopters of sustainable farming systems, and a matched household using conventional farming systems, were interviewed during the same week. Some paired sets were relatives, others best friends or the neighbor who was the loyal opposition. Findings statistically significant demonstrated that households using sustainable farming systems had older cars, were less likely to have central air conditioning, and employed older tractors than their paired household using conventional practices (Salamon et al., 1997, 1998). Sustainable farm households are proud of how they 'recycle' equipment, furnish from flea markets, and are self-sufficient for organic meat and/or produce. By making these choices sustainable households live counter to the consuming patterns in their immediate community.

The consuming society temptations are always a challenge to resist. In one small town the big event of the year was the opening of a MacDonald's restaurant. Local youth were overjoyed that their town finally had joined the consuming society and they could 'be like everyone else.' Sustainable farming systems, as a production choice, have not been considered in the context of being a choice about consumption. These farm households represent a counter-dialogue about rural production, consumption, and by implication community sustainability. Sustainable-adopter farm-households march to a different drummer, particularly the men.

In other settings, defining new standards of taste and validating new rural objects as art have been liberating for women, as in the emergent art scene of Southwest (Mullin, 2001). Consumption focused on the home and garden for this reason may be another indicator of a more flexible identity for women and/or the increasing flexibility of rural gender relations. Women in the USA have tended to take the lead as consumers and patrons of modern art. Gender relations in changing farm households provide a rich area for future study given how much of consumption theory describes feminine choices in taste and material objects (Ritzer, 2001). Are farm and rural women more susceptible than men to the coercion of suburban aesthetics for home and lifestyle?

Rural sociology has widely debated whether a rural/urban continuum exist, expressed as differences between the two types of space and place (see Bell, 1992). Whether consumption theory illuminates a further erosion of rural and urban differences is ripe for consideration in the relatively unexamined context of rural homes and household possessions. Sociology of agriculture researchers have concentrated on linking scale or the form of production and the quality of life in rural communities, since Goldschmidt's famous research on the industrialization of agriculture, in the 1940s (1978). Given the pervasiveness of a consumer culture perhaps the field must reframe how questions are asked – to connect forms and locales of consumption and the quality of life in rural communities. The resurgence of farmers' markets, for example, is driven by consumption demands of non-producing households. Farmers' markets represent a weekly ritual that allows the celebration of community on a regular basis (Kemmis 1995). Farmers' markets simultaneously help preserve smaller operations. In Santa Fe the community experimented successfully with a weekly indoor winter farmers' market during 2002–3. A suggestion exists that farmers' markets, which foster consumption, can revitalize moribund rural places and sustain local small-scale agriculture (Lyson et al., 1995). These are intriguing questions.

Among farm households it is crucial for us to examine whether heightened levels of consumption threaten the survival of mid-sized family farms, who historically have persisted by curtailing consumption in hard times. Similarly, changing gender roles in the USA have made daughters logical successors to the farm. Will this gender role transformation among farmers have implications for the feminine-driven suburbanized consumption on farms?

The challenge is for suburbanizing rural communities to use consumption preferences in ways that allow newcomers and oldtimers alike to preserve a sense of community and local vitality in a fast changing world. How consumption can be harnessed to build community among new and old rural households, should be a rich area for future research.

NOTE

1 Research about farming on the fringe derives from a field study of farmers and suburbanites in Kane County, Illinois, 40 miles west of Chicago funded by a 1996–98 grant from the Illinois Council for Food and Agricultural Research (C-FAR) Project 97–108. Research from a 1992–94 field study of Illinois farmers adopting sustainable farming systems versus paired farmers in the same community using conventional farming systems was supported by grant number LWF 62-01-03113 from the North Central Region Sustainable Agriculture and Research Program.

REFERENCES

Adams, Jane H. (1988) 'The decoupling of farm and household: Differential consequences of capitalistic development on Southern Illinois and Third World family farms', *Comparative Studies in Society and History* 30(3): 453–82.

Adams, Jane H. (1994) *The Transformation of Rural Life. Southern Illinois 1890–1990.* Chapel Hill, NC: University of North Carolina Press.

Appadurai, Arjun (1986) *The Social Life of Things: Commodities in Cultural Perspective.* Cambridge, UK: Cambridge University Press.

Babcock, Richard W. with Stevens, Lauren (1996) *Old Barns in the New World: Reconstructing History.* Lee, MA: Berkshire House Publishers.

Barlett, Peggy F (1989) 'Industrial agriculture', in Stuart Plattner (ed.), *Economic Agriculture.* Stanford, CA: Stanford University Press. pp. 253–91.

Barlett, Peggy F. (1993) *American Dreams, Rural Realities: Family Farms in Crisis.* Chapel Hill, NC: University of North Carolina Press.

Baudrillard, Jean (1983) *Simulations.* New York: Semiotext(e).

Baudrillard, Jean (1998) [1970] *The Consumer Society: Myths and Structures.* London: Sage.

Beale, Calvin L. (1989) 'Significant recent trends in the demography of farm people', *Proceedings of the Philadelphia Society for Promoting Agriculture* 1987–88: 36–39.

Bell, Michael M. (1994) *Childerley: Nature and Morality in a Country Village.* Chicago: University of Chicago Press.

Bell, Michael M. (1992) 'The fruit of difference: The rural-urban continuum as a system of identity', *Rural Sociology*, 57 (1): 65–82.

Bennett, John W. in association with Kohl, Seena B. and Binion, Geraldine (1982) *Of Time and the Enterprise.* Minneapolis, MN: University of Minnesota Press.

Bourdieu, Pierre (1984) *Distinction: A Social Critique of the Judgement of Taste.* Tr. Richard Nice. Cambridge, MS: Harvard University Press.

Chevalier, Sophie (1998) 'From woolen carpet to grass carpet: bridging house and garden in an English suburb', in Daniel Miller (ed.), *Material Cultures: Why Some Things Matter*, Chicago: University of Chicago Press. pp. 47–71.

Collier, Jane F. (1986) 'From Mary to modern woman: The material basis of Marianismo and its transformation in a Spanish village', *American Ethnologist*, 13: 100–7.

Collier, Jane F. (1997) *From Duty to Desire: Remaking Families in a Spanish Village.* Princeton, NJ: Princeton University Press.

Creed, Gerald W. (2000) 'Family values' and domestic economies', *Annual Review of Anthropology*, 29: 329–55.

Csikszentmihalyi, Mihaly and Rochberg-Halton, Eugene (1981) *The Meaning of Things: Domestic Symbols and the Self.* Cambridge: Cambridge University Press.

Di Leonardo, Micaela (1987) 'The female world of cards and holidays: Women, families and the work of kinship', *Signs*, 12: 440–53.

Doherty, William J. (1997) *The Intentional Family: Simple Rituals to Strengthen Family Ties.* New York, NY: Avon Books.

Doherty, William J. (2000) *Take Back Your Kids: Confident Parenting in Turbulent Times.* South Bend, IN: Sorin Books.

Dubbink, David (1984) 'I'll have my town medium-rural, please', *American Planning Association Journal*, 50 (4): 406–418.

Dudley, Kathryn Marie (2000) *Debt and Dispossession: Farm Loss in America's Heartland.* Chicago: University of Chicago Press.

Duncan, Cynthia M. (1999) *Worlds Apart: Why Poverty Exists in Rural America.* New Haven, CT: Yale University Press.

Fink, Deborah (1986) *Open Country, Iowa: Rural Women, Tradition, and Change.* Chapel Hill, NC: University of North Carolina Press.

Firey, Walter (1946) *Social Aspects to Land Use Planning in the Country-City Fringe: The Case of Flint, Michigan.* Michigan Agricultural Experiment Station. Special Bulletin 339.

Fitchen, Janet M. (1991) *Endangered Spaces, Enduring Places.* Boulder, CO: Westview Press.

Fuguitt, Glenn V., Brown, David L. and Beale, Calvin L. (1989) *Rural and Small Town America.* New York: Russell Sage Foundation.

Gasson, Ruth M. and Errington, Andrew (1993)*The Farm Family Business.* Wallingford, UK.: ... CAB International.

Giddens, Anthony (1990) *The Consequences of Modernity.* Stanford, CA: Stanford University Press.

Goldman, Robert and Dickens, David R. (1983) 'The selling of rural America', *Rural Sociology* 48: 585–606.

Goldschmidt, Walter (1978). *As You Sow.* 2nd ed. Montclair, NJ: Allanheld, Osmun.

Halle, David. 1993. *Inside Culture: Art and Class in the American Home.* Chicago: University of Chicago Press.

Harris, Daniel (2000) *Cute, Quaint, Hungry and Romantic: The Aesthetics of Consumerism.* New York: Basic Books.

Hatch, Elvin J. (1979) *Biography of a Small Town.* New York: Columbia University Press.

Hatch, Elvin J. (1992) *Respectable Lives: Social Standing in Rural New Zealand.* Berkeley, CA: University of California Press.

Hearn, Jeff and Roseneil, Sasha (eds) (1999) *Consuming Cultures: Power and Resistance.* London, UK: Macmillian.

Herbers, John (1986) *The New Heartland: America's Flight Beyond the Suburbs and How It is Changing Our Future.* New York: Times Books.

Huxtable, Ada Louise (1997) *The Unreal America: Architecture and Illusion.* New York: The New Press.

Johnson, Kenneth M. and Beale, Calvin L. (1998) 'The rural rebound', *The Wilson Quarterly*, 22 (2): 16–27.

Kammen, Michael (1999) *American Culture American Tastes: Social Change and the 20th Century.* New York: Alfred A. Knopf.

Kemmis, Daniel (1995) *The Good City and the Good Life: Renewing the Sense of Community.* Boston: Houghton Mifflin.

Kohl, Seena B. (1976) *Working Together: Women and Family in Southwestern Saskatchewan.* Toronto: Holt, Rinehart and Winston of Canada.

Kroma, Margaret (2002) 'Gender and agricultural imagery: Pesticide advertisements in the 21st century agricultural transition', *Culture and Agriculture*, 24: 2–13.

Kunstler, James Howard (1993) *The Geography of Nowhere: The Rise and Decline of America's Man-Made Landscape.* New York, NY: Simon & Schuster.

Lawrence-Zúñiga, Denise (1999) 'Suburbanizing rural lifestyles through house form in Southern Portugal', in Donna Birdwell-Pheasant and Denise Lawrence-Zúñiga (eds), *House Life: Space, Place and Family in Europe.* New York: Berg. pp. 157–175.

Lobao, Linda and Meyer, Katherine (2001) 'The great agricultural transition: Crisis, change, and social consequences of twentieth century US farming', *Annual Review of Sociology*, 27: 103–24.

Löfgren, Orvar (1999) *On Holiday: A History of Vacationing.* Berkeley, CA: University of California Press.

Lyson, Thomas A. and Falk, William W. (eds) (1993) *Forgetten Places: Uneven Development in Rural America.* Lawrence, KS: University of Kansas Press.

Lyson, Thomas A., Gillespie, Gilbert W. and Duncan Hilchey (1995), 'Farmer's markets and the local community: Bridging the formal and informal economy', *American Journal of Alternative Agriculture*, 10: 108–113.

Miller, Laura J. (1995) 'Family togetherness and the suburban ideal', *Sociological Forum*, 10 : 393–418.

Mullin, Molly H. (2001) *Culture in the Marketplace: Gender, Art, and Value in the American Southwest.* Durham, NC: Duke University Press.

Netting, Robert McC. (1984) 'Smallholders, householders, freeholders: why the family farm works well worldwide', in Richard R. Wilk (ed.), *The Household Economy: The Domestic Mode of Production Reconsidered.* Berkeley, CA: University of California Press. pp. 221–44.

Newby, Howard (1980) 'The rural sociology of advanced capitalistic societies', in Frederick H. Buttel and Howard Newby (eds), *The Rural Sociology of Advanced Societies.* Montclair, NJ: Allanheld, Osmun. pp. 1–30.

Olson, Richard K. and Thomas A. Lyson (eds) (1999) *Under the Blade: The Conversion of Agricultural Landscapes.* Boulder, CO: Westview Press.

Palen, J. John (1995) *The Suburbs.* New York: McGraw Hill.

Pollan, Michael (1998) 'The chain saws of landscape salvation.' *New York Times*, 2 April.

Riley, Robert (2002) Personal communication.

Ritzer, George (2000) *The McDonaldization of Society: New Century Edition.* Thousand Oaks, CA: Sage.

Ritzer, George (2001) *Explorations in the Sociology of Consumption.* London: Sage.

Salamon, Sonya (1992) *Prairie Patrimony: Family, Farm and Community in the Midwest.* Chapel Hill, NC: University of North Carolina Press.

Salamon, Sonya (2003) *Newcomers to Old Towns: Suburbanization of the Heartland.* Chicago: University of Chicago Press.

Salamon, Sonya, Farnsworth, Richard G., Bullock, Donald L. and Yusuf, Raji (1997) 'Family factors affecting adoption of sustainable farming systems', *Journal of Soil and Water Conservation*, 52 (2): 15–21.

Salamon, Sonya, Farnsworth, Richard G., and Bullock, Donald L. (1998) 'Family, Community, and Sustainability in Agriculture', in Gerard E. D'Souza and Tesfa G. Gebremedhin, (eds), *Sustainability in Agriculture and Rural Development.* Aldershot, U.K.: Ashcroft Publishing Co. pp. 85–102.

Salamon, Sonya, McGuire, Christine J., and Basic, Marni (nd). 'Cornfields to Culs de Sac: Invasion-Succession in Exurban Chicago.'

Salamon, Sonya and Tornatore, Jane B. (1994) 'Territory contested through property in a Midwestern post-agricultural community', *Rural Sociology*, 59 (4): 636–654.

Sheehy, Colleen J. (1998). *The Flamingo in the Garden: American Yard Art and the Vernacular Landscape.* New York: Garland.

Snipp, C. Matthew (1989) *American Indians: The First in this Land.* New York: Russell Sage Foundation.

Urry, John (1995) *Consuming Places.* London: Routledge.

Wilk, Richard R. (1989a) 'Decision making and resource flows within the household: beyond the black box', in Richard R. Wilk (ed.), *The Household Economy: The Domestic Mode of Production Reconsidered.* Berkeley, CA: University of California Press. pp. 23–52.

Wilk, Richard R. (1989b) 'Houses as consumer goods: social processes and allocation decisions', in Henry J. Rutz and Benjamin S. Orlove (eds), *The Social Economy of Consumption. Monographs in Economic Anthropology*, No. 6. Lanham, MD: University Press of America. pp. 297–322.

Wilk, Richard R. (2000) 'Consuming America', in Mary Margert Overbey and Kathyrn Marie Dudley (eds), *Anthropology and Middle Class Working Families: A Research Agenda.* Arlington, VA: American Anthropological Association.

24

Consumption culture: the case of food

Mara Miele

INTRODUCTION

In recent decades the attention to forms of consumption has grown in the social sciences (Miller, 1995: 1–57), especially among those authors involved in the debate about the definition of the 'postmodern' condition. The term 'culture of consumption' has been used by Featherstone (1991) to emphasize the world of consumer goods and to show that the structural principles of this world are of ever-increasing importance to the understanding of current changes in Western societies. More specifically, over the past two decades increasing attention has been paid by social scientists to the meanings, beliefs and social structures giving shape to food consumption (Lupton, 1996; Warde, 1999). This chapter explores some of the main perspectives that provide insights for understanding the contemporary consumption practices of Western societies, practices that are specific to *affluent societies* (Fine, 1995). In particular, it focuses on practices of food consumption. Therefore the perspectives discussed here are culturally specific and address the different aspects that affect changes in food preferences, eating habits and cooking repertoires of individuals who are in the position of having good access to food and plenty of choice.

In reviewing the main perspectives in contemporary studies of consumption this chapter addresses the distinction between a conceptualization of consumers as relatively individualized actors, those whose behaviour involves the selection of distinct goods, and a conceptualization of consumers as part of 'a muddier world of embedded, inter-dependent practices and habits, probably better explicable in terms of background notions such as comfort, convenience, security and normality' (Shove and Warde, 1997). Ultimately, this reflection on food consumption practices hopes to contribute to the debate on agro-food studies where, according to Goodman, '[C]onsumption has been neglected, under-theorized, treated as an exogenous structural category, and granted 'agency' only in the economistic, abstract terms of demand' (Goodman and DuPuis, 2002: 10 in Goodman, 2002: 272).

The body of literature relevant to understanding the changes in food practices is vast, and ranges from the specific area of 'the sociology of food'[1] to broader cultural studies.[2] Nevertheless, this chapter aims at discussing only two dimensions of food consumption practices. First, the broadening of food consumption tastes (that is, the *omnivorousness thesis*) which can be opposed to a focus on hierarchical consumption (that is, the *distinction thesis*). Secondly, the apparent contradiction between the persistence of traditional or un-reflexive food consumption (that is, *ordinary consumption*), and the growing number of cases of reflexive food consumption as suggested by the emergence of a plurality of food subcultures.

Having assessed some broad theoretical elements, the second part of the chapter focuses on empirical studies of consumer attitudes and shopping behaviour towards locally produced, and organic products. These cases[3] have often been interpreted in agro-food studies as more 'reflexive' consumption practices. Such practices, it has been argued, make it possible for producers of these product types to increase the number of alternative food chains. But a growing number of empirical studies of consumption of both organic and local products show that consumers use different strategies for coping with the mismatch between the desire to consume healthy, environmental and animal-friendly produced products and their wish to consume food according to aesthetic preferences (taste) in line with time and money constraints. Such studies suggest a growing *hybridity* in food consumption practices, that is, they suggest that consumer willingness to adopt healthier food consumption styles and environmentally and socially responsible purchasing practices is often negotiated with other desires and considerations, such as a concern for design and visual presentation of everyday products, or other contingencies that affect the full experience of the products, like the places of shopping and the services associated with them. It is this ensemble of factors, the chapter argues, that largely influences the success or failure of alternative food chains and markets.

THE SOCIOLOGY OF CONSUMPTION

From the mid-1980s the sociology of consumption has emerged as a specific field of inquiry within sociological studies. The seminal works of Baudrillard, Bourdieu, Castells, de Certeau and Maffesoli have inspired a blossoming literature that, in the following decades, has developed more actively in the Scandinavian countries and in the United Kingdom than in France, where it originated. According to Gronow and Warde (2001), its emergence coincided with the maturation of mass consumption in the north and Western Europe;[4] in the ongoing debate about the postmodern condition, mass consumption was re-evaluated (Featherstone, 1995).

Two key issues characterize the past two decades of studies in the sociology of consumption: while the influential work of Bourdieu still addresses the hierarchical character of consumption, much of the class-based account of consumption has been increasingly questioned and the theory of hierarchical consumption has been criticized from several angles. New social formations (e.g. Maffesoli's *affectual tribes*[5]) or the practices of more 'individualized' and 'self-reflexive' actors, moved by personal motivations and the search for a self-identity, have been acknowledged as now having a more important role than class belonging (Bauman, 1992; Beck, 1994 et al.). Warde (1997: 8–11) has defined this change as a passage from Bourdieu's conceptualization of consumption as *habitus* to consumption as a domain of *freedom* as conceived by postmodern theorists such as Bauman (1988, 1992), Giddens (1991) and Beck (1992).

Secondly, analysis of consumption has started to acknowledge the role of emotions, dreams, desires and pleasure. This is especially evident in the work of Featherstone (1991, 1995) and his theory of the aestheticization of everyday life in contemporary societies. This aestheticization process involves several aspects. In particular, it requires the growth and development of 'artistic subcultures' which seek to break down the barriers between Art (in the traditionally understood sense) and 'everyday life'. Here Featherstone proposes the example of The Bloomsbury Group who took the position that the greatest 'goods' in life were those of personal emotions, affectations and aesthetic enjoyment. Secondly, the aestheticization process refers to the rapid flow of signs and images which saturate the fabric of 'everyday life'. This dimension is evident in the work of Collins (2004) on the '*dream world*' of department stores, theme parks and carnivalesque events. It emphasizes the creative potential of mass culture and the apparent collapse of the boundary between high and popular culture.

Thus, the debate in the sociology of consumption over the past two decades has been dominated firstly by the theme of consumption as communication and secondly by its contribution to the creation of individual identity in a context where a 'consumer ethics' has replaced a previously existing 'work ethics' (Bauman, 1988: 75). Consumption now becomes more than just the pursuit of use-values or a claim to social prestige: it is also deeply associated with the sense of self and personality, and a growing number of studies (e.g. Shields, 1992) support the thesis that in contemporary Western societies a person is likely to be defined in terms of lifestyle or form of belonging to a group or, better, a postmodern *tribe* (see below) rather than in terms of either class belonging or personal qualities.

In this approach identity is defined as an interminable project, involving not only crucial life-choices and decisions, but, equally, their translation into a narrative, a life-story. Giddens illustrates this point very clearly:

> In the post-traditional order of modernity, against the backdrop of new forms of mediated experience, self-identity becomes a reflexively organized endeavour. The reflexive project of the self, which consists in the sustaining of coherent, yet continuously revised, biographical narratives, takes place in the context of multiple choice as filtered through abstract systems. In modern social life, the notion of lifestyle takes on a particular significance. The more tradition loses its hold, and the more daily life is reconstituted in terms of the dialectical play of the local and the global, the more individuals are forced to negotiate lifestyle choices among a diversity of options. (Giddens, 1991: 5, cited in Gabriel and Lang, 1995: 86)

According to this interpretation, individuals are increasingly obliged to *choose their identities*

(Warde, 1994: 878). This being the case, it has been suggested (for instance by Bauman) that consumer choice may become a major source of personal anxiety, since the individual is now responsible for his or her choices, so for his or her mistakes. This 'production model of the self' (Munro, 1996) implies that the acquisition of goods and services has become central to personal psychological well-being. It is no longer just that certain special objects give people a sense of security and satisfaction; rather, it implies that attempts at personal self-development and self-growth increasingly entail *constant* consumption. To the extent that people can, relatively freely, re-design their selves by purchasing new outfits and forming new associations then a high level of demand for new, or rather different, goods is likely to pertain.

However, more recently Alan Warde (1996) has criticized those approaches that stress only the glamorous, self-identifying aspects of consumer culture. He has argued that other sources of identity, particularly that associated with national, ethnic, occupational and kin groups, remain strong. These forms of identity are not dependent upon shared patterns of commercial consumption. Thus, he suggests that the 'production view' of the self overemphasizes the role of cultural products (particularly media outputs and icons of fashion) at the expense of the variety of practices that create and sustain social relations of kinship, friendship and association.

Consistent with this criticism, Gronow and Warde (2001), while welcoming the new post-modern approaches for the greater insights that they provide for our understanding of a significant number of current consumption practices, at the same time underline that all these studies have concentrated on what they define as 'extra-ordinary' consumption:

> The theories of consumption inherited from the last decade of scholarly inquiry have particular emphases on choice and freedom, taste and lifestyle, identity and differentiation, image and appearance, transgression and carnival. However, these considerations left out a good deal of the substantive field of consumption. Those actions which required little reflection, which communicate little social messages, which play no role in distinction and which do not excite much passion or emotion, were typically ignored. (2001: 3–4)

They suggest that the sociology of consumption has constructed an 'unbalanced and partial account' by focusing excessively on musical taste, clothing fashions, private purchase of houses and vehicles, and the attendance at 'high' cultural performances like theatre and museums. At the same time this debate has neglected practices of 'everyday food consumption' or the 'use of water and electricity, organization of domestic interiors and listening to the radio' (2001: 3–4).

In short, they propose to re-insert a concern with materiality into the studies of consumption practices.

THE DE-HIERARCHICALIZATION OF TASTES: DISTINCTION AND OMNIVOROUSNESS

Food and drink have been widely used as symbols of social position and status in sociological analyses and thus have had a high profile in consumption studies. The classic study of conspicuous consumption and status symbols by Thorstein Veblen in 1898 (Veblen, 1934) can be considered the first example and has been recognized as a model for the sociological analysis of consumer behaviour (Gusfield, 1994: 80). Bourdieu has more recently reinforced the view that food consumption accords with social hierarchies of various kinds in his study of distinctions between social classes in France. Yet, Bourdieu's strong emphasis on clear distinctions in taste and lifestyle based on class has more recently been criticized by several authors, both in the field of cultural studies (among others, Lamont, 1992), and from within the sociology of food (Mennel and Murcott, 1992; Warde, 1994, 1997, 1999). These studies all question Bourdieu's emphasis on a 'hierarchicalization' of tastes.

The diminishing role of class in explaining food consumption is underlined, first, in studies that point to the evolution of food consumption models towards a common standard (Mennel, 1985; Montanari, 1996). These historical analyses of the evolution of food consumption in Europe suggest that patterns of food consumption tend to run across rather than with the grain of class. The diminished role of class is also evident in studies of contemporary societies that underline a growing fragmentation in the world of food consumption, brought about by the growing sophistication of food tastes (omnivorousness, cosmopolitanism and so on – Warde, 2000).

As an illustration of the first approach we can cite Mennel (1985), who argues that social contrasts in food consumption have diminished during the later twentieth century. In his study he suggests that contrasts between classes especially, but also between regions, seasons and so forth, are less prominent. Mennel interprets the evolution of the attitude towards food in European societies from the Middle Ages as a long-term process of taming the appetite:

> In the Middle Ages there were great inequalities in the social distribution of nourishment, but in all social ranks there was an oscillating pattern of eating related to the insecurity of life in general, and food supply in particular. ... During the eighteenth century [with the general improvement of the food supply] social distinction came to be expressed more through the quality and refinement of cooking than through sheer quantitative stuffing. The change towards restraint of appetite began to be expressed in medical opinion in France and England .... In the nineteenth century the virtue of moderation and disdain of gluttony were increasingly stressed by bourgeois gastronomes, as a concern with obesity as a result of overeating began to be felt in well-to-do circles. ... The problem and the fear of fatness gradually spread down the social scale, 'slimming' becoming a prominent concern in the popular press in the twentieth century. The social standards of expected self control over appetite have developed so that they make much greater demands on individual people than formerly, and the growing incidence of eating disorders like anorexia and bulimia afflicting a minority appears to be related to these changes in social standard of the majority. (1992: 49)

We can see here that social distinctions associated with class hierarchy are seemingly diminishing over the longer term. There is also evidence over the shorter term of this diminishing hierarchy. In an analysis of contemporary food consumption trends, Warde (1997) also questions Bourdieu's assumption that 'differences in taste' are used to draw distinctions from 'others'. The relevance that Bourdieu (like Veblen, and Fine and Leopold, 1993) ascribes to social class and social distinction as determining consumption practices has left under-investigated other social groupings, particularly associations based on gender and generation, but also ethnic, local and national differences. By acknowledging that 'much individual consumption behaviour is contextually determined' Warde questions the theory of the survival of a social hierarchy of taste and wonders whether 'distinction in consumption can be mapped onto class structure or indeed onto any system of social positions' (1996: 307).

Warde underlines that one of the main features of affluent European contemporary societies is the enormous increase in the variety of the commodities and services in the field of food (as illustrated by Mennel, 1985). He believes this phenomenon has important consequences: in particular, it becomes hard to read the signs of social and aesthetic classification when there are too many cultural items on display, because the proliferation of variety makes aesthetic judgement and the detection of a cultural hierarchy more difficult (Warde, 1997). In a study on the diffusion of ethnic restaurants in the UK, Warde identifies three main consumer attitudes or strategies towards food which address the different social meanings of ethnic cuisine: omnivorousness, cosmopolitanism and distinction. In the sociology of culture the term cultural omnivorousness has been used by Peterson (1992) in order to describe the process by which people develop an appreciation and knowledge of an increasingly large number of cultural genres in different fields, for example, music[6] or food. Cultural omnivorousness has been interpreted as a move away from snobbish claims of exclusivity on the basis of an appreciation of high culture. It represents an attitude of denial and illegitimacy of any possibility of 'distinctions' in Bourdieu's sense. As Warde points out:

> Bourdieu presumes that cultural and social hierarchies coincide. Thus food tastes position people in social locations that reflect the social hierarchy. However, it is difficult to establish empirically whether there is a commonly acknowledged hierarchy of taste, for plurality of practices need not necessarily imply a relationship of superiority and inferiority. It is not entirely satisfactory to conclude that merely because privileged persons engage in particular cultural practices exclusive to them that other groups recognize any special cultural merit in those practices, even while participation is a marker of social privilege. ... One problem in making that presumption today with respect to ethnic cuisine is that both omnivorousness and cosmopolitanism might themselves be marks of social distinction. It is possible that having a wide knowledge is itself the current way to express high social position, exert social closure and operate effective cultural exclusion of others who lack this form of cultural capital. (2000)

Thus, as Warde and Martens (2000: 226) emphasize, class differences have not disappeared altogether: 'The professional and managerial classes are thronging to ethnic cuisine restaurants, while poorer, working class, older, provincial people are not. Familiarity with ethnic cuisine is a mark of refinement' (Warde and Martens, 2000: 226).

IDENTITY, REFLEXIVITY AND FOOD

Postmodern approaches conceptualize consumption as a process in which a purchaser of an item is actively engaged in trying to create and maintain *a sense of identity* through the display of purchased goods. As writers such as Baudrillard suggest, contemporary consumers do not buy items of clothing, food, body decoration, or a style of entertainment, for instance, in order to express an already existing sense of who they are. Rather, people create a sense of who they are through what they consume (Bocock, 1993: 67). Mike Featherstone has incorporated some postmodern insights into conceptualization of consumption and has suggested that:

> The term 'life-style' is currently in vogue. While the term has a more restricted sociological meaning in reference to the distinctive style of life of specific status groups, within contemporary consumer culture it connotes individuality, self-expression, and a stylistic self-consciousness. One's body, clothes, speech, leisure pastimes, eating and drinking preferences, home, car, choice of holidays, etc. ... are to be regarded as indicators of the *individuality* of taste and sense of style of the owner/ consumer. In contrast to the designation of the 1950s as an era of grey conformism, a time of mass consumption, ... we are now moving towards a society without fixed status groups in which the adoption of styles of life (manifest in choice of clothes, leisure activities, consumer goods ...) which are fixed to specific groups have been surpassed. (1991: 83; emphasis added)

Yet, we should not assume from such statements that individuals are simply increasingly and relentlessly *individualized* for there is also evidence that new social groupings come into being on the back of postmodernist consumption trends. An influential commentator on this aspect of consumption is Maffesoli, who interprets the passage from modernity to postmodernity as entailing a movement from individualism to collectivism, and from rationality to emotionality. Thus, Maffesoli situates the (postmodern) individual within new social groupings such as networks of friends, acquaintances and the small groups of people we encounter at different times during the day. According to Shields, Maffesoli's 'postmodern tribes' or 'affectual tribes' are more than a residual category of social life, they are the central feature and key social fact of our experience of everyday living:

> Between the time one might leave one's family or intimates in the morning and the time when one returns, each person enters into a series of group situations each of which has some degree of self-consciousness and stability. ... friends at the office, coffee 'klatches', associations of hobbyists, 'Neighbourhood Watch' community policing, and single-issue pressure groups are all examples of neo-tribes. (Shields, 1992: IX)

These groupings provide a strong tactile, embodied sense of being together. They are regarded as *neo*-tribes because they exist in an urban world where relationships are transitory, hence their identifications are temporary as people necessarily move on and make new attachments (Featherstone, 1995: 120). The 'affectual tribes' described by Maffesoli are different from traditional tribes described in anthropology because they are characterized by weaker ties, are temporary formations, are spontaneously chosen and are not mutually exclusive.

Case study: Slow Food

A good example of such neo-tribalism within the food arena is Slow Food, a movement that combines local organizations with broad conceptual aims. As Miele and Murdoch (2002) explain, this organization emerged out of the cultures of food that surround regional cuisines in Italy. It was established in 1986 in Bra, a small town in the Piedmont region, by a group of food writers and chefs. The immediate motivation was growing concern about the potential impact of McDonald's on local food cultures. The first Italian McDonald's had opened the previous year in Trentino Alto Adige, a region in the North-East. It was quickly followed by a second in Rome. This latter restaurant, because of its location in the famous Piazza di Spagna, gave rise to a series of protests (see Resca and Gianola, 1998 for a full account) and these protests provided the spur for the founding of Slow Food. As Renato Sardo, the director of Slow Food International, put it:

> There was a lot of public debate at the time [1986] about standardisation, the McDonaldisation, if you will, of the world. Up until then, any opposition was split in two. On the one hand there were the gastronomes, whose focus was fixed entirely on the pleasure of food. The other tradition was a Marxist one, which was about the methods of food production and their social and historical implications. Carlo Petrini, Slow Food's president, wanted to merge the two debates to provide a way forward. (quoted in *Observer Food Monthly*, 11 November 2001, 'Slowly does')

The movement's founders were concerned that the arrival of McDonald's would threaten not the growing up-market restaurants frequented by the middle/upper-class city dwellers, but local *osterie* and *trattorie*, the kinds of places that serve local dishes and which have traditionally been frequented by people of all classes. Because, in the Italian context, traditional eateries retain a close connection to local food production systems, Slow Food argued that their protection required the general promotion of local and typical foods.

In taking forward these concerns, Slow Food was established on the basis of a local structure, coordinated by a central headquarters in Bra. The local branches effectively engage in a range of activities aimed at strengthening local cuisines and attract people from very different backgrounds but with a common passion for food. These branches were initially established in all the Italian regions (and were called *condotte*) but soon began to spread to other European countries and then further afield (where they are called 'convivia'). In 1989 Slow Food was formally launched as an international movement. In that year representatives from twenty countries attended a meeting in Paris and agreed both an international structure and a manifesto. The manifesto asserted 'a firm defence of quiet material pleasure' and stated 'Our aim is to rediscover the richness and aromas of local cuisines to fight the standardisation of Fast Food.' The movement thus began to establish itself outside Italy, and as Miele and Murdoch (2002) note, convivia exist in 40 countries and the movement has around 80,000 members.

Every Slow Food group in Italy (*condotte*) usually consists of a group of promoters and a more or less stable, but flexible, group of friends, supporters and participants. They meet periodically in order to organize theme dinners, food and wine tours, tasting courses in connection with local restaurants, eno-gastronomic tourism, local food conventions, taste labs in schools and other sites, and a wide range of initiatives connected with the discovery or promotion of local products (for example, buying groups). Since it began to identify the importance of local cuisines in maintaining food diversity, Slow Food has also become aware of the problems faced by the producers and processors of typical products. It has therefore begun to play a more direct role in the protection and promotion of such products under an initiative called the 'Ark of Taste'. The Ark was launched in the late 1990s to 'save from extinction' such typical foods as cured meats, cheeses, cereals, vegetables and local breads (Petrini et al., 2001). As part of this project, Slow Food has begun a major 'census' of quality small-scale agro-industrial production and has encouraged Slow Food *osterie* and *trattorie* (that is, those listed in *Osterie d'Italia*) to include the products in their dishes. The aim is to 'protect the small purveyors of fine food from the deluge of industrial standardization'. To assist these small purveyors the movement has launched another organizational tier at the local level (called 'presidia'). These local Ark groups are charged with providing various types of practical assistance to small producers of typical products (for example, organizing commercial workshops, identifying new marketing channels).[7]

The Slow Food movement provides an example of what Featherstone has described as the process of aestheticization of everyday life where the knowledge about traditional cuisines and typical products is perceived and practised by its members as a form of art. It can be interpreted as an example of a new 'affectual tribe' of food lovers – brought together by a common passion for local and diverse foods and concerns about the 'extinction' of such foods. As one member of this tribe, Alberto Capatti (1999: 4) puts it, 'food is a cultural heritage and should be consumed as such'. Thus, for Slow Food an aesthetic appreciation of food requires an appreciation of the temporal flow of food from the past into the present. 'Slow Food', in Capatti's view, 'is profoundly linked to the values of the past. The preservation of typical products, the protection of species from genetic manipulation, the cultivation of memory and taste education – these are all aspects of this passion of ours for time' (1999: 5). Slow Food, as underlined by Capatti's words, also provides a good example of current de-hierarchicalization of food tastes and increasing omnivorousness: the celebration of diversity in regional cuisines, often consisting of 'poor dishes', such as the florentine zolfino beans soup or 'low status' ingredients, such as pork fat (e.g. *lardo di Colonnata*) promoted by the many local tribes (convivia), is a direct challenge to a priori classifications of 'refined' or 'superior' foods or dishes described by Bourdieu as fixed symbols of status used to reinforce class distinction (1984).

BETWEEN REFLEXIVITY AND ORDINARY CONSUMPTION, EXAMPLES OF FARMERS' MARKETS AND ORGANIC CONSUMPTION RE-LINKING WITH RURAL PRODUCTION

The example of Slow Food and other such new food movements suggests that in recent times there has been an enormous increase in moral discourses about the symbolic meanings of food (see also Mennell et al., 1992). An area that has generated a great deal of consumer concerns and lifestyle attention is the complex of ideas and values associated with terms such as organic, fairly traded, animal friendly and typical foods (Murdoch and Miele, 2004a, 2004b). The common feature of these ideas and values is the belief that the mass marketing of foods has led to a sameness and tastelessness that not only has diminished 'authenticity' and connection with specific localities through industrial production, but also exploits farm animals and producers in Third World countries.

It has been argued that these new moral and ethical concerns in industrialized countries have been reflected in the emergence of the alternative agro-food chains (Arce and Marsden, 1993; Ilbery and Kneafsey, 1999; Murdoch and Miele, 1999) and in an increase in the market share of organically produced, typical and fair trade products (DuPuis, 2000). An example of this position can be found in Sage, who argues:

> Recovering a sense of morality within the food and agriculture sector is arguably one of the important emerging characteristics of alternative food networks. One need not subscribe to claims for a 'widespread cultural rebellion over industrialized food production methods' to recognize the considerable unease, particularly amongst higher-income consumers, with the standardized mode of supply that engendered such fears over food safety. (2003: 49)

It is thought such values and concerns have inspired the birth and/or the growing popularity of a number of contemporary social movements such as the vegetarian associations, boycott campaigns, animal welfare groups, fair trade movements and buying clubs[8] for collective shopping of locally, ethically produced and fairly traded goods. However, while these social movements make widely supported ethical statements about animal rights, environmental sustainability, social responsibility and fair condition of exchange[9] and are increasingly successful in enrolling in their discourse a larger group of supporters,[10] they seem to be less successful when they make calls for more reflexive and ethically informed shopping behaviour to the broader public. This observation, which highlights a need to balance the reflexive and routine aspects of consumption, can be illustrated by reference to two examples: farmers' markets and organic consumption.

Case study: farmers' market

The recent popularity of direct selling outlets such as farmers' markets has been frequently associated with a desire to buy organic and/or locally produced products, an interest in supporting local economies and the possibility of a face-to-face relationship between consumers and producers. Farmers' markets have been acknowledged to 'hold the potential for a challenging of conventional production, retail and consumption patterns by alternatives which embrace discourses of the local, environmental awareness and direct contact between producer and consumer' (Holloway and Kneafsey in Sage, 2003: 49). Yet the situation can often be more complex than such statements suggest. For instance, La Trobe (2001) has recently conducted a study of customers attending the Stour Valley Farmers' Markets (Kent) in the UK in order to identify the reasons for using such markets. The results of this study point to a different set of issues than the ones exemplified in the position of Holloway, Kneafsey and Sage. According to La Trobe,

> the most commonly given response, quoted by 61% of the customers interviewed, when asked why they had attended the market was '*curiosity*' or '*for something to do*'. This response is likely to stem from the novelty value attached to farmers' markets in the UK. These markets, characterized by direct selling between producers and consumers, have only returned recently to the UK, with the first one being established in Bath in 1997. Surveys of customers at farmers' markets in America have shown that one of the prime motivations for shopping at these outlets is because of the *quality and freshness* of the produce. (2001: 185; emphasis added)

Thus the novelty of the experience is a key element of consumption practice. The results of the Stour Valley study also point to the relevance of freshness and quality while shopping for fruit and vegetables by the vast majority of the consumers in the sample (57 per cent), followed by a concern for price and value (14 per cent).[11] The interviews with the market stallholders underline the importance of these elements:

> the *quality and freshness* of the produce is of the utmost importance when selling their goods at the market. There is no point in taking along inferior quality produce as it will not sell, particularly when customers have become so accustomed to the quality of supermarket produce. For superior freshness, the produce on sale at farmers' markets is harvested either the evening before, for root vegetables etc., or on the morning of the market. (La Trobe, 2001: 186; emphasis added)

Only a minority of the other responses of the consumers surveyed referred to concerns of a social or ethical nature – these included the desire to buy local produce (6 per cent), the willingness to support the market (3 per cent), the intention to support the local producers (3 per cent) and the desire to buy seasonal products (0.5 per cent). In addition, cheaper products of at least equal quality to supermarket goods are thought to be an additional benefit offered by this type of outlet.[12]

Moreover, in addition to the relevance of the 'conventional' product characteristics, the social atmosphere is also regarded as a key character of the farmers' markets. In a study conducted in the United States it emerged that the vast majority (75 per cent) of customers arrived at the farmers' markets in groups to shop, compared with only 16 per cent arriving at supermarkets in such groups. In general, customers were far more likely to engage in a social encounter in the farmers' market than they were in the supermarket

(Sommer et al., 1981 in La Trobe, 2001: 186). These findings seem to indicate the emergence of a 'neo-tribe' of farmers' markets dwellers and sellers formed around the collective experience of shopping for fresh fruit and vegetables.

Another interesting consideration emerging from this study is that while private and public institutions promote farmers' markets in order to contrast the adverse environmental effects (for example, food miles and use of preservatives in order to ensure long shelf-life in supermarkets) and adverse social aspects (for example, price squeeze applied to producers by retailers), the success or the viability of such ventures does not seem to rely (principally) upon a growing number of consumers acting upon an ethically informed choice. Rather, the markets survive by attracting a growing number of ordinary consumers interested in novelty, freshness, quality and in the opportunity for shopping with friends in a friendly atmosphere. These research findings are more consistent with theorization of 'ordinary consumption' or increasing 'omnivorousness' than with the theory of 'reflexive consumption' and indicate the emergence of a creolized practice where the *immaterial* elements of the leisure shopping time and ethical or social considerations are coupled with the *materiality* of the conventional qualities of the produce (freshness, range etc.). They also problematize theories of short food supply chains which see such chains as bringing about 'fundamentally different type of relationship [between consumers and producers] than that found in conventional food supply chains' (Sage, 2003: 49). Indeed, both the consumers' participation in and the producers' organization of these types of alternative agro-food networks is not necessarily alternative to participation in, or cooperation with, conventional agro-food networks. As reported in La Trobe's study: 'The supermarkets themselves are even "jumping on the bandwagon" and holding farmers' markets in their car park or foyer' (La Trobe, 2001: 183). The interest of supermarkets in this kind of cooperation can be interpreted as an attempt to attract a newly formed *tribe* of consumers or, more simply, as an effort to broaden their services by adding the 'atmosphere' created by the farmers' market. In either case, for the consumers, shopping at the farmers' market and shopping at the supermarket are not mutually exclusive, but, rather, complementary experiences.

Case study: organic consumption

Similar issues, pointing to the growing complexity or complementarities between routines and reflexivity in consumption practices, are apparent also in some recent studies of consumer behaviour towards organic foods. Halkier (2001), in her study of young Danish consumers of organic products, challenges the common assumption that consumer practices are 'greened' as a result of reflection or that they are 'not greened' as a result of the routine character of consumption practice (Halkier, 2001: 32). Halkier also criticizes research that divides consumers into a continuum of segments from light-green to dark-green based on the argument that while consumers begin to incorporate environmental considerations into their practices via reflected choice, these choices over time become silent and non-reversible habits. Thus Halkier underlines the complex mixture of routinization and reflexivity that characterizes organic food practices as exemplified in some of her consumer interviews:

> Morten, a twenty year old shop assistant and vegetarian, includes a large number of different environmental friendly consumption practices in his consumption. To him, buying organic foodstuffs, non-harmfull washing powders and saving water have become natural habits that he does not reflect upon any more. He simply does not reach out for cauliflower from Italy outside the Danish season, when he is out shopping: '*I just think ... then, it's just become such a natural part of my day ... it does not occur to me to buy the other goods.*' At the same time, he often introduces normative discussions in all parts of his social network about environmental considerations in consumption and what the individual ought to do: '*Ok, at least they always get a kick, if they have bought milk that's not organic, such things that are so easy to get hold of. In that way, we talk about it ...*'. Thus Morten's [environmentally friendly consumption practices] apparently are characterized by a mixture of solitary tacit knowledge and social active reflexivity. (2001: 33; emphasis added)

Another example of complex practice is identifiable in the interview of another young consumer:

> twenty years old Anders studies biology and comes from a family where organic foodstuffs were consumed routinely. He buys and uses organic goods regularly, if quality, price difference and effect on environmental problems are in a reasonable relation to each other. This he believes to be the case for oat flakes, milk and eggs, but not for meat (because of the price) and not for foreign citrus fruits (because of the environmental effect of transport). Thus, he negotiates situationally his familiar habit of using organic goods: '*With daily goods, you ought to do something, if you can afford it.*' (2001: 34)

Halkier emphasizes that any interpretation of such consumption practices as driven solely by reflexive orientations to the self may be misleading. She indicates that even committed green consumers balance their ideas about 'environmentally correct consumption' and 'healthy foods' with more mundane consideration such as convenience, price, conventional quality and taste. Halkier's study is situated within the approach described by Gronow and Warde (2001) as 'ordinary consumption'. In particular, she seems to share their view that contemporary studies of consumption not only overemphasize the role of reflexivity but, by neglecting a large part of everyday ordinary consumption, fail to acknowledge the role that routines, constraints and preferences play in most consumption practices. Halkier thus proposes the concept of 'creolized practices' and argues that 'the sharp distinction between the routinized and reflexive aspects of consumption practices could begin to be overcome' (Halkier, 2001: 27).

These empirical studies point to a mixing of the routine and reflexive aspects of consumption. Moreover, they show that consumers do not interpret the normative claim for environmental consideration in consumption as an obligatory moral rule; rather, they negotiate environmental claims in relation to different consumption practices and consumption situations. They also show that consumers balance individualization against memberships of new consumer groupings of 'neo-tribes' where people come together as a result of expressive, emotional and aesthetic concerns and implicate their consumption practices within wider normative frameworks.

CONCLUSIONS

This chapter has aimed at critically examining some of the key approaches to food consumption: from Bourdieu's analysis that underlines the role of class in determining hierarchies of tastes as symbols of socio-economic statuses to postmodern approaches theorizing the end of fixed status groups and the increasing fragmentation of cultures. The more recent theorization of growing 'omnivorousness', 'cosmopolitanism' and 'distinction' represents a key conceptualization of contemporary trends in food consumption in affluent societies in relation to the increasing abundance of food and improved possibility for choosing between different foods, different qualities, different cuisines. While this work points to growing consumer interest in novelty and to the overcoming of 'old' hierarchies of tastes, nevertheless it does not exclude the possibility of the creation of new forms of social distinctions through the display of competence about a wide range of genres in foods.

More attention has then been dedicated to the criticism of the latter postmodern approaches and it has been suggested that they place a little too much emphasis on de-traditionalization, complexity and symbolic meanings of consumption and fail to acknowledge the role of habits, routines and materiality of food. Postmodern positions such as those of Bauman (1992), Beck (1992) and Giddens (1991), conceptualize consumption as a domain of freedom and disembeddedness from traditional social groupings (primarily class) and as an activity aimed at creating a sense of (temporary and changing) identity.

Here it has also been suggested that while processes of individualization are undoubtedly occurring, new modes of socialization are also appearing in the food sector. Maffesoli's conceptualization of postmodern tribes,[13] for its stress on weak and temporary ties that keep together its members, has been used to describe the emergence of new groups of 'food lovers' or 'food critics' and the process of re-embedding of these critical or individualized postmodern consumers in different and non-mutually exclusive types of social grouping.

We have seen that theories of alternative food networks frequently argue that reflexivity is the defining characteristic of 'new' food consumers. Yet, the studies of farmers' markets dwellers and organic consumers in Denmark quoted in the previous sections indicate that such new forms of consumption retain some surprisingly traditional features. In fact, such studies indicate that no clear distinction between routine and reflexive actions can be drawn in the food consumption field. This highlights the importance of seeing consumer practices as 'hybrid' in nature: they combine a whole range of considerations, some of which are deeply thought and dearly held, others of which are unthought and unexamined. As Gronow and Warde (2001: 226) conclude, consumption in modern societies can at the same time be both reflexive *as well as* habitual. In line with their view, it is perhaps time that we paid more attention to the unexamined, routinized aspects of food consumption, for a focus on the reflexive alternatives may mislead us into thinking that food consumption has become a greater realm of freedom than it actually has.

Such conceptualization might improve the understanding of the conditions for environmentally significant consumption practices and provide a platform for an assessment of possibilities and strategies for promoting sustainability in the food chains.

NOTES

1 Fine, 1995; Lupton, 1996; Mennel, 1985; Mennel et al., 1992; Montanari, 1996; Warde, 1996, 1997, 1999, 2000.

2 Bocock, 1993; Bourdieu, 1984; Campbell, 1995; Chaney, 1996; Douglas, 1979; Featherstone, 1991; Franklin, 1999; Gabriel and Lang, 1995; Gronow and Warde, 2001; Lamont, 1992; Maffesoli, 1992.

3 As well as GMO boycotts, animal-friendly produced products, fair trade and so on.

4 According to the same authors, unprecedented private affluence, represented by high average of personal consumption, fuelled increasing household expenditure on items like leisure, clothing, domestic technologies and automobiles, but most countries were also experiencing growing difficulties in sustaining the welfare state through the fiscal system. Simultaneously, both left-centred governments (see Tony Blair's 'New Labour' philosophy in the UK and the Olive Tree coalition 'Third Way' in Italy) and more clearly right-wing parties (e.g. in Spain) shared a renewed faith in the 'market' as the best mechanism for meeting social demands of goods and services. One outcome was the ideological legitimization of the principle of consumer sovereignty – that people should be permitted, some would say obliged, to determine for themselves the means for the satisfaction of their wants without external guidance or regulation.

5 Maffesoli takes neo-tribalism to refer to the intense, episodic social attachments which take place in the midst of the generalized neutrality and transparency that characterizes postmodern society. 'Tribal groupings', writes Maffesoli (1991: 12), 'cohere on the basis of their own minor values, and ... attract and collide with each other in an endless dance, forming themselves into a constellation whose vague boundaries are perfectly fluid.' Maffesoli is referring to the momentary coherence of people in 'emotional communities' such as art festivals, theatre auditoriums and so on, but also the symbolic gathering situated around brand-names such as Nike, Apple Macintosh, Calvin Klein ... where consumers recognize tribal status but retain an essential nomadic existence (Rojek, 1995: 151–152).

6 Warde (2000) quotes the study of Peterson and Kern (1996) which shows that all groups in the population professed to have a knowledge of a greater range of types of music in 1993 than in 1983, with those with more highbrow tastes having extended their repertoires most markedly.

7 A closely linked initiative is the 'Slow Food Award', first given in October 2000 in Bologna. The award was given to biologists, fishermen and small-scale entrepreneurs whose work helps defend the world's biodiversity. And, in a conscious emulation of McDonald's, Slow Food has established a 'Slow University' which aims to spread good practice in relation to the growing, processing, preparation and consumption of typical products.

8 Such as the Italian GAS and the numerous Box schemes in the UK and in the rest of Western Europe.

9 See Henson and Harper, 2001; Miele and Parisi, 2001.

10 See, for example, the growing number of members of vegetarian associations in the EU, the number of supporters of animal rights/welfare associations, and the number of participants in GAS (Box schemes) in Italy and so forth.

11 See La Trobe, 2001: 188 table 1.

12 Another piece of recent research aimed at studying the price of organic products in a farmers' market and the price of equivalent products in supermarkets showed that 'a number of organic vegetables available at the farmers' markets were actually cheaper than their equivalent non-organic products being sold in the nearby supermarkets. Only a few items had an organic supermarket equivalent which were shown to be between 20% and 170% more expensive than in the farmers' market' (La Trobe, 2001: 186).

13 Most often employed for describing youth groups formed around fashion items or music trends.

REFERENCES

Arce, A. and Marsden, T. (1993) 'The social construction of international food: a new research agenda', *Economic Geography*, 69 (3): 291–311.

Bauman, Z. (1988) *Freedom*. Milton Keynes: Open University Press.

Bauman, Z. (1992) *Intimation of Postmodernity*. London: Sage.

Beck, U. (1992) *Risk Society: Towards a New Modernity*. London and Thousand Oaks, CA: Sage.

Bocock, R. (1993) *Consumption*. London and New York: Routledge.

Bourdieu, P. (1984) *Distinction: A Social Critique of the Judgement of Taste*. Cambridge, MA: Harvard University Press.

Capatti, A. (1999) 'The traces left by time', *Slow*, 17: 4–6.

Campbell, C. (1995) 'The sociology of consumption', in Miller, D. (ed.), *Acknowledging Consumption, a Review of New Studies*. London and New York: Routledge.

Chaney, D. (1996) *Lifestyles*. London and New York: Routledge.

Collins, C. (2004) 'I shop therefore I know that I am: The metaphysical basis of modern consumerism', in Karin M. Ekstrom and Helene Brembeck (eds), *Elusive Consumption*. Oxford: Berg. pp. 27–45.

Douglas, M. and Isherwood, B. (1979) *The World of Goods: Towards an Anthropology of Consumption*. Harmondsworth: Penguin.

DuPuis, E.M. (2000) 'Not in my body: rBGH and the rise of organic milk', *Agriculture and Human Values*, 17: 145–176.

Featherstone, M. (1991) *Consumer Culture and Postmodernism*. London: Sage.

Featherstone, M. (1995) *Undoing Culture*. London: Sage.

Fine, B. (1995) 'From political economy to consumption', in D. Miller (ed.), *Acknowledging Consumption: A Review of New Studies*. London and New York: Routledge. pp. 127–163.

Fine, B. and Leopold. E. (1993) *The World of Consumption*. Blackwell: Oxford.

Franklin, A. (1999) *Animals and Modern Cultures*. London: Sage.

Gabriel, Y. and Lang, T. (1995) *Unmanageable Consumer*. London: Sage.

Giddens, A. (1991) *Modernity and Self-Identity: Self and Society in the Late Modern Age*. Cambridge: Polity Press.

Goodman, D. (2002) 'Rethinking food production-consumption: integrative perspective', *Sociologia Ruralis*, 42 (4): 271–278.

Goodman, D. and DuPuis, E.M. (2002) 'Knowing food and growing food: beyond the production consumption debate in the sociology of agriculture', *Sociologia Ruralis*, 42 (1): 6–23.

Gronow, J. and Warde, A. (2001) (eds), *Ordinary Consumption*. London: Routledge.

Gusfield, J.R. (1994) 'Nature's body and the metaphors of food', in M. Lamont and M. Fourrier (eds), *Cultivating Differences: Symbolic Boundaries and the Making of Inequality*. Chicago: University Chicago Press. pp. 74–103.

Halkier, B. (2001) 'Routinisation or reflexivity? Consumers and normative claims for environmental consideration', in J. Gronow and A. Warde (eds), *Ordinary Consumption*. London: Routledge. pp. 25–45.

Hinrichs, C.C., Gillespie, G.W. and Feenstra, G.W. (2004) 'Social learning and innovation at retailing farmers' markets', *Sociologia Ruralis*, 69 (1): 31–58.

Ilbery, B. and Kneafsey, M. (1999) 'Niche market and regional specialty food in Europe', *Environment and Planning A*, 31: 2207–2222.

Lamont, M. (1992) *Money, Morals and Manners*. Chicago and London: The University of Chicago Press.

La Trobe, H. (2001) 'Farmers' markets: consuming local rural produce', *International Journal of Consumers Studies*, 25 (3): 181–192.

Lupton, D. (1996) *Food, the Body and the Self*. London: Sage.

Maffesoli, M. (1992) *The Time of Tribes*. London: Sage.

Mennel, S. (1985) *All Manners of Food: Eating and Taste in England and France from the Middle Ages to the Present*. Oxford: Blackwell.

Mennel, S. (1992) 'On the civilizing appetite', in M. Featherstone, M. Hepworth and B. Turner (eds), *The Body: Social Process and Cultural Theory*. London: Sage. pp. 126–156.

Mennell, S., Murcott, A. and Otterloo, A.H. (1992) The sociology of food: eating, diet and culture', *Current Sociology*, 40 (2): 1–125.

Miele, M. (2001) 'Changing passions for food in Europe', in H. Buller and K. Hoggart (eds), *Agricultural Transformation, Food and the Environment*. Basingstoke: Ashgate. pp. 29–51.

Miele, M. and Murdoch, J. (2002) 'The practical aesthetics of traditional cuisines: Slow Food in Tuscany', *Sociologia Ruralis*, 42 (4): 312–328.

Miller, D. (ed.) (1995) *Acknowledging Consumption*. London and New York: Routledge.

Montanari, M. (1996) *The Culture of Food* (trans. C. Ipsen). Rome and Bari: Laterza.

Munro, R. (1996) 'The consumption view of self: extension, exchange and identity', in Edgell, S., Hetherington, K. and Warde, A. (eds), *Consumption Matters: The Production and Experience of Consumption*. Oxford: Blackwell Publishers. pp. 248–274.

Murdoch, J. and Miele, M. (1999) 'Back to nature: changing "world of production" in the food sector', *Sociologia Ruralis*, 39 (4): 465–483.

Murdoch, J. and Miele, M. (2004a) 'A new aesthetic of food? Relational reflexivity in the "alternative" food movement', in M. Harvey, A. McMeekin and A. Warde (eds), *Qualities of Food*. Manchester: Manchester University Press.

Murdoch, J. and Miele, M. (2004b) 'Culinary networks and cultural connections: a conventions perspective', in A. Amin and N. Thrift (eds), *The Cultural Economy Reader*. Oxford: Blackwell.

Peterson, R.A. and Kern R. (1996) 'Changing highbrow taste: from snob to omnivore', *American Sociological Review*, 61: 900–907.

Peterson, R.A. (1992) 'Understanding audience segmentation: from elite and mass to omnivore and univore', *Poetics*, 21: 243–258.

Petrini, C. with Watson, B. and Slow Food Editore (2001) *Slow Food, Collected Thoughts on Taste, Tradition, and the Honest Pleasures of Food*. White River Junction, VT: Chelsea Green Publishing Company.

Resca, M. and Gianola, R. (1998) *McDonald's: una storia Italiana*. Varese: Baldini & Castoldi.

Rojek, C. (1995) *Decentring Leisure: Rethinking Leisure Theory*. London: Sage.

Sage, C. (2003) 'Social embeddedness and relations of regard: alternative "good food" networks in south-west Ireland', *Journal of Rural Studies*, 19 (1): 47–60.

Shields, R. (1992) *Lifestyle Shopping*. London: Routledge.

Shove, E. and Warde, A. (1997) 'Noticing inconspicuous consumption'. Paper presented at the ESF TERM Programme workshop on Consumption, Everyday Life and Sustainability, at Lancaster University, April.

Veblen, T. (1934 [1899]) *The Theory of the Leisure Class: An Economic Study of Institutions*. London: George Allen and Unwin.

Warde, A. (1994) 'Consumption, identity-formation and uncertainty', *Sociology*, 28 (4): 877–898.

Warde, A. (1996) 'Afterword: the future of sociology of consumption', in S. Edgell, K. Hetherington and A. Warde (eds), *Consumption Matters: The Production and Experience of Consumption*. Oxford: Blackwell. pp. 302–313.

Warde, A. (1997) *Consumption, Food and Taste: Culinary Antinomies and Commodity Culture*. London: Sage.

Warde, A. (2000) 'Eating globally: cultural flows and the spread of ethnic restaurants'. University of Lancaster.

Warde, A. (2000) 'Eating globally: cultural flows and the spread of ethnic restaurants', in D. Kalb, M. van der Land, R. Staring, B. van Steenbergen and N. Wilterdink (eds), *The Ends of Globalization: Bringing Society Back In*. Boulder, CO: Rowman & Littlefield. pp. 299–316.

Warde, A. and Martens, L. (2000) *Eating Out: Social Differentiation, Consumption and Pleasure*. Cambridge: Cambridge University Press.

25

Tourism, consumption and rurality

David Crouch

INTRODUCTION

At the centre of this chapter is a concern to establish ways in which meanings of rurality may be understood as constructed and constituted through and in terms of tourism. In part this concern seeks to accommodate consideration of how rurality, in Britain in particular, has been transformed through processes of commodification and consumption in terms of tourism; but also of the ways in which ideas of rurality inform what happens in tourism. The power of commodification is considered in relation to the power of individuals to act in processes of consumption. Familiar emphasis on the power of commodification in constituting what rurality means is shifted to a consideration of the constitution of rurality through the encounter the individual makes as a tourist consumer. The tourist is problematized as active, engaged in a complex encounter with space, ideas and desire.

A discussion is advanced on embodied practice and performance as a focus of making sense of this human constitution of contemporary ruralities. The discussion revolves around reference to the content of British rurality, but the conceptual debate is by no means so restricted. Indeed, it is argued that a series of related debates is invigorating the ways in which rurality, tourism and consumption may be understood. The discussion of how rurality is constituted in terms of meanings is developed with a consideration of how consumption may complicate interpretation of issues of identity and power in relation to the contemporary rural and, in turn, to contemporary culture and society.

It is familiar to point to the increasing significance of tourism in the rural economy, and tourism's agencies as producer, generator and power for change. Processual connections between tourism and rurality can be considered in terms of the accelerating commodification of tourism and of rural areas and the tourism industry's commercial framing of what rurality is, or how it may be imagined. The ways in which the rural has been constructed and constituted in terms of tourism has tended to be explained through metaphor, representations and its contexts. These have often been considered in terms of the tourism industry's marketing and packaging of destinations, events and activities, familiarly positioned through the language and imagery of landscape, distinctive cultures, nature and wilderness (Cloke and Perkins, 1998; Ringer, 1998). Tourism has been theorized as a powerful component of the restructuring of the contemporary economic, social and cultural world and the economy, society and culture of the rural is no exception (Lash and Urry, 1994).

Thus commodification has come to be understood as increasingly significant in understanding rurality, in, for example, the construction of rural 'attractions' as commodities and commodity forms. Discourses on tourism and the rural have prioritized the power of commodification, marketing and of mediated cultural semiotics in the production of 'the rural' for tourism and have emphasized the power of visual cues in the gaze-consumption of the rural (Urry, 2002), as well as the power of the diverse components of the tourism industry to appropriate space and the meanings of space (Britton, 1991). Thus Urry argued that the power of the commodity of rural space is embedded in the visual components of tourism sites, and the meaning of the rural is conveyed in tourism through the visual components of landscape, nature and wilderness: 'the rural' is each of these.

However, an alternative discourse on rurality emerges from work on nature and the countryside in relation to practices. With Macnaghten, Urry argues that alternative ideologies compete with the commodification of the rural that has been emphasized in the literature, for example, in tourism (1995; see also Clarke et al., 1994; Macnaghten and Urry, 1998). Yet to argue an alternative ideology may oversimplify the complexity of consumption and the rural, not least in tourism. In this chapter the power of the industry to reconstruct the rural is considered through the work the tourist does as a consumer. Moreover, a complexity of cultural mediators informs the production of rurality that tourists may consume. It is with an outline of the dynamic of these frameworks that the chapter begins.

The main orientation of this chapter is provided through a critical argument that develops recent work in terms of the diversities and complexities of processes of consumption, and therefrom, of tourism and of rurality. This discourse seeks to reposition the familiar debate concerning the rural as powerfully mediated by and constructed in dominant cultural reference points. This chapter considers ways of theorizing the dynamic character of the rural through processes of tourism consumption. Thus green politics is considered in terms of its enactment of ideas of the rural; tourist destinations are considered from the perspective of a performance of spaces and cultures not only as ideas or products but as self-actualization, identity work and active reflexive and embodied consumption.

Whilst to acknowledge performance recognizes the power of contexts and frameworks in framing ruralities, it does not situate these as detached from the dynamic processes of subjectivity and reflexivity of the individual consumer. Thus the consumer, or rather the individual enacting complex processes of consumption, becomes a significant actor in constructing and making sense of the rural. A distinction is drawn between the land use changes and new imagery and metaphors of the rural that have resulted from processes of commodification and the ways in which the rural has become refigured in contemporary culture, where meanings, metaphor, new 'products', marketing and practice interact. In developing this discourse attention is given to the consumption of space, material and visual culture. In addition possible consequences for this refiguring of the rural in terms of power and politics are considered.

TOURISM COMMODIFICATION AND CONSUMPTION SHIFTS IN THE UK

Visiting rural spaces and cultures continues to be framed by cultural worlds in which individuals live. Television, literature, film and artwork interpenetrate other fields of knowledge of the imagined, material and sensuous character of cultures and the physical contours of geography outside cities and towns – a shorthand for the complexity and paradox around which rurality may be understood. A range of cultural contexts mediates the aesthetic of contemporary life, shapes the aesthetic values and meanings that individuals make, and seek, in what they consume (Urry, 1995). One of these numerous influences is the diverse content of the tourism 'industry' – a complexity of agents and agencies, private and public, lacking in coherence. That 'industry' works within and alongside these other cultural worlds, drawing upon their fragments to present the rural as material objects for profit and enjoyment and to influence its future, for example as sustainable environments. In the process, the material and imaginative content of the rural is multiply refigured. Ideas of authenticity and of 'the new' combine in hybrid forms. There is no singular story or version of the rural but numerous 'rurals' that are seen to serve different objectives (Crouch, 1992, 1994, 1997). This complexity may be referred to as 'postmodern' but certainly reflects contested ground and the complexity of interests as well as the increasing diversities of tourism consumption.

Ideas of the traditional character of farming countryside may be combined with contemporary industrial and commercial design in theme parks and centre parks. Stories from preindustrial land ownership are communicated through a designed display of heritage as 'play'. Alton Towers theme park, constructed in the 1980s, is located in a historic rural landed estate in the English Midlands, although the house itself has been demolished. Its new content of rides and events bears little relationship to the landed character of the place. Ironically 'the rural' is significant in the ways in which attractions are promoted, and thereby draw upon pre-commodification imagery. Historic landscapes and cultures are conveyed through contemporary cultural activities such as in the re-enactment of historic events. Global sporting events are framed in reference to traditional cultural forms; wilderness countryside is used as a backdrop in the promotion of, for example, designer walking products. Old and new hotels and small towns are marketed with reference to their rural location and rural stories, and theme parks can be promoted to evoke childhood stories based in 'old countryside' (Phillips, in Crouch, 1999). Health-promoting holidays are combined with an appeal to the surrounding countryside, associating images of the healthy body with the outdoors (Macnaghten and Urry, 2001).

These developments and promotions of tourism increasingly engage other cultural forms. Traditional art and both natural and built heritage associated with ideas of 'the rural' are interposed with the content of TV and film (Mordue, 2001). 'The rural' thus becomes 'concept' in a process of theming and contemporary commodification. Resisting such contemporary refiguring of the rural, interest groups appeal to the avoidance of this kind of 'rural makeover' by insisting on the unadorned character of the rural (Clarke et al., 1994; Urry, 1995). The Eden Project in the far south-west of Britain combines an ecological (rural) mission with the re-use of a disused quarry. Here flora from distinct regions of the globe are displayed in highly stylized glass domes in the middle of rural Cornwall. Other kinds of tourism enterprise refute any effort to re-engage references to 'the rural' and utilize the availability of space as their *raison d'être*, while it is the journey from where the visitors live that creates the idea of 'going away', often a crucial component of how tourism is understood (Crouch, 2001a; Rojek, 1993). However, at the same time the distinctive character of many of these tourism developments, sites and events is undifferentiated between rural and urban areas as the design used to wrap contemporary tourism sites becomes universal. This further complicates a decisive attention to what 'the rural' is and what it is not, both in terms of imaginative and material geographies. This is complicated further when the experience of the individual, momentarily as tourist, is considered.

During the past decade several authors argued for a shift from theorizing the rural from a contextual and representational focus to one that attends to the individual subject in interaction with contexts (Crouch, 1992; Philo, 1992). The rural emerges as multiple, diverse, less constituted according to the ways in which its various features may be represented or experienced as social distinctions but more in terms of the ways in which individuals, through their actions, make sense of it. Philo (1992) argued further that the elucidation of such diverse and multiple rurals was consequent upon rural geographies' neglect of particular social groupings or identities and their distinctive consumption of the rural as rural areas, features and ideas.

More recently the notion of the rural as 'made sense' through a consideration of its particular leisure and tourism consumption has been a subject of intellectual curiosity (Coleman and Crang, 2002; Crang, 1999; Crouch, 1997, 1999). The power of consumption, as multiplicities of practices, in elucidating the rural has tended to be ignored by dominant readings of the rural that have prioritized dwelling in terms of communities and settlements (Crouch, 1997). These practices contradicted prevailing ideas of the rural because those ideas conceptualize tourism in particular as 'other'; as distillations of the rural from 'outside'; as tourists, marketeers and developers. The significance of tourism developments in rural areas to appropriate ideas of the rural from a wide cultural context suggests instead their power in making sense of rurality in contemporary Britain. Indeed to consider rurality as a process of cultural framing requires influences such as these to be considered as predominant in what the rural currently signifies, and is, in terms of economy and society (Lash and Urry, 1994). Curiously, the cultural mediation of the rural that most significantly plays on its traditional, classic, cultural construction is less the processes of traditional rural interests and is more closely identified with tourism and leisure.

Thus 'the rural' as landscape, peace, nature and quiet playground remains powerfully embedded in tourist and leisure representations, from The Grand Tour and its landscape appropriation to the move, in the past 150 years to securing sites for National Parks, in the United States and the UK (Inglis, 2000; Macnaghten and Urry, 1998). Yet these representations profoundly re-imagine the contexts to which they may refer. In drawing upon literature and other arts representations they have framed the rural in terms of its commodity for dwelling, economy and much else, through for example the reconstitution of its cultural capital. Images of the rural, such as continuity and community, have been reworked into products that offer, and are projected to deliver, these qualities during a short period of visit by the tourist. Rurality is marketed as tourism products through which contemporary popular knowledges are projected framed in imagined pre-touristic cultural contexts. Tourism and the cultural work of framing the rural is itself a traditional practice. Constable may have been born into a farming and land-owning family, over which his versions may be fought in claims of integrity over landscape, but he engaged in an international cultural circuit. Wordsworth may have been 'a local' to the Lake District, but his vision, too, transcended the local (Urry, 1995). The Pont Aven painters of northern France and others of the late nineteenth century, framing contemporary ideas of what the rural really is/was, were tourists, and their imagery was used in the commercial communication of the rural as place and culture (Lubbren, 2001).

The profound appropriation of rural artefacts, locations, events and even cultures by tourism as an industry may suggest that any notion of a 'prior rural' as pre-touristic has been profoundly reshaped and reconstituted thereby. But of course this has not been done through the exclusive action of an industry. Yet the increased circulation of multiple imagery in cultural mediation, including the work of the tourism industry, may complicate and agitate the available polysemy of rurality. Therefrom, the potential polysemy of the rural is at once both increased, through the accelerated re-mixing of its constituents, and decreased through their increasing stylization. However, in the section that follows, the character of consumption processes is explored and the distinctive ways in which representations and contexts work are discussed with a view to establishing the complex interactions of influences and practices in the production of the contemporary rural.

REFIGURING THE RURAL IN TOURIST PRACTICE AND PERFORMANCE

In this section key conceptual interventions concerning consumption are oriented towards the work of the tourist in the constitution of the rural. These are the ideas of consumption as an active process (Miller, 1998); of tourists practising space, metaphorically and materially (Crouch, 1999, 2001a), in an embodied way; and doing performance, performing space, performing the rural (Edensor, 2001). These provide ground from which to reconsider the ways in which the rural is made sense of, and through which to assess and critique the ways in which the rural is reconstituted through the multiple actions of cultural mediators, including those of the tourism industry. Prevailing interpretations of countryside, landscape and nature, components of rurality, have frequently been concentrated in the following way. The power of the culturally mediated versions of the rural through tourism and other forms is articulated through their sign-value. The tourist specializes in 'the gaze' upon the cultural artefacts he or she encounters. Thus the prefigured significance of what the rural is becomes conveyed. The power of landscape as a visual commodity is of crucial importance and whether in visual promotion (for example, tourism brochures, advertising and television) or in the visual culture of particular sites, it conveys meaning and significance (Crouch and Lubbren, 2003; Macnaghten and Urry, 1998; Urry, 1990, 2002).

However, the rural may be less the subject of its projected sign-value than constituted in a more complex process of lay knowledge. In his discussion on the constitution of social worlds and of the self John Shotter (1993) points to the importance of actions in which individuals engage themselves in relation to others, and the pathways of those actions in enabling them, inter-subjectively, to construct and negotiate their worlds. Thus individuals participate in the world through a practical ontology through doing, making accessible understanding of how the rural – as lay geographical knowledge – may be constituted. From such a perspective it becomes possible to deconstruct the work and power of visual prefigured codes and identify how they are consumed. Individual subjects may draw upon both mediated significations and contexts of their social/cultural practice in a process of constituting lay knowledge; spaces encountered during the course of doing emerging as crucial components in remembering (Radley, 1990). Memory can be taken forward, reworked, in further practice (Crang, 2001).

Ways in which individuals act are profoundly bodily in character. Their action in relation to rurality, as an idea received and constituted in practice, may be no exception. Such a claim is pursued through ideas of embodied practice and of performance and performativities. It is argued that these perspectives enable a more critical exploration of the power of pregiven constructions of the rural (such as tourism marketing and other cultural mediation) and thereby of the interactions between tourism, consumption and the rural. Of course these various components are not polarizations but all engage in complex uneven temporal circuits of action that will be summarized towards the end of this chapter.

An individual doing the rural as a tourist may unsettle the power of parameters of 'given semiotics' available to the tourist, semiotics that become reflexively consumed; their significance refigured from their possible intentions set by the industry (Lash and Urry, 1994). Tourism as space and culture is constructed not only through the tourism industry but also through much wider frameworks of culture, and these add to the disruption of tourism-specific contexts. Moreover, other realms of what rurality may mean are part-constituted through the framing of the rural in relation to its projected consumption by tourists. Thus the tourist does not act in a tourism 'bubble' but, like any other individual consuming the rural, in multiple contexts and areas of life that overflow the limits of what the rural as touristic may be constructed to mean. So, to return for a moment to Shotter, the individual, here as tourist, has to 'make sense' of the rural through complex significations. Also, that sense is made through practice and performance.

Consumption is complex. Consumption, rather than being a linear process of utilization, constituted in and from the act of purchase, is a reflexive process of making sense of things in the act of consumption that extends significantly beyond the act of purchase (Miller, 1999). Consumption is about making sense, about negotiating and awarding cultural value. It may be argued further that consumption extends to anticipation, through ways in which desires are constructed and constituted, and projected forwards to the event (of a visit) itself. Consumption concerns the 'making sense' of objects, artefacts, spaces, that extend further beyond the market commodified – the rural incompletely understood as framed in commodification. Holiday brochures may seek to deliver value and significance, but their import to the 'consumer' may be very different. The issue, then, becomes one of identifying processes through which the commodified, as idea and materiality, is engaged in what the individual does. The idea of embodied practice offers a component of how products, objects or spaces may be made sense of, informed by so-called 'non-representational' theory. Its emphasis is on the individual as mind/body construction, as not merely thinking – or gazing – but as engaged in the world simultaneously bodily. The individual thinks and does, moves and engages the world practically and thereby imaginatively, and in relation to material objects, spaces and other people. The individual is surrounded by spaces and does not merely act as onlooker. Key informing ideas of non-representational work emerge from recent reworking of the ideas of Merleau-Ponty (1962). Visiting a tract, or site, the tourist/visitor is not merely presented with a 'view', but engages the space in which he or she finds themselves in a nuanced and complex way.

Touch, a feeling of surrounding space, sight, smell, hearing and taste are worked interactively. However, this mode of embodied practice does not operate merely as a gathering device but is worked through the way the individual uses his or her body expressively – turning, touching, feeling, moving on, dwelling, engulfed by the space around the self. Paul Cloke and Harvey Perkins (1998) explored dimensions of the individual acting, and enacting space bodily in a discussion of white-water rafting. The experience of rafting is contextualized through a series of coded images. Through what the individual as embodied human being does he or she engages the world through feeling; as connector as well as receiver. White-water rafting is prefigured in tourism promotion, the rural signified in represented iconography in connection with the drama of wilderness and the idea of action. The rural is experienced as bodily movement and instant response to material artefacts whose value in the experience is engaged mentally too, and through developed competences and in relation to technology (Lorimer, 2003). Things, artefacts, views and surrounding spaces become signified through how the individual feels, and how he or she feels about them.

The character of the surrounding world is engaged, and constituted expressively. Radley (1995) points to the significance of expressivity in the way individuals do things, and thereby the way in which performance is felt in a relationship with others, objects and spaces, and things are given significance. Macnaghten and Urry (1998) identified the power of painting and poetry in the formation of an idea of rurality as landscape, nature and wilderness that positions the individual in relation to ideas of how the individual consumes space. In terms of embodied practice or bodily action the significance of poetics works also from the power of the expressive individual. In acting in a poetic way, places, events and things are given value. Soile Veijola and Eva Jokinen (1994) explored the enactment of the bodily sense through a discourse on the body and the beach, not merely as body inscribed with conventions, but active. The body has been conceptualized as object of the gaze, and in terms of tourism consumption this has often meant adornment for the purpose of projecting status, image and character. The body emerges as a site of enactment and imagination; being playful and going beyond what is evidently 'there' in an outward, rational sense as prescribed in brochures and other visitors' guides, as Birkeland explored in her discussion of the experience of the North Cape in Norway (Birkeland, 1999; de Certeau, 1984).

Furthermore, actions of the individual as a consuming tourist are frequently done in relation to other people, that is, inter-subjectively. Therefore the character of events, experiences and sites is additionally attributed through what people do in relation to others (Crossley, 1995). For tourists this may include both other tourists and people who are at that moment not tourists, who may be locals and so on as explored in the encounter and experience of different people with island destinations in Scandinavia (Crouch, 2001b). The social anthropologist Tim Ingold has used ideas of embodied practice in thinking through how individuals may thereby relate to the world through the practical things they do and how they do them, which Harré has termed 'the feeling of doing' (Harre, 1993; Ingold, 2000). Ingold develops his notion of doing in terms of 'dwelling' (Ingold, 2000). He distinguishes between ideas for things, space and so on, as prefigured and determinate, and the motor of 'dwelling' that sustains the present and future, from which contemplation and new possibilities of reconfiguring the world, in tensions and flows, can occur. The rural in tourist locations is less proscribed than encountered, actively, as is intimately considered by Anne Game (1991) in terms of Bondi Beach.

Discussion on lived experience as performance concerns the working of protocols, competences and engagement – pregiven codes, habitually repeated, conservative, working to cultural givens (Dewesbury, 2000; Tulloch, 2000). Yet it can also be potentially disruptive and unsettling, or at least have the potential of openness in refiguring space and the self in relation to those protocols (Carlsen, 1996). Individual's actions are done in relation to the self or to others, 'performed' for others and the self – including self-regulation and negotiation inter-subjectively.

This potentially unsettling character of performance centres on performativity. This awkward term seeks to identify the potential of ordinary life actions, incidental, not focused around intentions and competences. It is likely that the tourist is less working to programmed menus of the rural than adjusting his or her experience along the way, in the event. Thereby the individual can move their life forward, discover new potential in things and in what they do, as 'becoming', the possibility of making something different. Whilst 'becoming' may concern profound rearrangement of life, through numerous momentary acts that may themselves be significant (Dewesbury, 2000), individuals do not only seek to move life forward, they have the potential both to secure where they are in life and what things mean to them, and to change it: 'holding on' and 'going further', becoming something new. Temporally being a tourist can assist in such a quest. Whilst tourism promotional imagery may use the rural as a template of 'escape', or 'finding the past', the individual encounters it also on his or her own terms. Visiting Disney-rural and vacation use of summer cottages in the Lakes region of the United States offer an opportunity for a mixing of the value of what the rural is (Miller, 1999; Warren, 1999).

The expressive character in performativity is especially significant in 'becoming' (Malbon, 1999; Radley, 1995; Thrift, 1997). Tourism in the countryside may be thought through dance, an event that has both possibilities of becoming through individual enactment but also is constituted through protocols (Nash, 2000). The combination of what is expected of the individual, what the individual expects, and what he or she discovers suggests the character of tension that surrounds the encounters with space the individual performs. Dance is one practice and performance in which the tourist may well find him- or herself involved, exemplified in summer music club culture in the Mediterranean and at rock concerts in the British countryside. However, the reach of performance arguably extends to all tourist activities and embraces profoundly what it is to be a tourist. Anne Game (1991) discusses how, in a process of being a visitor to Bondi Beach and to the English Pennines, she constructs her own 'material semiotics' of places through the bodily encounters that she makes. Acknowledging the significance of body-practices it is appropriate to argue that these become embodied semiotics (Crouch, 2001a). These insights may be directed to how rurality may be understood. In exercising performativity the individual inscribes his or her own distinctive character on the artefacts, sites, actions and events, and can figure the rural. Of course it is important to note that these actions and making geographical knowledge do not happen in isolation of contexts; they may embellish, rework, or contradict them. Cultural backgrounds, contexts and the visual and material culture of rural tourist sites may be worked further and in hybrid ways.

Performativity has a reconfiguring, or reconstitutive potential that can contribute to modulating life and discovering the new, the unexpected, in ways that may reconfigure the self, enabling the unexpected (Grosz, 1999). Immediately the potential of such a perspective for understanding tourism is considerable, given the ways in which tourism can be presented and contextualized in terms of its potential for breaking free, and the exotic. Edensor (1998) has explored the performance of tourism and its spaces and identified the diverse ways in which different groups of visitors at the Taj Mahal make their way around the site, amble and queue and make sense of their experience and the spaces they occupy. These individuals are not heroic, over-riding cultural contexts, mediations and circumstances of life, but are in a constant encounter with these, too. The tourist interacts with these contexts in making sense of what space, artefacts, location is. In terms of rurality, the tourist, the individual as visitor, plays an active role in its constitution as meaning, knowledge, value. For the tourist rurality is constituted in the process of encounter.

How prefigured ideas of rurality may work through the individual's encounter is problematic. Embodied practice is engaged in the flow of reflexive thinking the individual makes (Lash and Urry, 1994) to constitute the character of their knowledge. In part at least, individuals produce their own geographical knowledge through what they do and think (Crouch, 2001a), in non-linear relation to contexts, relations that involve multiple routes rather than acting as polarities of 'context' and 'practice'. Performatively the individual may make diverse sense of the encounter with the imagination, ideas and materiality of the rural. Paradoxically, the individual can in the encounter discover the unexpected in the rural and use it to hold on to the meanings, values and identities of the rural they already feel (Crouch, 2001a, 2003), evident in vacationing with a mobile home or 'caravan' in France, the United States or the UK. There is a process of dynamic tension as different significations of the rural may be used as reassurances, to 'hold on' and as opportunities, to 'go further'.

Of course, new products and images of the rural and new events that are commercially wrapped, contextualized, in particular cultural ideas of what signifies the rural, are actively significant in their contribution to framing what rurality means, how it is understood (Crouch, 2000). Stereotypes of 'Ayre's Rock', Donegal in western Ireland or the site of Little Big Horn, revamped in new interpretations, may build on new versions of old imagery, their significance contested but also plural and hybrid (Crouch, 1994). Yet to infer from their investment power that they are predominant in remaking the rural for even the tourist would be to make the process of consumption simplistic. A framework for figuring through the power of commodification in the reconstitution of consumption is sketched in the following section.

TOURIST CONSUMPTION, IDENTITY PROCESSES AND THE POWER TO CONSTITUTE RURALITY

Different prefigured meanings of the rural may be useful resources in this process but meanings may be discovered in the process of enactment. Urry (1995) identified a distinctive framing of the rural as a significant component in the mobilization of ideology and identity in terms of 'green issues'. This version of the rural is constructed ideologically with reference to an (imagined) rural outside such 'modernizations' as tourism commodification and its presumed exploitation. This repositioning of contemporary representations of the rural is itself a motor to empower self-identity through competing claims of what the rural is or ought to be. It has a touristic character, too, in its claims on particular kinds of activity and landscape that it deems appropriate. Yet performance and its reflexive potential need not only operate from prefigured alternatives through the conscious enactment of ideologies but may be discovered and worked in and through embodied practice and the performance itself. Furthermore, what was imagined and anticipated as secure or challenging may in the encounter become the reverse, or simply something different. Old identities of the rural, tradition, continuity, reassuring archetypes and authenticity may not be encountered in the same way. Identities may be progressed or complicated in the process, repositioning the rural in the way lives are made significant and in terms of contemporary culture.

Burkitt (1999) has engaged the embodied character of this process of practical ontology and its emergent ontological knowledge and its potential role in refiguring the self and negotiating identity. Through the practical act of living in the world the individual reconfigures their identity and their identity in relation to others. It is this capacity of performativity to be productive that characterizes 'becoming', the possibility of adjusting understandings, relationships and self-actualization. It produces feeling, changes the intensity and pitch of how things are signified, what things mean. The several strands of what individuals do as a tourist are negotiated in the performance 'made sense' as body-thinking beings (Burkitt, 1999). This process of negotiating leaves room for the individual to take projected idealizations of the rural where he or she likes. Harvey distinguished 'routine' and transformative action (1996). Yet in the apparently routine there can be the transformative. Domosh (1998) suggests that routine practices can generate their own micro-politics. This can be developed in terms of performance in the negotiation of values and relations with the world and others, in what particular material and imaginative encounters with the rural mean. Thus performances can be significantly in excess of prefigured meanings, frameworks of consumption and the anticipated realization.

The power of the tourism industry to appropriate and to commodify rurality may be taken as tourism consumption's critical influence in the contemporary reconstitution of rurality and its position in contemporary culture. However, the ways in which tourists consume the rural constitute an important feature of this process and complicate an analysis of tourism, power and rurality. Thinking through tourist practice and performance offers fresh insights into the role of rurality in contemporary culture by re-engaging theory into the concrete times and places of doing tourism as a tourist and their dynamic micro-politics. Temporal moments of being a tourist need to be engaged within complex flows of representation and commodification, practice, performance and knowledge. Urry (1995) suggests that individuals may use tourism or leisure to re-engage their identities in contemporary life. Apprehending tourism consumption critically through a perspective of the embodied tourist performing and practising space problematizes the power of values and attitudes that may be projected in the commodification of the rural. The power of tourism as an industry to reconstruct the rural through its ability to appropriate meanings and restyle the rural is only part of the story. Indeed, the structure–agency framework for thinking of rurality and tourism is complicated by being non-linear, and consumers influence producers. Moreover, as Urry (2003) has suggested, there is no hierarchy of influences and connections. Indeed, there is large potential for conjoining the debate on complexity through thinking practice and performance, important loose components of flow in the circulation of influenced and influences, in making sense of the rural.

There is no reason to assert that ideas of distinctive 'rural' cultures, present or past, their cultures and their locations, are fixed and proscribed as 'givens'. The tourist arrives at his or her sense of the rural through a complex process. Encounters can refigure. Moreover encounters may be significant in refiguring individuals' identities, and where individuals understand the rural to relate to their lives, and how ideas of the rural are given purchase in contemporary culture. The power of the tourism industry to transform the materiality of the rural is difficult to contradict. Its power to reconfigure what the rural means, culturally, in contemporary culture is something altogether different, partial, and more complex. Contexts, representations of events, activities or places, and the role of mediators in their construction remain powerful. Indeed, there is potential for cultural mediators in tourism to rework their approach to figuring the value of rural locations as Crouch and McCabe (2003) discuss in terms of Goa as tourism commodity or sustainable rural space. Yet cultural contexts work in relation to the individual and vice versa, and in a way that incorporates the bodily/mental character of the individual (Crossley, 1995; Nash, 2000; Tulloch, 2000).

The individual operates in relation to the cultures in which he or she lives and finds him/herself. Doing tourism by being a tourist plays a powerful role in this process. Equally, however, in order to make sense of the working of contexts in tourism it is appropriate to consider them in relation to, rather than in isolation from, the individuals who, amongst other things, enact them. Commodified versions of landscape, nature, industry, other people's lives and activities, sites and 'destinations' are inflected in the encounter. This includes a potential responsible dialogue that seeks to exploit neither rural spaces nor the individual as consumer (Crouch and McCabe, 2003).

Moreover, many individuals settled in rural areas have been influenced both by the cultural mediations of the rural, including those used in tourism industry, and by their own tourist practices over the years, and knowledge of those practices passed on between generations. Moreover, individual tourists are not merely the object of tourist promotion, but have emerged as tourists through their own experience, in their family or with friends, as they grew up. These various components are not polarizations but all of these are engaged in complex uneven temporal circuits of action (Crang, 2001). The consumption aesthetics of the rural in contemporary culture may inform what rurality means to the individual tourist. However, the insights from embodied practice, ontological knowledge and performance suggest that the construction of the rural in its consumption is much more than this. Tourism and other cultural mediations do not act on a *tabula rasa*: they work in amongst other ideas, desires and knowledges the individual has, and may seek to practise their ideas of the rural. In practice and performance those ideas can be refigured. The individual as tourist can no longer be adequately understood as 'other', but as a significant actor in the process of making the rural and as a producer of rurality. It is argued that these perspectives enable a more critical exploration of the power of pregiven constructions of the rural (such as tourism marketing) and thereby of the interactions between tourism, consumption and the rural. To consider rurality through a perspective of process, and from the position of the individual, consumer, tourist, provides a new framework through which contemporary ruralities may be explored further. Rurality is not so much constituted as a kaleidoscope of culturally mediated images. These images are part of the encounter the tourist makes, through which rurality is made sense.

BIBLIOGRAPHY

Birkeland I. (1999) The mytho-poetic in northern travel. In D. Crouch (ed.), *Leisure/Tourism Geographies*. London: Routledge. pp. 17–33.

Britton S. (1991) Tourism, capital and place: towards a critical geography of tourism. *Environment and Planning D: Society and Space*, 9: 451–478.

Burkitt I. (1999) *Bodies of Thought: Embodiment, Identity and Modernity*. London: Sage.

Carlsen M. (1996) *Performance: A Critical Introduction*. London: Routledge.

Clarke G. et al. (1994) *Leisure Landscapes*. London: Council for the Protection of Rural England. pp. 36–53.

Cloke P. and Perkins H.S. (1998) Cracking the Canyon with the awesome foursome. *Environment and Planning D: Society and Space*, 16: 185–218.

Coleman S. and Crang M. (eds) (2002) *Tourism: Between Place and Performance*. Oxford: Berghan.

Crang M. (1997) Picturing practices: research through the tourist gaze. *Progress in Human Geography*, 21 (3): 359–373.

Crang M. (1999) Knowing, tourism and practices of vision. In D. Crouch (ed.), *Leisure/Tourism Geographies*. London: Routledge. pp. 238–256.

Crang M. (2001) Rhythms of the city: temporalised space and motion. In J. May and N. Thrift (eds), *Time/Space: Geographies of Temporality*. London: Routledge. pp. 187–207.

Crossley N. (1995) Merleau-Ponty, the elusory body and carnal sociology. *Body and Society*, 1: 43–61.

Crouch D. (1992) Popular culture and what we make of the rural. *Journal of Rural Studies*, 8 (3).

Crouch D. (1994) Home, escape and identity. *Journal of Sustainable Tourism*, 2 (1): 93–101.

Crouch D. (1997) Others in the rural: leisure practices and geographical knowledge. In P. Milbourne (ed.), *Revealing Rural 'Others'*. London: Pinter. pp. 189–216.

Crouch D. (1999) The intimacy and expansion of space. In D. Crouch (ed.), *Leisure/Tourism Geographies*. London: Routledge. pp. 257–276.

Crouch D. (2000) Leisure and consumption. In V. Gardiner and H. Matthews (eds), *Changing Geographies of the United Kingdom*. London: Routledge.

Crouch D. (2001a) Spatialities and the feeling of doing. *Social and Cultural Geography*, 2 (1): 61–75.

Crouch D. (2001b) Tourist encounters. *Tourist Studies*, 1 (3): 253–270.

Crouch D. (2003) Spacing, performativity and becoming. *Environment and Planning A*, 35: 1945–1960.

Crouch D. and Lubbren N. (eds) (2003) *Visual Culture and Tourism*. Oxford: Berg.

Crouch D. and McCabe S. (2003) In R. Dowling and D. Fennel Wallingford (eds), *Ecotourism Policy and Planning*. UK: CABI. pp. 77–98.

De Certeau M. (1984) *The Practice of Everyday Life*. Berkeley, CA: University of California Press.

Dewesbury J-D. (2000) Performativity and the event. *Environment and Planning D: Society and Space*, 18: 473–496.

Domosh M. (1998) Those gorgeous incongruities: polite politics and public space on the streets of nineteenth-century New York. *Annals of the Association of American Geographers*, 88: 209–226.

Edensor T. (1998) *Tourists at the Taj*. London: Routledge.

Edensor T. (2001) Performing tourism, staging tourism: (re)producing tourist space and practice. *Tourist Studies*, 1 (1): 59–82.

Featherstone M. (1995) *Undoing Culture – Globalization, Postmodernism and Identity*. London: Sage.

Game A. (1991) *Undoing the Social: Towards a Deconstructive Sociology*. Buckingham: Open University Press.

Grosz E. (1999) Thinking the New: of futures yet unthought. In E. Grosz (ed.), *Becomings: Explorations in Time, Memory and Futures*. Ithaca, NY: Cornell University Press.

Harre R. (1993) *The Discursive Mind*. Cambridge, Polity Press.

Hughes G. (1998) In G. Ringer (ed.), *Destinations: Cultural Landscapes of Tourism*. London: Routledge.

Inglis F. (2000) *The Delicious History of the Holiday*. London: Routledge.

Ingold T. (2000) *The Perception of the Environment: Essays in Livelihood, Dwelling and Skill*. London: Routledge.

Lash S. and Urry J. (1994) *The Economies of Signs and Space*. London: Sage.

Lorimer H. (2003) In C. Waterton and B. Szersinski (eds), *Performing Nature*. London: Blackwell. pp. 130–144.

Lubbren N. (2001) *Rural Artists' colonies in Europe, 1870–1910*. Manchester: Manchester University Press.

Macnaghten P. and Urry J. (1998) *Contested Natures*. London: Routledge.

Malbon B. (1999) *Clubbing: Dancing, Ecstasy, Vitality*. London: Routledge.

Merleau-Ponty M. (1962) *The Phenomenology of Perception*. London: Routledge.

Miller D. (ed.) (1998) *Material Culture: Why Some Things Matter*. London: Routledge.

Mordue T. (2001) XXX *Tourist Studies*, 1 (3): XIXX.

Nash C. (2000) Performativity in practice: some recent work in cultural geography. *Progress in Human Geography*, 24 (4): 653–664.

Philo C. (1992) Neglected rural geographies. *Journal of Rural Studies*, 8 (3):

Radley A. (1990) Artefacts, memory and a sense of the past. In D. Middleton and D. Edwards (eds), *Collective Remembering*. London: Sage.

Radley A. (1995) The elusory body and social constructionist theory. *Body and Society*, 1 (2): 3–23.

Ringer G. (ed.) (1998) *Destinations: Cultural Landscapes of Tourism*. London: Routledge.

Rojek C. (1993) *Ways of Escape*. London: Routledge.

Shotter J. (1993) *The Politics of Everyday Life*. Cambridge: Polity Press.

Thrift N. (1997) The still point: resistance, expressive embodiment and dance. In S. Pile and M. Keith (eds), *Geographies of Resistance*. London: Routledge. pp. 124–154.

Tulloch J. (2000) *Performing Culture*. London: Sage. p. 14.

Urry J. (1995) *Consuming Places*. London: Routledge.

Urry J. (1999) Sensing leisure spaces. In D. Crouch (ed.), *Leisure/Tourism Geographies*. London: Routledge.

Urry J. (2002) *The Tourist Gaze*, 2nd edn. London: Sage.

Urry J. (2003) *Global Complexity*. Cambridge: Polity Press.

Veijola S. and Jokinen E. (1994) The body in tourism. *Theory and Society*, 11: 125–151.

Warren S. (1999) Cultural contestation at Disneyland Paris. In D. Crouch (ed.), *Leisure/Tourism Geographies*. London: Routledge. pp. 109–126.

Williams D. and Kaltenbom B. (1999) Leisure places and modernity. In D. Crouch (ed.), *Leisure/Tourism Geographies*. London: Routledge. pp. 214–230.

26

Gender and sexuality in rural communities

Jo Little

INTRODUCTION

The discussion of identity in rural studies, whether it be in the examination of the lifestyles of particular groups or in the conceptualization of the notion of identity itself, has tended to take as a starting point the work of the early 1990s and in particular Chris Philo's 1992 article on 'neglected rural geographies'. The recognition, at this time, that rural social scientists had ignored the experiences of many of those living in the countryside and failed to acknowledge the different rural realities that existed beyond mainstream, taken-for-granted, versions of rural living, prompted an interest in how particular 'identities' engaged with and contributed to the rural community. Detail on the everyday lives of elderly people, young people, people of colour etc. was accompanied by discussions of how dominant constructions of rurality served to 'other' the values, beliefs and needs of marginalized groups.

While some commentators have bracketed women with such marginalized others, the interest in rural gender studies predates this more general concern for rural identities. The inclusion of women amongst Philo's 'neglected others' may have encouraged a welcome commitment to learning more about rural women's lives, but the idea that gender inequalities are embedded within rural social relations and that understanding the nature of and the reasons for such inequalities is a necessary task for rural researchers has been growing in momentum (albeit sporadically) since the 1970s.

The purpose of this chapter is to examine critically the development and contribution of work on gender to rural studies. In so doing the chapter highlights key phases and 'moments' in the application of gender perspectives within rural research as well as documenting particular findings in the study or rural gender identities. What is especially important here, and what distinguishes the study of women as a group from other neglected identities, is the sense in which the focus on gender challenges dominant perspectives in rural studies in a reconfiguring of ideas of power and inequality in the rural community. The chapter includes an examination of sexuality within this broader consideration of rural gender identities. It is important that sexuality is not subsumed within the issue of gender and the intention here is to look specifically at work on sexuality as a particular direction within the wider interest in gender. As is argued towards the end of the chapter, however, the relationship between the social and the sexual is highly significant in terms of understanding gender identity and is an area that is emerging in very contemporary work as critical to future rural social scientific research.

THE ADOPTION OF FEMINIST THEORY IN RURAL STUDIES

This chapter is largely about the progress that has been made in developing gender perspectives within rural studies and so it is concerned with research and writing that has provided evidence of gender difference, often within a discussion of women's lives and experiences, or debated the notion of gender inequality (or both). Before turning to this work and its contribution to rural studies, however, it is worth making some more general observations about the research context within which it emerged; to note in particular the rather traditional development of theoretical perspectives within rural studies which was manifest in an early (and, some would argue, continuing) reluctance of rural social scientists to acknowledge or investigate gender difference or to engage with feminist theory.

The charge that rural studies has been slow to respond to theoretical developments in the social and spatial sciences (Phillips, 1998) holds particular significance in the case of work on gender and feminist perspectives. As has been noted elsewhere (Cloke, 1997; Little, 2002a), the emphasis on land use and description that characterized early rural studies in geography ensured that the kinds of theoretical debates taking place elsewhere in the discipline concerning, for example, spatial science and humanism, failed to influence the nature and direction of rural geographical research. In rural sociology at this time the rural community was a much more significant focus for research but, again, remained remarkably untheorized. Theoretical debate was reserved, so it seemed, for agricultural production and, in particular, the adoption of Marxist approaches.

By the 1980s the theoretical weakness of rural studies had been recognized as a problem, especially given the major changes taking place in rural economies and societies of Western capitalist countries. A need to explain the restructuring of rural areas prompted the adoption and discussion of political economic approaches, initially in the theorization of agricultural change but, increasingly, in the discussion of the broader transformation of rural economies (see Marsden, 1998). Within such approaches little attention was paid to the lives of individuals and where the social changes associated with rural restructuring were discussed, it was generally through reference to social class. Any mention of gender at this time was in individual (and largely peripheral) accounts of women's roles (as was the case with the earlier 'community studies') and not in the application of feminist perspectives broadly across rural studies. It was not until the late 1980s that any challenge to the androcentric direction of existing theory was mounted in attempts to provide a feminist reading of social and economic change in rural communities.

The 1990s provided a more sympathetic theoretical context for the inclusion of work on gender and the development of feminist approaches to rural studies. As noted, the concern with difference and with the detail of individuals' lives in rural communities has invited a direct focus on the experiences of rural women and this has stimulated an interest in the gender differences encapsulated in those experiences. Furthermore, the recognition of gay men and lesbians as 'significant others' (see Bell and Valentine, 1995) has enriched and broadened not only the understanding of difference but also the theoretical and conceptual importance of gender and feminist approaches in explanations of difference.

While such engagement with notions of gender and sexual difference is important, rural studies remain somewhat cautious in the acceptance of theoretical approaches that prioritize these differences. As argued below, work on gender identity in a rural context has much to offer, both empirically and theoretically, the development of gender approaches within the social sciences more generally. But rather than leading debate or participating in its early introduction, rural social scientists have tended to follow the lead given elsewhere. For example, work on the body and on the construction of masculinities and feminities has only recently found its way into rural studies, despite being broadly adopted in the social sciences generally. Moreover, there is little evidence, as yet, of mainstream rural studies being significantly altered by an interest in gender and feminist approaches. In other words, there is still a sense of work on gender operating at the margins of rural studies rather than being incorporated more broadly into the centre.

Having outlined the reluctance of rural studies to fully embrace work on gender and sexuality or to provide effective support for the development of feminist theory, the remainder of this chapter will focus more positively on the progress that *has* been made in discussion of rural gender and sexual identities. The following section will briefly outline early work on women's roles in agriculture and the rural community, looking, in particular, at arguments surrounding the traditional nature of gender roles and relations underpinning such work. These arguments and the contribution they have made to understanding rural gender inequality are well established in the literature, however. Thus most attention is given in what follows to more recent explorations of rural gender identities, of different constructions of rural masculinities and femininities and of the role of sexual identity in the performance of gender in rural communities.

Gender and agriculture

The area of rural studies in which work on gender first became established in its own right was agriculture. During the 1970s and 1980s social scientists in the UK, Europe and the United States began to document the differing roles of men and women involved in agriculture and, in particular, the different responsibilities of members of the family farm household (Gasson, 1981; Sachs, 1983; Symes and Marsden, 1983). Attention was drawn to the gender division of labour within both the productive and the reproductive spheres and to the lack of recognition given to many of the tasks traditionally carried out by 'farm wives'. Women's contribution, it was argued, was central to the survival of the family farm and yet was rarely recognized in conventional analyses of agricultural production.

A number of studies were published in the United States, the UK and Europe recording the range of tasks performed by women in the everyday running of the farm, tasks that became downgraded once established as 'women's work' and seen as peripheral to the real agricultural business (see Repassy, 1991; Sachs, 1983). As well as providing useful detail on the gender division of agricultural labour, such work was couched within an important theoretical debate concerning the relationship between production and reproduction. It was argued (see, for example, Whatmore, 1990) that the lack of recognition given to women's agricultural labour was part of the broader undervaluing of women's work and its association with the reproductive activity of the household. Women were seen largely as domestic workers and their involvement in both the reproduction of the household and the productive work on the farm was controlled by patriarchal power relations. Research argued that a feminist analysis of agriculture demonstrated the patriarchal nature of not only the labour process but also the ownership of the farm business and rights over property. It also showed, particularly in the United States, how women's contribution changed with the growing use of technology and the development of capitalist farming methods (see Fink, 1991; Friedland, 1991; Shaver, 1991).

Studies of gender roles and patriarchal gender relations within agriculture have continued to contribute to an understanding of contemporary farming lifestyles and economies (see Almas and Haugen, 1991; Alston, 1990; Teather, 1996). During the 1980s, as debate focused on the transformation of capitalist agriculture, research showed how women's participation in farm diversification became crucial to the survival of many farm businesses. Again, it was argued that such women's contribution through activities such as bed and breakfast and farm tourism was peripheral to the real economic activity on the farm, even where the income generated was necessary to the survival of the farm itself (Bouquet, 1984; Bouquet and Winter, 1987; Gasson and Winter, 1992).

Interest in gender and agriculture has developed since the early work on women's roles and power relations on the farm to include a variety of themes and approaches. Some of this recent work on constructions of masculinity and femininity within agriculture is considered below in a discussion of more contemporary theoretical discussions of rural gender and sexual identity. In the United States and Australia in particular, a body of work has developed on gender and sustainable agriculture. Such work, some of it adopting an eco-feminist perspective, has focused on the involvement of women in environmentally friendly and organic agriculture in terms of both work on the farm (see Peter et al., 2000; Sachs, 1996) and in the wider organizational networks which support farmers in these areas (Liepins, 1998). Other work has remained more traditional, however, and continues to draw attention to the undervaluing of women's agricultural labour and the gender division of labour on the farm. Some of this research has attempted to put a contemporary spin on more conventional studies of women and agriculture in a development of debates on patriarchy within the farm household and in farming organizations. Bennett (2001, 2005), for example, has revisited questions of decision-making within the family farm enterprise, arguing that 'patriarchal webs' can be identified through which control of women's decision-making is exercised by senior members of the extended farm family household. Shortall (2002) has applied theories of the state to the discussion of farm women's organizations and to policies surrounding the allocation of state funding to farm businesses. Studies such as these demonstrate the continuing need to pay attention to gendered power relations within agriculture with reference to both established and more recent theoretical debates.

Gender relations and the rural community

Work on gender in the context of the wider rural community was rather more sporadic in its early days and while the various 'community studies' so central to the writing of rural sociologists in the 1960s and 1970s often included specific reference to women's lives in the village, there was little sustained consideration of gender divisions in rural society. Despite this lack of a coherent body of work on gender, however, the 1970s saw the publication of an article that I believe was not only central to early debates on women's role in the rural community, but continued to influence ideas about gender and rurality over the next couple of decades if not beyond. This was a chapter by Davidoff, L'Esperance and Newby entitled 'Landscapes with figures: home and community in English society', published in a book edited by Mitchell and Oakley (1976).

In this article Davidoff et al. went beyond a description of men and women's lives in the rural community in the development of a theoretical framework for the conceptualization of rural gender relations. These authors argued that rural women's role was shaped by a powerful domestic ideology and that their centrality to the family was mirrored by their role as 'lynchpins' of the rural community. While women's position at the heart of the rural family and community was seen as vital to the continued existence and symbiosis of both, that position was ascribed and maintained by patriarchal gender relations. Davidoff et al.'s chapter was neither long nor particularly complicated. It did, however, succeed in making a number of very important arguments. First, that women's and men's roles in the rural community continued to be structured by, and reinforced, a powerful set of patriarchal gender relations in which men, regardless of their position within the agricultural community, controlled women's roles within both the household and the village. Secondly, Davidoff et al.'s work specifically located gender relations as part of the social and cultural construction of the rural community. Thirdly, and related to this second point, they maintained that there was an identifiable *rural* ideology at work and that the beliefs, assumptions and values surrounding the rural community shaped gender relations in an particular way.

It was the conventional and traditional nature of rural gender relations that Davidoff et al. saw as so important. These have played, and continue to play, a very pivotal role in the notion of the rural idyll. The association between the family and the rural community has always been very strong; the sense of friendliness, honesty, supportiveness, tradition etc. celebrated as part of the rural community being frequently linked to the survival of the family. Davidoff et al. linked this centrality of the family directly to the gender relations encapsulated within it showing how women's domestic role underpinned not only the rural family but also the rural community. On both a practical level in the tasks ascribed to women in a 'voluntary' capacity, and an ideological level, the rural community drew on and sustained women's primary roles of wives and mothers.

> The underlying theme of 'home' was also the quest for an organic community; small, self-sufficient and sharply differentiated from the outside world. Like the village community it was seen as a living entity, inevitably compared to the functional organs of a body, harmoniously related parts of a mutually beneficial division of labour. The male head of this natural hierarchy, like the country squire, took care of and protected his dependants. (Davidoff et al., 1976: 152)

Following the publication of Davidoff et al.'s article, more work directly focusing on women and the rural community began to emerge in the UK, the United States and Australia. While such work was often rather descriptive, mainly outlining the practical constraints imposed on women's roles by rurality (particularly in relation to problems of accessibility and the poor service provision experienced by many rural communities) (see, for example, Little, 1986; Stebbing, 1984) much was underpinned by a recognition of the importance of rural gender relations to the detail of men and women's day-to-day lives within the rural community (Dempsey, 1987, 1992; Poiner, 1990). Thus, work showed how rural organizations, labour markets and even spaces, for example, reinforced a traditional domestic division of labour and an ideology of family that ensured the continuation of patriarchal power relations within the village (see Middleton, 1986; Stebbing, 1984).

The more conventional nature of rural women's role was illustrated in work on the labour market and access to employment. Research published in the early 1990s claimed that while women's participation in paid work nationally had been growing rapidly over the past decades, in rural areas it remained relatively low (Little, 1991a; Rural Development Commission, 1991). Moreover, women were seen to be disadvantaged in relation to both the sorts of work they did and the choices available to them. So rural women who were in employment tended to work in jobs that were less well paid and more insecure than those done by women living in urban areas. They were also more likely to have interrupted and fragmented work histories and to work in jobs for which they were over-qualified. The research argued that while some of these employment characteristics were a result of the practical constraints operating on the rural labour market and on women's access to paid jobs outside the immediate locality, rural women's involvement in paid work was also strongly influenced by attitudes towards the relationship between the family, women's employment and their child-caring role (see Little, 1991b), attitudes that were still deeply conservative in rural communities. Similar conclusions were identified by other research on rural women's employment in other countries (see Dempsey, 1987).

Into the 1990s research and writing on rural women focused more directly on the gender and rural ideologies behind their daily lives. Such work sought to develop understanding of the traditional and domestic assumptions behind women's roles in discussion of what Hughes (1997) termed the 'cultures of rural womanhood'. Increasingly, the influences of feminist geography and feminist studies more generally began to make their mark as rural social scientists started to recognize the varying experiences of individual men and women within the dominant patterns of rural gender inequality. Studies turned from debates on the broad and general directions of patriarchal gender relations to the hybrid, relational and situated nature of different rural gendered identities. It is to those more nuanced recent directions in the study of rural gender and sexuality that this chapter now moves.

RURAL GENDER IDENTITIES

In discussing rural gender identities this section turns to more contemporary writing by rural social scientists. Some of this work focuses on difference in the experience of gender – not simply difference between urban and rural and men and women but within these categories – while some also incorporates issues of sexuality and the relationship between gender and sexual identity. Central to the ideas presented here is that gender identity is not fixed or given but is made and re-made constantly, in response to countless different national and local sets of meanings and practices. While this work develops considerably beyond the directions of the 1970s and 1980s discussed above, it owes much to earlier attempts to illuminate gender inequality and to identify gendered power relations as significant to the social construction of the rural community.

Studies of gender in relation to the community and labour market revealed that while differences clearly existed between the experiences of men and women in rural areas, they could also be seen in the day-to-day lives of groups and individuals within those overarching categories. It had always been recognized in, for example, studies of access to transport, that there were differences in the extent to which different women experienced the constraints of their gender roles, depending on, for example, class position. Now, however, what was being suggested (in line with feminist debate elsewhere) was that the whole experience of being a man or a woman varied according to these differences.

One evident difference which was seen to transform the experience of rurality was age and some of those involved in research on rural youth argued that the gender identities of young women in particular varied significantly from their mothers. Hughes (1997), in her discussion of rural familial ideologies, noted the greater flexibility allowed to younger women in terms of the performance of their domestic roles and questioned the extent to which 'cultures of rural womanhood' retained their significance for younger women. In a more recent study, Little and Morris (2002) drew attention to changing patterns of paid work amongst rural women suggesting that traditional attitudes towards rural women's role in the labour market may be changing. Leyshon (2002), in research on rural youth in South-West England, found that young girls felt marginalized in relation to the dominant feminine identities available to them in the spaces of the village.

Beyond examining some of the key characteristics of rural gender identities and showing the variation between different groups and individuals, there have been few attempts to explore the significance of the notion of gender identity to the study of rural gender or to a broader theoretical understanding of social relations in rural communities. The kinds of debates seen elsewhere in geography and the social sciences concerning the concept of gender identity and the tension between the fluid and shifting nature of the idea of gender as a form of identity and the more fixed and predictable characteristics of gender as a political category were not confronted in a rural context. Thus there has been little questioning, within the adoption of the concept of gender identity in rural studies, of the validity of continuing to refer to 'rural women' or 'rural men' as a specific, definable group. In other words, whether or not the recognition of difference, so central to the study of gender identity, had challenged the basic, taken for granted categories of rural gender divisions such that they were no longer relevant.

In discussing rural women's involvement in paid work, I have briefly highlighted elsewhere the dilemmas surrounding the conceptualization of gender identity together with the problems of continuing to speak of 'rural women' and 'rural men' without heeding the differences within those over-arching categories (Little, 1997).

> Underlying the observations and assumptions [about rural women's identity] is a sympathy with the notion of multiple identities and a firm belief that women do not share a single gender identity. Having made this point, however, certain aspects of gender identity may be shared by women in particular places at particular times. A recognition of difference – 'the surrendering (as Gibson-Graham [1996: 213] put it) of epistemological claims about women's shared identity' – should not mean we reject any possibility of commonality between women or that we throw out the feminist project which is to challenge the relative powerlessness of women as women. (Little, 1997: 142)

Ironically, although debates around 'otherness' and exclusion have helped to raise awareness of the importance of gender divisions, it is in the consideration of marginalization itself that the differences within genders become particularly apparent. Rather than seeing rural women *per se* as marginalized or excluded from dominant constructions of rurality, it may be more appropriate to conceptualize particular elements of their identity as 'other' to such constructions. Thus, as I argue elsewhere, while it may not be particularly helpful to label rural women as a group as marginal, aspects of the gendered identities of some rural women may be shared or common.

The notion of the country woman is still an attractive one but not one, perhaps, which encapsulates, necessarily, the entire identities of many rural women. While there may be no single construction of a 'country woman' or 'country man', the work of many of those currently researching and writing on gender and rurality seems, both implicitly and explicitly, to support the idea of certain dominant and hegemonic characteristics within masculine and feminine rural gender identities. It is to an examination of these characteristics as part of the broader constructions of rural masculinities and femininities that recent work on gender identity has turned. A growing body of work is emerging in which dominant but also shifting and contested forms of, in particular, rural masculinities, are explored. Emphasis has been placed, in such writing, on the links between hegemonic forms of masculinity and the relationship between gender identity, nature and the land, as I shall go on to discuss below.

Masculinity, femininity and rurality

Much of the writing that has started to look at rural masculinity has focused on agriculture and on male farmers. The dominant theme in this work has been the role of the physical environment in the construction of hegemonic rural masculinities. The masculinity of farmers, it is argued, revolves around classic expectations of physical strength and fitness. 'Real' farmers (men and women) are tough enough to cope with extreme weather conditions and hostile landscapes and their success measured in terms of their ability to 'control' the environment rather than letting it control them (see Bryant, 1999; Liepins, 1998; Saugeres, 2002). They work in conditions in which the comforts of the body are denied – getting dirty and uncomfortable and doing without sleep are seen, similarly, as markers of a good farmer.

> The farming occupation is an example of a traditional, hegemonic masculinity because physical strength and manual work have been important characteristics of farming. The height of masculinity can be reached when men have to overcome nature in order to make a living. (Brandth, 1995: 125)

Hostility to women farmers has been shown to be related to doubts surrounding their physical capacities and the perceived strength of their bodies (see Liepins, 2000; Saugeres, 2002).

Liepins (2000) has illustrated how these hegemonic forms of masculinity are reflected in representations of farmers and the farming industry. Drawing on images contained within the Australian and New Zealand farming press, she notes the overwhelming dominance of poses that reveal and emphasize farmers' physique and, in particular, the strength of their bodies. She writes:

> Photographs ... build on a sense of active (often battling) masculinity articulated by men who exhibit 'roughness' and strength. Pictures of drenching, dagging and shearing sheep, or branding and moving cattle, are common: men are seen as active participants in agriculture, performing dirty work in muddied clothes in rugged outdoor conditions (including steep slopes, barren or rough backgrounds, and harsh weather). Moreover, common arrangements of bodies create a sense of strength and ruggedness. Men are shown holding equipment or stock with legs splayed or arms bared; veins are accentuated and muscles are tensed. (2000: 614)

Similar images have been identified in representations of masculinity in forestry by Norwegian researchers Brandth and Haugen (1998, 2000). Lumberjacks, like farmers, are generally depicted as tough, rugged and fit, their bodies able to cope with the harsh environment and the physical demands of their job. Emphasis is placed, within such representations, on the remote wilderness conditions within which the lumberjacks work and the suggestion that only particular kinds of men can survive the demands placed upon them by such an extreme version of 'nature'.

A number of studies have explored the relationship between the traditional construction of masculinity amongst farmers, with its emphasis on physical strength and power, and the increasing use of technology in agriculture – technology that can theoretically relieve farmers of many of the tasks dependent on bodily strength (see Peter et al., 2000). Brandth (1995: 126), in her discussion of gender images in tractor advertisements, argues that tractors and other large, powerful pieces of agricultural machinery are 'symbols of a type of masculinity that is often communicated through the male body, its muscles and strength'. Such machines may save farmers from some heavy manual work but they reinforce rather than subvert the traditional masculine images of power and potency. Adverts used to sell tractors and other agricultural machinery are directed to men farmers in what Brandth suggests is a 'mutual process of construction':

> The users, the men farmers, give the tractor a gender, and the tractor makes the farmers into real men. (1995: 128)

In addition, understanding these machines and becoming competent mechanics in order to use and repair them has become an important aspect of masculinity in farming (see Saugeres, 1998).

Interestingly, several authors mention the importance of risk, in relation both to the environment and to business decisions, as a feature of farming/rural masculinities (see Bryant, 1999; Ni Laoire, 2002). The dangers of the environment are seen as matched, increasingly, by those of the market place and the 'real' farmer marked not only by the ability to survive harsh climates and landscapes but also to make tough and risky business decisions. Machinery and, in particular, sophisticated and up-to-date technology are an acceptable way of reducing risks in farming; the fact that they too require some sort of mastery is indicative of the development of new and shifting forms of masculinity within agriculture.

Those writing in this area have shown how ideas about rural masculinity are underpinned by conceptual binaries in the construction and representation of gender identities and, in particular, the relationship between masculinity and femininity. Rose (1993) discusses the ways in which 'land' and 'nature' are conceptualized as feminine in their association with fertility and reproduction. Culture and science have the ability to control feminine nature and are conceptualized as masculine. Thus, as Saugeres sums up:

> men's domination and mastery over women is parallel to men's domination over nature and animals ... (G)ender differences are constructed and articulated around a system of dual opposition. In this system what is constructed as feminine is devalued compared to what is constructed as masculine. (2002: 375–6)

Challenges or alternatives to the dominant hegemonic masculinities found within agriculture are emerging, it is suggested, in response not only to increasing use of computer technology but also to new attitudes to nature and the environment. Peter et al. (2000) explore changing forms of masculinity in the context of the transition to sustainable agriculture in the Midwest of the United States. They show how farmers belonging to a sustainable farming group, although still identifying with traditional forms of masculinity and gender relations, were more open to change and to a more 'dialogic conception of masculinity'. These farmers appeared to accept a less polarized view of masculinity in which flexibility did not threaten their identities as men. Significantly, farmers belonging to the sustainable agriculture group exhibited greater willingness to give up social and environmental control on their farms; they were more likely to acknowledge the uncertainties of the weather and the market and the possibilities of making mistakes. More flexible forms of farming masculinity also had implications for attitudes towards gender roles and were found to be linked to a greater involvement and recognition of farming women.

There is evidence that the forestry industry is also becoming characterized by new forms of masculinity in which the traditional attributes of physical toughness and strength associated exclusively with natural/outdoors settings are being added to, if not replaced, by administrative and managerial control in the boardroom. Such shifts, as Brandth and Haugen (2000) point out, may not signal a reversal in dominant attitudes towards gender relations in forestry but do appear to open the way for greater involvement by women in various roles within the forestry industry.

Control of nature is also one of the dominant features of rural masculinity beyond agriculture. While there are relatively few accounts of the construction of non-farming masculinities, those that do exist show similar links between masculine identity, nature and control of the environment. Woodward, in her study of military training, demonstrates how the countryside is incorporated into the construction of what she terms the 'warrior hero' as the dominant military construction of masculinity. Conquering the fear that they feel in relation to the landscape during their training is seen as an essential step in the creation of a soldier, as is the ability to carry on as they become 'exhausted, cold, wet, hungry and injured'.

> The sheer physical challenge of the route marches and mountain running is presented as a test of one's manhood. The warrior hero must be fit enough to conquer landscapes; indeed he is literally made in the landscape of the Army's training areas. (Woodward, 2000: 651)

Woodward goes on to point out in her account of the construction of the 'warrior hero' that failure in the physical aspects of military training is defined in feminine terms. Those who cannot complete particular tasks in the allotted time or way are branded 'fairies' or 'girl guides'.

Although studies of rural masculinity are relatively few, they far outnumber those of rural femininity. The bias in terms of studies of gender relations towards a focus on women has not been repeated in research and writing on sexual identity. Morris and Evans (2001) provide one of the few discussions of rural femininity in their examination of the gendered representations of farm life in the agricultural press. Their study explores Connell's (1987) claims that the mass media present dualistic representations of hegemonic masculinity and emphasized femininity, arguing that, when applied to representations of women in the farming press, such a conceptualization is limited. In particular it fails to accommodate the complexity of representations of femininity in agriculture which tend, so Morris and Evans argue, to emphasize the 'women achievers', particularly those running their own businesses as part of the wider farm enterprise. They do concede, however, that the

> celebration of business success of farm women thinly disguises a commodification of women's reproductive roles within farm households ... [and that], at a deep level, a dualistic construction of gender identities has proved remarkably resilient. (2001: 388)

Little (2003) has also made reference to constructions of rural femininity in her work on the embodiment of heterosexual gender identities. She notes the continuing currency of traditional notions of rural women's sexual identities, in particular the *de*-sexualization of women's feminine identities and the contradiction between true country women and aggressive, sexy femininity. These ideas are returned to below in discussion of work on the rural body.

While, as noted above, the village as a site for the construction and performance of gender relations has attracted some attention, there has been very little direct reference to constructions of either masculinity or femininity within the social setting of the rural community. The exception is the discussion of homosexual masculine and feminine identities as described below. Campbell (2000), Kraack (1999) and Leyshon (forthcoming) have all considered rural masculinity within the social space of the village pub and have all shown how hegemonic forms of masculinity are reproduced in the performance of drinking, being used to gain acceptance by some and to marginalize others. While in these studies the authors stress that the kinds of masculinity that are performed in the village pub are not specifically or exclusively *rural*, they do note the importance of the pub as a highly important rural space where young men learn and reproduce these forms of masculine identity.

The various discussions of gender identity in both agricultural and non-agricultural settings have provided important insights into the performance of rural masculinities and femininities. They have drawn attention to highly stereotypical and conventional constructions and representations of masculine and feminine identities in the rural environment and community. Despite this, however, very few studies have developed these findings in discussion of sexual identity *per se* or the nature of sexual relationships amongst those living in the rural community. This is due in part to a continuing theoretical separation between gender and sexual identity but, more specifically, to a more general failure to conceptualize heterosexuality. Thus, gender has been seen as disembodied and de-eroticized and the construction and performance of particular sexual roles (and imaginings) have not been seen as informing broader notions of gender identity. Elsewhere (see Little, 2003) I have discussed how this failure to make links between rural gender identities and sexual identities relates to a wider reluctance by feminist theory to unpack the notion of heterosexuality and to fully explore the 'normalization' of heterosex within contemporary Western society. In recognition of this problem, Richardson writes:

> most of the conceptual frameworks we use to theorize human relations rely implicitly upon a naturalized heterosexuality – where (hetero)sexuality tends either to be ignored in the analysis or is hidden from view, being treated as an unquestioned paradigm. (1996: 1)

Heterosexuality's naturalization means that it is rarely acknowledged as a sexuality, as a sexual category or identification – and where it is, it is seen as a monolithic category.

Recently feminist geographers have drawn on post-structuralist writings to destabilize the simplistic division of gender and sex, challenging the notion of heterosexuality as 'normal' – and arguing that it is not a natural product of the biological urge to reproduce but rather is socially produced and maintained. Feminists such as Rich (1980) have talked about the dominance of heterosexuality as the product of the institutionalization of a range of social practices, rituals and laws, maintaining that such institution has closed other avenues of sexual expression and impelled people to accept confining and oppressive sexual identities.

Other feminist theorists, notably Butler (1990), also attack the naturalization of heterosexuality showing how heterosexual identities are constructed and constantly reconstructed through repeated performance.

There is insufficient space here for a lengthy theoretical discussion of sexual identity and the naturalization of heterosexuality in the maintenance of gendered power relations through the relationship between the sexual and the social. What is important here is to acknowledge the relevance of debates on the nature of heterosexuality itself for the construction and performance of gendered identities in rural communities. The observed importance of the family and of conventional versions of masculinity and femininity in rural areas needs to be located within a wider engagement with the notion of heterosexuality. Hubbard (2000) has argued that space and place are extremely important in terms of the range of ways in which heterosexuality is played out and while he uses these claims to support a focus on the 'scary' heterosexualities of spaces of prostitution, they can also be mobilized in the study of the more conventional sexualization of the heterosexual spaces of the village. In drawing attention to the lack of theoretical discussion around heterosexuality as a form of sexuality, theorists have noted that the same criticism cannot be levelled at homosexuality. It seems that the conceptualization of sexual identity has focused mainly on debates around the 'deviance' of homosexual identity and same-sex relationships. This is also the case in rural studies and the attention that has been given to sexual identity specifically has tended to be confined to exploring the nature and significance of homosexuality in rural areas. The next section will examine the progress that has been made by such work in understanding the relationship between gay and lesbian identities and rurality.

Homosexuality and rurality

To some extent the study of homosexuality in rural areas has parallels with the development of work on gender and rurality more generally in terms of an initial focus on the day-to-day experiences of lesbian and gay rural residents later supported by broader, more theoretically informed, debate on the nature and construction of rural homosexual identity. Work in the mid-1990s started to challenge the academic neglect of rural homosexuality in examining the lives and experiences of gay men and lesbians living in the countryside. Such work was an attempt to draw attention to the (perceived) absence of gays and lesbians in villages and to document the exclusion of those who did live there. It told of the practical difficulties facing homosexual people in rural areas; in particular the lack of support services, health provision and information for rural gays and lesbians, especially young men and women growing up in the countryside, and also of the marginality of homosexual lifestyles within mainstream lay discourses and understandings of the rural (see Bell and Valentine, 1995; D'Augelli and Hart, 1987; Kramer, 1995; Valentine, 1997). This work was stimulated and informed by the discussion of neglected rural geographies and the growing interest in the relationship between place and identity that helped to frame rural studies in the 1990s.

While many accounts of gay and lesbian rural lifestyles emphasize the more negative characteristics, particularly the lack of acceptance afforded to homosexuality by the dominant society and culture of rurality, some present the rural in a more positive light. Smith and Holt (2002), for example, have discussed the concentration of lesbian women in the Pennine village of Hebdon Bridge and noted the broad acceptance of their lifestyles by other, heterosexual, residents, and the strength that this gives the lesbian women in their daily lives.

In another study Valentine (1997: 111) describes the creation of 'lesbian lands' in the United States in the 1970s; separatist rural communities. Here women lived on communal farms in an effort to gain the 'freedom to articulate a lesbian identity, to create new ways of living and to work out new ways of relating to the environment'. As Valentine recognizes, however, these attempts by lesbian women to create their own versions of a rural idyll were ultimately unsuccessful, due in part at least to the boundaries and exclusions created by the desire for unity and common ways of rural living. Some very different positive gay rural experiences are recounted by Will Fellows (1996) in his stories of 'farm boys' living in rural/agricultural parts of the Midwest of the United States. Fellows notes the sexual freedoms associated with aspects of living and working on a farm, and discusses the gay rural farm worker as the object of attraction to gay men from the city. The opportunities that farm work provided for the development of strong, fit and tanned masculine bodies was seen, by some, as a positive attribute in the rural gay scene.

Bell (2000) develops these insights into a broader discussion of the construction of the rural (and particularly nature) as a space for certain forms of same-sex activity. Within a theoretical framework in which the urban/rural divide is 'meshed in complex and distinct ways' with the dichotomies of masculine/feminine and homosexual/heterosexual, Bell explores different meanings and practices of 'rural gay masculinity'. He draws, in his discussion on various different representations of gay masculinity, from the stereotypical 'rural sodomite' presented by some Hollywood movies, to the idyllic Edenic or Arcadian images of erotic literature and the search for deep masculinity in nature by movements such as the Gay Fairies and the mythopoetic men's movement (see Bonnett, 1996). In discussing these examples and the lived realities of the men involved within them, Bell shows how the consideration of homosexuality in rural areas needs to recognize the varying discursive and material constructions of both 'homosexual' and 'rural' and also to acknowledge the 'interplay between the symbolic and the experiential' (Bell, 2003: 179). In so doing, as he warns, we must look very closely at conceptualizations of gay and lesbian sexual identity and be wary of the dangers of essentializing such identities and their relationship with rurality.

Rural embodiment

In his consideration of the homosexual practices of rural gay men (or gay men in rural spaces), Bell touches on another under-researched area of rural studies, namely embodiment. He suggests that focusing on the embodied experience of the rural can reintroduce a material dimension to 'run alongside the emphasis on representation found in some work on rural others'. As Bell notes, the social sciences have witnessed a 'widespread rediscovery' of the body and an initial concern by those involved in gender and sexuality has developed into a much more general interest in embodiment across very diverse subject areas. Rural researchers have, however, been relatively slow to respond to this reawakening of interest and have only very recently begun to engage with either theoretical or empirical material on the body.

The absence of work on rural embodiment is somewhat surprising, as Little and Leyshon (2003) observe, since the rural provides an excellent context for the exploration of certain characteristics and tensions in the relationship between place and the performance of sexual identity. Some reference has been made to the body in discussions of rural masculinity; as noted above, such discussions have identified the importance of stereotypical representations of the male body as muscular and strong in hegemonic rural masculinities. Some of this work has commented on the sense in which the body has been seen as one side of a dualism (in binary opposition to the mind), being associated with nature, emotion and the feminine, and thus in need of control by masculine science and rationality. These ideas have been touched upon in the work of eco-feminists and developed by rural social scientists in the examination of women's role in environmental movements and in more sustainable forms of agriculture (Liepins, 1998).

Discussions of the relationship between sexuality and the body in the context of nature and the environment have not been taken up to any great extent in relation to the rural society and community. A few studies have commented on the rural body as largely absent from or disguised within rural spaces and Little (2003) has linked this absence with hegemonic heterosexuality in rural communities. Using the example of nude calendars produced by rural villagers, Little shows how the rural body is generally represented as 'unsexy'. The more aggressive and pouting sexuality usually incorporated in nude calendar images is replaced in these rural examples, so she argues, with a jocular and coy nudity that suggests that the rural body is an object of fun rather than desire. Little builds on these observations in further discussions of rural heterosexuality in which she stresses the conventional nature of heterosexual identities in rural areas and the construction of heterosexual identities as very benign and 'non-sexual'.

Leyshon (2002) also talks about the conventional constructions of rural sexuality and the body in his examination of youth identity. He draws on a series of interviews with young women living in rural communities in discussions of the rural body and shows how they regard the kinds of sexual identities available to them in rural areas as very traditional and old-fashioned. These women claim to go outside the rural area to perform aspects of their sexual identities. Clearly the link here between embodiment and sexual identity of young women has a profoundly spatial (rural) aspect – the dominant image of the rural body for young women is not one that sits easily with the identities that they perform (and aspire to). As other authors writing on the body have done, Leyshon (2002) makes use of Butler's (1990) work on the creation and recreation of gender identity through the performance of daily activities, arguing that neither gender nor sexual identities are essential or fixed but rather are confirmed through repeated, stylized and understood performance.

CONCLUSION

Although gender differences within the rural landscape and community have not gone unnoticed in the post-war development of rural studies, it is only in the past twenty years or so that an explicit focus on gender and sexuality has become an accepted part of mainstream rural research and publishing. This chapter has examined the development of work on gender from the early interest in rural gender roles and relations through to more contemporary studies of rural gender identities. Rather than provide a straight historical account of the different developments in the study of rural gender, the chapter has chosen to emphasize more recent (and I believe innovative) work on gender identity and the construction of rural masculinity and femininity. In so doing, however, it has also stressed the interconnectivity of different 'phases' of the development of rural gender studies and the continuing relevance of wider patriarchal power relations and hegemonic notions of masculinity and femininity to the more individual and fluid performance of gender identity.

The recent examination of rural masculinity and femininity has injected new vigour into the study of rural gender and sexuality. Although not significant, perhaps, in terms of volume, the emerging work on sexual identity and the rural body has been important in drawing in theoretical and conceptual ideas from beyond the rural sphere and in using these to highlight some interesting directions for future research. Rural studies has begun to recognize the relevance of work on gender identity beyond a need to identify the experiences of different (in particular, marginalized) identities in the countryside and new work emerging has shown the centrality of issues of sexual and gender identity to the cultural construction of rurality and to the assumptions and values inscribed in day-to-day social relations within the rural community. These more recent studies have varied in content and direction but are united in their call for further research to be undertaken on the formation and performance of rural gender identities.

As noted in very recent work, studies in rural gender identities have started to benefit from a move away from representation in focusing on performance and on the material practices through which gender and sexual identities are produced and sustained. Such work has included an emphasis on the body in discussing how the performance of gender relations is itself part of the production and reproduction of masculinity and femininity. There needs to be sustained attention in this area such that in the development of future research directions work on the performance of gender identity is extended beyond the discussion of sexuality *per se* and is used to inform studies of gender in the rural community more broadly. In other words, those familiar areas of gender inquiry – the labour market, the community and the rural economy – need to be revisited using more recent theoretical approaches to gender identity.

In extending the understanding of gender identities, rural social scientists also need to engage with two related areas of recent academic interest: senses and emotions. Work on the senses and on different ways of 'knowing' or experiencing the rural has been touched on in the past in different ways within a rural context – for example, in work on landscape perception – but there has been little sustained attempt to show how different senses may produce or inspire different ways of encountering rurality (but see Macnaghten and Urry, 2001). Clearly there is significant overlap here with work on the body and on how different bodily experiences of the rural may result from exploring different sensory connections. Similarly, the rural provides an important site for the examination of emotions. Again, work here is not entirely unheard of (see, for example, work on stress amongst farmers) but it needs to be brought into more mainstream rural research. Emotional responses and well-being are obviously relevant to much of what we study as rural social scientists – for example, social exclusion, marginalization and inequality within the rural community all incorporate elements of feeling and emotion while issues such as farm animal welfare and even the development of alternative food networks can be understood in the context of emotional response and need. Recently we have seen in research on rural fear and safety an attempt to focus specifically on emotions as a way of understanding behaviour and imaginings within the rural community (see Panelli et al., 2004). There is much scope for thedevelopment and extension of approaches used in such work to many other areas of rural research.

Throughout the development of rural gender studies has run the question of place – in particular, whether the characteristics and patterns being observed in the nature of gender inequality, gender relations and gender identities can be seen as specifically 'rural'. While early work on gender roles was keen to point out that rural women encountered particular problems due to the generally poor accessibility and the absence of many basic services in the countryside, later work on gender relations argued that rural women were disadvantaged as women and that rural gender relations were, like those elsewhere, a function of the operation of patriarchy and its relationship with capitalism. But in this work, and in subsequent examination of gender identity with its emphasis on individual and varying expressions of gender, there remains an underlying belief in the existence of an identifiable set of assumptions, attitudes, values and associations which shape *rural* gender in a particular way. Thus patriarchal gender relations may underpin the household division of labour in rural families just as in urban families, but they are given a particular resonance by the more traditional and conservative attitudes that exist in rural communities and by the effect these attitudes have had on the historical development of gender. More recently, the idea that rurality is a social construction has reinforced a belief in the specificity of rural gender relations and identities and encouraged arguments concerning the mutual constitution of gender and rurality. There is thus a strong defence of the importance of the relationship between gender and space in work on rural gender identities in terms of both the importance of the rural as a site for the performance of gender and also the role of rurality in the construction of gender.

REFERENCES

Almas, R. and Haugen, M. (1991) Norwegian gender roles in transition: the masculinisation hypothesis in the past and in the future. *Journal of Rural Studies*, 7: 79–84.

Alston, M. (1990) Feminism and farm women. *Australian Social Work*, 43: 23–27.

Bell, D. (2000) Farm boys and wild men: rurality, masculinity and homosexuality. *Rural Sociology*, 65: 547–561.

Bell, D. (2003) Homos in the heartland: male same-sex desire in rural USA. In P. Cloke (ed.), *Country Visions*. London: Pearson.

Bell, D. and Valentine, G. (1995) Queer country: rural lesbian and gay lives. *Journal of Rural Studies*, 11: 113–122.

Bennett, K. (2001) *Voicing Power: Women, Family Farming and Patriarchal Webs*. Working Paper 62, Centre for Rural Economy, University of Newcastle.

Bennett, K. (2005) Out of the ashes: women, work and rural economies. In J. Little and C. Morris (eds), *Critical Perspectives on Gender and Rurality*. London: Ashgate.

Bonnett, A. (1996) The New Primitives: identity, landscape and cultural appropriation in the mythopoetic men's movement. *Antipode*, 28: 273–291.

Bouquet, M. (1984) Women's work in rural south-west England. In N. Long (ed.), *Family and Work in Rural Society*. London: Tavistock.

Bouquet, M. and Winter, M. (eds) (1987) *Who From Their Labours Rest? Conflict and Practice in Rural Tourism*. Aldershot: Avebury.

Brandth, B. (1995) Rural masculinity in transition: gender images in tractor advertisements. *Journal of Rural Studies*, 11: 123–133.

Brandth, B. and Haugen, M. (1998) Breaking into a masculine discourse. Women and farm forestry. *Sociologia Ruralis*, 38: 427–442.

Brandth, B. and Haugen, M. (2000) From lumberjack to business manager: masculinity in the Norwegian forestry press. *Journal of Rural Studies*, 16: 343–356.

Bryant, L. (1999) The detraditionalisation of occupational identities in farming in South Australia. *Sociologia Ruralis*, 39: 236–261.

Butler, J. (1990) *Gender Trouble: Feminism and the Subversion of Identity*. London: Routledge.

Campbell, H. (2000) The glass phallus: pub(lic) masculinity and drinking in rural New Zealand. *Rural Sociology*, 65: 562–581.

Cloke, P. (1997) Country backwater to virtual village? Rural studies and 'the Cultural Turn'. *Journal of Rural Studies*, 13: 367–375.

Connell, R. (1987) *Gender and Power: Society, the Person and Sexual Politics*. Cambridge: Cambridge University Press.

D'Augelli, A. and Hart, M. (1987) Gay women, men and families in rural settings. *American Journal of Community Psychology*, 15: 79–93.

Davidoff, L., L'Esperance, J. and Newby, H. (1976) Landscape with figures: home and community in English society. In J. Mitchell and A. Oakley (eds), *The Rights and Wrongs of Women*. Harmondsworth: Penguin.

Dempsey, K. (1987) Economic inequality between men and women in an Australian rural community. *Australian and New Zealand Journal of Sociology*, 23: 358–374.

Dempsey, K. (1992) *A Man's Town: Inequality between Women and Men in Rural Australia*. Melbourne: Oxford University Press.

Fellows, W. (1996) *Farm Boys: Lives of Gay Men from the Rural Midwest*. Madison, WI: University of Wisconsin Press.

Fink, V.S. (1991) What work is real? Changing roles of farm and ranch wives in south-eastern Ohio. *Journal of Rural Studies*, 7: 17–22.

Friedland, W. (1991) Women and agriculture in the United States: a state of the art assessment. In W. Friedland, L. Busch, F. Buttel and A. Rudy (eds), *Towards a Political Economy of Agriculture*. Boulder, CO: Westview Press.
Gasson, R. (1981) Roles of women on farms. *Journal of Agricultural Economics*, 32/1: 11–20.
Gasson, R. and Winter, M. (1992) Gender relations and farm household pluriactivity. *Journal of Rural Studies*, 8: 387–397.
Gibson-Graham, J. (1996) *The End of Capitalism (as we knew it)*. Cambridge, MA: Blackwell.
Hubbard, P. (2000) Desire/disgust: mapping the moral contours of heterosexuality. *Progress in Human Geography*, 24: 191–217.
Hughes, A. (1997) Rurality and cultures of womanhood: domestic identities and moral order in rural life. In P. Cloke and J. Little (eds), *Contested Countryside Cultures: Otherness, Marginalisation and Rurality*. London: Routledge.
Kraack, A. (1999) It takes two to tango: the place of women in the construction of hegemonic masculinity in a student pub. In R. Law, H. Campbell and J. Dolan (eds), *Masculinities in Aotearoa/New Zealand*. Palmerston North: Dunmore Press.
Kramer, J. (1995) Bachelor farmers and spinsters: gay and lesbian identities and communities in rural North Dakota. In D. Bell and G. Valentine (eds), *Mapping Desire: Geographies of Sexualities*. London: Routledge.
Leyshon, M. (2002) Youth identity, culture and marginalisation in the countryside. Unpublished PhD thesis, Department of Geography, University of Exeter.
Leyshon, M. (2005) Making the pub the hub: power, drinking and rural youth. In J. Little and C. Morris (eds), *Critical Perspectives on Gender and Rurality*. London: Ashgate.
Liepins, R. (1998) Women of broad vision: nature and gender in the environmental activism of Australia's 'Women in Agriculture' movement. *Environment and Planning A*, 30: 1179–1196.
Liepins, R. (2000) Making men: the construction and representation of agriculture-based masculinities in Australia and New Zealand. *Rural Sociology*, 65: 605–620.
Little, J. (1986) Feminist perspectives in rural geography: an introduction. *Journal of Rural Studies*, 2: 1–8.
Little, J. (1987) Gender relations in rural areas: the importance of women's domestic role. *Journal of Rural Studies*, 3: 335–342.
Little, J. (1991a) Theoretical issues of women's non-agricultural employment in rural areas, with illustrations from the UK. *Journal of Rural Studies*, 7: 99–106.
Little, J. (1991b) Women in the rural labour market: a policy evaluation. In T. Champion and C. Watkins (1991) *People in the Countryside: Studies of Social Change in Rural Britain*. London: Paul Chapman.
Little, J. (1997) Employment, marginality and women's self-identity. In P. Cloke and J. Little (eds), *Contested Countryside Cultures: Otherness, Marginalisation and Rurality*. London: Routledge.
Little, J. (2002) *Gender and Rural Geography: Identity, Sexuality and Power in the Countryside*. London: Pearson.
Little, J. (2003) Riding the rural love train: rural community and the heterosexual family. *Sociologia Ruralis*, 43: 401–417.
Little, J. and Leyshon, M. (2003) Embodied rural geographies. *Progress in Human Geography*, 27: 257–272.
Little, J. and Morris, C. (2002) *The Role and Contribution of Women to Rural Economies*. Report to the Countryside Agency, Cheltenham.
Macnaghten, P. and Urry, J. (2001) *Bodies of Nature*. London: Sage.
Marsden, T. (1998) Economic perspectives. In B. Ilbery (ed.), *The Geography of Rural Change*. Harlow: Addison–Wesley–Longman.
Middleton, A. (1986) Marking boundaries: men's space and women's space in a Yorkshire village. In T. Bradley, P. Lowe and S. Wright (eds), *Deprivation and Welfare in Rural Areas*. Norwich: Geo Books.
Morris, C. and Evans, N. (2001) 'Cheesemakers are always women': gendered representations of farm life in the agricultural press. *Gender, Place and Culture*, 8: 375–390.
Ni Laoire, C. (2002) Masculinities and change in rural Ireland. *Irish Geography*, 35: 16–28.
Panelli, R., Little, J. and Kraack, A. (2004) A community issue? Women's feelings of safety and fear. *Gender, Place and Culture*, 11: 445–467.
Peter, G., Mayerfield Bell, M. and Jarnagin, S. (2000) Coming back across the fence: masculinity and the transition to sustainable agriculture. *Rural Sociology*, 65: 215–233.
Phillips, M. (1998) Social perspectives. In B. Ilbery (ed.), *The Geography of Rural Change*. Harlow: Addison–Wesley–Longman.
Philo, C. (1992) Neglected rural geographies: a review. *Journal of Rural Studies*, 8: 193–207.
Poiner, Gretchen (1990) *The Good Old Rule: Gender and Other Power Relationships in a Rural Community*. Sydney: Sydney University Press.
Repassy, H. (1991) Changing gender roles in Hungarian agriculture. *Journal of Rural Studies*, 7: 23–36.
Rich, A. (1980) Compulsory heterosexuality and lesbian existence. *Signs*, 5: 631–660.
Richardson, D. (ed.) (1996) *Theorising Heterosexuality: Telling it Straight*. Buckingham: Open University Press.
Rose, G. (1993) *Feminism and Geography: The Limits of Geographical Knowledge*. Cambridge: Polity Press.
Rural Development Commission (1991) *Women and Employment in Rural Areas*. Rural Development Commission Research Report 10. Salisbury, RDC.
Sachs, C. (1983) *Invisible Farmers: Women's Work in Agricultural Production*. Totowa, NJ: Rhinehart Allenheld.
Sachs, C. (1996) *Gendered Fields: Women, Agriculture and Environment*. Boulder, CO: Westfield Press.
Saugeres, L. (1998) Representations of femininity and masculinity: gender relations and identities amongst farm families in a French community. Unpublished PhD thesis, Manchester Metropolitan University.
Saugeres, L. (2002) The cultural representation of the farming landscape: masculinity, power and nature. *Journal of Rural Studies*, 18: 373–384.

Shaver, F.M. (1991) Women: work and the evolution of agriculture. *Journal of Rural Studies*, 7: 37–44.

Shortall, S. (2003) Politics, gender and the farmyard: political restructuring in Northern Ireland. In J. Little and C. Morris (eds), *Critical Perspectives on Gender and Rurality*. London: Ashgate.

Smith, D. and Holt, L. (2005) Lesbian migrants in the gentrified valley and 'other' geographies of rural gentrification. *Journal of Rural Studies*.

Stebbing, S. (1984) Women's roles and rural society. In T. Bradley and P. Lowe (eds), *Locality and Rurality: Economy and Society in Rural Regions*. Norwich: Geo Books.

Symes, D. and Marsden, T. (1983) Complementary roles and asymmetrical lives: farmers' wives in a large farm environment. *Sociologia Ruralis*, 23: 229–241.

Teather, E. (1996) Farm women in Canada, New Zealand and Australia redefine their rurality. *Journal of Rural Studies*, 12: 1–14.

Valentine, G. (1997) Making space: lesbian separatist communities in the United States. In P. Cloke and J. Little (eds), *Contested Countryside Cultures: Otherness, Marginality and Rurality*. London: Routledge.

Whatmore, S. (1990) *Farming Women: Gender, Work and Family Enterprise*. London: Macmillan.

Woodward, R. (2000) Warrior heroes and little green men: soldiers, military training and the construction of rural masculinities. *Rural Sociology*, 65: 640–657.

27

Rurality and racialized others: out of place in the countryside?

Paul Cloke

INTRODUCTION: ETHNICITY AND THE 'UNUSUAL' COUNTRYSIDE

A report by Raekha Prasad (2004) in the *Guardian* newspaper deployed an orthodox shock tactic to highlight the complex interconnections between ethnicity and rurality in the British countryside. Three people are pictured, posing against the background of a lake by a forest, and the banner asks 'What's unusual about this scene?' The answer to the conundrum lies neither in the unspoilt solitude of the place, which might be expected to display the crowded nature of the rural honeypot, nor in the absence of the high-technology boots, anoraks and rucksacks which typically denote the paraphernalia of 'serious' walkers enjoying a highly embodied and entirely acceptable cultural experience of the countryside. No, the representation of 'unusual' in this context is signified by the clear non-white ethnicity of the subjects of the photograph – their 'Asian-ness' is stereotypically suggested in clothing, headgear, the men's beards and so on – and the intended shock counterposes the expectation that the countryside is typically a white domain.

To acknowledge that the presence of ethnic minorities in rural scenes is 'unusual' is one thing, but Prasad's report proceeds to identify more sinister links between rurality and ethnicity via vox-pop testimonies from black and Asian Britons who have variously found themselves 'disowned by rural England' (p. 2). Andrea Levy, a black novelist, tells us that 'in the countryside I am so acutely aware of what I look like, not because people are hostile or unfriendly, but just because you are different. I always get the feeling when I walk into a country pub that everyone is looking at me, whether they are or not. You are glowing with colour' (p. 3). Benjamin Zephaniah, a black poet, talks of the time when he was staying on a friend's farm and went out for a jog within the boundaries of the farm. 'When I got back to his house, the place was surrounded by police, a helicopter circling above. "We have had reports of a suspicious jogger" the police said' (p. 2). Lemm Sissay, a black poet who was fostered and brought up in the country, tells of his experience in a countryside that he finds 'beautiful but incredibly damaged': 'Growing up, I used to ask myself why everyone had such a big issue with the colour of my skin. Men would shield their women from me, bars would go quiet. The incendiary racism that is in the country is never challenged' (p. 3).

These autobiographical notes reflect on a range of identity clashes between minority ethnicity and rurality – difference, suspicion, marginalization and racism all lurk behind the seemingly unusual presence of people of colour in British countryside settings. While such identity clashes are highly visible in city sites, it is almost as if ethnic others are typically rendered invisible by rurality, such that any moments of becoming visible represent unusual intrusions into the conventional cultural norms of rural life. In this chapter I want briefly to explore what it is about rurality and rural culture which sponsors this 'othering' of ethnic minority people, and to discuss the mechanisms of social and cultural regulation which position people of colour as 'out of place' in the countryside. Identities and subjectivities centred around ethnicity are by no means the sole axis of rural othering, but as this book portrays, both extreme and banal racisms represent very significant socio-cultural problematics in rural areas, and as with other axes of marginalization, rendering them highly visible is the first step to a more socially inclusive future.

RURALITY AND OTHERNESS

Rurality is a complex concept. At one level, we are often content to fall back on key characteristics which historically have been associated with rurality and rural life and which translate into objects of desire in contemporary society. As discussed in Chapter 2, the countryside is thus often viewed in terms of being dominated (either currently or in the recent past) by extensive land uses, being characterized by small-scale settlement that has strong affinity with the surrounding environment, and offering a close-knit and cohesive way of life.

At another level, however, we draw from these material characteristics a series of ideas and understandings about the meaningfulness of rurality. Somewhere deep down in our cultural psyche there appear to be longstanding handed-down precepts about what rurality represents, emphasizing the enabling power of nature to offer opportunities for lifestyle enhancement through the production and consumption of socially cohesive, happy and healthy living at a pace and quality that differs markedly from that of the city. Rurality has thus become cross-referenced with tranquillity, goodness, wholeness and problem-freedom, and at a more obviously political level it maps onto cartographies of identity, encapsulating for some a treasury of norms and values which both illustrate and shape what is valuable in a nation, a region or a locality.

Much has been written about the so-called 'rural idyll' over the years (see, for example, Bell, 1997; Bunce, 1994, 2003; Cloke and Milbourne, 1992; Halfacree, 1995; Little and Austin, 1996; Mingay, 1989; Short, 1991; Williams, 1973), both to confirm and to challenge these ideas of aesthetic pastoral landscapes acting as sites for humans working together in harmony and achieving both contentment and plenty. However, despite the overarching and sometimes stereotyping characteristics of the concept of 'idyll', the idea endures both in the direct representation of a range of contemporary cultural paraphernalia, and in reflexive and instinctive knowledges about the rural which are lived out in perception, attitude and practice. As Bunce (2003) explains,

> the values that sustain the rural idyll speak of a profound and universal human need for connection with land, nature and community, a psychology which, as people have become increasingly separated from these experiences, reflects the literal meaning of nostalgia; the sense of loss of home, of homesickness.

Rurality, then, not only represents spaces and values that satisfy basic psychological and spiritual needs, but also sponsors a cultural idyll which represents a natural and inevitable counterpoint to the rise of urban modernism.

This seemingly straightforward understanding of rurality needs to be tempered by a series of complicating acknowledgements about how rural metanarratives play out in real life, of which three will be summarized here. First, it is clear that rurality is not homogenous and that rural areas are *different*. In the English context, there are significant differences between the metropolitan ruralities of areas close to cities and the more peripheral ruralities of remoter areas. Equally, within the UK account has to be taken both of regional distinction and of the significance of nationhood in fashioning, for example, the ruralities of Wales and Scotland. Moving beyond Britain, the extensive ruralities of larger land continents exaggerate both materially and relatively the differences between rural areas. Outback Australia, the Canadian northern territories and the American Midwest, for example, can only loosely be treated spatially or conceptually as 'the same' as each other, and comparisons with, say, rural Berkshire are even more tenuous.

Secondly, rural areas are essentially dynamic. Far from being timeless, unchanging sites of nostalgia they are being reconstructed economically and recomposed socially by the globalized food industry, by the increasing mobility of production and people, and by the niched fragmentation of consumption and the commodification of place. Wilson (1992) argues that the end-product of rural dynamism is a blurring of the boundaries between the urban and the rural. Rural areas are becoming culturally urbanized through the all-pervading spread of urban-based mass media and other cultural output. Social trends of counter-urbanization have brought 'urban' people into rural areas, and out-of-town movements of factories and shopping malls have had similar impacts. In these and other ways, then, the city has moved out into the countryside. By contrast, the countryside has also, to an extent, been moving into the city, for example with the development of urban 'villages' and in the now pervasive trend of heightening the visibility and performance of urban nature. As I have written elsewhere, all this renders the rural–urban distinction indistinct:

> New regions focus on the hybrid relations between cities and surrounding areas; new information technologies permit the traversing of time-space obstacles; new forms of counterurbanisation result in spatial cross-dressing both by the arriving in and leaving of places. (Cloke, 2003: 2)

Thirdly, rural areas will often display dystopic characteristics in visible illustrations of the seamier side of rural life. Bell's (1997) account of cinematic depictions of small town America demonstrates how the 'horror' of rural life counterposes any narrative of idyll. The foot and mouth epidemic which consumed much of rural Britain during 2001 provided clear evidence of a non-idyllic countryside, as news media portrayed vivid images of the funeral pyres of culled livestock, and the curiously empty fields of farms deprived of their essential animality. Whether in the imaginative texts of film, or in the seemingly more mimetic (but equally imaginative) representation of news coverage, there is now evidence galore of rurality without idyll. Previous assumptions about rural areas as problem-free are challenged by a determined flow of suggestions that rural life is little different from that of the city in terms of crime, drug-addiction, poverty and other apparently urban problems. The material realities of 'living-in-the-idyll' are being seriously challenged by these depictions.

These characteristics of difference, dynamism and dystopia suggest that the rural–urban dualism has been largely overtaken by events. Rurality can no longer be regarded as a single space but is rather a multiplicity of social spaces that overlap the same geographical area (Mormont, 1990). However, while the marked opposition between the *geographical* spaces of urban and rural is being broken down, the imagined opposition between the *social* significances of urban and rural are being maintained and in some ways enhanced. Indeed, it is the social space of rurality – often fuelled by idyllistic concepts – which is the magnetic force that pulls together the category 'rural' or 'countryside' in the contemporary discourses of everyday life. And, as Halfacree (1993) has explained, the multiple meanings that constitute the social space of rurality are increasingly diverging from the geographical spaces of the rural. The symbols of rurality are becoming detached from their referential moorings, as socially constructed rural space is becoming increasingly detached from geographically functional rural space. Thus, despite our ever-more nuanced understanding of the different happenings in rural places, there is considerable scope for socially constructed significations of rurality to dominate both the territory of ideas and meanings about the rural, and the attitudes and practices which are played out in and from that territory.

One of the most important outcomes of debates about socially constructed rurality has been a deep concern about the cultural and political domination afforded by hegemonic ideas about rurality and rural people. Philo's (1992) intervention to highlight the neglected rural geographies hidden away in and by such hegemonic social constructions was seminal in the search for ways to give voice to rural 'others'. He emphasized that social constructions of life are dominated by white, male, middle-class narratives and pointed to the discursive power through which the all-embracing commonalities suggested by social constructions of the rural idyll serve in practice to exclude individuals and groups of people from a sense of belonging to, and in, the rural on the grounds of their race, ethnicity, gender, age, class and so on. Subsequent studies of rural 'others' (see Cloke and Little, 1997; Milbourne, 1997) have sought to identify the practices and devices that exclude particular individuals and groups in this way. For example, studies of poverty in rural areas (see, for example, Cloke, 1997; Cloke et al., 1995, 1997a) have connected the idyllized imagined geographies of rural lifestyles with the suggestion that poverty is being hidden or rejected *culturally*, both by decision-makers who refuse to accept that poverty can exist in idyllic rural areas, and by rural dwellers themselves who legitimize poverty as a disadvantage of rural life that is often compensated 'naturally' by the benefits of a rural setting. In this way, rurality can signify itself as a poverty-free zone, and socially constructed rural idylls both exacerbate and hide poverty in rural geographic space. Similarly, studies of rural homelessness (see Cloke et al., 2002) recognize that implicitly idyllized and romantic notions of rural living sponsor the idea that homelessness cannot occur in caring rural communities:

> The idea of a 'helpful' community, then, discursively constructs rural society, and problems such as homelessness within that society, in such a way that rurality and homelessness are further disconnected. (2002: 68)

A concern for 'other' people and practices of 'othering', then, has become of paramount importance in tracing the consequences of how cultural significations of rurality have impacts on and in rural life itself.

RACIALIZED OTHERS IN THE COUNTRYSIDE

> The countryside is popularly perceived as a 'white landscape' ... predominantly inhabited by white people, hiding both the growing living presence and the increasing recreational participation of people of colour. Thus in the language of 'white' England, ethnicity is rarely an issue associated with the countryside. Its whiteness is blinding to its presence in any other form than the 'non-white'. (Agyeman and Spooner, 1997: 197)

The terms race and ethnicity are often substituted for each other without care for more precise definition. In this chapter, 'race' (following Jackson, 2000) is taken to reflect a social construction that marks out human difference usually on the basis of distinct skin colour. By contrast, ethnicity is taken to be a more self-referential term by which particular groups linked by birth express the differences in their culture and lifestyle (Jackson, 1992). The reference made by Agyeman and Spooner to how the 'whiteness' of the countryside reflects on its other as 'non-white' reflects an assertion that race rather than ethnicity lies at the heart of processes and practices by which people of colour are othered in and by the rural.

There is now a strong body of work which has demonstrated how rural landscapes have served as key symbols of English national identity (Daniels, 1992, 1993; Kinsman, 1993, 1995). The countryside, it is argued, is more than the sum of its parts; it reflects core qualities of Englishness and is inhabited by those who are intent on upholding the rural and social fabric of the nation. The absence of 'non-white' people in countryside landscapes, then, has reinforced a collaboration between whiteness and national identity in the symbolism of rurality – a collaboration which, for the most part, remains unspoken, but which has been detailed by extreme right-wing political parties such as the National Front (Coates, 1993) and the British National Party (Roberts, 1992), for whom a green and pleasant land free from invading (non-white) aliens represents ideologically potent fascist imagery.

Given these strong connections between rurality and the symbolization of national identity, morality and social fabric, it is inevitable that social constructions of rurality will become embroiled, often unknowingly, in ideas of Englishness from which people of colour are excluded. Brace (2003) has argued that the mapping out of the nature of Englishness involves both 'the search for and valorization of untouched, unsullied places that epitomized the things about Englishness which were so valued', and 'constructing a moral geography of places which were signally un-English' (2003: 69). The vilification of problematic people and places is only made possible by its corollary – the celebration of unproblematic people and places, and rurality offers many opportunities for such celebration. However, the unsullied nature of rurality is predicated on a heritage that is assumed to be white and Anglo-Saxon, with other social and cultural groups being excluded because they potentially threaten the political narrative of 'acceptable' history and heritage. The glorious Englishness of the rural is founded on the relations and values of colonial history – a history told by white English historians who, as Agyeman points out, have written out the stories of people of colour from their hegemonic narratives:

> soldiers from North Africa used the Roman environment of the Borders. They were garrisoned on Hadrian's Wall. People from Asia were brought, often as whole villages, to Britain to work in the Yorkshire and Lancashire cotton mills. Many of our stately homes were financed, built and exotically landscaped through African–Caribbean slavery ... Has the presence of these and other people been routinely celebrated in visitor attraction and interpretation facilities, or has it been quietly and unceremoniously swept under the carpet? (1995: 5)

If rurality has been linked with racial purity, and the countryside has been viewed as the core of national identity in which country people somehow become an essence of England, then such symbolisms are founded on skewed histories and deliberate exclusions which are bound to fuel unthinking partialities amongst the included and fear and disenfranchisement amongst those excluded.

Explorations of racialized otherness in the countryside have been relatively few and far between. Most rural and social texts and commentaries acknowledge the othering of people of colour by turning to a particular contribution by the black photographer Ingrid Pollard (see Kinsman, 1995), whose series *Pastoral Interlude* comprises self-portraits and portraits of friends (also black) against rural backgrounds – a stone wall in front of an extensive landscape, a country churchyard, a stream. For example, a picture of a black person in a rural setting is intended to be unusual, disconcerting, an interlude to pastoral norms. The mood of the images is enriched by their captions:

> it's as if the Black experience is only ever lived within an urban environment. I thought I liked the Lake District; where I wandered lonely as a Black face in a sea of white. A visit to the countryside is always accompanied by a feeling of unease, dread.
>
> feeling I don't belong. Walks through leafy glades with a baseball bat by my side.
>
> a lot of what *made england great* is founded on the blood of slavery, the sweat of working people … an industrial *revolution* without the Atlantic Triangle.
>
> death is the bottom line. The owners of these fields; these trees and sheep want me off their *green and pleasant land*. No Trespass, they want me *dead*. A slow death through eyes that slide away from me. (Pollard, n.d.)

According to Kinsman (1993), the lone black figures in Pollard's images are wistful and resigned, while the angrily ironic captions protest the white ownership of land which seeks to evict, repatriate or even destroy black 'intruders', even though their labour was a crucial factor in the economic development that underpins histories of land ownership. Pollard thus connects the historic death of black slaves with what she regards as the slower 'death' of black people in contemporary Britain, suffering – through racialized surveillance from country folk – from unease, dread, fear. Pollard's work has become an iconic source, turned to by academics wishing to illustrate racialized otherness in the countryside, although as the Women in Geography Study Group (1997) argues, her work also reflects gendered otherness. While Pollard's excellent work is certainly worthy of respectful re-examination, there is a whiff here of white academics (myself included) often being content just to have one (token?) example to reflect the non-white other. More recently, other academic studies have sought to widen the base of evidence (see, for example, Spooner-Williams, 1997, reflecting on Agyeman, 1989, 1991, 1995; and Dabydeen, 1992), but the rural racialized other has certainly been given less attention than alternative othered groups.

The other major explorations of racialized otherness have come in a series of reports, for example: '*Keep Them in Birmingham*' (Jay, 1992), a challenge to racism in south-west England published by the Commission for Racial Equality; *Staring At Invisible Women* (Esuantsiwa Goldsmith and Makris, 1994), a report on the experiences of black and minority ethnic women in rural areas published by the National Alliance of Women's Organizations; and *Not in Norfolk* (Derbyshire, 1994), a report on the experiences of ethnic minority people living in Norfolk, published by the Norfolk and Norwich Racial Equality Council. A feature of each of these excellent reports is an emphasis on the invisibility of race and racism in rural settings. To some extent, the mediated impression that race issues are associated with urban areas means that their invisibility in rural areas is unsurprising. However, as the reports make clear, the popular myth that such invisibility is due to an absence of people of colour in the countryside is just a myth. There is a small, but significant, presence of people of colour living in and visiting the countryside, and the belief that no race issues exist there is due only to cultural invisibility. Many of Jay's respondents from local authorities, health authorities and other voluntary organizations felt that questions of racial equality were 'not applicable' in their areas of jurisdiction, both because they seriously underestimated the scale of ethnic minority presence, and because of a broad cultural denial of any 'problems with race'. He concluded that 'the commonest response to the project was one of indifference or even hostility; racial equality is evidently not part of the agenda' (1992: 43). Esuantsiwa Goldsmith and Makris experienced similar attitudes from statutory and non-statutory organizations which denied that race-related problems occurred in rural areas. However, their evidence from women of colour confirmed the presence of racism in the countryside:

> Racism was experienced by the women we consulted as a lack of acceptance into white society, and discrimination against minority culture, religion, colour and way of life through fear and ignorance of other people. (1994: 23)

Moreover, they concluded that women of colour suffered from a double dose of othering practices because rural communities tend to be dominated by patriarchal structures that prevent consultation with women and therefore render invisible their needs, views and aspirations. Women of colour are additionally disadvantaged because of the invisibility of issues connected with race. Derbyshire's findings add a further dimension to the invisibility of rural racism. She records the almost universal reaction from white respondents that 'there is no problem here' and suggests that the reasoning behind this response is because

> Those [people of colour] who are here are encountered individually and so constitute no real threat, or they are working within the context of restaurants and takeaways and in this sense 'know their place'. (1994: 21)

This reference to 'knowing their place' seems crucial to the understanding of racialized others in the countryside, because ethnic minorities not only find themselves positioned culturally as 'out of place' in the countryside, but also experience practices and attitudes which seek to purify that place should it be transgressed.

IN PLACE/OUT OF PLACE IN THE COUNTRYSIDE

The othering of ethnic minorities in rural areas needs to be understood against a broad-brush background of how idyllistic cultures of rurality serve to signify key facets of what rural life should be like (Cloke et al., 2000). As discussed in Chapter 32, Cresswell's (1996) discussion of how uses of space come to be constructed as 'appropriate' or 'inappropriate' is instructive here. He not only suggests that an established social order can be (re)produced by a naturalization of common sense, but also that boundaries will be erected between common sense notions of what is appropriate or inappropriate in particular places. He argues that there is therefore a 'taken-for-granted' doxa of what goes on in places, and that different places display different such doxa, or senses of the obvious. When such commonsense doxa are transgressed, things will seem out of place because of a lack of doxa conformity. Cresswell concludes that particular places are implicated in the construction and reproduction of particular ideologies, and that socially constructed meanings of place can become naturalized as they are taken for granted.

Cresswell's ideas fit well with Sibley's (1995) arguments about the geographies of social exclusion, in which spatial purification is seen as a key factor in the organization of social space:

> The anatomy of a purified environment is an expression of the values associated with strong feelings of objection, a heightened consciousness of difference and, thus, a fear of mixing at the disintegration of boundaries. (1995: 78)

So not only do we need to acknowledge a taken-for-grantedness about what is 'in place' or 'out of place', but we need to grasp how places assume a symbolic importance that constructs and reproduces a desire for order in which local environments require ordering and purification. It is by these means that spaces become implicated in the construction of deviancy. The purity of spaces reinforces their difference from other places and sponsors a policing of social and spatial boundaries. Speaking of New Age travellers, Sibley argues

> A rigid stereotype of place, the English countryside, throws up discrepant others ... These groups are other, they are folk-devils, and they transgress only because the countryside is defined as a stereotypical pure space which cannot accommodate difference. (1995: 107–108)

Although it is important to reiterate that rurality should not be viewed as any kind of self-defining or naturalistic category, we can use Cresswell's and Sibley's arguments to suggest a thesis about the othering of ethnic minorities in rural areas. First, the lived presence of ethnic minorities in the UK tends to be spatialized as 'in place' in urban environments. Secondly, rural areas are typically understood as signifying essentially white characteristics of Englishness. Taken together, the presence of non-white ethnic minorities in rural areas is not just 'unusual' (as discussed in the introduction to this chapter) but may be deemed as out of place, representing a transgression of the orthodoxies assured by socio-spatial expectations. We might therefore expect to find the countryside policed as a purified space in this respect, because a rejection of difference is deeply embedded in its social system.

Such expectations find echoes in studies of race and ethnicity in rural areas. Agyeman (1989), for example, reflects that for white people 'ethnicity' is viewed as being out of place in the countryside, and Spooner-Williams (1997) interprets white women's narratives of race in the village in terms of attempts by middle-class residents to purify their local space. Evidence on these issues needs to be treated with care, as personal statements of individual attitudes towards race and ethnicity in the countryside may well contrast with attitudes attributed to 'the village' collectively. In this way, white people can deny their own complicity in racist purification yet can sometimes acknowledge the existence of racism by displacing it onto other individuals or onto 'the village' more generally.

Rachel Spooner-Williams assesses these contrasting discourses of rural race in terms of the 'public' and 'private' faces of racism. She found that many white women claimed not to recognize race as an issue in rural settings – a denial which often assumes away the difference in others of colour so as to maintain the dominance of the presumed sameness of whiteness. The public discourse of 'we are all the same' is sometimes accompanied by displacement of racist attitudes to other (often older and longer-standing) village residents, but is mostly twinned with discourses of 'we all have every opportunity' and thereby with a blindness to racial difference and to the presence of prejudice and discrimination. She also found that many women of colour in rural settings become co-opted into these discourses through a desire to 'fit in'. The private face of rural race, however, strongly reflects a purification of village spaces in which white residents demand conformity to their perceived orthodoxies and exclude those who transgress these boundaries of what is in place in the village setting. Sometimes being 'out of place' seems to be a matter of constructions of cultural incompetence (Cloke, 1994; Cloke et al., 1997b) reflected in cultural distinctions relating to how to keep house, tend a garden or join in with village society. However, such constructions often represent the superstructure of a far deeper vessel of anxiety, discrimination and racism in rural settings, which collectively serve to bleach the cultural identity of people of colour, and to police the cultural boundaries of rural whiteness.

RURAL RACISMS

While it would be erroneous to assume all white rural residents to be racist, the evidence from available research nevertheless clearly points to the presence of rural racism which polices the 'purity' of rurality in a number of different ways, ranging from the subtle to the downright criminal. The purification of rural space varies in visibility – a village up in arms over proposals to establish a centre for asylum seekers in their vicinity seems to be 'acceptable' as a very public expression of outrage, while the miseries suffered by a black child in a village school will be more private yet equally reflect processes and practices of rural purification. Despite the risk of overgeneralization, a useful starting point for the identification of rural racism is therefore to identify a range of covert and overt practices and experiences by which people of colour are made to feel excluded, marginalized and othered in rural environments.

Covert racism comes in many forms, but at its heart lie mechanisms, assumptions, inflections and orthodoxies which serve to deny people of colour any distinct cultural identity in rural settings, presenting them with lifestyle choices involving a denial of ethnic identity so as to 'fit in', or a celebration of ethnic identity often leading to cultural isolation. Derbyshire's (1994) study in Norfolk demonstrates how the onus is always on ethnic minority people to fit in to rural society, and her interviews with people of colour illustrate how coping with life in white rural society requires compromise and self-denial:

> I have to make changes to fit in, for my own mental health. I'm not sure it's the way I should go. There is no platform for me ... I'm allowed to entertain but they are toning me down!
>
> If you don't fit in you pay the price. It's hard to have the confidence to fight back – if you are the one who suffers. You must try and fit in and suppress any feelings against this.
>
> Here I keep a large part of myself hidden. It's like I have two lives. (1994: 33)

Even where people of colour are content to strive for community acceptance, they face an inevitability that acceptance will usually only be partial. Esuantsiwa Goldsmith and Makris quote an Asian woman interviewee as suggesting that

> The attitude of white society in rural areas appears contradictory. They imply that if minority people conform and adopt white culture they will be accepted. But even if we do we are still regarded as different. (1994: 22)

This cultural isolation is often compounded by difficulties in receiving community support and a lack of information networks or links for ethnic minority residents in rural areas. These covert racisms are often cloaked in a façade of polite condescension, as rural residents will assume a public face of tolerance towards the 'strangers' in their midst. Jay illustrates this point through the life experiences of a black interviewee living in the rural south-west of England.

> For one man, the experience of being a black person in an almost totally white environment was that he encountered great ignorance and was regarded as a piece of 'exotica'. 'They treated me as someone who needed to be patronised; it was as though I had just stepped off the boat' ... (1992: 21)

While people of colour may experience such ignorance in all kinds of spatial settings, the white heat of the localized cauldron of rural society, where everybody knows everybody else, exacerbates the isolation and othering which results.

More overt forms of rural racism are also evident from these research studies. Narratives of straightforward discrimination in rural areas abound, and it is interesting to note in the following two examples drawn from Jay (1992) that discrimination often seems to result from transgression of the *imagined* spaces of rurality – what is and what is not 'in place' – which can be as much to do with the expectations of wider society as it is constructed by rural people themselves:

> a young black student working for a degree in institutional management [arrived] at a hotel in Cornwall to begin an industrial placement ... The hotel management was surprised to find that the trainee was black, and the following day he was asked to leave, since his colour 'might affect the trade' ... at a different Cornish hotel, a black woman who had just started work as a chambermaid was dismissed because members of a coach party staying there 'expressed virulent dislike at the idea of a black chambermaid attending their rooms'. (1992: 17)

Such obviously discriminating racisms were dealt with under race relations and employment legislation, but other practices of racialized purification in rural areas take the form of sustained harassment, which is more difficult to counteract both in legal terms and because of the costs of resultant exclusion in a small community. People from ethnic minorities who operate restaurants and shops in rural areas, for example, will often have to tread a precarious line between the abusive harassment attracted by their non-white identities. Derbyshire (1994) reports the story of one such takeaway owner in rural Norfolk:

> Children are the worst. They say so many rude words to you. 'You fucking foreigner' ... You go and talk to their parents. They don't take it seriously. They laugh at you or say the same thing ... People leave without paying. Do they do that to the English? ... Children threw a condom through [the] window. We sent the mother a solicitor's letter. It means she never comes anymore and she has told all in the neighbourhood not to come here for food ... Windows have been broken and the insurance won't pay anymore ... I'm stressed, I can't sleep. We are selling up and going after 15 years. (1994: 32)

Such violent harassment is not confined to a few isolated cases. Jay's (1992) research in Dorset and Somerset uncovered a series of incidents in which Asian families had been forced to leave their homes and businesses because of the hostility of other villagers, and from Devon and Cornwall he gathered reports of racial abuse and threatening and violent behaviour experienced by restaurant workers and owners.

These covert and overt examples of rural racism clearly illustrate some of the practices used to 'purify' the whiteness of rural space, when transgressions occur through what are constructed as the 'out of place' presence, activities or cultural differences of non-white others. Returning to the photograph in Prasad (2004), we might even suggest that part of 'what is unusual about this scene?' is the absence of any indication of stares, demands to 'fit in', cultural restrictions on ethnic identity, condescending politeness, discrimination, harassment or violence – those responses that usually accompany non-white 'transgressive' presence in the blinding whiteness of the countryside. At a time when the cultural politics of rurality appear to have focused sharply on developing a 'rural voice' to speak for the countryside against a range of perceived threats from the urban, there is an urgent need to evaluate that rural voice in terms of its potentially exclusive vision of rural affairs. Hegemonic speaking out on behalf of rural people inherently reinforces the processes and practices of exclusion which accompany hegemonic rurality. An alternative cultural politics emphasizing a countryside of difference seems long overdue but is essential if we are to break down the interior boundaries of rurality which both concrete difference and 'purify' otherness.

ACKNOWLEDGEMENTS

This chapter is a slightly amended version of P. Cloke (2004) 'Rurality and racialised others: out of place in the countryside', in N. Chakraborti and J. Garland (eds), *Rural Racism* (Cullompton, UK: Willan Publishing). It is reproduced by kind permission of the Editors and the publisher.

REFERENCES

Agyeman, J. (1989) 'Black-people, white landscape', *Town and Country Planning* 12: 336–338.

Agyeman, J. (1991) 'The multicultural city ecosystem', *Streetwise* 7: 21–24.

Agyeman, J. (1995) 'Environment, heritage and multiculturalism', *Interpretation* 1: 5–6.

Agyeman, J. and Spooner, R. (1997) 'Ethnicity and the rural environment', in P. Cloke and J. Little (eds), *Contested Countryside Cultures*. London: Routledge.

Bell, D. (1997) 'Anti-idyll: rural horror', in P. Cloke and J. Little (eds), *Contested Countryside Cultures*. London: Routledge.

Brace, C. (2003) 'Rural mappings', in P. Cloke (ed.), *Country Visions*. Harlow: Pearson.

Bunce, M. (1994) *The Countryside Ideal*. London: Routledge.

Bunce, M. (2003) 'Reproducing rural idylls', in P. Cloke (ed.), *Country Visions*. Harlow: Pearson.

Cloke, P. (1994) '(En)culturing political economy: a life in the day of a rural geographer', in P. Cloke, M. Doel, D. Matless, M. Phillips and N. Thrift (eds), *Writing the Rural: Five Cultural Geographies*. London: Paul Chapman.

Cloke, P. (1997) 'Poor country: marginalisation, poverty and rurality', in P. Cloke and J. Little (eds), *Contested Countryside Cultures*. London: Routledge.

Cloke, P. (2003) 'Knowing ruralities?', in P. Cloke (ed.), *Country Visions*. Harlow: Pearson.

Cloke, P. and Park, C. (1985) *Rural Resource Management*. London: Croom Helm.

Cloke, P., Goodwin, M., Milbourne, P. and Thomas, C. (1995) 'Deprivation, poverty and marginalisation in rural lifestyles in England and Wales', *Journal of Rural Studies*, 11: 351–366.

Cloke, P. and Little, J. (eds) (1997) *Contested Countryside Culures*. London: Routledge.

Cloke, P. and Milbourne, P. (1992) 'Deprivation and lifestyles in rural Wales: rurality and the cultural dimension', *Journal of Rural Studies*, 8: 359–371.

Cloke, P., Milbourne, P. and Thomas, C. (1997a) 'Living lives in different ways? Deprivation, marginalisation and changing lifestyles in rural England', *Transactions of the Institute of British Geographers*, NS 22: 210–230.

Cloke, P., Goodwin, M. and Milbourne, P. (1997b) 'Inside looking out, outside looking in: different experiences of cultural competence in rural lifestyles', in P. Boyle and K. Halfacree (eds), *Migration Into Rural Areas: Theories and Issues*. London: Wiley.

Cloke, P., Milbourne, P. and Widdowfield, R. (2000) 'Homelessness and rurality: "out-of-place" in purified space', *Environment and Planning D: Society and Space* 18: 715–735.

Cloke, P., Milbourne, P. and Widdowfield, R. (2002) *Rural Homelessness*. Bristol: Policy Press.

Coates, I. (1993) 'A cuckoo in the nest: the National Front and green ideology', in J. Holder, P. Lane, S. Eden, R. Reeve, U. Collier and K. Anderson (eds), *Perspectives On the Environment: Interdisciplinary Research Network On the Environment and Society*. Aldershot: Avebury.

Cresswell, T. (1996) *In Place, Out of Place: Geography, Ideology and Transgression*. Minneapolis, MN: University of Minnesota Press.

Dabydeen, D. (1992) *Disappearance*. London: Secker and Warburgh.

Daniels, S. (1992) 'Place and geographical imagination', *Geography* 77: 310–322.

Daniels, S. (1993) *Fields of Vision*. Cambridge: Polity Press.

Derbyshire, H. (1994) *Not in Norfolk: Tackling the Invisibility of Racism*. Norwich: Norwich and Norfolk Racial Equality Council.

Esuantsiwa Goldsmith, J. and Makris, M. (1994) *Staring At Invisible Women: Black and Minority Ethnic Women in Rural Areas*. London: National Alliance of Women's Organizations.

Halfacree, K. (1993) 'Locality and social representation: space, discourse and alternative definitions of the rural', *Journal of Rural Studies* 9: 1–15.

Halfacree, K. (1995) 'Talking about rurality: social representations of the rural as expressed by residents of six English parishes', *Journal of Rural Studies* 11: 1–20.

Jackson, P. (1992) 'The politics of the streets: a geography of Caribana', *Political Geography*, 11: 130–151.

Jackson, P. (2000) 'Race', in R. Johnston, D. Gregory, G. Pratt and M. Watts (eds), *Dictionary of Human Geography*. Oxford: Blackwell.

Jay, E. (1992) *'Keep Them in Birmingham': Challenging Racism in South-West England*. London: Commission For Racial Equality.

Kinsman, P. (1993) 'Landscapes of national non-identity: the landscape photography of Ingrid Pollard'. Working Paper 17, Department of Geography, University of Nottingham.

Kinsman, P. (1995) 'Landscape, race and national identity: the photography of Ingrid Pollard', *Area* 27: 300–310.

Little, J. and Austin, P. (1996) 'Women and the rural idyll', *Journal of Rural Studies* 12: 101–112.

Milbourne, P. (ed.) (1997) *Revealing Rural 'Others'*. London: Pinter.

Mingay, G. (ed.) (1989) *The Rural Idyll*. London: Routledge.

Mormont, M. (1990) 'Who is rural? Or how to be rural: towards a sociology of the rural', in T. Marsden, P. Lowe and S. Whatmore (eds), *Rural Restructuring*. London: David Fulton.

Philo, C. (1992) 'Neglected rural geographies: a review', *Journal of Rural Studies* 8: 193–207.

Pollard, I. (n.d.) *Monograph*. London: Autograph.

Prasad, R. (2004) 'Countryside retreat', *Guardian* 28 January, pp. 51–53.

Roberts, L. (1992) 'A rough guide to rurality: social issues and rural community development', *Talking Point* No. 137. Newcastle upon Tyne: Association of Community Workers.

Short, J. (1991) *Imagined Country*. London: Routledge.

Sibley, D. (1995) *Geographies of Exclusion: Society and Difference in the West*. London: Routledge.

Spooner-Williams, R. (1997) 'Interpreting cultural difference: articulations of "Race", Gender and Rurality in Britain and New Zealand/Aotearoa'. Unpublished PhD thesis, School of Geographical Sciences, University of Bristol.

Williams, R. (1973) *The Country and the City*. London: Chatto and Windus.

Wilson, A. (1992) *The Culture of Nature*. London: Routledge.

Women in Geography Study Group (1997) *Feminist Geographies: Exploration in Diversity and Difference*. Harlow: Longman.

28

Rural change and the production of otherness: the elderly in New Zealand

A. I. (Lex) Chalmers and Alun E. Joseph

INTRODUCTION

The invitation to reflect on views of the elderly[1] in contemporary rural studies provides us with an opportunity to review some of the significant trends in rural research and commentary over the past ten years. Over this decade we discern changes that relate directly to the lives of elderly people, changes in methodological emphases and (most importantly, perhaps) changes in the awareness of constructions of the elderly. Our own work is perhaps typical. It began by casting the elderly as a distinct category within a demographically based analysis of rural change in Waikato, migrated through explorations of the place of the elderly in the national and regional space economy, to arrive at deconstructions of the lives of the elderly and their individual reactions to social and political processes that affect them. This chapter uses this experience to characterize and extend research trends focusing on the rural elderly. While restrictive in some senses, this method permits us to be truly reflexive in our approach to a postmodernist rurality and diverse rural life-worlds. It also gives us the comfort of a situated analysis that acknowledges the specificity of context and the contingency of events (Rowles, 1988; Kearns and Joseph, 1997).

Figure 28.1 summarizes the generic interplay between the operational, conceptual and empirical elements of contemporary research relating to the rural elderly. The reshaping of rural space is engineered without direct reference to the worlds of the elderly, but the lives of the elderly in place are greatly affected by the changes put in place. Our evolving understanding of these changes over time has been governed by a temporal location relative to the evolving target of study (rurality and the place of the elderly in it), the state of rural studies and our reflexive relationship with both. Our current understanding of the rural elderly is anchored in a recognition of cumulative change in rural spaces and places over the past decade, informed by an expanding literature and fundamentally shaped by our sequential engagement with both. Inevitably there are time lags, such that contributions about a past reality are made to a literature base that has already moved on.

Structurally, the substantive discussion that follows is organized in three sections. We consider it to be important first to reflect on the (re)conceptualization of rural studies over the past decade. The broad trajectory is well known, but we consider our contexted review 'adds value' through explicit reference to commentaries on the elderly. We reflect particularly upon the enrichment of the normative and political economy approaches to geographies of the elderly that has flowed from engagement with the postmodernist challenge and the growing acceptance of qualitative and ethnographic research in rural studies.

We then move to consider theorizations of the processes and mechanisms of change in the rural sector and explore the constructions that underlie the characterization of the elderly. We work outwards from the local literature that informed our own work, with the intention of amplifying the conceptual developments reported in the first substantive section of the chapter. We are aware that

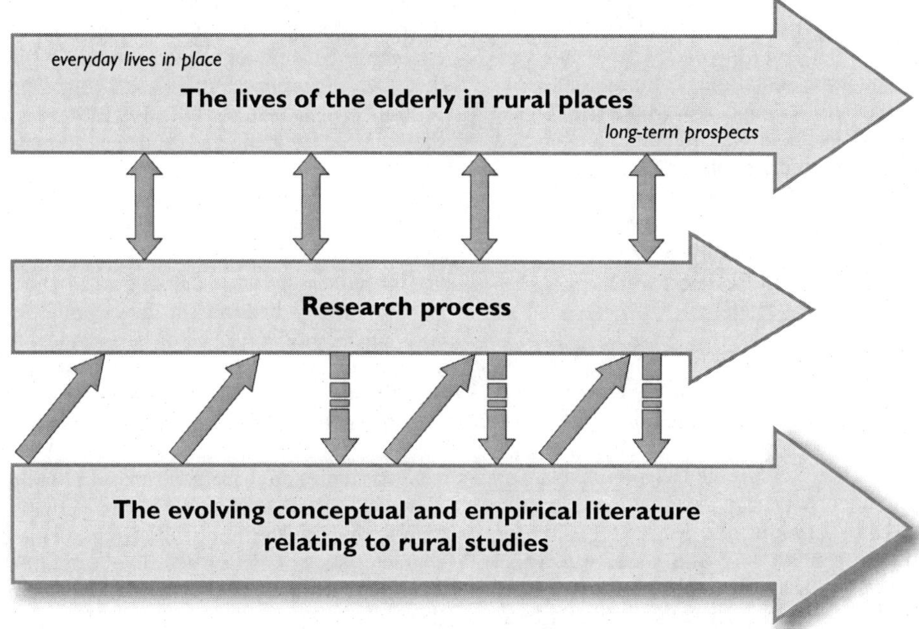

Figure 28.1 ***Dynamics of the research process embedded within conceptual and lived experiences***

such a review is selective, but we cite strategically from the literature in three related areas in which analysis of, and reporting on, the lived experience of the elderly has been central.

The first thematic review considers research that constructs the elderly as a demographic cohort defined by mandatory retirement and faced with the prospect of 'staying on' or relocating at a critical life stage. The literature on collective ageing-in-place versus movement to preferred 'spaces' or locations is substantial, and the interpretations of patterns of movement are necessarily general but nevertheless rich in implications of difference and otherness. The second theme we document describes work based on an awareness of the social and political dimensions of ageing and of the need to acknowledge the special requirements, in housing and health care for instance, of 'sector' groups such as the elderly. In this discussion we note the historical construction of entitlement by elderly people and cast this expectation against a political economy characterized by the withdrawal of the state and the downloading of social responsibilities to individuals and their families and communities. Finally, we explore research based on a qualitative, contextualized engagement with the lived experience of rural people. While the documentation of lives and lived experience often reveals diversity, the formal deconstruction of the category 'rural elderly' is less common.

The final substantive section of this chapter expands upon dominant and emerging themes in the construction of the elderly as other in rural places. We begin by asking how the rural elderly can be cast in the role of *discarded other*, when in many cases they are the embodiment of specific places. We note the related political economy view of the *elderly as consumers* rather than as producers, and point to the emergence of a *technology divide* between elderly consumers and their communities. We then consider some emergent ideas about able-ist constructions of place that cast the elderly as the *disabled other*. We round out our discussion of themes in the construction of the elderly as other by considering manifestations as *ethnic other* and *nostalgic other*. The latter leads us to question how constructions of place, and rural places in particular, are created. We point to some analyses of rural coverage in mass media (television and press), and review the placement of the elderly in this coverage.

In framing conclusions for the chapter, we consider the significance of recent theoretical and empirical progress for future studies of the elderly as other. We bring to this task an awareness of the need to (i) promote discussion about the lives of the elderly in rural places across the variety of economic, social and cultural contexts in which people experience ageing and (ii) connect our suggestions about research directions to new and important directions in rural studies.

BROAD CONCEPTIONS OF RURAL GEOGRAPHIES AND THE RURAL ELDERLY

When we began working in this area in the early 1990s, we reviewed the literature on the changing demography and geography of rural New Zealand. In our reading, we confronted a landscape largely devoid of theoretical markers, with the notable exception of structuralist work focusing on agricultural change and the local effects of state restructuring (for example, see Cloke, 1989; Britton et al., 1992; Fairweather, 1992; Moran et al., 1993). In most cases, it seemed that the literature on antipodean rurality had followed the international trend of data-rich but largely atheoretical case studies. There was published research on the changing populations of rural areas (Franklin, 1969; Heenan, 1979; Cant, 1980) and on the nature of rural places (Franklin, 1978; Mackay, 1984), but in contrast the number of studies using qualitative, socio-political or ethnographic approaches to life in rural places was modest (Pawson and Scott, 1992; Keating and Little, 1994; Wilson, 1994). We thus faced the challenge of designing a case study of the impact of economic and social restructuring on rural communities and their elderly residents without the benefit of baseline studies or a deep understanding of the processes of change that were, in retrospect, transforming the space-economy of New Zealand and its constituent rural places (see Le Heron and Pawson, 1996; Liepins and Bradshaw, 1999; Joseph et al., 2001). We turned to the international literature to supplement our understanding of unfolding events in New Zealand.

From his standpoint in the 1980s Krout (1988: 103) bemoaned 'the slow pace at which any systematic developments in thinking and writing about the experience of being old in rural places have occurred'. Our review of the international literature in the early 1990s initially confirmed Krout's assessment, but subsequently we encountered an emerging and constructive tension between methodologically distinct views of rural space and place. We saw that at times the contestation of 'rural' inhibited research on specific threads of rural change, but more frequently that it encouraged such work. Ten years on, we consider that the 'disturbance' of post-structuralism and postmodernism (Philo, 1992, 1993; Murdoch and Pratt, 1993, 1994) liberated our research agenda. This liberation at once confirmed our intuitive feeling that important things were happening in rural places and provided us with the means to link with an evolving community of ideas.

The refocusing on matters of place and space in rural studies began with the critique of rural community studies as being simplistic in their equating of 'community' with discrete ways of life, usually agrarian-based, in particular places (Williams, 1973; Lewis, 1979; Day and Murdoch, 1993).

Such studies often included an overly-simplistic view of spatiality (Newby, 1986), perhaps best represented by the notion that the sociological characteristics of a place could be 'read off' from its relative location on the rural–urban continuum (Halfacree, 1993). The critique of community studies as being excessively and simplistically place-oriented was often accompanied by the advancement of theorizations of space and society (Murdoch and Pratt, 1993).

The earliest and most prescriptive of these theorizations used the methods of political economy to construct models of socio-spatial relations that cast events in rural places as subject to the spatial relations of 'capital' and 'class', elevating them above 'community' as the focus of analysis (Murdoch and Pratt, 1993). Subsequent reactions to the excessive determinism of political economy models, especially among those seeking to build upon Giddens's (1984) theorization of locality-based structure and (human) agency, sought to re-assert the importance of place. The attempt to re-balance space and place remained evident in the restructuring approach, the broad thrust of which 'has been to treat the locality as constituted through the operation and intersection of general processes and relationships, such as spatial divisions of labour, which are inevitably realized and mediated in particular places' (Day and Murdoch, 1993: 88). What, though, of people in places?

To observers in the early 1990s it appeared that rural research generally paid only limited attention to the geographies of rural people (Philo, 1992), often choosing to 'privilege particular conceptions of reality over others' (Murdoch and Pratt, 1994: 84) by focusing on the narratives of the majority, or those who shape social constructions of 'reality'. It is clear that the elderly (and particularly those who are ill or frail) were not well served by narratives stressing, for example, the acceptability of trade-offs between increased travel and 'improved' (read 'centralized') service facilities. The creation of new opportunities across rural space at the expense of local facilities and services was full of implications for elderly people with 'small lives' circumscribed by the location of home and the strictures of declining mobility; for them, space collapses into the confines of place (Joseph and Martin-Matthews, 1993).

Like Harper (1987: 309), we believed (and still do) that 'place emerges from the barrenness of space through the meaning imposed on it by … people as the result of … experiences'. Moreover, the 'wide-angled lens' of postmodernism encouraged us to valorize the experience of ageing individuals and their restructured communities, to esteem 'the more specific "stories" that "other" people in "other" places tell themselves when seeking to make sense of their specific and situated existence' (Philo, 1992: 199). Postmodernist approaches also helped us build important bridges, because 'what a focus on localities can share with a shift toward postmodernism is a recognition of the potential significance of both the local and variety' (Murdoch and Pratt, 1993: 421). While not without its flaws (Little, 1999), postmodernism also helped us to re-insert 'community' into our analysis of the rural experience, as both a context for the description of human experience and as a unit of analysis in its own right. In particular, we took up the views of Massey (1991), as later extended by Kearns and Joseph (1997: 19), 'restructuring not only takes place in places but also takes place in the lives of individuals; … the summation of lives in place is surely "community", community being at once a concept rooted in geography and (a shifting) socio-cultural reality'. It is this sensitivity to shifting rural geographies and experiences (Cloke and Little, 1997) that led us to see, tentatively at first but then with more conviction, the elderly as 'other'.

SETTING THE STAGE: CREATING RURAL SPACE AND LOCATING THE ELDERLY

Demographic change and rural communities: the elderly as 'cohort'

The elderly in rural places, and elsewhere, are almost invariably characterized with reference to their chronological age. Thus, we consider here macro-constructions of the (rural) elderly based on chronological age that view 'the elderly' as a collective, or cohort, that can be divided into implicitly homogeneous sub-categories. Thus reference is often made to the 'young elderly' (aged 65 to 74, for instance) and the 'old elderly' (75 or older) (Joseph and Fuller, 1991). The elderly cohort is open-ended at the 'top' but defined at the 'bottom' by retirement, a social construct. Irrespective of whether an individual loses the capacity to be 'productive' at 40 or 80, legislated rights to state-sponsored welfare, generally a pension, and services at an arbitrary age (usually 65) has legitimized analysis of the distribution of this 'dependent' population group.

In the rural context, analysis of population structures in particular places or regions is often cast in comparative terms; proportions of population in the 65 or older age group are compared, with some populations being characterized as more 'aged' than others (Chalmers and Joseph, 1997; Bryant and Joseph, 2001; Rosenberg and Moore, 2001). The prime source of information for these descriptions of rural populations is the national quinquennial or decennial population census. Extensions of place-focused demographic analysis often feature consideration of migration, through which ageing in place governed by mortality patterns is modified by structured flows of older adults. Migration effects are sometimes estimated through direct analysis of flows or, more commonly, imputed through methods such as Cohort Survival Analysis (Chalmers and Joseph, 1997). Related behavioural studies probe motivations for migration (Lewis, 1998), especially at the point of retirement.

All these studies, explicitly or implicitly, *classify* the elderly and in one way or another *read meaning* from their presence (e.g., see Joseph and Cloutier, 1991). Thus, rural communities rich in amenities and services may report percentages well above national or regional norms in the 65 to 74 age group because of net in-migration. This is often characterized as good for communities; the young elderly bring income into communities and ask for little support (Joseph and Cloutier, 1991). For instance, in the North American literature dealing with 'Snowbirds' (McHugh, 2000: 84), these seasonal migrants to Florida and other climatically favourable destinations are depicted as having 'health and vigor, expression of autonomy and independence in ageing, financial wherewithal and a sense of adventure in exploring new places and people'. In contrast, high proportions of older, 'dependent' elderly people in rural communities are interpreted in the opposite way

(Joseph and Chalmers, 1998). In this manner, demographic analysis, itself based upon the social construction of retirement, becomes the basis for the social construction of implications for communities.

Political economy and rural places: implications for the elderly

Political economy adds the rich layer of history to constructions of the elderly based on chronological age. We suggest that by virtue of their 'experience in the world' the elderly have a temporally situated perspective on issues in political economy, but that this is not well recognized in macro-constructions of the elderly as a group within national economic policy settings. The elderly are not passive recipients of welfare transfers; through multiple lives lived, they have been actively involved in the construction of contemporary social and political systems. Major changes in these systems (principally those controlled by the state) were engineered in the last two decades of the twentieth century. These changes were generally made without the assent of groups increasingly dependent on state support for health and welfare. While Joseph and Chalmers (1999) document the resistance of elderly people threatened by change in social policy and local provisioning, we see the dominant story as one of systematic exclusion from policy debates.

Within the political economy framework, we suggest that national economies characterized as 'post-Fordist' (Amin, 1994) point to and produce particular views of the elderly. For example, a process euphemistically known as 'economic restructuring' was introduced in New Zealand by an incoming Labour government in 1984. The key elements of this process have been described as those of a 'Schumpeterian workfare economy' (Jessop, 1994: 263), the signature characteristics of which include an increasing structural competitiveness on the supply side, subordination of social policy to the needs of labour market flexibility, and promotion of productivist social policy over consideration of re-distributive welfare rights. Elderly people, who had contributed to an effective Keynesian welfare state in New Zealand since 1935, with state support implicitly available from 'the cradle to the grave', were faced with change, along with many other sector groups. The 'otherness' of the elderly was distinctive within the restructuring environment, however. The elderly were seen as consumers of resources (typically through funding for superannuation) and conspicuously lacking capacity to adapt (re-train, re-habilitate, re-locate) to new conditions. There is little evidence that the new forms of rural governance (Little, 2001) emerging in a post-Fordist environment will offer much new support for the rural elderly; the state has, in a sense, produced the definitive 'end game'.

Within the substantial literature on the social, political and economic restructuring of national economies in the late twentieth century there is an identifiable body of work focusing on the rural sector. Within this work, and for the purpose of our review of otherness later in the chapter, we identify two broad processes of change, one focused on the elderly as a sector group and the other on rural spaces and places within the national space-economy. The former features the weakening of state commitments to health and social support systems, and includes the privatization of caring services previously regarded as core components of the welfare state (Barnett, 1999). The latter embraces the transformation of rural communities as places in which change is experienced, and features the withdrawal of state support for employment in hitherto unproblematic rural infrastructure and industry (Le Heron and Pawson, 1996). The rural elderly live their lives at the intersection of these processes of change.

Ethnographies: making sense of the lives of the elderly

In contrast to the work done with aggregate cohort data and/or distant actors in the political economy, investigations of the lived experience of individual elderly people build outwards to an understanding of the ways that they are constructed as other in rural places. We use the term ethnography to describe a range of methods that seek to uncover the lived experience of the rural elderly; the methods range from qualitative data collection in surveys to participant-observation. Deconstructions of rural experience are not new, and there are interpretations of lives of the rural elderly that sit alongside the macro explanations offered in political economy approaches (e.g. see Harper, 1987). Central to these investigations is engagement with the concept of community. Liepins (2000) notes that community is constructed and maintained through a wide set of meanings and practices. We will argue below that, for the elderly, these *meanings* are richly textured and *practices* are deeply ritualized in everyday life.

The call from Philo (1992, 1993) to focus attention on neglected groups in rural studies was perhaps inevitable, given the opportunities created by the turbulence of new ideas in social science. Indeed, a number of scholars in rural studies followed personal research trajectories that replicate the examples reported in Cloke et al. (1994: 162) and summarized in Phillips (1998). Engagement with the life-worlds of neglected groups rekindled interest in ethnographic method. While there are reservations about method and practice in ethnographic research, Herbert (2000) and Hall and Kearns (2001) argue that ethnographic methods are complex but essential. Narratives of the elderly are essential to the recognition of life-worlds in rural places, especially as this relates to a broadening understanding of the impact of disability, at any age, on the lived experience of change. It is our view that the voices of elderly people help to bridge the mind–body binary that currently characterizes disability studies; no other socially constructed group has such access to lived experience.

CONTEMPORARY CONSTRUCTIONS OF THE ELDERLY AS OTHER IN RURAL PLACES

We have discussed in general terms the processes by which otherness is produced at the level of the reference group ('the elderly cohort', 'pensioners') and noted how these views might become inscribed in the lives of individual elderly people. The process of inscription flows in two directions. Flowing from the community towards the elderly are normative expectations of dependency and disability as people age. The reciprocal flow occurs when the condition and circumstances of particular frail elderly people are extrapolated falsely to the group at large. The distinction we point to is captured in the subtle separation of *the elderly* and *elderly people*; the former discourages consideration of individual experience, while the latter does not.

We have noted that the collective called 'the elderly' is often constructed in universal space. We focus now on constructions of the elderly and elderly people in *rural* places. We argue that while changes in the political fabric of society (like policy on retirement) affect all elderly people, changes in the rural sector take these effects to the local and specific; the rural setting produces subtle and particular forms of construction (Rowles, 1988). We draw on our New Zealand research experience to illustrate the multiplicity of otherness, working away from the macro constructions of political economy (the rural elderly as discarded other and as consumers) towards those generated by particular experiences in place (the rural elderly on the other side of the technological divide) and in the lives of rural people (the elderly as disabled other). We also recognize that there may be competing or overlapping sources of otherness, and this is illustrated with reference to the elderly as ethnic other. Throughout, we draw on the viewpoints expressed by elderly people themselves in 1992 (and these verbatims can be found in Chalmers and Joseph, 1998, and Joseph and Chalmers, 1998). This 'insider' perspective leads us to add a sixth perspective – the elderly as nostalgic other.

We make reference to our work in rural Waikato, and Tirau in particular (Figure 28.2). For nearly a century Waikato has been the centre of the New Zealand dairying industry, and pastoral agriculture has provided the economic base for dozens of rural communities like Tirau (population 705 in 2001). The changes of the last decade have had a significant effect on these communities, and have transformed lives and expectations of those who have lived (often for decades) within them.

The elderly as the discarded other in rural communities

A state-supported pension at age 65 is a feature of social policy in many countries. This universal benefit is the most fundamental driver of separation between those defined as 'elderly' and 'not elderly' in rural places, as it is more broadly. Those who receive such income support are cast as dependent upon society at large and dependants in their communities, even though they may have contributed to superannuation schemes throughout their working lives and to their community as taxpayers and volunteers after retirement (Joseph and Chalmers, 1998).

In New Zealand, and probably elsewhere, most elderly people living in rural communities are long-term, if not life-long, residents (Chalmers and Joseph, 1997). Active engagement, often in the agricultural workforce, gave these people income and standing in the places they live, but the act of retirement produced a significant change in profile in the community. The change in economic capacity, and transfer to 'beneficiary' status, separates men in particular from mainstream life in rural places. They stand out in daytime streetscapes dominated by the structured comings and goings of workers, women and

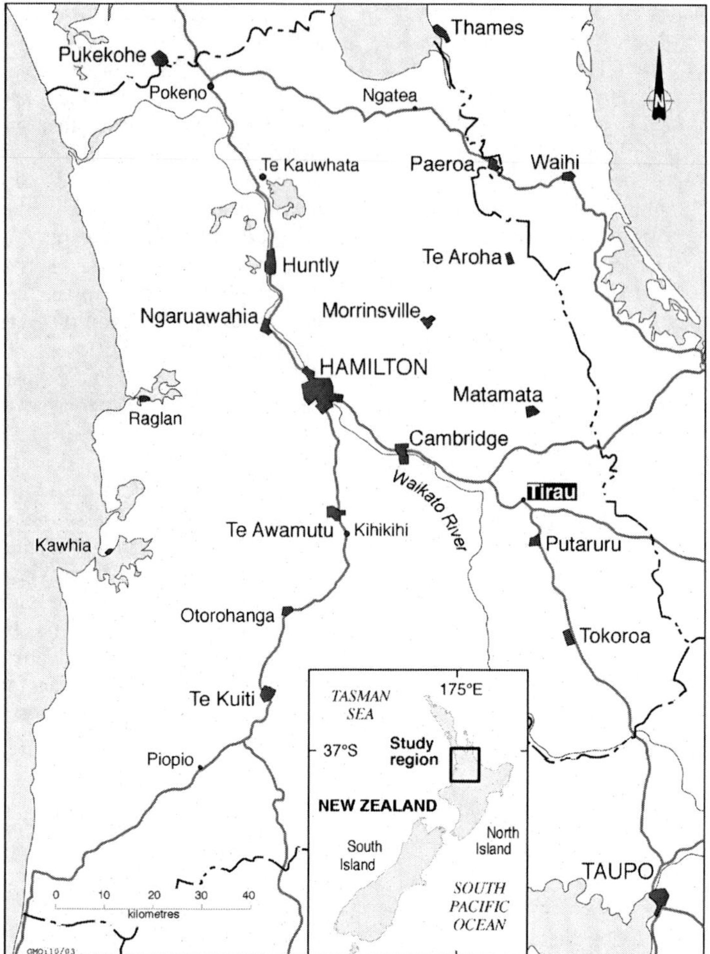

Figure 28.2 ***Rural Waikato and Tirau***

children. We saw this alienation from the 'mainstream', working population echoed in general resentment against younger people, '*Tirau used to be a caring community, but it is all young ones here now. I make myself known but they don't want to know,*' and in expressions of withdrawal, '*I'm no longer involved with the church; they are mainly young people there now.*'

Despite having departed the workforce, from the political economy perspective elderly people still maintain their status as consumers. Thus, the tendency to 'discard' the elderly may be selective because some elderly people may be conspicuous consumers. They are certainly seen in this light by those selling age-targeted accommodation and services (Joseph and Chalmers, 1996). In the rural context, 'retirement villages' are usually marketed to elderly newcomers drawn to rural communities by amenity and lifestyle considerations. These newcomers often bring with them equity derived from the sale of city properties, so at least in the short term they may be cast as 'consumers' rather than discarded as beneficiaries (Joseph and Cloutier, 1991).

The rural elderly as consumers

The consumption of goods and services related to the maintenance of lives in place provides the basis for aggregate manifestations of otherness. The elderly can, at once, be seen as a subset of consumers affected by broad changes in provisioning systems and as a target group for age-specific provisioning, and this is used as an organizing principle here.

The elderly in Tirau were outspoken about the withdrawal of services in the late 1980s. As one elderly resident put it, '*by taking services away from Tirau it's no longer attractive for older people to live here.*' Our interviews revealed that the (restructuring-generated) closure of local banks, post offices and shops had an effect on the elderly not found so pervasively in other groups (Joseph and Chalmers, 1995). Rural lives are sustainable only if increased mobility can compensate for this reduction in local services: elderly people in Tirau reported increasing concerns about their capacity to move freely within the community and a reluctance to drive significant distances, especially at night. They recognized that in their service-depleted communities, '*giving up driving would be dynamite*', and replete with implications for the maintenance of independence. As one elderly resident summed it up, '*I'd have to rely on others if I wanted to go out of Tirau. That takes away your independence, which is such an important asset.*'

Concerns about (long-term) accommodation are characteristic of all elderly people, but the production and consumption of accommodation creates distinctive environments for those in rural areas. As Laws (1993) notes, the creation of retirement villages for elderly people with resources became an enormous commercial enterprise in the last three decades of the twentieth century. 'Security' and 'accessibility' are features of environments designed for elderly living, and the benefits of intergenerational residential segregation (Harper, 1997) are rarely challenged. Our experience indicates that migration to amenity-rich environments is not uncommon, but local movement to state or community-supported retirement and care facilities (often in nearby peri-urban areas) is the expectation of the majority of rural elderly people (Joseph and Chalmers, 1996). As one of our Tirau respondents expressed it, '*When you get really old, you must leave the community and go into a home somewhere.*' Another made the link back to the absence of community-based support services, '*People could manage in their own homes if they could get help or assistance. They wouldn't need to go to the Old Folks' Home.*'

While new rural partnerships involving for-profit and non-profit organizations may be negotiated in the post-Fordist era, the universal influence of the state will play through into the lives of elderly people in rural places through the provision of services and accommodation (Jones and Little, 2000). Our work in Waikato (Joseph and Chalmers, 1999) emphasizes the importance of continued state support of accommodation for those elderly people with very limited financial resources, and brought home to us how short a step it is for some elderly people to slip from being consumers of services to being clients of the state.

The rural elderly on the 'other side' of the digital technology divide

Early in the last century, Fiske (1912) recognized the capacity of technology to transform rural communities. During the lifetime of those now over the age of 64, the widespread adoption of the telephone and increasing car ownership have been the transformational agents of major effect; rural space has been changed irrevocably by them.

In contrast to earlier, universalist changes driven by the state (for example, the development of state-run telecommunications and rural road improvement), those in rural place have recently been expected to adopt new technologies individually. Some, however, have been excluded by lack of resources or by the absence of connectivity to national networks. The question we now explore is the extent to which the latest technological transition, that driven by digital technology (Grimes, 2000; Malecki, 2003), has influenced the lives of the elderly in rural communities.

With respect to everyday lives, the two services on which the rural elderly are most dependent are postal and banking services. Both have been affected by 'restructuring' in the past ten years, and our work in Waikato indicates that the physical withdrawal of these services and their digital reconstitution has the effect of putting the elderly on the 'other side' of the digital divide. We noted the resistance of the elderly, who value face-to-face contact, to new technologies such as the ATM (Joseph and Chalmers, 1998), with many of our respondents agreeing with Argent and Rolley's (2000) assertion that local bank branch closures have left a 'hole' in the community. The feeling of exclusion flowing from the pervasive introduction of new technology in banking is captured elsewhere in New Zealand in a contribution in the *Southland Times*:

> Once again the rural areas and the seniors of our communities are being passed over in favour of the almighty dollar ... try explaining to a person of 80 years plus that they may no longer walk down to their local bank and withdraw money to go and buy their little bits and pieces – telebanking, EFTPOS [Electronic Fund Transfer at Point Of Sale] machines and ATM machines are just a foreign language to them. (Thornton, 1997)

Demands for familiarity with digital technology as a prerequisite for access to the day-to-day necessities of living in rural communities systematically discriminates against the elderly (Joseph and Chalmers, 1998). However, there is evidence in SeniorNet that some elderly people have found that 'going digital' greatly enriches lives in rural places, but for most there is only a remote chance that digital access to goods and information will compensate for the reduction of local retail outlets and services. We see the continued preference of the elderly for 'talk-back' radio, that makes few demands on technological competence (Roberts, 2001), as further indication of the desire to avoid the new and complex.

The rural elderly as disabled other

Harper (1997) explores the characteristics of individuals that define their experience in rural places. She distinguishes male and female experience, and points out that in the ageing process increasing physical and mental limitations determine individual lives. A focus on disability is instructive in our consideration of elderly people in rural areas. Hall and Kearns (2001) review contrasting approaches to disability, noting three waves of responses to the de-institutionalization of the 'disabled'. They point to access and service supply work with the physically challenged, and with the mentally sick, but note only limited research on the mentally challenged. The second wave of research, where 'disability was rethought as a socially constructed entity, rather than a biological certainty' (Hall and Kearns, 2001: 238), provides a useful critique of the able-ist view of the elderly. The progressive physical frailties of advancing years are well known, and mental conditions of dementia and Alzheimer's are part of common constructions of the elderly. As Harper (1997: 187) states, 'old people are defined by the ageing body, young people are not'; the ageing body is the sign, 'disability' is the signifier, and the process of exclusion is cued to it.

In rural environments, Milligan (1999: 230) notes that 'the difficulty of maintaining confidentiality in small towns and the stigma and social conspicuousness of those with MIH [Mental Ill Health] in rural as compared to urban environments' produces some exclusionary practices. More broadly speaking, pensioner housing and small residential care facilities are usually recognized as distinct and separate spaces in rural communities; the de-institutionalized parallel of an asylum without walls is the village within a village. In some cases, local resistance to the establishment of care facilities for elderly disabled people formalizes the exclusionary practice of rural communities. The consequences are clear – 'a lack of exposure in civic life may serve to inhibit public awareness of difference' (Hall and Kearns, 2001: 240).

Among the rural elderly people of the Waikato we found stoicism and a resistance to construction as a dependent or disabled person (Chalmers and Joseph, 1998; Joseph and Chalmers, 1998). There were accounts of both mutual and local support for daily living of elderly people. As one of our respondents noted in a matter-of-fact way, '*I help older people, I found two of them dead.*' Another noted, '*The district nurse keeps an eye on me so I can manage a bath.*' While care-within-the-community has been a keystone of disability services in the past decade, such services are not available in most rural areas and elderly people have been forced to look beyond their local community for such services (Joseph and Chalmers, 1996).

The rural elderly as ethnic other

Our awareness of Maori cultural practice, including language teaching, resource use and local governance, encouraged us to look at the construction of the elderly in rural places where ethnic diversity was a factor. Maori migrated to urban areas in significant numbers in the 1960s, but have returned to their rural *marae* (home setting) under the pressures of economic restructuring (Kearns and Reinken, 1994). While the increasing number of Maori in rural places has put pressure on local resources, the change in status with age is often the reverse of the experience in Pakeha (non-Maori) communities, and very distant from the theorization of ethnicity in rural places presented in Agyeman and Spooner (1997). Maori have respect for elders (*kaumatua*/men and *kuia*/women) for their knowledge of *tikanga Maori* (Maori protocol and practice), and their maintenance of *te reo* (language) during the urban diaspora. This respect has been carried through explicitly to the realm of political economy and is manifest in issues of governance (Little, 2001). Much rural planning requires consultation with Treaty of Waitangi partners under the 1991 Resource Management Act. *Kaumatua* and *kuia* are central in this process.

Three generalizations emerge from our exploration of New Zealand rural communities with significant Maori populations. In some of these communities, the language and cultural practice has been sustained by elderly people in the face of threats from a homogenized global culture imported from urban places. In some cases this resilience has had mixed benefits; the distrust of state-supported health services and the preference for traditional medicines and practitioners

(*Tohunga*) has not always had good effects. Hirini et al. (1999) note the continued patterns of chronic illness among (elderly) Maori and note a reliance upon 'providers such as community based tribal or Maori services and traditional Maori healers' (Hirini et al., 1999: 147). The second generalization is that emphasis on discrete nuclear families is less evident in some Maori communities, and community inclusiveness has valued roles for elderly people, especially on the *marae* (meeting/speaking place). The final generalization relates to tensions that arise in rural communities where ethnicity provides the basis for unequal access to resources (Scott et al., 2000).

The elderly as nostalgic other

The constructions of otherness we have reviewed are formalisms derived from lines of significant research on the elderly generally and on elderly people in rural places. Our field experience suggests that the layering of human experience in time and space is too complex to render completely in this way, and our final comments relate to the subtle narratives that contribute to contested rural living in Tirau. We build on the 'voices' of the elderly themselves, as they comment not on their own personal position, but on the position of older people in rural communities and on the changing shape of place.

Our interest is in the way the local community is represented. We read the comments of the elderly as 'nostalgia', and suggest that while the speakers identify with Tirau, there is an element of regret about the inevitability of change and, implicitly, the potential for progressive separation from community.

> A house is just a house, it's the people who make your lives.
>
> Tirau means a lot to me, a big lot. I know everybody. You can have a yarn with somebody at the letterbox.
>
> I go to the hotel every night for a couple of handles, and to meet my friends. Tirau is where I live and where I enjoy living. I feel like an insider. People ask my advice.
>
> Communities alter. People don't visit like they used to. I feel we don't have a community.

Nostalgia is both comforting and threatening at a personal level. The threat lies in romantic envisioning of the past, and the potential this has for separating the (elderly) viewer from contemporary rural living. Collective nostalgia about rural life reinforces some idyllic myths. Phillips et al. (2001) have drawn attention to the role of popular culture, and radio/television in particular, in the portrayal of the rural. Elderly people are not a part of the MTV generation, but form part of the 'Archers market' (BBC radio serial) for constructions of the rural, a market well supplied with a string of 'quality' British TV programmes internationally. These programmes fuel nostalgia for a golden era, a generation (30 years) or more removed from the present. Legitimate and accurate representations of former realities or not, this material provides the basis for contemporary (re)constructions of place by long-term rural residents, and the related contestation of time and place experience.

We suggest that nostalgic views of place framed by the media diminish the importance of lived experience, of 'situated nostalgia', in the contemplation of rural futures and new rural governance (Jones and Little, 2000). 'Restructuring' had impacts on the rural, and long-term residents understand these better than most. The views they offer may be dismissed as nostalgic, but they could also be taken as part of a prescription for sustainable rural communities which provides at least an implicit role for older people. As one of our respondents declared, '*people should give back to the community if they enjoy living here*.' However, it is also clear that there will continue to be competition for the right to imagine community futures.

CONCLUSION

We suggest that the conventional templates introduced at the beginning of the chapter have provided useful lenses for our re-consideration of the construction of the elderly as other in rural places. We note that, characteristically, many jurisdictions have broad policies in place to address issues associated with exclusion and marginalization of individuals in the rural sector. However, these policies often lack coherence; they simultaneously create and ameliorate otherness amongst the elderly as a collective and produce and reduce marginalization for individual elderly people. Thus elderly people can, depending on the policy reference point, be cast as dependants or consumers, service users or volunteer providers. While it has to be recognized that even the best-intentioned policy can do little to compensate for increasing frailty, acute illness or bereavement (Joseph and Chalmers, 1998), it can be argued that, in comparison with other groups, the elderly are systemically disadvantaged by the dominant construction of their otherness. For example, in New Zealand during the 1990s Maori gained improved access to resources for development, and women secured greater access to opportunities and resources. With respect to the elderly, these observations underscore the continuation of exclusion on the basis of a fundamental distinction: resources directed to the elderly are cast as a current cost rather than as an investment in the future. Agencies filling the gaps created by the 'hollowing out of the state' (Joseph and Chalmers, 1999: 165) are able to point to the long-term benefits of investment in community development for most excluded groups, but supportive measures for those characterized as elderly are more likely to be represented as 'entitlements'. New support is thus made less likely as communities focus on development potential (Joseph and Chalmers, 1998).

The bleak assessment of the (excluded) elderly as a drain on community futures imagined in terms of economic development feeds through to the media representations of rural communities. Sadly, this is nowhere better represented than in the small rural town in Waikato that has been the focus of our project. Between September 2000 and March 2001, Tirau received press and television coverage focusing on the change from a point 'in the 1980s [when] tiny Tirau was dying' to the current 'Tirau transformed into Waikato's boom town' (Taylor, 2001). The coverage lauds the action of new entrepreneurs who have drawn investment by 'branding' Tirau as a destination and developing iconic local features (a 'castle', a sheepdog-shaped information centre and a host of antique shops) along State Highway 1. The transformation bypasses the lived experience of those ageing-in-place in Tirau and is symptomatic of the general tendency for economic development and social capital to proceed on different trajectories (e.g., see Merrett, 2001).

Without specific attention to the lives of everyone in rural communities, survivability and sustainability are challenged. In rural places like Tirau, the debate about sustainability remains constructed largely in terms of the dominant hegemony. Measures of success remain couched in terms of entrepreneurial values and not lived experience. We argue that two sets of actions are important in the context of new research about the rural elderly: challenging constructions of otherness and reducing barriers to inclusive rural communities. Visibility in rural places promotes local awareness and valorization of the elderly as community members. Roles as retainers of community history and providers of experience are obvious pointers to the reconstruction of the elderly as *valued other* within rural places, and their roles as entrepreneurs and commercial leaders remain unexplored.

In terms of future directions for research, we see opportunities to (re)link considerations of the 'elderly as other' with complementary research interests in rural studies. While we share the reservations expressed by Rowles (1988) about comparative (urban–rural) studies, we see value in comparative research that preserves and valorizes the embeddedness of life-worlds in communities and the contingencies of personal circumstances. Given the progressive engagement with Philo's (1992) original agenda and the expansion of research on young rural lives (Panelli, 2002), studies immersed in overlapping intergenerational life-worlds seem both feasible and intellectually intriguing. In New Zealand, the distinctive nature of Maori beliefs and practices associated with ageing and the elderly offer scope for research that probes ethnic constructions of otherness.

We also see opportunities for breaking the confines imposed by the categorizations coded within this chapter. In our own work we are increasingly questioning the casting of elderly people as (only) consumers in the process of commodification that is increasingly permeating rurality in the twenty-first century (Cloke and Goodwin, 1992; Hoggart et al., 1995; van Dam et al., 2002). Older/elderly rural people seem to be involved as new-start entrepreneurs, facility operators and community promoters. This is not to say that we do not see continued utility in the categorization of otherness advanced earlier – recent work in rural mental health (Philo et al., 2003) and rural homelessness (Cloke et al., 2002) ensures that new implications for the (all too often invisible) rural elderly as other will continue to emerge.

NOTE

1 We use the term elderly as a loose expression generally describing individuals over the age of 64. Where we need to distinguish further culturally constructed categories in the substantive discussion that follows, we use appropriate or internationally recognized qualifiers.

REFERENCES

Agyeman, J and Spooner, R (1997) Ethnicity and the rural environment. In P Cloke and J Little (eds), *Contested Countryside Cultures*. London: Routledge. pp. 197–217.

Amin, A (ed.) (1994) *Post-Fordism: A Reader*. Oxford: Blackwell.

Argent, N and Rolley, F (2000) Financial exclusion in rural and remote New South Wales: a geography of bank branch rationalisation, 1981–1998. *Australian Geographical Studies* 38: 182–203.

Barnett, R (1999) Hollowing out the state? Some observations on the restructuring of hospital services in New Zealand. *Area* 31 (3): 259–270.

Britton, S, Le Heron, R and Pawson, E (1992) *Changing Places in New Zealand: A Geography of Restructuring*. Christchurch: New Zealand Geographical Society.

Bryant, C and Joseph A (2001) Canada's rural population: trends in space and implications in place. *Canadian Geographer* 45 (1): 132–137.

Cant, G (1980) Rural depopulation: patterns and processes. In Cant, G (ed.), *People and Planning in Rural Communities*. Studies in Rural Change, Number 4. Christchurch, New Zealand.

Chalmers, A and Joseph, A (1997) Population dynamics and settlement systems. *New Zealand Geographer* 53 (1): 14–21.

Chalmers, A and Joseph, A (1998) Rural change and the elderly in rural places: commentaries from New Zealand. *Journal of Rural Studies* 14 (2): 155–165.

Cloke, P (1989) State deregulation and New Zealand's agricultural sector. *Sociologia Ruralis* 29: 34–48.

Cloke, P and Goodwin, M (1992) Conceptualizing countryside change: from post-Fordism to rural structure coherence. *Transactions of the Institute of British Geographers*. NS 17: 321–336.

Cloke, P and Little, J (eds) (1997) *Contested Countryside Cultures: Otherness, Marginalisation and Rurality*. London: Routledge.

Cloke, P, Doel, M, Matless, D, Phillips, M and Thrift, N (1994) *Writing the Rural: Five Cultural Geographies*. London: Paul Chapman Publishing.

Cloke, P, Milbourne, P and Widdowfield, R (2002) *Rural Homelessness: Issues, Experiences and Policy Responses*. Bristol: Policy Press.

Day, G and Murdoch, J (1993) Locality and community: coming to terms with place. *Sociological Review* 41 (1): 82–111.

Fairweather, J (1992) *Agrarian Restructuring in New Zealand*. Research Report 213, Agribusiness and Economics Research Unit, Lincoln University.

Fiske, G (1912) *The Challenge of the Country*. New York: Association Press.

Franklin, H (1969) The age and sex structure of North Island communities. In J Forster (ed.), *Social Process in New Zealand*. Auckland: Longman Paul. pp. 49–76.

Franklin, H (1978) *Trade Growth and Anxiety: New Zealand Beyond the Welfare State*. Wellington: Methuen.

Giddens, A (1984) *The Constitution of Society: Outline of the Theory of Structuration*. Los Angeles, CA: University of California Press.

Grimes, S (2000) Rural areas in the information society: diminishing distance or increasing learning capacity? *Journal of Rural Studies* 16 (1): 13–21.

Halfacree, K (1993) Locality and social representation: space, discourse and alternative representations of the rural. *Journal of Rural Studies* 9 (1): 1–15.

Hall, E and Kearns, R (2001) Making space for the 'intellectual' in geographies of disability. *Health and Place* 7 (3): 237–246.

Harper, S (1987) A humanistic approach to the study of rural populations. *Journal of Rural Studies* 3 (4): 309–319.

Harper, S (1989) The British rural community: an overview of perspectives. *Journal of Rural Studies* 5 (2): 161–184.

Harper, S (1997) Contesting later life. In P Cloke and J Little (eds), *Contested Countryside Cultures: Otherness, Marginalisation and Rurality*. London: Routledge, pp. 180–196.

Heenan, B (1979) Internal migration: inventory and appraisal. In R Neville and C O'Neill (eds), *The Population of New Zealand: Interdisciplinary Perspectives*. Auckland: Longman Paul. pp. 60–88.

Herbert, S (2000) For ethnography. *Progress in Human Geography* 24 (4): 550–568.

Hirini, P, Flett, R, Long, N, Millar, M and MacDonald, C (1999) Health care needs for older Maori: a study of kaumatua and kuia. *Social Policy Journal of New Zealand* 13: 136–153.

Hoggart, K, Buller, H and Black, R (1995) *Rural Europe: Identity and Change*. London: Edward Arnold.

Jessop, B (1994) Post-Fordism and the state. In A Amin (ed.), *Post-Fordism: A Reader*. Oxford: Blackwell. pp. 251–279.

Jones, O and Little, J (2000) Rural challenge(s): partnership and new rural governance. *Journal of Rural Studies* 16 (2): 171–183.

Joseph, A and Chalmers, A (1995) Growing old in place: a view from rural New Zealand. *Health and Place* 1 (2): 79–90.

Joseph, A and Chalmers, A (1996) Restructuring long-term care and the geography of ageing: a view from rural New Zealand. *Social Science and Medicine* 42 (6): 887–896.

Joseph, A and Chalmers, A (1998) Coping with rural change: finding a place for the elderly in sustainable communities. *New Zealand Geographer* 54 (2): 28–36.

Joseph, A and Chalmers, A (1999) Residential and support services for older people in the Waikato, 1992–1997: privatization and emerging resistance. *Social Policy Journal of New Zealand* 13: 154–169.

Joseph, A and Cloutier, D (1991) Elderly migration and its implications for service provision in rural communities: an Ontario perspective. *Journal of Rural Studies* 7 (4): 433–444.

Joseph, A and Fuller, A (1991) Towards an integrative perspective on the housing, services and transportation implications of rural aging. *Canadian Journal on Aging* 10 (2): 127–148.

Joseph, A and Martin-Matthews, A (1993) Growing old in aging communities. *Journal of Canadian Studies* 28: 14–29.

Joseph, A, Lidgard, J and Bedford, R (2001) Dealing with ambiguity: on the interdependence of change in agriculture and rural communities. *New Zealand Geographer* 57 (1): 16–26.

Kearns, R and Joseph, A (1997) Restructuring health and rural communities in New Zealand. *Progress in Human Geography* 21 (1): 18–32.

Kearns, R and Reinken, J (1994) Out for the count? Questions concerning the population of the Hokianga. *New Zealand Population Review* 20: 19–30.

Keating, N and Little, H (1994) Getting into it: farm roles and careers of New Zealand farm women. *Rural Sociology* 59 (4): 720–736.

Krout, J (1988) The elderly in rural environments. *Journal of Rural Studies* 4 (2): 103–114.

Laws, G (1993) 'The land of old age': society's changing attitudes towards urban built environments for elderly people. *Annals of the Association of American Geographers* 83 (4): 672–693.

Le Heron, R and Pawson, E (1996) *Changing Places: New Zealand in the Nineties.* Auckland: Longman Paul.

Lewis, G (1979) *Rural Communities. Problems in Modern Geography.* Newton Abbot: David and Charles.

Lewis, G (1998) Rural migration and demographic change. In B Ilbery (ed.), *The Geography of Rural Change.* Harlow: Longman. pp. 131–160.

Liepens, R (2000) Exploring rurality through 'community': discourses, practices and spaces shaping Australian and New Zealand rural 'communities'. *Journal of Rural Studies* 16 (1): 83–89.

Liepins, R and Bradshaw, B (1999) Neo-liberal agricultural discourse in New Zealand: economy, culture and politics linked. *Sociologia Ruralis* 39 (4): 563–582.

Little, J (1999) Otherness, representation and the rural construction of rurality. *Progress in Human Geography* 23 (3): 437–442.

Little, J (2001) New rural governance? *Progress in Human Geography* 25 (1): 97–102.

McHugh, K (2000) Inside, outside, upside down, backward, forward, round and round: a case for ethnographic studies in migration. *Progress in Human Geography* 24 (1): 71–90.

Mckay, P (1984) *Rural Health and Related Social Issues.* Northland Community College, Kaikohe.

Malecki, E (2003) Digital development in rural area: potentials and pitfalls. *Journal of Rural Studies* 19 (2): 201–214.

Massey, D (1991) The political place of locality studies. *Environment and Planning A* 23 (2): 267.

Merrett, C (2001) Declining social capital and non-profit organizations: consequences for small towns after welfare reform. *Urban Geography* 22 (5): 407–423.

Milligan, C (1999) Without these walls: a geography of mental ill-health in a rural environment. In R Butler and H Parr (eds), *Mind and Body Spaces.* London: Routledge. pp. 221–239.

Moran, W, Blunden, G and Greenwood, J (1993) The role of family farming in agrarian change. *Progress in Human Geography* 17: 22–42.

Murdoch, J and Pratt, A (1993) Rural studies: modernism, postmodernism and the 'post rural'. *Journal of Rural Studies* 9 (4): 411–427.

Murdoch, J and Pratt, A (1994) Rural studies of power and the power of rural studies: a reply to Philo. *Journal of Rural Studies* 10 (1): 83–87.

Newby, H (1986) Locality and rurality: the restructuring of rural social relations. *Regional Studies* 20 (3): 209–215.

Panelli, R (2002) Young rural lives: strategies beyond diversity. *Journal of Rural Studies* 18 (2): 113–122.

Pawson, E and Scott, G (1992) The regional consequences of economic restructuring: the West Coast, New Zealand (1984–1991). *Journal of Rural Studies* 8 (4): 373–386.

Phillips, M (1998) Social perspectives. In B Ilbery (ed.), *The Geography of Rural Change.* London: Addison–Wesley–Longman.

Phillips, M, Fish, R and Agg, J (2001) Putting together ruralities: towards a symbolic analysis of rurality in the British mass media. *Journal of Rural Studies* 17 (1): 1–27.

Philo, C (1992) Neglected rural geographies: a review. *Journal of Rural Studies* 8 (2): 193–207.

Philo, C (1993) Post-modern rural geography? A reply to Murdoch and Pratt. *Journal of Rural Studies* 9 (4): 429–436.

Philo, C, Parr, H and Burns, N (2003) Rural madness: a geographical reading and critique of the rural mental health literature. *Journal of Rural Studies* 19 (2): 259–281.

Roberts, P (2001) Electronic media and the ties that bind. *Generations* 25 (2): 96–98.

Rosenberg, M and Moore, E (2001) Canada's elderly population: the challenges of diversity. *The Canadian Geographer* 45 (1): 145–150.

Rowles, G (1988) What's rural about rural ageing? An Appalachian perspective. *Journal of Rural Studies* 4 (2): 115–124.

Scott, K, Park, J and Cocklin, C (2000) From 'sustainable rural communities' to 'social sustainability': giving voice to diversity in Managakahia Valley, New Zealand. *Journal of Rural Studies* 16 (4): 433–446.

Taylor, K (2001) Tirau transformed into Waikato's boom town. *New Zealand Herald* Monday, 26 March, p. A8.

Thornton, A (1997) Elderly disadvantaged by bank closures. *Southland Times*, Edition 1, 22 April, p. 4.

van Dam, F, Heins, S and Elbersen, B (2002) Lay discourses of the rural unstated and revealed preferences for rural living. Some evidence of the existence of a rural idyll in the Netherlands. *Journal of Rural Studies* 18 (4): 461–476.

Williams, R (1973) *The Country and the City.* London: Chatto and Windus.

Wilson, O (1994) They changed the rules: farm family responses to agricultural de-regulation. *New Zealand Geographer* 50 (1): 3–13.

H Exclusion

29

Inclusions/exclusions in rural space

David Sibley

In this chapter, I will be reflecting on ways in which rural spaces are bounded through fear and anxiety and how rural communities, through political mobilization and legislation, attempt to exclude those who are deemed not to belong. I will not be attempting, however, to generalize about the power relations that secure the spaces of the dominant groups in rural societies and marginalize and exclude the weak. The geographies of inclusion and exclusion are rooted in particular histories and space-economies and it is difficult to avoid what might be deemed a parochial or, at best, a localized account of the issues. This is not to say that there are no theoretical cues with wide application relating, for example, to racism or to psychoanalytical concepts of fear but these have to be connected to the histories and geographies of the rural as they are configured in particular states and regions. Thus, rather than presenting a postmodern picture of the social and cultural spaces of the rural according to which, as Jameson (1991) has suggested, the world can be represented as one of differences rather than divisions, in the plane of geographical diversity rather than of historical succession, I will suggest that difference and division cannot be disconnected. I would follow Soja (1993: 113), who has argued that 'we must be open right from the start to the possibility of being modern and postmodern at the same time ... Analyzing and interpreting ... the simultaneous and uneven development of modernity and postmodernity is crucial to making practical and political sense of the contemporary world.' It is evident, for example, that during the period of high modernity when capitalist states were attempting to obliterate difference through practices of assimilation and incorporation, they were at the same time romanticizing and celebrating ethnic and racialized difference.[1] Similarly, today, we could argue that the rhetoric and practice of 'social inclusion' is simply incorporation in another guise, a process of fitting 'the excluded' into positions in the mainstream economy where they can be more efficiently exploited at the same time that some of these excluded groups are featured in seemingly positive and progressive representations of multicultural societies. Thus, the central theme of this essay is to understand how progressive states or 'cultures of progress',[2] to use characteristically modernist concepts, respond to difference, specifically in rural locales.

MARGINS, EXCLUSION AND RESISTANCE IN HISTORICAL PERSPECTIVE

Many of the ideas that have provided a foundation for recent thinking on exclusion/inclusion originated in urban research. W.E.B. DuBois, for example, in *The Philadelphia Negro* (1899), recognized the importance of boundary processes in the emergence of black settlement in North American cities, involving both processes of spatial exclusion operating in the housing market through the discriminatory practices of landlords, and self-exclusion which he attributed to well-justified fears of racism and common bonds provided by churches and other black institutions.

The bounding of black space, however, DuBois also attributed to discrimination against black workers in the job market, confining them to the poorest areas of the city. Thirty years later, Edith Abbott and her co-workers showed similar insight in their account of the marginalization of the Latino/ Latina and black populations of Chicago (Abbott, 1936). These writers had a good appreciation of the importance of power relations in the construction of social space (more so than most of their contemporaries in the Chicago School of Sociology[3]) although, in other respects, their theorizations of space were limited. There was, in fact, little that I would recognize as progress in thinking about issues of spatial exclusion until the appearance of work on the social psychology of space and psychogeographies,[4] starting with Richard Sennett's *The Uses of Disorder* (1970). Sennett's work is important because it was the first to make a case for connecting arguments about the development of the self, drawing on the writing of Erik Erikson (1970); about the gratifications and anxieties of groups, and the construction of social space.[5] Specifically, Sennett suggested a connection between a 'purified identity', associated with adolescence in 'normal' development, and purified communities/spaces, which had a low tolerance for social difference or spatial heterogeneity, that is, they were exclusionary social spaces. He then argued that the North American suburb was the archetypal purified social space. There were similar ideas emerging in social anthropology, following Mary Douglas's *Purity and Danger* (1966), which described taboos and purification rituals in tribal societies in rural Africa, and in sociology, particularly Basil Bernstein's thesis on the structuring and regulation of education which was readily translatable into the opposing spatial processes of purification and mixing (Bernstein, 1971). Exclusion was again central to Bernstein's argument, specifically, the rejection of knowledge which was judged to be polluting and transgressive in the context of highly structured, strongly classified, hierarchical systems.

These theoretical accounts of the production of exclusive and exclusionary groups and spaces which appeared in the late 1960s and 1970s were innovative and they have clearly influenced thinking in human geography and sociology on questions of boundary construction, inclusion and exclusion (Sibley, 1981, 1995; Cresswell, 1996; Hetherington, 1998). Sennett's early work certainly could be criticized for its employment of rather sweeping social and spatial categories, like 'the purified suburb', and for binary thinking which was a feature of all three authors' theorizing. Although I would argue that this can be quite creative, it might also be seen as a shortcoming in that it erases the 'spaces in-between' and ambiguous others, a problem recognized by Ed Soja (1996) but also considerably earlier by Gunnar Olsson (1980). Before studies from the mid-1990s onwards, which have been concerned specifically with issues of inclusion and exclusion in rural spaces, I can find no applications of Sennett, Douglas or Bernstein to rural questions in developed societies. There are, however, two significant engagements with rural societies, both by black women, which raise a similar point to Sennett's critique of North American cities. The first is Ingrid Pollard's photographic project on English rural landscapes in which she demonstrates the dominant whiteness of place by photographing her black African self in landscapes that are saturated with symbols of white national identity – and echoes of colonialism and imperialism. The second is bell hooks's observation on growing up in a small town in the southern states of the United States, where black space was a refuge and where she reinhabited 'a location where black folks associated whiteness with the terrible, the terrifying, the terrorizing' (hooks, 1992: 170). Such fears, and the violence of segregation in the rural American South, are aspects of black geographies that have rarely been articulated in the academic literature and this neglect is itself a disturbing exclusion.[6]

Pollard and hooks are important because they challenge white hegemony, not least, in authorship, but also because they signal resistance to physical and psychic exclusion and to dominant white categorizations of spaces and difference. It is necessary to think more about what exclusion means for those who are subject to it and who may resist both the process and the idea. This involves moving beyond those political and academic discourses of social exclusion which underpin western European and North American policy initiatives to consider 'other voices' and the experience of those on the margins. In this vein, it is worth noting a neglected commentary on the concept of marginality by Janice Perlman (1976: 72). Her concern was the welfare of the inhabitants of *favelas*, the squatter settlements in and around Rio de Janeiro, but her argument has much wider significance. As she observed: 'Paradoxically, the characteristic way to handle the dread of the masses is to profess a desire to integrate them into the very system which is producing the social and economic situation called "marginal". Highly partisan social labeling is thus attributed to both squatters and squatter settlements and it is transmitted through the socialization process.' Perlman recognized that we have to think critically about concepts such as 'margin', 'marginality' and 'exclusion' and ask, first, whether or not groups are, from their perspective, marginalized or excluded and, secondly, what the social and economic consequences are for those categorized in this way.

THE PRODUCTION OF EXCLUSIONARY RURAL SPACE

A few examples may serve to demonstrate the difficulty of generalizing about processes of exclusion in rural areas.[7] (1) Along the lightly populated north bank of the Red River in Oklahoma, the east–west highway connects small settlements with nothing but a store, a gas station and the trailer homes of native North Americans. In a North American context, this is clearly a racialized landscape of poverty, one that is a widespread and persistent feature of the rural United States. (2) Close to the Hungarian border with the Ukraine and Romania, around Nyíregyhaza, villages that were once populated by Hungarian farmers are now entirely Roma settlements, lacking most basic services. As the Hungarians have moved to the larger towns, the poorest of the Hungarian Roma have replaced them (Ládanyi and Szelényi, 1996). (3) In coastal Talamanca province in Costa Rica, the principal region of banana production by multinational corporations (Chiquita, Dole, Delmonte), the indigenous (Bribri) and the black population of island Caribbean origin are the poorest in the country, with living standards markedly lower than those of the farmers, primarily of European origin in the coffee and dairying regions of the central highlands. The rural geography of Costa Rica is strongly racialized and there are clear connections between income, power, race and space.

These three cases of rural marginality have common elements, principally racism, but each has to be understood also in terms of particular histories – in the case of Native Americans in Oklahoma, relocation in the nineteenth century as part of a wider pattern of oppression; for the Hungarian Roma, economic decline in rural Hungary following the end of the communist regime and the redundancy of the Roma in the labour force during the shift to a market economy; in Costa Rica, the operations of the United Fruit Company since the late-nineteenth century and current exploitation of plantation workers in the global economy, including Nicaraguan migrant workers over the border in Panama, and so on. One other common element is that ethnic difference has long been characteristic of these rural regions. Although there is segregation at the local level in all three cases and distanciation and closure shape the social geography of these regions, ethnic and racialized difference are unremarkable features of the rural landscape. There are currently no pressures to exclude any of these populations from rural space (although there were pressures from white farmers to expel black workers from Costa Rica early in the twentieth century). As in many rural economies, these racialized minorities are exploited but they are also an acknowledged part of the scene. Their labour is either a necessary part of the capitalist production process[8] or, as in the case of the Hungarian Roma, they constitute a residual labour force and an 'other' for workers in the formal economy. As Dwyer and Jones (2000) have remarked in relation to racialized difference in the United States, distancing – both social and spatial – occurs at both local and regional levels but racialized minorities are not wished away in imagined geographies of the rural, except perhaps in some fascist imaginings of rural Hungary.

To someone from outside the United Kingdom and some other European states, the above observations may seem commonplace – international migrations and internal movements of people have made rural spaces spaces of difference, where these differences are often reflections of a racialized division of labour. In the United Kingdom, however, and particularly in England, rural space has often been represented as not only ethnically homogeneous but as a depository of core values. Thus, rural England has been particularly difficult terrain for those who are seen not to subscribe to the core values of 'the rural community', a term that has been taken also to represent the values of the nation. I am talking here not about the ways in which rural society may be defined objectively but how it has been imagined by those who have an interest in rejecting diversity. This problem requires a rather different theoretical perspective to the questions of marginality indicated by the previous examples. Particularly, we have to understand how feelings about others translate into imaginary geographies and how these imaginary geographies influence policy and practice. One way into this is to think about some psychoanalytical interpretations of socio-spatial relations.

PSYCHOGEOGRAPHIES OF RURAL ENGLAND

Psychoanalytical readings of place and space are problematic. Gordon (2001) has suggested that the farther we get away from the clinical situation of analysis the more difficult psychoanalytical theorizing becomes: 'the further one moves from the individual patient', he argues, 'the less purchase psychoanalytical ideas can have. Outside of the therapeutic encounter, anything and everything can be true, psychoanalytically speaking. But if everything is true, then nothing can be false and therefore nothing can be true' (2001: 27). However, it is not unusual in interpretative social science to use metaphor to illuminate problems, notwithstanding the impossibility of testing claims in a scientific sense, and psychoanalysis constitutes a rich source of metaphors for describing problems at scales beyond the individual. It is also the case that psychoanalysts themselves have considered both collectivities, in group analysis, and spaces, from the home to global scales of interest,[9] in their work. Rural spaces, as they are imagined and as these imaginings translate into practice, seem to me to be entirely appropriate objects for this kind of analysis.

We might first think of rural England as an imagined community held together by historical narratives which features only certain actors, generally those with some power, like landowners, and which recognizes distinct boundaries of the rural that are under threat from the incursions of various others. Can we represent this problem as one concerning deep-seated anxieties about social relations and space using the language of psychoanalysis and do we gain anything by doing so?

Drawing on Kleinian object relations theory,[10] Gianna Williams (1997) has described patients in her practice, where she works primarily with adolescents, who develop 'no entry defences'. They have a dread of being invaded. What she means is that these patients have projected their anxieties onto others who threaten to invade the self, to return the pain to where it originated. In extreme cases, the anxiety of invasion becomes overwhelming. This might be manifested in eating disorders (primarily anorexia), but 'a defensive rejection of input [is] often not confined to food intake but [extends] at times so widely that it might be referred to as a "no entry" system of defences' (Williams, 1997: 927). As Jane Desmarais (2001: 82–83) has suggested, 'it has become possible to see that anorexic behaviour permeates all levels of our social structure. Anorexia may not only arise from an individual's difficulties; it may also occur as a social and cultural symptom of a common condition called Modern Living.' She goes on to suggest that this wider problem is the pathology associated with impairments in 'taking from another'. In other words, we could describe this as a problem of coping with difference. Fears are projected onto groups and individuals who threaten the boundaries of the self through movement, contact or touch. Thus, people with difficulties in establishing relationships with those marked as different have their anxieties reinforced. The erection of 'no entry' defences raises the stakes and some kind of intervention, usually psychotherapy, is needed to work towards a more porous self which does not resist so much the presence and introjection of others.

The way in which such resistance to difference has been expressed collectively in rural England is, first, through the definition of a community from which sources of anxiety have been removed – these anxieties are projected elsewhere – and, second, by sharpening the boundary between the country and the city. The city, in this imagined geography, is cosmopolitan and a source of others who are potentially transgressive. Rural space loses its ambiguous and anomalous elements. Power lines, gravel-pits, quarries and chicken-processing factories disappear and there are substantial chunks of rural England, like the old coal-mining areas of south Yorkshire and south-west Durham, which would not fit the vision. This imagined rural space is essentially agricultural, but lacking in class division and ethnic difference. Those who are not white and not culturally sedentary, particularly Traveller–Gypsies, similarly melt away. This particular disappearing act involves romanticizing nomads as an integral element of rural society in the past, when they provided services to farmers and blended nicely into the landscape. These Gypsies are 'real'. Defined in this way, Gypsies figure in a mythical rural community without appearing out-of-place. This representation provides the standard for judging those nomads who currently live and move through the countryside and, predictably, they are generally judged to be 'not-real', inauthentic, out-of-place. The emphasis on a history of belonging and on 'roots' particularly isolates the nomadic. As Morley (2000: 33) comments: 'The over-valuation of home and roots has as its necessary correlative the suspicion of mobility.' Through these selective inclusions and omissions, rural space is purified and a boundary between the pure and impure is created, sustained by a set of what Williams describes as 'no entry' defences. This psychogeography is best expressed in the discourses of rejection associated with conservative and regressive rural social movements, such as the English Countryside Alliance. The Alliance's website (www.countryside-alliance.org.uk) conveys the clear message that the social space of the countryside embraces 'a way of life', that rural England 'has a voice' and that the state should not meddle in the lives of 'country people'. There is a clear sense of ownership in the propaganda of the organization, expressed in the urging of the chief executive to 'listen to *your* countryside' (www.epolitix.com, 2000). This countryside is manifestly not pluralistic and the sentiments of the Alliance convey all the problems associated with the idea of community, whether rural or urban.

The idea of proprietorship and shared values which 'community' embodies has clear exclusionary implications. As Iris Young (1990: 235) puts it: 'if existing together with others in relations of mutual understanding and reciprocity is the goal, then it is understandable that we exclude and avoid those whom we do not or cannot identify'. Similarly, Zygmunt Bauman (cited by Robins, 1996: 33) has argued that strangers serve to demonstrate the problematic nature of community: '[S]trangers bring the outside in and, in so doing, they seem to disturb the resonance between physical and psychic distance, the sought after co-ordination between moral and topological closeness, the staying together of friends and the remoteness of enemies.' Like all communities, however, this rural English community which is identified through a combination of landscape and a particular way of life is imaginary. The 'strangers' may be a part of another, pluralistic vision of rural society, but it is one which is denied by the dominant collectivity. Redundant coal-miners, chicken factory workers, New Age Travellers, Gypsies, for example, may all be present but they do not qualify for membership of this restricted community of country folk, those who understand the ways of the countryside.

This vision of the Countryside Alliance echoes some of the feelings about home described in Sigmund Freud's 1919 essay on the uncanny, or the unsettling (*Das Unheimliche*), (Freud, 1919, 1985). Freud's observations include several references to negative sentiments expressed by people in rural Swabia, such as 'Is it still *heimlich* to you in your country where strangers are felling your woods?' and 'That which comes from afar ... assuredly does not live quite *heimelig (heimatlich)* [at home] or *freundnachbarlich* [in a neighbourly way] among the people'. Freud's main point, however, was that feelings of comfort gained from being at home among your own kind, could never remove feelings of insecurity, anxieties about invasion and transgression. The greater the concern with security, the more insecure the individual or group becomes. Thus, the *heimlich* became the *unheimlich* and the urge to strengthen the no entry defences becomes even greater. The similarity between rural Swabia in the early twentieth century and the Countryside Alliance's vision of rural England at the end of the twentieth century could suggest that the latter group is looking backwards, having difficulty in coming to terms with the reality of the modern/postmodern countryside. However, their anxious response to difference is reflected also in modern forms of governance and legislative responses to various 'others' in rural space. I would argue that there is no necessary progress in terms of embracing difference. To quote Morley:

> The pervasive assumption of a natural – or originary – world in which people are (or in happier days, were) rooted in their own proper soils or territory is described by Lisa Malkki as the 'metaphysics of sedentarism' – through which a culture is equated with a people and that people with a particular geographical place or territory. Clearly, the assumptions of homogeneity and purity built into this model are at odds with the actual state of flux of peoples and cultures in many parts of the contemporary world. *However, this racialized and territorialized model of culture continues to have enormous influence in popular consciousness.* (2000: 39; emphasis added)

SEDENTARISM, MOVEMENT AND RURAL SPACE

Geographies are often created by what Cooper (1998: 4) calls 'flash-point events'. These would include the moral panics of the 1960s charted by Stanley Cohen (1972). The episodes that Cohen selected to demonstrate the dynamics of a moral panic were essentially spatial and transgressive. The youth cultures whose behaviour scandalized respectable society in the 1960s had an impact because their rituals were performed in places where they 'did not belong', particularly the family spaces of English seaside resorts. The alarmist way in which the presence of Mods and Rockers in seaside resorts during public holidays was reported in some newspapers created a short-lived boundary separating the normal and the deviant. At the end of the twentieth century, the co-presence of mobile minorities, travelling in and stopping in rural areas in England, and other groups trying to maintain what they regard as the purity of the countryside, had created similar flashpoints, moral panics and boundary issues. There is a long history of tensions between settled people and nomads but in developed societies, as I have argued above, there is now a general assumption that the population will be sedentary. Thus, movement itself becomes deviant unless it is the movement of otherwise sedentary people for leisure and work purposes. This has been reflected in legislation designed to criminalize the movement of particular pariah groups, people who are pariahs partly because they are mobile or nomadic. Although legislation like the public order sections of the Criminal Justice and Public Order Act, which became effective in England and Wales in 1994, is about territorial control generally, the groups targeted, like Gypsies, New Age Travellers, hunt saboteurs and environmental protesters, are the ones most likely to disturb the cherished vision of the rural scene as one of stability and harmony, with a social structure that safeguards the countryside and produces its rightful custodians. Hunt saboteurs, for example, challenge the fundamental assumption that only the country folk championed by the Countryside Alliance know how to manage the countryside. Furthermore, hunts, as Cooper (1998: 10) argues, 'might be seen as an integral state structure since they help to reproduce dominant social norms and relationships'. Thus, the state may support rural practices like hunting if the party in power recognizes the dominant groups in rural areas as a part of its constituency.

The connection between these aspects of governance and moral panics is that the latter may influence the law. This is demonstrable in relation to the Criminal Justice and Public Order Act because New Age Travellers and environmental protesters were specifically targeted during parliamentary debates on the legislation. Some members of parliament, mostly on the right, also projected their anxieties onto some less likely others, generally located in cities by the antagonists, including transgressive football supporters (Sibley, 1997). Debates echoed popular discourses, in tabloid and local newspapers, for example, which have consistently employed alarmist metaphors of movement (invasion, flood, etc.) and the language of 'dirt' to create negative images of all nomadic groups – Gypsies, Irish and Scottish Travellers and New Age Travellers.

New Age Travellers had provoked panics over several decades, particularly in relation to their festivals and assemblies at sacred sites, such as Stonehenge. Hetherington (2000) documents the panics surrounding the Peace Convoy (the first ascription for New Age Travellers, current in the 1960s), who were subject to police violence on several occasions, and more recent episodes when New Age Travellers were represented as a more general threat to 'the rural community' and their presence contributed to the institution of more general movement controls, culminating in the 1994 Act. The combined effect of moral panics and what Cooper (1998) has termed 'excessive governance'[11] has been a periodic redefinition of the rural in terms of who does and who does not belong.

BEARS, TRADITION AND MODERNITY

Kevin Hetherington (2000: 170) suggests that 'what is significant in the ambivalence of the stranger is the way he or she disrupts the routinized cognitive ordering of everyday life that makes it a source of familiarity, this thinking-as-usual'. Thus, we could argue that 'strangers', like New Age Travellers, create a consciousness of a rural community that is defined in opposition to various 'others' who threaten to disturb the unity and social cohesion of that community. The boundary of the rural is, thus, contingent on the perception of a potentially transgressive 'other' but this boundary and the community which it encloses are recognized only by those occupants of rural space who feel threatened. These relationships between boundary, community and other have varying manifestations, as my second example demonstrates.

In Slovenia, a state that has been rapidly modernizing since the collapse of socialist Yugoslavia, large mammals, particularly bears, wolves and lynx, have an important place in society in the sense that responses to these animals serve to define different communities, which can be distinguished roughly as 'traditional' and 'modern'. The largest Slovenian mammal, the brown bear, has had a particularly important role in constructing rural geographies which map onto these broad social categories. The animal has recently entered public discourse partly as a result of an isolated event. On 24 February 2000, there was an attack on an elderly man by a female bear near the village of Kot pri Ribnici, 40 kilometres from the capital, Ljubljana. The man was badly injured but he survived. According to Krystufek and Griffiths (2003: 132), 'National opinion [about the risks posed by bears and the need for conservation] was now clearly divided, with the urban population advocating further bear conservation whilst the rural people, especially from the brown bear area, opposed this.' The brown bear had already been politicized in the previous year, however, through the granting of an exceptional permit for the killing of an animal 25 kilometres south of Ljubljana, on the grounds that it represented a significant threat to property and to human life. It had been alleged that the politician whose party formed the national government at that time wanted the bear removed so that he would feel more secure about building a house in the area. These two isolated events made the placing of brown bears a contentious political issue – essentially a polarized argument about the inclusion or exclusion of the species. Two aspects of the problem warrant some elaboration.

First, a core area was established in 1966 by the parliament of the Socialist Republic of Slovenia to provide protection for the brown bear. This was quite a progressive move, anticipating Europe-wide mammalian conservation measures. Clearly, the boundary of the core area, which stretches from the Julian Alps, where it connects with Austrian brown bear habitats, to Croatia, had to be policed. It created an inside and an outside and, thus, the possibility of transgression. It also created expectations about the sanctity of ecological relations within the core zone although these expectations were primarily those of the educated urban middle class. Conversely, for some farmers, the core zone represents an unwarranted restriction on hunting. The creation of a protected area for bears could be seen as a key element in the generation of a moral panic following the attack at Kot pri Ribnici. The brown bear became a potentially transgressive animal and the possibility that it might cross a boundary served to make it more dangerous. We might argue that containment or encapsulation rendered it decidedly *unheimlich*. Fear was amplified by the repeated use of one photograph of a fierce animal on its hind legs, facing the camera, in reports in the yellow (tabloid) press, which generally supported the anti-bear conservation lobby. The facts of the Kot pri Ribnici attack, where the victim inadvertently walked between a female bear and her cubs, all of whom were hidden in the forest undergrowth – it was an exceptional event – were disregarded. Other stories, particularly a newspaper report in May 2000 that a brown bear attacked a school bus outside the core area when, in fact, the bus had collided with the bear, served to fuel the panic. It was amplified further by two other associations. One was the coupling of bears and the Slovenian Roma (Gypsies) as serious social problems by a member of the Slovenian parliament, suggesting a more generalized fear of difference, and the other was a comparison made between bears and collaborators during the German/Italian occupation in the Second World War.[12] The brown bear could also be seen to bring into being or, at least, increase consciousness, of two communities – one rural and traditional and the other modern and urban. In this sense, the bear served as the stranger who disrupts the taken-for-granted routines of everyday life.

DECONSTRUCTING SELF, OTHER AND COMMUNITY

Both these case studies concern imagined communities which, I have argued, are important politically because they bring into being binary oppositions and self–other tensions that inform policy and practice. Critical research, however, should not be concerned solely with documenting and making sense of such conflicts but should also present alternative views of socio-spatial relations which might serve to weaken boundaries and encourage mutual understanding and integration, in other words, it should be concerned with social justice. Walkerdine and Lucey (1989, cited in Lucey and Reay, 2000: 153) have similarly suggested that what we should be doing is 'un-splitting things, integrating them, considering them to be part of each other', although these authors also think that this might 'muddy the waters', presumably in the sense that patterns of inequality and exploitation may become more opaque. Lucey and Reay's own work is interesting in that they demonstrate that in an inner-city context, race and class do not map onto conventional self–other categorizations, which tend to reflect middle-class fears. In their work with children in inner-London, they show that it is those who are poor, and often racialized, who have the strongest sense of personal and group boundaries – expressed in anxieties about others in their neighbourhoods. Living on declining local authority housing estates, children from low-income families are the ones most likely to feel threatened by drug dealers or vandalism, for example. Middle-class children, by contrast, showed a greater sense of 'mastery' of the city and, in most respects, were more secure than their working-class peers. Thus, Lucey and Reay challenge orthodox moral geographies.

Similar analyses of rural communities and of rural–urban conflicts are needed in order to counter the stereotyped views of urban populations voiced by groups like the Countryside Alliance in the UK, and of rural populations who are assumed by both rural and urban interest groups to constitute 'the rural community'. This requires an effort to capture the world-views of disparate groups, something which is most likely to be achieved through ethnography and, particularly, through long-term involvement with people. The researcher should be involved as a sympathetic observer if not a participant in their lives. Such research is, by definition, local and time-consuming and may not produce much in the way of conventional outputs during the time-frames used in academic accounting procedures. However, it is the research mode which is most likely to uncover key aspects of identity and relationships to place.

The complex nature of identity within groups living in rural areas, in terms of origins, views of self and other, and aspirations, is suggested in Lowe and Shaw's study of New Age Travellers in England (Lowe and Shaw, 1993). These authors show, for example, that the conception of New Age Travellers as a disaffected urban minority which has 'invaded' rural space is hardly the whole truth. The following extracts from their research demonstrate the inappropriateness of widely used social and spatial categorizations:

> I come from a farm cottage on a hill on the Lancashire–Yorkshire border, so whichever side of the hill you went down, they thought you came from t'other side, because nobody came from the middle, like. So I've always felt a bit alienated in England. This is how I choose to live my alienation. Travellers aren't a totality, like. It depends which Travellers you start looking for. There are people out there who are typical Travellers but I'm not one. I've never been off the island [mainland Britain], me. Went to the Isle of Man once, when me mam was feeling rich, and a seagull shat on her hat. She didn't like it. It were a new hat, special. That's a true story that. My father was a coalminer. They've shut those pits down now ... I've got a brother. Only one. He's a funny bugger, a Catholic. He went to fight for the British Army in Northern Ireland. They don't come any stranger than that ... I didn't want to do any of that. (1993: 104–105)

> I'm a city boy but during the summer you get that vibe to get out of London. Bedlam was the second biggest sound system at Castlemorton [a New Age Traveller festival site], plus our lot found the site ... We'd been to Lechlade, played there, done a party on Stroud, went and chilled out in Wales for a bit and then went looking for the Avon Free Festival. We knew it was one of the biggest things England had seen for a long time. (1993: 171)

While the second story conforms to the geography of the mobile urban raver constructed by legislators in the British parliament in the 1990s, the first is quite different. It hints at the uncertain sense of self of the respondent as well as the complexity of social relations and relationships to place which contribute to identity. It is this kind of ethnographic material which is needed to challenge simplistic notions of belonging and not-belonging in rural space and to defuse moral panics. I would argue that such methods serve to support an anarchistic project on socio-spatial relations, one that might produce a boundary-free and more just countryside.

CONCLUSION

In this chapter I have tried to suggest that generalization about processes of exclusion and inclusion in rural areas is difficult and, in most respects, inappropriate. The social composition of rural spaces and symbolic role of the rural in narratives of belonging and not-belonging are too variable and rooted in particular histories to warrant much theorizing across nation-states and cultures. In the social sciences there is, however, a history of attention to difference which provides some pointers to key research issues and which suggests directions for future research on rural exclusion.

Early work on racism and economic exploitation in North American cities but, more particularly, the social psychological perspective on the construction of social space in cities developed by Richard Sennett in the 1970s, has applications in regard to issues of exclusion and inclusion in rural areas. I have suggested that we can develop more sensitive analyses of boundary construction and spatial purification by drawing on certain aspects of psychoanalysis, particularly Kleinian object relations theory. If the central problem is anxiety about social difference, movement and transgression, then we need to combine psychoanalytical perspectives with ethnographies that focus, first, on questions of identity and place, and, secondly, on feelings about others. The ultimate objective of this kind of research should be to gain a clearer understanding of processes of exclusion and to provide the knowledge that might be used in the production of more socially just geographies. This objective does not necessarily require inclusion, however, if inclusion means incorporation and continued domination. The maintenance or encouragement of group autonomy, which has been a consistent theme of humanistic anarchism, also provides a route to social justice but it is one which may conflict with the objective of social inclusion. Thus, the politics of rural exclusion and inclusion remain fraught and problematic.

NOTES

1 Romanticization is suggested by the use of the term 'real', as in 'real Eskimos' (Inuit), who were presumed to have an authentic relationship with the natural world. Housing programmes for the Inuit in Canada, Alaska and Greenland during the 1960s and 1970s, which involved relocation in planned settlements, were designed to remove them from nature. This was justified with the argument that hunter–gatherer economies were no longer sustainable but this was a convenient way of removing indigenous peoples from their land to facilitate oil and gas exploration. The intended modernization process made them less than real from the perspective of the state and they became 'a social problem'.

2 'Culture of progress' was a term used by John Berger to characterize modern capitalist and socialist states which are bound to the idea of future development – capital accumulation, social change, and so on – as opposed to 'cultures of survival', like peasant cultures, which are bound to annual cycles of food production and repeated acts of survival (see Berger, 1979).

3 For a discussion of differences in the treatment of ethnicity in early urban sociology, see Sibley (1995).

4 The differences between social psychology and psychoanalysis lie primarily in their respective claims to scientific status. While Sigmund Freud initially regarded psychoanalysis as scientific, following his early training in neurology, the field has developed as a humanistic, hermeneutic discipline. Psychology, by contrast, projects itself as a science and there are very few psychologists who would accept the theories or methods of psychoanalysis and psychotherapy.

5 There are clear difficulties in generalizing about groups and about social space on the basis of theories that were developed initially in relation to the self – it is impossible to escape doing some violence to reality – but these problems are probably no more serious than generalizations about group attitudes and behaviour in sociology and political science. It is just as reasonable to do empirical analyses in psychogeography at the scales of interest of, say, urban sociology. See also note 9.

6 Apart from the work of hooks and Pollard, it is worth noting the pioneering work of Bill Bunge on black communities in cities in the United States in which black people articulated their own geographies (see Bunge, 1971).

7 These three cases are all ones of which I have some, rather limited, knowledge but I have not done research on any of them. Impressions gained from looking and talking to people in these regions have been enough to convince me of the complexity and peculiarity of rural poverty and exclusion. In the case of Costa Rica, I have benefited from conversations with researchers at the Asociacion Nacional para Asuntos Indígenas (ANAI), San Jose, and in Hungary with János Ladányi, Budapest University of Economics.

8 The significance of rural areas in the exploitation of labour within developed societies is demonstrated in David Harvey's account of the deaths of workers in a chicken processing plant in Hamlet, North Carolina, USA. As he argues, 'Those living in relatively geographically isolated rural towns of this sort are … easy prey for an industry seeking a cheap, unorganised and easily disciplined labour force' (Harvey, 1996: 335).

9 Such work by psychoanalysts would include Hannah Segal's writing on the Cold War and the nuclear threat (Segal, 1987, 1995), Ian Parker on local exchange trading systems (Parker, 1997) and Julia Kristeva on international migration (Kristeva, 1988).

10 Melanie Klein shifted the emphasis of psychoanalysis from the father (in Freud's work) to the mother. Object relations, as she developed it in relation to her clinical work with young children, is concerned with the development of the self through the internalization (introjection) of others, initially the mother, in terms of pain and pleasure ('good' and 'bad' objects) and the projection of good and bad onto others. These are processes involved in boundary formation and socialization and, I would argue, in the formation of social space.

11 Davina Cooper is concerned with what may be termed inappropriate activities for governance, which may be the provenance of both state institutions and private bodies. My argument here is that panics may prompt legislation that is unnecessary, where issues may be better resolved by negotiations between interested parties. Why, for example, should raves, an example of a youth activity controlled by the law in England and Wales since 1994, be treated differently to, say, clubbing, when both, if necessary, could be dealt with under legislation dealing with public nuisance? (See Cooper, 1998: 5–6.)

12 Collaboration with the occupying German forces during the Second World War is still an issue in Slovenia. The continued existence of ruined villages which are the sites of Second World War massacres serves as a potent reminder of this conflict.

REFERENCES

Abbott, E (ed.) (1936) *The Tenements of Chicago, 1908–1935*. Chicago: Chicago University Press.

Berger, J (1979) *Pig Earth*. London: Writers and Readers Publishing Cooperative.

Bernstein, B (1971) *Class, Codes and Control: Volume 1*. St Albans: Paladin.

Bunge, W (1971) *Fitzgerald*. Cambridge, MA: Schenkman.

Cohen, S (1972) *Folk Devils and Moral Panics*. London: MacGibbon and Kee.

Cooper, D (1998) *Governing Out of Order*. London: Rivers Oram Press.

Cresswell, T (1996) *In Place/Out of Place*. Minneapolis, MN: University of Minnesota Press.

Demarais, J (2001) 'Anorexic and passive resistance: a literary case study', *Soundings*, 18: 82–93.

Douglas, M (1966) *Purity and Danger*. London: Routledge and Kegan Paul.

DuBois, WEB (1899) *The Philadelphia Negro: A Social Study*. New York: Benjamin Blom.

Dwyer, O and Jones, JP III (2000) 'White socio-spatial epistemology', *Social and Cultural Geography*, 1 (2): 197–208.

Erikson, E (1970) *Childhood and Society*. Harmondsworth: Penguin.

Freud, S (1919/1985) 'The 'uncanny', *The Penguin Freud Library*, 14: 339–376.

Gordon, P (2001) 'Psychoanalysis and racism: the politics of defeat', *Race and Class*, 42 (4): 17–34.

Harvey, D (1996) *Justice, Nature and the Geography of Difference*. Oxford: Blackwell.

Hetherington, K (1998) *Expressions of Identity*. London: Sage.

Hetherington, K (2000) *New Age Travellers*. London: Cassell.

hooks, b (1992) *Black Looks*. London: Turnaround.

Jameson, F (1991) *Postmodernism, or The Cultural Logic of Late Capitalism*, Durham, NC: Duke University Press.

Kristeva, J (1988) *Strangers to Ourselves*. Hemel Hempstead: Harvester Wheatsheaf.

Kryštufek, B and Griffiths, H (2003) 'Anatomy of a human: brown bear conflict–case study from Slovenia in 1999–2000', in Kryštufek, B, Flajšman B and Griffiths, H (eds), *Living with Bears: a large European Carnivore in a shrinking world*. Ljubljana: Ecological Forum, pp. 127–153.

Ladányi, J and Szelényi, I (1996) 'Class, ethnicity and ecological change in post-communist Hungary and at the turn of the millenium Budapest'. Unpublished ms.

Lowe, R and Shaw, W (1993) *Travellers*. London: Fourth Estate.

Lucey, H and Reay, D (2000) 'Social class and the psyche', *Soundings*, 15: 139–154.

Morley, D (2000) *Home Territories*. London: Routledge.

Olsson, G (1980) *Bird in Egg*. London: Pion.

Parker, I (1997) *Psychoanalytic Culture*. London: Sage.

Perlman, J (1976) *The Myth of Marginality*. Berkeley, CA: University of California Press.

Robins, K (1996) *Into the Image*. London: Routledge.

Segal, H (1987) 'Review of "Against the state of nuclear terror"', *Free Associations*, 9: 137–142.

Segal, H (1995) 'From Hiroshima to the Gulf War and after: a psychoanalytic perspective', in A Elliott and S Frosh (eds), *Psychoanalysis in Contexts*. London: Routledge.

Sennett, R (1970) *The Uses of Disorder*. Harmondsworth: Penguin.

Sibley, D (1981) *Outsiders in Urban Societies*. Oxford: Blackwell.

Sibley, D (1995) *Geographies of Exclusion*. London: Routledge.

Sibley, D (1997) 'Endangering the sacred: nomads, youth cultures and the English countryside', in P Cloke and J Little (eds), *Contested Countryside Cultures*. London: Routledge. pp. 218–231.

Soja, E (1993) 'Postmodern geographies and the critique of historicism', in JP Jones, III, W Natter and T Schatzki (eds), *Postmodern Contentions*. New York: Guilford Press. pp. 113–136.

Soja, E (1996) *Thirdspace*. Oxford: Blackwell.

Walkerdine, V and Lucey, H (1989) *Democracy in the Kitchen*. London: Virago.

Williams, G (1997) 'Reflections on some dynamics of eating disorders: "no entry" defences and foreign bodies', *International Journal of Psychoanalysis*, 78: 927–941.

Young, I (1990) *Justice and the Politics of Difference*. Princeton, NJ: Princeton University Press.

30

Rural poverty

Ann R. Tickamyer

INTRODUCTION

The majority of the world's poor are found in rural areas, including a substantial portion of the impoverished populations of the richest nations (World Bank, 2000). Yet, increasingly, the rural poverty in their midst is ignored by the postindustrial nations, rendered either invisible or irrelevant to the urban preoccupations of most analysts and policy-makers. In the rest of the world, rural poverty is the dominant mode of deprivation; in the postindustrial nations of the global north, while it no longer represents the majority of the poor, it remains disproportionately large and among the most desperate forms of economic hardship, but one that generates relatively little academic, public, or political interest. Social and economic policy both within and across nations largely neglects rural poverty and directly or indirectly supports efforts to turn the rural poor into the urban proletariat and underclass. This belies an assumption that rural and urban poor are only tenuously connected.

In fact, in the era of accelerated globalization, transnational linkages between the rural poor and the rural and urban poor dominate in particular locales and labour markets, creating economic regions that transcend national boundaries. Economic migrants, guest workers and refugees cross borders with and without proper documentation to supply cheap labour to both rural and urban industries as part of the accelerating transnational flow of capital, goods and labour. The effort to escape poverty, injustice or lack of opportunity in one location often results in finding new forms in another and inextricably connects the two places.

The United States (US) and the nations of the European Union (EU) are prime examples of both poverty in the midst of great affluence and the hidden nature of many of these pockets of poverty. Rural poverty is particularly difficult to uncover in these settings for both material and ideological reasons. The concentration of populations, policy-makers, scholars and the media in urban centres shifts attention away from rural regions and permits both neglect and romanticization of rural life and livelihoods (Cloke, 1995; Tickamyer et al., 2004). In the US poverty is conflated with the problems of inner city ghettos and their racial and ethnic minorities. In the UK and continental Europe, national and EU-wide concern with issues of social exclusion turns attention away from spatial dimensions of poverty to vulnerable populations defined on the basis of social and demographic characteristics (Atkinson, 1998; Barnes et al., 2002; Geddes and Benington, 2001; Hauser and Becker, 2003; Ott and Wagner, 1997; Philo, 1995).

The US serves as an instructive example of the neglect of rural poverty in the developed nations, as well as the often obscured but integral links between rural and urban poor both intra and internationally. The US is not unique in these spatial disparities, but the contradiction embodied in being the pre-eminent global power and the world's wealthiest nation combined with high levels of persistent poverty and spatial inequality underscores the importance of addressing the conditions and sources of poverty in the rural periphery. This article examines rural poverty in the developed world, focusing on the sources of poverty in peripheral regions and the ways these are connected within and across national borders, using the US as the primary example. This approach is guided by the assumption that it is not possible to understand forms of poverty and inequality without attention to the historically specific patterns of development and linkages between different social institutions that create local culture and society. In other words, no one theory or even combination of theories can explain all rural poverty; rather, it requires a mix of theory and attention to unique features of place. Where appropriate, comparisons are drawn to comparable conditions in European nations, and the discussion concludes with a brief comparison of the state of knowledge about rural poverty in the US and Europe.

RURAL POVERTY IN THE UNITED STATES

For most of the twentieth century, rural poverty rates far exceeded urban poverty figures. As late as 1959, the percentage of poor people residing in nonmetropolitan areas was more than double the poverty rates for metropolitan areas, the jurisdictions that contain large cities and their suburban fringe (US Department of Agriculture, Economic Research Service, 2000).[1] Even central city poverty, the urban slums and ghettos most often identified as the epitome of impoverished and degraded living conditions, did not compete with rural poverty until the last third of the century. In the mid-1970s, rural and central city poverty rates converged for the first time, and central city poverty overtook rural poverty by the end of the decade (RSS Task Force, 1993). Subsequently, rural poverty rates represented by nonmetropolitan counties continued to far exceed overall urban metropolitan areas, but have generally remained somewhat lower than central city figures, although the exact relationship between the two varies by time and place (US Census Bureau, 2000).

Depending on the time span examined, poverty rates can be viewed as either remarkably stable or part of a long-term downward trend. Poverty rates declined rapidly in the years following the Second World War until the1970s. After that, changes in poverty rates levelled off and even reversed. For almost two decades they fluctuated from year to year with changes in the economy, but remained high for both rural and urban locations. In 1989 almost 16 per cent of nonmetro and 13 per cent of metro residents were classified as poor, using the official measures.[2] This represented approximately 9 million poor people living in rural areas. Then in the 1990s, poverty rates resumed their historic decline. By the end of the twentieth century, all poverty rates had decreased substantially. In 1999, rural poverty was at 14.2 per cent, 3 percentage points higher than the urban rate of 11.2 per cent, but 4 points lower than central city poverty (Lichter and Jensen, 2002; US Department of Agriculture, Economic Research Service, 2000). At 7 million persons this was 2 million fewer people than the previous decade and was the lowest recorded level since 1979. This decline accompanied the longest economic expansion in post-Second World War US history. Future directions undoubtedly are tied to changes in local, national and global economies. Continued improvement or even maintenance of lower levels appeared doubtful in the face of widening recession beginning early in the new century. Longer-term trends are not yet obvious, although they are clearly heavily dependent on economic conditions.

High rates of rural poverty are even more striking if the spotlight is placed on spatial distribution of poverty – the places that have extremely high concentrations of poor people. The most common way to distinguish rural and urban spaces, using metro/nonmetro county classifications, masks great variation within these boundaries. Because population densities permit finer spatial discrimination, and because social indicators of urban welfare carry great political weight, the differences between central city (often inclusive of but not necessarily synonymous with urban ghettos) poverty rates and those found in the remainder of counties designated as within metro areas are routinely reported. However, there are large differences within and between nonmetropolitan areas as well. Rural areas vary in their proximity to metro areas, population densities and history of economic development and activity, and each of these has relevance for poverty levels. Nonmetropolitan counties range from those located on the urban fringe that serve as bedroom communities for affluent commuters to jobs in nearby urban centres to the most remote and sparsely populated wilderness areas. They include a large number of counties that have been designated as persistently poor – nonmetro counties that have had excessively high poverty rates over long periods of time – 20 per cent or more for each decennial Census since 1960 (Nord, 1997).

Persistent poverty counties are primarily located in the rural South and West of the US, and typically have distinctive economic, sociocultural and racial-ethnic histories and identities. The plantation South or Blackbelt (including the Mississippi Delta and Southeastern coastal plains), the Appalachian and Ozark highlands,

Indian reservations in the West and Southwest, and the colonias of the Southwestern borderlands, consistently exhibit the highest poverty rates, in many cases encompassing more than half the population (Nord, 1997). These places also have high rates of unemployment, low levels of human and social capital, lack of investment and infrastructure development, poor or non-existent service provision, and frequently inefficient or corrupt political and administrative institutions. Even as a homogenizing commercial culture submerges social and demographic differences, these 'forgotten places' remain distinct, mired in poverty with their populations exposed to high risk of economic deprivation and all its correlates. Specific sources and conditions vary, as do the primary groups at risk, but regardless of whether they are the descendants of African slaves located in the Blackbelt of the former slave-holding states, conquered and colonized Native Americans isolated in Indian reservations, both recent and longstanding Hispanic residents and migrant workers, or the old settlers and more recent 'ethnic' whites of rural Appalachia, they share common experiences of deprivation, exploitation, discrimination and subordination to powerful elites (Billings and Blee, 2000; Duncan, 1992, 1996, 1999; Eller, 1982; Fitchen 1981; Hyland and Timberlake, 1993; Lyson and Falk, 1993; Pickering, 2000; Seltzer, 1985; Snipp, 1996; Snipp et al., 1993).

The existence and persistence of large pockets of poverty located in rural hinterlands and often associated with distinctive racial and ethnic groups undermine common stereotypes about the rural poor that contribute to their obscurity and neglect. To the extent that it is considered at all, rural poverty is seen as less threatening than urban poverty, associated with hard luck and hard times rather than bad values and dissolute lifestyles. Popular attachment to romanticized views of rural landscapes and communities and imputed agrarian values of hard work and self-reliance create a more benign image of rural life and poverty that has little to do with its realities, but compares favourably to equally distorted stereotypes of the malignancy of urban social problems. The outcome is a highly nostalgic and romanticized view of rural life 'that glosses over rural poverty and that personifies its forms in the guise of white working class families who are seen as less threatening to dominant social values and elites' (Tickamyer et al., 2004). The reality is much more complex.

To a limited extent these stereotypes have a basis in social fact. Historically, the rural poor have been more likely to be white, to be found in traditional, married couple households, to have lower rates of extra-marital child-bearing, to have fairly substantial rates of formal labour force attachment, and to have lower levels of welfare use (RSS Task Force, 1993). However, these differences are relative, belying the diversity that exists in rural America. In rural areas just as in urban centres, women, single-headed households, children, and race and ethnic minorities are over-represented, with growing convergence between the demographic characteristics of urban and rural poor (Albrecht et al., 2000, Lichter and Jensen, 2000; McLaughlin et al., 1999). Women are at greater risk for poverty than men or their urban counterparts and continue to be particularly vulnerable if they occupy multiple risk categories. Race and ethnic minorities endure some of the deepest and most grinding forms of poverty, as is evident from their over-representation in persistent poverty counties in the South and Southwest. Even in the 'whitest' pockets of persistent rural poverty, such as are found in rural Appalachia, there is much greater race and ethnic variation than is often credited or documented (Billings and Blee, 2000). The demographics of rural poverty can only be explained in conjunction with theories of poverty.

EXPLANATIONS FOR RURAL POVERTY

Numerous theories have been offered to explain rural poverty. They tend to fall into two groups corresponding to the distinction between poverty of persons and poverty of place, or those that focus on individual characteristics and those that target institutional and structural conditions. Even within these broad categories, there is great variation (and confusion) between those that emphasize the correlates and characteristics of poverty, whether personal or spatial, while others provide theoretical exegesis. For example, poverty is often explained by reference to demographic characteristics such as gender, race, ethnicity and marital status, or more theoretically, by culture of poverty models that purport to explain why these correlates of poverty status persist. Similarly, the spatial distribution of poverty is often attributed to characteristics of poor places such as high unemployment, low levels of human and social capital, and the prevalence of particular forms of economic activity, or to theories of development and dependency that also attempt to explain the sources and distribution of these characteristics of rural poverty.

These explanations are often posed as mutually exclusive or polarized models that must be embraced to the exclusion of alternative approaches. While this has had great political success in driving policy debates and political campaigns, it has much less academic currency (Tickamyer and Duncan, 1990). The characteristics of poor persons and places are very real, and may provide a good way to predict the odds of being poor, but they have little explanatory power. Being female does not explain poverty, but the dynamics of patriarchal gender relations supplies the necessary understanding for women's heightened vulnerability. Similarly, being a member of a racial or ethnic minority may make it more likely to be poor, but its real significance is as a marker of ongoing social relations of subordination and discrimination that create that probability. High unemployment rates generate high poverty, but the reasons behind particular patterns of employment, such as deindustrialization, economic restructuring, uneven development, or globalization have much greater theoretical power.

To complicate matters more, the diversity that exists within rural places and spaces in the social, demographic, political and economic development of specific groups and regions means that no one theoretical approach will fit all cases. The 'new poverty' of the agrarian heartland differs sharply from the persistent poverty of portions of Appalachia or the deep South. Deindustrialization, economic restructuring and capital consolidation of formerly highly industrialized rural extractive industries have different sources and outcomes than found in locations lacking a history of industrial production and that have little more than subsistence economies and dependence on transfer payments of one sort or another. Thus understanding rural poverty requires sorting through a myriad of proximate and underlying causes, correlates, theories and explanations to emerge with a richer more nuanced understanding of the causes and consequences of different forms of rural poverty. Understanding particular instances of rural poverty requires placing these theories in both spatial and temporal contexts.

INDIVIDUAL EXPLANATIONS

There are numerous forms of individual explanations for rural poverty, generally involving some variant of individual characteristics combined with explanations for the prevalence of these traits among the poor. As indicated above, these generally begin by enumerating specific demographic traits associated with high poverty levels, focusing on race, gender, marital status, education and labour force attachment. Female-headed households and their children, often black or brown, who possess few marketable skills and little labour force experience, are especially vulnerable to poverty, and these traits are elevated from the status of correlates to proximate causes. In turn these are given theoretical coherence by explanations that purport to show why these traits are so prevalent among the poor.

Culture of poverty

Apart from low-level empirical generalization that provides coherence to these descriptive findings, such as demographic and human capital models, the most popular theory used to tie these together, and the one with the greatest political weight, is found in variants of the culture of poverty. These vary from crude prejudices about the inherent laziness, lack of initiative and disregard for mainstream social norms and values found among some groups (almost *sui generis*) to more sophisticated formulations that posit initial misfortune, regardless of source, that is subsequently enshrined in originally adaptive but ultimately maladaptive behaviours and attitudes. The resulting 'culture of poverty' responds to economic hardship by reproducing irresponsible behaviours and beliefs. Thus the real or attributed prevalence of marital instability, family disorganization, extra-marital child-bearing and substance abuse is combined with attributions of character flaws that include lack of work ethic, hedonism, fatalism and inability to defer gratification to form a constellation of dysfunctional coping mechanisms used by poor people that directly contravene twentieth-century middle-class moralism and practice. These are hypothesized to form a culture (or more accurately a subculture) that explains how poverty is entrenched and perpetuated from one generation to the next. More recently, conservative analysts added welfare dependency to the list of both primary causes and consequences of the culture of poverty, perpetuating the circular logic embedded in this concept. Whatever the initial source of poverty in this explanation, the almost inevitable development of a culture of poverty ensures that it will trap its victims in a vicious cycle that is difficult if not impossible to exit.

It should be noted that culture of poverty theories are less frequently applied to rural poverty, and this is a measure of the somewhat ambiguous status of the rural poor. In fact, because of its often highly pejorative application and its association with racial stigma, compared to the more benign view of rural poverty described previously, the rural poor are often explicitly or implicitly excused from its provisions. Nevertheless, as members of the larger class of poor people, they come under its embrace. Furthermore, when the focus is refined to refer not to rural poverty in general but to specific groups and pockets of rural poor, the interpretation shifts. Southern blacks and white Appalachians, for example, are widely treated as textbook examples of the culture of poverty (Duncan, 1999). In the latter case, views of the entire region are conflated with culture of poverty traits (Billings and Blee, 2000; Billings and Tickamyer, 1991).

Crude culture of poverty theories carry little weight in academic circles – in fact there is a large literature devoted primarily to debunking them – but they continue to have enormous impact politically. In a highly individualized culture, public opinion finds them congenial explanations for individual and group failure to prosper and a way to blame the victim rather than social conditions that could be ameliorated through public policy. This model also underpins the 'war on welfare' that gained ground in policy circles in the 1980s and subsequently became the basis of new public policy in the 1990s (Handler, 1995; Schram, 1995; Tickamyer et al., 2000). The notion that welfare use breeds dependency and the constellation of behaviours collectively labelled the culture of poverty became the driving force behind the political campaign to 'reform welfare as we know it' by dismantling the collection of Depression era New Deal and 1960s War on Poverty programmes that constituted the US safety net during the second half of the twentieth century. The success of this effort has had significant consequences for rural poverty and the rural poor.

STRUCTURAL EXPLANATIONS

Modernization

Structural theories take a number of forms, beginning with the dominant modernization paradigm that formed the basis of post-war development theory and practice to numerous variants of Marxist and neo-Marxist models of dependency, under- and uneven development that replaced it later in the century and culminating with the current dominance of globalization theories. The commonly expressed idea that pockets of rural poverty were the forgotten places, the people left behind in the prosperity that came from industrial development and urbanization, directly expressed the notion that rural poverty was the result of the failure of economic development to unfold or to 'take off' in the linear fashion that characterized economic development in early modernization theories (Rostow, 1960). Although based on the highly developed and still enormously influential tenets of neo-classical economics, the crude applications of this approach were in many ways equivalent to the overly simplistic culture of poverty models, but in this case applied to regions and places rather than to individuals. Poor rural places suffered from inefficient and unproductive economies that failed to grow and thrive in a modernizing world, because of traditional values and practices that prevented successful adoption of new technologies and new institutions. Although initially most prominently applied to explain the impoverished state of the majority of the world's non-industrialized and industrializing nations following the Second World War, this model was also coopted to explain internal development problems.

Again, Appalachia provides the quintessential example for this approach. Despite its long history of industrial production in resource extraction (mainly coal, but also timber and other minerals) and steel, textile and furniture manufacturing industries, Appalachian poverty was widely viewed by both scholars and policymakers as the outcome of a traditional local culture bypassed by the modern world and lacking the institutions and resources of a dynamic economy to compete in a highly industrialized and urbanized society. Other poor rural areas shared in this model to greater or lesser degrees, depending on the specifics of their economies and local history. In most cases, it was a distorted and partial view that failed to recognize the realities of local history and economic development.

Development paradigms

The failure of places such as Appalachia to properly conform to the development model promoted by modernization theory created a search for alternative explanations. Here too the theories that gained currency to explain widespread, persistent and in some cases deepening poverty among developing nations were adapted for domestic use. The rise of dependency theory, world systems theory and other variants of 'the development of underdevelopment' undermined the notion that the problems of poverty were inherent in a stunted development process, hindered by adherence to traditional culture. Instead, poverty was posited to be the result of inequalities created and promoted by social relations of exploitation, domination and discrimination that enabled rich regions, nations, and interests, whether spearheaded by the state or by capital, to plunder natural resource-rich but otherwise poor regions for their wealth. In the process they distorted or suppressed local institutions and social relations, stunting their ability to grow and prosper. Numerous variants of these theories abound, but in the US for a long time the most popular approach used an internal colonial model to depict poor areas such as Appalachia, the deep South, Indian reservations, and Southwestern colonias as internal colonies, exploited and impoverished by powerful external global, national and regional corporate interests (RSS Task Force, 1993).

The model fits better in some places than others. There can be no doubt that the conquest of the original inhabitants of the Americas, the appropriation of their land and livelihoods, and their resettlement and restriction to reservations often located on remote and seemingly unproductive territories follows a classic model of the exercise of colonial power. On the other hand, in other areas, the template does not fit quite so well. There have been numerous efforts to depict Appalachia as an internal colony exploited by external elites, but in fact the perpetuation of inequality and exploitation was also widely and successfully practiced by local, indigenous elites (Billings and Blee, 2000; Duncan, 1999). Similarly, the extreme poverty and discrimination of the Mississippi Delta can be largely attributed to the success of local elites in preventing external influences, whether social, economic, or political, from gaining any foothold in the region. It is only in very recent times that a combination of federal legal interventions, reverse migration of African Americans returning to their places of origin with new resources at their disposal, and new economic development in tourism and gaming industries has broken the stranglehold of the white planter class on local conditions (Duncan, 1999). These elites not only thoroughly dominated local politics and exploited the large underclass, both white and black, but also used their power to actively discourage external investment in the area to maintain their positions of power and privilege. Other areas of persistent rural poverty have similarly complicated histories and patterns of underdevelopment (Lyson and Falk, 1993; Tickamyer et al., 2004).

Other structural approaches variously emphasize the operation of class, markets, political process, or social institutions, but they share a common concern with the social structures that create inequality and depress opportunity through institutions that dominate and exploit. The concrete outcomes of these processes are the creation of economies that cannot produce adequate employment or income for workers in the region, industries that cannot compete in regional, national or global markets, crumbling or non-existent infrastructure that cannot attract new investment, livelihood practices that cannot sustain an adequate standard of living, and stocks of human and social capital that cannot compete in a global economy.

Economic change, restructuring and globalization

The notion that poor rural places have been left behind, frozen in an impoverished past, is countered by the nature of the ongoing processes that buffet them and that contribute to their current state. Economic restructuring in a variety of guises plays a major role in creating new poverty or perpetuating old forms. Here again, the specific forms and processes differ, but the outcomes are similar. The growth of highly efficient, capital-intensive technologies, whether in agriculture or in mining, in resource extraction or in manufacturing, have hit rural areas especially hard. Lacking diversified economies for the most part, the loss of jobs entailed in these processes, especially of high wage work, can devastate communities and even whole regions. The growth of service sector jobs, especially in rural places where these tend to be low-end positions in retail, tourism, or care work, does not substitute for declining old line industries. Thus the agriculture sector and its workers have suffered destructive losses in jobs and population from the explosive growth of corporate agriculture with its capital- and technology-intensive methods. The long-term trend is exacerbated by periodic crises of overproduction and deflation of land values, such as occurred in the US farm crisis of the 1980s. Similarly, capital-intensive mining operations have resulted in the precipitous decline of relatively high-wage, if hazardous, employment. Depletion of natural resources and environmental regulation such as found in the Northwest timber industry and coastal fisheries have similar impacts. Deindustrialization found in peripheral rustbelt communities of the upper Midwest and the Southern mill towns and branch assembly plants, where the high costs of manufacturing have made industrial production non-competitive in the world economy, further depresses rural regions. Finally, the culmination and acceleration of all these trends in the internationalization and integration of global markets, where products and labour must compete not just locally, regionally or nationally, but in a global market place where there is always some location with access to cheaper labour, less state regulation, or greater subsidies and incentives to relocate, make rural areas continually vulnerable to economic insecurity.

MID-LEVEL AND META-THEORIES

Intermediate between micro- and macro-level explanations for poverty are theories that attempt to link these levels of analysis. Theories that emphasize the relationships between processes that operate at different levels – such as those that examine the connections between gender, poverty and work; or race, discrimination and employment; or the networks that create opportunity for individuals in particular places – serve as examples of efforts to connect people and places and the dynamics of individual opportunity and action with structural forms and processes. Finally, there have been efforts at deconstructing dominant discourses that prevail in scholarly and particularly important, policy debate about poverty and its causes. While these rarely address rural poverty directly, they have important implications for understanding prevailing patterns of poverty and inequality that are relevant by implication or extension to rural policy debates.

Gender and patriarchy

Feminist theories have been among the most influential sources of analyses of the high incidence of women's poverty and exploitation, both spatially and temporally. It is no accident that women and children have the highest rates of poverty in both core and peripheral regions and nations, although the specific processes vary by place as well as how they intersect with other sources of inequality such as class, race and sexual identity. At their most global, feminist theories of patriarchy have attempted to describe and disentangle the complex patterns of gendered privilege and exploitation that create greater risk of poverty for women and their children. More immediately, they focus on women's disadvantage in the labour market and the household and the intersections of the two realms.

Women remain responsible for the majority of reproductive labour while increasingly required to seek forms of waged labour as well. Whether they are the sole sources of support in growing numbers of female-headed households or their wages are part of a household's income packaging as male claim to a family wage disappears in the post-Fordist era, women's waged labour is both necessary and normative. The outcome is a double or even triple day of work combining formal, informal and household labour. Although the evidence is not definitive, there is much speculation that there is a growing informal sector in advanced industrial and postindustrial nations, analogous to those found in postcolonial economies and that lack of formal employment opportunities increases women's informal economic activity (Tickamyer and Wood, 2003). Regardless of where it is located or how it is compensated, women's work both directly and indirectly supplies much of the cheap labour that fuels rural industrialization, first in depressed areas of the core's rural regions and then increasingly in industrial enterprise zones established in rural areas of peripheral regions and nations. In both places, women's unpaid work subsidizes the reproduction of the household and the labour force. Their paid labour supplies necessary income, but also is the major source of labour power in rural areas (Dickinson and Schaeffer, 2001).

In rural areas in the US, women's traditional concentration in unpaid reproductive, informal and family labour results in work histories that are not competitive for jobs that provide a living wage or benefits, even in the rare event that such employment is available in rural places (Tickamyer and Henderson, 2003). Ironically, and somewhat contradictorily, women's poverty has been exacerbated by both industrialization and deindustrialization. Women, like other rural residents, benefited from the relocation of manufacturing industries to rural areas, seeking cheap land and labour. Even poorly paid work in manufacturing can supply economic stability to both individual workers and communities. However, in those places where rural industrialization created relatively stable jobs, women were typically relegated to the least desirable and lowest paid jobs (Fink, 1998). Whatever advantage was to be gained from steady employment has disappeared as capital moves offshore or to border lands to find a cheaper and even less demanding labour force, generally, young unmarried women, exporting patterns of rural poverty and exploitation from the global north to the southern periphery.

Race and ethnicity

These also play prominent roles in explanations of some forms of rural poverty. Race and ethnic minorities are often concentrated in rural regions that share histories of extreme class stratification and political disenfranchisement created by both legal and extra-legal oppression, exploitation and discrimination, although these are place- and population-specific and vary dramatically in their forms if not their outcomes. The specific dynamics often are inadequately researched (Snipp et al., 1993). While slavery, the plantation system and their legacies both in the rural South and their connections to urban ghettos have received great scrutiny, other forms of racial and ethnic subjugation in rural America do not garner similar attention. Thus the history of the reservation system for subjugating American Indians, ranging in their early history as virtual concentration camps to their current status of rural ghettos, while well documented, nevertheless remains under-studied (Pickering, 2000; Snipp, 1996; Snipp et al., 1993). Similarly, the enormously complex history of diverse Hispanic populations in the US, both native and migrant, is not systematically scrutinized or well understood. Connections to rural poverty run the gamut from expropriation of Hispanic communal lands in the Southwest to the more recent development of colonias of impoverished rural residents (Maril, 1989; Montejano, 1987; Saenz and Ballejos, 1993). Migrant farm workers, both native and foreign-born, documented and undocumented, are among the poorest and most exploited of Hispanic groups and demonstrate the complex connections that link the rural poor of the global north and south (Tardanico and Rosenberg, 2000). The removal of the meat packing industry and other food processing plants from urban centres to rural areas has also contributed to the growth of new populations of Hispanic workers in rural areas that formerly had little ethnic diversity. In all these cases, it is not that good research is entirely lacking, rather it is case-based, episodic and frequently not focused on or integrated into an understanding of both the structures and dynamics of rural poverty.

Political power and social capital

Many of these approaches explicitly or implicitly refer to individual and group position in the labour market, seeing poverty as the outcome of lack of jobs and employment opportunities. Labour market factors are not sufficient to explain rural poverty, however, especially if the focus is on place-based poverty. In many poor rural places, individual deficits in human capital that create labour market disadvantage are reinforced and exacerbated by lack of political power, lack of social capital and lack of institutional support for the most disadvantaged groups in the community. The public sector is absent or controlled by a small economic elite, and it is dedicated to their interests. The inadequate public sector is accompanied by exclusion from community institutions that range from the schools and churches to stores and credit sources. This pattern reinforces economic inequality, suppressing opportunities for both individual mobility and community and regional development. The limited amount of social capital that can be mobilized by poor persons tends to be family-based, failing to provide the larger linkages that would give access to larger institutions and opportunities, thus only reinforcing disadvantage (Duncan, 1999, 2001).

Again Appalachia and Southern plantation regions such as the Mississippi Delta serve as quintessential, although not unique examples. In both regions, the history of violent enforcement of a politics of exclusion has kept power in the hands of a local elite, has kept new investment from penetrating local economies or providing competition for control of resources, and has kept both local and external institutions from mobilizing opposition. Alternatively, in places where historically, local institutions are less class based and biased, such as rural New England, rural poverty is less virulent and persistent. Possibilities for individual mobility transcend the limited assistance possible from family sources, and civic engagement, social capital, public infrastructure and investment are available to assist the poor as well as the power elite. Not coincidentally, these places tend to be more demographically homogeneous, and class relations are less overlaid with historical divisions by race, ethnicity or labour unrest (Duncan, 1999).

Current preoccupation with social capital as an engine of development, poverty alleviation and the spread of democratic institutions takes numerous forms (Saegert et al., 2001). On one level, it merely serves as an extension of individual resources, extending notions of human capital to include the less tangible connections and networks that link people to opportunities beyond their immediate circles. At another level, it provides a means of conceptualizing and measuring community-based resources and ways to link individuals to those assets. However, wide variation in both the measurement and conceptualization of this concept and a tendency to conflate it with other fundamental conceptual tools, ranging from social class to civic engagement and community make it a promising but underdeveloped tool for analysing rural poverty.

Rural–urban links

Often forgotten in the 'forgotten places' view of rural poverty are the connections to urban poverty. Historically, rural and urban poverty and inequality are inextricably connected as migration streams flow back and forth between rural and urban places, depending on a variety of push and pull factors ranging from local political, economic and even climatic conditions to the vagaries of the global economy. This is especially true of the agricultural sector where the long pattern of increasing concentration, industrialization and corporate ownership has disrupted rural livelihoods. This historical trend further impoverishes already poor residents and creates new categories of rural poor who then migrate to urban areas seeking alternative employment. Thus the urban black ghettos of the northern industrial centres were populated by the displaced sharecroppers and agricultural workers of the plantation South and their descendants after the mechanization of Southern agriculture following the Second World War. These groups maintain ties to their rural roots, returning home in good times to contribute to family and communities who need their help and in bad as a refuge from hard times (Duncan, 1999; Falk, 2003; Stack, 1996). The farm crisis of the 1980s in the rural Midwestern grain fields created new poverty among family farmers, many of whom lost their farms and livelihoods and ultimately migrated to urban centres in search of new lives and livelihoods.

While the general flow is from rural to urban centres, it is not always unidirectional, as the example of African-American movement demonstrates. Appalachian coal miners and subsistence farmers also migrated to urban industrial centres, especially in the automotive industry, but often maintained close ties to places of origin and travelled back and forth depending on the relative state of the urban versus rural economies. Downturns in the industrial sector or increased demand for coal have led to reversing migration streams, often transferring poverty from one location to another. The ongoing demand for cheap labour in both farm and food processing industries pulls migrant workers into rural areas, often recruited through intermediaries in urban centres. Thus the meat packing industry of rural Iowa recruits from the barrios of Los Angeles with little regard for whether these are native or migrant, documented or undocumented workers (Fink, 1998). These jobs are difficult, dangerous and poorly paid, keeping already poor people poor, providing little avenue for mobility. Finally, on a more macro level, the operation of the global economy shifts populations as capital shifts investments and bases of operations, seeking cheap inputs and large returns. The result is to enrich some and impoverish many others. The wholesale movement of the textile and garment industries to peripheral regions in the world economy means economic hardship for rural Southern textile communities, long centred on this industry.

Migration and globalization

Increasingly, transnational movements of capital, goods and labour characterize the links between the rural poor of the periphery and both the rural and urban poor of the core. Many of the examples already alluded to exemplify this trend. Migrant farm workers and food processing workers from Mexico and Central America supply a large proportion of the labour force in the US agricultural sector. Their remittances in turn support the rural villages and families who remain behind. The loss of rural industry to peripheral regions with cheaper labour and lax regulation, whether in peripheral regions of the US or in the industrializing nations of Asia or Latin America, impoverishes abandoned communities and motivates new rounds of migration.

The new global order redistributes populations as well as goods and wealth. This movement is accelerated by the trade agreements that constitute the new global order, such as the North American Free Trade Agreement (NAFTA). NAFTA 'binds together Mexico, the United States and Canada in a neo-liberal regime of denationalized, reciprocal regulation of trade, investment and associated affairs' (Tardanico and Rosenberg, 2000: 3). Although the movement of people remains formally restricted, the opening of borders inevitably rearranges populations and materially affects the fortunes of sending and receiving groups with profound consequences for the wealth and well-being of different regions and sectors in the participating nations.

The NAFTA example illustrates this process and the resulting emergence of new regions from the reorganization of economic relations that integrate disparate populations. In this case, especially noteworthy is the juxtaposition of the southern states of two of the three signatories – the US and Mexico (Tardanico and Rosenberg, 2000). These linked regions are part of the new global division of labour that restructures the material fortunes of places, regions and nations. While the two souths have very different locations in core and periphery and divergent development trajectories, they share a number of characteristics including histories of high poverty rates, repressive labour practices, exploitative race relations and dependence on primary commodity production that make comparison of the development outcomes of free trade and globalization particularly instructive (Tardanico and Rosenberg, 2000). The impacts on inequality and poverty of different groups and regions within each country is as important as the differences in each nation's rank in the global world order (Sernau, 2001). Rural peoples and places are especially susceptible to increasing impoverishment and inequality in both rich and poor nations alike.

Policy and development discourse

A number of analysts have scrutinized both domestic poverty and international development policy as discourse, examining the practices used to define, marginalize and stigmatize the poor and ultimately contributing to the construction of policies that recreate and reinforce existing power relations (Escobar, 1995; Schram, 1995). This approach uses textual deconstruction techniques to decode the discursive practices that underlie how poverty has been conceptualized and politicized. The representation of social reality becomes the primary means by which new understandings are forged and old forms of inequality are reproduced. Scrutiny of the historical and situational context provides insight into how the issues are formulated, the way problems are defined and the tools and policy instruments selected to guide interventions. Discursive practices – the words and texts and theories applied to a given issue – are representative of not only power relations that disadvantage certain groups and places, but also the currency that creates them. In Escobar's words, development is

> a historically singular experience, the creation of a domain of thought and action, by analyzing the characteristics and interrelations of the three axes that define it: the forms of knowledge that refer to it and through which it comes into being and is elaborated into objects, concepts, theories, and the like; the system of power that regulates its practice; and the forms of subjectivity fostered by the discourse, those through which people come to recognize themselves as developed or underdeveloped. (1995: 10)

Escobar's analysis applies to the construction of the theory and the reality of the 'third world', 'underdevelopment' and the poverty of developing nations. While the mechanisms by which these meanings were created, enforced and ultimately employed to further impoverish the subjects of this discourse are necessarily specific by time and place, similar analyses have been applied to the poor of developed nations, but in very few cases have these specifically examined rural poverty. In the US the subjects have focused on the poverty establishment, welfare dependency, welfare reform and the feminization of poverty (Fraser, 1990; Fraser and Gordon, 1994; Schram, 1995). In Britain, Cloke et al. (1995a, 2002) analyse how concepts of deprivation and disadvantage are juxtaposed with an idyllic view of rural life to construct distorted notions of rural poverty and homelessness, but this is a rare example of analysis focused specifically on *rural* poverty in a Western nation.

RURAL POVERTY POLICY AND THE RESTRUCTURED WELFARE STATE

Thus explanations for the persistence of rural poverty are not merely academic but have significant implications for the forms devised for policy interventions and programmes designed to ameliorate economic distress. Decisions to target individuals or structural conditions, to focus on people or place-based policies, to use positive or negative sanctions, rest partly on understanding what causes poverty and therefore where leverage can be effectively applied. Alternatively, policy can deliberately or inadvertently create or increase hardship for particular persons and places. In the US a highly individualistic interpretation prevails and one that tends to be punitive toward the poor. However, the virtual absence of a rural poverty discourse or even a more general view of rural life, means that poverty policy overlooks the conditions of poor people and places outside the metropolitan core. There has been little systematic rural economic development policy and few efforts to tailor more general safety net programmes to the particular problems of the rural poor. In the absence of formal rural development policies, a mixture of industrial and agricultural policies affect rural fortunes and livelihoods, but largely by default and to the advantage of large corporate interests rather than struggling families and communities (Buttel et al., 1993; O'Connor, 1992). Similarly, safety net programmes have either ignored spatial variation or disadvantage rural peoples and places, more by default than by design (Tickamyer et al., 2004).

Rural development policy

There is little independent rural development policy in the US. Instead, there are agricultural policies, primarily focused on creating more efficient agricultural production and providing price supports for particular commodities and their producers, and there are a handful of infrastructure and industrial development programmes that target persistently poor regions. Commodity programmes are designed to benefit large-scale corporate agriculture and have been accused of hastening the demise of small producers and family farmers. They certainly do nothing to boost the fortunes of poor rural peoples and communities. Efforts have been made to end or scale back these programmes, especially as they come head to head with pressures to end subsidies and support free trade as part of globalization. However, currently there is little evidence of any serious movement in this direction. In fact, the most recent policy legislation restores and enlarges subsidies that had been previously cut back or eliminated. Infrastructure investment projects such as the Appalachian Regional Commission, the Delta Commission and more recent enterprise and empowerment zones, while targeted at poor places, also have little direct impact on rural poverty, although they have selectively improved roads, hospitals, and similar basic economic development infrastructure. Furthermore, with few exceptions (cf. Glasmeier and Fuelhart, 1999), there has been little systematic evaluation to assess their impacts. For both types of programmes, effects on rural communities, residents and their living conditions have been indirect, haphazard and often inadvertent, and there has been little interest among policymakers in creating more general rural development policies. In fact, every effort to create a broader policy for the agricultural sector, beginning in the depression era, has met with little support and often strong opposition, even though programmes were generally modest and based on the widely accepted modernization paradigm that prevailed in economic development circles (O'Connor, 1992).

Safety net programmes

The provision of social welfare programmes in the US historically has been organized by gender, race and spatial factors (Gordon, 1990; Tickamyer, 1995–96). For example, the single most effective programme in reducing poverty, the social security programme, first implemented as part of safety net provision in the Depression era New Deal legislation, excluded farm workers, effectively eliminating large numbers of the poorest rural residents, including many racial and ethnic minorities. Later, in the 1960s, War on Poverty expansions of New Deal programmes[3] extended access and created new forms of assistance. However, they still had a limited focus primarily on female-headed households and their children, again disadvantaging rural households who were more likely to be in traditional families and to be part of the working poor.

Even though the War on Poverty was prompted in part by disturbing images of rural poverty, specifically the hardship and deprivation clearly visible in Appalachian and deep South landscapes, their primary focus quickly turned to urban ghettos, especially in the wake of the civil unrest and urban disturbances of the era. Problems specific to rural residents receded in public interest and few efforts were made to tailor safety net programmes to the specific conditions of the rural poor. Not only did the demographics of the rural poor disqualify them from some forms of interventions but they also suffered from greater isolation, lack of transportation and information that limited access to services. Furthermore, while eligibility was determined by national policies, benefit levels for most programmes were set by states. The regions that include the largest concentrations of persistent rural poverty coincided with states with the lowest benefit levels, the most punitive policies and often the most corrupt and discriminatory local administrations. Finally, in many places, the rural poor are more reluctant to use safety net services even when eligible. This reluctance has been attributed to greater stigma attached to their use in rural areas, whether from cultural reasons or the lack of privacy in small rural communities. Whatever the reason, the outcome is less access and less use of social welfare safety net programmes among the rural poor.

The restructured welfare state

Growing dissatisfaction with the contours of the welfare state and its safety net programmes surfaced and gained strength among both the political right and left. Conservative analysts argued that poverty and its correlates were the outcome of an overly generous welfare system that encouraged welfare dependency and other antisocial behaviour. More progressive critics pointed to the inconsistencies, inefficiencies and inequities embedded in welfare bureaucracies that undermined their effectiveness as well as individual initiative and work effort. Both views conflated poverty and welfare dependence and played on popular prejudices and stereotypes about the poor that equated poverty with deviance and deprivation.

The movement to restructure the welfare state gained momentum in the 1990s, culminating in the successful passage with bipartisan support of the Personal Responsibility and Work Opportunity Reconciliation Act (PRWORA) in 1996. This legislation ended the 60-year history of New Deal and War on Poverty entitlements, eliminating the primary programme of cash assistance, and put new limits on eligibility for and receipt of assistance.[4] The purpose was to move welfare recipients off assistance into work, either by carrot or by stick, incentives or sanctions, and to devolve specific authority for safety net programmes to state and local jurisdictions. Programmes varied dramatically from state to state and in some states from county to county, reflecting different priorities, politics and opinions about the causes and remedies of poverty. However, the overall approach and the dominant position in most states was to view poverty as a national and urban problem, responding to the greater numbers and visibility of the urban poor.

Even though devolution implies that welfare policy theoretically is responsive to local needs, the invisibility and lack of knowledge of the specific problems and issues of the rural poor meant that few programmes acknowledged and planned with their needs in mind (Tickamyer et al., 2002). Yet the rural poor face specific obstacles to moving from welfare to work that frequently differ from urban populations. While these vary by place, they include lack of employment opportunities, lack of living wage jobs with benefits, lack of transportation, lack of childcare, lack of human and social capital, lack of local institutional and organizational capacity, and general lack of infrastructure investment and support that could supply these missing ingredients. Mirroring the lack of interest in rural poverty in policy circles, there has been relatively little formal scrutiny of outcomes of welfare reform in rural communities. With few exceptions (see Weber et al., 2002 for one of the most comprehensive collections of research on the subject), studies that document impacts have ignored rural peoples and places or have only looked crudely at metro–nonmetro differences, ignoring the variation that exists within and between rural locations. Overall, there is good reason to expect that the rural poor may have a particularly rough time meeting the requirements of welfare reform, but as yet this is largely speculative.

IS THE US UNIQUE?

US exceptionalism is often assumed in a variety of historical and developmental contexts, so the question arises of whether conditions and sources of rural poverty are also a uniquely American phenomenon. Obviously, the specific social and spatial landscapes are necessarily unique as are the historical, social, political and cultural configurations that construct the American experience. However, it can be argued that, in a number of significant ways, there are at least as many similarities as differences, and rural poverty in the US is just one manifestation of rural poverty in postindustrial nations. Areas of known similarity include the relative invisibility and neglect of rural poverty with concomitant concern with urban and aspatial perspectives; the tendency to view rural poverty through an ideological and idealized lens; the importance of economic restructuring, de-industrialization, global competition and the changing nature of natural resource extraction and livelihoods in rural poverty. Additionally, many other areas require further investigation, such as the influence of public policy and the shape of the welfare state on the rural poor.

The invisibility of rural poverty in postindustrial Europe is a case in point, witnessed by the lack of interest in specifically rural poverty in the major studies and assessments of current poverty in European nations. For example, reports from waves of the European Community Household Panel (ECHP), a major effort to collect and analyse comparable longitudinal data across the members of the European Union, emphasize multi-dimensional, dynamic, life course analyses, but with few exceptions provide little subnational information by type of place (Barnes et al., 2002). Regional differences can be inferred in part from national differences. Thus poverty rates vary widely across nations (from 9 per cent in Denmark to 29 per cent in Portugal in the first wave of the ECHP), with poorer, less developed countries with higher levels of agriculture employment displaying high rates of poverty (Barnes et al., 2002; Van den Bosch et al., 1997). Within-country regional differences in the UK (Cloke et al., 1995b; McCormick and Philo, 1995) and parts of Europe (Barnes et al., 2002) also may partly reflect variation in rural industrial development, similar to distressed regions in the US. Thus parts of southern Italy, the former East Germany and regional divisions in Britain present suggestive profiles of poverty and exclusion.

European concern with 'social exclusion' bears a limited resemblance to the American focus on aspatial social characteristics of the poor. Accounts of disproportionate poverty among children, elderly, disabled, women, single-parent households, substance abusers, unemployed youth, immigrant and ethnic minorities and crowded housing developments echo similar concerns and findings in the US (Barnes et al., 2002; Geddes and Benington, 2001; Hauser and Becker, 2003; Ott and Wagner, 1997). However, there is greater emphasis on the impediments to full participation in society and the ways these are manifested. Poverty itself is defined to emphasize this social and relational aspect of poverty. Thus the European Commission Council statement of 19 December 1984 defines poverty as 'persons whose resources (material, cultural and social) are so limited as to exclude them from the minimum acceptable way of life in the member states in which they live' (as quoted in Atkinson, 1998: 2). Furthermore, the extent to which social exclusion or its manifestations are spatially correlated or distributed intranationally is not given a high priority in these analyses, and the implicit assumption of national and urban priority appears to dominate European scholarly accounts as in US analyses as well.

The focus on social exclusion may be related to an ideological dimension of the invisibility of rural poverty. In the UK, one of the few countries with extensive research on rural poverty and lifestyles, the idealization of the rural countryside has served to obscure the existence of poverty and hardship in its midst. Paul Cloke and his colleagues (Cloke, 1995; Cloke et al., 1995a, 1995b; Milbourne, 1997; Philo, 1995; Woodward, 1996), based on extensive investigation via the Rural Lifestyles Project, argue that the rural 'idyll' or idealization of rural landscapes and lifestyles denies the existence of serious deprivation or marginalized populations in the British countryside, despite substantial evidence of serious need. It is not clear whether this discourse also prevails in other European nations. However, it is suggestive of the power of rural romanticism that in the two places with relatively extensive rural poverty research records – the US and the UK – discursive practices serve to obscure the extent and reality of rural poverty, by distortion in the US and denial in the UK (Cloke, 1995).

Debates on the causes of exclusion also share similarities to US accounts, centring on the dynamics of post-Fordist economic restructuring and globalized capitalism versus the deficiencies and failures of government policies and welfare systems (Geddes and Benington, 2001). Geddes and Bennington (2001: 18) argue that the highest rates of poverty are found in European regions with a large proportion of the population employed in agriculture, but the largest numbers of poor people are found in more industrialized countries that have experienced de-industrialization and high unemployment. Certainly, industrial and agricultural restructuring is implicated in poverty, including rural poverty in all postindustrial nations.

Similarly, arguments have been mounted that state welfare policies have a major role in perpetuating poverty and exclusion (compared to fear of dependency in the US). Similar to the US, some blame welfare policies as the source of the problem, rather than the correlate, consequence or solution to poverty. Welfare programmes are seen to reinforce cultures of exclusion and inhibit initiative, while providing inadequate and fragmented services that are far removed from community contribution or participation (Geddes and Benington, 2001). Others see the dismantling of the welfare state as implicated in problems of poverty, especially poverty in isolated rural areas (Meert, 2000).

Generalizations are not easy to make given the wide variation across developed countries in the types of welfare system, ranging from those that fit Esping-Anderson's (1991) standard categories of corporatist, social democratic and liberal to 'rudimentary welfare states' with minimal welfare provision (Millar and Middleton, 2002). However, it should be no surprise that some of the highest rates of poverty are found in liberal or neo-liberal systems such as the US and the UK as well as in the rudimentary welfare states (that is, in the less developed nations of the EC). By definition, the latter will have fairly high rates of rural poverty. The relationship of type of welfare regime to the spatial distribution of poverty requires greater scrutiny, although accounts of poverty policy in both the US and the UK (Cloke, 1995, 1997;

Cloke et al., 1995a, 1995b; Milbourne, 1997) are highly suggestive of the value of this inquiry.

CONCLUSION

The processes described are not unique to the US but common to many nations, both in the core and the periphery. It would be presumptuous and inaccurate to say that rural poverty in the developing nations is strictly analogous to counterparts in the core, but in an increasingly interconnected and globalized economy, many of the same processes can be predicted or are already under way. The parallels described by Tardanico and Rosenberg (2000) between southern regions of Mexico and the US under NAFTA, despite huge differences in their national economies and polities, point to the direction that development may take under globalization. Similarly, concerns about decentralized rural programmes in Mexico mirror the questions about capacity that ensue from devolution in the US. 'A particular problem of decentralized rural development programs is the typically limited capacity of local governments to improve policy outcomes (Fox and Aranda, 2000: 189). 'Poverty or development' are not two parallel universes but rather interrelated processes whose intersection enriches some, impoverishes many, and which are particularly problematic for rural people and places. Future research must scrutinize these connections, assuming the difficult task of documenting historically specific conditions while analysing the increasingly complex processes that link people and places across physical and cultural borders.

NOTES

1 US counties are classified as metropolitan or nonmetropolitan, depending on population size and proximity to a large population centre. The county is the basic administrative unit used to collect and report official statistics and social indicators, and it is the geo-political area for which the most consistent and comparable data are available. Strictly speaking, however, the metro/nonmetro designation is not synonymous with urban and rural. It is possible to have extensive rural areas in metropolitan counties and some nonmetro counties may be integrated into urban economies but not yet captured and classified as part of the metro area.

2 The US government has calculated an official measure of poverty based on income poverty since the mid-1950s. The poverty threshold uses a formula to determine a minimum subsistence level based on cost of living data, family size and composition that is adjusted annually for inflation. Persons residing in families with incomes below the appropriate figure are classified as poor. In 2001, the poverty line was approximately $9,000 for a single individual and $18,000 for a family of four. This compares with median income in 2000 of almost $51,000 overall ($41,000 in rural areas; $62,000 for a family of four and $21,000 for a single individual; US Census, 2002). The establishment of an absolute poverty threshold contrasts with the approach used by European Community nations to measure income poverty, which typically uses some variation on half the average income per member nation (Atkinson, 1998).

3 The term 'War on Poverty' was coined by President Lyndon Johnson in his inaugural speech in 1964 and became a primary focus of his domestic agenda to build 'the Great Society'. However, the original idea to mobilize vast resources in 'a comprehensive assault on poverty' originated in the Kennedy administration (Katz, 1996: 262). War on Poverty programmes entailed numerous extensions and expansions of existing programmes and the creation of many new services to assist the poor, improve health, nutrition and welfare, and extend access to safety net provisions to previously omitted groups. Additionally, it marked the mobilization of numerous grass roots social action efforts, community organizations and civil rights campaigns that materially affected US politics and policies (Katz, 1996).

4 The PRWORA legislation eliminated the federally mandated Aid to Families with Dependent Children (AFDC), a programme that entitled anyone who was eligible to aid and substituted Temporary Assistance to Needy Families (TANF), a time-limited programme with a maximum eligibility of five years. It was funded with block grants to states, and an important component of the restructuring was to devolve major responsibility to states to design and implement programmes in response to local needs and knowledge. This included authority to determine length of eligibility which varied across states from two years to the full 60 months maximum allowed in the authorizing legislation.

REFERENCES

Albrecht, Don, Albrecht, Carol and Albrecht, Stan (2000) 'Poverty in nonmetropolitan America: impacts of industrial, employment, and family structure variables', *Rural Sociology* 65 (1): 87–103.

Atkinson, A.B. (1998) *Poverty in Europe*. Oxford: Blackwell.

Barnes, Matt, Heady, Christopher, Middleton, Sue, Millar, Jane, Papadopoulos, Fotis, Room, Graham and Tsakloglu, Panos (2002) *Poverty and Social Exclusion in Europe*. Cheltenham: Edward Elgar.

Billings, Dwight B. and Blee, Kathleen M. (2000) *The Road to Poverty: The Making of Wealth and Hardship in Appalachia.* New York: Cambridge University Press.

Billings, Dwight B. and Tickamyer, Ann R. (1991) 'Development and underdevelopment: the politics of Appalachian development', in T.A. Lyson and W.W. Falk (eds), *Forgotten Places: Uneven Development and the Underclass in Rural America*. Lawrence, KS: University of Kansas Press. pp. 7–29.

Buttel, Frederick et al. (1993) 'The state, rural policy, and rural poverty', in Rural Sociological Society Task Force on Persistent Rural Poverty (ed.), *Persistent Poverty in Rural America*. Boulder, CO: Westview Press. pp. 292–326.

Cloke, Paul (1995) 'Rural poverty and the welfare state: a discursive transformation in Britain and the USA', *Environment and Planning A* 27: 1001–1016.

Cloke, Paul. (1997) 'Poor country: marginalisation, poverty and rurality', in Paul Cloke and Jo Little (eds), *Contested Countryside Cultures: Otherness, Marginalisation and Rurality*. London and New York: Routledge. pp. 252–271.

Cloke, Paul, Goodwin, Mark, Milbourne, Paul and Thomas, Chris (1995a) 'Deprivation, poverty and marginalization in rural lifestyles in England and Wales', *Journal of Rural Studies* 11 (4): 351–365.

Cloke, Paul, Milbourne, Paul and Thomas, Chris (1995b) 'Poverty in the countryside: out of sight and out of mind', in C. Philo (ed.), *Off the Map: The Social Geography of Poverty in the UK*. London: CPAG. p. 83.

Cloke, Paul, Milbourne, Paul and Widdowfield, Rebekah. (2002) *Rural Homelessness: Issues, Experiences and Policy Responses*. Bristol: The Policy Press.

Dickinson, Tory and Schaeffer, Robert K. (2001) *Fast Forward: Work, Gender, and Protest in a Changing World*. Lanham, MD: Rowman and Littlefield.

Duncan, Cynthia M. (ed.) (1992) *Rural Poverty in America*. New York: Auburn House.

Duncan, Cynthia M. (1996) 'Understanding persistent poverty: social class context in rural communities', *Rural Sociology* 61: 103–124.

Duncan, Cynthia M. (1999) *Worlds Apart: Why Poverty Persists in Rural America*. New Haven, CT: Yale University Press.

Duncan, Cynthia M. (2001) 'Social capital in America's poor rural communities', in Susan Saegert, J. Phillip Thompson and Mark R. Warren (eds), *Social Capital and Poor Communities*. New York: Russell Sage Foundation.

Eller, Ronald D. (1982) *Miners, Millhands, and Mountaineers: Industrialization of the Appalachian South, 1880–1930*. Knoxville, TN: University of Tennessee Press.

Escobar, Arturo (1995) *Encountering Development: The Making and Unmaking of the Third World*. Princeton, NJ: Princeton University Press.

Esping-Anderson, G. (1991) *The Three Welfare States of Capitalism*. Cambridge: Polity Press.

Falk, William W. (2003) 'Sense of place and rural restructuring: lessons from the Low Country', in William Falk, Michael Schulman and Ann Tickamyer (eds), *Communities of Work: Rural Restructuring in Local and Global Context*. Athens, OH: Ohio University Press.

Fink, Deborah (1998) *Cutting into the Meat Packing Line: Workers and Changes in the Rural Midwest*. Chapel Hill, NC: North Carolina University Press.

Fitchen, Janet M. (1981) *Rural Poverty in America: A Case Study*. Boulder, CO: Westview Press.

Fox, Jonathan and Aranda, Josefina (2000) 'Politics of decentralized rural poverty programs: local government and community participation in Oaxaca', in Richard Tardanico and Mark B. Rosenberg (eds), *Poverty or Development*. New York: Routledge. pp. 179–196.

Fraser, Nancy (1990) 'Struggle over needs: outline of a socialist-feminist critical theory of late-capitalist political culture', in Linda Gordon (ed.), *Women the State, and Welfare*. Madison, WI: University of Wisconsin Press. pp. 199–225.

Fraser, Nancy and Gordon, Linda (1994) 'A genealogy of dependency: tracing a keyword of the US welfare state', *Signs* 19 (Winter): 328–329.

Geddes, Mike and Benington, John (eds) (2001) *Local Partnerships and Social Exclusion in the European Union: New Forms of Local Social Governance*. London and New York: Routledge.

Glasmeier, Amy and Fuelhart, Kurtis (1999) 'Building on past experiences: creating a new future for distressed counties'. A Report to the Appalachian Regional Commission. Retrieved 25 May 2002. (http://www.arc.gov/research/ building.htm)

Gordon, Linda (ed.) (1990) *Women, the State and Welfare*. Madison, WI: University of Wisconsin Press.

Handler, Joel F. (1995) *The Poverty of Welfare Reform*. New Haven, CT: Yale University Press.

Hauser, Richard and Becker, Irene (eds) (2003) *Reporting on Income Distribution and Poverty: Perspectives from a German and a European Point of View*. Berlin: Springer-Verlag.

Hyland, Stanley and Timberlake, Michael (1993) 'The Mississippi Delta: change or continued trouble?', in T.A. Lyson and W.W. Falk (eds), *Forgotten Places: Uneven Development in Rural America*. Lawrence, KS: University of Kansas. pp. 76–101.

Katz, Michael (1996) *In the Shadow of the Poor House*, rev edn. New York: Basic Books.

Lichter, Daniel T. and Jensen, Leif (2002) 'Rural America in transition: poverty and welfare at the turn of the twenty-first century', in B.A. Weber, G.J. Duncan and L.E. Whitener (eds), *Rural Dimensions of Welfare Reform Welfare, Food Assistance, and Poverty in Rural America*. Kalamazo, MI: UpJohn Institute.

Lyson, Thomas A. and Falk, William W. (eds) (1993) *Forgotten Places: Uneven Development and the Loss of Opportunity in Rural America*. Lawrence, KS: University Press of Kansas.

Maril, Robert Lee (1989) *Poorest of Americans: The Mexican Americans of the Lower Rio Grande Valley in Texas*. Notre Dame, IN: University of Notre Dame Press.

McCormick, James and Philo, Chris (1995) 'Where is poverty? The hidden geography of poverty in the United Kingdom', in Paul Milbourne (ed.), *Revealing Rural 'Others': Representation, Power and Identity in the British Countryside*. London and Washington: Pinter. pp. 1–22.

McLaughlin, Diane, Gurdner, Erica and Lichter, Daniel (1999) 'Economic restructuring and changing prevalence of female-headed families in America. *Rural Sociology* 64: 394–416.

Meert, Henk (2000) 'Rural community life and the importance of reciprocal survival strategies', *Sociologia Ruralis* 40: 319–338.

Milbourne, Paul (1997) 'Hidden from view: poverty and marginalization in rural Britain', in Paul Milbourne (ed.), *Revealing Rural 'Others': Representation, Power and Identity in the British Countryside*. London and Washington: Pinter. pp. 89–116.

Millar, Jane and Middleton, Sue (2002) 'Introduction' in Barnes et al. *Poverty and Social Exclusion in Europe*. Cheltenham: Edgar Elgar.

Montejano, David (1987) *Anglos and Mexicans in the Making of Texas, 1836–1986*. Austin, TX: University of Texas Press.

Nord, Mark (1997) 'Overcoming persistent poverty – and sinking into it: income trends in persistent poverty and other high poverty rural counties, 1989–94', *Rural Development Perspectives* 12 (3): 2–10.

O'Connor, Alice (1992) 'Modernization and the rural poor: some lessons from history', in Cynthia M. Duncan (ed.), *Rural Poverty in America*. Westport, CT: Auburn House. pp. 215–233.

Ott, Notbuga and Wagner, Gert G. (eds) (1997) *Income Inequality and Poverty in Eastern and Western Europe*. Heidelberg: Physica-Verlag.

Philo, Chris (ed.) (1995) *Off the Map: The Social Geography of Poverty in the UK*. London: CPAG.

Pickering, Kathleen (2000) 'Alternative economic strategies in low-income rural communities: TANF, labour migration, and the case of the Pine Ridge Indian Reservation', *Rural Sociology* 56 (4): 148–167.

Rostow, W.W. (1960) *The Stages of Economic Growth: A Non-Communist Manifesto*. Cambridge: Cambridge University Press.

RSS (Rural Sociological Society) Task Force on Persistent Rural Poverty (1993) *Persistent Poverty in Rural America*. Boulder, CO: Westview Press.

Saegert, Susan, Thompson, J. Phillip and Warren, Mark R. (eds) (2001) *Social Capital and Poor Communities*. New York: Russell Sage Foundation.

Saenz, Rogelio and Ballejos, Marie (1993) 'Industrial development and persistent poverty in the Lower Rio Grande Valley' in Thomas A. Lyson and William W. Falk (eds), *Forgotten Places: Uneven Development in Rural America*. Lawerence, KS: University Press of Kansas, pp. 102–124.

Schram, Sanford (1995) *Words of Welfare: The Poverty of Social Science and the Social Science of Poverty*. Minneapolis, MN: University of Minnesota Press.

Seltzer, Curtis (1985) *Fire in the Hole: Miners and Managers in the American Coal Industry*. Lexington, KY: University Press of Kentucky.

Sernau, Scott (2001) *Worlds Apart: Social Inequalities in a New Century*. Thousand Oaks, CA: Pine Forge Press.

Snipp, C. Matthew (1996) 'Understanding race and ethnicity in rural America', *Rural Sociology* 61 (1): 125–141.

Snipp, C. Matthew et al. (1993) 'Persistent rural poverty and racial and ethnic minorities', in Rural Sociological Society Task Force on Persistent Rural Poverty (ed.), *Persistent Poverty in Rural America*. Boulder, CO: Westview Press. pp. 173–199.

Stack, Carol B. (1996) *Call to Home: African Americans Reclaim the Rural South*. New York: Basic Books.

Tardanico, Richard and Rosenberg, Mark B. (eds) (2000) *Poverty or Development*. New York: Routledge.

Tickamyer, Ann R. Henderson, Debra A., White, Julie Anne and Tadlock, Barry L. (2000) 'Voices of welfare reform: bureaucratic rationality versus the perceptions of welfare participants'. *Affilia* 15: 173–192.

Tickamyer, Ann R. Wood, Teresa A. (2003) 'The social and economic context of informal work', in William Falk, Michael Schulman, Ann Tickamyer (eds), *Communities of Work: Rural Restructuring in Local and Global Context*. Athens, OH: Ohio University Press.

Tickamyer, Ann R. (1995–96) 'Public and private lives: social and spatial dimensions of women's poverty and welfare policy in the United States', *Kentucky Law Journal* 84: 721–744.

Tickamyer, Ann R. and Duncan, Cynthia M. (1990) 'Poverty in rural America', *Annual Review of Sociology* 16: 67–86.

Tickamyer, Ann R. and Henderson, Debra (2003) 'Rural women: new roles for a new century', in D. Brown and L. Swanson (eds), *Challenges for Rural Women in the Twenty First Century*. Philadelphia, PA: Pennsylvania State University Press.

Tickamyer, Ann R., Duncan, Cynthia M. and Heffernan, Kara. (2004) 'Poverty and inequality in rural America', in Gwendolyn Mink and Alice O'Connor (eds), *Poverty in the United States: An Encyclopedia of History, Politics, and Policy*. Santa Barbara, CA: ABC-CLIO.

Tickamyer, Ann R., White, Julie, Tadlock, Barry and Henderson, Debra (2002) 'Where all the counties are above average', in B.A. Weber, G.J. Duncan and L.E. Whitener (eds), *Rural Dimensions of Welfare Reform Welfare, Food Assistance, and Poverty in Rural America*. Kalamazo, MI: UpJohn Institute.

US Census Bureau (2000) 'Income 2000'. Retrieved 25 May 2002. (http://www.census.gov/hhes/www/income.00.html)

US Department of Agriculture, Economic Research Service (2000) 'Briefing room: rural income, poverty, and welfare'. Retrieved 28 January 2001. (http://www.ers.usda.gov/Briefing/ruralpoverty)

Van den Bosch, Karel, De Lathouwer, Lieve and Deleeck, Herman (1997) 'Poverty and social security transfers – results for seven countries and regions in the EC', in Notbuga Ott and Gert G. Wagner (eds), *Income Inequality and Poverty in Eastern and Western Europe*. Heidelberg: Physica-Verlag. pp. 91–124.

Weber, Bruce A., Duncan, Greg J., and Whitener, Leslie A. (eds) (2002) *Rural Dimensions of Welfare Reform*. Kalamazoo, MI: Upjohn Insitute.

Woodward, Rachel (1996) '"Deprivation" and "the rural": an investigation into contradictory discourses', *Journal of Rural Studies* 12: 55–67.

World Bank (2000) *World Development Report 2000/2001: Attacking Poverty*. Oxford: Oxford University Press.

31

Rural housing and homelessness

Paul Milbourne

INTRODUCTION

Housing and homelessness represent key constituents of rural social change and welfare. The presence of particular forms of housing, its suitability and its accessibility all play a part in shaping the social composition of particular rural places. Rural housing connects with broader economic and socio-cultural processes in rural areas. Inadequate housing conditions and the abandonment of housing in rural areas are indicative of broader problems concerning marginal rural economies, rural depopulation and rural poverty. By contrast, gentrified rural housing markets are symplomatic recent processes of rural re-population and the 'capture' of rural spaces by middle-class groups. Within such processes, rural housing is constructed as a desirable positional good that carries a great deal of cultural capital. Housing has also been drawn into a series of political, social and cultural conflicts in rural areas as different groups seek to project their own identities and promote particular meanings of rurality through housing.

Another important aspect of rural housing and homelessness relates to the theme of social welfare. Over recent years, academic and policy discussions of rural housing have largely been concerned with issues of affordability and need. The main focus of these discussions has been on the uneven and unequal forms of access linked to particular systems of housing in rural areas, and the failure of these systems to accommodate a broad range of social groups. While much of this attention has focused on the needs of groups located within rural housing systems, the last few years have also witnessed attempts by rural researchers to consider the needs of the homeless – a group that is effectively locked out of formal housing systems.

This chapter is concerned with making connections between housing and homelessness and these broader social, cultural, economic, political and welfare rural contexts. The chapter provides a comprehensive and critical review of academic approaches to and evidence on rural housing and homelessness in advanced capitalist countries. In doing this, a key intention is to take stock of current academic understandings of rural housing and homelessness and suggest ways that these subjects can be approached more critically by rural researchers. The geographical coverage of the chapter can be seen to reflect that of the academic literature on rural housing and homelessness, with its main focus being on academic writings on these subjects in the United States and the UK. The structure of the chapter also reflects dominant academic approaches to housing and homelessness in rural areas, with these two subjects tending to be researched separately – housing as a discrete sub-field of rural studies, while homelessness is usually discussed in relation to rural poverty.

The body of the chapter is structured around three main sections. The first focuses on rural housing and considers systems of rural housing production and regulation, connections between rural housing, society and culture, and key rural housing welfare issues. The second section of the chapter then moves on to explore homelessness in rural areas and, in particular, focuses on the scale, nature, experiences and welfare contexts of rural homelessness. The chapter finishes with a critical assessment of the state of academic understandings of rural housing and homelessness, and the identification of ways that new theoretical research agendas can be developed to make connections between these two subjects.

RURAL HOUSING

Writing more than twenty years ago about the state of rural housing research in Britain, Phillips and Williams (1982: 3) referred to 'a general neglect of rural housing issues, as any of the standard books on housing would reveal'. Unfortunately, there remains an element of truth in this statement in relation to rural housing research and writing over the past couple of decades. Today, one might use the term 'marginalized' rather than 'neglected' to describe the state of rural housing literature in the UK. There is no doubt that the volume of research on rural housing issues has increased over recent years and attempts have been made to provide more critical and theoretically informed understandings of rural housing, but the cutting edge of housing research continues to remain focused on urban spaces. As discussed in the introduction, however, the marginalization of rural housing as a research area in the UK is much less pronounced than that in most other countries; possibly with the exception of the United States, where there is a developed literature on rural homelessness. In these other countries, it is easier to describe rural housing as a 'neglected' research area, with very little published material evident, and much of what has been published being characterized by weakly developed theoretical analytical frameworks.

Notwithstanding this general marginalization of housing within rural studies, it is possible to highlight three main themes within the rural housing literature. The first concerns structures and regulatory processes bound up with rural housing, involving private markets and the role of the state as housing provider. The second theme connects housing with social and cultural change in rural areas, and considers the role of housing within rural demographic processes. Welfare represents the third key theme discussed in this section, and here attention is given to linkages between rural poverty and housing conditions, and broader issues of housing need in rural areas. I will consider each of these themes in turn.

Rural housing structures and regulatory processes

One of the most significant findings that cuts across rural housing studies undertaken in different countries concerns the peculiarities of rural housing structures. The picture that emerges is one in which the private sector occupies a much more dominant position in providing housing in rural areas compared with the situation in urban spaces. With a wide range of statistical evidence published on housing tenure, this picture is relatively easy to demonstrate for different countries. In England, for example, the private sector accounts for the tenure of 84 per cent of rural households, but 77 per cent of those in urban areas (Countryside Agency, 2003). By far the largest proportion of this private housing is provided through private ownership, with ownership rates generally higher in rural areas. In the United States, the national rate of ownership is 68 per cent compared with 76 per cent in non-metropolitan areas (Housing Assistance Council, 2002); the 2001 Census New Zealand records a home ownership rate of 76 per cent in its rural areas compared with 67 per cent in urban areas; and owner-occupation accounts for 80 per cent of rural households in Australia but only 69 per cent of households in major urban areas (Tonts et al., 2001).

While relatively little research has examined the reasons underlying the increased dominance of private property ownership in different rural areas, it is possible to highlight three main explanations. The first is that this rural–urban ownership differential reflects different social structures in rural and urban areas, with higher proportions of more affluent groups living in and moving to rural areas. However, as will be discussed later in this chapter, though this point certainly holds true in Britain, it is more problematic in relation to other countries. A second explanation for high rural ownership levels concerns particular cultural attributes that are seen to be connected with rural living, most notably those concerning self-reliant forms of community living. For example, commenting on the situation in Ireland, Finnerty et al. (2002: 133) argue that 'the almost exclusive attachment to ownership in rural areas has virtually eliminated rental options to the point that they are rarely considered to be realistic accommodation strategies even by households who cannot afford to buy'.

A third explanation involves addressing the rural tenure situation not from the perspective of high levels of property ownership but from that of low levels of social housing provision. It is clear that the state (and associated agencies) has played a relatively insignificant role in providing social housing in rural areas. In France, for example, Auclair and Vanoni (2003) point out that the social housing sector provides one-quarter of the country's new housing, but only 8 per cent in rural areas. The reasons for this limited intervention of the state in rural housing markets vary from one country to another. However, it is possible to identify three key factors. The first concerns the practical difficulties faced by agencies in providing units of social housing in small and isolated rural communities (Hoggart, 2003). The second factor relates to the cultural politics of rural areas in particular countries, which often lead to proposed schemes of social housing being resisted by local communities (see Auclair and Vanoni, 2003; Cloke et al., 2002; Yarwood, 2003). Structures of local rural welfare provision represent the third factor. Writing about the situation in New Zealand, Davey and Kearns (1994) suggest rural housing provision tends to be detached from national welfare programmes, while Newby (1979), argues that historical local political structures in rural England have been largely responsible for low levels of social provision in the countryside (see also Rogers, 1976 and Milbourne, 1998). Newby highlights how rural councils dominated by larger farmers and landowners were often reluctant to use newly acquired powers to construct units of social housing in the inter-war and immediate post-war periods, as they had a vested interest in maintaining their workforce within systems of tied accommodation.

More recently, further work has been undertaken on impacts of welfare restructuring on the provision of social housing in rural areas. With 'new right' central government administrations in power in different countries over the 1980s and 1990s, significant attempts were made to reduce levels of national welfare expenditure and these attempts fed through to social housing programmes in rural areas. In the United States, the Housing Assistance Council (2002) highlights major cut-backs in federal government programmes of low- and moderate-income housing schemes in rural areas over recent years, with the number of units of affordable housing provided by one particular government scheme of housing assistance, the Section 515 Rural Housing Program, falling from almost 12,000 in 1994 to less than 2,000 units in 2002. A similar situation has been reported in the UK, where housing policies introduced in the 1980s by the Thatcher governments brought about dramatic cut-backs in new social housing building programmes, relegated the roles of local authorities from housing providers to enablers, and gave many social housing tenants the right to purchase their properties at dramatic discounts. Drawing on regulationist accounts of welfare restructuring, Milbourne (1998) examines the impacts of these policies in rural areas. A key point that emerges from this work is that while there has been a general residualization of the social housing sector in the UK, this process has been more pronounced in rural areas, where historically lower levels of social stock have been reduced at much faster rates than in the cities. Milbourne (1998) also highlights important processes of *social residualization* linked to social housing restructuring in rural areas – as more affluent tenants have purchased the better quality social housing stock, thus widening the gap between social rental tenants and those able to enter owner-occupation – and processes of *spatial residualization*, as reductions in the stock of the social housing have been greatest in smaller villages.

Research in Britain has also focused on new structures of local social housing provision in rural areas that have emerged over recent years. Work has traced the development of new partnership arrangements that have drawn in a broader range of agencies to provide social housing in the countryside. Studies have highlighted the increasing complexity of social housing provision in local rural spaces, as the previous activities of one agency, local government, have been replaced by partnerships involving, *inter alia*, local government, housing associations, landowners, building companies and financial institutions (see Milbourne, 1998; Cloke et al., 2001; Yarwood, 2003). However, what these studies also show is the continued influential position of local government within new partnerships of social housing provision. As Milbourne comments in relation to housing partnerships in two areas of rural Wales, 'the local housing authority maintains a powerful position as both enabler and (indirect) provider of social housing opportunities' (1998: 183).

In addition to examining the distinctive structures of rural housing, academic attention has been given to the processes through which housing systems are regulated in rural areas. Unfortunately, much of this work has focused on regulatory processes within the context of rural Britain, although it is possible to tease out some broader conclusions from these studies. The bulk of research on the regulation of rural housing has focused on the linkages between housing and planning systems. In the 1970s and early 1980s, much of the UK-based academic work on rural housing was concerned with the ways in which housing was bound up with settlement planning in rural areas (see Cherry, 1976). While this focus remains evident in more recent rural planning texts (for example, Gilg, 1996), the period since the mid-1980s has witnessed a shift in focus to the more negative impacts of planning systems on the provision of housing for particular groups of the rural population. The planning system has been accused of restricting the provision of new properties generally in rural areas, and the supply of affordable housing more particularly, with the operation of key settlement policies channelling new residential development into towns and larger villages.

While Sibley (1995) has provided a broad-ranging critique of the UK planning system, arguing that it has been developed to protect middle-class constructions of the countryside as a predominantly natural space free of 'disorderly development', others have explored the more specific relationships between the planning system and the provision of affordable housing in rural areas. Rogers (1985), for example, has examined the ways in which local government in rural areas has attempted to introduce so-called 'local needs' housing policies through the planning system. Work by Shucksmith (1981, 1990a) has critically evaluated the impacts of such planning policies on local housing need, with his work in the English Lake District indicating that such planning approaches to dealing with housing need represent rather blunt policy instruments. More recently, academic attention has been given to other central government planning responses to housing problems in rural Britain, with evaluations undertaken of new schemes of low-cost housing provision for local groups in housing need on so-called 'exceptional sites' where such developments would not normally be allowed (see Gallent, 1997; Williams et al., 1991; Yarwood, 2003).

More limited attention has been directed to the operations of different fractions of private capital within rural housing markets. Drawing on ideas of domestic property classes (see Saunders, 1984), efforts have been made to examine the different interests bound up with what has been termed the private property class – those agents involved in the provision of new housing and services concerned with its supply and distribution – in rural Britain (see Shucksmith, 1990a, b and Milbourne, 1997). Four main groups within the private property class can be identified: industrial capital (property developers); commercial capital (exchange professionals); finance capital (banks and building societies); and landed capital (landowners). Shucksmith (1990) highlights potential conflicts within the private capital class in relation to housing in rural areas. He suggests that industrial capital will tend to favour an unregulated housing market and high levels of land release, that commercial and finance capital will support highly regulated housing markets since they lead to property price rises and the attraction of wealthier house purchasers, while the landed capital class contains oppositional interests between those who own land in development and protected areas.

The activities of these fractions of private capital have also been explored within the context of rural housing markets in Wales (Milbourne, 1997, 1999). This work points to the complexities associated with the relationships between these private property classes, as proposed by Shucksmith (1990a, b), and the importance of remaining sensitive to the specificities of different types of rural housing market when considering the interests of members of the private property class. More generally, these studies provide an insight into the role and influence of property developers, financial institutions and exchange professionals within local rural housing markets. It also indicates that property developers in these rural Welsh study areas are generally local and small-scale companies involved more in the development of existing properties than the construction of new housing. As such, they play a relatively insignificant role in the regulation of housing compared with the increased significance of larger property developers reported within urban studies (see, for example, Barlow and Savage, 1986).

Similarly, key findings from studies of the regulatory roles of financial institutions and exchange professionals in urban areas would appear to be at odds with those emerging from research in rural Wales (Milbourne, 1997, 1999). Little evidence was revealed of discriminatory lending policies – the so-called 'redlining policies' (see Boddy, 1980), whereby mortgage advances in particular residential areas are prohibited regardless of the individual circumstances of potential mortgagees. Rather, local managers of these financial organizations pointed to the more limited spatial segregation within and internal heterogeneity of rural housing markets that made the operation of such spatially discriminatory policies unworkable in rural areas. However, these local financial institutions remained suspicious of recent interventions within private housing markets by planning authorities to meet local housing need.

Research on exchange professionals in the 1990s indicates important shifts in the structure and roles of the estate agency sector in the national housing market (see Beaverstock et al., 1992). Linked to the deregulation of financial services in Britain, estate agencies were given the power to broaden their activities from selling properties to offering a wider range of financial services to purchasers. This led to an increased number of takeovers of smaller firms by financial institutions and larger estate agencies over the late 1980s and early 1990s, and the increased significance of particular agencies in the housing market. However, research in rural Wales indicates that these processes of restructuring have impacted only marginally on the local estate agency sector, which remains dominated by long-established, small firms providing a narrow range of services to customers (see Milbourne, 1997, 1999). In addition, most of the local agencies involved in the study were seen to act 'more as passive intermediaries than active agents of change; generally reflecting rather than determining the changing structure of the local market' (Milbourne, 1999: 224–5).

Rural housing, society and culture

Alongside consideration of rural housing structures and processes of regulation, academic attention has been given to the connections between housing and demographic change in rural areas. In previous decades, the presence of abandoned properties in states of considerable disrepair in rural areas came to represent an extremely visual indicator of rural population decline. More recently, processes of rural re-population that have been reported in most advanced capitalist countries have also impacted on housing systems in rural areas. While not denying the continuation of population out-movement from particular rural areas, it has been the connections between rural population growth and housing that have occupied the attention not only of housing researchers but also of those scholars interested in exploring social change in the countryside.

Across a number of studies that have been undertaken on population growth in rural areas, the nature of rural housing is identified as a significant factor in attracting in new groups. However, the exact role played by housing varies significantly between countries. The contrast is most stark between the UK and, perhaps, the Netherlands (see Heins, 2003), and the other countries included within this review. In relation to the latter group of countries, which includes the United States, Australia, New Zealand, and other European countries, several studies have pointed to the affordability of rural housing (relative to that in cities and major towns) as being an important factor behind population movements to rural areas. As Auclair and Vanoni (2003: 79) note in relation to France, 'decisions to move to the countryside are frequently driven by the knowledge that cheap housing is readily available outside major urban centres ...'. Indeed, in most countries, statistical evidence indicates that rural housing tends to be much less expensive than equivalent properties in urban areas (see Tonts et al., 2001; Housing Assistance Council, 2002). This lower price of rural housing has meant that it has facilitated movements of lower-income groups to rural areas. Beer and Maude (2002), for example, highlight how thousands of income-support recipients relocate on an annual basis from cities to nonmetropolitan areas of Australia, stating that 'housing affordability is the single most important factor influencing decisions by income-support recipients to move away from Sydney and Adelaide' and that 'cheaper and better housing in nonmetropolitan areas facilitates their moves' (2002: ix). In addition, studies in the United States have indicated that the increased presence of particular forms of low-cost housing, such as 'manufactured housing', in nonmetropolitan areas has enabled lower income groups to relocate from cities (see Belden and Weiner, 1997).

These connections between rural housing and population change contrast strongly with those reported in the British countryside. Here, rural housing is also seen as a key attractor of in-moving groups, but for different reasons and for different in-moving groups. Rural housing in Britain needs to be viewed less as a low-cost accommodation option and more as a positional good that carries a great deal of social status and cultural capital. In general terms, the average price of housing in the countryside generally, and those rural areas within commuting distances of key employment centres, is higher than the national mean (see Countryside Agency, 2003). It is possible to point to two factors that help explain this peculiar relationship between housing and social change in rural Britain. The first concerns the restricted supply of housing in rural areas, linked to the heavy regulation of land for development. The second factor relates to the demand for housing in the countryside. The domination of idyllic constructions of rural living within British society has meant that the 'house in the country' has become a much sought after commodity, particularly amongst the middle-classes (see Thrift, 1989; Cloke et al., 1998; Murdoch, 1995; Phillips, 2003). In combination, these two factors have led to increased levels of competition for housing in rural areas, and particularly in those parts of the countryside that are constructed as attractive in terms of their proximity to urban labour markets or in relation to other symbolic landscape features.

While I will discuss the welfare consequences of this competition for rural housing in the next section of this chapter, it is important to note that several studies have revealed important housing conflicts bound up with these demographic and social processes. Perhaps the first study to explore this issue was undertaken by Newby et al. (1978) in their research on property, paternalism and power in the English rural region of East Anglia. Their work highlighted how the in-movement of new groups into these rural spaces had resulted in new alliances between established (farmer and agricultural worker) groups and the formation of new cultures of localism. Thus emerged ideas of locals and newcomers and socio-cultural differences and conflicts between the two groups that have remained dominant within academic and policy discussions of rural housing conflicts ever since. However, their work on social change and conflict in this part of rural England did go on to complicate such simplistic categories of 'local' and 'newcomer' by pointing to the ways that housing was able to generate new alliances between established farmers and newcomer middle-class groups to contest proposals for new schemes of social housing provision.

Over subsequent years, other writers have attempted to provide more sophisticated accounts of class-based conflicts involving rural housing. Cloke and Thrift (1989, 1990), for example, have questioned the usefulness of the 'local' and 'newcomer' categories in explaining rural (housing) conflicts, arguing that more sophisticated accounts of rural conflicts are required based around deeper understandings of the roles of different socio-economic classes. In relation to rural housing, they suggest that conflicts are less bound up with clashes between new and established groups, and more concerned with middle- and working-class tensions. Furthermore, they suggest that emerging intra-class conflicts between different fractions of the middle classes are often played out through the arena of housing. Such arguments have also been proposed by Marsden and Murdoch (1994) in relation to the impacts of the middle classes in rural areas of South-East England, by Bell (1994) in his ethnographic study of a village in Hampshire, England, and by Cloke et al. (1997) in the context of socio-cultural change and conflict in different parts of the English and Welsh countryside.

These (socio-economic) class-based analyses of rural housing conflicts have been given a further 'housing twist' by Milbourne (1997) and Shucksmith (1990a, b) through their utilization of theories of domestic property classes. In addition to focusing on the (competing) interests of different fractions of private capital, ideas of domestic property classes also extend to the consideration of the associational interests of what are termed the property owning and non-owning classes. According to Saunders (1984), the property-owning class has a vested interest in ensuring the capital accumulation of its property and so will tend to favour a restricted provision of housing supply. This interest, it is argued, is shared across the spectrum of owners, including outright owners and mortgagees, and in-moving and more established groups. By contrast, the non-owning class is not affected directly by property price fluctuations, and so will tend to possess different attitudes towards the housing market, and new property development more particularly. Thus, the focus for analysing housing conflicts shifts from those groups defined in relation to length of residence or socio-economic status, to those occupying different housing positions.

While recognizing the value that domestic property class theory brings to accounts of housing conflicts in rural areas, Milbourne (1997) paints a more complex picture of the interests of the property-owning and non-owning classes. Based on a study of two housing markets in rural Wales, he demonstrates that housing tenure does play an important part in explaining attitudes towards new housing development in these areas, with a higher proportion of owners opposing new housing than non-owners. However, other findings from the study indicate different interests within these two property classes – between outright owners and mortgagees, and social and private rental tenants – with age-based factors appearing to be of particular significance in influencing attitudes towards housing. Furthermore, perceptions of local housing need in these areas cut across conventional class divisions, whether defined in relation to occupations or housing positions. Such findings would appear to confirm those of earlier studies that point to the importance of bringing together socio-economic classes, housing classes and lifestyle defences in understandings of housing conflicts in rural areas (see Short et al., 1986, 1987).

Finally in this section on housing, society and culture in rural areas, further attention needs to be given to the connections between housing and culture in understanding rural housing conflicts. In a British context, perhaps the clearest example of such cultural conflicts surrounding housing is found in areas of rural Wales, where Welsh remains a significant everyday language, and where domestic property has taken on a symbolic role in the contestation of social, cultural and linguistic changes. Over the 1980s, holiday cottages owned by English people became a key symbolic issue and sometimes literally a burning issue within the arson campaigns of Meibion Glyndwr, a 'direct action' Welsh nationalist movement. As Cloke et al. (1997: 26) have noted, 'the image of the holiday cottage ablaze has become a potent icon of resistance and cultural separateness within parts of the Welsh countryside'. More recently, Cymdeithas Yr Iaith Gymraeg (the Welsh Language Society) has positioned the operations of the rural housing market centrally within its language campaigns. These campaigns tend to conflate cultural and economic issues bound up with rural housing; with in-moving groups constructed not only as more affluent than many households within Welsh-speaking rural communities but as threatening the future of the Welsh language and its cultural attributes.

Rural housing welfare

A third significant theme that runs through international rural housing literature is that of housing disadvantage and welfare. Here attempts have been made to link issues of housing need to broader discussions on poverty and social exclusion in rural areas. While the definition of housing need and the nature of the linkages between housing, poverty and exclusion vary across different studies, two important aspects of rural housing need can be identified. The first concerns the juxtaposition of rural poverty and inadequate housing conditions, while the second focuses on housing needs and the costs associated with accessing housing in rural areas. Each of these aspects will be discussed in this section of the chapter.

Rural housing research undertaken in the different countries included within this review points to the gradual improvement of rural housing conditions over the period of the second half of the twentieth and the early years of the present century. Prior to this, it is generally accepted that the state of much of the rural housing stock was inadequate and unacceptable in policy terms, with strong linkages evident between the poor state of this housing and the poverty experienced by its occupants. In an earlier review of rural housing research in the UK, Rogers (1985) states that the physical conditions of housing in the countryside preoccupied academic researchers until the 1960s. In addition, the Housing Assistance Council (2002) reports the findings of a 1934 government survey which indicated that 56 per cent of farm households were without indoor water, more than two-thirds lacked electricity or kitchen sinks, and only half of all farm buildings were in sound structural condition. Over recent decades, though, key policy interventions, processes of rural re-population and improvements in rural economies have dramatically reduced the number of households living in such inadequate properties. Returning to the situation in the United States, the latest figures reveal that only two per cent of non-metropolitan households are living in what are termed situations of 'severe housing inadequacy' (Housing Assistance Council, 2002). Similarly, a major study of rural living in England undertaken in the early 1990s revealed that almost all households had access to mains electricity and water supplies, with only a small minority – 8 per cent – reporting structural defects with their properties (see Cloke et al., 1994).

Notwithstanding these general improvements in the condition of rural housing over recent decades, several academic studies have continued to focus on the subject for three main reasons. First, while only affecting a minority of the rural population, official statistics indicate that rural housing conditions are considerably worse than those found in urban areas. For example, in France, Auclair and Vanoni (2003) state that the proportion of properties lacking one or more basic amenities in remoter rural areas is almost double the national French average. Official statistics point to a similar situation in the United States, where 'rural homes comprise a little over one-fifth of the nation's occupied housing units, [but] they account for over 30 per cent of units without adequate plumbing' (Housing Assistance Council, 2002: 30). A second reason for continuing to focus on housing conditions concerns the presence of high levels of poor conditions in particular rural areas. Research by Harvey (1994) pointed to in excess of 10 per cent of rural households lacking water, an indoor toilet and a bath/shower in a case study area in rural Ireland (see also O'Shea, 1996), and Tonts et al. (2001) and Econsult et al. (1989) have highlighted the inadequate state of the housing stock in remoter rural parts of Australia, where lower property prices combine with poorer quality accommodation. In the United States, spatial concentrations of poor housing conditions have been identified in the South which generally correspond with areas of persistent rural poverty.

A third reason relates to the social composition of rural households living in inadequate housing conditions. Milbourne (2004) reports that rural poverty studies conducted in both the UK and the United States indicate that inadequate housing conditions tend to be associated with low-income and poor groups. In addition, US government figures highlight that non-white and Hispanic households in rural areas are almost three times more likely to be living in sub-standard housing than their white counterparts (Housing Assistance Council, 2002). Other studies in the United States point to other groups of the rural population that experience particularly severe problems with the physical state of their accommodation. Peck, for example, discusses the poor housing conditions experienced by farm-workers, who 'too often ... live in shacks, barns or chicken coops; along riverbeds; in hand-dug caves; or amid unsafe electrical wiring, raw sewage, and polluted water' (1999: 84), while research by Martinez et al. (1999) indicates the severe housing problems faced by other marginal members of the agricultural labour force. They focus on border colonias, which they define as 'the hundreds of quasi-rural communities along the US–Mexican border ... characterized by extreme poverty and severely substandard housing conditions' (1999: 49). Populated by impoverished immigrant farm-workers, the situations of housing in these colonias, they suggest, are not dissimilar to those found in squatter settlements in less developed parts of the world:

> Colonia communities are concentrated pockets of poverty, characterized by dirt roads without adequate drainage control and substandard, overcrowded housing. This housing often is constructed out of 'found' materials such as cardboard, wood palettes, and corrugated metal sheets. It is also not uncommon to find people living in abandoned buses or tar paper shacks. Colonia residents often lack access to potable drinking water, use illegal cesspools or septic tanks, and/or dispose of waste directly into open trenches. (1999: 51)

While the inadequate state of the rural housing stock can be viewed as a surrogate of housing disadvantage and poverty in rural areas, issues of housing needs have generally been explored in more direct terms by rural housing researchers over the past couple of decades. Within this body of work, attention has been given to the failure of welfare policies to meet the needs of particular social groups in rural areas and the impacts of recent processes of rural re-population on housing access. In the UK, such concerns with rural housing needs can be traced back 25 years, with the work of Larkin pointing to the nature of housing problems in the rural county of Dorset in England:

> There were young families living with their parents who had been on housing waiting lists for about ten years, whilst their children grew up in more and more cramped surroundings. There were parents who had ended up walking the streets with their children as a result of the homelessness policies being operated by councils in that area. There were families with young children who were living in winter let accommodation, gaining a brief respite of independence before eviction and a return to over-crowding and perhaps a caravan site over the summer. There were families living in rural slums, usually isolated privately rented cottages with no basic amenities. (1979: 71)

Around this same period, those researchers interested in poverty and deprivation in rural areas began to focus on the housing components of these problems (see McLaughlin, 1986; Phillips and Williams, 1984; Shaw, 1979; Walker, 1978), highlighting how housing represents an integral component of multiple deprivation affecting certain households in rural areas. Such issues were then taken forward by more specific studies of housing competition and needs in the countryside (Dunn et al., 1981; Phillips and Williams, 1982).

More recently, rural housing researchers have been preoccupied with setting out the scale and profile of housing needs in rural areas through the utilization of measures of housing affordability. This has been particularly true in the UK and the United States, where most of this work has taken place at the academic–policy interface. In the UK, national assessments of rural housing need have been made using a variety of methods. Clark (1990) aggregated findings from a large number of local surveys undertaken by rural community councils in England, which enabled him to estimate that about 377,000 households living in rural England were in housing need in the late 1980s. Others have utilized broader surveys of rural living to highlight the scale and nature of rural housing needs in selected rural localities. For example, Cloke et al. (1994), from a survey of 3,000 households in 12 rural localities in England in 1990–91, report that 68 per cent of respondents pointed to local groups that were encountering difficulties in securing housing, with young people identified as the main group experiencing housing need (see also Rugg and Jones, 2000). Cloke et al. (1994) also explored the nature of these rural housing problems through commentaries provided by respondents. What emerges from this research is a mix of supply- and demand-side housing pressures, linked to processes of middle-class in-movements and reduced rental housing opportunities, and similar sets of housing problems being faced by young people in rural areas to those identified by Larkin (1979) 15 years previously.

Rural housing needs in England have been examined in more sophisticated ways over the past few years in an effort to provide more comprehensive and spatially sensitive assessments of housing affordability. Undertaken by government rather than academic researchers, local income and property selling price data have been combined to provide new accounts of housing need in rural areas. In overall terms, rural housing emerges as less affordable than that in urban areas (recording an income to mortgage cost index of 4.94 for 2002 compared to 4.66 in urban areas). However, the geographies of housing affordability in rural England appear to be extremely complex, with a considerable regional and local variability in affordability scores (see Countryside Agency, 2003).

Rural housing affordability has also formed the focus of a large volume of academic and policy-based studies in the United States. Using 'official' indicators of housing need, the Housing Assistance Council (2002) calculate that 5.5 million households – one-quarter of all nonmetropolitan households – are 'cost-burdened', (paying more than 30 per cent of their incomes on housing costs) with 2.4 million of these paying more than half their incomes towards the costs of housing. Furthermore, most of these 'cost- burdened' rural households are characterized by low incomes and about one-third are living in the rental sector. As with the situation in England, rural housing affordability is characterized by a great deal of spatial variation in the United States. Problems of affordability appear to be much more pronounced along the West Coast (in the states of California, Oregon and Washington), 'amenity' states, such as Colorado's 'ski counties', and areas of high and persistent poverty, for example in Native American reservations and the Appalachian mountains. It is in these latter 'poor places' that housing problems are particularly acute:

> In these areas, incomes are so low that many residents cannot afford housing even though costs are much lower than the national average. When incomes and housing prices are both depressed in communities such as these, the quality of housing is also low. (2002: 29)

The scale of such housing problems in parts of nonmetropolitan America has led Belden and Weiner (1999) to call for rural housing welfare to be taken more seriously by academic and policy communities. In an edited text, the authors bring together a broad range of case study material on housing need in rural America. While brevity prevents a broad review of the studies included within this text, it is worth pointing to particular themes. Spatial concentrations of rural housing problems are discussed by McCray, who uses evidence drawn from studies of housing need in the rural South. She suggests that rural housing problems in the South are characterized by a complex mix of factors concerning affordability and conditions which themselves impact on rural community living:

> Both housing quality and affordability issues negatively influence the 'attachment to place' of ill-housed residents. As a result, strong emotional bonds to community and family are diminished ... Housing problems link families and communities in systematic social networks. (1999: 46, 47)

Chapters by Martinez et al. (1999) and Peck (1999) explore housing need amongst farm workers in nonmetropolitan areas of America. Both highlight the severe housing difficulties faced by this group, as inadequate housing conditions combine with low wage levels and rising rates of unemployment associated with the agricultural sector. Krofta et al. (1999) focus their analysis of rural housing problems on the rural Midwest, an area commonly constructed as the 'heartland' of the United States but within which, over recent decades, processes of economic restructuring have created new sets of rural housing problems and 'the number and proportion of very poor and homeless households have increased' (1999: 72). Finally, attention is given to the hidden financial costs associated with manufactured housing (typically consisting of trailers or mobile homes) which represents

16 per cent of the nonmetropolitan housing stock (see Straus, 1999; and also Housing Assistance Council, 2002).

RURAL HOMELESSNESS

The preceding discussion of rural housing has clearly pointed to the existence of a series of important housing problems linked to housing conditions and access, and has indicated connections between housing, poverty and social exclusion in rural areas. It is these latter linkages that form the focus of the second substantive section of this chapter. Alongside consideration of the problems faced by groups and individuals within rural housing markets, more limited attention has also been given to the homeless, a group that is effectively locked out of formal housing systems. Compared with the volume of work on rural housing issues, the number of studies of homelessness in rural areas is relatively small. In most of the countries covered in this review there has not been any serious academic coverage of rural homelessness.[1] By far the bulk of work on the subject has been completed in the United States, where rural homelessness has received much more academic scrutiny than general rural housing issues. The main reason for this is that homelessness has been studied as a component of rural poverty, a subject that has been widely researched in the United States. Other studies of rural homelessness have been recently completed in the UK, although the focus of mainstream research remains fixed on rural housing issues. Notwithstanding this spatial imbalance in the academic coverage of rural homelessness, recent studies have provided theoretically informed and empirically rich accounts of the complex connections between homelessness and rurality. It is these accounts that form the focus of this section of the chapter. More specifically, attention will be given to the invisibilities of rural homelessness, its statistical extent and profile, the nature and experiences of homelessness in rural areas, and issues of welfare governance bound up with rural homelessness.

Hidden forms of homelessness in rural areas

Dominant academic, policy and lay discourses of homelessness in advanced capitalist countries have tended to juxtapose particular forms and spaces of homelessness. Considerable attention has been given to literal or roofless homeless groups and their visible concentrations within large cities. Evidence of such socio-spatial linkages can be seen in a range of recent texts on homelessness, as the following extract from a key book on homelessness in the United States, UK and Canada illustrates:

> During the 1980s, like other visitors to American cities and to the South Bank in London, I was profoundly disturbed by encounters with people who sleep in doorways or live in cardboard containers. These scenes provided tangible evidence of public policy failure. (Daly, 1996: xiii)

Consequently, as Aron and Fitchen (1996) have commented, much of what is known about homelessness is based on studies conducted in urban spaces. Indeed, it has only been over the past few years that a broader range of forms and spaces of homelessness have begun to be acknowledged within research and policy communities. This is the case with homelessness in rural areas, which has been neglected not only by researchers of homelessness but also by those working on rural housing and rural welfare issues.

Two key factors can be identified for this neglect of rural homelessness as a research arena. The first relates to the reduced visibilities of homelessness in rural areas and the associated methodological problems of researching hidden groups of the population. The rural homeless remain hidden not only due to the dispersed settlement structure of rural areas but through the lack of services and facilities for homeless groups, such as hostels, shelters and drop-in centres, in these spaces. Arguably, it is the presence of such sites in urban areas that provides homelessness with its visibility (see Stover, 1999; Cloke et al., 2000a, c). While the absence of such sites/sights does not indicate the absence of homeless people in rural areas, it does make their identification more problematic in research terms. As Aron and Fitchen (1996: 81) have commented in relation to homelessness in rural America, '[rural communities] do have woods, campgrounds, and remote hills and valleys where literally homeless people may be found, albeit with difficulty'. Linked to this factor, the nature of homelessness in rural areas also acts to cloak its visibility. As the previous quotation indicates, the literal homeless are often less conspicuous when present in hidden spaces. Moreover, it has been claimed that literal forms of homelessness constitute only a small proportion of homelessness in rural areas:

> Literal homelessness often is episodic, whereas the condition of being without permanent adequate housing

> usually is longer term. The rural homeless typically move from one extremely substandard, overcrowded, and/or cost-burdened housing situation to another, often doubling or tripling up with friends or relatives. While housed in these precarious situations, the rural homeless do not meet the predominant interpretation of literal homelessness. They are, however, without permanent adequate homes. (Stover, 1999: 76)

As such, the small number of researchers that have focused on homelessness in rural areas have called for broader definitions of homelessness to be acknowledged. In particular, Fitchen (1992) has argued for consideration to be given not only to those who are roofless or living in emergency accommodation, but to those poor households that are living in precarious housing situations, what she describes in the title of one of her papers as those 'living on the edge of homelessness'.

Alongside issues bound up with the nature and visibilities of rural homelessness, we can also point to powerful cultural constructions of rural spaces and rurality that act to deny any legitimacy to homelessness in rural areas. In some countries, such as the UK, it has been claimed that dominant ideas of rurality have become associated with ideas of problem-free forms of rural living (see Cloke and Milbourne, 1992). In others rural areas are often constructed as spaces of community, belonging and home, not least through the strong associations that exist between rural spaces and national identities. While these constructions of rurality are predominantly (re) produced and (re)circulated beyond rural spaces, it is also the case that they permeate the everyday lives of rural people. As Paul Cloke, Rebekah Widdowfield and I argue:

> rurality and homelessness are discursively non-coupled ... [through] the socio-cultural barriers that exist within the practices, thoughts and discourses of rural dwellers themselves, leading them to deny that homelessness exists in their place. (2002: 66)

Recent work on rural homelessness in England by the three of us has provided detailed discussions of the complex relationships between homelessness and rurality. We have drawn on sociological and cultural geographical literatures to claim that homelessness is often rendered 'out of place' in rural spaces that are constructed in privileged and purified terms (see Cresswell, 1996; Takahashi, 1998; and Sibley, 1995). This work also provides a range of case study material on cultural denials of homelessness in rural areas, denials that are linked to middle-class groups moving to the English countryside in search of 'idyllic' rural lifestyles, and the attitudes of local elites who create difficulties for those welfare agencies dealing with homeless groups in rural areas.

The scale and profile of homelessness in rural areas

Assessing the extent of homelessness generally and that in rural areas in particular is fraught with difficulty. In most countries, there is relatively little available statistical information on homelessness, and what does exist is usually restricted to counts of the literal homeless or rough sleepers. These statistics tend to be based on snapshot counts of rough sleepers in known sites of the homeless located in towns and cities, and thus, provide only a partial and geographically biased account of the scale of the problem. It is therefore not surprising to learn that rough sleeper counts in the United States and UK consistently point to more significant concentrations of homeless groups in urban spaces. In the United States, for example, the 1990 Population Census rough sleeper count recorded a level of homelessness in urban areas of 17.5 per 10,000 residents compared with 1.2 per 10,000 in rural areas. While individual studies of homelessness in the United States, using more sensitive approaches to measuring rough sleeping, have shown higher incidences of rural homelessness, it remains the case that levels of literal homelessness in rural areas fall below those recorded in towns and cities. For example, Aron and Fitchen (1996) report that research in the state of Kentucky using data supplied by a broad range of homeless and welfare agencies revealed a level of street/literal homelessness in rural areas of 13 per 10,000 residents but 22 per 10,000 in urban parts of the state.

In the UK, researchers have access to another source of official data on the incidence and geographies of homelessness. Since 1977, central government has collected statistics on homelessness as part of welfare policies aimed at dealing with particular forms of homelessness and specific groups of homeless people. These statistics provide an extremely useful, although still partial, source of information on the shifting incidence of homelessness at different spatial scales. Official homelessness statistics have been used to highlight the scale and profile of homelessness in rural areas over the 1990s (Cloke et al., 2001a; see also Lambert et al., 1992, for an earlier analysis of these data).

Cloke, 2001a indicate that homelessness in rural areas accounted for 14.4 per cent of all officially recorded homelessness in England in

1996. When standardized against population totals, the rate of homelessness in rural areas stood at 3.5 per 10,000 households compared with 5.7 per 10,000 households in urban areas. However, rural homelessness was shown to have risen in proportional terms over the 1990s from 11.8 per cent in 1992 to 14.4 per cent in 1996. More recent analyses have indicated that the rural proportion of the homelessness total has continued to rise, with rural homelessness accounting for 18.8 per cent of England's homelessness in 2000–01 (Countryside Agency, 2003). In addition to this overall proportional increase in the rate of rural homelessness, analyses of official data reveal how absolute levels of homelessness in the remoter rural areas rose over the 1990s (Cloke et al., 2001a). Thus, while urban areas continue to record higher standardized rates of homelessness, rural homelessness has taken on an increased significance in England over recent years.

Research in the UK and the United States has pointed to the spatial unevenness of homelessness in rural areas. In the US context, work by Aron and Fitchen (1996) has indicated that highest rates of homelessness tend to be found in those remote rural areas associated with declining primary economic activities and persistent poverty. However, they also identify rising levels of homelessness in more accessible rural areas that have been subject to in-migration pressures and processes of housing gentrification. In England, research also has pointed to the spatial complexities of rural homelessness. While rural homelessness levels are generally higher in southern regions of the country than in the North, within individual regions a great deal of variation is evident in relation to levels of rural homelessness (see Cloke et al., 2001a).

Attention has also been given to the statistical profile of homelessness in rural areas. In England, official statistics indicate strong similarities between the profile and causes of homelessness in rural and urban areas, with these similarities linked to the narrow definition of homelessness utilized by central government. Households with or expecting children account for the bulk of homeless cases in rural areas (comprising slightly less than three-quarters of the total in 1996), with the remaining cases consisting of households considered vulnerable on account of age, disability, domestic violence and other reasons. In relation to the key causes of homelessness in rural areas, four factors account for almost 90 per cent of cases: the loss of rented accommodation; the refusal of family or friends to continue to provide housing; the breakdown of a personal relationship; and the inability to maintain mortgage payments. While these four factors explained the bulk of homelessness cases in urban areas, housing factors appear to play a more important part in causing homelessness in the countryside (see Cloke et al., 2001a). Strong connections between rural housing structures and homelessness have also been identified in the US, with Fitchen's (1992) research pointing to the ways that inequitable rural housing systems not only lead to people becoming homeless but create additional problems for these people once they become homeless.

Profiles of rural homelessness have also been produced in the United States, although at the sub-national level. The National Coalition for the Homeless (1999), for example, reports key findings from research by the US Department of Agriculture (1996: 1), which indicates that the rural homeless are 'more likely to be white, female, married, currently working, homeless for the first time, and homeless for a shorter period of time' than their urban counterparts. In addition, a study by Vissing (1996) in the state of Kentucky showed that families and single mothers constitute the most significant homeless group in rural areas. Other state-focused studies point to migrant agricultural workers and Native Americans as important groupings of the rural homeless, and higher levels of homelessness linked to domestic violence in rural areas (see Housing Assistance Council, 2002). However, what emerges from these US studies, and from the research in England, is that the profile and causes of homelessness in rural areas are characterized by a great deal of geographical specificity – linked to the different socio-cultural and economic contexts of particular rural spaces and places.

The nature and experiences of rural homelessness

While not wanting to downplay the significance of statistical assessments of the scale and nature of homelessness in rural areas, perhaps the most interesting material on rural homelessness has emerged from local studies that have explored the experiences of being homeless in different rural spaces (Fitchen, 1991, 1992 in the United States and Cloke et al., 2003 in rural England). Utilizing qualitative methodologies, these studies have sought to expose the nature of homelessness in rural areas through interviews with welfare agency workers and encounters with homeless people in local rural spaces. What emerge from this work are rich narratives on the types of obstacles – physical and cultural – that homeless people and welfare agencies are faced with in rural areas and the innovative ways that these obstacles are overcome.

In the United States, Fitchen (1991, 1992) has focused on homeless people in rural parts of upstate New York. Adopting a broad definition of homelessness to include those groups and individuals living 'on the edge of homelessness', Fitchen provides a detailed analysis of the characteristics of rural housing problems and homelessness in this part of the country and the ways that homeless people adopt coping strategies to deal with their homelessness. In relation to the former issue, Aron and Fitchen (1996) highlight five housing-related problems that can lead to situations of homelessness in upstate New York: housing supply–demand imbalances, whereby larger numbers of poor households are searching for fewer affordable properties; falling levels of owner-occupation; the rigid regulation of housing construction and land-use; increasing levels of rent burdens; and insecure tenancies. Turning to the coping strategies of homeless people, Aron and Fitchen suggest that rather than make use of shelters and hostels (assuming that they are present), most of the rural poor adopt 'three common strategies to fend off literal homelessness' (1996: 83). First, they will move in with family members on a temporary basis. A second course of action involves moving into inadequate, unaffordable or unsafe forms of accommodation that might prevent literal homelessness but lead to other types of problem, such as rent arrears or health problems. Third, Aron and Fitchen point to frequent movements from and to different properties as a means of avoiding literal homelessness in rural areas, stressing that:

> poor people with unsatisfactory housing tend to move suddenly, shifting from one rented small-town apartment to another, from one village to another, from trailer park to trailer park, from village to trailer park and back to village and back to country. (1996: 84)

Similar themes emerge from recent research on rural homelessness undertaken in England (Cloke et al., 2003). Based on interviews/encounters with homeless people, anonymized case notes provided by local welfare agencies, and formal interviews with welfare agency workers in two case study areas, this work highlights a number of key experiences of being homeless in rural spaces that are worthy of discussion here. The first concerns the ways that dominant constructions of homelessness as roofless feed into understandings of homelessness in rural areas. Not only does the (hidden) nature of rural homelessness allow it to be denied any legitimacy by incoming middle-class residents and local elite groups, those people experiencing forms of homelessness in rural areas may not even recognize their own situations as homelessness.

A second theme that emerges from this English research concerns the out-of-place nature of rural homelessness and the ways that homeless people cope with their out-of-placedness in rural spaces. Drawing on the writings of Cresswell (1996), it is suggested that dominant constructs of rural spaces (as spaces of home, community, safety) have become associated with particular social groups and modes of behaviour within these spaces (see Cloke et al., 2001c, 2002, 2003). However, once constructed in these terms it is necessary to protect rural space from actual or potential transgressions by perceived 'other' groups and activities. Sibley (1995) highlights the ways in which 'pure' spaces, such as the countryside, are able to be policed effectively through the increased visibilities of 'others' within these spaces. Such perspectives have been applied to studies of homelessness in rural England (see Cloke et al., 2002). It has been shown how 'inappropriate' groups, such as travellers, and activities, for example a proposed drug and alcohol rehabilitation centre, in different villages, are excluded from rural spaces by groups of monied and empowered residents. More particularly, examples are provided that illustrate how the increased visibilities of homeless people in village spaces often lead to the accentuation of difference and deviance, and the active, though informal, policing of this deviance. A local church representative in one case study village makes the following comment:

> if you're homeless … in [name of village] it sticks out … if someone's milling around and going round the same places … it's more exposing for that person. It identifies and accentuates their sense of failure and lack of worth. (Cloke et al., 2001a: 86)

In other cases, particular local cultures of rurality act to position homelessness as very much out-of-place in village contexts. For example 'Pete', a young man made homeless as a consequence of a family relationship breakdown, feels compelled to move to a proximate town after his attempts to beg on the village street lead to him being ridiculed by local people.

A third theme that can be identified from this English work relates to the coping tactics employed by homeless people in rural areas. A key concern here is the way that the homeless deal with their increased visibilities in rural spaces, as well as the limited provision of welfare services in these spaces. Attention is given to the issue of movement, with homeless people portrayed as being engaged in complex patterns of movement within rural spaces and also between rural and urban spaces (see Cloke et al., 2000a, 2001c, 2003). In some cases, movement represents a key constituent of camouflage for homeless people in rural areas, with attempts made to seek out the less visible sites of the countryside located away from the types of village spaces mentioned earlier. In other cases, the restricted provision of services for the homeless in rural areas pushes homeless people into making decisions about 'making do' locally or moving to places of increased provision. In the former case, homeless people may be forced to rely on inappropriate and unaffordable forms of accommodation, some of which may lead to future situations of homelessness. Movements to other places to secure welfare services may well involve temporary visits to particular sites or more permanent relocations, with these movements involving short and long distances. However, most involve movements from rural locations to towns and cities, which not only perpetuate the reduced visibilities of homelessness in rural areas, but can create additional problems for homeless people during periods of intense personal vulnerability.

Rural homelessness and welfare governance

Recent work on rural homelessness has also focused on the ways that homeless people are dealt with by welfare agencies. However, while much of this work has explored the particular local responses to rural homelessness and the practical difficulties encountered by welfare agencies in meeting the needs of homeless people in rural areas, attempts have also been made to make connections between national homelessness policies, local initiatives aimed at dealing with rural homelessness and theoretical literatures on the shifting nature of welfare governance. It is these latter connections that I want to focus on within this section of the chapter (although see Cloke et al., 2001, 2002, for a discussion of practical responses to rural homelessness in the UK). Of particular interest here are the different scales of welfare governance and the emergence of more complex local systems of welfare delivery. In the context of England, the development of a national homelessness policy from the 1970s onwards has been explored, with particular attention paid to important shifts in homelessness policy that occurred in the 1990s (see Cloke et al., 2001b). Running through these policy developments, it is argued, has been a desire on the part of central government to develop national responses to homelessness, thus ensuring that homeless people are dealt with consistently by different local authorities. The 1996 Housing Act, though, attempted to reduce the role played by the central and local states in responding to homelessness. It produced a more restrictive official definition of homelessness, reclassified homelessness as an element of housing need, and shifted policy responses to homelessness into the private housing sector.

The impacts of this Act have been examined in rural case study areas in South-West England (Cloke et al., 2001b). Key findings that emerge from this work include key spatial assumptions bound up with this national legislation and the importance of local agencies in interpreting and mediating national policy. In relation to the former point, it was claimed by rural local authorities that the 1996 Act was produced in response to homelessness policy actions in urban areas. In terms of local responses to this legislation by rural authorities, it was shown that these authorities had been able to work within this new national policy context but to change particular policies so that they continued to deal with homeless groups in similar ways. As such, the desire of central government to provide consistent and, in the case of the 1996 Act, more restrictive, policy responses to homelessness would appear to have been thwarted by the actions of local government (see also Mullins and Niner, 1999). These actions lend support to broader critiques of the UK welfare system which have stressed how central policy is mediated in different ways at the local delivery level (see Cochrane, 1994; Stoker and Mossberger, 1995).

In addition to these kinds of vertical linkage between central and local policies, Stoker and Mossberger point to the importance of horizontal connections as 'circumstances and actors create the conditions for specific alliances and particular ways forward in different localities' (1995: 220). These types of horizontal connection are also examined in relation to the delivery of welfare services for homeless groups in rural areas (see Cloke et al., 2000b, 2001b). Drawing on recent discussions of the increasing prominence of systems of governance in delivering welfare (and other public) services, attention is given to the development of and power relations associated with partnerships of support for homeless people in their case study areas. The picture that emerges from this work is rather complex and supports the idea of an increasing spatial differentiation of welfare provision for homeless groups, with different clusters of actors coming together for different sets of reasons in different local spaces. Furthermore, local government emerges as a (still) powerful player within these new partnerships of welfare support for homeless people in their study areas, with these partnerships operating within the 'shadow of hierarchy' (Jessop, 1997). However, what this work also demonstrates is that these new partnership arrangements are also able to challenge existing power relations associated with local homelessness welfare delivery and to develop more constructive local policy discourses of rural homelessness.

CONCLUSION

This chapter has demonstrated that a large volume of academic work now exists on rural housing and homelessness, covering a broad range of important themes. Rural researchers have focused critical attention on rural housing structures and regulatory processes, the position of housing within broader processes of socio-cultural change and the welfare components of housing in rural areas. Extending work on rural poverty, issues of housing welfare have been explored in relation to rural homelessness, with coverage given to the natures, scales, experiences and welfare policy contexts of homelessness in rural spaces. Within this body of work, important attempts have been made by rural researchers to position the subjects of rural housing and homelessness within critical theoretical contexts linked to political economy and regulationist approaches, welfare restructuring, economic and domestic property classes and ideas of socio-cultural spatialization. Other researchers have been content to provide more descriptive and policy-relevant accounts of housing and homelessness in rural areas. While there is clearly room for both approaches within studies of rural housing and homelessness, it has arguably been the latter that has dominated academic coverage of these subjects over the past few decades. As such, there remain important gaps in our knowledge of rural housing and homelessness and there exists a need to develop new theoretically based agendas for research on these subjects. In the remaining parts of this concluding section I want to suggest four areas of research that, collectively, will add to academic understandings of rural housing and homelessness.

The first area involves making connections between housing and homelessness and broader socio-cultural, economic and political structures and processes in rural areas. To date, these studies have been undertaken of the roles played by housing in relation to population change and socio-cultural shifts and conflicts in rural areas, and on the linkages between agricultural political economies and rural housing. However, the position of housing within these wider socio-cultural, political, policy and economic contexts requires further attention from rural researchers. In particular, more needs to be known about the complex relationships between the production/regulation of housing and regional/local rural economies, the significance of housing within processes of in- and out-movement in rural areas, how rural housing is socially constructed by different groups in rural spaces, and the cultural and political representations of rural housing need and homelessness.

Second, there remains a need to explore the interconnections between the different components of the rural housing system. Much of the academic work to date has focused on examining individual components of this system – relating to production, regulation, consumption, welfare and so on – with relatively little attention given to how these different components are connected. For example, we need to provide more sophisticated accounts of the relative importance of structural, regulatory and consumption-based factors in shaping particular rural housing systems. Likewise, it would be useful to investigate how the nature of housing needs connects with processes of housing production and regulation, housing consumption, as well as broader processes of economic restructuring.

A third area of work is required that can make meaningful connections between housing and homelessness in rural areas. The chapter has pointed to an unhelpful division that exists within the rural literature between a well-established rural housing genre and a more recent body of work on rural homelessness that has developed more from studies of rural welfare than rural housing. Most researchers of housing welfare in rural areas have stopped well short of engaging with issues of homelessness, restricting their scope to needy groups positioned within formal housing systems. Recent writings on rural homelessness in the UK and the United States, though, have pointed to some interesting links between the make-up of housing systems and homelessness in rural areas. Emerging findings from this work would appear to indicate that housing structures play a more significant role in causing and perpetuating homelessness in rural spaces than in the city. Similarly, it is clear that many policy and political actors involved in rural affairs continue to focus their interventionist welfare strategies around discourses of housing need rather than homelessness. Indeed, while the presence of poverty in rural areas is now widely recognized by rural policy actors at national and local levels, it would seem that homelessness and homeless people remain positioned as out of place in a wide range of rural contexts.

Fourth, further attention needs to be directed towards the complex geographies of rural housing and homelessness. It is possible to identify two dimensions of this spatial complexity. The first relates to the uneven coverage given to the subjects of rural housing and homelessness by rural researchers in different countries. Much of the published academic material on these topics has been focused on the UK and the United States, with a significant knowledge gap evident in relation to the nature of rural housing and homelessness in other advanced capitalist countries. It is unclear whether this neglect relates to the particular features of housing and homelessness in these countries, to the nature of rural studies within them, or to a combination of these factors. What is clear, however, is that the task of producing an international review of rural housing and homelessness literatures has been made more complicated by the limited coverage given to these subjects outside of the United States and the UK. While a couple of comparative texts are available (see Gallent et al., 2003; Cloke and Milbourne, 2006), there exists a need for much more attention to be given to these international dimensions of rural housing and homelessness to ensure that the structural, social and welfare themes discussed in this chapter receive wider geographical coverage.

It is also the case that rural housing systems are characterized by a great deal of spatial unevenness within individual countries. Whether such geographical variation is viewed in terms of differences between remote/marginal and accessible/peri-urban rural areas, high and low price housing markets or different socio-cultural, economic and political spatial contexts, it is clear that geography plays a crucial part in understandings of housing and homelessness in rural areas. While national systems of housing regulation and welfare have been developed in most countries to ensure the consistent delivery of housing and housing welfare services, the local context remains important in shaping the structures, distribution and consumption of rural housing and the provision of social support for those experiencing housing need and homelessness in rural areas. Consequently, rural researchers need to remain sensitive towards these local geographies of rural housing and homelessness.

NOTE

1 An edited book on rural homelessness by Cloke and Milbourne (2006) provides a comparative account of rural homelessness in North America, Europe, Australia and New Zealand.

REFERENCES

Aron, J and Fitchen, J (1996) Rural homelessness: a synopsis, in J Baumohl (ed.), *Homelessness in Rural America*. Phoenix, AZ: Oryx Press.

Auclair, E and Vanoni, D (2003) France, in N Gallent, M Shucksmith and M Tewdwr-Jones (eds), *Housing in the European Countryside: Rural Pressures and Policy in Western Europe*. London: Routledge.

Barlow, and Savage, (1986) The politics of growth: cleavage and conflict in a Tory heartland, *Capital and Class*, 30: 156–182.

Beaverstock, J, Leyshon, A, Rutherford, T and Thrift, N (1992) Moving houses: the geographical reorganization of the estate agency industry in England and Wales in the 1980s, *Transactions of the Institute of British Geographers*, 17 (2): 166–182.

Beer, A and Maude, A (2002) *Local and Regional Economic Development Agencies in Australia*. Adelaide: LGASA.

Bell, M (1994) *Nature and Mortality in a Country Village*. Chicago: University of Chicago Press.

Belden, J and Weiner, R (eds) (1999) *Housing in Rural America*. Thousand Oaks, CA: Sage.

Boddy, M (1980) *Building Societies*

Cherry, G (ed.) (1976) *Rural Planning Problems*. London: Leonard Hill.

Clark, D (1990) *Affordable Rural Housing*. Cirencester: ACRE.

Cloke, P and Milbourne, P (eds) (2005) *International Perspectives on Rural Homelessness*. London: Routledge.

Cloke, P and Thrift, N (1987) Intra-class conflict in rural areas, *Journal of Rural Studies*, 3: 321–334.

Cloke, P and Thrift, N (1990) Class, change and conflict in rural areas, in T Marsden, P Lowe and S Whatmore (eds), *Rural Restructuring*. London: David Fulton.

Cloke, P, Goodwin, M and Milbourne, P (1997) *Rural Wales: Community and Marginalization*. Cardiff: University of Wales Press.

Cloke, P, Goodwin, M and Milbourne, P (1998) Cultural change and conflict in rural Wales: competing constructs of identity. *Environment and Planning A*, 30: 463–480.

Cloke, P, Milbourne, P and Thomas, C (1994) *Lifestyles in Rural England*. London: Rural Development Commission.

Cloke, P, Milbourne, P and Widdowfield, R (1999) Homelessness in rural areas: an invisible issue?, in T Kennett and A Marsh (eds), *Homelessness: Exploring the New Terrain*. Bristol: The Policy Press.

Cloke, P, Milbourne, P and Widdowfield, R (2000a) The hidden and emerging spaces of rural homelessness, *Environment and Planning A*, 32: 77–90.

Cloke, P, Milbourne, P and Widdowfield, R (2000b) Partnership and policy networks in rural local governance, *Public Administration*, 78 (1): 111–133.

Cloke, P, Milbourne, P and Widdowfield, R (2000c) Homelessness and rurality: 'out of place' in purified space, *Environment and Planning D: Society and Space*, 18 (6): 715–735.

Cloke, P, Milbourne, P and Widdowfield, R (2001a) The geographies of homelessness in rural England, *Regional Studies*, 35 (1): 23–37.

Cloke, P, Milbourne, P and Widdowfield, R (2001b) The local spaces of welfare provision: responding to homelessness in rural England, *Political Geography*, 20 (4): 493–512.

Cloke, P, Milbourne, P and Widdowfield, R (2001c) Homelessness and rurality: exploring connections in local spaces of rural England, *Sociologia Ruralis*, 41 (4): 438–453.

Cloke, P, Milbourne, P and Widdowfield, R (2002) *Rural Homelessness: Issues, Experiences and Policy Responses*. Bristol: The Policy Press.

Cloke, P, Milbourne, P and Widdowfield, R (2003) The complex mobilities of homeless people in rural England, *Geoforum*, 34 (1): 21–35.

Cloke, P, Phillips, M and Thrift, N (1998) Class, colonisation and lifestyle strategies in Gower, in P Boyle and K Halfacree (eds), *Migration Into Rural Areas: Theories and Issues*. Chichester: Wiley.

Cochrane, A (1994) Restructuring the local welfare state, in R Burrows and B Loader (eds), *Towards a Post-Fordist Welfare State?*, London: Routledge.

Countryside Agency (2003) *The State of the Countryside*. London: The Countryside Agency.

Cresswell, T (1996) *In Place/Out of Place: Geography, Ideology and Transgression*. Minneapolis, MN: University of Minnesota Press.

Daly, G (1996) *Homeless: Policies, Strategies and Lives on the Street*. London: Routledge.

Davey, J and Kearns, R (1994) Special needs versus the 'level playing field': recent developments in housing policy for indigenous people in New Zealand. *Journal of Rural Studies*, 10 (1): 73–82.

Dunn, M, Rawson, M and Rogers, A (1981) *Rural Housing: Competition and Choice*. London: Allen and Unwin.

Econsult (1989) *Rural Centres Housing Study*. Canberra: Australian Government Publishing Services.

Finnerty, J, Guerin, D and O'Connell, C (2002) 'Ireland', in N Gallent, M Shucksmith and M Tewdwr-Jones (eds), *Housing in the European Countryside: Rural Pressures and Policy in Western Europe*. London: Routledge.

Fitchen, J (1991) Homelessness in rural places: perspectives from upstate New York, *Urban Anthropology*, 20 (3): 177–210.

Fitchen, J (1992) On the edge of homelessness: rural poverty and housing insecurity, *Rural Sociology*, 57 (2): 173–193.

Gallent, N (1997) Planning for affordable rural housing in England and Wales, *Housing Studies*, 12 (1): 127–137.

Gallent, N, Shucksmith, M and Tewdwr-Jones, M (eds) (2003) *Housing in the European Countryside: Rural Pressures and Policy in Western Europe*. London: Routledge.

Gilg, A (1996) *Countryside Planning*. London: Routledge.

Harvey, B (1994) *Combating Exclusion: Lessons from the Third EU Poverty Programme in Ireland, 1989–1994*. Dublin: Combat Poverty Agency.

Hillier, J, Fisher, C and Tonts, M (2002) *Rural Housing, Regional Development and Policy Integration: An Evaluation of Alternative Policy Responses to Regional Disadvantage*. Australian Housing and Research Institute.

Hoggart, K (2003) England in N Gallent, M Shucksmith and M Tewdwr-Jones (eds), *Housing in the European Countryside*. London: Routledge.

Housing Assistance Council (2002) *Taking Stock: Rural People, Poverty and Housing at the Turn of the 21st Century*. Washington, DC: Housing Assistance Council.

Jessop, B (1997) Capitalism and its future: remarks on regulation, government and governance, *Review of International Political Economy*, 4 (3): 561–581.

Krofta, J, Crull, S and Cook, C (1999) Affordable housing in the rural Midwest, in J Belden and R Weiner (eds), *Housing in Rural America*. Thousand Oaks, CA: Sage.

Lambert, C, Jeffers, S, Burton, P and Bramley, G (1992) *Homelessness in Rural Areas*. London: Rural Development Commission.

Larkin, A (1979) 'Rural housing and housing needs', in J Shaw (ed.), *Rural Deprivation and Planning*. Norwich: GeoBooks.

Marsden, P and Murdoch, J (1994) *Reconstructing Rurality*. London: UCL Press.

Martinez, Z, Kamasaki, C and Dabir, S (1999) The border colonias: a framework for change, in J Belden and R Weiner (eds), *Housing in Rural America*. Thousand Oaks, CA: Sage.

McCray, J (1999) Affordable housing in the rural South, in J Belden and R Weiner (eds), *Housing in Rural America*. Thousand Oaks, CA: Sage.

McLaughlin, B (1986) The rhetoric and reality of rural deprivation, *Journal of Rural Studies*, 2 (4): 291–308.

Milbourne, P (1997) Housing conflict and domestic property classes in rural Wales. *Environment and Planning A*, 29: 43–62.

Milbourne, P (1998) Local responses to central state social housing restructuring in rural areas, *Journal of Rural Studies*, 14 (2): 167–184.

Milbourne, P (1999) Changing operations? Building society and estate agency activities in rural housing markets, *Housing Studies*, 14 (2): 211–227.

Milbourne, P (2004) *Rural Poverty: Marginalisation and Exclusion in Britain and America*. London: Routledge.

Mullins, D and Nimer, P (1999) A prize of citizenship? Changing access to social housing, in A Marsh and D Mullin (eds), *Housing and Social Policy*. Buckingham: Open University Press.

Murdoch, J (1995) Middle-class territory?: some remarks on the use of class analysis in rural studies, *Environment and Planning A*, 27 (8): 1193–1336.

National Coalition for the Homeless (1999) *Rural Homelessness*. NCH Fact Sheet 13, The National Coalition for the Homeless (www.nch.ari.net/rural.html).

Newby, H (1979) *Green and Pleasant Land?* London: Hutchinson.

Newby, H, Bell, C, Rose, D and Saunders, P (1978) *Property, Paternalism and Power*. London: Hutchinson.

O'Shea, E (1996) Rural poverty and social services provision, in C Curtin, T Haase and H Tovey (eds), *Poverty in Rural Ireland: A Political Economy Perspective*. Dublin: Oak Tree Press.

Peck, S (1999) Many harvests of shame: housing for farmworkers, in J Belden and R Weiner (eds), *Housing in Rural America*. Thousand Oaks, CA: Sage.

Phillips, D and Williams (1982) *Rural Housing and the Public Sector*. Aldershot: Gower.

Phillips, D and Williams (1984) *Rural Britain: A Social Geography*. Oxford: Blackwell.

Rogers, A (1976) Rural housing, in G Cherry (ed.), *Rural Planning Problems*. London: Leonard Hill.

Rogers, A (1985a) Rural housing: an issue in search of a focus, *Journal of Agricultural Economics*, 36 (1): 87–89.

Rogers, A (1985b) Local claims on rural housing, *Town Planning Review*, 56: 367–380.

Rugg, J and Jones, A (2000) *Getting a Job, Finding a Home: Capturing the Dynamic of Rural Youth Transitions*. York: Joseph Rowntree Foundation.

Saunders, P (1984) Beyond housing classes: the sociological significance of private property rights in the means of consumption, *International Journal of Urban and Regional Research*, 8: 202–227.

Shaw, J (ed.) (1979) *Rural Deprivation and Planning*. Norwich: GeoBooks.

Short, J, Witt, J and Fleming, S (1986) *Housebuilding, Planning and Community Action: The Production and Negotiation of the Built Environment*. London: Routledge.

Short, J et al. (1987) Conflict and compromise in the built environment: house building in central Berkshire, *Transactions of the Institute of British Geographers*, 12: 29–42.

Shucksmith, M (1981) *No Homes for Locals*. Farnborough: Gower.

Shucksmith, M (1990a) *Housebuilding in Britain's Countryside*. London: Routledge.

Shucksmith, M (1990b) A theoretical perspective on rural housing, *Sociologia Ruralis*, 30: 210–229.

Sibley, D (1995) *Geographies of Exclusion*. London: Routledge.

Stoker, G and Mossberger, K (1995) The post-Fordist local state: the dynamics of its development, in J Stewart and G Stoker (eds), *Local Government in the 1990s*. London: Macmillan. pp. 210–227.

Stover, M (1999) The hidden homeless, in J Belden and R Weiner (eds), *Housing in Rural America*. Thousand Oaks, CA: Sage.

Strauss, L R (1999) Credit and capital needs for affordable rural housing, in J Belden and R Weiner (eds), *Housing in Rural America*. Thousand Oaks, CA: Sage.

Takahasi, L (1998) *Homelessness, AIDS and Stigmatization*. Oxford: Oxford University Press.

Thrift, N (1989) Images of social change, in C Hamnett, L McDowell and P Sarre (eds), *The Changing Social Structure*. London: Sage.

Tonts, M, Fisher, C, Owens, R and Hillier, J (2001) *Rural Housing, Regional Development and Policy Integration*. Australian Housing and Urban Research Institute.

US Department of Agriculture (1996) *Rural Homelessness: Focussing on the Needs of the Rural Homeless*. Washington, DC: US Department of Agriculture (Rural Economic and Community Development).

Vissing, Y (1996) *Out of Sight, Out of Mind: Homeless Children and Families in Small-Town America*. Lexington, KY: University Press of Kentucky.

Walker, A (ed.) (1978) *Rural Poverty*. London: CPAG.

Williams, G, Bell, P and Russell, L (1991) *Evaluating the Low Cost Rural Housing Initiative*. London: HMSO.

Yarwood (2002) Parish councils, partnership and governance: the development of exceptions housing in Malvern Hills District, England, *Journal of Rural Studies*, 18 (3): 275–291.

Part 3

NEW RURAL RELATIONS

32

Rurality and otherness

Paul Cloke

INTRODUCTION: RURAL OTHERS AND OTHER RURALS

There can be few more quoted interventions in the recent history of rural studies than Chris Philo's (1992) essay on how rural geography (and by implication other rural disciplines) has neglected myriad non-mainstream windows onto the rural world. His argument, crafted within a review of Colin Ward's (1990) book *Child in the Country*, gave critical prominence to the ways in which conventional accounts of rural life and rural change have tended to view the social and cultural characteristics of rurality through an implicit (and sometimes explicit) lens of typically white, male, middle-class narratives. He suggests:

> there remains a danger of portraying British rural people … as all being 'Mr Averages', as being men in employment, earning enough to live, white and probably English, straight and somehow without sexuality, able in body and sound in mind, and devoid of any other quirks of (say) religious belief or political affiliation. (Philo, 1992: 200)

To some extent, Philo's intervention has been embraced as a reminder of a series of 'forgotten' items on the rural research agenda, and as such has affirmed the continuing development of studies of women, ethnicity, age, sexuality and the like in rural settings. However the significance of his critique also lies in recognizing the discursive power with which hegemonic representations of rural culture can act as an exclusionary device, serving to marginalize particular individuals, groups of people and even places on the grounds that the practice and performance of gender, age, sexuality, disability and so on is somehow other to the rural norm.

Over the past decade or so, research in rural studies has sought to bring revelation to the exclusionary qualities embedded within social and cultural constructions of rurality. Two books, *Contested Countryside Cultures* (Cloke and Little, 1997) and *Revealing Rural 'Others'* (Milbourne, 1997) illustrate some of the research being carried out in the UK context to unravel the 'Mr Average' approach, but in general there remains a suspicion that the pursuit of neglected rural geographies has taken the form of marking out particular facets of rural difference – for example, gender, sexuality and 'alternative' lifestyles – as the 'obvious' outworkings of new polyvocal narratives of the rural. This privileging of particular identities of otherness has been problematic on a number of counts. First, there is a danger that we trivialize vital issues such as gender, race and sexuality by labelling them as 'other', often in a relativist framework which can be less concerned with a commitment to emancipatory social practice than with a commitment to the political empowering of pleasure. Secondly, there is a risk that this narrow focus on what constitutes 'otherness' in rural settings will restrict the selection of rural subjects for research, as the richness of rural representation and myth involves some places and people with more seductive appeal than others. So far as *places* are concerned:

> Idyll-ised landscapes and places are favoured for study at the expense of less glamorous subjects and things which are woven within landscape tapestries. 'Ordinary' other places can become shadowed out by the privileging of special landscapes, with the result that the 'messiness' of rural space is sometimes lost. (Cloke and Little, 1997: 11)

A similar argument can be made about the privileging of key *social* identities of otherness:

> An off-the-hip listing of otherness – gender, sexuality, race, age, disability, alternativeness, and so on – has emerged which both risks the oversimplification of these very important and complex domains of difference, and serves to ignore other othernesses – those which are less easily categorised and perhaps hybrid, those which are less glamorous, those traits of otherness which may be a partial and transitory aspect of people's identity but which will often not be used to categorise the whole self and so on – which attribute to positionings and identities that are important in the messiness of rural populations and their lifestyles. (Cloke and Little, 1997: 11)

Thirdly, a focus on certain 'obvious' others in rural settings can often lack the depth of sustained, empathetic and contextualized research necessary to 'do justice' to the individuals, groups and places concerned. Any tendency to 'flit in and flit out' of the lives of rural others has important implications in terms of practising research as tourism or even voyeurism of the subjects concerned.

EXPLORING IDEAS OF OTHERNESS

In part, then, the difficulties of researching rural others may be methodological, but it is also clear that the engagement of rural studies with issues of otherness has been hampered by conceptual imprecision about what constitutes otherness, and how it might be theorized critically so as to avoid an over-strict ordering of the other. There are a number of issues here which contextualize any appreciation of the rural 'other'. Initially it is important to recognize the relationship between self and other in identity formation, by which to represent otherness is to gaze into a mirror of the self. The self-gaze on 'our' world is often tunnel-visioned, casting a self-referencing frame on the people and places around us. As Shurmer-Smith and Hannam (1994) point out, the 'automatic' and 'obvious' nature of this self-referencing gaze often defies efforts to think and behave beyond our own representations of identity. They suggest that 'as soon as we start to think about people who are not ourselves, we lapse into the language of "othering" … those who are not like "me" can start to slide into homogeneous mass of difference from "me" exactly the same as each other' (1994: 89). Identity, then, is crucially bound up in the self–other relationship, and identity formation will often relegate others to a more or less tolerated periphery, marginalized by individualist or cosy group politics, and often rendered invisible in the canvas we use to impose knowledgeable order on a hugely variegated world:

> Identity is … associated with processes of self-recognition, belonging and identification with others. Identity is also a way whereby we create forms of distinction between ourselves and those who we see as being like us and those who we see as different. We generally do this by creating divisions between those with whom we identify and those with whom we do not. Identity, therefore, is how we do membership and how we include or exclude others from membership of a particular identification. (Hetherington, 2000: 92)

These divisions and boundaries between the self and others become formed through a series of cultural representations in which people, places and things frequently elide so as to naturalize the context for selfhood (Sibley, 1995). It is therefore important to recognize how rural settings somehow serve to naturalize particular kinds of self-identity and, by implication, other-identity.

Recent studies of rural poverty and rural homelessness illustrate this point. Researchers (see Cloke et al., 1995; Fabes et al., 1983; Woodward, 1996) have consistently argued that socio-cultural barriers exist within the practices, thoughts and discourses of rural dwellers themselves which lead them to deny the presence of serious social problems such as poverty in rural places. In particular, cultural constructions of rurality which associate rural England with some kind of pastoral idyll have been shown to exert a pervasive yet obfuscatory influence over people's ability to recognize the existence of poverty and homelessness in the midst of that idyll. However, it is not just concealment which is embedded in discourses of rurality. Drawing on the work of Cresswell (1996), it is possible to see how the use of rural spaces becomes constructed as appropriate or inappropriate, as the established social order is reproduced by a kind of naturalization of common sense. Such naturalizations become concreted into rural society by the erection of boundaries between common sense notions of appropriate or inappropriate behaviour in particular places. When transgressions of appropriate behaviour occur, they seem 'out-of-place':

> The occurrence of 'out-of-place' phenomena leads people to question behaviour and define what is and is not appropriate for a particular setting … although 'out-of-place' is logically secondary to 'in-place', it may come first existentially. That is to say, we may have to experience some geographical transgression before we realise that a boundary even existed. (Cresswell, 1996: 22)

The self–other relationship, then, becomes embedded in taken-for-granted meanings about place, which are both socially and historically constructed, and are essentially implicated in the creation and maintenance of ideological beliefs. Such embedded boundaries are often neither dormant nor merely attitudinal. Sibley has shown how the built environment assumes a symbolic and practical significance which reinforces the desire for order in which the environment itself becomes ordered and purified and boundaries become 'policed'.

> In this way, space is implicated in the construction of deviancy. Pure spaces expose difference and facilitate the policing of boundaries – the exclusionary practices of the institutions of the capitalist state are supported by individual preference for purity and order ... A rejection of difference is embedded in the social system. (1995: 86–7)

Recent research on rural homelessness (Cloke et al., 2000, 2002) has shown how the rural 'self' relates to the homeless 'other'. Homelessness is assumed to be out-of-place in rural space, representing a key transgression of orthodox socio-spatial expectations. Equally, rural areas are variously purified of homelessness as boundaries are policed and the rejection of difference is embedded in the social system (see Milbourne, Chapter 31 in this volume).

The self–other relationship tells us much about how processes of identity formation, both individual and corporate, lead to a subjugation of the other, which can take the form of the policing of socio-spatial boundaries and the purification of space. However, it is exploration of the 'same'–'other' distinction which has led rural researchers to question how processes by which order is imposed on things have tended to distinguish between an other that can be recognized and known, and some other other which is unordered or unconforming to taken-for-granted patterns by which society distributes its understanding. In other words, one way of dealing with Philo's thesis of neglected rural studies is to move out in obvious directions from Mr Average, say to his wife, daughter, neighbour, elderly acquaintance, local shopkeeper and so on, recognizing the importance of gender, age, ethnicity and sexuality amongst others in the awareness of otherness. I have pointed out above some of the dangers in this approach, but what is occurring here is that only easily recognizable and understandable territories of otherness are being explored. As Philo suggests, this view of otherness comes from being locked into the 'thought-prison of "the same"' (1997: 22) – whereby the appreciation of otherness is restricted to that which fits with conventional patterns of ordering. From this viewpoint it seems impossible to get beyond the same to that other which is both present and absent, everyday but foreign, commonplace yet unexpected, unknown and not looked for.

Doel (1994) has characterized this distinction as between the 'Other of the Same' and the 'Other of the Other', arguing that with the best will in the world, researchers seeking to reach out to other people, practices and places succumb to inherent conventional impulses which accentuate the Other of the Same, and thereby end up reaffirming the patterns of order which serve to obscure the differences offered by the Other of the Other. Doel's differentiation between the Other of the Same and the Other of the Other is thus:

> While the former belongs to the theoretical-practice, falling under its justification and influence, the latter does not. More precisely, in the former difference and otherness are *appropriated* to the same, while in the latter they are *ex-appropriated*. Writing about otherness can therefore follow one of two transgressions ... *to the right*: appropriation, overcoding, territorialization, accumulation and capture; *to the left*: exappropriation, decoding, deterritorialization, expenditure and flight. (1994: 1042; emphasis in the original)

Any critical review of rural studies (see, for example, Philo, 1997) would have to conclude that there is a strong tendency to cling tenaciously to the cartographies of the Other of the Same, and in so doing rurality, rural people and rural places have effectively been simplified, and rendered predictable. The Other of the Other remains elusive.

ENCOUNTERING RURAL OTHERS

In the remainder of this chapter I discuss a range of studies which, in different ways, have attempted to explore beyond the Other of the Same in order to connect with the Other of the Other in rural settings. Before doing so, however, it is salient to acknowledge that there are no easy pathways for such exploration. In some senses it may be considered that the Other of the Other can only be encountered theoretically, through the affirmation of that which is other inherent in postmodernism, post-structuralism or deconstruction. However, I find it fruitful to agree with Philo's call for 'attentive engagements with all manner of substantive "other rurals" present in the total historical-geographical record' (1997: 40) – a call for narratives of how particular rural others actively produce other rurals not only imaginatively but also practically. This need for grounded engagement allows no side-stepping of crucial issues about how encounters with others can avoid conventional patterns of ordering that merely reinforce the Other of the Same.

Perhaps the most obvious starting point to encounters with others is to practise a rural studies that is more open to the circumstances and voices of 'other' people in 'other' places. This 'giving voice' to others has been aided by the increasingly significant adoption of qualitative and ethnographic research methodologies, and is able to contribute strongly to the idea of a polyphonic rural understanding. However laudable this aim might be, the idea of 'giving voice' has itself been subject to critical scrutiny. In part, critical concern has focused on the argument that merely giving voice to others does not ensure that issues of power in the rural area are understood. As Murdoch and Pratt (1993: 422) argue, 'simply "giving voice" to "others" by no means guarantees that we will uncover the relations which lead to marginalization or neglect'. Equally there are concerns about the ways in which by 'giving voice' to others, social scientists risk subordinating those other voices to the orthodox codes and conventions of academic discourse, essentially walking a tightrope between rewriting narratives of the rural from other perspectives and inscribing a colonial textuality within conventional common sense narratives (Barnett, 1997; Clayton, 2003).

The situated engagement with others seems to depend upon two seemingly different but often interconnected research impulses. The first is to stand back and listen – to wait for the other to interrupt our patterns of logic and expectation. Doel's (1994) account of the Other of the Other culminates in a plea that researchers remain alert to the counter-intuitive and often surprising call of the Other, which he views not as a stationary presence waiting to be discovered, but something which will always be arriving in ways that are 'alien to the intellectually competitive academy where most of us work, distinct from the sanitized, secularized, materialistic social spaces from which most of us hail and into which most of us have been socialized' (Philo, 1997: 25). The second impulse is to engage in active partnerships in the construction of new interactive knowledges and relationships. The wider development of postcolonial ethics of connection in social science has raised new possibilities for such partnerships. As Rose has argued:

> At the margins, within the domain of the 'other', one knows that the world, life and people express themselves with rich and interactive presences that are invisible from the viewpoint of deformed power, except perhaps, as disorder and blockage. The dismantling of this oppressive and damaging pole is a necessary step in moving towards dialogue ... we must embrace noisy and unruly processes capable of finding dialogue with the peoples of the world and with the world itself. (1999: 179)

So too with the peoples of the rural world and with the rural world itself. It is not a case of conversion of the other, of converting them into our world. Instead it is a commitment to enter into and in some cases dwell in the world of the other, making it not simply a workplace, but a place of residence. Of course there are myriad questions about how this might be achieved, each of which has a bearing on the setting of boundaries of professional/personal practice, and our performance in and amongst those boundaries (see Cloke, 2004). But dialogue arising from the embrace of noisy and unruly processes will undoubtedly involve new forms of commitment to, and even conversion for, the other.

EMBRACING RURAL OTHERS

Any attempt to establish ideal type illustrations of how rural research has forged encounters with rural others and other rurals risks overstating the achievements of these strategies of standing back and listening, and pursuing partnership dialogue. Without question, researchers working in these areas make modest claims about their ability to break out of the thought prison of the same. However, in what follows I briefly discuss four examples of such engagements. Many such examples are available (see, for example, the collections edited by Cloke, 2003; Cloke and Little, 1997; Milbourne, 1997; the contribution of Halfacree, 1996, 2003; and many of the chapters in this Handbook). What follows then are indicative examples of the potential for exploring rural otherness in research settings.

Rural mental health

A recent research programme undertaken by Chris Philo, Hester Parr and Nicola Burns (Philo et al., 2003; Parr et al., 2004) has drawn on in-depth interviews with over 100 users of psychiatric services in rural Scotland to investigate the complex dynamics of inclusion and exclusion of people experiencing mental health issues in different rural settings. Their innovative and critically committed study makes valuable contributions to the field of rural mental health in at least two main ways. First, they demonstrate how a plurality of rural spaces is formed by different overlapping surfaces of rurality – physical, demographic, economic, social and cultural. The covariation of these spaces produces and reproduces very significant differences (some subtle, some dramatic) between rural places, and it is in these differences that an openness emerges for a range of different, and often unpredictable, mental health outcomes which 'topple out of the circumstances of particular rural people and rural places' (Philo et al., 2003: 278).

Some of these different surfaces refract interesting light on rural others and other rurals in this context. The absence of anonymity in small places presents real difficulties in maintaining personal confidentiality and avoiding 'gossip', with the result that individual conduct, movement and service use tend to be visible and identified in potentially stigmatic ways. Cross-cutting demographic spaces of 'locals' and newcomers', and economic spaces of 'affluent' and impoverished', often serve to jumble up what can seem from the outside to be the 'comfortable' nature of the countryside, and are in turn overlain by cycles of particular economic pressure which are linked with the mediation of physiological distress both at home and in the workplace. Given that rural mental health can be linked to poverty, inadequate employment and the everyday struggles of 'getting by', the socio-spatial impacts of mental health problems will similarly be jumbled up in these overlapping layers of surface distinctions.

Although the presence of a rural 'community' supposedly offers networks of natural support in rural areas, the close-built social networks of some rural places drive some individuals away from seeking such informal or professional support. At the same time, these individuals can be subject to the strictures of rural cultural spaces which can, for example, require residents to conform to cultural expectations of self-sufficiency, resourcefulness and problem-freedom. Such conformity often takes the form of fatalism, derived from longstanding experience of hardship and sometimes framed within religious beliefs which accentuate a predestined existence that can be practised through fatalistic attitudes towards death and illness.

In these kinds of ways Philo et al. skilfully unravel the multiple surfaces of rural spatiality and demonstrate how somewhat unordered and certainly uncomfortable mental health outcomes can be found amongst the jumbled crevices and joints of these various surfaces. The second major contribution of their study is to listen to the voices of users of mental health services and discover a range of relations and practices by which people with mental health problems become variously 'included' and/or 'excluded' in their rural settings, with these lines of inclusion and exclusion often being blurred by the overlay of temporary and often superficial inclusionary moments. Most of their interviewees reported times of caring acceptance and understanding of their circumstances, suggesting at least a broad tolerance of mental health problems in the Scottish Highlands. Rural living can be replete with random and systematic acts of kindness from neighbours, family and friends, forming a platform of practical assistance and broader sympathetic concern for people experiencing mental health problems. However, Parr, Philo and Burns also detect more dutiful and even invasive forms of concern within the various and sometimes ambiguous practices of 'popping in', 'asking after' and 'sharing concern'. They conclude:

> To feel included in a small rural place can be an uneasy social state. The very workings of socially proximate but spatially dispersed communities are predicated on ambiguous social interactions which sometimes happen at a distance (asking after someone, phoning, and writing). For those bound into dense but distributed networks, illness can cement or disrupt senses of community embeddedness. (2004: 406)

By contrast to these practices and relations of inclusion, these authors found that knowledge of an individual's mental ill-health often brings negative impacts on local social relations in rural settings. Public forms of stigmatizing and 'differencing' were common, whether in the cessation or changed nature of conversation, or in the avoidance of interaction both through conscious behavioural tactics and subconscious body-language. Being ignored in the public spaces of a small rural place is shown by the authors to be a powerful act of rejection. Equally the cessation or changed nature of visits to the private space of home cements the experience of exclusionary social relations. Again, however, those interviewed in this study told a 'complex and untidy story of inclusion and exclusion, tolerance and rejection, articulated through reflection on the micro-spatial interactions of social life in spatially dispersed but socially closed communities' (Parr et al., 2004: 414). In bringing this complexity and untidiness to broader attention, the researchers go some considerable way towards teasing out new and surprising forms of how rural others can construct other rurals.

Rural children

The second illustration of how aspects of rural otherness can be explored is drawn from Owain Jones's groundbreaking work on rural childhood and children (Jones, 1997, 1999, 2000, 2001). He begins his intervention in the familiar territory of recognizing how children's worlds in rural spaces are often structured according to discourses that portray childhood in the countryside as a form of ideal. Constructing rural areas as good and appropriate places in which to bring up children can be linked with migratory patterns by which adults may move to a rural setting in order to give their children a country childhood. Drawing on the narratives of country childhood idylls to be found in a range of literature and poetry, Jones suggests:

> Country childhoods are seen powerfully in terms of a synthesis of innocence, wildness, play, adventure, the companionship of other children, contact with nature, agricultural spaces and practices, healthiness, spatial freedom, and freedom from adult surveillance. (1997: 162)

His scrutiny of these texts, however, has led him to suggest ways in which the otherness of childhood is exposed in and through the affinity between children and disordered rural spaces. Just as parents will choose particular disciplines, spaces and lifestyles for their children's activities in the light of particular understandings of childhood, so these discourses of childhood are penetrated and shaped by the embodied lives, practices and relations of children.

Jones recognizes the disordered spaces of rural childhood as markers of 'becoming children' in the striated fabric of modern society. Such spaces represent fractures in hegemonic spatial understandings from which upwellings of otherness can flow, rendering adult-oriented spaces disordered, facilitating a softening of rigid scales and processes, and conjuring up pauses and between-spaces in which the conflicts and pressures of disciplined order can abate. Disordered spaces offer children an opportunity to unleash potential to become themselves, or to become other than their normal regulated selves.

Consider, for example, the following extract from Dylan Thomas's autobiography, *Portrait of the Artist as a Young Dog* in which it is the *dilapidation* of his uncle's farmyard which makes it so attractive to him:

> The ramshackle out houses had tumbling, rotten roofs, jagged holes in their sides, broken shutters, and peeling whitewash; rusty screws nipped out from the gangling, crooked boards. … There was nowhere like that farmyard in all the slapdash country, nowhere so poor and grand and dirty as that square of mud and rubbish and bad wood and falling stone, where a bucketful of old bedraggled hens scratched and laid small eggs. (1965: 12)

Here, then, is a scene which differs significantly from the idyllized countryside of romantically nurtured and architecturally convincing cottages, pristine farmyards, manicured landscapes and the like. Instead the ramshackle, rusty and rubbish-strewn space offers another imagination of rural others and other rurals, an imagination which perhaps remains rather romanticized in an Arcadian sort of manner, yet reflects an unkempt and disordered space where adults do not, or cannot, bring a purity of order to the spaces they control, and where the otherness of children is less constrained. In such space, the child can be constructed as something other than innocent, living out that otherness to adult ordering and expectations.

Jones's work is significant because it blurs the distinction between the real embodied lives of children and the imagined constructed ideas of childhood. He challenges more commonplace views on the otherness of children – that is, as other to adults – and understands the disordering of rural spaces by children not in terms of innocence which is soon to be lost, but rather as a deterritorialization in which children can become other. He highlights the other rural spaces constructed by children, and demonstrates how childhood becomes associated with places and spaces which are beyond or outside adult control and ordering, where the fabric of Mr Average's adult world has become scrambled and torn, and where spatial order is disrupted, folded, made other. As Jones makes clear, childhood is itself a highly differentiated category, varying across gender, age, culture, class and family structure as well as individual disposition. Practices and relations of scrambling and disruption will be equally differentiated, although not necessarily along these lines of identity. Indeed one of the most significant aspects of Jones's recognition and naming of these neglected spatialities is that the seemingly disordered spaces of (albeit differentiated) children can be regarded as territories for becoming other, not in terms of fixed categories, but rather as constantly ebbing and flowing contested constructions and deconstructions of spacetimes and timespaces.

Asylum seekers and rural centres

A third illustration of encounters with rural others can be found in Phil Hubbard's (2004) account of the opposition to plans to accommodate asylum seekers in a number of rural locales in England. In this case the other – that is, asylum seekers – is present only in virtual, imagined form, yet it is clearly defined in terms of an other that is recognizable in relation to the rural self. Hubbard's authoritative and insightful study therefore raises important questions about the ability of rural society and rural space to accommodate otherness, and demonstrates how the differentiation between self and other can be formed around firm anxieties about the what-is-absent rather than the what-is-present.

The research centres on proposals to locate reception centres for asylum seekers on ex-military sites in rural areas is Nottinghamshire (Newton) and Oxfordshire (Bicester), each to house some 750 applicants for asylum. In each case a local action group was formed within a few days of the story being leaked out in the national press, and over a short period of time these groups began to marshall arguments against the centres on the grounds that they were in the interests neither of the local community nor the asylum seekers themselves, thus distancing themselves from any potential allegations of racism, but revealing fundamental anxieties about individual and collective identities.

Hubbard's study draws on a discourse analysis of letters of protest by local residents and suggests that 'local opposition to the asylum centre was based on predominantly negative constructions of asylum seekers as an other group' (2005: 8). Thus asylum seekers were characterized as non-productive and personally culpable for their current circumstances, and these bedrock subjectivities provided a firm foundation on which to build further edifices of the identity of asylum seekers as burdensome, threatening and illegal. To permit such others a place in these rural areas would be to bring stigma by association. Two fundamental sets of assumptions seem to have been at work here. First, regardless of the varied backgrounds of asylum seekers (and especially that many would be white Europeans), protesters assumed that asylum seekers would be unable to assimilate with the law-abiding, hard-working, self-sufficient and ordered nature of rural society. Secondly, it was assumed that asylum seekers would integrate more successfully in urban areas, where there are higher concentrations of both social services, and non-white residents.

These taken-for-granted associations between rurality and ethnicity were matched by other understandings of rural space, notably concerning sexuality. The heterosexual order of rurality is contrasted with the promiscuity of the city, and bringing in 'outsiders' to rural places translates into a perceived threat to this sexual order. Hubbard thus reports that:

> the identification of asylum seekers as potential rapists was a recurring motif used to justify the anti-accommodation centre protest, with many objectors suggesting that 'dumping' hundreds of young single men in a rural locale was not merely 'inviting vandalism and theft' but also 'rape, sexual and physical assault, lewd conduct and worse'. (2005: 14)

Such deep-seated fears reveal how even the *potential* presence of others in rural spaces provokes highly structured borders between what is in place and out of place, and confirms that such boundaries range widely between the body, the family, the community and the nation.

While these expressions of opposition can be understood against a very specific framework of cultural imaginaries relating to English rurality that draw out very specific fears about the different identities and subjectivities of others, it is clear (as Hubbard himself emphasizes) that the resultant discourses of protest nevertheless share similarities with equivalent outpourings of opposition from urban residents. Hubbard's study therefore raises specific and important questions about both the cultural construction of rurality and about the ways in which the borders of rural orthodoxy are policed so as to secure a purification of rural space. The next step could be to seek to hear the views of successful asylum seekers who have settled in rural areas, so as to destabilize these all-too-certain identifications of otherness, and to open out perhaps more uncomfortable, and certainly less ordered boundaries between same and other in rural spaces.

Animals and the co-construction of rural places

As a final example of research that has acknowledged and encountered rural others who have constructed other rurals, it is worth emphasizing that the notion of otherness can stretch well beyond the social domain to include the relational presence of animals in rural spaces (see Jones, Chapter 13 in this Handbook). Recent research in rural New Zealand (Cloke and Perkins, 2005) has focused on how the presence and performance of cetaceans – whales and dolphins – has been responsible for the co-constitution of the booming tourist town of Kaikoura. Until the late 1980s Kaikoura was a sleepy fishing village of less than 2,000 inhabitants, known only for its locally caught lobsters, and a series of rusting relics from the whaling era. By 2002, the population of the town had nearly doubled, and it was receiving nearly 1.5 million tourists per annum. Studies of rural otherness in Kaikoura could, of course, focus on incomers or on transient tourists, each bringing with them distinctions of identity and subjectivity capable of transforming the nature of the rural space concerned. However, this transformation of a small coastal backwater into an international travel destination can be traced to the dramatic development of a high-profile eco-tourism industry in the town, and this in turn depends on a plentiful supply of majestic and playful otherness in the form of cetaceans in nearby waters.

The rise of eco-tourism in Kaikoura has been spearheaded by the emergence of two lead-brand companies – Whale Watch® and Dolphin Encounter® – the former an enterprise led by the local Maori tribe (Ngai Tahu) which runs three large boats with a capacity to take some 65,000 tourists per annum out to view the local whale population, and the latter a firm offering opportunities to 70 or so people per day to swim with dolphins. Both Whale Watch® and Dolphin Encounter® are strongly featured in the contemporary bibles of travel such as *Lonely Planet* and *Rough Guide*, and have quickly assumed the status of an essential stop-off on the backpacker trail of New Zealand. Several more offshore spin-off opportunities have followed, including plane and helicopter flights to view whales from the air, swimming with seals and shark encounters, and Kaikoura itself has become swamped with cetacean iconography throughout its range of accommodation, restaurant and retail services.

Given these circumstances, it is clear that the development of Kaikoura has been co-constituted by the presence of cetaceans; without the whales and dolphins, the town would not be the place it is today. Its popularity with tourists reflects the enormous cultural appeal of cetaceans which exert very considerable charisma, and offer relational encounters often understood in terms of being the thrill of a lifetime. This study shows how the whales and dolphins are the subject of a whole range of anticipatory knowledges and expectations by tourists, who will have been exposed to powerful image constructions by different media characterizing cetaceans in terms of characteristic embodied manoeuvres such as the sperm whale's surface 'blow' and majestic flick of the flute when diving, and the acrobatics of dolphins performing their leaps, jumps, flips and somersaults. As Cloke and Perkins suggest:

> The lure of Kaikoura, then, is the place-experience of encounter with these most special animals and the experience-performance of getting in amongst the whales and dolphins in their own world, seeing them perform their trademark manoeuvres, metaphorically or literally being close enough to reach out and touch them. These trophy moments will provide unforgettable experiences and open up the possibility of more liminal and imaginative encounters as a result of being part of the relational assemblage co-constituted by the cetaceans themselves. (2005: 15)

To some extent, it is difficult to examine human existence with animals in a manner that respects their otherness and difference, although some attempts have been made to examine these co-constitutive place performances in terms of the cetaceans and not just the humans. The sustained attempts in New Zealand to bring scientific study to bear on the impacts of eco-tourism on cetaceans have been inconclusive, partly because the history of close exposure of cetaceans to human activity has meant that they have become habituated to human proximity, and have developed adaptive responses which may not necessarily be helpful. There is some evidence that cetaceans are incurring low-level but long-term stress from eco-tourism encounters which could impact on reproductive and immune systems and perhaps even increase rates of mortality. However, it might equally be argued that there are glimpses here of relational achievements in which cetaceans do convey their otherness on something like their own terms. Overall the presence and performative capacities of cetaceans can be regarded as co-constituting Kaikoura as an intelligible place. Although the enabling skills of tour operators are crucial to the overall assemblage, it is in fact the presence of cetaceans which folds this network around itself. The very nature of this appealing and charismatic 'other rural' of Kaikoura is, then, dependent on the cetacean others which in this case have been so influential in bringing together the layers of rural space.

CONCLUSION

In some ways these four illustrations serve to reinforce some of the conceptual and methodological difficulties rehearsed in the first part of this chapter. There does seem to be a deep-seated privileging of the Other of the Same as opposed to the Other of the Other in some of these examples, and yet each indicates territories of complexity and untidiness which suggest surprising as well as intuitive senses of otherness. Thus Philo et al. not only identify those experiencing mental health problems as a significant rural other but, more than that, demonstrate that mental health issues in rural areas, and our understanding of them, can be opened up to surprising, cross-cutting senses of otherness. Their account shows how those experiencing mental ill-health can be the subject of socially and culturally regulated boundaries, rendering such illness as out of place in rural spaces, but they go well beyond the obvious levels of generalization here and open out highly differential niches for sometimes highly differentiated experiences of mental ill-health, and in doing so they prove sensitive to at least some of the call of the Other of the Other.

Jones's study of the otherness of childhood again explores beyond the obvious boundaries of children's otherness to the adult orthodoxy. In identifying the unruly and disordered spaces of rural childhood, he highlights fractures in the normal understandings of rurality from which otherness upwells and overflows in the becoming-children of childhood. He demonstrates how rural space can both imaginatively and substantively be reterritorialized in the immanent performances of childhood. Hubbard, too, emphasizes potential points of rupture in the ordered spaces of the rural. His analysis of the fear expressed by rural residents of absent and unknown others clearly shows how the threat of otherness can be made concrete by creating a counter-identity for others based on their supposed transgression of the self. To some extent, such processes of othering are unsurprising, representing a clear case of the other of the same, but the example points sharply towards the necessity to deconstruct the 'Self of the Self' in order to be more open to the 'Other of the Other'.

Finally, the example of the role of the cetaceans in co-constituting the place of Kaikoura reinforces both the mystery and the importance of continuing to explore otherness in rural studies (although some care to challenge the precise 'rural' nature of this illustration). Although to some extent the cetacean otherness of whales and dolphins has been tamed and colonized by human tourism in Kaikoura, the role of cetaceans is recognizably crucial to the nature of the place, even though their character, identity and language remains stubbornly other to our attempts to appropriate and capture them to our same. Although we try to convert 'them' into 'our' world, we have in fact taken small steps to enter into and dwell (albeit briefly) in their world. It is these tentative steps towards embracing the unintelligible and sometimes unruly world of others that ultimately will bring us closer to those neglected and new othernesses which hithertofore have been passed over in our emphasis on the colonizing order of rural space and rural society.

REFERENCES

Barnett C (1997) '"Sing along with the common people": politics postcolonialism and other figures'. *Environment and Planning D: Society and Space*, 15: 137–54.

Clayton D (2003) 'Critical, imperial and colonial geographies'. In K Anderson, M Domosh, S Pile and N Thrift (eds), *Handbook of Cultural Geography*. London: Sage. pp. 354–368.

Cloke P (ed.) (2003) *Country Visions*. Harlow: Pearson.

Cloke P (2004) 'Exploring boundaries of professional/personal practice and action: being and becoming in Khayelisha Township, Cape Town'. In D Fuller and R Kitchin (eds), *Radical Theory/Critical Praxis: Making a Difference Beyond the Academy?* Praxis (e)Press. pp. 92–102.

Cloke P and Little J (eds) (1997) *Contested Countryside Cultures*. London: Routledge.

Cloke P and Perkins H (2005) Cetacean performance and tourism in Kaikoura, New Zealand. *Environment and Planning D: Society and Space*, 23.

Cloke P, Milbourne P and Thomas C (1995) 'Poverty in the countryside: out of sight and out of mind', in C Philo (ed.), *Off the Map: The Social Geography of Poverty in the UK*. London: Child Poverty Action Group.

Cloke P, Milbourne P and Widdowfield R (2000) 'Homelessness and rurality: "out-of-place" in purified space?' *Environment and Planning D: Society and Space*, 18: 715–35.

Cloke P, Milbourne P and Widdowfield R (2002) *Rural Homelessness: Issues, Experiences and Policy Responses*. Bristol: The Policy Press.

Cresswell T (1996) *In Place, Out of Place: Geography, Ideology and Transgression*. Minneapolis, MN: University of Minnesota Press.

Doel M (1994) Deconstruction on the move: from libidinal economy to liminal materialism. *Environment and Planning A*, 26: 1041–1059.

Fabes R, Worsley L and Howard M (1983) *The Myth of the Rural Idyll*. Leicester: Child Poverty Action Group.

Halfacree K (1996) Out of place in the country: travellers and 'rural idyll'. *Antipode*, 28: 42–72.

Halfacree K (2003) 'Landscapes of rurality: rural others/other rurals'. In I Robertson and P Richards (eds), *Studying Cultural Landscapes*. London: Arnold. pp. 141–169.

Hetherington K (2000) *New Age Travellers: Vanloads of Uproarious Humanity*. London: Cassell.

Hubbard P (2005) '"Inappropriate and incongruous": opposition to asylum centres in the English Countryside'. *Journal of Rural Studies*, 21: 3–18.

Jones O (1997) 'Little figures, big shadows: country childhood stories'. In P Cloke and J Little (eds), *Contested Countryside Cultures*. London: Routledge. pp. 158–179.

Jones O (1999) 'Tomboy tales: the rural, nature and the gender of childhood'. *Gender, Place and Culture*, 6: 117–136.

Jones O (2000) 'Melting geography: purity, disorder, childhood and space'. In S Holloway and G Valentine (eds), *Children's Geography: Living, Playing, Learning.* London: Routledge.

Jones O (2001) '"Before the dark of reason": some ethical and epistemological considerations on the otherness of childhood'. *Ethics, Place and Environment*, 4: 173–178.

Milbourne P (ed.) (1997) *Revealing Rural Others.* London: Pinter.

Murdoch J and Pratt J (1993) Rural studies: modernism, postmodernism and the 'post-rural'. *Journal of Rural Studies*, 9: 411–428.

Parr H, Philo C and Burns N (2004) 'Social geographies of rural mental health: experiencing inclusions and exclusions'. *Transactions of the Institute of British Geographers*, NS29: 401–419.

Philo C (1992) 'Neglected rural geographies: a review'. *Journal of Rural Studies*, 8: 193–207.

Philo C (1997) 'Of other rurals?' In P Cloke and J Little (eds), *Contested Countryside Cultures.* London: Routledge. pp. 19–50.

Philo P, Parr H and Burns N (2003) Rural madness: a geographical reading and critique of the rural mental health literature. *Journal of Rural Studies*, 19: 259–281.

Rose D (1999) 'Indigenous ecologies and an ethic of connection'. In N Low (ed.), *Global Ethics and Environment.* London: Routledge. pp. 175–187.

Shurmer-Smith P and Hannam K (1994) *Worlds of Desire Realms of Power: A Cultural Geography.* London: Arnold.

Sibley D (1995) *Geographies of Exclusion: Society and Difference in the West.* London: Routledge.

Thomas D (1965) *Portrait of the Artist as a Young Dog* London: Dent.

Ward C (1990) *The Child in the Country.* London: Bedford Square Press.

Woodward R (1996) '"Deprivation and the rural", an investigation into contradictory discourse', *Journal of Rural Studies*, 12: 55–68.

33

Political articulation: the modalities of new critical politics of rural citizenship

Michael Woods

INTRODUCTION

On 12 August 1999, veteran anti-militarist campaigner turned farmer José Bové led a group of his supporters in 'dismantling' a new McDonald's restaurant under construction in the southern French town of Millau. Ten months later, a crowd of 100,000 descended on Millau in carnival mood for Bové's trial, celebrating his new-found iconic status. For the cosmopolitan gathering of well-wishers, Bové and his actions symbolized many things. Most overtly, they were a strike against globalization, with campaigners from around the world called as defence witnesses in an attempt to invert the trial into a tribunal on globalization. At another level, the targeting of McDonald's symbolized Bové's crusade against *malbouffe* – 'rotten food' – earning plaudits from environmentalists and 'real food' advocates. For French nationalists, meanwhile, the attack was a patriotic act of anti-Americanism. Yet, in many ways Bové's was a fundamentally *rural* protest. The protesters he led came from the *Confédération paysanne*, a radical farmers' union that he had helped found to represent smallholders. The provocation for the protest came from US trade restrictions on the import of Roquefort cheese, a staple product of the Millau region, combined with discontent at the power of multinational agro-food corporations to undermine locally rooted traditions of food production. And Bové's rapid elevation to national folk-hero in France was attributed to his ability to appeal to romantic notions of the rural idyll among the urban middle classes. At the heart of Bové's anti-globalization protest was a rural citizen fighting for his rights.

Bové is not alone. Protests by rural campaigners have become an increasingly prominent feature on the political landscape of countries across the developed world. In March 2001, farming and countryside groups organized a Rally for Rural America in Washington, DC, inviting comparisons with the 'tractorcade' occupation of The Mall by the American Agricultural Movement in the 1970s. In London, a quarter of a million protesters joined the Countryside Rally in 1997 and the Countryside March in 1998, prompted by parliamentary efforts to ban fox-hunting; whilst in France the *Chasse, Pêche, Nature et Tradition* (Hunting, Fishing, Nature and Tradition) party won seven seats in the 1999 elections to the European Parliament. Increasingly militant farmers in the UK have engaged in blockades of ports, dairies, supermarket distribution depots and oil refineries, borrowing tactics employed by their French counterparts. At a more local level, numerous small-scale protests against housing developments, new roads, quarries, windfarms, waste disposal sites and so on have flared up against the perceived threat to 'rural character'.

At the same time, many more rural residents have become politically engaged in a less confrontational manner through involvement in community action. Grassroots initiatives have sought to respond to rural change by providing services, developing facilities and promoting community regeneration through the pooling together of individuals' resources within the community. Partnerships are often formed with local and outside agencies to draw down external funds for community projects. In Kielder on the English–Scottish border, villagers formed a development trust to run local services when the Forestry Commission withdrew their paternalistic presence; whilst residents of Chemainus, on Vancouver Island, responded to the closure of the major employer, a sawmill, by creating murals of the community's history and turning the town into a tourist attraction. Volunteers in East Fairfield, Vermont, have converted the old village school into a community centre and rural health clinic; meanwhile, in Booneville, Mississippi, a residents' initiative to provide a children's play area has grown into an ambitious project refurbishing the small town's main street. Each of these initiatives – and the thousands more like them – have involved individuals assuming a share of responsibility for the governance of their own community.

The common thread that unites the rural protests with the community development initiatives is citizenship. Both are expressions of a dynamic within rural societies that is re-defining the relationships between individuals, communities and the state. As rural residents come to articulate themselves as rural *citizens* – through the proactive performance of rights and responsibilities – so relations of power are shifted, and conflicts arise over representations of place and rurality. In these ways, there is emerging a new critical politics of rural citizenship. Critical, that is, both in the way that the mobilization of rural citizenship challenges established rural social and political structures, and in the sense in that the form and conditions of the new rural citizenship are still in a process of becoming, and thus are still open to contestation and critique. This chapter explores the dimensions of this new critical politics, treating citizenship both as an object of inquiry and as a lens through which changes in rural societies may be analysed. First, the chapter discusses the concept of 'citizenship' and its application in rural society, identifying the emergence of the 'new rural citizenship' with wider processes of social, economic and political change. Second, the chapter examines the two main contexts in which rural citizenship has been mobilized – the devolution of governmental responsibilities to 'active citizens' in the context of state restructuring, on the one hand, and the activation of protest movements claiming and defending the rights of rural citizens in the context of rural restructuring and changing rural–urban relations, on the other. These sections demonstrate the application of the concept of citizenship as an analytical device in understanding rural change, but also begin to raise questions about citizenship and the critical politics of rural citizenship itself. Thus, third, the chapter proposes an agenda for the analysis of the critical politics of rural citizenship, focusing on issues of identity, leadership and accountability.

CITIZENSHIP AND RURALITY

Citizenship codifies the relationship between the individual and the state. At one level, citizenship is a mark of belonging. When configured at a national scale this is a legal status and is tightly policed as such; however, citizenship is most commonly performed at a local scale, within particular communities, where citizenship is defined through custom and practice. The respective *de jure* and *de facto* citizenships empowered at these scales are not necessarily concomitant and the disjuncture between the two has become a focus of research exploring inequalities in the mobilization of different residents' social, economic and political entitlements within localities (see Smith, 1989).

At a second level, therefore, citizenship is about the receipt of rights and the performance of responsibilities. Many of the rights and responsibilities afforded by national citizenship are legally stipulated and enforced, although some exist by means of tacit or implicit understandings. In contrast, the rights and responsibilities involved in the practice of *local citizenships* are essentially contingent on local culture and are open to re-negotiation and contestation. Furthermore, as Marshall (1950, 1964) postulated, the rights and responsibilities of citizenship are not set to any absolute standard, but are the product of a dynamic process of social development. As such, citizenship is always politicized and contested.

The ongoing dynamic of citizenship is observed in the recent recognition within urban studies of a *new politics of citizenship*. This has two dimensions. The first refers to the testing and expansion of local citizenship rights as cities respond to their increasingly diverse and multicultural communities and to the pressures of globalization (see Dunn et al., 2001; Gorman, 2000; Menahem, 2001); whilst the second dimension is concerned with the reworking of citizen responsibilities as urban governance is restructured (Lowndes, 1995; Raco and Imrie, 2000). This includes not just the transference of responsibilities to citizens as part of a strategy of 'rolling back' the state, but also citizens' perceptions that the state, at both a local and a national level, has failed to deliver or is impotent to deliver the outcomes that they desire, and that citizens must act to find their own solutions. Arguably, both of these dimensions apply equally to rural societies as to urban, yet the analysis of citizenship in a rural context has been much more limited. To a large extent, the explicit employment of concepts of citizenship and citizens' rights in rural studies has been focused on issues involving disputes over legal rights, such as access to land, hunting and travellers (see Bromley and Hodge, 1990; Halfacree, 1996a; Parker, 1996, 1999a; Parker and Ravenscroft, 2001; Ravenscroft, 1998). However, as Parker (2002) demonstrates, the potential use of citizenship as a concept in rural research is far greater, extending into areas where notions of citizenship have already been implicitly evoked in research without necessarily being described in the language of citizenship.

Parker (2002) employs citizenship as a framework for exploring the processes and consequences of contemporary rural change, engaging with citizenship both as a normative and as an analytical concept. Thus, starting from the familiar ground of land rights, Parker demonstrates how conceptions of citizenship that emphasize the rights of citizens have been drawn upon normatively within disputes over land management and amenity use. Thus, for example, political debates surrounding legislation to extend public access to open countryside in the United Kingdom can be represented as a fundamental conflict between those who perceive the countryside as a shared national treasure and therefore position a right of access – the so-called 'right to roam' – as part of their citizen's rights, and those who oppose the legislation on grounds of defending the rights of property owners (Parker, 2002; Parker and Ravenscroft, 2001). Similar appeals to citizen's rights can be observed in conflicts over hunting in both North America and Europe, where action by the state to control or regulate hunting is positioned as an infringement of, or attack on, the rights of rural citizens. One explicit example of this representation is found in Texas, where pro-hunting campaigners sponsor a political action committee, Citizens for the Preservation of Rural Lifestyle, 'dedicated to protecting our God given rights as rural Texans from the misplaced intrusions of Urban Sprawl' (CPRL, 2001). By adopting the self-description of 'citizens' campaigners exert their 'right' to protest and to be listened to, and turn disputes over land use, development, environmental impact, lifestyle and service provision into battles to protect or promote claimed 'rights'.

The enrolment of the rhetoric of citizenship into rural conflicts in this way highlights the contingent and contested nature of citizenship rights (Parker, 2002). Citizenship is both defined and enforced downwards by the state, and claimed upwards by individuals and groups. The state, in its various guises, can act in some circumstances to guarantee or enable citizens' rights, but in others to constrain citizenship. Rights and responsibilities need to be negotiated in a transaction that Parker and Ravenscroft (2001) argue is analogous to the exchange of capital and debt in the economic 'gift' relationship. Resolution of these tensions and dynamics does not occur within a vacuum, but is shaped through efforts by individuals and agencies to define both their own identities, and those of the spaces that they inhabit. As such, the (re)definition of rural citizenship is inseparably entwined with the reconstitution of rural space.

Furthermore, as Parker observes, 'citizenship can be viewed as a cultural phenomenon in addition to a legal status or bundle of rights associated to the citizen' (2002: 6), such that, 'the ethnic or ethical community of citizens plays a strong part in determining their "citizenship" in terms of day-to-day visible (and more opaque) community politics' (p. 6). From certain perspectives, the promotion of community-based 'cultural citizenship' can be an empowering force helping to create a more inclusive, participatory rural society. From others, however, it can be an exclusionary force, denying rights to those who 'do not fit' into the imagined community. In different ways, the gated communities of rural America (see Phillips, 2000) and the imposition of legal controls on the movement of New Age travellers in rural Britain (see Halfacree, 1996b; Parker, 1999a), both represent manifestations of an exclusionary rural citizenship in practice.

'Rural citizenship' is hence a complex creature, at once both a normative strategy and a mode of interpretation; both an object of inquiry and an explanatory concept; both a source of empowerment and a vehicle for exclusion.

THE PARADOX OF RURAL CITIZENSHIP

For some the very idea of a 'rural citizenship' will grate with the embeddedness of long-standing cultural associations between citizenship and urbanity. The words *citizen* and *citizenship* share a common etymological root with *city* in the Latin *civitas*, and from its earliest formulation the concept of citizenship was perceived as an urban phenomenon. In the city-states of Classical

Greece, citizenship entitled participation in the fora of face-to-face decision-making that urban life permitted. In the medieval period, reference to citizens was used to distinguish city dwellers from their rural counterparts, emphasizing the political and mercantile rights of chartered boroughs. Later, with the rise of modernity, citizenship stood for the nurturing within urban centres of a progressive, liberal, participatory political and social society. As the leading philosopher of citizenship, Jean Jacques Rousseau, remarked: 'houses make a town, but citizens a city' (Rousseau, 1993: 192).

Rousseau's distinction was intended as a critique of the loose usage of the word citizen, by equating not just citizenship with the city, but also the city with a body politic rather than with a geographical entity. In the development of this idea by Weber (1958), urbanization was represented as a means by which individuals were transformed into citizens, liberated from historic ties of loyalty and tradition and opened to rational ways of organizing and governing society. For Weber, the emergence of the modern Western city was characterized by the establishment of an oath-based citizenship based on the principle of equal rights, where all citizens are subject to the same laws and same tribunals, and the ability for all citizens to participate in political debates and decision-making through a vibrant civil society. Fundamentally, it is this oath-based community which defines the nature of the city and its civil society, and which is translated into the discursive distillations of charters, plans, rules and regulations, and the institutions of civic government and civil society (see also Lowndes, 1995; Poulle and Gorgeu, 1997). From this formulation emerges the understanding of citizenship as a balance of rights and responsibilities – citizens are defined by the rights they enjoy, but equally they have responsibilities to the collectivity of the city.

In this model, neither urbanity nor citizenship are necessarily restricted to large concentrations of population – the geographical city – but it was such locales which provided the conditions for the emergence of these phenomena. In contrast, the traditional social formations of rural areas were presented as the antithesis of citizenship. Rural societies in European nations – and to some extent their colonial settlements – were characterized by entrenched social hierarchies, not by equality. They lacked the institutions of civil society, such communal institutions as did exist, notably the church, served to reinforce the social order not to facilitate democratic participation. Moreover, the economic dominance of agriculture, commonly with a concentration of land ownership and a tied labour force, underscored particularistic relations in which rights were enjoyed only by the few and responsibilities were to individuals – the employer or the landowner – not to a collective. At the same time, the fundamental inequalities were cloaked by a discourse of an organic community which promoted community solidarity and loyalty and a reverence of tradition. Rural people and rural cultures were valorized as the guarantors of ancient tribal or national values and histories, such that, as Ramet comments in the context of Serbia, 'the countryside traditionally teaches one what it is to be a good *national*, while the city teaches one what it is to be a good *citizen*' (Ramet, 1996: 72).

There were exceptions. In Switzerland citizenship was until the nineteenth century associated more with the rural cantons than with the towns. In North America, the pioneer small towns of New England are perceived to have embodied much of Weber's urbanity (Mattson, 1997; de Tocqueville, 1946; see also Lowndes, 1995), yet their nascent civic society equally served to distinguish the incorporated townships from their unincorporated rural hinterlands. Attempts to build a more widespread rural citizenship through the American progressive movement in the early twentieth century failed as a political project. By and large, however, the ideas of Rousseau, Weber and others informed a discursive separation of urban and rural, with each perceived as separate political fields, with different dynamics of power, and requiring different modes of government.

However, in the late twentieth century this discursive topography was undermined by processes of social and economic restructuring. As has been extensively documented elsewhere (Marsden et al., 1993; Mormont, 1990; Woods, 1997, 2003), changes in agriculture, shifts in rural economic and employment patterns, increased mobility, counter-urbanization, demographic evolution, advances in technology and communication, cultural homogenization, better standards of education, the expansion of leisure and tourism, and growing environmental and consumer awareness, have all contributed to a refashioning of the spatial and social character of rural areas – and have provoked a several-fold political response. First, competition has emerged within rural localities for influence over the local power structure and government, displacing previously hegemonic elites (Woods, 1997, 1998a). Secondly, conflicts have developed at national, regional and local scales concerned with the meaning of 'rural' and the regulation of rurality (Mormont, 1987, 1990; Woods, 1998b). Thirdly, the restructuring of the countryside has required the establishment of new mechanisms and institutions for the governance and regulation of rural space, and even of new territories of governance (Goodwin, 1998; Morquay, 2001; Woods and Goodwin, 2003). Fourthly, as external actors have become increasingly involved in aspects of rural governance, so elements within rural society have been mobilized to defend symbols of 'traditional' rurality from external interference (Mormont, 1990; Woods, 2003, 2004).

CITIZENSHIP AND GOVERNMENTALITY

The changes wrought through processes of rural restructuring, described above, were paralleled in the late twentieth century by a restructuring of the role and scope of the state. Whilst not explicitly concerned with the rural, state restructuring has impacted upon the way in which rural areas are governed, not least by reconceptualizing citizenship. The modes of government dominant in the mid-twentieth century had tended to embody Marshallean notions of citizenship which emphasized rights above responsibilities. In particular, Marshall (1964) had argued that in the twentieth century the *social rights* of citizens had gained prominence over political and civil rights in the priorities of government. Social rights defined citizens' entitlement to public health and education services, social security and state protection from poverty and deprivation. These rights were provided for in broadly social-democratic states, such as the UK and the Scandinavian countries, by the centrally governed 'welfare state' and by considerable state intervention in economic development (manifest, for example, in the English Rural Development Commission). In more liberal regimes, including the United States, Canada and Australia, Esping-Anderson (1990) suggests that social rights were subjugated to individual liberty, with state provision of welfare mobilized only where welfare entitlements could not be met through the market (see also Lowndes, 1995). Notably, in the United States at least, the exceptional cases of state intervention often related to rural areas, where the state assumed an interest in infrastructure provision and, on occasion, experimented with large-scale interventionist projects such as the Tennessee Valley Authority.

The rise of the New Right in the 1980s promoted a new ideology of the minimal state in which the demands of Marshall's social citizenship were identified as a key factor in the fiscal crisis of government (King, 1987). The New Right re-fashioned citizenship by prioritizing rights of individual liberty above social rights and by emphasizing the responsibilities of citizens as well as their rights. In particular, two new models of citizenship emerged.

First, the idea of the *consumer-citizen* was advanced, emphasizing the citizen's *right to choose* (for example between service providers), and portraying the relationship between the citizen and the public services as equivalent to that of consumers and businesses (see Urry, 1995, 2000). In a rural context, this model has been applied not just to issues of service provision, but also to questions around the commodification of the countryside and demands for recreation rights (see Parker, 2002). Moreover, the logic of consumer-citizenship has been adopted by radical campaigners through the use of consumer choice and consumer protest tactics (e.g. boycotts) to intervene in food, agricultural and environmental policy networks (see Parker, 1999b, 2002). Changing relationships between producers and consumers have contributed to this trend, particularly as the power of transnational corporations within rural economic structures has become increasingly recognized, making such corporations targets for protest activity alongside state agencies.

Secondly, the renewed emphasis on the responsibilities of citizens has been articulated through the concept of *active citizenship*. Here, citizenship is perceived not as something that is passively received from the state, but as something that must be actively performed by individuals through participation in governance and sharing responsibility for the defence of citizenship rights. Benevolence and mutual co-operation are thus positioned as items of self-interest with citizens mobilized to participate in community action. The model of active citizenship was most explicitly promoted by the British government of John Major in the early 1990s, becoming introduced into rural policy across a range of arenas including education (Ribchester and Edwards, 1999), crime prevention (Yarwood and Edwards, 1995), public access (Parker, 1999c) and community development (Edwards, 1998). However, it has resonance with the practices of community organizing within the American Rural LISC programme (Gittell and Vidal, 1998) and the Clinton–Gore Empowerment Zone/Enterprise Communities Initiative (Aigner et al., 2001); as well as in Australian rural entrepreneurship initiatives (Herbert-Cheshire and Lawrence, 2001).

Governmentality and citizen action

These political developments have been further theorized as forming part of a shift in governmentality – or the way in which government renders society governable (Foucault, 1991; Rose, 1993). Following Rose (1996a, 1996b), it has been argued that advanced Western democracies have experienced a shift in the regime of governmentality from 'managed liberalism', characterized by state intervention in the social sphere and centralized state planning, to a new governmental rationality of 'governing through communities' which,

> does not seek to govern through 'society'; but through the regulated choices of individual citizens, now construed as subjects of choices and aspirations to self-actualization and self-fulfilment. Individuals are to be governed through their freedom, but neither as isolated atoms of classical political economy, nor as citizens of society, but as members of heterogeneous communities of allegiance, as 'community' emerges as a new way of conceptualizing and administering moral relations amongst persons. (Rose, 1996a: 41)

The manifestation of this shift in governmentality has been observed in British rural policy by Murdoch (1997) and in an Australian context by Herbert-Cheshire (2000). Murdoch identifies the components of the integrated rural policy introduced in the first 'Rural White Paper' for England in 1995 as embodying a strategy of governing through communities, in contrast to a previous emphasis on national scale planning (e.g., see Murdoch and Ward, 1997). For example, Murdoch notes that from the start the paper places limits on the legitimate scope of state intervention, highlighting the diversity of the countryside, the limited ability of government to foster local character and questioning the appropriateness of central government to initiate policies that are in keeping with local circumstances. Instead, the document stresses the importance of rural people 'helping themselves' through engagement in their own communities. By appealing to the notion that 'people in the countryside have always needed to take responsibility for looking after themselves and each other' (DoE, 1995: 16, quoted by Murdoch, 1997: 114), the paper seeks to translate a perceived characteristic of traditional rural societies into a duty of citizenship – thus fostering an *active rural citizenship* in which rural residents assume a share of responsibility for finding solutions to rural problems across a range of policy areas through collective community-scale action.

Parallels to this policy shift are identified by Herbert-Cheshire (2000) in the output of the Queensland 'Positive Rural Futures' conference of 1998, which she examines as an example of the wider emphasis within Australian rural development strategy on self-help and bottom-up, community-based initiatives. Herbert-Cheshire records that, as in the English Rural White Paper, the conference placed limits on the scope of government and sought to justify the strategy of self-help by emphasizing 'the *rights* of rural people to seek their own solutions to problems and, correspondingly, the "empowering" effect of enabling them to do so' (Herbert-Cheshire, 2000: 210; emphasis in the original). In analysing these policies through the lens of governmentality, Herbert-Cheshire is not only able to present a more sophisticated reading of their significance than previous functional explanations, but also identifies a number of issues for further research. These include, how the notions of self-governing individuals and communities are constructed in political discourse; the political rationalities that are used to justify levels of (non)intervention; and the discourses, forms and outcomes of empowerment at a local level. Whilst Herbert-Cheshire attempts to address these questions with regard to the Queensland case, they may equally be adopted as concerns through which a more critical and theoretically engaged research agenda could be advanced to take forward the often rather empiricist body of work on rural community action. Paying attention to the concept of citizenship in these contexts – and to the way its use relates to strategies of governmentality – will help to reveal not just the mechanisms and outputs of community development initiatives, but also the social and power relations that lie behind them, thus providing a greater insight into their longer-term significance for rural societies.

Citizenship rights and rural protest

The promotion of a proactive rural citizenship described above has emphasized the enactment of citizens' responsibilities. At the same time, however, a more reactive strand of citizenship has been mobilized through campaigns and protests which have evoked citizens' rights. This new wave of rural protest has gathered momentum in reaction to the disruptive effects of the social and economic restructuring of rural areas. As Mormont (1987, 1990) observed, processes of restructuring have produced multiple understandings of rural space mapped over the same territory, such that 'if what could be termed a rural question exists it no longer concerns issues of agriculture or of a particular aspect of living conditions in a rural environment, but questions concerning the specific functions of rural space and the type of development to encourage within it' (1987: 562). As such, the unifying feature between a diverse and extensive set of protests variously focused on social, economic, political or environmental objectives is a common interest in defending or promoting a particular representation of rurality. In this way, it can be argued that there is emerging a loosely cohered 'rural movement' (of many sometimes antagonistic elements), which displays the characteristics of a new social movement in informally mobilizing a loose network of diverse organizations and individuals around the common interest of rural identity (Woods, 2003).

Rural protest and campaign groups rarely draw explicitly on the language of citizenship to describe their activities, yet the motivation for their mobilization often reflects concerns with citizens' rights. First, whilst the neo-liberal state has sought to redefine citizenship by reducing the emphasis on social rights, citizens themselves have reacted by mobilizing in defence of perceived social entitlements, including public education and health services, subsidised transport and postal services and, more loosely, amenities such as village shops and street-lighting. The reactive defence of public services is not exclusive to rural areas, but it often carries a greater pertinence in a rural context where deregulation has exposed services to market forces that deem their delivery uneconomic in remote and sparsely populated regions. As Gray and Lawrence (2001) observe in Australia, a 'domino effect' of collapsing services 'accelerates as the most basic services disappear and the towns suffer the tyranny of centralized bureaucratic decision-making. Basically, "critical mass" formulations come into effect. Services are removed when governments [or private corporations] deem the level of demand is below that for "efficient" delivery' (2001: 104).

Furthermore, many of the services and amenities threatened by deregulation and rationalization are also central to discursive constructs of rural community, such that their withdrawal is seen as undermining the rural experience. Mormont (1987), for example, describes how the proposed closure of a village school was to become an issue around which 'rural struggles' could be mobilized, 'insofar as inhabitants not only felt deprived of a service to which they considered they were entitled, but also of a local institution with which they could identify' (1987: 564).

Citizen action in support of rural public services is not restricted to the defence of existing service levels, but may also claim an entitlement to additional state intervention in response to perceived new needs resulting from rural restructuring. For instance, a perceived heightened threat from crime in rural localities has generated demands for more policing as a right of citizenship (see Yarwood and Gardner, 2000). Similarly, discourses of national citizenship have been employed to justify expectations of equality in service provision between rural and urban localities. Such expectations have informed the agenda of the conventionally sectorally organized rural lobby groups in the United States which succeeded in identifying rural development objectives with infrastructural development towards a uniform provision of health, education, water, electricity, transport and so on (see Browne, 2001; Lapping et al., 1989). Demands for urban levels of services in rural locations have also been advanced at a local level by recent in-migrants, used to street-lighting or sidewalks or traffic speed controls, who consequently campaign to introduce such measures in their new homes – a dimension of counter-urbanization identified in the UK by Newby et al. (1978) as early as the 1970s.

Secondly, some rural protests have claimed a right of protection by the state, not in a contemporary context from military threats, but from adverse economic pressures. This is particularly notable with respect to agriculture where the cumulative effects of trade liberalization, globalization, farm modernization and productivist expansion, food and animal welfare scares and changing consumer practices has pushed many smaller and more marginal farmers in to precarious positions financially and sparked a sometimes militant response. Campaigns by groups such as the National Family Farm Coalition and the Farm Defenders in the USA, and the direct-action protests of Farmers for Action in the UK and the *confédération paysanne* in France, may illustrate the complexity of modern rural politics in targeting supermarkets, food processors and oil companies alongside governments, but their strategy is underpinned by an expectation that the state has a *duty* to intervene to support farmers. Often, this expectation is discursively paired with an assertion that farmers have historically discharged their civic responsibility by feeding the nation. Significantly, both the discourses and the demands of such agricultural protest movements tend to be scaled on the nation-state. There is little in the way of cooperation or collaboration between farmers' organizations in different countries to defend their collective interests against the actions of the World Trade Organization, or multinational food processing and retailing corporations, or industrial agriculture. Rather, demands tend to be for protectionist policies to privilege home-grown foodstuffs, whilst simultaneously and paradoxically also demanding that national governments act to guarantee their own export markets.

Thirdly, rural campaigners have emphasized the civil rights that restrict the legitimate power of the state over individuals. Actions by governments to regulate planning and development, to introduce conservation measures, and, most notably, to control or prohibit forms of hunting, have been portrayed as attacks on individual liberty. In the UK, the Countryside Alliance's campaign against legislation to ban the hunting of wild mammals with hounds, has been in part articulated through the language of liberty and freedom. The organization adopted the motto 'Life and Liberty' and one group of long-distance marchers to the Countryside Rally in London in July 1997 followed a banner inscribed with words from the Welsh national anthem – 'For freedom they shed their blood' (Woods, 1998c, 2004). Moreover, the proposed legislation was presented not just as an attack on the rights of hunters, but also as a precedent for a broader erosion of citizenship rights. In doing so, the Countryside Alliance subverted the conventional association of citizenship with the city by positioning the countryside as the repository of a more elemental citizenship with an intrinsic commitment to liberty born from tradition and evasion of polluting foreign influences.

This representation echoes the distinction drawn by the American right between 'natural citizens' and 'fourth amendment citizens' – the former being born into citizenship of the US state of their birth, but, according to the militia movement, having no obligation to the federal US government or its laws; and the latter being citizens who received their citizenship through the Bill of Rights (Dyer, 1998). 'Fourth amendment citizens' therefore include all immigrants to the United States accepted into citizenship, but also, it is argued, the majority of 'natural citizens' who have consented to the federal government's authority by completing census forms, registering births, applying for driving licences, paying taxes, and so on. The militia movement therefore seeks a revival of the perceived 'superior' condition of 'natural citizenship' by encouraging citizens to rescind their 'contractual relations' with federal government and resisting attempts by the federal government to impose what is regarded as its illegitimate authority. In this discourse, rural areas are positioned both as, again, the repository of the values of 'natural citizenship' and as spaces within which the self-sufficient communities required by 'natural citizenship' might be planted. At the same time, the perceived failure of the federal government to adequately address problems of economic depression and poverty in rural America has provided fertile ground for the ideas of the militant right. As Dyer (1998) remarks (see also Kimmel and Ferber, 2000; Stock, 1996):

> Antigovernment behavior in rural America is becoming increasingly tied to the idea of the sovereignty of the individual – the belief that citizens can take certain steps to legally remove themselves from the authority of the current federal government. The idea is that if the federal government won't help rural America, then rural America will simply govern itself by ignoring federal authority. (Dyer, 1998: 174)

The American militia movement represents an extreme reaction to rural social, economic and cultural change, but it also highlights the political volatility of rural regions. As historical power structures and governmental regimes that emphasized social order and stability have been undermined, so new forms of political organization have been sought. Central to this process has been the dissemination of the discourse of citizenship within rural societies with its message of empowerment for individuals and communities. However, whilst this has on the one hand permitted a partial withdrawal of the state by 'empowering' citizens and communities to assume greater responsibility for their own self-governance, it has also 'empowered' rural citizens to mobilize in opposition to the actions of external agents, including the state, and in defence of their communities. These contrasting routes offer different strategies for 'community resistance' to forces of globalization, social modernization and economic liberalism which, left unchecked, might bring into question the continuing viability of some rural communities and rural practices (see Gray and Lawrence, 2001; also Gray, 1991; Pawson and Scott, 1992). The processes by which any one strategy is adopted by a community or a group of rural citizens clearly forms an issue for future research that is as yet little explored. Furthermore, by adopting citizenship as a lens through which both rural protests and rural community action might be observed, attention is drawn to common questions about the discursive settings of such developments, and about the micro-politics that pervade their constitution and implementation.

TOWARDS A RESEARCH AGENDA

The new critical politics of citizenship are already reshaping the political relations of rural societies, both in terms of internal power structures and systems of governance, and in terms of external relations with the state, corporations and urban society. The emergence of the new critical politics requires researchers to adopt new frameworks for the analysis of power and politics in rural society. In particular, research needs to explore how the new relations of rural citizenship are constituted and how political action by rural citizens is mobilized. Some of the issues to be addressed are shared with research on the new urban citizenship and new urban governance, including issues about accountability and democracy. Others have a particular significance within rural society. For example, the articulations of place identity that constitute the spaces of dependence (Cox, 1998) for rural citizenship are arguably more complex than those that exist for urban citizenship, especially if 'rurality' itself is represented as a basis for citizen identity. Equally, issues of leadership in rural citizenship must be examined against the historic backdrop of paternalism and twentieth-century models of community organization that have followed a different trajectory to patterns of political participation in urban societies. This final section of the chapter discusses these issues further, indicating the elements that might be drawn into a new research agenda.

Place identity and rural citizenship

It was noted earlier that citizenship is a reciprocal relationship between the individual and the state that is performed through the exchange of rights and responsibilities within specific communities (see also Faulks, 2000). Communities need not necessarily be geographical places, but in the context of citizenship they most commonly are. In legalistic terms, citizenship is ascribed by a state with a territorial expression, and a citizen's location inside or outside that territory will determine the immediate meaning of that citizenship (for example, entitlement to health care or to diplomatic assistance). In terms of everyday practice, rights and responsibilities tend to be enacted within geographical communities of work or residence. Ironically, however, the rise of the 'new rural citizenship' has coincided with the weakening of traditional rural community interactions. From this perspective, the assertion of the citizenship discourse can also be read as an attempt to reconnect individuals to their geographical communities. However, the communities that are revived in this way are very different to the rural communities of the past. Idyllized discourses of rurality may play a significant role in influencing their new identity, but they are being remade in a world where essentially 'urban' cultures and lifestyle patterns prevail.

French rural researchers François Poulle and Yves Gorgeu (1997) described this as the creation of an '*urbanité rurale*' (rural urbanity), which they see as directly complicit in the construction of rural citizenship. Returning to Weber's understanding of citizenship within his model of the modern Western city, Poulle and Gorgeu identify similar features in the reconstituted countryside. Noting that in Weber's conception urbanity was never intended as a property only of towns, they label this phenomenon *l'urbanité rurale*, and contend that rural development initiatives and new forms of governance which encourage local residents to participate in the decision-making life of their community are forging a new rural citizenship equivalent to that traditionally associated with cities (Morquay, 2001). The *urbanité rurale* thesis identifies a parallel for the oaths of confederation of the Weberian city in the *intercommunualité* accords by which French communes agree to collaborate across a defined territory to achieve objectives aimed at the enhancement of that territory to a specifically defined model. Local citizens are hence enrolled into a particular development trajectory through consent to a particular representation of their community. Transposed to Anglophone countries, a similar process can be identified as occurring through local community plans and surveys, cooperative agreements, village appraisals and planning for real exercises, where local residents are enrolled in strategic discussion about the physical and discursive shape of a territory and thus in identifying priorities for the future governance of the territory.

The concept of *l'urbanité rurale* has a particular resonance with Schroeder's (1980) thesis that the character of American small towns is 'related to tenuous, informal agreements among a plurality of residents about what is most important to the town' (Norris-Baker, 1999: 253), and that as towns are faced with threats from social and economic change, these 'covenants' serve to mobilize citizens to act in defence of aspects of the community which they deem as worthy of protection.

A related analysis is contained within the concept of endogenous development advanced by Ray (1998a, 1999a, 1999b). Endogenous development is here positioned as a characteristic of the era of reflexive modernity in which understandings of place and identity are produced through dialogues between the local and the extra-local (Giddens, 1994; Ray, 1999a; see also Bonanno, 2000). It therefore looks both inwards to engage residents in the cultivation of a community's own 'development repertoire', including the local ownership of resources and local choice over the deployment of resources, and outwards to establish rights to act (Ray, 1999b). As in the concept of *l'urbanité rurale*, the model of endogenous development involves the redefinition of a rural territory through the negotiation of a development trajectory which draws in local citizens to perform roles of problem identification, target-setting and project implementation. The principle of endogeneity demands active citizenship as a condition of its operation. Ray (1998a, 2000; see also Shucksmith, 2000), particularly applies the concept to the European Union's LEADER programme, which is at least in theory based on the principle of 'bottom-up development' – although the depth of citizen engagement might be questioned. Equally, however, features of endogenous development might be identified in grass-roots ecosystem management movements in the western United States (Swanson, 2001), in the Australian Landcare programme (Lockie, 1999; Sobels et al., 2001), and in British, North American and Australian small town regeneration initiatives.

Both the endogenous development and *l'urbanité rurale* concepts focus on citizen engagement at a local scale – although both also allow for the rescaling of territories of governance as part of the process of reconstructing the identity of communities (exemplified by *intercommunualité* arrangements and LEADER group territories respectively – Buller, 2000; Poulle and Gorgeu, 1997; Ray, 1998b). However, the scale at which citizenship is performed has become an increasingly complex question as classical assumptions about the enactment of citizenship within communities of everyday interaction cannot be sustained. Active citizenship and rural protest may connect individuals to particular geographical communities, but these need not necessarily be the traditional units of village, valley, township and so on. The construction of new scales and communities of political identification in the reconstituted rural forms one avenue for future research. Moreover, it could be argued that the types of citizen engagement described above simply constitute a local citizenship that just happens to take place within a 'rural' area. The mobilization of a truly 'rural citizenship', practised at the scale of the amorphous and ambiguous entity known as 'the rural', is far more difficult. This remains the major obstacle faced by the nascent, and as yet fractious, rural movement (Woods, 2003) – but within that movement there are groups which claim to represent rural interests as a whole, and which therefore need to engage individuals in political action at that scale as *rural citizens*. Significantly, the most notable of these, the British Countryside Alliance, ran an exercise on its website in early 2002 asking its members to comment on the definition and classification of 'rural' and 'rural areas', and the characteristics of 'being rural'.

Mobilization and leadership

Active citizenship may be practised within communities, but it essentially involves the mobilization of individual citizens. Previous paternalistic structures of rural leadership relied upon the active participation in governing and policy-making processes of an elite few, whose assumption of leadership status often resulted from family tradition or from expectations which followed from their position as significant landowners or in visible occupations such as postmasters, bank managers and clergy (Grant, 1977a, 1977b; Morel, 1977; Newby et al., 1978; Woods, 1997). Newer forms of governance have necessitated the enrolment into leadership roles of a far larger number of rural residents – and have relied on the discourse of 'active citizenship' to persuade people that they have a responsibility to participate. Nevertheless, empirical evidence suggests that in many rural areas there is a shortage of citizens willing to take on more responsible leadership roles; and that many new community 'leaders' feel ill-equipped to engage effectively with more experienced political actors and professionals (see, for example, Edwards and Woods, 2004; Edwards et al., 2000, 2003).

In some places the recognition that new rural community leaders might not emerge organically, but that active citizens might need to be trained in leadership skills, has been addressed through schemes explicitly aimed at 'building community leaders'. Leadership development initiatives operate in a number of parts of the United States, many under the umbrella of the W.K. Kellogg Foundation's Rural America Initiative. These include the Environmental Partnerships in Communities (EPIC) leadership training programme in Vermont, which involves a ten-week series of evening meetings and weekend retreats, with participants nominated by existing leaders in rural communities (Richardson, 2000). In the workshops, 'participants role play, practice coping with tensions over local issues (such as the siting of a landfill), learn how to access resources, learn how to give a television interview, write a press release, and persuade a reporter to write about their work' (Richardson, 2000: 112–113). A similar model has been developed in Australia, for example in the Queensland Building Rural Leaders programme. Funded by the Queensland government as a training scheme, the programme involves a stand-alone introductory module on rural community change, leadership styles, personality and communication skills, and five follow-up modules on goal-setting and personal growth, team-building, strategic planning, creative thinking, business development and media and presentation skill enhancement (Herbert-Cheshire and Lawrence, 2001).

Significantly, the syllabi of both the Vermont and the Queensland schemes emphasize business, communications and campaigning skills as the necessary attributes of a 'rural leader'. In this way, it may be argued that the leadership development programmes are engaged in a reconstitution of rural leadership, in which the role model is no longer the squire, but the animateur or the entrepreneur. Leadership is hence envisaged not as a paternalistic or coercive relationship, but rather as an act of *facilitation*. As Richardson opines,

> strategic rural leadership is collaborative. It involves a diversity of people, opinions, and perspectives and is grounded in an understanding of the physical and cultural environments of the local communities. Such leadership can understand, at least to some extent, both the component parts and the whole system of the rural community in its context, local and global. (2000: 87–88)

The promotion of a new model of rural leadership provides a number of opportunities for research, not just in an evaluative manner, but also to explore the remoulding of power relations within a community that may result, the dynamics of enrolling both endogenous and exogenous resources (and, indeed, the role of both endogenous and exogenous leaders), and the motivations, self-definition and self-reflection of those individuals who become drawn into leadership roles in this way. The concept of citizenship can help to provide insight on questions such as whether the practice of self-awareness is fundamental to the practice of citizenship (Touraine, 1992). Self-awareness in this sense involves the realization by the individual of their particular social, economic and political circumstance, and the subsequent self-identification and articulation of their best interests, goals and objectives. For example, Deléage (2001) identifies the formation of a sustainable agriculture movement in Brittany as an example of the construction of a 'new citizenship' as it requires the self-conscious detachment of farmers from old corporatist alignments and their participation in a new activist network based on the articulation of a new community of interest.

A further issue arises in the extent to which 'new' rural leaders seek to engage with the established structures and institutions of rural government. In some places, local councils and other elements of elected local government provide a vehicle through which new leadership groups emerge and function. In others, traditional local government is perceived to lack the capacity or the legitimacy to act and the emergence of the new rural citizenship has been identified with the development of non-governmental community associations, partnerships and campaign groups. In most localities the nascent structure of facilitative leadership involves a combination of both formal, established, local government bodies and civil society groups – indeed, Welch (2002) contends that the legitimacy of rural local government now depends on its engagement with voluntary and community sector actors. The processes by which such local regimes of practice are constituted, the decisions made by individual citizens about their own personal pathways of participation, and the factors that produce different amalgams of state and civil society activity in different localities are all key areas for future research.

Representation and democracy

Concepts such as active citizenship are full of democratic promise, implying a redistribution of power back from the state to the people. Initiatives of community engagement, self-government, partnership and leadership development are all promoted with the rhetoric of empowerment and self-determination. Citizens' protests, whether over farming or fuel prices, hunting or school closures, have been heralded as part of a new direct democracy where 'people power' bypasses supposedly moribund elected representatives. Yet, all involve a retreat from two traditional cornerstones of citizenship in liberal democracies – the principle of one citizen, one vote, and the accountability of political leaders to citizens through the ballot box.

Untangling the new relations of power, authority and representation thrown up by the 'new rural citizenship' hence forms a third potential avenue for research. First, more needs to be known about the social and demographic characteristics of those 'active citizens' who have assumed leadership roles in rural community governance and in rural protest movements. Anecdotal evidence suggests that whilst there has undoubtedly been a broadening of participation, not all parts of a community are equally engaged. Certainly, the kinds of skills and personal resources identified as desired attributes for 'new' rural leaders – such as communication and organizational skills – tend to favour people with particular educational and occupational backgrounds. Where 'leaders' are self-declared, questions might be asked about how they subsequently establish their authority to lead within a community. Conversely, studies in Missouri and southern England (O'Brien et al., 1998; Woods, 1998d) have both shown that elite networks remain a key feature of rural community leadership, becoming used for the recruitment of new 'leaders' and thus further reinforcing social biases. Frequently, an emphasis is placed on 'consensual leadership', yet little is known about how decision-making processes actually operate in these contexts and how leaders, whether they emerge by self-volunteering or by appointment, are accountable to the communities they purport to represent or act on behalf of.

Secondly, if 'active citizens' are drawn from only particular sections of a community, issues arise over the representations of the community they promote and the consequences for policy formulation. Proposals adopted as a result of citizen engagement are sometimes opposed by other local citizens who have rationalized the priorities for the community differently – but in these cases, the existence of conflict points to at least two mobilizations of active citizenship. More interesting, perhaps, are the silences that prevail when excluded groups are unable to challenge dominant representations of place, either because they cannot mobilize the resources to compete, or because they have been defined as outside the concerned local citizenry. Cloke et al. (2000), for example, demonstrate how a collaborative initiative to respond to homelessness in Taunton, England, privileged the representation of citizens' interests advanced by local business, and excluded the voices of local homeless people.

CONCLUSION

The emergence of a new critical politics of rural citizenship has undermined many conventional academic representations of rural politics and society. Assumptions that rural areas are effectively devoid of political conflict and therefore there is little for political research to study, and models of rural communities as stable, hierarchically structured, paternalistic societies are no longer sustainable in an age when instances of rural community action and rural protests are highly visible across the developed world. As the rural political canvas is being redrawn, the opportunities for research are considerable. More needs to be known about why these changes in rural political practice have come about, and how they have been activated in particular localities; about the policy contexts in which they have been developed; about the mobilization of rural actors, either in protest movements or in community action initiatives; and about issues of place, power, leadership, representation and accountability, among others.

Research on these themes has the potential to draw on a number of theoretical perspectives. First, theories of citizenship can inform analysis of the constitution of citizenship and identity in a rural context, the rights and responsibilities that are attributed to citizenship, and the spaces and scales through which citizenship is performed. In particular, insights can be gained from recent work on citizenship in a cosmopolitan world (Isin, 2002) and on 'multi-layered' citizenship which is constituted through multiple group identities (Yuval-Davis, 1999, 2000). Secondly, theories of governmentality have already been employed to analyse the normative use of citizenship in the redistribution of responsibilities between the state, communities and individuals (Herbert-Cheshire, 2000; Murdoch, 1997) and there is scope for this work to be extended. Thirdly, research on citizen mobilization in rural protest could be positioned within the conceptual framework of work on social movements and resistance more broadly (Woods, 2003).

By engaging with these broader theoretical perspectives, a framework can be developed for the analysis of the new critical politics of rural citizenship that is fully sensitized to the particular historical, cultural, social and economic dynamics that frame the contemporary rural experience. Such an approach would inform rural studies as a whole, as the concerns of the new critical politics of rural citizenship are central to so many elements of rural life. The shifting of responsibility for the governance of rural communities to 'active citizens', for example, has implications for the provision of public services in rural areas and for strategies of economic development and community regeneration. The mobilization of rural citizens in protest events, meanwhile, will influence policy decisions with respect to agriculture, conservation, infrastructure provision and rural culture, as well as to the broader canvas of rural–urban relations. An understanding of the dynamics of rural citizenship is hence fundamental to the study of rural societies in the new century.

REFERENCES

Aigner, S.M., Flora, C.B. and Hernandez, J.M. (2001) The premise and promise of citizenship and civil society for renewing democracies and empowering sustainable communities. *Sociological Inquiry*, 71 (4): 493–507.

Bonanno, A. (2000) The crisis of representation: the limits of liberal democracy in the global era. *Journal of Rural Studies*, 16 (3): 305–323.

Bromley, D. and Hodge, I. (1990) Private property rights and presumptive entitlements: reconsidering the premises of rural policy. *European Review of Agricultural Economics*, 17: 197–214.

Browne, W.P. (2001) Rural failure: the linkage between policy and lobbies. *Policy Studies Journal*, 29 (1): 108–117.

Buller, H. (2000) Re-creating rural territories: LEADER in France. *Sociologia Ruralis*, 40 (2): 190–199.

Cloke, P., Milbourne, P. and Widdowfield, R. (2000) Partnership and policy networks in rural local governance: homelessness in Taunton. *Public Administration*, 78 (1): 111–133.

Cox, K.R. (1998) Spaces of dependence, spaces of engagement and the politics of scale, or: looking for local politics. *Political Geography*, 17 (1): 1–24.

CPRL (2001) Citizens for the Preservation of Rural Lifestyle website; <http://www.cprl.com>. Accessed 8 November 2001.

Deléage, E. (2001) New forms of rural citizenship: the example of the Sustainable Agriculture Network. Paper presented to the European Congress for Rural Sociology, Dijon, France, September.

DoE (1995) *Rural England*. London: Department for the Environment.

Dunn, K., Thompson, S., Hanna, B., Murphy, P. and Burnley, I. (2001) Multicultural policy within local government in Australia. *Urban Studies*, 38 (13): 2477–2494.

Dyer, J. (1998) *Harvest of Rage*. Boulder, CO: Westview Press.

Edwards, B. (1998) Charting the discourse of community action: perspectives from practice in rural Wales. *Journal of Rural Studies*, 14 (1): 63–78.

Edwards, B. and Woods, M. (2004) Mobilising the local: community, participation and governance, in L. Holloway and M. Kneafsey (eds), *Geographies of Rural Cultures and Societies*. Aldershot: Ashgate.

Edwards, B., Goodwin, M., Pemberton, S. and Woods, M. (2000) *Partnership Working in Rural Regeneration*. Bristol: The Policy Press.

Edwards, B., Goodwin, M. and Woods, M. (2003) Citizenship, community and participation in small towns: a case study of regeneration partnerships, in R. Imrie and M. Raco (eds), *Urban Renaissance: New Labour, Community and Urban Policy*. Bristol: The Policy Press. pp. 181–204.

Esping-Anderson, G. (1990) *The Three Worlds of Welfare Capitalism*. Cambridge: Polity Press.

Faulks, K. (2000) *Citizenship*. London: Routledge.

Foucault, M. (1991) Governmentality, in G. Burchell, C. Gordon and P. Miller (eds), *The Foucault Effect: Studies in Governmentality*. London: Harvester Wheatsheaf.

Giddens, A. (1994) Living in a post-traditional society, in A. Beck, A. Giddens and S. Lash (eds), *Reflexive Modernisation: Politics, Tradition and Aesthetics in the Modern Social Order*. Cambridge: Polity Press.

Gittell, R. and Vidal, A. (1998) *Community Organizing: Building Social Capital as a Development Strategy*. Thousand Oaks, CA: Sage.

Goodwin, M. (1998) The governance of rural areas: some emerging research issues and agendas. *Journal of Rural Studies*, 14 (1): 5–12.

Gorman, A. (2000) Otherness and citizenship: towards a politics of the plural community, in D. Bell and A. Haddour (eds), *City Visions*. Harlow: Pearson.

Grant, W. (1977a) *Independent Local Politics in England and Wales*. Farnborough: Saxon House.

Grant, W. (1977b) *The Role Perceptions of Rural Councillors*. University of Warwick, Department of Politics, Working Paper No. 12. Coventry: University of Warwick.

Gray, I. (1991) *Politics in Place: Social Power Relations in an Australian Country Town*. Cambridge: Cambridge University Press.

Gray, I. and Lawrence, G. (2001) *A Future for Regional Australia: Escaping Global Misfortune*. Cambridge: Cambridge University Press.

Halfacree, K. (1996a) Displacing the rural idyll, in C. Watkins (ed.), *Rights of Way: Policy, Culture and Management*. London: Cassell.

Halfacree, K. (1996b) Out of place in the countryside: travellers and the 'rural idyll'. *Antipode*, 29: 42–71.

Herbert-Cheshire, L. (2000) Contemporary strategies for rural community development in Australia: a governmentality perspective. *Journal of Rural Studies*, 16: 203–215.

Herbert-Cheshire, L. and Lawrence, G. (2001) Creating the active citizen: fostering entrepreneurship in rural Australia. Paper presented to the European Congress for Rural Sociology, Dijon, France, September.

Isin, E. (2002) *Being Political: Genealogies of Citizenship*. Minneapolis, MN: University of Minnesota Press.

Kimmel, M. and Ferber, A.L. (2000) White men are this nation: right-wing militias and the restoration of rural American masculinity. *Rural Sociology*, 65 (4): 582–604.

King, D.S. (1987) *The New Right: Politics, Markets and Citizenship*. Basingstoke: Macmillan.

Lapping, M.B., Daniels, T.L. and Keller, J.W. (1989) *Rural Planning and Development in the United States*. New York: Guilford Press.

Lockie, S. (1999) The state, rural environments and globalisation: 'action at a distance' via the Australian Landcare program. *Environment and Planning A*, 31 (4): 597–611.

Lowndes, V. (1995) Citizenship and urban politics, in D. Judge, G. Stoker and H. Wolman (eds), *Theories of Urban Politics*. London and Thousand Oaks, CA: Sage. pp. 160–180.

Marsden, T., Murdoch, J., Lowe, P., Munton, R. and Flynn, A. (1993) *Constructing the Countryside*. London: UCL Press.

Marshall, T. (1950) *Citizenship and Social Class*. Cambridge: Cambridge University Press.

Marshall, T. (1964) *Class, Citizenship and Social Development*. Chicago, IL: University of Chicago Press.

Mattson, G.A. (1997) Redefining the American small town: community governance. *Journal of Rural Studies*, 13 (1): 121–130.

Menahem, G. (2001) Urban civic worlds: a conceptual and empirical exploration. *Local Government Studies*, 27 (1): 37–60.

Morel, A. (1977) Power and ideology in the village community of Picardy: past and present, in R. Forster and O. Ranum (eds), *Rural Society in France*. Baltimore, MD: Johns Hopkins University Press. pp. 107–125.

Mormont, M. (1987) The emergence of rural struggles and their ideological effects. *International Journal of Urban and Regional Research*, 7 (4): 559–575.

Mormont, M. (1990) Who is rural? Or, how to be rural: towards a sociology of the rural, in T. Marsden, P. Lowe and S. Whatmore (eds), *Rural Restructuring: Global Processes and Their Responses*. London: David Fulton.

Morquay, P. (2001) L'invention de nouveaux territories: une urbanité rurale, in J-P. Deffontaines and J-P Prod'homme (eds), *Territoires et acteurs du développement local: des nouveaux lieux de démocratie*. La Tour d'Aigues: Éditions de l'Aube. pp. 133–142.

Murdoch, J. (1997) The shifting territory of government: some insights from the Rural White Paper, *Area*, 29: 109–118.

Murdoch, J. and Ward, N. (1997) Governmentality and territoriality: the statistical manufacture of Britain's national farm. *Political Geography*, 16 (4): 307–324.

Newby, H., Saunders, P., Bell, C. and Rose, D. (1978) *Property, Paternalism and Power*. London: Hutchinson.

Norris-Baker, C. (1999) Aging on the old frontier and the new: a behavior setting approach to the declining small towns of the Midwest. *Environment and Behavior*, 31 (2): 240–258.

O'Brien, D.J., Raedeke, A. and Hassinger, E.W. (1998) The social networks of leaders in more or less viable communities six years later: a research note. *Rural Sociology*, 63 (1): 109–127.

Parker, G. (1996) ELMs disease: stewardship, corporatism and citizenship in the English countryside. *Journal of Rural Studies*, 12 (4): 399–411.

Parker, G. (1999a) Rights, the environment and Part V of the Criminal Justice and Public Order Act 1994. *Area*, 31: 75–80.

Parker, G. (1999b) The role of the consumer-citizen in environmental protest in the 1990s. *Space and Polity*, 3 (1): 67–83.

Parker, G. (1999c) Rights, symbolic violence and the micro-politics of the rural: the case of the Parish Paths Partnership Scheme. *Environment and Planning A*, 31: 1207–1222.

Parker, G. (2002) *Citizenships, Contingency and the Countryside: Rights, Culture, Land and the Environment*. London: Routledge.

Parker, G. and Ravenscroft, N. (2001) Land, rights and the gift: the Countryside and Rights of Way Act 2000 and the negotiation of citizenship. *Sociologia Ruralis*, 41 (3): 381–398.

Pawson, E. and Scott, G. (1992) Community responses on the West Coast, in S. Britton, R. Le Heron and E. Pawson (eds), *Changing Places in New Zealand*. Christchurch, NZ: New Zealand Geographical Society.

Phillips, M. (2000) Landscapes of defence, exclusivity and leisure: the rise of rural private communities in North Carolina, in J.R. Gold and G. Revill (eds), *Landscapes of Defence*. London: Prentice Hall.

Poulle, F. and Gorgeu, Y. (1997) *Essai sur l'urbanité rurale: cinq territoires ruraux, leurs serments et leurs modes de gouvernement*. Paris: Syros.

Raco, M. and Imrie, R. (2000) Governmentality and rights and responsibilities in urban policy. *Environment and Planning A*, 32 (12): 2187–2204.

Ramet, S. (1996) Nationalism and the idiocy of the countryside: the case of Serbia. *Ethnic and Racial Studies*, 19 (1): 70–87.

Ravenscroft, N. (1998) Rights, citizenship and access to the countryside. *Space and Polity*, 2 (1): 33–48.

Ray, C. (1998a) Culture, intellectual property and territorial rural development. *Sociologia Ruralis*, 38 (1): 3–20.

Ray, C. (1998b) Territory, structures and interpretation – two case studies of the European Union's LEADER I Programme. *Journal of Rural Studies*, 14 (1): 79–88.

Ray, C. (1999a) Endogenous development in the era of reflexive modernity. *Journal of Rural Studies*, 15 (3): 257–267.

Ray, C. (1999b) Towards a meta-framework of endogenous development: repertoires, paths, democracy and rights. *Sociologia Ruralis*, 39 (4): 521–537.

Ray, C. (2000) The EU LEADER programme: rural development laboratory. *Sociologia Ruralis*, 40 (2): 163–171.

Ribchester, C. and Edwards, B. (1999) The centre and the local: policy and practice in rural education provision. *Journal of Rural Studies*, 15 (1): 49–63.

Richardson, J. (2000) *Partnerships in Communities: Re-weaving the Fabric of Rural America*. Washington, DC: Island Press.

Rose, N. (1993) Government, authority and expertise in advanced liberalism. *Economy and Society*, 22: 283–299.

Rose, N. (1996a) Governing 'advanced liberal democracies', in A. Barry, T. Osborne and N. Rose (eds), *Foucault and Political Reason*. London: UCL Press. pp. 37–64.

Rose, N. (1996b) The death of the social? Re-figuring the territory of government. *Economy and Society*, 25 (3): 327–356.

Rousseau, J-J. (1993) *The Social Contract and Discourses*. London: Dent.

Schroeder, F. (1980) Types of American towns and how to read them. *Southern Quarterly*, 19: 104–135.

Shucksmith, M. (2000) Endogenous development, social capital and social inclusion: perspectives from LEADER in the UK. *Sociologia Ruralis*, 40 (2): 208–218.

Smith, S. (1989) Society, space and citizenship: human geography for the new times? *Transactions of the Institute of British Geographers*, 14: 144–156.

Sobels, J., Curtis, A. and Lockie, S. (2001) The role of Landcare group networks in rural Australia: exploring the contribution of social capital. *Journal of Rural Studies*, 17 (3): 265–276.

Stock, C.M. (1996) *Rural Radicals: Righteous Rage in the American Grain*. Ithaca, NY: Cornell University Press.

Swanson, L.E. (2001) Rural policy and direct local participation: democracy, inclusiveness, collective agency and locality-based policy. *Rural Sociology*, 66 (1): 1–21.

Tocqueville, A. de (1946 [1835]) *Democracy in America*. London: Oxford University Press.

Touraine, A. (1992) *Critique de la modernité*. Paris: Fayard.

Urry, J. (1995) *Consuming Places*. London: Routledge.

Urry, J. (2000) *Sociology beyond Societies: Mobilities for the Twenty-first Century*. London: Routledge.

Weber, M. (1958) *The City*. Glencoe, IL: Free Press.

Welch, R. (2002) Legitimacy of rural local government in the new governance environment. *Journal of Rural Studies*, 18: 443–459.

Woods, M. (1997) Discourses of power and rurality: local politics in Somerset in the twentieth century. *Political Geography*, 16 (6): 453–478.

Woods, M. (1998a) Advocating rurality? The repositioning of rural local government. *Journal of Rural Studies*, 14 (1): 13–26.

Woods, M. (1998b) Researching rural conflicts: hunting, local politics and actor-networks. *Journal of Rural Studies*, 14 (3): 321–340.

Woods, M. (1998c) The people of England speak? Nation, rurality and countryside protest. Paper presented to the Rural Economy and Society Study Group Conference, Aberystwyth, September.

Woods, M. (1998d) Rethinking elites: networks, space and local politics. *Environment and Planning A*, 30: 2101–2119.

Woods, M. (2003) Deconstructing rural protest: the emergence of a new social movement. *Journal of Rural Studies*, 19: 309–325.

Woods, M. (2004) Politics and protest in the contemporary countryside, in L. Holloway and M. Kneafsey (eds), *Geographies of Rural Cultures and Societies*. Aldershot: Ashgate.

Woods, M. and Goodwin, M. (2003) Applying the rural: governance and policy in rural areas, in P. Cloke (ed.), *Country Visions*. London: Pearson.

Yarwood, R. and Edwards, B. (1995) Voluntary action in rural areas: the case of Neighbourhood Watch. *Journal of Rural Studies*, 11 (4): 447–461.

Yarwood, R. and Gardner, G. (2000) Fear of crime, cultural threat and countryside. *Area*, 32 (4): 403–412.

Yuval-Davis, N. (1999) The multi-layered citizen: citizenship at the Age of 'Glocalization', *International Feminist Journal of Politics*, 1: 119–136.

Yuval-Davies, N. (2000) Citizenship, territoriality and the gendered construction of difference, in E. Isin (ed.), *Democracy, Citizenship and the Global City*. London: Routledge.

34

New rural social movements and agroecology

Eduardo Sevilla Guzmán and Joan Martinez-Alier

There are new movements emerging in the world in defence of agricultural policies favourable to traditional agroecological methods. The agroecological antagonism to neo-liberal globalization is described here mainly with reference to networks in Latin America (because of our own direct knowledge and participation in them), but it is a worldwide phenomenon, as shown by movements in India also described here. These movements have been born out of local resistance to seed multinationals, the degradation of ecosystems and the threats to livelihoods because of agricultural modernization. They also oppose subsidized exports of agricultural surpluses from the US and EU. These movements are based on ancient knowledge of farming systems and also on the innovations of low input agriculture. The main actors are not neo-rural postmodern organic farmers (as they might exist in the United States and Europe) but spokesmen for large rural populations, sometimes peasants, sometimes landless labourers (as the MST in Brazil). Such movements are interpreted in this chapter in the wider context of a world movement of dissidence formed by a network of networks. By 'agroecology' we refer here to a collective practice of agriculture which explicitly considers not only economic and social aspects (income, employment) but also environmental and ecological aspects (pollution, soil conservation, cycles of nutrients, energy flow). Therefore there is a link between agroecology as a practice and the science of agroecology (Altieri, 1987; Gliessman, 1998). Agroecology in our view promotes the endogenous potential of agriculture, relying on traditional peasant knowledge, though being also open to innovations that help sustainability (Sevilla Guzmán and Woodgate, 1997).

THE RISE IN LATIN AMERICA OF RURAL ARTICULATION OF DISSIDENCE AGAINST NEO-LIBERAL GLOBALIZATION

The usual explanation for the disappearance of the active agricultural population in the process of economic development is that, as agricultural productivity increases, production cannot increase *pari passu* because of a low demand for agricultural produce as a whole. Therefore, the active agricultural population decreases not only in relative but also in absolute terms, and indeed this has been the path of development – in Britain even before the First World War, in Spain since the 1960s, not yet in India. Now, however, agricultural productivity is not well calculated: nothing is deducted from the value of production on account of chemical pollution and genetic erosion, and the inputs are valued too cheaply because fossil energy is too cheap, and because unsustainable use is made of soils and some fertilizers. What the ecologically correct prices should be is unknown; the important point is that the ecological critique of the economics of agriculture opens up a large space for neo-Narodnik argument, a space that is being increasingly taken up around the world. Issues such as biodiversity conservation, threats from pesticides and energy saving are transformed into local arguments for improvements in the conditions of life and for cultural survival of peasants. Such arguments have become widespread in new networks such as the Via Campesina (the Peasant Way), which has instituted an international Peasant's Day, the 17th April, the anniversary of the massacre of

19 members of the Movement of the Landless (MST) in 1996 in El Dorado, Parà, Brazil.

The convergence of those that, at the beginning of the 1980s, were called 'revolutionary peasant unions', took place in Managua in December 1981 during the 'Continental Conference of Agrarian Reform and Peasant Movements'. There an interaction was initiated which would lead to the birth of the Continental Peasants Movement in Latin America. The different Latin American organizations (with a small European representation) thus became aware of the similarities in both their means of struggle and their ideological evolution. Such is the case of the Andalucian SOC – Sindicato de Obreros del Campo[1] (land labourers union) – and the Brazilian MST, legalized in 1984, but at work in an embryonic state in Rio Grande do Sul since 1978 (cf. Navarro, 1996; De Medeiros, 1999; Mançano Fernández, 2000; Wizniewsky, 2001). This process of convergence between indigenous and peasant organizations became more consolidated on the South American continent through the formal organization of the Latin American Congress of Peasant Organizations (CLOC) in 1994 in Peru. We would point out here that there was an interaction between the MST (as a proto-organization) and other groups in the first half of the 1980s, which became more intense in the 1990s. These first interactions involved productive experiments of an agroecological nature (Sevilla Guzmán, 1999) and the creation of the first European committees in support of the Mexican Neo-Zapatism and the MST and then those that developed around the SOC.

Probably the next step in this process of confluence of independent peasant organizations took place on 14/15 November 1984, with the Latin American Conference of Independent Peasant Organizations, organized in Mexico by the Coordinadora Nacional Plan de Ayala. Here the Peasant Confederation of Peru, the National Federation of Peasant Organizations of Ecuador, the Independent Peasant Movement of the Dominican Republic, the National Confederation of Peasant Workers Union of France, the Union of Rural Workers and the recently founded MST of Brazil exchanged experiences.

The MST started in the south of Brazil and has spread to the whole country. It has withstood violent armed repression in Paranà, Parà and other states. Its tactics consist in occupation, settlement and immediate cultivation of large properties. Some of the MST leaders also belong to the Workers' party, though the MST is more to the left. Other spaces of confluence in the dissidence process include the international exchange events convoked by the MST of Brazil in 1985 and by the FENOCI of Ecuador in 1986. In Ecuador in 1987 the First Andean Exchange Workshop of Peasant Indigenous Organizations was held. In Central America, in 1987, the COCENTRA was created and, in 1989, ASOCODE. In October of that same year indigenous and peasant organizations of the Andean region and the MST of Brazil named their continental campaign '500 years of indigenous, black and popular resistance' in Bogota, Colombia. Three continental conferences were held, as well as several meetings coordinated by different Latin American countries, with the assistance of European rural (or so-called peasant) organizations.

THE ZAPATISTA MOVEMENT AS ONE CREATOR OF THE ANTAGONISTIC RURAL DISCOURSE

The key social actor, along with the MST, in the configuration of antagonistic rural praxis and discourse was the Neo-Zapatista Movement of Chiapas. Mexican peasant agriculture was and is under increasing threat because of food imports from the United States, which increased under the NAFTA free trade treaty between the US, Canada and Mexico. Eco-Zapatism was overdue in Mexico. In the early 1990s, Guillermo Bonfil had published his deeply moving account of vanishing indigenous Mexico (Bonfil Batalla, 1998). It has now become general knowledge in Mexico that indigenous cultures and biodiversity go together (Toledo, 1996, 2000). Biodiversity is valuable even when it has no market. The Chiapas rebellion came into the open against the NAFTA on the day it became operative (1 January 1994), helping to make indigenous peasantry a political subject.

Neo-Zapatism came to signify, in 1994, a reaction against the attacks on Mexican peasant agriculture and a real incentive towards the convergence and coordination of the movements that question economic globalization and neoliberalism at world level, as well as the progressive consolidation of a new antagonistic discourse. In fact, the Zapatista movement made possible the introduction of socio-cultural diversity into the worldwide anti-neo-liberal movement's discourse (when this was in its gestation period); that is to say, the enormous diversity of subjects, territories, resources, traditions and realities that the world was made up of at the end of the twentieth century.

In an attempt to come up with a synthesis, the characteristics of Neo-Zapatism, an age-old and at the same time new social movement, are the following:

1. The acceptance of a historical continuance between its processes of collective social action and those developed by those ethnic groups which through multiple processes throughout 500 years have put up resistance to colonization and oppression generated by the expansion of the European socio-cultural identity.
2. The attribution to economic globalization and neo-liberalism in present times, of the historical oppression suffered by the indigenous communities. Specifically the foreseeable impact of the NAFTA on the indigenous communities of Chiapas, which added to their resistance to the eviction of their communities and to the subordination to the interests of the timber companies and landowners.
3. This struggle against exclusion does not end with their confrontation with the modernizing socio-economic system. They are also fighting for the recognition of the Native Indians in the Mexican constitution. The diversity of the ethnic groups which make up their movement has led them to defend the recognition of differences: 'We want a world where all worlds fit in'.
4. They demand a democracy unadulterated by external or internal mismanagement, corruption and distortion of the true participation of people. To this effect, they are Mexican patriots who oppose the 'foreign domination of North American imperialism'. Moreover, they aim to make a true democratic change to the political organization so that 'those that are in charge also have to obey'.

From the depths of the Lacandona forest, the EZLN and Subcomandante Marcos developed an 'informational strategy' to fulfil the establishment of an 'autonomous communication' to reach public opinion and to generate a process of confluence with all the groups that are excluded from the modernizing socio-economic system. With this, they not only developed a way of defending themselves with the spoken word ('We only take up arms to make a statement'), but also aimed to generate networks of dissidence to the socio-economic and cultural oppression which they suffer.

This was how the Zapatista movement, through its 'autonomous communication' made contact with the, then incipient, economic anti-globalization social movements, holding debates which took place in the context of the campaign of '50 years are enough', against the half-century of existence of global financial institutions (the International Monetary Fund, the World Bank). Demonstrations took place in different places throughout the world, culminating in the alternative forum 'The Other Voices of the Planet' which developed in Madrid in the autumn of 1994. Continuing with its dynamics of resistance and informational struggle, the EZLN called, in Spain in the summer of 1997, the Second Intergalactic Conference against Neo-liberalism and for Humanity, by means of an itinerant celebration throughout various towns and cities that had as its driving force local Zapatista committees. In Andalucia the militant members of the SOC played a central role in the organizational infrastructure of the congress, especially in the closing acts which took place in El Indiano, a large farm which was acquired after many years of struggle involving occupations and imprisonments. This was one of the agroecological experiences that the cooperatives of the SOC carried out as a 'place for reflection and sociopolitical and productive practice' (Sevilla Guzmán, 1999; Guzmán Casado et al., 2000).

THE IMPACT OF THE FTAA

The biggest and most devastating impact that, in the short term, the economic globalization process is having on peasant and family-run agriculture is caused by the policies of the freeing of international agricultural trade (Rosset, 1998) coupled with the subsidies to exports in the United States (and the EU). In this sense, the NAFTA must be contemplated within a global strategy that intends to configurate a 'Free Trade Area in America' (FTAA). It intended to deregulate the market, services and investments throughout both American continents in such a way that the multinationals had the right to use natural resources indiscriminately. Dorval Brunelle (2001) illustrated the repercussions of this deregulation with a Mexican example: 'The Mexican government had to pay 16.7 million dollars to the Californian firm Metalclad Corp., because a Mexican municipality would not authorize the installation of a hazardous waste dump against which the local population had been mobilized.' The approval of the FTAA meant the gradual elimination of any type of tariff. Therefore, products coming from the United States and Canada had free access and were exempt from custom and non-custom restrictions.

Likewise the FTAA would mean unrestricted access to bidding and contracts for public sector supply. Local companies were left in the hands of the multinational market to carry out activities linked to water and energy provision in the urban economies of Latin American countries. The third requirement of this amplification of trade centred on the patents over life and intellectual property, leaving in the hands of the multinational corporations the provision of seeds, as well as the technological packets linked to the agriculture that industrialized farming, introduced throughout the Latin American area, requires.

Thus an antagonism towards the FTAA emerged and it is still mounting. It appeared in the form of antagonism towards the FTAA from the American trade union movement and the social movements crystallizing in the appearance of a Continental Social Alliance (CSA). In fact, this process began in the ministerial meeting in Denver in 1995. The trade union movement of the 35 countries of the Americas, including Cuba, with the support of the Pan-American Regional Employment Organization (PREO) – the continental wing of the International Confederation of Free Unions – organized a parallel conference to express their mistrust of the FTAA. The following year, the American Union Movement assembled in the Colombian city of Cartagena to elaborate a document reflecting on this subject and to put pressure on government representatives. The process continued that same year in Brazil. During the meeting of the presidents of the member countries of Mercosur, where 'both the first central trade union of the USA, the AFL–CIO and the ORIT sent representatives to offer support to their South American counterparts who had reached an agreement to celebrate an international day of struggle for the workers of Mercosur'. However, it was in Belo Horizonte in 1987 that the first convergence between the American trade union movement and the civic organizations against free trade occurred. These have since worked together on an alternative project to the FTAA, The decision was taken to create a Continental Social Alliance (CSA) which would face up to the FTAA, elaborating, in a participative way, concrete and viable alternatives.

In 1998, the five existing national coalitions against free trade[2] called for the first Summit of the American People. This took place in Santiago de Chile from the 14th to the 17th of April, parallel to the 'second summit' of the leaders of the 'American States'. Environmental and feminist associations as well as several associations of alternative American social movements responded to that call. There a programmatic document was produced of great relevance to the configuration of alternatives to global neo-liberalism, 'Alternatives for America: towards an agreement between the people of the continent'. In this document it was established that:

> Trade and investments should not be an end in themselves but a means capable of guiding us towards a fair and long-lasting development. It is fundamental that citizens exercise their rights in the formulation and evaluation of the social and economic policies of the continent. The central objectives of such policies should be the promotion of economic sovereignty, the collective well-being and the reduction of inequalities on all levels. (Brunelle, 2001)

The fact that the Latin American Congress of Peasant Organizations (CLOC) was involved with this dynamic, representing of the Peasant Movement of Latin America and the Caribbean, is relevant to our line of argument.

Antagonism towards globalization in the American continent should be analysed in the much wider context of global dissidence. Here the Movement against Maastricht and Economic Globalization (MAM) and the confluence against the Multi-lateral Investment Agreement (MIA) developed parallel and confluent dynamics. In effect, from 1990 to 1995 multiple European social movements joined forces by incorporating in to their ideas and debates calls for a struggle against the rapidly developing 'Europe of Capital'. Hence, feminist, ecological, pacifist and Third World groups and all the collectives committed to the fight against poverty, with ethical and solidarity ideals, joined together, consolidating the MAM. On the other hand, the confluence against the MIA acquired special relevance in Canada, France, the United States, central Scandinavian countries and several countries from the periphery such as Malaysia, the Philippines, India and Brazil. Friends of the Earth and Le Monde Diplomatique conducted vigorous campaigns against the MIA. The joining together of these two fronts of economic anti-globalization began to interfere with the plans of global neo-liberalism. This forced a delay in the signing of the MIA, at the heart of the OECD, in Paris in October 1998, through the configuration of Global Action of the People.

THE CONSOLIDATION OF THE ANTAGONIST NETWORK AGAINST NEO-LIBERAL GLOBALIZATION: GLOBAL ACTION OF THE PEOPLE

Since the First Intergalactic Conference against Neo-liberalism and for Humanity, which took place in the Lacandona forest in the summer of 1996, and the Second Intergalactic Conference, which took place in Spain, the processes of confluence have quickened, leading to the creation of Global Action of the People (AGP) against free trade. This group was the first coordinator, on a world level, against economic globalization and neo-liberalism. In Geneva, at the beginning of 1998, the very first meeting of the AGP was attended by some three hundred activists from all over the world.

> There were representatives from the Southern periphery, of the indigenous people that inhabit the most recondite places on the planet, that suffer a threat to their habitats and territories as a result of the unstoppable expansion of globalization (the Maoris of New Zealand, the CONAIE of Ecuador, the Mayan indians, the Ogonis of Nigeria ...); also the peasant movements of those places on the planet where there still exist important contingents of population living in the traditional rural world (Nepal, India ...), as well as new peasant movements that are fighting for access to community ownership of land (MST of Brazil). There were also representatives of those metropolitan movements fighting against the consequences of the so-called plans of structural adjustment of the [International Monetary Fund] and the [World Bank] that urban populations are suffering (eg. the teachers' movement in Buenos Aires, or the movements from the slums of Mexico City). Also represented were the new workers' organizations (many of them clandestine due to repression) of the maquila industries in Central American countries, and even organizations representing people with specific problems as such is the case with certain Afro-American communities in Caribbean countries. (Fernandez Durán and Sevilla Guzmán, 1999: 365)

The dissident groups from countries of the centre of the system were also diversely represented:

> In Geneva the French unemployed movement, as well as certain organizations on the European network against unemployment, precariousness and social exclusion, attended. North American organizations that work with the homeless, also Food Not Bombs. New organizations in defence of part-time workers or those threatened by privatization processes. The squatter movement and the self-managed social centres from different European countries. In fact the meeting in Geneva was organised thanks to the active participation of the squatter movement of the city. Some direct action organizations from the ecological environment, amongst which the movement Reclaim the Street from Great Britain stood out. As well as the different groups and networks that attempt to unveil the consequences that the Maastricht treaty had on the population of the European Union countries. (1999: 366)

Why is it that such diverse social groups join forces to fight against free trade? This question can only be answered in the context of the debates that the different groups have carried out in order to identify the nature of globalization, subject to the command of the profit logic of multinational companies. The transnational joining of states, in the form of their international institutions – fundamentally the IMF, World Bank and World Trade Organization, is coactively imposing economic policies that openly impact negatively on both human work and natural resources. The large multinational corporations have been studied since the early 1990s by different social collectives and networks that have witnessed how pacifist, feminist and ecological claims have been seemingly incorporated into sales campaigns as slogans. At the same time these very same transnationals use the workforce from the periphery through the relationships they maintain with their production lines and affiliated suppliers. They exploit precariousness and child labour, impose a total absence of social benefits and a union prohibition, amongst other human rights transgressions, as well as paying wages so low that workers are unable to feed their families.

In a similar way, the dissidence against economic globalization came to the conclusion that neo-liberal politics mean a growing degradation of natural resources, revealing the commercial, financial and speculative mechanisms which pull down thousands of hectares of forest, transforming this land for the growth of crop or tree plantations, forcing indigenous groups, whose livelihood depended on the forest, to move. The uprooting of mangroves around the Tropics to the benefit of shrimp exports became an international scandal. Also, attention started to be drawn to the human and environmental damage caused by the obligation to pay external debts (emphasized by the Jubilee 2000 campaigns).

THE EMERGENCE OF AGROECOLOGY FROM THE ANTAGONISM PROCESSES TOWARDS NEO-LIBERALISM AND GLOBALIZATION

In the past few decades there have been various productive experiences that show the emergence of a new management model of natural resources, based on local knowledge and its merging with modern technologies. Many of these recreate, in some aspects, historical forms of socio-economic organization linked to socio-cultural identity.

Conventional agricultural science would not hesitate in labelling such experiences as a new paradigm of anti-modern rural development. Such experiences are dispersed worldwide (Pretty, 1995). They are born from processes of resistance in the interstices of agricultural modernization and they offer a list of productive and social strategies.

There are two social spaces where such 'productive dissidence' towards agricultural modernization can be found, according to Victor Manuel Toledo. They are 'focal points of civilizatory resistance'. The first, which he refers to as 'postmodern', is made up of 'a polychrome range of social and countercultural movements'. The second social space is located on certain 'islands of pre-modernity or pre-industriality',

> those enclaves of the planet where western civilization did not or still has not managed to impose its values, practices, corporations and modern actions. They are predominantly, although not exclusively, rural, in countries such as India, China, Egypt, Indonesia, Peru or Mexico, where the presence of various indigenous populations (made up of peasants, fishermen, shepherds and craftsmen) confirm the presence of civilizatory models different to those originated in Europe. These do not constitute immaculate archaisms, but contemporary syntheses or forms of resistance born from the encounters that have taken place in the last few centuries between the expansive force of western civilization and the ever present forces of the 'peoples without history'. (Toledo, 2000: 53)

THE EXPERIENCE OF INDIA

Elements in the movement for agroecology in the south are the collective defence of agrobiodiversity, food security and the in situ conservation or co-evolution of plant genetic resources. Thus in Mexico, beyond the neo-Zapatism born in Chiapas, a wider movement has risen since 2002 called 'En Defensa del Maíz', against maize imports from the United States. In India, as Kothari puts it (1997: 51), a single species of rice (*Oryza sativa*) collected from the wild some time in the distant past, has diversified into approximately 50,000 varieties as a result of a combination of evolutionary/habitat influences and the innovative skills of farmers. This contribution to genetic diversity is a fact that the modern seed industry conveniently sidesteps, and that the consumers of industrialized countries have ignored until recently. Mexican peasants never thought of patenting or instituting other types of intellectual property rights on the varieties of maize that have been used by the commercial seed industry.

Agricultural biopiracy is a topic which the Food and Agriculture Organization of the United Nations (FAO) has been discussing for some twenty years under the name of Farmers' Rights. Even some governments say that

> if a company takes a seed from a farmer's field, adds a gene and patents the resulting seed for sale at a profit [or otherwise 'improves' the seed by traditional methods of crossing, and then protects it under the UPOV rules], there is no reason the initial seed should be free. They also say patents ignore the contributions by indigenous peoples, who often are the true discoverers of useful plants and animals, or of farmers who improve plants over the generations. The negotiation run by the Food and Agriculture Organization [on Farmers' Rights] is weighing whether to compensate traditional farmers for work on improving crops and maintaining different varieties. (Pollack, 1999)

But, then, who wants the Third World farmers to continue growing and locally freely sharing or selling their own low-yielding, low-input seeds? From the point of view of international capitalism, replacing their seeds by commercial seeds would be more conducive to economic growth. Should not traditional seeds be forbidden on grounds of lack of sanitary or yield guarantees?

There is then a growing alarm in southern countries which are centres of agricultural biodiversity, or close neighbours to them, because of the disappearance of traditional farming. This new awareness, which goes totally against the grain of development economics, is helped by the social and cultural distance between the seed companies (often multinationals) and the local peasants and farmers. While conservation of 'wild' biodiversity in 'national parks' is seen often as a 'northern' idea imposed on the south (as to some extent is really the case), the conservation of in situ agricultural biodiversity was for many years left aside by the large wilderness northern organizations. It was pushed instead by specific NGOs such as RAFI and GRAIN, also by southern scientists and by southern groups who developed pro-peasant ideologies.

There are deliberate attempts in India by groups and individual farmers to revive agricultural diversity. In the Hemval Ghati of the Garhwal Himalaya, some farmers under the banner of the Beej Bachao Andolan (Save the Seed Movement) have been travelling in the region collecting seeds of a large diversity of crops. Many farmers grow high-input high-yield varieties for the market but also other varieties for their own families. An important issue is not only to promote the survival of many varieties of the main crops (wheat and rice) but also to keep alive other food crops that have not been subject to 'Green Revolution' seed substitution – like bajra, ramdana and jowar, and also pulses in general. In the south of the country, the somewhat grandly named 'seed satyagraha' of the Karnataka Rajya Raitha Sangha (KRRS), became well known in the early 1990s.[3]

Monsanto has used the loopholes in legislation or in effective regulation to introduce transgenic crops outside the United States. Thus, there is a debate in some parts of India against the introduction of Bt cotton (that is, cotton seeds into which the bacillus thurigiensis has been genetically engineered to act as an insecticide). In Andhra Pradesh, the farmers' movement APRS uprooted and burned two crop sites in 1998, and alerted the state parliament and government to ban further field sites, while in Karnataka the leader of the farmers' movement KRRS transparently called on the company to reveal the exact locations of its field tests of transgenic Bt cotton. Monsanto has been more successful elsewhere. There was little opposition in Argentina to transgenic soybeans (Pengue, 2000).

In India, on 30 November 1999, the first day of the WTO conference in Seattle, several thousand farmers gathered in Bangalore at the Mahatma Gandhi statue in the park. They issued a 'Quit India' notice to Monsanto, and they warned the prestigious Indian Institute of Science not to collaborate with Monsanto in research. The company was urged to leave the country or face non-violent direct action against its activities and installations. Agribusiness had already been warned with the destruction of Cargill facilities in one district back in 1993. The KRRS leaders have travelled around the world, being much involved in the anti-neo-liberal dissidence against the WTO because the new regulations on international trade bring in their wake the enforcement of property rights on commercial seeds, which unjustly do not recognize the original raw material and knowledge, while preventing farmers' local gifts or sale of such commercial seeds. In 2001 the KRRS was still trying to prevent the wholesale introduction of transgenic Bt cotton in India.

Also in India, Navdanya is a large network of farmers, environmentalists, scientists and concerned individuals which is working in different parts of the country to collect and store crop varieties, evaluate and select those with good performance, and encourage their reuse in the fields (Kothari, 1997: 60–61), certainly a more participatory strategy than that of ex situ cold storage. What other name but 'ecological neo-Narodnism' can be given to such initiatives? Reality is contradictory, and movements against Cargill and Monsanto are combined in India with movements for subsidized industrial fertilizers. However, who would have thought twenty years ago that praise for organic agriculture would be expressed not by professional ethnoecologists or agroecologists or by northern neo-rural environmentalists but by real farmers from India in international trade meetings? This is not homespun oriental wisdom combating northern agricultural technology, it is not identity politics only. On the contrary, it must be interpreted as part of an international worldwide trend with solid foundations in agroecology.

Should there be a rush in southern countries to impose intellectual property rights on crop varieties, animal races and medicinal knowledge? In India, Anil Gupta has long confronted this question with a pioneering large-scale ground-level effort to document the local communities' knowledge regarding old and innovative resource uses in the form of local registers. The objectives are manifold: the exchange of ideas between communities, the revitalization of local knowledge systems and the building up of local pride in such systems, and the protection against intellectual 'piracy' by outsiders (Kothari, 1998: 105). The protection arises because prior registration and publication would stop patenting. As Anil Gupta (1996) has said repeatedly, if somebody is to patent some properties of neem, why not ourselves, Indian farmers and scientists? The main thrust of his work, however, has been to enhance local pride in the existing processes of conservation and innovation, and to stop outside advantage being taken gratis from this work.

TOWARDS A LATIN AMERICAN AGROECOLOGICAL MOVEMENT

There is no space here (and we lack sufficient knowledge) for a review of other similar movements in countries in Asia and Africa. We shall now very briefly review some Latin American agroecological movements. In South America productive dissidence to agricultural modernization can be found in the south of Brazil, in the states of Paraná, Santa Catarina and Rio Grande do Sul and extending through Misiones up to the historical region of Gran Chaco, from the north of Argentina and Paraguay as far up as the south of Bolivia.

In Argentina, probably the most relevant agroecological experience that has so far emerged takes place in the province of Misiones.[4] Here, a peasant agroecological movement has brought together a range of productive experiences based on the 'improvement of the traditional, productive diversification, specialization in some sectors and the strengthening of production for family consumption'. Such experiences emphasize the transformation of production and the search for new markets in *ferias francas de Misiones* (fairs). With reference to the creation of one of these fairs, one of the organizers said, 'We didn't invent Ferias Francas, we are recreating an age-old experience …'. In this province, 27 fairs take place every week of the year, in which more than 2,000 farmers take part in order to sell their produce directly to customers (Carballo, 2000). Probably the most beneficial work, agroecologically speaking, that is carried out in Misiones is that of the Organic Farming Network of Misiones.

Experiences with agroecological initiatives can also be found in north Santafesino,[5] and in all of Gran Chaco. In the past few years a network of farmers and NGOs has taken shape, exchanging experiences (some with more than 20 years of experience, as is the case of INCUPO) and coordinating actions generating training courses for technicians and producers in agroecology.

In the north of the province of Santa Fe an 'agroecological week of the Santa Fe province' has been developed. Since 1998, in the city of Rosario, there have existed 'urban communitarian ecological food gardens' on *villas miseria*, which provide 'local health centres' with medicinal plants rescued from Toba knowledge (Martinez Sarasola, 1992: 441–476).

If the agroecological movement is significant in the north of Argentina it is more so in Brazil, especially in the states of Paraná (with the fundamental action of AS-PTA), Santa Catarina (with the official support of EPAGRI) and, above all, in Rio Grande do Sul where EMATER (the state organism for agricultural extension) adopted agroecology as its official policy (until 2002), declaring that the state is 'free of transgenics'. There is in Brazil today the strongest movement in the world for land reform, the MST (the Movement of the Landless), whose social origins are in Rio Grande do Sul. In 1999 the MST declared itself against transgenic crops, and in January 2001 the MST, together with Rafael Alegria and other leaders of Via Campesina, and with José Bové of the French Confederation Paysanne, became the media stars of the Porto Alegre World Social Forum when they symbolically destroyed some Monsanto experimental fields in the village of Nao-me-toques. The context was the prohibition of transgenic soybeans in Rio Grande do Sul by the state government. Even if the valiant attitude of the government and judiciary in Rio Grande do Sul against transgenic crops would finally fail because of federal overruling, it has served to propel the MST in an ecological direction. The transgenic issue has sparked off a general discussion on agricultural technology inside the MST.

The rural–urban link of the Brazilian experiences of Rio Grande do Sul is especially relevant in Porto Alegre, where a few days a week entire streets fill up with market stalls, where many cooperatives establish 'agroecological socialization links' with consumers (Caporal, 1998; Costabeber, 1998; Caporal and Costabeber, 2001). However, the Brazilian agroecological phenomenon is much more widespread, as hundreds of productive agroecological experiences can be found throughout the country (Canuto, 1998).

Similarly, in the states of Jalisco (Morales Hernández, 1999) and Michoacan (Toledo, 1991) in Mexico, there exist several experiences that through social collective action organize their production and marketing to face up to conventional markets. Likewise, in Chile, the excellent work of CET (previously in Santiago, now in Temuco), with its ramifications throughout the country and even throughout the rest of Latin America through CLADES (with its magazine *Agroecology and Development*), provide good examples of agroecological experiences, and which acquire special significance in the Mapuche territory. Also, in Colombia, a Red de Custodios de Semillas (seed wardens) exists which is composed of farmers who exchange experiences, reinforcing a recuperation of local peasant knowledge. Quite a few such alternative management proposals also have a strong indigenous content.

In the land reforms of the past 50 years, the highland peasantries of the central Andes fought against the modernization of the haciendas, which sought to get rid of them; they stayed put, and increased their holdings. There are more established communities and more community (pasture) land in the Andes now than 30 or 40 years ago. This bothers the neo-liberals. The peasantry has not yet decreased in numbers, despite migration, but now the birth rate is coming down. Will Quechua and Aymara communities survive as such? Only 40 years ago, integration and acculturation was the destiny traced for them by local modernizers (such as Galo Plaza in Ecuador) and by the US political-anthropological establishment. Their resistance today would be helped by improvement in the terms of trade for their production, if subsidized imports of agricultural products from the United States and Europe were stopped, if they could get subsidies (in the form of payments for Farmer's Rights, for instance, and subsidies for the use of solar energy), and if they could exercise organized political pressure for this purpose. We see explicitly for the first time in the Andes and also in Mesoamerica an agroecological pride which provides a foundation for an alternative development or, as Arturo Escobar would put it, for an alternative to development. If not this, what then? Should Andean peasants, with low-yielding agriculture, give up farming and livestock raising as the economy grows, give up their communities and their languages? Should then some of their grandchildren, as the economy grows still more, come back in small numbers as subsidized mountain caretakers, making music and dancing as Indians for the tourists? In the final analysis, in situ agricultural biodiversity and local food security could be assured as part of a movement which would put a much higher value also on the preservation of cultural diversity. This is what PRATEC in Peru, founded by the dissident agronomist Eduardo Grillo, tried to do, building upon the work by agronomists from remote provinces, such as Oscar Blanco who long defended cultivated species such as quinua and many tubers (the 'lost crops of the Incas') against the onslaught of imported subsidized wheat. PRATEC is romantic and extremist, but the subject it puts on the table is realistic and down-to-earth. It is not their fault that it is not considered worth the attention of multilateral banks or even of universities (Apffel-Marglin and PRATEC, 1998). In the University of San Simón de Cochabamba in Bolivia there is an Agricultural Institute (AGRUCO) which is reviving Andean peasantry agroecology (Delgado, 2002; Tapia Ponce, 2002; likewise Stephan Rist, 2002 in the University of Berne).

Farmers and peasants from the movements and experiences discussed in this chapter, from in Argentina, Brazil, Bolivia, Mexico, Chile and Colombia, met in December 1998 in Pereira, Colombia, and established a declaration of principles, as members of the Agroecological Movement of Latin America and the Caribbean (MAELA). In this declaration they expressed their 'opposition to the neo-liberal model … for its degradation of nature and society'. At the same time they established, as a right of their local organizations, the 'management and control of natural resources … without dependence on external input (agrochemical and transgenic), for the biological reproduction of their cultures', underlining its 'support of the promotion, exchange and diffusion of local experiences of civil resistance and the creation of alternatives in the use and conservation of local varieties' (MAELA, 2000). They also expressed their 'solidarity with the MST of Brazil, the peasant movements of Bolivia, the Mapuches of Chile, the indigenous peasants of Chiapas', amongst other groups, as an example of international peasantry.

A BRIEF CONCLUSION

In this chapter we have reviewed several movements in countries of the south based on an explicit agroecological awareness. These movements are very different from the small neo-rural postmodern organic farming movements of the United States and Europe. We are still far from being able to provide a complete taxonomy of such movements in the south, and in fact nobody seems yet able to provide a whole picture. So, this chapter gives some detailed information on some cases but only a very brief (superficial, and second-hand) view of other cases. However, there are some undoubted developments: a new network such as Via Campesina has arisen; many agronomists now write theses and books on agroecology based on peasant knowledge; the debates on Farmers' Rights, biopiracy, in situ coevolution of agricultural biodiversity reach public opinion. The agricultural policies of the United States and the European Union (protectionism against some imports, large export subsidies for many other products undermining world peasant agriculture) are under attack. There is a confluence of views from peasants groups in the south and from some circles in the European Union against such policies. In Europe this is characterized as the *Agrarwende*, against subsidized exports but in favour of subsidies to farmers based on the multifunctionality of agriculture. In the south, subsidies to agroecological peasants would be even more justified on grounds of in situ biodiversity conservation and coevolution, energy efficiency, food security, cultural conservation. Such a policy of subsidies would require an international agreement, perhaps based on a notion of paying back an ecological debt from north to south for so many cases of biopiracy.

Under the discussion on agroecology lurks a large question that is still outside the political and economic agenda. Has the march of agriculture in the past 150 years in Western countries been wrong? What is the agronomic advice that should be given not only in Peru or Mexico, but even more in India, in China? Should they preserve their peasantries or should they get rid of them in the process of modernization, development and urbanization? Compared to 100 years ago, and because of population growth, the number of peasants in the world has increased considerably; therefore such questions are indeed relevant.

In summary, agricultural policy should balance environmental, economic, social, cultural values at different geographical and time scales. In some interpretations, modern agriculture is characterized by lower energy efficiency, genetic and soil erosion, ground and water pollution. From another point of view, in the language of economics, modern agriculture achieves increased productivity. Another, non-equivalent, description of agricultural development will emphasize loss of indigenous cultures and knowledge. There is here a clash of scientific perspectives, also a clash of values. How to integrate the different points of view? How to decide on an agricultural policy in the presence of such opposite, legitimate points of view? The role of the rural social scientist that we have adopted is to study experiences of peasant agroecological movements and extract theoretical principles for two purposes: first, to help design participatory strategies of local development, second, to intervene in the policy discussions at higher levels on the role of agriculture in today's world. The worldwide peasant agroecological movement is now an actor in these debates, as seen very clearly in the World Social Forums of Porto Alegre both in 2001 and 2002. The wider scene is the worldwide movement against neo-liberal globalization in all its aspects (financial, trade, environment, politics), a network of networks, in which agrarian movements are just one actor. Perhaps an unexpected one.

NOTES

1 This land labourers union (SOC) was, in fact, the expression in the 1980s of the final stage of a peasant movement led by the land labourers or peasants without land who demonstrated a huge potential and capacity for struggle in Southern Spain for more than 100 years. In the 1980s there was discontent with almost total mechanization of work: coinciding with a grave industrial crisis, this meant that land labourers had little or no opportunities of alternative work. With their wish to look for new alternatives which would surpass the traditional claim to land, the SOC moved towards new social movements in general, and towards the ecological movement in particular. From the interaction of their activism with their productive experience, there emerged a clear approach of ecological management of natural resources, similar to organic farming but disagreeing with some of its styles of farming. The aspect which they most dislike is the emphasis on healthy eating and the commercial interest which organic farming shows, in contrast to the social aspects.

In its fight for land the SOC had had access to land on several farms. Some of these lands were obtained through continuous occupations and evictions which led to frequent imprisonments, and others through renting or purchase. There was always union pressure and support from the more progressive sectors of the church and the university, as well as some socio-economic and cultural institutions. This meant that, in the first half of the 1980s, the SOC was accompanied by different non-peasant groups in their many actions. These ranged from peaceful demonstrations and marches looking for support from the villages and cities on their itineraries to 'symbolic' occupations of land or other more problematic temporary take-overs of local government buildings, airports or even the Andalusian Parliament building. The ISEC (Institute for Sociology and Peasant Studies, The University of Cordoba, Spain) of the University of Cordoba has collaborated with SOC since it was founded in 1978.

2 The Alliance for Responsible Trade (ART) of the United States; Common Frontiers of Canada; the Red Mexicana de Acción contra el Libre Cambio (RMALC); the Quebec network for Continental Integration; the Red Chilena por la Iniciativa de los Pueblos (RCHIP, which is presently called the Alianza Chilena por la Iniciativa de los Pueblos, ALCIR).

3 Cf. the letter from M.D. Nanjundaswamy, 'Farmers and Dunkel Draft', *Economic and Political Weekly*, 26 June 1993, and the emailed newsletter of the KRRS. Also, Anil Gupta (1998), esp. last chapters, for a description of the KRRS up to the mid-1990s.

4 Our knowledge of this experience is due to our unforgettable friend 'el coya Cametti', with whom we shared an enriching experience in the Maestria del ISEC in la Rábida.

5 In spite of the grave social situation, disturbances, environmental degradation and the progressive depopulation of North Santafesino, there exists a wide nucleus of institutions and independent technicians which, for some years now, have made great efforts in the search for an alternative development. Many producers of the region share these ideals and some years ago started to make changes using agroecological practices. There now exists an inter-institutional articulation whose first success was the excellent diagnosis of the Chaco Argentino (1999) which was carried out by the Chaco Argentina Agroforestal Network financed by the Secretary of Natural Resources of the Argentinian central government. Incupo and Fundapaz participated in this diagnosis, thus potentiating the constitution of a Santafesina Agroforestal Board. This group developed many experiences in North Santafesino, including: (a) rotative pasture trials on forested low-lands in Vera, where FundaPaz, INTA, MAGIC Vera participated from 1994 to 1997; (b) forestry and pasture management experiences with

small producers of the Cuña Boscosa Santafesina developed by FundaPaz from 1992 to 2000; (c) development of means of protection against overpasturing of cattle, with the participation of CATIE of Costa Rica; (d) experiences with forestal plantations by means of intercalating the cultivation of local tree species by FundaPaz in 1995 and 1996; (e) recuperation of soil and impoverished natural pastures with producers of the La Cabral area, on the part of PSA, INTA San Cristóbal, with similar experiences in San Manuel – La Sarita, on the part of PSA; (f) selective thinning out of woodland with the production of ecological charcoal and the management of natural pastureland with rotative pasture in San Cristóbal on the part of INTA San Cristóbal; (g) forestry management experiences with small producers in the Colonia Piloto Villa Guillermina area, on the part of the PSA from 1997 to 2000; (h) agroforestry management experiences in the north-east Santafesino on the part of Pastoral Social de Rafaela, Incupo and FundaPaz from 1995 to 2000; (i) cataloguing of the native woodland flora of the province of Santa Fe, carried out by G.D. Marino and J.F. Pensiero, for the Subsecretary of Culture – the provincial government of Santa Fe; (j) introduction of subtropical pastures in the rainforests of the Chaco, on the part of a team which was coordinated by G.D. Marino of the university Department of Agricultural Sciences; (k) gathering of floral information of young *quebrachales* of the Cuña Boscosa Santafesina, FACA and FundaPaz; (l) cataloguing of flora and bird life in the province of Santa Fe, by members of the aforementioned university department; (m) experience of InCuPo moving from the monocrop cultivation of sugarcane to the productive recuperation of the Tacuarendí area from 1995 to 2000; (n) sustainable management of the 'coast and islands' ecosystem through diversified agroforestal production in the Romang area developed by the InCuPo from 1993 to 2000; (o) trials in the harvesting and storing of water for human consumption, of which the Pastoral Social de Rafaella was in charge from 1996 to 2000. The information gathered from these experiences led to the institutional articulation of the Agroforestal Board Santafesina, which is committed to working for and combining efforts in the preservation of the natural environments of the region, contributing with ideas and activities for productive and demographic recuperation of an agroecological nature.

REFERENCES

Altieri, M.A. (1987) *Agroecology: The Scientific Basis of Alternative Agriculture*. Boulder, CO: Westview Press.

Apffel-Marglin, F. and PRATEC (1998) *The Spirit of Regeneration: Andean Culture Confronting Western Notions of Development*. London: Zed.

Bonfil Batalla, G. (1998) *México profundo: Una civilización negada*. Mexico: Grijalbo.

Brunelle, D. (2001) 'Una alianza social desafía a Washington: Estados Unidos quiere un mercado hemisférico bajo su control' in *Le Monde Diplomatique*. April, Argentina: Edicción Cono Sur.

Canuto, J.C. (1998) 'Agricultura ecológica en Brasil. Perspectivas socioecológicas'. Doctoral thesis, ISEC, University of Cordoba, Spain.

Caporal, F.R. (1998) 'La extensión agraria del sector público ante los desafíos del desarrollo sostenible'. Doctoral thesis, ISEC, University of Cordoba, Spain.

Caporal, F.R. and Costabeber, J.A. (2001) *Sustentabilidade e Cidadania*. Porto Alegre: Programa de Formaçao Técnico-Social da EMATER/RS.

Carballo, C. (2000) 'Las ferias francas de Misiones. Actores y desafos de un proceso de desarrollo local'. Working Paper No. 9. Buenos Aires, Centro de Estudios y Promoción Agraria (CEPA). Mimeo.

Costabeber, J.A. (1998) 'Acción colectiva y procesos de transición agroecológica en Rio Grande do Sul, Brasil'. Doctoral thesis, ISEC, University of Cordoba, Spain.

Delgado Burgoa, F. (2002) *Estrategias de autodesarrollo y gestión sostenible del territorio en ecosistemas de montaña*. La Paz: Plural editores.

De Medeiros, L.S. and Leite, S. (1999) *A formaçao dos assemtamentos rurais no Brasil. Processos sociais e políticas públicas*. Rio de Janeiro: Editora da Universidade.

Fernández Durán, R. and Sevilla Guzmán, E. (1999) 'La resistencia contra la globalización económica y el neoliberalismo', in T. Ricaldi Arévalo (ed.), *Una nueva mirada a la ecología humana*. Cochabamba, Bolivia: CESU-UMSS/UNESCO.

Fernández Durán, R., Etxezarrata, M. and Sáez, M. (2001) *Globalización capitalista: Luchas y resistencias*. Bilbao: Virus.

Figueroa Zapata, M. (2000) 'Andalucía, una experiencia particular', in *Transgénicos: Biotecnología en el agro*. La Plata: The University of La Plata.

Gliessman, S.R. (1997) *Agroecology: Ecological Processes in Sustainable Agriculture*. Chelsea: Ann Arbor Press.

Gupta, Anil (1996) 'Social and ethical dimensions of ecological economics', in R. Costanza et al. (eds), *Going Down to Earth: Practical Applications of Ecological Economics*. Washington, DC: Island Press/ISEE.

Gupta, Anil (1998) *Postcolonial Development: Agriculture in the Making of Modern India*. Durham, NC: Duke University Press.

Guzmán Casado, G., González de Molina, M. and Sevilla Guzmán, E. (2000) *Introducción a la agroecología como desarrollo rural sostenible*. Madrid: Mundi-Prensa.

Kothari, A. (1997) *Understanding Biodiversity. Life, Sustainability and Equity*. Hyderabad: Orient Longman.

MAELA (2000) *Perspectivas del movimiento groecológico latinoamericano en el nuevo milenio*. Cochabamba, Bolivia: AGRUCO.

Mançano Fernández, B. (2000) *A formaçao do MST no Brasil*. Petrópolis: Editora Vozes.

Martinez-Alier, J. (1992) *De la economía ecológica al ecologismo popular*. Barcelona: Icaria.

Martínez-Alier, J. and Guha, R. (1997) *Varieties of Environmentalism*. London: Earthscan.

Martínez Sarasola, (1992) *Nuestros paisanos los indios*. Buenos Aires: Emecé.

Martins Carvalho, H. (2000) 'A emancipaçao do movimento no movimento de emancipaçao social continuada (reply to Zander Navarro) in Boaventura de Sousa Santos. (ed.), *Reinventando a emancipação social*. Lisbon and Sâo Paul.

Morales Hernández, J. (1999) 'La articulación entre potencial endógeno y entrono externo, en el diseño de estrategias de agricultura sustentable para la comunidad de Juanacatlán, Jalisco, México'. Doctoral thesis, ISEC, University of Cordoba, Spain.

Navarro, Z. (1996) 'Política protesto e cidadania no campo. As lutas sociais dos colonos e dos trabalhadores rurais no Rio Grando do Sul'. Río Grande do Sul: Federal University of Rio Grande do Sul.

Pengue, Walter A. (2000) *Cultivos transgénicos: ¿hacia donde vamos*? Buenos Aires: Lugar Editorial/ UNESCO.

Pollack, A. (1999) 'Biological products raise genetic ownership issues', *New York Times*, 26 November.

Pretty, N.J. (1995) *Regenerating Agriculture*. London: Earthscan.

Rist, S. (2002) *Si estamos de buen corazóm, siempre hay producción*. La Paz: Plura ediciones.

Rosset, P. (1998) *Mitos de la revolución verde*. Oakland: Food First, EE.UU.

Sevilla Guzmán, E. (1999) 'Asentamientos rurales y agroecología en Andalucía' in *Cuadernos Africa, América Latina*. No. 35. Special Issue on North-South Relations.

Sevilla Guzmán, E. (2002) 'Agroecología y desarrollo rural sustentable: una propuesta desde Latinoamérica', in S. Sarandón (ed.), *Agroecología. El camino hacia una agricultura sustentable*. La Plata: Ediciones Científicas Americanas.

Sevilla Guzmán, E. and Woodgate, G. (1997) 'Sustainable rural development: from industrial agriculture to agroecology', in Michael Redclift and Graham Woodgate (eds), *The International Handbook of Environmental Sociology*. Cheltenham: Edward Elgar.

Tapia Ponce, N. (2002) *Agroecología y agricultura campesina sostenible en los Andes bolivianos*. La Paz: Plural editores.

Toledo, V.M. (1991) 'La resistencia ecológica del campesinado mexicano (en memoria de Angel Palerm)', *Ecología Política*, No. 1.

Toledo, V.M. (1996)

Toledo, V.M. (2000) *La paz en Chiapas: luchas indígenas y modernidad alternativa*. Tlaxpana: Ediciones Quinto Sol.

Wizniewsky, J.G. (2001) 'Los asentamientos de reforma agraria y la perspectiva de la agricultura sostenible: los casos de Hulha y Piratini, Rio Grande do Sul, Brasil'. Doctoral thesis, ISEC, University of Cordoba, Spain.

35

Performing rurality

Tim Edensor

INTRODUCTION

In this chapter, the ways in which people are predisposed to carry out unquestioned and habitual practices in rural settings is explored by using the metaphor of performance. Moreover, the ways in which the materialities and meanings of rural space are reproduced, consolidated and contested, along with the identities of those who dwell and move within them, can also be considered by examining how rurality is staged so as to accommodate particular enactions. It is through the relationship between the array of characters playing out particular roles, and the spaces in which they perform, that ruralities are routinely produced. Thus common sense understandings of who appropriately belongs in rural space, what (kinds of) countryside symbolizes and what actions are fitting can be grounded in performances and stagings.

Usefully, performance foregrounds identities (of spaces and individuals) as continually in process as actors rehearse and repeat conventions about what to do in specific settings. We will see that whilst the rural is an assemblage of differently connected and constituted spaces, attempts to fix the identity of space, place and rural subjectivities through performance by different groups testify to the desire for fixity and certitude in conditions of continual social and cultural flux. For it is especially the rural realm which is assigned significance as that which remains the same in a changing world, and repetitive performances may reassuringly convey the illusion of stasis. This is why the rural has become such a powerful signifier of national identity (Edensor, 2002). However, as Macnaghten and Urry (1998) have pointed out, so contested has the countryside become as the venue for a huge variety of practices that desires for stasis are apt to be thwarted by the coincidence of contesting performances across the countryside and on particular stages.

By exploring particular kinds of scripts, roles, forms of stage management, choreography, improvisation and reflexivity, I will investigate some varieties of rural performance. Examples will be primarily grounded in the United Kingdom, but references to other contexts will occasionally be provided to indicate the widespread applicability of notions about rural performance and staging. Different rural performances are enacted on different stages by different actors: at village greens, farm-life centres, heritage attractions, grouse moors, mountains, long-distance footpaths and farmyards, and in rural spaces identified as 'wilderness' (Rothenberg, 1995). Performances depend for their coherence on being performed in such particular settings where they reinforce group and placial identities. In fact, recognizable environments around which enactions can be mobilized – around props, with fellow actors – are essential to the sustenance of performance. Thus different 'natural' stages accommodate performances and are (re)produced by them. Performances are socially and spatially regulated to varying extents. Stages might be carefully managed, and enactions can be tightly choreographed or closely directed. Moreover, performances might be scrutinized by fellow performers to minimize any diversions from conventions. However, unconventional performances can undermine attempts to fix meaning and action on stage. Alternatively, the stage's boundary might be blurred, be cluttered with other actors playing different roles, and thus may facilitate more improvisatory performances.

As I have said, performances are increasingly acted out by competing actors on the same stages. For instance, there is competition between adventure tourists, ramblers, hunters and farmers on the mountains of Britain, each group possessing contesting ideas about what activities are appropriate to these domains. Implicit in much of these enactions and stage management are assumptions and allegations about those who properly belong on these stages and those who lack performative competence. For example, in a rather clichéd account, townies are accused of being insufficiently aware of 'country ways' and 'nature', in their sentimentalized, effete views about farming and conservation. On the other hand, country folk may be derided as rural throwbacks who cannot competently perform according to 'modern' norms, norms that middle class 'incomers' to rural places competently wield as cultural capital in political struggles against 'development', commonly supported by their adversaries, these long-standing residents (Pahl, 1970; Newby, 1980). More pertinently perhaps, the British countryside continues to be represented as a stage populated by a cast of white characters across film, television and advertising spheres, implicitly excluding black characters and reifying the rural as a white realm (Agyeman and Spooner, 1997).

Recent discussions of performance have revolved around the theoretical contributions of Judith Butler and Erving Goffman. Both provide very rich concepts with which to explore social performance and yet I think both limit the ways in which performance can be used as an analytical tool. Butler uses the notion of performativity to identify how *gender* and *sex* categories are reproduced by actions, disavowing essentialist notions that they are pre-discursive or 'natural'. By adopting a gamut of attributes assigned as 'feminine', girls and women continually 'perform' gender, a repetitive iteration which seems to produce unambiguously gendered bodies (Butler, 1993: 9). Butler effectively foregrounds identity as an ongoing process but unfortunately distinguishes between *performance* and *performativity*. The former is characterized as self-conscious and deliberate, the latter is reiterative and unreflexive, a dualism which neglects the blurred boundaries between purposive and unreflexive actions. Seemingly reflexive performances may become 'second nature' to the habituated actor, and similarly, unfamiliar surroundings may provoke acute self-awareness of iterative performances where none had previously been experienced. Perhaps Bourdieu's (1984) notion of habitus can mediate between these positions since it is constituted through a *practical* reflexivity in which embodied know-how modulates unforeseen events.

Goffman insists that social life is inherently dramatic and that we invariably play particular roles. In 'front-stage' situations, driven by an urge for 'impression management' (1959), we *strategically* enact performances which are devised to achieve particular goals. However, this insistence on the instrumentality of role-playing – especially on the front stage – conjures up a continually self-reflexive individual, intentionally communicating values to an audience. This captures some modes of performance but does not consider a host of unreflexive, habitual enactions. It is my purpose to explore here how performance can be understood as both deliberately devised and habitual, but rather this should be grasped as an interweaving of conscious and unaware modalities, part of the flow of ongoing existence. This becomes particularly central to any discussion of rural performance since the rural is often constructed as that stage upon which the reflexive self can carry out work towards self-realization, that space in which the 'real' self can be attained. Accordingly, pre-existing techniques, performative dispositions and discourses surround rural activities, often as 'second nature'.

Addressing the division between reflexivity and non-reflection, Merleau-Ponty argues that intellectual faculties are secondary attributes which are 'rooted in practical and pre-reflective habits and skills' (Crossley, 2001: 62). Using the metaphor of football to explain that much action is practical and engaged rather than contemplative, Merleau-Ponty argues that performing football depends on a contextual awareness that incorporates the skills and know-how needed to read and play the game within the relational space of the players' interactions (2001: 74–79). A purely self-aware consciousness would minimize performative effectiveness, but players' shared assumptions, skill and use of space epitomizes a practical reflexive performance. The rural performances discussed here possess this capacity for 'playing the game' whether they use rural stages only periodically, even when they have clearly been devised to stage the rural as commodity, but especially when they are the everyday performances of country dwellers. All are shaped by forms of 'second nature' although such dispositions may be incomprehensible to outsiders who cannot immediately immerse themselves in an unfamiliar field.

These themes of spatial specificity, regulation and reflexivity re-emerge in the following attempt to distinguish the kinds of performance that express and transmit rural identity. There are many rural performances and stagings that could be explored. There are the dramatized rituals of grouse and pheasant shoots, the performances of rural folk customs such as Morris Dancing and well dressing, folk music festivals, and popular film and television dramas. Here, however, I restrict myself to looking at forms of performance that occur in the United Kingdom although I have chosen three areas which will resonate in other contexts. I will look at the enactions of leisure-seeking city dwellers who usually enter the countryside for a limited period, the increasing touristic staging of the rural and, finally, the everyday performances of 'country folk'.

CITY FOLK PERFORM RURALITY

It is ironic that in Europe and North America, dwelling and moving within the rural is conceived as the antithesis of performance for many city dwellers, who promote excursions to the countryside for pleasure as precisely the means to escape from the urban binds that confine the self to 'false' dispositions and stifled bodily sensations. Csordas describes the body in the city as 'primarily a performing self of appearance, display and impression management' (1994: 2), whereas it is a part of common sense understanding that the body in the countryside does not need to adhere to the 'falsities' inherent in these practices. Binary understandings that declare the urban to be the realm of fakeness, over-socialization and alienation in contradistinction to the unmediated physical and mental release from urban convention found in the rural, emerged during the romantic era when a distinct set of engagements with the countryside were mooted (Bunce, 1994). Yet what becomes apparent is that this quest for liberation from urban constraints equally depends upon the adoption of diverse performative procedures to attain 'peak experiences', or at least a heightened sensory awareness of one's body-self and surroundings.

As I have mentioned, performative techniques and technologies are mobilized in particular settings. Thus, when city dwellers enter certain kinds of rural stage, they are usually informed by pre-existing discursive, practical, embodied norms which help to guide their performative orientations. These collective performative norms constitute a 'discrete concretization of cultural assumptions' (Carlson, 1996: 16) which tend to reinforce social and cultural norms about what to do in specific theatres. Accordingly, forms of rural space are (re)produced by distinct kinds of performance, having their symbolic and practical values consolidated by the ways in which actors dramatize their allegiance to place. Thus both individual and group identities and imagined geographies are performed and thereby (re)produced. For instance, besides inscribing paths and signs in rural space, reaffirming their obviousness as routes – and producing patterns of erosion – walkers delineate particular kinds of landscape as suitable for (particular kinds of) walking.

Like all social performances, culturally specific ways of acting in rural theatres are organized around which clothes, styles of movement, modes of looking, photographing and recording, expressing delight, communicating meaning and sharing experiences are deemed to be appropriate in particular contexts. Initially, particular enactions are learnt so that the necessary competence is acquired, and the suitability of the performance is also likely to be subject to the disciplinary gaze of co-participants and onlookers. Through such socially constituted approaches to being and acting in rural contexts, urbanities gradually lose self-consciousness and self-monitoring as they become more grounded in 'common sense' and unreflexive assumptions about how and where to walk, how to 'appreciate' and comment upon beauty, how to climb, run, ski or relax. The consolidation of embodied practical norms is thus forged through internalizing performative norms, being subject to external surveillance and the re-iterating of such enactions when in particular kinds of space. Communal conventions are thereby continually rendered consistent so that they appear 'obvious' to participants who are usually unaware of their cultural locatedness unless challenged by alternative performances.

In addition to the performative norms rehearsed and then ingested by visitors to the countryside, a host of technologies and procedures enfold actors into rural pursuits, mediating the experience of the rural through their 'proper' use and application, simultaneously confining and extending experience and human capacities (Macnaghten and Urry, 2000: 2). Judith Adler (1989) has shown how travel programmes, brochures, accounts and guidebooks are 'a means of preparation, aid, documentation and vicarious participation' for tourists. By following the 'norms, technologies, institutional arrangements and mythologies' (1989: 1371) which are instantiated in particular places and tours, tourists continually reconstruct tourism and tourist space. If we focus specifically on visitors to rural areas as members of coach parties, rambling clubs, birdwatching groups or members of conservation bodies, we can similarly identify the technologies utilized by such groups: the guided tours, walking poles, boots (see Michael, 2000), rucksacks, clothing, binoculars and identification booklets. Such technologies are competently and habitually utilized so as to make them agents in the performance. As such they restrict certain kinds of alternative actions through their properties but also extend human capacities and possible actions. For instance, binoculars enable the identification of birds, especially when in the hands of an experienced birdwatcher who is able to quickly focus upon a moving or distant species and rapidly identify it, perhaps also by using a bird identification handbook. Thus the feasibility of identifying birds and adding them into the 'twitcher's' checklist becomes part of a sequence of activities: spot bird, focus upon it, identify it and list the sighting. The techniques and technologies cited here enable the furtherance of this kind of birdwatching as performance but necessarily restrict other performative modalities that are possible. For instance, a focus on small, moving birds restricts awareness of multiple other features in the landscape, perhaps ignores numbers of birds elsewhere, foregrounding as it often does the centrality of the rare. Yet the shared code and enactions are rarely questioned by participants and the tools of their pursuit tie them into normative performances.

Ways of acting in the rural are laid out in authoritative books which detail how to conduct oneself as a climber, walker and birdwatcher (for instance, see *Bill Oddie's Introduction to Birdwatching*, 2002) which sustain the performative norms of groups entering rural domains for pleasure. Equally, appropriate actions are upheld by directors and choreographers in the form of group leaders, who are usually well rehearsed in forms of knowledge, ability to utilize technologies and the physical skills necessary to negotiate rural space as walking terrain, rock face or nature reserve, and thus are exemplary performers. These coordinating experts lead by example and monitor the performances of others in their orbit of influence, embodying cultural capital through bodily techniques and experience gained through previous achievements. In climbing, for instance, excursions are planned, choreographed and directed by route leaders. Moreover, these experienced climbers determine the roles members of climbing parties should adopt (Beedie, 2003), assess whether participants are sufficiently equipped, fit or adept to meet climbing challenges, monitor food intake, maintain a pace appropriate to a route march, and disseminate and teach skills such as ropework and compass reading. The authoritative regimes installed by these 'directors' stabilize the relations between equipment and clothing, bodily maintenance, manoeuvres and collective enactions which sustain performing climbing. This seems to militate against achieving the escape from social conventions and bodily restrictions that many climbers seek.

There are then, a host of performative conventions informed by forms of 'common sense' and praxis which shape how city dwellers conduct their recreation on rural realms, permeated with discursive, experiential, technical and moral rules governing 'appropriate' conduct. However, in a postmodern, post-productive countryside, the quantity and range of enactions have multiplied. For instance, a growing number of adventure sports use the rural as a theatre for feats of daring and excitement. Particular stages are venues for the various adherents of these activities but these may be the same stages used by walkers, fell-runners, mountain-bikers, birdwatchers, botanists and hunters who may all enact distinctly different practices. Accordingly, contesting notions of how specific rural stages ought to be utilized increasingly inform conflicts between different actors. For instance, in recent years, there have been heated debates between walkers and naturalists and skiing enthusiasts about how the Cairngorms area of Highland Scotland should be used. These activities and their proponents wield distinct cultural values variously based around different notions of conservation, land use, productivity, beauty, individuality and stewardship.

Not only is this contestation about preferred performative norms exchanged between the adherents of different pursuits but also occurs within particular spheres of activity. For example, I have argued that devotees of different walking styles enunciate their own values whilst impugning those of other walkers. Typically, such contests focus upon whether walking should be conceived as sport or relaxing pleasure, whether it is better to walk solitarily or accompanied, how far and fast to walk, whether to follow the 'beaten track' or make one's own path through nature (Edensor, 2000). Mobilizing their own variants of status and common sense, such performers also implicitly reconstruct rural stages in distinct ways as well. In a similar vein, Lewis (2000) highlights how different climbing performances assert that only they achieve the aim of *bringing one closer to nature and oneself.*

In the light of this contention, I think that it can still be maintained that a key theme in all rural leisure pursuits is the quest for an apparently unmediated exchange between self and 'nature', a quest to 'feel' that inheres in a host of activities from naturism (Bell and Holliday, 2000) to more recently developed action sports in New Zealand (Cloke and Perkins, 1998). They vary, however, in the extent to which they seek contingency and manage risk as a means to promote 'instinctive' responses, and the degree to which they aspire to experience a sensual and unmediated nature.

Despite my insistence upon the prevalence of (proliferating) performative conventions for urbanites seeking an 'authentic' self in the 'authentic' realm of the rural, it is important that performance is not conceived as static, no matter how rigorously regulated space and action might be through self-surveillance and that of other participants. In addition – and this is especially pertinent to rural space – the brute force of incidental and institutionalized qualities can interrupt any progress. For the affordances of the countryside, the unregulated and unexpected effects of climate, terrain, animals and plants are apt to impact upon the body so as to jar it from its performative normalcy. Amongst a host of potential disruptions, nettles and thistles sting, insects bite, frisky horses frighten, muddy paths may be slippery and occasion tumbles, downpours drench and powerful smells such as silage disturb thoughts of a rural idyll. Thus the material qualities of the rural are likely to act back upon the walker whose early sauntering and visual delight is replaced by fatigue, pain and an acute awareness of gradient, surface and obstacle. What this also confirms is that the performances described above are never merely visual but involve a diverse sensual encounter with the rural, depending upon the degree of temporal and physical immersion, and drawing upon tactile, auditory and olfactory senses in an engagement with space and materiality.

STAGING THE RURAL

In theatrical terms, tourism encourages the production of distinct kinds of stage and is an activity which sustains a host of competing performative norms (Edensor, 2001). Local urban tourism strategies increasingly seek to compete by advertising their distinct charms. Whilst they often attempt to create unique niches, certain themes and enactions recur. As it becomes a post-productionist space, the countryside has similarly become a realm of diverse tourist attractions as farms diversify their sources of income and other rural entrepreneurs, landholders and politicians seek to identify potentially marketable buildings, rituals and customs, landscapes, histories and signs of 'tradition'. Thus particular rural regions are apt to promote generic landscapes and attractions – national parks, for instance – as well as key symbolic sites and events, along with more homely and apparently mundane spaces of rural life.

The stage management of touristic spaces particularly tries to produce affective, sensual and mediatized experiences – within a format of 'edutainment'. This partly involves the production of 'themed' spaces. Typically identified as encoded shopping malls, festival marketplaces, cultural quarters and waterfront attractions, such attractions are proliferating and include a wide range of rural tourist sites. The extension of these themed spaces into the countryside includes themed pubs and restaurants, manor houses and mansions which hold medieval banquets, the dwellings of rural workers and remnants of industry such as mills and water wheels, farm life centres, rural museums, heritage sites and audio-visual displays. Rural identity is often produced in 'indigenous', 'traditional' or 'folkloric' customs where tourists collect signs of local or national distinctiveness.

In highly encoded spaces, stage managers attempt to 'create and control a cultural as well as a physical environment' (Freitag, 1994: 541), through strict aesthetic monitoring and the inclusion of clear visual and sonic cues. Through the use of 'sceneography' (Gottdiener, 1997: 73), the tourist gaze is directed to particular attractions and commodities and away from 'extraneous chaotic elements', effectively reducing 'visual and functional forms to a few key images' (Rojek, 1995: 62), often media-based in origin. Accordingly, in such spaces, highly selective aspects of the countryside are turned into experiential commodities. The selection of particular forms of information, objects worthy of the tourist gaze and self-evident routes can restrict alternative enactions and viewpoints, reinforcing practical performative norms about how to behave, what to look at and where to go.

To exemplify this dramatic production of rurality, one may consider a small area of southern Shropshire, a predominantly rural English county, which is replete with rural attractions. This one area contains a golf centre, a small breeds and owl centre, a vineyard, two themed gardens, a family farm and a cheese dairy. These productions collectively stitch together an intertextual landscape in which both productive, conservation and leisure landscapes are cultivated, and the breeding of animals and manufacture of food and drink can be witnessed 'backstage'. These rural spaces are staged by managers, landscapers, farmers and gardeners, animal carers and falconers, cheese-makers and wine-makers, along with other employees. Moreover, at all of these attractions there are various levels of participation. At the animal and farm centres visitors are encouraged to handle animals, to 'collect eggs, feed pigs and calves, hand milk a cow and cuddle kittens and chicks'. Similarly, directed performances are found at the vineyard where 'tutored tastings' can be arranged. While the cheese centre is primarily devoted to display, visitors are encouraged to consume the produce manufactured on site. At the gardens, preferred routes enable those conventional forms of strolling and gazing which reinforce the use and meanings of such spaces, and golf decorum instantiates the normative performances of players.

Thus a range of performances are inculcated and played out in highly staged settings of rurality. The dramatic thematizing of places through physical makeovers, selective narration and imagery, and advertising is complemented by the production of particular kinaesthetic effects (see Pine and Gilmore, 1999), typically through the installation of rather smooth, built-in affordances. Of course, such stagings, in aestheticizing and regulating performances, are likely to erase any reference to conflict and oppression, masking the issues of power and domination which may inhere in their local performance and participation, but imprint them with different traces of power.

Another way in which sites can be dramatically contextualized, produced as theatrical spaces, is through capitalizing on a nexus between media and place. The production of film and television dramas in identifiable geographical settings has given rise to a proliferation of tourist sights. Morley and Robins argue that 'the "memory banks" of our time are in some part built out of the materials supplied by the television and film industries' (1995: 90). And this is also the case with the cinematic portrayal of the rural (Mitman, 1999). Here the intertwining of everyday television drama and tourism reinforces a network that constitutes a thoroughly dramatized landscape. These television dramas, mapped onto the distinctive landscapes in which they are set, produce a theatrical signature through which the scenery can be familiarized, associated with characters, episodes and props. As the series of televisual signifiers condenses, the network of associations between theatrical rural spaces provides extensive opportunities for re-envisaging the dramatic conventions of these series, TV settings and literary settings. Accordingly, the county of Yorkshire alone possesses regions advertised for their association with the television series *Last of the Summer Wine*, *Heartbeat* (see Mordue, 1999) and *All Creatures Great and Small*, all of which recycle rather nostalgic, bucolic versions of country life stocked with quaint eccentrics.

In association with this mediatization of the rural, a species of touristic performative space, the audio-visual display, is proliferating whether supplementary to existing attractions or purpose-built. Often organized around dramatized histories or mythological fantasies, such productions merge with and recycle media images and narratives. Dramatizing the rural on screen and in fiction involves the citation of a stock of rural characters and settings that can be mobilized in the imagining of multiple ruralities. For instance, in Glencoe in the Scottish Highlands, using audio-visual displays and animatronic representations, Highland Mystery World energizes the ancient mysteries of the region right before your eyes. It brings to life the myths, legends and folklore that rest so richly within Scottish history. Here you will embark upon a journey back in time to the world of bogles, kelpies and fachns. Experience the Highlands of yesteryear. Enter the mysterious stone circle and meet visitors from long ago who will tell you amazing tales of Scotland's legendary past. Experience the riddles of the serpents' cavern where the Blue Man of the Minch and other legends come alive just for you. Feel for yourself the sensations, sights, sounds and smells of ancient Scotland (http://www.alltravelscotland).

And in Penrith, on the England/Scotland border, the Rheged centre has a huge cinema presentation in which we are invited to journey back in time to 'meet Celtic Warriors, delve into Arthurian legend, ride with fearsome Border Reivers and come face to face with a larger than life William Wordsworth' (htttp//www.rheged.com). At a larger scale, the Smoky Mountain National Park in Tennessee, USA, is filtered through Dollywood, an amusement park which articulates the distinctive, 'down home' folksiness of rural life through the character of country music star Dolly Parton. Alongside themed 'white knuckle' rides and musical attractions, guests can visit simulacra of an old country church, a schoolhouse and Dolly's childhood home, as well as trying out 'traditional' country crafts.

Such productions minimize the potential for improvisation and attempt to eradicate ambiguity, with key personnel – the stage managers, choreographers, directors, costume designers and stage-hands – coordinating the sequence of events and training participants in accordance with the 'goals, constraints, resources, conventions and technologies of particular culture-producing groups and their audiences' (Spillman, 1997: 8). Alternative, contesting and ironic performances can never be ruled out, however.

Nevertheless, such is the profusion of staged and performed rural spaces that the diversity of themes can become bewildering. This was particularly evident at the Cheshire (agricultural) Show of 2002. Featuring a cast of thousands, there was no single production but a multiply staged countryside in which hundreds of stages (stalls, displays, show arenas, demonstrations, play areas and information centres) vied with each other for the attention of the large crowds who wandered around the extensive site. What was clear was that in contradistinction to earlier shows, where the emphasis was on the display of livestock and farming techniques and apparatus, at the Cheshire Show a more post-productivist inclination featured consumption and leisure pursuits. The proportion of farm machinery and livestock has dwindled. Instead there are a welter of crafts and foodstuffs on display. Mounds of bread, jams and honey, cheeses, cakes and biscuits are loaded onto rustic tables, and particularly, the growing importance of specialized and organic produce implies a lifestyle that avows a close connection with the countryside. Such foodstuffs are dramatized, polished and carefully displayed, pieces are offered to taste, and render a facsimile of a market at variance to supermarket shopping against which it is defined. Moreover, there are demonstrations of how to prepare and cook these more 'authentic' products.

Craft products that imply an imagined country lifestyle are promoted, producing the countryside as the site for well-hewn walking sticks, shooting sticks, plants, wooden furniture for the garden, fencing, decorative ironwork, wood-carvings, paintings of animals and rural scenes, and a host of other commodities. The elements of rural lifestyle are closely allied to magazines such as *Cheshire Life*, which feature and blur the boundaries between rural and suburban. Not only do such typically 'hand-crafted' goods posit a vestigial world at variance to urban mass production but they also imply a cast of craft workers who produce them in workshops and sheds. Indeed, craft-making displays feature at a number of sites, featuring flower-arranging and the like. In addition to these lifestyle accoutrements, there are several clothes shops peddling not the high fashions and fripperies of the high street but sturdy, sensible and practical wear – the boots, the waxed jackets and the tweeds which appropriately clothe the 'traditional' rural body and imply distinct leisure and practical activities.

A different craft ethos and aesthetic is also consolidated in display areas in which old farm machinery, traction engines and fairground attractions are the provenance of skilful enthusiasts. That these highly polished specimens are located at a county show suggests that the countryside is the last refuge of specialized engineering knowledge, connoting a nostalgia for an era of skilled work and closeness between 'man' (inevitably) and machine, and staging the past as utopian. Other forms of expertise with animals also foreground the rural as the space in which traditional and largely unsentimental knowledge about animals is mobilized. Competencies of breeding, rearing, handling and training animals are all on display. Rare breeds of rabbits and chickens are displayed as exotic products and well-tended beasts, the more traditional dressage and show-jumping events stage the command of horses, and the highly choreographed manoeuvres through which dog owners cajole their pets reveal a masterly competence. Besides these skills, a somewhat more recent emphasis on conservation-related proficiency adds a dimension that constructs the rural as that realm which is subject to the stewardship of those knowledgeable in conserving and managing nature. Accordingly, the relevant skills of fence-laying, dry-stone walling and pond-making are complemented with more archaic activities such as bodging, which again reinforce the staging of the countryside as reiterative in that rurality requires the ongoing performance of skilful action for its sustenance. There are clear dissonances here with the stalls of agribusiness and agro-chemical sales teams which foreground intensive production.

Further tensions may be found in the proliferation of rural leisure pursuits, as exemplified by the many displays advocating membership of a host of activities in addition to the 'traditional' sporting pursuits of the country 'sportsman', shooting, hunting and fishing. The decentring of these formerly 'appropriate' activities comes with other shooting pastimes such as cross-bow marksmanship, clay pigeon shooting and archery, but more importantly, by displays promoting cycle cross, paintballing, hang-gliding, orienteering and abseiling. The countryside here, then, is the stage for an expanding range of leisure activities, open to both rural and urban dwellers to extend and display their skills and competencies, experiences which (re)construct the rural as (constituted by different forms of preferred) space.

As a multiply-staged event, the Cheshire Show could be analysed in much greater detail along a number of axes of enquiry. Here, my intention has been to show the different ways in which the countryside is staged and performed in contesting and contrasting ways, multiplying the ways in which it is envisioned and utilized as a distinct realm by different actors.

EVERYDAY PERFORMANCES: POPULAR COMPETENCIES, EMBODIED HABITS AND SYNCHRONIZED ENACTIONS

Despite a geographical focus upon the representational, symbolic and purposive lineaments of spatial identity, the most grounded, situational relationship between people and space occurs within the mundane sphere of the everyday. Accordingly, the ways of dwelling, working, socializing and relaxing in familiar space can be considered as largely unreflexive habits, quotidian performances that tether people to place, producing serial sensations via daily tasks, pleasures and routines. In the countryside, as elsewhere, distinct structures of feeling are wrought through a feel for the tasks at hand and for the environment in which they are performed, as repetitive interaction with tools, space, humans and other animals is carried out. Habitual, unreflexive performances produce *becoming* subjects through embodied, affective and relational (to other people, to spaces and to objects) practices in a world-in-process (Nash, 2000: 655), but continuity in quotidian performance also serves to provide spatial consistency, a sense of *being* in space. To perform in this fashion is to comfortably dwell in a homely environment, using 'common sense', and unreflexive, sensuous interaction with familiar space. Drawn out over time and in consistently traversed and inhabited space, the habitual performative is consolidated in the successive enactions of daily routine, and becomes irrevocably social through the synchronic and coordinated actions of fellow inhabitants, producing a collective 'non-cognitive dwellingness' (Macnaghten and Urry, 2000: 7).

These everyday spaces of domesticity and work have been identified as 'taskscapes' by Ingold and Kurttila (2000), which accommodate an 'unnoticed framework of practices and concerns' in which we dwell as 'habituated body subjects' (2000: 90–91). Space thus not only is understood and experienced cognitively but is approached with what Crouch (1999) calls 'lay geographical knowledge', a participatory disposition in which the influences of representations and semiotics are melded with sensual, unreflexive, practical knowledge to produce a mundane way of 'being-in-the-world'. Interestingly, Ingold and Kurttila (2000) illustrate the embeddedness of the taskscape with reference to a rural space, the north of Finland. They particularly detail how the Sami inhabitants 'understand' and experience the weather sensuously, in contradistinction to climatologists who 'scientifically' *record climate*. This everyday knowing is a flexible skill which adapts, for instance, to new forms of transport such as the snowmobile and motorbikes, and is thus an apprehension which depends upon a multi-sensory awareness that facilitates spatial orientation and coordinates activity, an immersed, space-making practice which embeds identity.

As in other performances, these quotidian enactions are variously constrained and enabled by the materialities or affordances of space – as 'a concrete and sensuous concatenation of material forces' (Wylie, 2002: 251) – which mesh with inhabitants' bodily practices to become embedded over time. The surfaces, textures, temperatures, atmospheres, smells, sounds, contours, gradients and pathways of places encourage humans – given the limitations and advantages of their normative physical abilities – to follow particular courses of action, and produce an everyday practical orientation towards the taskscape. This is complemented by a sensual apprehension of the textures of turf, hay and soil, the smells of beasts and vegetation, and the sounds of animals and machinery. Particular rural environments, therefore, greatly influence the kinds of habitual performances that persist. Such performances in place are highly contextual and therefore innumerable, but to briefly highlight the forms of common sense, practical disposition and sensuous knowledge that inform mundane rural performances, I draw upon a weekly column by Charles Allan which appeared for several years in the *Glasgow Herald* newspaper, and has been reproduced as a series of volumes entitled *A Farmer's Diary*, recording the everyday events at a mixed farm in Aberdeenshire.

The column is clearly addressed to farmers and, although full of light-hearted anecdotes about friends, family and neighbours, is principally concerned with the trials and tribulations of modern farming. Daily farming life is portrayed as a series of practical considerations and projections that produces a disposition shaped by the uncertainties of farming, the pitfalls, moments of good fortune, upturns and downturns. It is also marked by strong diurnal and seasonal routines, such as sowing, harvesting and breeding animals. There are long passages about how crop conditions are assessed, potential yields, creating and applying silage and fertilizers, buying and managing livestock, where and how to seek medical and chemical and mechanical advice, and a whole array of techniques and know-how. Thus there is a habitual practical connection with a network of 'experts' but, in addition, there is recourse to local lore, an intersubjectively constituted form of common sense: '[T]hey say hereabout that for every day you delay sowing your barley after April Fool's Day you lose a hundredweight of yield' (1995: 40). This situated knowledge emerges from conversations with 'experts' and other farmers, refers back to 'traditional' forms of farming and is thus ongoing, contested and contingent.

There is also, however, a necessity to become embroiled in new bureaucratic imperatives emerging out of Common European Agricultural policies which forces farmers to adapt to policies based around set-aside land, price-setting, conservation, aesthetic management and diversification, and an assortment of procedures involving form-filling. This, Allan contends, takes farmers away from that which they know about and competently perform and is thus anathema to the farmer's 'feel for the game'.

In short, these are the daily preoccupations of one who is concerned with the productive potential of land. Accordingly, there is none of the romantic performances of leisure-seeking urbanites and little recourse to lyrical rural depictions, aesthetic judgements being confined to machinery and livestock. The stage here is assessed in terms of its productivity, a point which Allan makes explicit in one column: '[T]o a farmer, beauty in the land is order. It is the triumph of his choice of flora and fauna over that which would have occurred in his absence' (1995: 63).

The practices identified above are popular rural *competencies* that facilitate the running of one's life. Farming requires a second nature about where, how and when to plant and tend crops, maintain soil fertility and where to go for advice. As well as demanding an intimate geographical knowledge, these tasks also require a practical ability to carry them out with a minimum of fuss. Such competence embraces a knowledge of locality, a geography of practical action within one's taskscapes which incorporates regular sites, routes and regions, as well as hedgerows, farm yards, barns, streams, fences and pastures as constituents of 'activity spaces' (Massey, 1995). Integral to the unconscious ways in which identity is performed, these culturally located unreflexive enactions help to constitute a sense of rural belonging. The extent to which this is grounded in the rural might be evident when country folk move to the city and require different practical resources to accomplish everyday tasks, at variance to their own practical habits, skills that need to be learned before they become 'second nature'. An unfamiliarity with cultural contexts can reveal the situated nature of such forms of embodied knowledge. We can identify 'the small differences in style, of speech or behaviour, of someone who has learned our ways yet was not bred in them' (Williams, 1961: 42). The discomfort urbanites are apt to feel when confronted by a herd of bullocks or barbed wire fences testify to this situated habituation. As elsewhere, in the rural sphere there is normative etiquette which instantiates which forms of conduct are appropriate in particular contexts, embodied habits evident in ways of walking, sitting and being convivial which constitute shared worlds of meaning and action.

Space-making practices depend on collective rituals whether daily, weekly or annually. Simultaneous quotidian performances in the pursuit of work, leisure and reproduction compose distinct kinds of cultural rhythmicity or social pulse, and the seasonal routines of farming consolidate time-geographies which shape the ways in which people's trajectories separate and intersect in regular ways. The detraditionalization of temporal and spatial social patterns is perhaps less evident for farmers who remain tethered to daily, weekly and seasonal imperatives. Similarly, the fixtures of rural stages – the local shop, pub, sports ground and village hall – provide enduring sites around which routines are performed and communally coordinated as 'place ballets' (Seamon, 1979).

Clearly, class, ethnic and gendered forms of habitus intersect with rural dispositions. For instance, there are distinctive forms of playing, socializing, home-making and interacting with nature that do not feature in the farmer's account, and the quotidian lives of farm workers, farmers' wives, shepherds and gamekeepers would articulate a different range of routines, practical concerns and sensations. To focus briefly on gender, I will give a short account of the display provided by the Young Farmers Association at the Cheshire Show, more specifically of the particular gendered attributes and associated activities foregrounded. A photographic and artefactual display emphasized the accomplishment of distinct forms of rural cultural capital. Male members were represented as acquiring competence and gaining prizes in farm handyman skills and stock-judging, and in sports such as tug o' war and rugby. Females, on the other hand, were rewarded for their cookery, handicraft and homecraft skills. Although not entirely organized around these activities – there were awards for photography and debating as well – these specific and separate gendered activities give some indication of the ways in which rural masculinities and femininities might be performed and collectively esteemed as resolutely heterosexual, foregrounding forms of skill and physical capital. Again, whilst normative, such performances can be challenged by alternative versions of sexuality (see Bell, 2003) and gender, for instance, in the counter-cultural rural performances espoused by communards (see Berger, 1981; Pepper, 1991) and 'New Age travellers' (Sibley, 1997).

Everyday, habitual performances are constituted by an array of techniques and technologies, practical, embodied codes which guide what to do in particular settings. And where these are communally shared, they help to achieve a working consensus about what are appropriate and inappropriate enactions, come to constitute conventions about 'appropriate' ways of behaving which consolidate ways of inhabiting the rural. These unreflexive 'good' habits mean that it may be difficult for townies to 'pass' as rural. Through everyday habitual performance 'a consistency is given to the self which allows for the end of doubt' (Harrison, 2000: 503), as habit is internalized and ingrained through interaction with others. Habits organize life for individuals, linking them to groups so that 'cultural community is often established by people together tackling the world around them with familiar manoeuvres' (Frykman and Löfgren, 1996: 10–11). These shared habits strengthen affective and cognitive links, consolidate skills and doxa, and minimize unnecessary reflection every time a decision is required.

These shared forms of practical enactions, everyday knowledge and embodied approaches to quotidian problems form mundane choreographies which are forged by doing things rather than thinking about them. In this light, Bourdieu refers to 'automatic gestures or the apparently insignificant techniques of the body – ways of walking or blowing one's nose, ways of eating or talking' which 'engage the most fundamental principles of construction and evaluation of the social world' (1984: 466). Crucially though, these and other habits are not static. The habitus is the practical basis for action, but consists of 'forms of competence, skill and multi-track dispositions' rather than 'fixed and mechanical blueprints' (Crossley, 2001: 110). Habits are therefore full of flexible skills which can operate in an improvisatory fashion within a known field but may flounder outside. Nevertheless, the popular sayings, 'old habits die hard' and 'you can't teach an old dog new tricks' are partly accurate, since the familiar social world consists of enduring contexts and habits which depend upon each other.

CONCLUSION

Performance is an interactive and contingent process which succeeds according to the skill of the actors, the context within which it is performed and the way in which it is interpreted by an audience. Even the most delineated social performances must be re-enacted in (even slightly) different conditions, inferring that actors must continually strive for consistency and fixity of meaning (Schieffelin, 1998). In order to retain their validity, performative norms need to be continually enacted, whether these are the conventions of urbanites visiting the country, staged displays or the unreflexive habits of everyday rural life.

Yet the countryside, like other realms, is increasingly full of diverse performances which spark competing notions about what actions are 'appropriate', 'competent' and 'normal', and also may produce a reflexive awareness of the habitual performances which are so integral to individual and group identities. The prevalence of alternative, rebellious, 'post-tourist', cynical or involuntary performances can reveal the previously unseen bases which underpin conventional enactions. As Judith Butler has pointed out, a 'forced reiteration of norms' (1993: 94) can provide a template from which to deviate. Although the rural performances discussed here are primarily British and all are Western, forms of understanding the rural inform a wider diversity of performances than can be covered here. Nevertheless, emerging out of globalization, the proliferation of multiple, simultaneous enactions on rural stages – and a variety of different stages – means that people constantly confront other actors and practices which may contradict and challenge cherished, embodied and unreflexive ways of doing things. By this means, habits are brought to the surface, a revelation which produces conventions as 'something on which one must take up a stance', creating the dilemma of 'whether to kick the habit or to stick tenaciously to it' (Frykman and Löfgren, 1996: 14). Accordingly, creolized performances emerge as people utilize an expanded range of cultural resources, they 'improvise local performances from (re)collected pasts, drawing on foreign media, symbols and languages'

(Clifford, 1988: 15). Patterns of performance are therefore becoming more 'varied, differentiated and de-differentiated' (Rojek, 2000: 9).

Through the examples chosen, I have tried to show that rural performance can range from reflexive, strategic self-presentation to unreflexively embodied practice. They can be spectacular commercial presentations, conventional reflexive ways of experiencing the countryside, carnivalesque celebrations or sedimented in the unreflexive, embodied dispositions of everyday life.

I want to conclude by considering the changing performative contours of the British countryside and the extent to which time-honoured performances are becoming reconfigured through confrontation with external influences. To show the persistence and contestation of a countryside performance with a currently high profile I will briefly focus upon foxhunting, which can be figured as exemplifying numerous performances.

Foxhunting has long been a dramatic event, with its hunt followers and urban visitors enthusiastic about witnessing the spectacle, with its costumes, extravagant movement and the vibrant sounds of thundering hooves and barking hounds. The competencies – the skill of the riders and houndsmen and the blowing of horns – the specific roles – the field master, the whippers-in and kennel men – and the rituals – the blooding of new recruits to the chase, the hunt ball – all contribute to the dramatic shape of the hunt. Implicit in this production are the embodied rural skills which participants mobilize and the 'traditional strictures' about dress and riding etiquette, although these are 'invented traditions' (Hobsbawm and Ranger, 1983) from the nineteenth century. The entwinement of hunting and its aesthetics with a dominant vision of rural England is enshrined in normative representations in the shape of the numerous paintings and reproductions which fill country pubs and homes.

Foxhunting has more latterly become an activity that symbolizes country lore, an embodied pleasure, an emblem of those 'traditional' rural ways typically misunderstood by an 'urban political elite'. As Franklin (2001) points out, hunting of all kinds tends to involve an embodied, skilful, sensual engagement with the rural so that hunting terrain may be considered as a taskscape, a realm in which hunters mobilize a highly attuned, cultivated disposition for sensing prey. In this sense, hunting is a performance that involves a deep affinity with an unsentimentalized nature, but one that is inherently unpredictable, contributing to its embodied and aestheticized pleasures. This, Franklin asserts, makes it very different from the more detached touristic practices of urban visitors.

Nevertheless, for many urban dwellers and rural residents, and particularly campaigners against the sport, foxhunting is an irredeemably brutal, sadistic, class-bound pursuit which reflects the atavistic nature of many rural dwellers who are unenlightened about new ethical regimes concerning animal rights. It is thus highly symbolic of the archaic, outdated patterns of country living from an era in which a conservation ethos was not prominent. The opponents of hunting tend to conceive the countryside as a stage for cruelty-free pursuits bound up with leisure, aesthetic appreciation and conservation. It is evident then, that at a time when the distinction between the rural and urban is more blurred than ever before, hunting is being staged as that which distinguishes an entrenched divide between rural and urban dwellers, thereby producing imaginary rural and urban realms and actors. Defiantly defended and performed by its adherents, it is vilified and sabotaged by anti-hunt campaigners, who mobilise and advocate performative tactics, disposition and appearance.

Most strikingly, the controversy about hunting has generated a gigantic staging of the rural in the urban in the shape of the Countryside Alliance march in London in September 2002. This occasion, pre-eminently a huge demonstration against the threat to foxhunting from future government legislation, produced a self-conscious performing of rural identity apparent amongst the 400,000 marchers who descended upon the city for the day. Described by the *Guardian*'s correspondent as 'the corduroy clad hordes' (Brockes, 2002), the almost wholly white participants of all ages donned an array of cloth caps, deerstalkers, body warmers, rugby shirts, tweeds and waxed jackets, articles of clothing at variance to those worn by urban populations. Amongst other markers of rural belonging, the distinction was further underlined by the sounds of hunting horns and a spirit which the marchers believed distinguished them from ill-behaved demonstrators and contemporary urban dwellers: 'It's such a lovely atmosphere,' said Philip Bushill-Matthews, the Conservative MEP for the West Midlands. 'And you know there'll be no litter,' said his wife. 'We come from the small villages where people dispose of their rubbish cleanly' (Brockes, 2002).

REFERENCES

Adler, J. (1989) 'Origins of sightseeing', *Annals of Tourism Research*, 16: 7–29.

Agyeman, J. and Spooner, R. (1997) 'Ethnicity and the rural environment', in P. Cloke and J. Little (eds), *Contested Countryside Cultures*. London: Routledge.

Allan, C. (1995) *A Farmer's Diary: Volume Three.* Buchan, Aberdeenshire: Ardo Publishing.

Beedie, P. (2003) 'Mountain guiding and adventure tourism: reflections on the choreography of the experience', *Leisure Studies*, 22: 147–167.

Bell, D. (2003) 'Homos in the heartland: male same-sex desire in rural USA', in P. Cloke (ed.), *Country Visions.* London: Pearson.

Bell, D. and Holliday, R. (2000) 'Naked as nature intended', *Body and Society*, 6: 127–140.

Berger, B. (1981) *The Survival of a Counter-Culture: Ideological Work and Everyday Life Among Rural Communards.* Berkeley, CA: University of California Press.

Bourdieu, P. (1984) *Distinction.* London: Routledge.

Brockes, E. (2002) 'The day cross country came to town: time out in London for 400,000 rural marchers', *Guardian*, 23 September.

Bunce, M. (1994) *The Countryside Ideal: Anglo-American Images of Landscape.* London: Routledge.

Butler, J. (1993) *Bodies That Matter: The Discursive Limits of Sex.* London: Routledge.

Carlson, M. (1996) *Performance: A Critical Introduction.* London: Routledge.

Clifford, J. (1988) *The Predicament of Culture.* Cambridge, MA: Harvard University Press.

Cloke, P. and Perkins, P. (1998) '"Cracking the Canyon with the Awesome Foursome": representations of adventure tourism in New Zealand', *Environment and Planning D: Society and Space*, 16: 185–218.

Crossley, N. (2001) *The Social Body: Habit, Identity and Desire.* London: Sage.

Crouch, D. (1999) 'Introduction: encounters in leisure/tourism', in D. Crouch (ed.), *Leisure/Tourism Geographies: Practices and Geographical Knowledge.* London: Routledge.

Csordas, T. (ed.) (1994) *Embodiment and Experience.* Cambridge: Cambridge University Press.

Edensor, T. (2000) 'Walking in the British countryside', *Body and Society*, 6 (3–4): 81–106.

Edensor, T. (2001) 'Performing tourism, staging tourism: (re)producing tourist space and practice', *Tourist Studies*, 1: 59–82.

Edensor, T. (2002) *National Identity, Popular Culture and Everyday Life.* Oxford: Berg.

Franklin, A. (2001) 'Neo-Darwinian leisures: the body and nature: hunting and angling in modernity', *Body and Society*, 7 (4): 57–76.

Freitag, T. (1994) 'Enclave tourist development: for whom the benefits roll?', *Annals of Tourism Research*, 21: 538–554.

Frykman, J. and Löfgren, O. (eds) (1996) 'Introduction', in *Forces of Habit: Exploring Everyday Culture.* Lund: Lund University Press.

Goffman, E. (1959) *The Presentation of Self in Everyday Life.* New York: Doubleday.

Gottdiener, M. (1997) *The Theming of America.* Oxford: Westview Press.

Harrison, P. (2000) 'Making sense: embodiment and the sensibilities of the everyday', *Environment and Planning D: Society and Space*, 18: 497–517.

Hobsbawm, E. and Ranger, T. (eds) (1983) *The Invention of Tradition.* Oxford: Blackwell.

Ingold, T. and Kurttila, T. (2000) 'Perceiving the environment in Finnish Lapland', *Body and Society*, 6 (3/4): 183–196.

Lewis, N. (2000) 'The climbing body: nature and the experience of modernity', *Body and Society*, 6 (3/4): 58–106.

Macnaghten, P. and Urry, J. (1998) *Contested Natures.* London: Sage.

Macnaghten, P. and Urry, J. (2000) 'Bodies of nature: introduction', in *Body and Society*, 6 (3/4): 1–11.

Massey, D. (1995) 'The conceptualisation of place', in D. Massey and P. Jess (eds), *A Place in the World? Places, Cultures and Globalisation.* Oxford: Oxford University Press.

Michael, M. (2000) 'These boots are made for walking ...: mundane technology, the body and human–environment relations', *Body and Society*, 6 (3–4): 107–126.

Mitman, G. (1999) *Reel Nature: America's Romance with Wildlife on Film.* London: Harvard University Press.

Mordue, T. (1999) 'Heartbeat Country: conflicting values, coinciding visions', *Environment and Planning A*, 31: 629–646.

Morley, D. and Robins, K. (1995) *Spaces of Identity.* London: Routledge.

Nash, C. (2000) 'Performativity in practice: some recent work in cultural geography', *Progress in Human Geography*, 24 (4): 653–664.

Newby, H. (1980) *Green and Pleasant Land? Social Change in Rural England.* London: Hutchinson.

Oddie, B. (2002) *Bill Oddie's Introduction to Birdwatching.* London: New Holland Publishers.

Pahl, R. (1970) *Readings in Urban Sociology.* Oxford: Pergamon Press.

Pepper, D. (1991) *Communes and the Green Vision.* London: Merlin Press.

Pine, J. and Gilmore, J. (1999) *The Experience Economy.* Boston, MA: Harvard Business School Press.

Rojek, C. (1995) *Decentring Leisure.* London: Sage.

Rojek, C. (2000) *Leisure and Culture.* London: Macmillan.

Rothenberg, D. (ed.) (1995) *Wild Ideas.* Minneapolis, MN: Minnesota University Press.

Schieffelin, E. (1998) 'Problematising performance', in F. Hughes-Freeland (ed.), *Ritual, Performance, Media.* London: Routledge.

Seamon, D. (1979) *A Geography of the Lifeworld.* London: Croom Helm.

Sibley, D. (1997) 'Endangering the sacred: nomads, youth cultures and the English countryside', in P. Cloke and J. Little (eds), *Contested Countryside Cultures.* London: Routledge.

Spillman, L. (1997) *Nation and Commemoration: Creating National Identities in the United States and Australia.* Cambridge: Cambridge University Press.

Williams, R. (1961) *The Long Revolution*, London: Chatto and Windus.

Wylie, J. (2002) 'Becoming-icy: Scott and Amundsen's South Polar voyages, 1910–1913', *Cultural Geographies*, 9: 249–265.

Index